55th American Association of Swine Veterinarians Annual Meeting 2024

AA005487

Nashville, Tennessee, USA
24 - 27 February 2024

ISBN: 978-1-7138-9240-3

Printed from e-media with permission by:

Curran Associates, Inc.
57 Morehouse Lane
Red Hook, NY 12571

Some format issues inherent in the e-media version may also appear in this print version.

American Association of Swine Veterinarians
830 26th Street
Perry, IA 50220-2328
USA

Phone: 515-465-5255
Fax: 515-465-3832

aasv@aasv.org

Additional copies of this publication are available from:

Curran Associates, Inc.
57 Morehouse Lane
Red Hook, NY 12571 USA
Phone: 845-758-0400
Fax: 845-758-2633
Email: curran@proceedings.com
Web: www.proceedings.com

Proceedings of the 55th AASV Annual Meeting

February 24-27, 2024
Nashville, Tennessee

Leading AASV

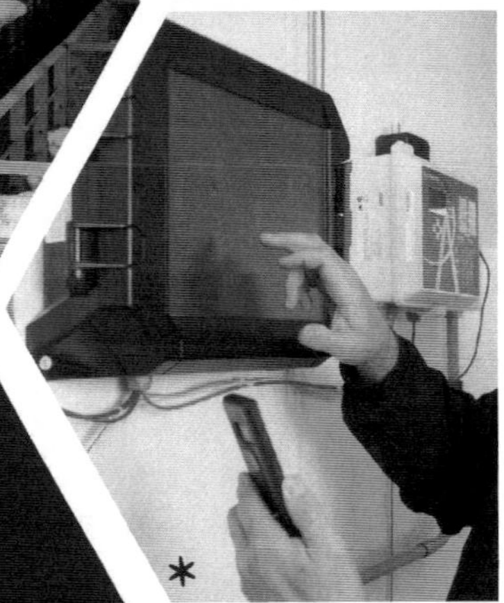

into the Future

AMERICAN ASSOCIATION SWINE VETERINARIANS

aasv.org/annmtg

* Photo courtesy of the National Pork Board, Des Moines, Iowa, USA

Table of Contents

Howard Dunne Memorial Lecture

Alex Hogg Memorial Lecture

Sunday: Research Topics

Sunday Session #1: Student Seminar

Sunday Session #2: Industrial Partners

Sunday Session #3: Industrial Partners

Sunday Session #4: Industrial Partners

Poster Session: Veterinary Students

Poster Session: Research Topics

Poster Session: Industrial Partners

Monday: Leading AASV into the Future

Monday #1: Disease Elimination: Theory to Move us Forward

Monday #2: Sustaining the Farm in the Face of Evolution

Monday #3: Immunology Toolbox: Today and Tomorrow

Tuesday: Driving Demand and Protecting the Product

Howard Dunne Memorial Lecture

Next Generation Swine Veterinarians: Who are we and where are we going?

Joel Nerem, DVM

In remembrance and honor of
Howard W. Dunne

Dr. Howard W. Dunne received his veterinary degree from Iowa State University in 1941. He was in engaged in private practice for one year and industry for four years before entering graduate school at Michigan State University. After receiving his PhD in 1951, he spent two years as deputy chief of veterinary microbiology at the Chemical Corps Biological Laboratories in Camp Detrick, MD. In 1953, he joined the faculty of Pennsylvania State University, where he headed the veterinary science research program and the large animal diagnostic laboratory. Dr. Dunne was internationally known for his work on swine diseases, particularly hog cholera and the swine reproductive viruses then referred to as the SMEDI syndrome. Dr. Dunne authored more than 90 scientific publications, and is perhaps best known for initiating and editing the textbook *Diseases of Swine*, which is still widely considered the "bible" of swine disease.

In 1971, the American Veterinary Medical Association presented Dr. Dunne with the XII International Veterinary Congress Prize to recognize his contributions to the international understanding of veterinary medicine. At the 1974 Congress of the International Pig Veterinary Society (IPVS), Dr. Dunne was elected president of the 1976 IPVS Congress, to be held in Ames, Iowa. His sudden death at the age of 61 in September of 1974 saddened the swine veterinary community, and the 1976 IPVS Congress was dedicated to his memory.

At the suggestion of Dr. Allen Leman, a financial gift made by the 1976 IPVS Congress organizers to the fledgling American Association of Swine Practitioners was combined with additional association funds to endow this prestigious annual lecture in remembrance and honor of Dr. Howard W. Dunne.

HOWARD DUNNE MEMORIAL LECTURERS

1978	Michael Muirhead	1990	Stanley Curtis	2002	Tim Blackwell	2014	Daryl Olsen
1979	W. J. (Bill) Smith	1991	Roy Schultz	2003	Lonnie King	2015	Gregory Stevenson
1980	N.C. Nielsen	1992	Joseph Connor	2004	Robert Desrosiers	2016	John Harding
1981	Allen Leman	1993	Camille Moore	2005	Scott Dee	2017	Jeff Zimmerman
1982	Michael Wilson	1994	Robert Morrison	2006	Gregg BeVier	2018	William R. DuBois
1983	Edward Bohl	1995	Rodney Johnson	2007	Tom Burkgren	2019	John T. Waddell
1984	Richard Ross	1996	Paul Yeske	2008	Timothy J. Loula	2020	Bret D. Marsh
1985	Kent Kislingbury	1997	Barbara Straw	2009	John U. Thomson	2021	Jerome O. Geiger
1986	Ralph Vinson	1998	Mike Daniel	2010	James F. Lowe	2022	Angela Baysinger
1987	Gary Dial	1999	Larry Rueff	2011	Locke A. Karriker	2023	Egan Brockhoff
1988	Thomas Alexander	2000	David Reeves	2012	R. Dean Boyd		
1989	Steven Henry	2001	Max Rodibaugh	2013	Craig Rowles		

https://doi.org/10.54846/am2024/1

Next Generation Swine Veterinarians: Who are we and where are we going?

Joel Nerem, DVM
Pipestone Veterinary Services, Pipestone, Minnesota

Introduction

Swine veterinarians have played a crucial role in the success of the global swine industry.[1] Their training, expertise, and innovation have contributed to the success of pig farmers and integrated production companies for decades. Many great practitioners have paved the way for our generation of swine veterinarians. We stand on their shoulders – we have big shoes to fill – and there is much yet to be accomplished.

Today's swine industry faces many challenges. PRRS virus remains our biggest health nemesis.[2] African swine fever is now present in more countries and swine producing regions than ever.[1] Markets are tough. Consumers and the public are making more demands on farming practices through support of regulation, such as California's proposition 12, a demand for sustainability, and a myriad of supply chain preferences. These impact areas such as antibiotic use and housing types. One could argue that there has never been a more challenging time to be a pig farmer or a swine veterinarian.

All these challenges in front of us create a confusing playing field. What should we focus on? What should our priorities be? How should we move forward? I think the answers to these questions are found in the same approach that our veterinary predecessors utilized. When in doubt, do what is right for:

1. The pig,
2. The farmer, and
3. The consumer

Where the interests of these three converge is the "sweet spot" for high impact by this generation's swine vets. To accomplish much, we will need to not only focus on the target, but we will also need to apply "Next Generation" solutions to the challenges in front of us.[1-3]

Focus on the pig

Health

Improving health is the greatest impact veterinarians can have on the pig (and the farmer's farm). It is at the center of our wheelhouse. If we are not focusing on health, we should probably not call ourselves veterinarians. By improving the health of pigs, we have a tremendous impact on animal well-being and the farm's profitability. Health should be our number one priority. Improving it is the right thing to do. While developing disease prevention and elimination strategies, vaccination schedules, diagnosing disease, and prescribing appropriate treatments are all critically important, to take the US swine herd's health to the next level, we need to do some things differently:

1. Better understand the cost of disease

Every disease costs the farm. It is a tax that must be paid on every group of pigs. We need to better quantify the tax each pathogen assesses to the farmer/producer. Having this knowledge will better inform our strategy decisions and conversations with pig owners for implementing necessary changes.

2. Implement Next Generation Biosecurity

Next Generation Biosecurity at the breed-to-wean farm has been well defined.[2,3] It involves (1) well executed industry standard mechanical transmission interventions, (2) air filtration to block aerosol transmission, and (3) year-round use of an appropriate feed mitigant. All breeding herds should have these.

As an industry we must dramatically improve wean-to-market biosecurity.[4] For many farms it is almost non-existent. This must change. In general, we know what to do – we just aren't doing it. Understanding the cost-benefit of disease and biosecurity interventions will assist in these discussions.

3. Better farm and production system design

Our swine veterinarian predecessors helped transition pig farming from farrow-to-finish site production to multi-site and age segregated systems. This contributed to improved health and performance. We need to press for similar system design changes, specifically:

a. Move sow farms out of pig dense regions. Breeding herds in pig dense regions get infected more often and send infected pigs everywhere. This multiplies disease problems. Poorly located sow farms create broken production systems.[2]
b. Flow growing pig sites all in/all out.[5] Endemically infected and continuous flow growing pig sites perpetuate disease, increase the need for antibiotics, and create a disease risk to neighbors. All in/all out management gives every group of pigs a clean start and the opportunity to achieve their genetic potential and optimizes profits to farmers.

4. Disease elimination

We continue to live with swine diseases in the US that can be eliminated.[1] Occasionally we add some to the list. Rarely, do we ever remove one from the list. It's time to start getting rid of some:

- **Farm by farm:** Lots of pathogens can be eliminated at the sow farm without needing depopulation and repopulations. PRRS, *Mycoplasma hyopneumoniae*, and the enteric coronaviruses (PEDV, TGEV, etc) come to mind. If we are reluctant to eliminate these costly pathogens from breeding herds, then, likely, we haven't implemented the correct biosecurity to prevent reintroduction, or our farm is in the wrong location.
- **National disease elimination:** I don't think we are ready to eliminate PRRS nationally, but we do have the tools, strategies, and expertise to eliminate *Mycoplasma hyopneumoniae* and the enteric coronaviruses. What we lack is the will to do it. I think it is time to take some diseases off the national list and these are where I think we should start. We are starting to organize as an industry to see such bold goals possible. The US Swine Health Improvement Plan (SHIP) could be an avenue for this effort.

Welfare

Like health, pig welfare fits squarely in the wheelhouse of the swine vet.[6] We should be the champions of swine well-being. We took an oath. It should be our passion. And while improving pig health is the single biggest way we can improve pig welfare, there are other priorities we should focus on. Some thoughts on swine welfare include:

- **Improving animal welfare should improve pig performance.** Therefore, we should embrace the topic and continually seek to improve it.
- **The public (our customers) view animal welfare as a high priority.** They even approve ballot initiatives that they think improve pig welfare, even when they do not. While some of these initiatives may be misguided, veterinarians have the responsibility (and opportunity) to ensure that the pig's welfare is improved as new requirements and production practices are applied.
- **Never walk past an issue.** Always embrace it as a teaching opportunity for caregivers and a way to make a pig's life better.
- **Who are your swine welfare experts?** Are you collaborating with them? Are we innovating in this space? I have had the privilege to work with several champions of swine welfare in my career: Dr. Angela Baysinger, Dr. Carissa Odland, and Dr. Cara Haden. Who is on your animal welfare expert panel list? How are they helping to improve animal welfare on your farms?

Focus on the farmer

Obviously, most everything we do to improve the health, well-being, and performance of the pig we are also doing to help the farmer. While the current hog market conditions in North America, and much of the world, present a threat to every farm's profitability and future, the threat of a foreign animal disease (FAD) incursion and our ability to respond also presents a real existential threat to every pig farming operation in North America. Our reliance on international exports for up to 30% of our USA production cements this reality. As a result, we need to be champions for FAD preparedness. This means several things:

- Every farm site we work with should have a Secure Pork Supply plan.
- Every site should have an accurate premises ID with an address that directs one to the farm site.
- Every farmer/producer should have 30 days of swine movement data available at all times – we have the tools available: AgView, RabApp, spreadsheets, etc.
- US Swine Health Improvement Plan[7] will likely be codified by USDA soon. Its current focus is FADs. I think it is currently the single best organized effort in FAD preparedness. I think all producers should enroll and that we should encourage and facilitate this.

To accomplish these goals, we need to work collaboratively with industry, academia, government, and the meat packing industry to address this concern. Each group has a unique role and unique strengths, and the effort should create synergies between these groups and efficiencies for farmers.

One of the most critical components to FAD preparedness is traceability. The swine industry is an industry on wheels and our ability to quickly track animal movements will be critical to any FAD response. It will also be necessary to maintain the confidence of our export customers and facilitate a quicker return to normal sales. For these reasons, we need a modern and "next generation" traceability system with 100% industry compliance.

What is a "Next Generation" traceability system? It needs to be:

- Real time. This system should capture movements as they happen. It shouldn't report what was supposed to happen.
- Accurate. Captures geolocation for both origin and destination as it happens.
- Easy to do. Think Uber for pigs.
- Mandatory. All pigs moved anywhere in the USA are required to be traced.
- Inexpensive.

What are the benefits of a "Next Generation" traceability system? They include:

- Rapid and accurate ability to trace movements from infected sites.
- Capability to regionalize and compartmentalize for purposes of trade.
- Greater surety to State Animal Health officials for managing risks of inter-state trade.
- Improve supply chain and attribute verification, which benefits retailers and consumers.
- Further distinguish USA swine industry as the best in the world.

Focus on the consumer

The consumer might seem to be an odd priority to think about for a group of swine veterinarians. After all, the farmer is our customer, right? Ultimately, the consumer is the farmer's customer and in recent years has made significant demands on the swine industry.[8-10] From simple preferences to state regulations, the public is changing the way producers must raise pigs and run their farms. Examples of this include:

- Certified Organic
- Open pen gestation
- Ractopamine-free
- No Antibiotics Ever
- Antibiotic Use Tracking (San Francisco)
- Prop 12
- Common Swine Industry Audit
- Gene Edit technology

This list is most likely to get longer and not shorter. We need to accept this reality and embrace consumer choice as an opportunity for the farmer and the veterinarian. Just like early adoption of new technologies provides an opportunity to capture more profit, early adoption of meeting new consumer preferences can also result in revenue opportunities for farmers. Because many of these consumer preferences are directly related to pig health and welfare, there is significant need and opportunity for veterinarians to get engaged and steer how these consumer demands are met – for the benefit of the pig and the farmer.

Discussion

We are blessed to be part of a profession that provides so many opportunities and rewards. We face many challenges. There is so much that needs to be done. This generation of swine vets owes much to those that came before us. We also possess the skills, technology, and "Next Generation" solutions to take our industry and clients to a new level of health, performance, and profitability. Therefore, the questions are now:

1. Will we do it?
2. Who will step up?
3. Are you prepared to serve your customer?

My encouragement to all swine veterinarians:

- **Be a Next Generation veterinarian.** Learn about and advocate for pathogen elimination – on the farm, regionally, and nationally. Promote FAD preparedness, industry-wide traceability, and US SHIP participation. Advocate for our industry to the consumer.
- **Practice Next Generation veterinary medicine.** Create new farm/system designs and eliminate pathogens in order to take your clients and the US swine herd to a new level of health. Design and implement new traceability tools that are real-time, accurate, and easy to use. Bring solutions to farmers as they seek to address evolving consumer preferences and always be an advocate for the pig.

References

1. Spronk G, Havas K, Dee S. Will swine veterinarians lead by meeting the next generation needs of our industry? *JAVMA*.2023. **doi.org/10.2460/javma.22.10.0480**

2. Havas K, Brands L, Cochrane R, Spronk G, Nerem J, Dee SA. An assessment of enhanced biosecurity interventions and their impact on porcine reproductive and respiratory syndrome virus outbreaks within a managed group of farrow-to-wean farms, 2020-2021. *Frontiers in Veterinary Science* section *Veterinary Humanities and Social Sciences.* 2022. **doi:10.3389/fvets.2022.952383**

3. Dee S, Brands, L, Nerem J, Schelkopf A, Spronk G, Kikuti M, Corzo C, Havas K. Improvements in swine herd biosecurity reduce the incidence of porcine reproductive and respiratory syndrome virus. *JAVMA* (accepted for publication).

4. Schelkopf A. Wean to market biosecurity standards: Creating, assessing towards, and benchmarking within a given standard for wean to market biosecurity standards. Proceedings of the AD Leman Swine Conference, p.54, St. Paul, Minnesota. 2023.

5. Dee SA and Joo HS. Prevention of the spread of PRRS virus in endemically infected pig populations by nursery depopulation. *Vet Rec*. 1994. 135:6-9.

6. Johnson AK, Rault JL, Marchant JN, Baxter EM, O'Driscoll KM Improving young pig welfare on-farm: The Five Domains Model. *J Anim Sci.* 2022 Jun 1;100(6):skac164. **doi: 10.1093/jas/skac164**

7. Holck T. How can US SHIP advance the control the control and elimination efforts of swine diseases? Proceedings of the AD Leman Swine Conference, p.32, St. Paul, Minnesota. 2023.

8. Weber GM, Hoban TJ, Kendall PA, Bull LS . Consumer concerns about modern technology in agriculture: considerations for undergraduate and graduate teaching. J Anim Sci. 1995 Sep;73(9):2727-32. **doi: 10.2527/1995.7392727x**

9. Harlander SK. The evolution of modern agriculture and its future with biotechnology. J Am Coll Nutr. 2002 Jun;21(3 Suppl):161S-165S. **doi: 10.1080/07315724.2002.10719260**

10. Niamir-Fuller M. Towards sustainability in the extensive and intensive livestock sectors. Rev Sci Tech. 2016 Nov;35(2):371-387. **doi: 10.20506/rst.35.2.2531**

Alex Hogg Memorial Lecture

Past, present, and future challenges for the swine veterinary profession

Chris J. Rademacher, DVM

In remembrance and honor of Alexander Hogg

Alexander Hogg
1920-2006

A native of Scotland, Alex (pronounced "Alec") Hogg immigrated to the United States with his family when he was six years old. After reaching adulthood, he served in the US Navy during World War II before entering veterinary school at Kansas State University. He received his DVM degree in 1950, and was a mixed animal practitioner in Iowa for twenty years. Dr. Hogg returned to college – this time, Iowa State University – to complete a Master's Degree in 1972. He spent the following years serving and educating swine veterinarians and producers in his roles as Nebraska Swine Extension veterinarian and professor at the University of Nebraska-Lincoln. Upon "retirement" from the university in 1990, Alex worked as a technical services consultant for MVP Laboratories. Ever an enthusiastic learner, he graduated from the University of Illinois Executive Veterinary Program in 1995.

An active and longstanding member of the AASV, Dr. Hogg joined what was then the American Association of Swine Practitioners shortly after it was formed. He served as secretary/treasurer of the association from 1974 to 1976, and was elected president in 1979.

The many awards and honors bestowed upon him over the years included several presented by the AASV: The Howard Dunne Memorial Award in 1985, the Meritorious Service Award in 1996, and in 2005, the first-ever AASV Foundation Heritage Award for a lifetime of achievement in veterinary medicine.

ALEX HOGG MEMORIAL LECTURERS

2011 Joseph F. Connor
2012 Lisa Tokach
2013 John T. Waddell
2014 Mark J. Engle
2015 C. Scanlon Daniels
2016 Peggy Anne Hawkins
2017 Matthew Turner
2018 Rodger Main
2019 Deborah Murray
2020 William L. Hollis
2021 Jeremy Pittman
2022 James A. Kober
2023 Attila Farkas

https://doi.org/10.54846/am2024/2

Past, present, and future challenges for the swine veterinary profession

Chris J. Rademacher, DVM
Iowa State University, Ames, Iowa

Introduction

It is with the most humble appreciation that I am able to deliver the 18th Alex Hogg Memorial Lecture this year. My sincerest thanks to Dr. Angela Baysinger and the 2024 AASV Annual Meeting Planning Committee for this great honor. I will tell you from the outset that I am unworthy of this honor, but will do my very best to share some of what I have learned and what I see as some of the challenges facing our profession now and in the future.

Although I never had the pleasure to really get to know Alex, he was certainly renowned within our association and was someone I admired and whose career I watched as I contemplated my own career path. Alex's story is one truly defined as a "lifelong learner." Alex immigrated through Ellis Island with his family when he was 6 years old, went to business college and then signed up for military duty during WWII. After the war, he came back and pursued his passion for agriculture by getting his DVM degree in 1950 from Kansas State University. He proceeded to practice in a small mixed-animal clinic for 20 years before going back to get a master's degree in veterinary pathology from Iowa State. He then became a Professor and Extension Veterinarian at the University of Nebraska-Lincoln and served in that capacity until his retirement in 1990. When most retirees pick up golf or travel, Alex served as a Technical Service Veterinarian for MVP labs and penned Hogg Sense for the next 11 years as well as participating in EVP when he was 75!! He will always continue to be the gold standard in what "lifelong learning" should look like. The thing I've always appreciated about Alex's lectures and presentations is just how applied they were. I have always thought that the role of Extension is one of a "translator." We take fantastic and technical research performed at universities and then translate it into language that producers and caretakers can understand and take home and apply in their daily operations. Concurrently we also report back what they are seeing to the researchers for further investigation. I have always thought that Alex was a master at this. I am a proud recipient of a Hogg Scholarship (a huge thank you to Mary Lou for all that she has done for the AASV!!) and am always trying to walk in Alex's footsteps, although those shoes are rather large to fill.

While many of the former Hogg Lecture themes have centered around "lifelong learning," Dr. Baysinger has allowed me some latitude to discuss some of the current and future challenges that face our profession. Most of the previous lectures given by my very stimulating and thought-provoking colleagues were designed to make you pontificate on the pearls of wisdom they acquired during their professional careers. My approach is a little different in that I am a data person. My goal is to present some information to you and get you to think about some of the current challenges we face and how we need to adapt in the face of our evolving profession.

Swine practice retention survey

Recently, a peculiar trend in swine veterinary practice began to emerge. Practitioners and AASV board members started to hear reports of early career swine veterinarians who were leaving swine practice to pursue careers in companion animal medicine. Swine veterinarians have historically enjoyed a salary advantage over their companion animal colleagues. That has recently changed with the venture capital investment into companion animal hospitals and clinics. This infusion of external capital has led to the ability for those entities to offer higher compensation packages with significant sign-on bonuses as well as student loan repayment plans in exchange for multi-year commitments. In discussion with other livestock veterinary groups, they have reported similar findings.

In response to these reports, the AASV Early Career Committee (ECC) commissioned a survey to collect more information on this emerging issue. The initial goal was to reach out to those ex-AASV members who had left the swine industry entirely and see if they would be willing to share some of those reasons why they left (phase 1). After development of the questions for the survey, there was further discussion to expand the survey to include a second phase. This phase would include veterinarians who left practice for other positions within the swine industry, such as technical service, academic, or governmental positions. The committee was instrumental in identifying a list of veterinarians who would be eligible to be surveyed for each phase.

The AASV Board of Directors approved the survey questions as well as the funding to complete both phases. The Iowa State Center for Survey and Statistical Methods (CSSM) was engaged to perform the surveys and compile the aggregated data, in order to maintain confidentiality of the participants. Surveys were completed during the summer and fall of 2023.

Phase 1

For phase 1, we were able to receive feedback from 11 veterinarians who had left the swine industry. The majority of the respondents had less than 10 years of experience. The age of the respondents ranged from 27 to 41 years of age. There were 4 males and 7 females who participated in this survey. When analyzing what type of swine practice the participants had most recently been employed in, 7 of the 11 respondents most recently worked for a swine production company. The majority of the veterinarians who had left the swine industry either left for mixed practice (n = 4) or companion animal practice (n = 5), whether they were privately owned practices or corporately owned entities.

When looking at potential reasons for making their career change, there were some enlightening responses. In terms of salary, only 4 of the respondents reported receiving significant pay increases as part of the reason for the career change. When evaluating the extent of the pay increase, the range reported was between 10 and 30%, with more of the responses weighted towards the 30% or more end of the range. When inquiring whether debt retirement was part of the compensation package, none of the respondents indicated that debt forgiveness was part of their compensation package. About half of the respondents (n = 6) also received a sign-on bonus ranging from 5 to 20% of their salary. In terms of travel, all respondents had indicated that travel, both within the state and out of state, was a significant part of their jobs while working within the swine industry. All the respondents also indicated that they travel significantly less with their current position than with their previous position.

One of the open-ended questions dealt with how the respondents defined "work-life balance." All of them referenced how their previous job had affected their definition. It is not possible to report all the answers here, but several themes began to emerge. One of the most prominent responses dealt with the issue of really being able to "disconnect" or "not feel like they are on call or have to be thinking about work 24/7" with their responsibilities within their swine practice role. Several mentioned that after transitioning to companion animal practice, while their daytime schedules were rigorous, they did not have to take anything home with them. Another theme was to have flexibility to be able to handle family commitments or emergencies. Almost all the respondents (n = 9 out of 11) felt like their work/life balance had improved since their career change by leaving the traditional swine practice.

The other open-ended question that was revealing inquired about some of the contributing factors that led to their decision to exit swine veterinary practice. What was revealed was that there were some factors that are typical to any industry and then some that were very specific to swine practice. Some of those reasons surrounded the relationship with their supervisors, lack of advancement opportunities, gender-related challenges, compensation compared to companion animal colleagues and the culture that existed in the organization that they worked for. These are common in many other industries as well. Some of the reasons that were more unique to swine practice dealt with the solitary nature of their work, lack of interaction with their colleagues, excessive time spent traveling to locations as well as duties that were assigned that didn't necessarily align with their training (ie, compliance-related activities). There were a couple of responses related to COVID-related challenges that were challenging for many swine veterinarians during that period.

Phase 2

For phase 2, we were able to receive feedback from 12 veterinarians (n = 12) who had left swine practice but were still involved in other facets of the swine industry. The majority of the respondents (n = 8) had 10 or more years of experience. The current age of this group ranged from 29 to 43 years of age with most over 35. That made this demographic slightly older and more experienced than the phase 1 group. There were 4 males and 8 females who participated in this phase of the survey, so the gender breakout was similar to phase 1. The majority of this group came from a production veterinary medicine background (n = 8). The types of jobs that this group left swine veterinary practice for tended to be technical service (n = 7) for pharmaceutical, genetic or other animal health related companies. In evaluating responses that may have explained reasons for leaving swine practice, compensation was one that was highlighted. Half of the respondents reported receiving an increase in salary with the move from practice to an allied industry. The salary increases ranged from 5 to 30% or more. Only 2 respondents received a sign-on bonus and only 1 veterinarian reported receiving a debt forgiveness package.

The definitions of work/life balance were very similar between both phases 1 and 2. Common themes were being able to disconnect from work when the workday is done and not feeling like they are having to be on call or be thinking about work when they are home. As in phase 1, almost all respondents reported having a better work/life balance after leaving swine veterinary practice and transitioning into their new roles. A majority of the comments indicated the flexibility they felt they had in their new roles, in comparison to their roles in swine practice. This was a repeated theme throughout the survey for this phase. Another couple of themes that were prominent were the increase in PTO (paid time off) and improved workplace culture with their job change. Another benefit that was reported was that the health care plan was often improved as well.

For phase 2, there were a couple of open-ended questions that dealt with reasons why they left practice and what it would take for them to consider taking a job back in veterinary practice. The reasons they left really had to do with the work/life balance and flexibility that their new jobs afforded them compared to their careers in swine practice. Improvements in benefits and a difference in workplace culture were noted as well. Some additional items mentioned included looking for new challenges, less stress, more mentorship, and feeling valued. In terms of what items would have to change in order for them to consider returning to a career in swine practice, it really fit into four main categories: comparable salary, increasing the number of veterinarians/technicians to share the workload, improved culture where veterinarians feel valued, and an improved work/life balance.

Future of veterinary students and swine education

In looking at estimates of future veterinary supply, the number of companion animal veterinarians is projected to grow by more than 20%, from 80,000 in 2022 to approximately 98,000 by 2030.[1] Contributors to this growth include the continuation of historical increases in class sizes and the addition of three new schools graduating their first classes in 2023, 2024 and 2025. On top of that 20% growth, there are at least 10 new veterinary schools in various stages of development. Even with those estimates, there is an expectation of needing nearly 41,000 additional veterinarians to meet the burgeoning growth of companion animal healthcare by 2030.[2] Even counting new grads over the next 10 years, a shortage of 15,000 veterinarians will likely occur in 2030. In the meantime, it is not surprising that companion animal clinics, especially those with the financial backing of large corporations, are going to compete with the already limited supply of veterinarians with very competitive salary and bonus structures as well as large debt-relief packages to help current students with the enormous student debt coming out of schools. Corporate groups employ approximately 40% of practicing veterinarians. Brakke Consulting predicts 25% of veterinary practices will be part of a corporate group in 2023.[3]

Another challenge for the swine industry is that the swine veterinary education options are dwindling across the veterinary schools across the US. The AASV Collegiate Activities Committee commissioned a 21-question survey that was sent to 39 veterinary colleges across North America and the Caribbean. What they discovered is that there is a continuing decline of swine-related activities within veterinary colleges. In terms of faculty positions, the number of faculty engaged in swine activities was low to extremely low for the majority of institutions, as shown by the median FTEs (0, 0.2, 0.3, and 1 for extension, service, teaching, and research, respectively).[4] Two-thirds (18/29 that responded) of veterinary education institutions did not track or require specific swine competencies to meet graduation requirements, even though a tracking system for competencies was in place. It is going to be difficult to continue to recruit new veterinarians into swine practice if the majority of schools are not even invested in swine faculty positions to help generate interest.

Compensation will be another area that will have to evaluated, especially with the intense competition for a finite number of new graduates. The AVMA annually publishes a report on the economic state of the veterinary profession which reports on factors such as compensation, debt, labor markets, and practice characteristics. The report is free to AVMA members and can be found at **https://ebusiness.avma.org/ProductCatalog/product.aspx?ID=2094.** Figure 1 shows the mean starting salary in private practice that is broken out across different practice sectors. You can see that starting in 2016, the companion animal sector (either exclusive or predominant) began to exceed the starting salaries for food animal practitioners (exclusive or predominant). Those figures really escalate beginning in 2021-2022 to the point where the mean starting salaries peak at around $120,000 for companion animal veterinarians.[5] In Figure 2, we superimposed the mean salary for veterinarians who are 1 year from graduating from veterinary school (ie, starting salary) from the last three AASV salary surveys (2017, 2020, 2023) onto the graph provided in the AVMA report. You can see that we are generally in line (but last) with other food animal practitioners, but have recently fallen significantly behind our companion animal colleagues.

Tools of the future for veterinary practitioners

According to the survey, it is pretty clear that travel was a major concern for those who left swine veterinary practice. With the declining number of swine practicing veterinarians and the geographical distribution of our swine production locations, it is becoming clear that we are going to have to think differently about what swine practice is going to look like moving forward. This will allow us as an association to begin to investigate tools, technology and programs that allow veterinarians to more efficiently use their time to use their skills to improve health and animal well-being.

Telehealth is an over-arching term that encompasses the use of technology to remotely deliver health information or education. Telemedicine is a more specific term that is defined by the AVMA as "use of medical information exchanged from one site to another via electronic communications regarding a patient's clinical health status."[6] There are federal and state statutes that regulate telemedicine. The AVMA policy is in general agreement with current federal and state guidelines, which allow for the use of telemedicine within a previously established veterinarian-client-patient relationship (VCPR). The federal VCPR definition (and that of most states) does not allow for establishment of a VCPR through electronic means. The AASV came out with a position statement in April of 2023 that supports the AVMA position on telemedicine and encourages swine veterinarians to continue to explore the uses of emerging technologies to "improve the standard of care and animal welfare, improve biosecurity, and bolster foreign animal disease preparedness." As technologies continue to improve and broadband internet coverage expands, these tools should help address some of the time and travel requirements mentioned by the veterinarians in the survey.

Speaking of foreign animal preparedness, one of the areas of concern if we do encounter a foreign animal disease (FAD) is having the veterinary resources to be able to collect samples from farms that are in quarantine or control areas. A collaborative effort between the AASV, National Pork Board, United States Department of Agriculture, and Iowa State University Swine Medicine Education Center has developed the Certified Swine Sample Collector (CSSC) program. This program allows Category II veterinarians

Figure 1: Mean starting salaries in private practice.

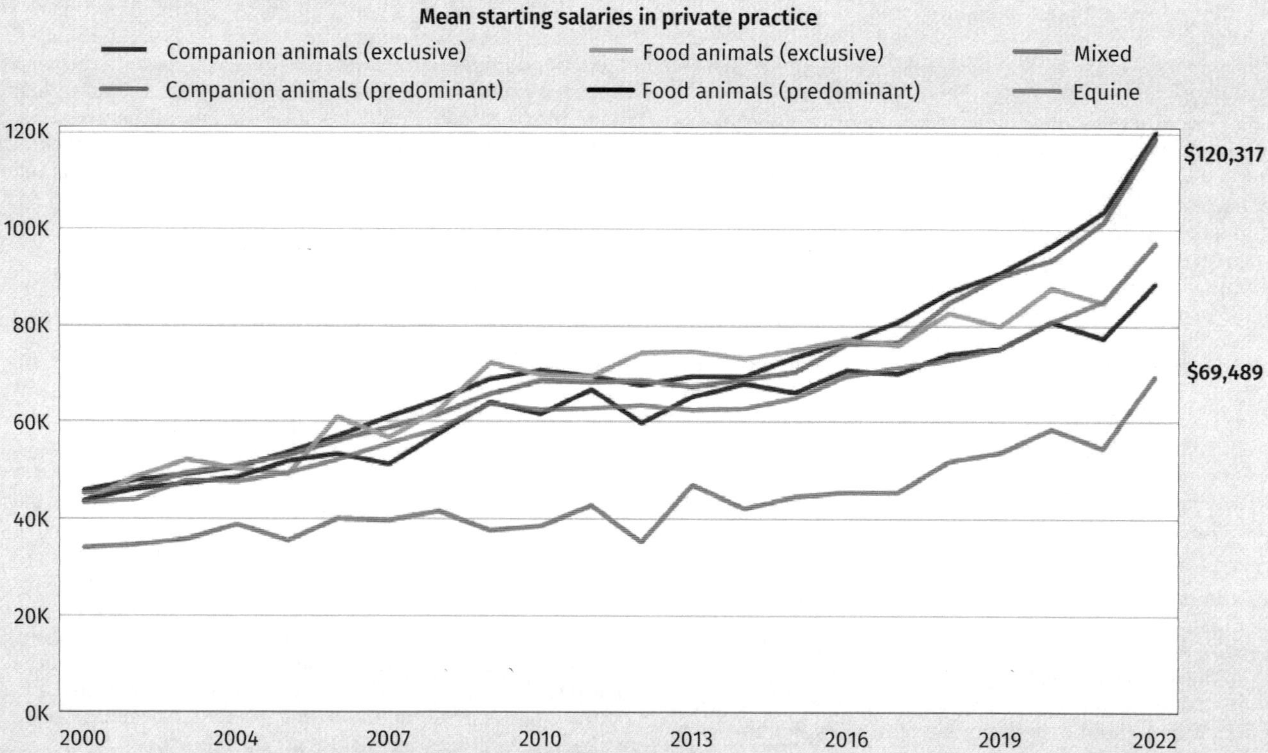

Figure 2: Mean starting salaries in private practice with the starting salaries from the AASV salary surveys (2017-2023) superimposed on the AVMA report.

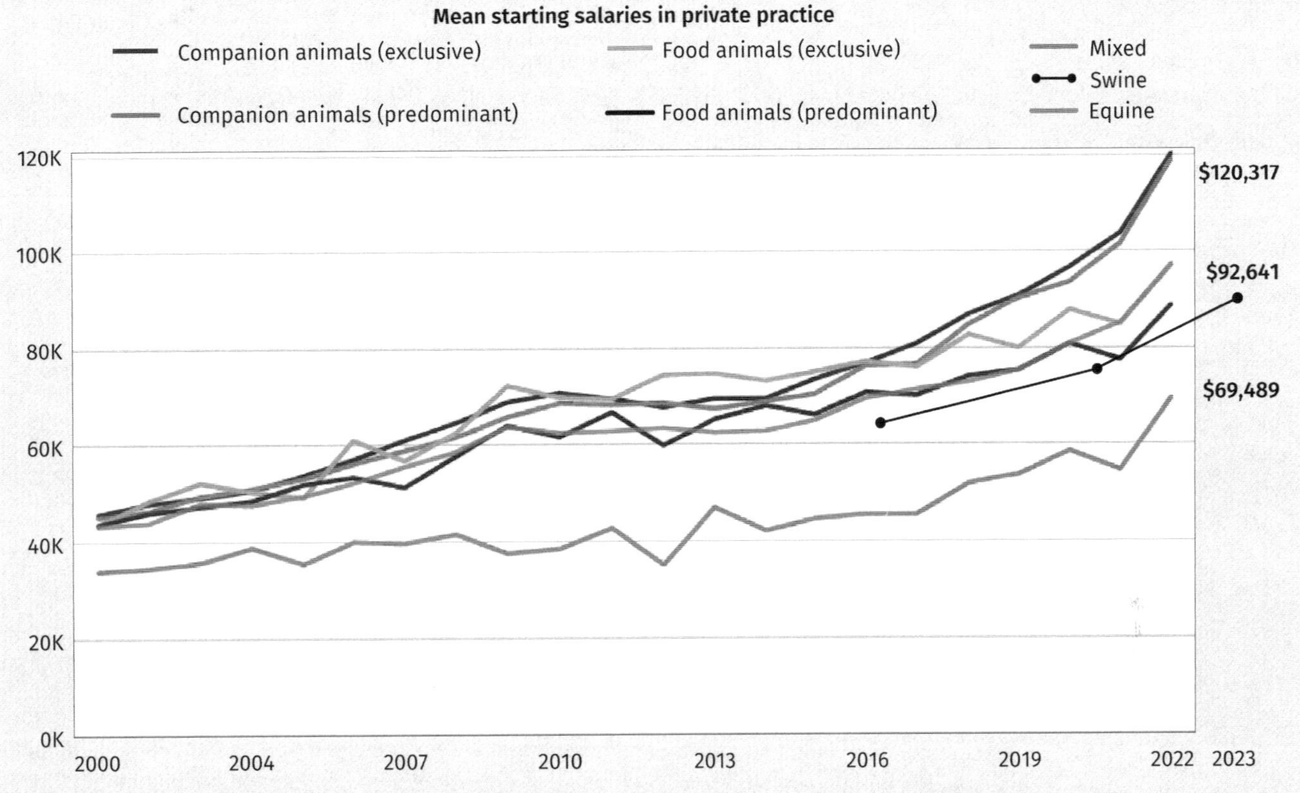

to train and certify on-farm staff to collect and submit various sample types in advance of an FAD outbreak. This is highly encouraged and, in my opinion, one of the most important FAD preparedness activities that we should be doing. Please contact the State Animal Health Official (SAHO) in your state for more information. The training materials are located at **www.securepork.org**. One additional side benefit of assembling these training materials and training/certifying farm staff addresses some of the feedback received during the swine veterinary retention survey: namely, veterinarians having to travel long distances to perform mundane tasks such as bleeding pigs and taking routine surveillance samples. Having caretakers on the farms (underneath a valid VCPR and certified in CSSC program) trained to collect these samples may be a way to better leverage the swine veterinarian's time to focus on more problem-solving activities.

Mental health and well-being will continue to gain importance in veterinary and corporate practice moving forward. The AVMA study also evaluated veterinary well-being. It surveyed veterinarians about potentially leaving the profession and found that their primary reasons for considering leaving involved mental health, poor work/life balance and the number of work hours. These factors represented 65% of the respondents' reasons for job dissatisfaction.[5] Tools such as soft skills training in personality profiling and conflict resolution can be important for young veterinarians beginning their careers as they must learn to deal with their clients and supervisors, which can be just as important as working with their animals. Many progressive clinics have hired corporate psychologists or mental health first aid officers to help their veterinarians address and deal with these issues early to avoid the burnout that can occur.

Closing

This is still the greatest veterinary sector to practice in!! I began my career with the idea that I was going to be a large animal practitioner, but that all changed when I attended the Leman Conference in 1994!! What I saw was a small group of intelligent, hardworking, passionate veterinarians who I would call my friends and colleagues from the moment I met them and now to this day. I wanted to join this small, intimate group and now when I come to the AASV annual meeting, it feels like an extended family reunion, rather than just a continuing education event. The one constant in life is change and we have seen that both swine producers and veterinarians are very adaptable to change. Whether it's moving sows from pens into stalls and back into pens again or moving weaning age from 35 days to 15 days and now back to 24 days, we are very adaptable to change!! The current challenges are no different, but it will take a different mindset and way of thinking than how we have thought about swine practice in the past. It's up to us!

Acknowledgments

I want to thank all those who have always been willing to share and collaborate, especially to those who are new in our profession. I can profoundly remember how important that was to me as a student and young veterinarian. I also want to thank all of those who have mentored me along the way; in particular, Drs. Gary Dial, Brad Freking, Steve Pollmann, Barry Wiseman, Pat Halbur, and Rodger Main. I also want to thank Drs. Locke Karriker and Justin Brown for their assistance in preparing this manuscript and presentation. Also, my wife Melissa, my daughters Isabella and Alexandra as well as my son Aidan, who have given me all that is precious and special in my life and allow me to do what I love to do.

References

1. Carlson R. AVMA Is Committed to Addressing the Industry's Challenges. *Today's Veterinary Business.* October/November 2023. Accessed Nov 3, 2023. **https://todaysveterinarybusiness.com/avma-veterinarians-1023**

2. Mars Veterinary. Tackling the Veterinary Professional Shortage. *Mars Veterinary.* Accessed Nov 3, 2023. **https://www.marsveterinary.com/tackling-the-veterinary-professional-shortage/**

3. OTTO Veterinary. (2022). 2022 Veterinary Stats. *OTTO Veterinary.* Accessed Nov 3, 2023. **https://otto.vet/2022-veterinary-stats**

4. Brown, Justin T., et al. "Swine Medicine Education: A Survey of North American and Caribbean Veterinary Colleges Curricula." *Journal of Veterinary Medical Education* (2023): e20230086.

5. American Veterinary Medical Association. Report on the Economic State of the Veterinary Profession. Schaumburg (IL): American Veterinary Medical Association; 2023. **http://www.avma.org/report-on-veterinary-economics**

6. American Veterinary Medical Association. "Telemedicine (Policy)." AVMA, **www.avma.org/resources-tools/avma-policies/telemedicine**. Accessed November 5, 2023

https://doi.org/10.54846/am2024/3

FFN titers in sows and piglets following homologous inactivated PRRSV administration to sows

L. Britton[1], BS; J. S. Pittman[2], DVM, MS, DABVP; T. Fangman[3], DVM, MS, DABVP; D. Nolan[3], DVM
[1]Department of Animal Sciences, North Carolina State University, Raleigh, North Carolina; [2]Smithfield Hog Production, Warsaw, North Carolina; [3]Huvepharma, Peachtree City, Georgia

Introduction

The control of porcine reproductive and respiratory syndrome virus (PRRSV) relies on early diagnosis and monitoring, biosecurity practices, herd management, and immunization.[1] Immunization strategies often center around commercial modified-live vaccines (MLV) that have shown some efficacy and may reduce the duration of viral shedding but may be unpredictable and are potentially dependent on the homology to the field virus.[2] Endemic outbreaks often lead to delayed herd cleanups and extended closure periods as vaccines are developed to combat specific strains. As PRRSV is constantly evolving, a major problem is the inability to cover genetically diverse field strains with MLV vaccines.[3] The heterogeneity of field PRRSV strains leads to increased risk with use of MLV, as transmission of vaccine virus and reversion to virulence is possible as well as the inability of the MLV to cover the field strain.[4] Autogenous inactivated vaccines (AIV) could provide more homologous protection with a specific contemporary strain. Studies have demonstrated increased homologous neutralizing antibody titers in sows and their piglets when a PRRSV AIV vaccine was used.[6,7] This study was designed to determine if a homologous PRRSV AIV would elicit a measurable booster of neutralizing antibody (measured as fluorescent focus neutralization: FFN) in PRRSV exposed dams, and if a higher level of FFN antibodies were transferred to their piglets.

Materials and methods

The farm where this study was conducted was a previously PRRSV naive 10,000-sow farrow-to-wean commercial farm that experienced a PRRSV (RFLP 1-4-3) outbreak (Week 0). The farm was a recent high health startup after herd depopulation (60 weeks pre-outbreak) with no known disease presence. The study started 19 weeks post-outbreak (wpo), with time to baseline production (TTBP) being achieved at 29 wpo.[5] Table 1 consists of farm performance averages 30 weeks prior to the outbreak (Period 1), the 30-week outbreak period until TTBP was achieved (Period 2), and the 30 weeks following (Period 3).

PRRSV (RFLP 1-4-3) from positive serum was used to create live virus inoculum (LVI), the PRRSV AIV, and was propagated on cell culture for the homologous FFN assay used in this study. The farm was closed to replacement animals and 1 ml LVI (10^3 TCID50) was given intramuscularly to the entire breeding herd at 6 and 13 wpo to ensure uniform exposure and immunity.[5] The farm was not exposed to other PRRSV vaccines or viruses during the study and did not experience any new disease throughout the duration of the study. For this evaluation, 90 females, in one breeding group, 8 weeks pre-farrow (wpf) were blocked into 3 parity groups: 30 gilts, 30 parity 1, and 30 parity 2/3 (split evenly as 15 parity 2 and 3 each). Within the block, sows were randomly allocated into 3 treatment groups: control (CNT) were unvaccinated, treatment 1 (TX1) received the AIV at 3 wpf, and treatment 2 (TX2) received the AIV at 8 and 3 wpf. At each vaccination event, enrolled dams received a 2cc dose of Autogenous 1-4-3 PRRSV vaccine. Please see Table 2 for treatment specifications and the distribution of animals across treatments.

Sows were bled at: 8, 6, 3 and 1 wpf. Serum samples were sent to South Dakota State VMDL for PRRSV PCR and homologous FFN testing. The FFN assay utilized in this evaluation was developed at South Dakota State VMDL to measure functional virus neutralizing antibodies specific to a known 1-4-3 challenge not immediately correlated to protection but believed to be indicative of a protective response. In the farrowing house, no pig movement was allowed as was consistent with current on farm protocol for PRRSV elimination. At 2 - 3 days of age (doa) 4 piglets from each sow were ear tagged and blood was individually collected via jugular venipuncture. At 21 - 24 doa blood was individually collected from the same pigs. All blood was spun down, serum poured off and saved in a -80 ultralow freezer within 24 hours. Serum was submitted for PRRSV PCR (pooled by litter) and 1 piglet per litter with both serum samples was randomly selected for PRRSV FFN. Sows in the study that remained in the herd at 20, 26, 48 and 56 weeks post final vaccination (1 wpf) were sampled for duration of FFN antibody.

For examinations of differences in antibodies by treatments, the Wilcoxon rank sum test was applied (rank-based analysis). The titer values were transformed with log base 2 (log_2 scale) to

Table 1: Herd performance 30 weeks prior to PRRSV break (period 1), during the break (period 2), and for 30 weeks after baseline production was reached (period 3). ANOVA was used with Student-Newman-Keuls multiple range test. Superscripts with differing letters are significantly different within column.

Period	AVG TB	AVG LB	% SB	% MUM	% PWM	PWMFY	ABORT
1	15.1 ± 0.1[a]	13.9 ± 0.1[a]	5.8 ± 0.3[a]	2.3 ± 0.2[a]	15.3 ± 0.5[a]	26.8 ± 0.2[a]	5.1 ± 0.4[a]
2	15.8 ± 0.1[b]	12.4 ± 0.3[b]	8.6 ± 0.8[b]	12.1 ± 1.8[b]	29.2 ± 5.6[b]	21.3 ± 0.9[b]	9.9 ± 2.5[a]
3	16.4 ± 0.1[c]	15.0 ± 0.1[c]	6.2 ± 0.2[a]	2.4 ± 0.1[a]	15.2 ± 0.3[a]	28.4 ± 0.3[c]	6.6 ± 0.7[a]

AVG TB: average total born; AVG LB: average liveborn; % SB: average % stillborn; % MUM: average % mummies; % LOSS: average of % pig losses; % PWM: average % pre-weaning mortality; PWMFY: average of pigs weaned/mated female/year; Abort: avg abortions.

Table 2: Treatment specifications

Treatment block	Control	Treatment 1	Treatment 2
Pre-farrow vaccination	No vaccination	AIV 3 WPF	AIV 8 and 3 WPF
Gilts	10	10	10
Parity 1	10	10	10
Parity 2	5	5	5
Parity 3	5	5	5
Total number enrolled	30	30	30

AIV: autogenous inactivated vaccines; WPF: weeks pre-farrow

preserve the ranking. Where all three treatments were compared to one another, the Bonferroni correction was applied within each time point for the three tests. Where antibody levels were compared between mothers and piglets to determine associations, the Spearman correlation was applied to the raw values. To examine the association between maternal antibody levels and the change in antibody levels over time in piglets, the difference for the piglets was calculated on the log^2 scale with an offset of 1 where 0 values occurred. To examine potential association between survival rate of piglets and the mother's antibodies, the Spearman correlation was again applied, this time comparing maternal antibodies to the proportion of piglets born alive. To examine potential association between treatment and successful farrowing, the extension to the Fisher's exact test was applied. Where a significant association was found, the ordinary Fisher's exact test was used to compare between the two groups which were most disparate.

Results

All three farm production periods (prior, during, and after PRRSV break) had significant differences ($P \leq .05$) observed in totalborn, liveborn pigs, and pigs weaned per mated female per year. There was statistical significance for period 2 compared to periods 1 and 3 for stillborns, mummies, and pre-weaning mortality. No sera collected from the study animals tested PRRSV PCR positive throughout the study, but all study sows were exposed to the on-farm 1-4-3 isolate prior to the study. No difference in FFN titer was detected between treatment groups at the start of the study (8 wpf). There was a significant difference between FFN titers of TX2 compared to CNT at 6, 3 and 1 wpf. There was also a significant difference between FFN titers of TX2 compared to TX1 at 6 wpf, but not at 3 and 1 wpf. Comparing TX2 piglet FFN titers to the other treatments at 24 doa, there was significant difference ($P = .007$) with TX2 having higher titers than TX1 or CNT. There was a significant correlation between the piglets' titer at 2-3 (r=0.767, $P < .001$) and 21-24 doa (r=0.774, $P < .001$) with the sows' 1 wpf titer (r=0.767, $P < .001$). Please see Table 3 for mean FFN titer results.

Table 3: Results reported here are mean FFN titer values transformed with log base 2. The Wilcoxon rank sum test analyzed the rank of the titer. Superscripts with differing letters are significantly different within column.

	Sows, weeks pre-farrow				Pigs, day of age	
Week	8	6	3**	1^	3	24*
Control	8^a	8^a	6^a	4^a	32^a	4^a
Treatment 1	8^a	8^a	8^{ab}	8^{ab}	32^a	4^a
Treatment 2	8^a	16^a	16^b	16^b	32^a	8^b

* At 24 days of age pigs in treatment 2 had significantly higher titers than non-treatment 2 ($P = .007$)

** At 3 weeks pre-farrow: sows in treatment 2 had significantly higher titers than controls ($P = .036$)

^ At 1 week pre-farrow: sows in treatment 2 had significantly higher titers than controls ($P = .036$)

Discussion

A 2-dose pre-farrow vaccination program with a homologous PRRSV AIV demonstrated statistically higher mean FFN values compared to unvaccinated controls. While not statistically significant, it is interesting to note that the TX2 sow titers increased, TX1 sow titers remained consistent and CNT sow titers declined through the second half of gestation leading up to farrowing. This work supports findings by others[6,7] that homologous PRRSV AIV can boost previous homologous immunity and can be transferred to piglets. The FFN results would suggest that all suckling pigs received good colostrum antibodies to PRRSV from their respective mothers (median 32 for all groups). Unfortunately, in this study, demonstration of reduced litter viremia could not be evaluated as all litters were negative for PRRSV at both sample points. This is likely due to the herd already achieving PRRSV stability 30 weeks after the outbreak (11 weeks into the study), classifying the farm as positive stable undergoing elimination (II-B) and transitioning to provisionally negative (III) as negative replacement animals remain negative.[8] Monthly wean pig sera monitoring for PRRSV began testing negative at 28 wpo and continued to test negative for all subsequent sampling events.

Limitations of using AIV is the prolonged process from virus isolation to vaccine manufacturing and implementation. While this may be a lengthy process, it could be beneficial for herds that have extended closures or encounter more difficulty in returning to stabilization. Further work should be conducted with homologous PRRSV AIV with increased sample sizes to determine if there is repeatability and if the transfer of higher maternal FFN to piglets is protective and reduces viremia in a PRRSV unstable population as has been demonstrated in other challenge studies.[6,7] More work should also be done to determine if higher FFN titers in piglets translates to reduced viremia within the herd, better protection, and how post weaning performance may be improved in a population with PRRSV. There is additional opportunity to have closely homologous AIV vaccines available for farms that break with similar isolates and to determine if AIV is useful in stabilizing farms that do not undergo any elimination protocol. The future application of FFN testing could be used to evaluate the efficacy of vaccination on dams in eliciting an immune response as well as monitoring pigs throughout the suckling and growing phase on the potential protection received through colostrum. FFN testing could also be used as a management strategy through use as a monitoring tool to determine a farm's homologous immune status over time throughout a closure.

References

1. Pileri E, Mateu E. 2016. Review on the transmission porcine reproductive and respiratory syndrome virus between pigs and farms and impact on vaccination. Vet Res (47) 108.

2. Chae C. Commercial PRRS Modified-Live Virus Vaccines. Vaccines (Basel). 2021 Feb 22;9(2):185. **doi: 10.3390/vaccines9020185**

3. Meng XJ. Heterogeneity of porcine reproductive and respiratory syndrome virus: implications for current vaccine efficacy and future vaccine development. Vet Microbiol. 2000 Jun 12;74(4):309-29. **doi: 10.1016/s0378-1135(00)00196-6**

4. Renukaradhya GJ, Meng XJ, Calvert JG, Roof M, Lager KM. Inactivated and subunit vaccines against porcine reproductive and respiratory syndrome: Current status and future direction. Vaccine. 2015 Jun 17;33(27):3065-72. **doi: 10.1016/j.vaccine.2015.04.102**

5. Linhares DC, Cano JP, Torremorell M, Morrison RB; Comparison of Time to PRRSv-Stability and Production Losses between Two Exposure Programs to Control PRRSv in Sow Herds; Prev Vet Med. 2014 Jun 2. pii: S0167-5877(14)00196-2. **doi: 10.1016/j.prevetmed.2014.05.010**

6. Geldhof MF, Van Breedam W, De Jong E, Lopez Rodriguez A, Karniychuk UU, Vanhee M, Van Doorsselaere J, Maes D, Nauwynck HJ. 2013. Antibody response and maternal immunity upon boosting PRRSV-immune sows with experimental farm-specific and commercial PRRSV vaccines. Vet Microbiol. 167(3-4):260-71.

7. Kick AR, Wolfe ZC, Amaral AF, Cortes LM, Almond GW, Crisci E, Gauger PC, Pittman J, Käser T. 2021. Maternal Autogenous Inactivated Virus Vaccination Boosts Immunity to PRRSV in Piglets. Vaccines (Basel). 9(2):106.

8. Holtkamp DJ, Polson DD, Torremorell M, Morrison B, Classen DM, Becton L, Henry S, Rodibaugh MT, Rowland RR, Snelson H, Straw B, Yeske P, Zimmerman J. Terminology for classifying swine herds by porcine reproductive and respiratory syndrome virus status. *J Swine Health Prod.* 2011;19(1):44–56.

https://doi.org/10.54846/am2024/4

Field experiences with a novel PRRS live virus vaccine based on the naturally nonpathogenic G16X strain from 2021 to 2023 in Mexico

J. H. Lara[1], DVM, MSc; A. Carranza[1], DVM; F. Zuckermann[2], DVM, PhD; D. Sarfati[1], DVM, SPAP; J. Torres[1], DVM, MBA; B. Lozano [1], DVM
[1]Laboratorio Avi-Mex, S.A. de C.V., [2]Pathobiology Department, University of Illinois at Urbana-Champaign, Urbana, Illinois

Introduction

The efficacy of a novel PRRS live virus vaccine based on the naturally nonpathogenic G16X strain was evaluated in 19 commercial swine operations located in different regions of Mexico, all of which have a very high prevalence of PRRSV.

Methods

Eight operations are in Mexico's Northwest, six in the West, one in the Central region and two in the Eastern part of the country. All the farms enrolled in the study were already using some type of commercial PRRS modified live virus (MLV) vaccine to control PRRSV prior to switching to the G16X live virus vaccine. The main parameter to assess vaccine efficacy was the average pig mortality rate for 6 months prior to beginning the use of the novel G16X vaccine (historical mortality rate), as compared to the mortality rate observed afterwards. Other production parameters were also selectively measured. A total of 219,230 piglets were involved in these field trials. In 12 farms, the G16X vaccine was administered intranasally (IN) to piglets on the third day of age (DA). In 6 farms, the vaccine was administered either intramuscularly (IM) or IN (after weaning) between the 3rd and the 4th week of age (WA). In one farm, the vaccine was administered IM at 10th WA (fattening period). The mortality rates in each farm at either 10th WA or the end of the production cycle were recorded and compared to the respective historical mortality rate. Serum samples from 20 piglets per farm were collected every 3 weeks to detect the PRRS-specific antibodies by ELISA and the presence of PRRSV ORF5 RNA by RT-qPCR. PCR positive samples from all farms were sequenced to determine the type of PRRSV present.

Results

Observation of pigs for 5 hours after G16X vaccination revealed no adverse clinical effects. ELISA tests were positive for PRRSV antibodies with increasing S/P ratios observed over time at 3 and 6 weeks postvaccination and thereafter, indicating effective immunization and exposure to field virus. Serum samples obtained 3 weeks postvaccination were positive for G16X ORF5 sequence, exhibiting its characteristic 1-6-2 RFLP pattern, indicating replication of the vaccine virus. Exposure to field PRRSV was confirmed in all the participating farms at ≥ 6 WA by a positive ORF5 RT-qPCR with a sequence distinct from G16X. Combined, 30 different field viruses with 21 distinct RFLP patterns and < 88% nucleotide or amino acid sequence homology to the G16X virus were detected. As shown in the Table 1, in all farms there was a reduction in the historical mortality rate ranging from 36 to 55% after initiating use of the novel G16X vaccine.

Discussion

The nonpathogenic vaccine strain G16X proved to be a safe and much more effective immunoprophylactic tool than the MLV vaccines currently being used to reduce production loses in farms with a high prevalence of PRRSV located in various regions of Mexico.

Table 1: Mortality rates before and after initiating use of G16X vaccine

Groups	Historical mortality (6 months)	Mortality G16X	Difference
3rd DA (IN) Mortality until 10th week	13.62%[a]	6.77%[b]	6.85%
Weaning (IN or IM) Mortality until end of the cycle	14.55%[a]	7.98%[b]	6.57%
Fattening (IM) Mortality until the end of the cycle	6.54%[a]	2.35%[b]	4.19%

[a,b] Different letter means statistically significant at $P < .05$ between historical and G16X mortality of each group by Chi-squared
DA = days of age; IN = intranasal; IM = intramuscular

References

1. H. Lara, D. Sarfati, J.A. Torres, J. Gonzalez, D. Herrera, B. Lozano, F.A. Zuckermann. Field trial experience with an intranasal nonpathogenic novel vaccine against PRRS. Conference of Research Workers in Animal Diseases, 2023; 158.

2. H. Lara, D. Sarfati, J.A. Torres, J. Gonzalez, D. Herrera, B. Lozano, F.A. Zuckermann. Field trial experience with an intranasal nonpathogenic novel vaccine against PRRS. NAPRRS/NC229 ICSVD. 2022; 67.

3. Mendoza P.R., Montes R., Escobedo M. Lara P.J.H, Sarfati M.D., Torres A., Lozano B. Resultados productivos y económicos en una granja porcina comercial de ciclo completo, en el estado de Querétaro, ante la estrategia de vacunación, con la cepa activa naturalmente apatógena G16X de PRRS y acciones complementarias de manejo. LV Congreso Nacional AMVEC 2023. 127.

4. Carranza G.A., Pacheco RRR, Borrego E.J.L., Castro G.R., Torres I.J.A., Sarfati M.D., Lara P.J.H., Lozano D.B. Uso de una vacuna activa apatógena para la reducción de mortalidad y daños productivos provocados por VPRRS en la etapa de engorda en una granja comercial en el estado de Sonora, México. Congreso Nacional AMVEC 2023. 157.

https://doi.org/10.54846/am2024/5

What is behind tongue tips sampling and other welfare-friendly postmortem samples for accurately detecting PRRSV?

M. Kikuti, PhD; C.M. Mellini, DVM; X. Yue, PhD, M. Culhane, DVM, PhD; C. A. Corzo, DVM, PhD
Dept of Veterinary Population Medicine, University of Minnesota, Saint Paul, Minnesota

Introduction

Specimens collected from dead pigs are a welfare-friendly and possibly cost-effective active surveillance method with comparable detection to ante mortem specimens for PRRSV detection. Processing fluids has emerged as a sensitive tool for PRRS monitoring[1] suggesting the use of tissues as viable specimens. Welfare-friendly ante-mortem specimens such as oral, nasal, and rectal swabs have been proposed for PRRS diagnosis and can be collected from dead pigs; however, sample collection demands piglet restrain. Tongue tips exudate have recently gained attention as a viable postmortem specimen for PRRS monitoring.[2] We aimed to determine whether postmortem sampling of suckling piglets allows for the detection of PRRSV and determining the sensitivity of each postmortem specimen studied.

Methods

Two farrow-to-wean farms undergoing a wild-type PRRS outbreak and that had decided to pursue elimination were conveniently selected. Samples were collected at each farm at approximately 8 and 20 weeks after the outbreak detection. Postmortem sampling of piglets found dead or euthanized according to farms' regular animal care protocols was conducted. Postmortem specimens consisted of individual sterile swabs of the nasal, oral, and rectal cavities, tongue tips, and blood collected directly from the heart. All specimens were stored and tested individually for PRRSV by rRT-PCR. Additionally, we collected blood via jugular venipuncture from 30 and 60 live piglets during the first and second sampling points, respectively. Sera from live piglets were tested for PRRSV by rRT-PCR in pools of five and if positive pools was then tested individually. Sera from live piglets were used to estimate the within-farm PRRS prevalence using the current industry monitoring sampling approach. The sensitivity and agreement of tongue tips and oral, nasal, and rectal swabs were calculated using the postmortem intracardiac sera as the assumed gold standard.

Results

PRRSV was detected by rRT-PCR in 79% of live piglets' sera on farm 1 visit 1 (F1V1), 10% on farm 1 visit 2 (F1V2), 63% on farm 2 visit 1 (F2V1), and 0% on farm 2 visit 2 (F2V2). Overall, oral swab sensitivity when compared to the same dead animal's serum was 90.9%, 84.8% for nasal swabs, 75.8% for rectal swabs, and 97.0% for tongue tips. Compared to sera, agreement ranged from 70.2% for tongue tips (kappa 0.45, $P < 0.001$) to 89.6% for oral and nasal swabs (kappa 0.77, $P < 0.001$).

Discussion

Postmortem sampling of piglets to detect PRRSV is a welfare-friendly alternative. Any of the assessed specimen types were capable of identifying PRRSV on F1V1, F1V2, and F2V1. PRRSV was identified in F2V2 through serum and tongue tips, possibly by due to environmental contamination. Although this might be an issue when trying to assess within-herd prevalence, having highly sensitive detection methods, even those that detect environmental contamination, might represent an improvement in detecting presence of the virus on a farm.

References

1. Vilalta et al. Effect of litter aggregation and pooling on detection of porcine reproductive and respiratory virus in piglet processing fluids. Journal of Veterinary Diagnostic Investigation. 2019. p. 625-8.

2. Baliellas et al. Porcine Reproductive and Respiratory Syndrome Surveillance in breeding Herds and Nurseries Using Tongue Tips from Dead Animals. Veterinary Sciences. 2021;8(11):259.

https://doi.org/10.54846/am2024/6

Evaluation of PRRSV vertical transmission using stillborn tongue tip fluids sampling

I. Machado, DVM, MS; P. Li, DVM, MS; A. P. Poeta Silva, DVM, MS, PhD; E. S. Magalhaes, DVM, MS, PhD; O. H. Osemeke, DVM, MS, PhD; S. Jayaraman; D. C. A. Moraes, DVM, MS; M. Mil-Homens, DVM, MS; G. A. Cezar, DVM, MS; T. Petznick, DVM, MS; G. S. Silva, DVM, MS, PhD; D. C. L. Linhares, DVM, MBA, PhD
VDPAM, Iowa State University, Ames, Iowa

Introduction

Porcine reproductive and respiratory syndrome virus (PRRSV) transmission to piglets may come directly from sows or littermates. Although ~85% of the breeding herd is in the gestation phase, there is no practical and reliable method for PRRSV surveillance validated for that group. While processing fluid is a practical, affordable, and sensitive approach for screening suckling pigs for PRRSV, it does not answer whether the PRRSV came from sows, other pigs, or contaminated fomites. Conversely, tongue tip fluids (TTF) is a risk-based sample as it targets dead pigs.[1] It has been documented that stillborn pigs are more likely to test PRRSV-positive than live newborn littermates.[2] The objective of the study was to evaluate PRRSV vertical transmission by using stillborn TTF to detect PRRSV RNA by RT-qPCR.

Methods

Two PRRSV-positive unstable breeding herds were recruited for this study. Within 24 hours of farrowing, sow tonsil scraping, stillborn serum, and stillborn TTF within the litter were collected. Samples were individually tested for PRRSV RNA by RT-qPCR, and the probability of PRRSV detection was compared between sample types, litters, and individual animals.

Results

A total of 66 sows and their stillborn piglets were tested within 24 hours post-farrow. Thirty-three litters (50%) had at least one PRRSV-positive result of tonsil scraping, stillborn TTF, or stillborn serum samples. There were 19 sows (29%) PCR-positive on tonsil scraping, of which five litters had at least one stillborn TTF positive, and four had at least one stillborn serum positive (Table 1). Overall, 19 litters (29%) had at least one TTF-positive, and 11 litters (17%) had at least one serum sample positive (Figure 1). Regarding stillborn TTF and serum samples correlation at an animal-level comparison, except for one stillborn pig, all serum-positive stillborn piglets also tested positive for TTF, while 14 stillborn piglets were PRRSV-positive on TTF with negative results for serum samples, and 94 stillborn piglets were negative on both TTF and serum sample (Table 2).

Table 1: Litter-level sow tonsil scraping and stillborn serum results

		Stillborn serum samples		
		+	–	
Sow tonsil scraping samples	+	4	15	19
	–	7	40	47
		11	55	66

Table 2: Animal-level stillborn tongue tip fluids and stillborn serum results

		Stillborn serum samples		
		+	–	
Stillborn tongue tip fluids	+	11	14	25
	–	1	94	95
		12	108	120

Figure 1: Litter-level positivity illustration for the three sample types.

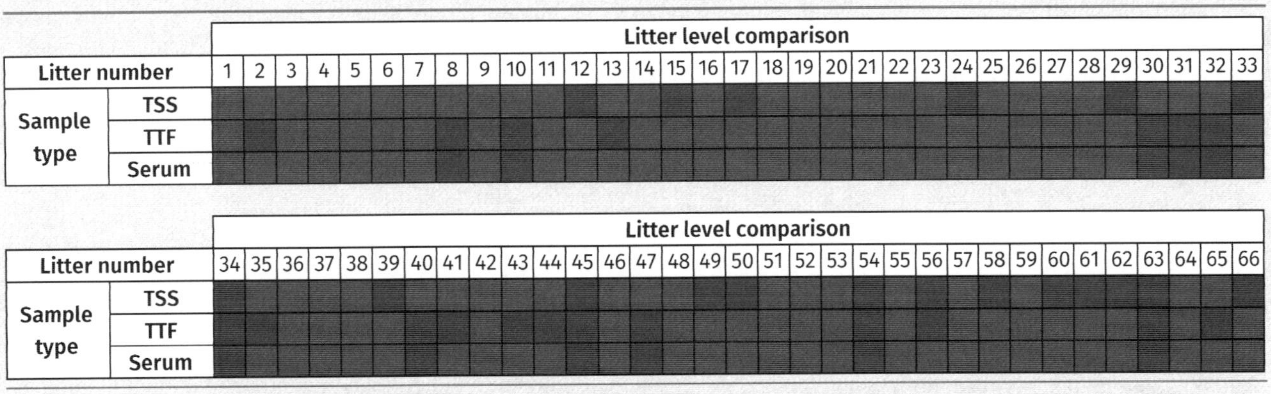

TSS = tonsil scraping samples. TTF = tongue tip fluids. Green color cells = PRRSV-RNA negative results. Red color cells = PRRSV-RNA positive results.

Discussion

TTFs from stillborn piglets are a proxy of the gestating population PRRSV status. Stillborn TTFs had a greater positivity than matching stillborn sera, possibly picking up viral RNA from systemically infected stillborns who were no longer viremic. Additionally, while tonsil sampling can effectively detect PRRSV RNA in pigs infected for a long period,[3] this study revealed several instances of sows having PCR-negative tonsil scraping giving birth to PRRSV-positive piglets. This demonstrates that relying solely on sow tonsil scraping for PRRSV detection may not reflect the offspring's status. To optimize cost and time, collecting multiple tongue tips in the same bag can be employed,[1] similar to the approach used for processing fluids.

References

1 Machado, I. F., Magalhães, E. S., Poeta Silva, A. P. S., Moraes, D. C., Cezar, G., Mil-Homens, M. P., ... & Linhares, D. C. (2022). Porcine reproductive and respiratory syndrome virus RNA detection in tongue tips from dead animals. *Frontiers in Veterinary Science*, 9, 993442. **doi.org/10.3389/fvets.2022.993442**

2. Jansen, T., Waddell, J., Linhares, D., Machado, I., LeFevre, C. (2023). Comparison of tongue tips, serum, and processing fluids for PRRSV monitoring in neonates. *54th AASV Annual Meeting*. 2023. **doi.org/10.54846/am2023/26**

3. Walker, H.L., Bowman, A.S., Ferreira, J.B., Nelson, S.W., Pairis-Garcia, M. and Arruda, A.G., 2021. Tonsil scrapings for porcine reproductive and respiratory syndrome virus detection in growing pigs under field conditions. Journal of Swine Health and Production, 29(2), pp.72-80. Available at: **https://www.aasv.org/shap/issues/v29n2/v29n2p72.html**

https://doi.org/10.54846/am2024/7

Have we seen this PRRSV before? Where? When? SDRS PRRSV BLAST: An informative tool to support swine veterinarians and producers

Srijita Chandra[1]; Guilherme Cezar[1]; Kinath Rupasinghe[1]; Michael Zeller[1]; Rodger Main[1]; Albert Rovira[2]; Franco Matias-Ferreyra[3]; Angela Pillatzki[4]; Melanie Prarat[5]; Craig Bowen[6]; Daniel C L Linhares[1]; Mengdi Huai[1]; Giovani Trevisan[1]
[1]Iowa State University, Ames, Iowa; [2]University of Minnesota, Saint Paul, Minnesota; [3]Kansas State University, Manhattan, Kansas; [4]South Dakota State University, Brookings, South Dakota; [5]Ohio Animal Disease and Diagnostic Laboratory, Reynoldsburg, Ohio; [6]Purdue University, West Lafayette, Indiana

Introduction

The Swine Disease Reporting System (SDRS, **https://www.fieldepi.org/sdrs**) is a collaborative initiative from six veterinary diagnostic laboratories (VDL): Iowa State University, University of Minnesota, Kansas State, South Dakota State University, Ohio state-based VDL, and Purdue University. The motto of the SDRS team has been to implement innovative technology to monitor swine health diagnostic data and report their findings to veterinarians, producers, and researchers. The team maintains a private sequence database containing over 100,000 open reading frame-5 (ORF-5) of the porcine reproductive and respiratory syndrome virus (PRRSV). The objective of the current work was to monitor the detection of new sequences and develop an open-source web-capability allowing veterinarians and producers to compare their ORF-5 sequences with those in the SDRS database. The second object was to identify new sequences entering the SDRS database from 2010 to August 2023.

Methodology

A one-of-a-kind implementation of a Basic Local Alignment Search Tool (BLAST) specific to the PRRSV, named the SDRS PRRSV BLAST, was implemented following the widely used NCBI BLAST,[1] identifies local similarities between the query ORF-5 nucleotide sequence and the database sequences and calculates the statistical significance of the matches using the Smith-Waterman heuristic algorithm.[2] Two major use cases handled by the tool are:

- Veterinarians and stakeholders can look up their own PRRSV ORF-5 sequence and identify when and where similar sequences have been seen before on a real-time basis.
- New sequences entering the SDRS containing ≤ 6 ambiguities and ≥ 597 base pairs are scanned, and sequences having nucleotide identity < 95% are detected and flagged as new, creating the capability to report trends of PRRSV evolution through years and months/seasons.

The data flow inside the process is simple and intuitive (Figure 1). The web application users can copy and paste their ORF-5 sequence from their diagnostic report into the search box on the website. Once the tool produces output for the input query, relevant details such as closest matching sequence IDs and their similarity percentages are displayed on the browser along with metadata of the matching sequence IDs such as received date, state, and lineage.

Figure 1: A flowchart diagram representing how the SDRS PRRSV BLAST tool works with the SDRS database to provide updated similarity information.

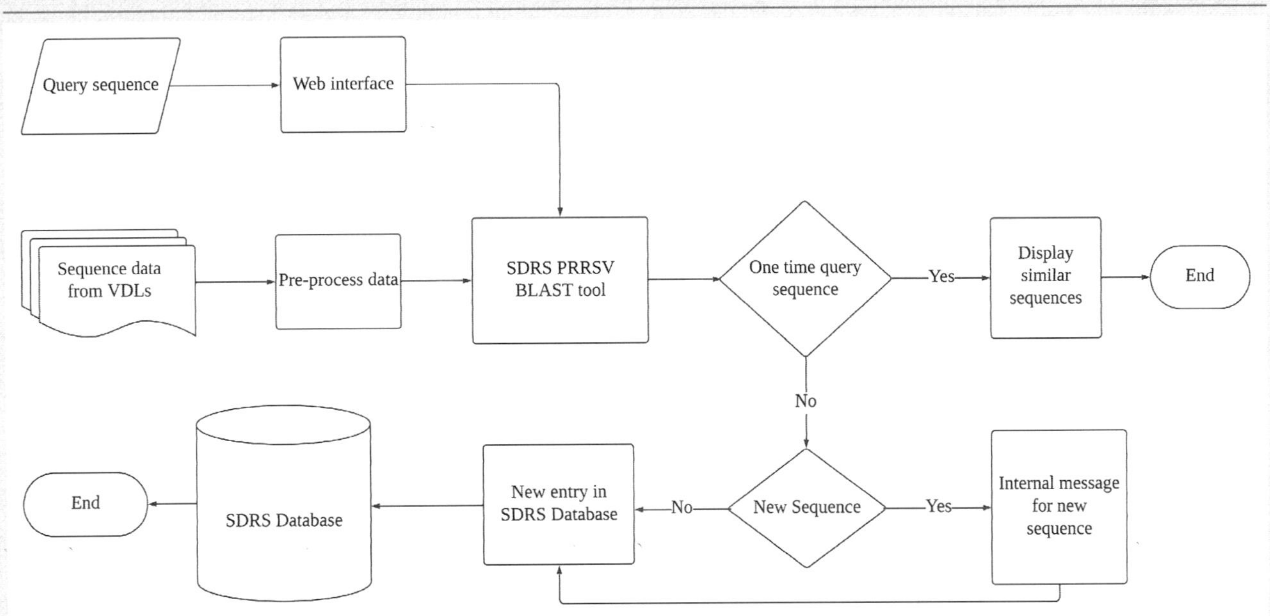

The process was customized for sequences entering the SDRS database. All sequences received by SDRS go through the tool to record the highest similarity before being added to the SDRS database. While building the tool, all available sequences until December 2009 (~20,000 sequences) were first added to the database. These formed the base data for carrying out BLAST data analysis. Every sequence entered in the SDRS by receive date since 2010 (~89,000 sequences) was passed through the tool to find similar sequences. The ID and the percentage similarity of the most similar sequence present in the SDRS database were recorded. The individual sequence was then added to the SDRS database along with the similarity details so that the newly added sequence would also provide a search basis for the next sequence. The described approach is a convenient way to update the database as the next sequence being searched is also compared to the last added sequence. The latter cuts down on the search time for new sequences while simultaneously reducing the reporting time. The database is updated daily depending on the incoming sequences.

The recorded percentage similarity is an important parameter to identify if a sequence should be flagged as new. Sequences having nucleotide identity < 95%, containing ≤ 6 ambiguities and ≥ 597 base pairs were flagged as new sequences.

Result

More than 89,000 sequence data entering the SDRS database since 2010 were observed, and 131 of these were unique sequences having < 95% nucleotide identity with sequences present in the SDRS database. These were considered new sequences. The large number of new sequences detected in 2010 and 2011 could be attributed to fewer sequences in the SDRS database for comparison with the incoming sequences (Figure 2). A decrease in sequences from unknown farm types indicated improvements made in VDLs to capture and record submission data through the years (Figure 2). The PRRSV epidemic years for the PRRSV Lineage 1A RFLP 1-7-4 in 2014 and 2015, and 2021 for the L1C.5 (also known as L1C variant RFLP 1-4-4) had the lowest number of new sequences detected. From the 131 new sequences, 45.8% (n 60) were recovered from sites with no farm type information provided, 37.4% (n 49) from grow-finish farms, 16.8% (n 22) from adult/sow farm types (Figure 2).

Iowa is the state with the leading number of new sequences detected (Figure 3), which could be in conjunction with the state being the leading hog producer state in the country. Minnesota, Indiana, and Illinois are other high pork producer states and followed behind closely in the number of new PRRSV sequence detections. Additionally, detection trends through the years were identified with the number of new sequence detections increased in specific months, with higher numbers being detected during December (n 15) and January (n 14). The numbers go down significantly in the warmer months. From the 131 sequences identified as new 43.51% (n 57) were unclassified lineages (unknown), 9.92% (n 13) as Lineage 5A, 6.1% (n 8) as L1H, 6.1% (n 8) as Lineage 1B. These were followed by 5.34% (n 7) sequences classified as Lineage 1A, 4.58% (n 6) as Lineage 1C-Others, 3.81% (n 5) each as Lineage 1F, Lineage L3 and Lineage 9A, 3.05% (n 4) as Lineage 1E, 2.29% (n 3) as Lineage 1D, 1.53% (n 2) as Lineage L2 and L6 each. Only one sequence (0.76%) from the Lineages 1C.1, 1C.2, 8C, 8D, 8E and 9C were detected since 2010.

Figure 2: Farm types where the new sequences have been detected since 2010. In the last 5 years, 61.11% (22 of 36) of new sequences detected come from grow-finish farms.

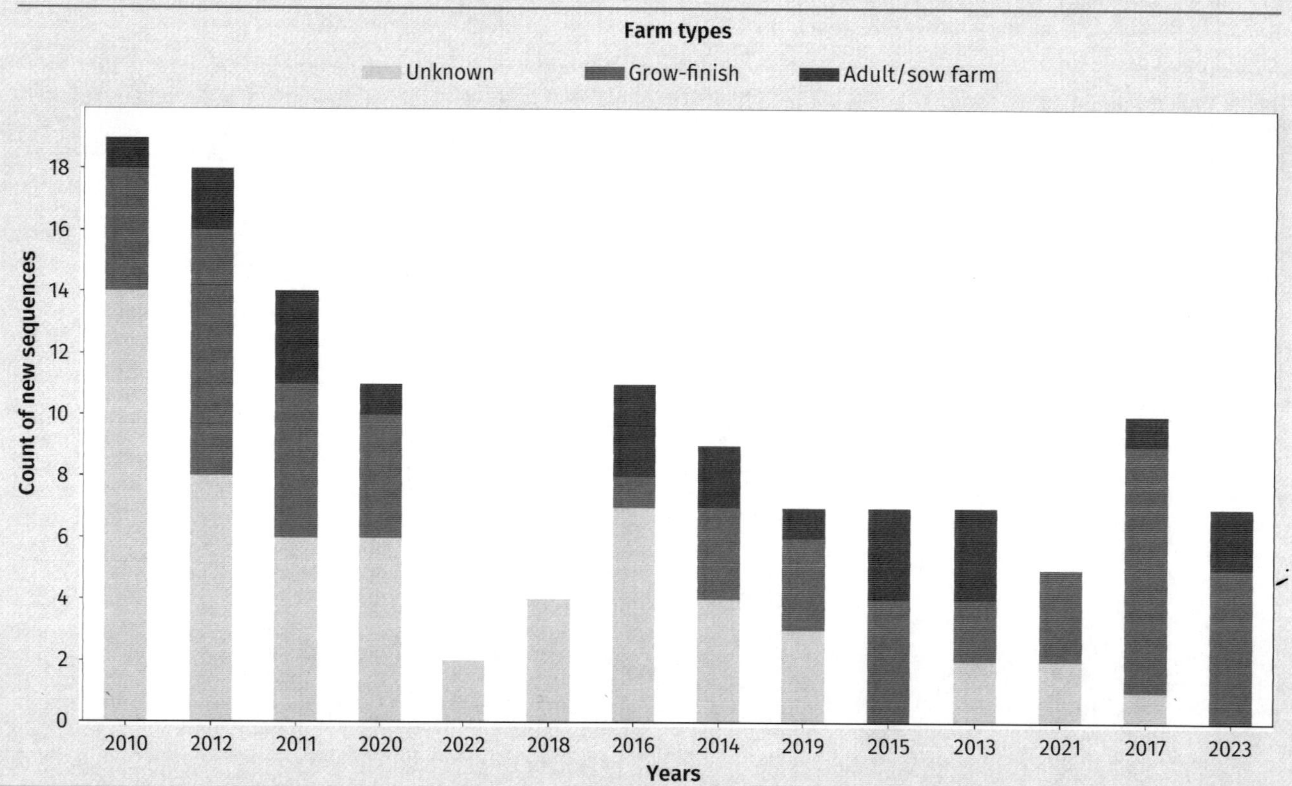

Figure 3: State wise distribution of new sequence count from 2010 till August 2023.

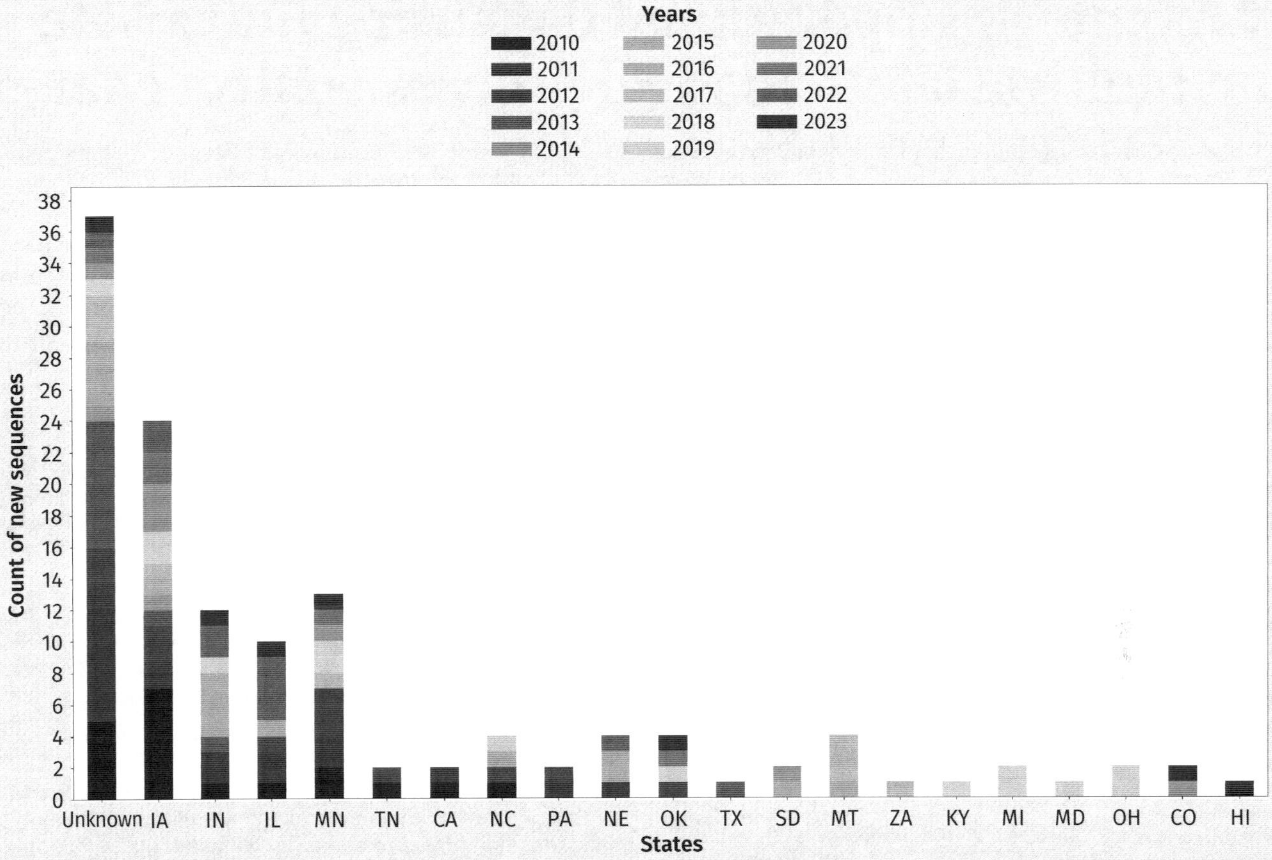

Discussion

The SDRS PRRSV BLAST tool was successfully implemented and has been made available on the SDRS website (**https://www.fieldepi.org/sdrs**) to any veterinarians, producers, or stakeholders who want to query their sequence. The developed tool helps stakeholders easily access spatiotemporal information on newly detected sequences. Stakeholders can position their sequences to those shared by this large collaborative network infrastructure to identify if the query sequence has been seen before, where, and when. Further, we BLAST any incoming sequence to check its novelty in the field. Any prospective generated information is incorporated into the SDRS monthly reports to inform the stakeholders about the detection of new sequences and revealed trends. Moreover, the SDRS is readily available to assist in interpreting the tool's findings or related information. Still, additional work is needed to identify the clinical relevance of new sequences and their spread pattern in the swine industry.

References

1. Altschul SF, Gish W, Miller W, Myers EW, Lipman DJ. Basic local alignment search tool. J Mol Biol. 1990 Oct 5;215(3):403-10. **doi: 10.1016/S0022-2836(05)80360-2**

2. Smith TF, Waterman MS. Identification of common molecular subsequences. J Mol Biol. 1981 Mar 25;147(1):195-7. **doi: 10.1016/0022-2836(81)90087-5**

3. Zeller MA, Arendsee ZW, Smith GJD, Anderson TK. classLog: Logistic regression for the classification of genetic sequences. bioRxiv, 2022.08.15.503907; **doi: https://doi.org/10.1101/2022.08.15.503907**

4. Yim-Im W, Anderson TK, Paploski IAD, VanderWaal K, Gauger P, Krueger K, Shi M, Main R, Zhang J. Microbiol Spectr. 2023 Nov 7:e0291623. **doi: 10.1128/spectrum.02916-23**

https://doi.org/10.54846/am2024/8

Early detect PRRSV outbreaks in breeding herds by monitoring operational data using univariate and multivariate statistical process control charts

Mafalda Pedro Mil-Homens[1]; Swaminathan Jayaraman[1]; Kinath Rupasinghe[1]; Chong Wang[1]; Fernanda Dorea[2]; Giovani Trevisan[1]; Daniel Linhares[1]; Derald Holtkamp[1]; Gustavo Silva[1]
[1]Iowa State University, Iowa; [2]National Veterinary Institute, Uppsala, Sweden

Introduction

Porcine reproductive and respiratory syndrome virus (PRRSV) has a big impact on swine production worldwide,[1-3] with losses of around $664 million annually in the United States.[4] PRRSV clinical signs, such as respiratory distress, lethargy, anorexia, abortions, stillborn, weak piglets and mummies ("porcine reproductive and respiratory syndrome (PRRS)| Iowa State University" 2023),[5] affect the productivity indicators. Data obtained from livestock production processes mirrors the biological nature of these processes.[6] With that, despite not being part of the swine production systems' routine,[7] monitoring production data deviations was shown to be helpful in understanding production patterns, the potential introduction of pathogens, such as PRRSV, and the impact this pathogen's introduction has on production.[8,9,11]

Previous studies have employed statistical process control (SPC) in swine, dairy, and cattle production and demonstrated utility in detecting deviations associated with subclinical and clinical diseases in livestock production, showcasing their effectiveness in monitoring production patterns and identifying potential signs of pathogen introduction. Notably, using SPC charts has proven beneficial in scenarios such as monitoring feeding levels in cows during the dry period and monitoring calves' water intake to detect disease signals.[12,13]

This study used two different SPC charts, the cumulative sum (CUSUM) and the multivariate cumulative sum (MCUSUM). CUSUM charts have aided in monitoring deviations in different contexts, such as evaluating water consumption to assess changes in pig health and well-being.[14] MCUSUM is helpful when data with interconnected indicators is analyzed, and the purpose is to identify shifts in the mean of those indicators collectively.[15] These charts have been reported as useful for monitoring early detection of bovine illness.[16]

While univariate SPC charts are the most used in monitoring swine industry productivity indicators,[9,11,17,18], they face limitations in aggregating information from multiple variables and detecting deviations in correlated variables.

To address these challenges, this study explores the application of multivariate control charts allowing a comprehensive overview of the process's behavior by enabling information aggregation. In addition, they are more effective at detecting deviations because they consider the structure of correlations between variables.[19] Therefore, this study aimed to assess the performance of univariate and multivariate control charts in detecting deviations associated with PRRSV outbreaks, utilizing a combination of productivity, diagnostic, and electronic sow feed data for a more integrated and nuanced approach to disease surveillance.

Methods

Study design

This retrospective observational study was conducted in swine breeding herds in the United States. Sixteen swine breeding herds were included from March 2022 to May 2023, totaling 14 PRRSV outbreaks. Processing fluids were collected weekly and tested by quantitative PRRSV reverse transcriptase polymerase chain reaction (RT-qPCR) to define a PRRSV outbreak, and the prospective PRRSV status for each herd was determined based on these results.

Eligibility criteria

The eligibility criteria for selecting herds included: **a)** availability of weekly PRRS status; **b)** willingness of producer to share weekly production data; **c)** continuous breed-to-wean operation; **d)** herd having at least 21 weeks with no PRRS outbreak to create baseline.

Data and outcomes

Three different data sources were obtained and consolidated in this study. **1)** The weekly sow productivity indicators were obtained from the keeping recording systems used by the study participants (PigKnows© and Maximus©). **2)** The feed data was obtained from Gestal© and Maximus©, and received daily or weekly depending on the farm's system. **3)** The diagnostic results were obtained from the Iowa State University Veterinary Diagnostic Laboratory (ISU-VDL).

The sow productivity indicators assessed in this study were the pre-weaning mortality rate, number of abortions, number of dead sows, and percentage of neonatal losses (sum of stillborn and mummies). The feed data, indicator was the weekly number of off-feed events (the number of times the feeding machine did not read feed consumption in 24 hours for a specific sow). Processing fluids were collected weekly in all farms to determine the PRRSV status. At the VDL, the samples were aggregated in a weekly pool and tested by RT-qPCR. A positive result in the RT-PCR was considered an outbreak.

The outcomes were the percentage of early detection, and time to detect PRRSV outbreak for each productivity indicator. Early detection was defined as the identifying outbreak before changing the PRRSV status based on processing fluids results by RT-qPCR testing. The time to detect disease was defined as the interval between the week the model detected a deviation, and the week the farm reported a change in PRRSV status based on processing fluids results by RT-qPCR testing.

Statistical analysis and model performance

Two statistical process control methods were used: CUSUM and MCUSUM. The CUSUM model was built with a decision interval of 5 (number of standard errors of the summary statistics at which the cumulative sum is out of control), a shift detection of 1 (measured in standard errors of the summary statistics), a baseline of 21 weeks for the cumulative sum as default in the package qcc() from R. The MCUSUM model was built using the mult.chart() function from package MSQC in R. A target value of h = 3.5, an α = 0.05, and control limits of k = 0.5[15] were used with a baseline of 21 weeks. The performance was calculated by comparing the models' alarms (deviations of the indicators from the control limits) with the results from the RT-qPCR testing using processing fluids. The accuracy, false positives and false negatives were calculated using 2 x 2 tables.

Results

In the univariate model, the weekly number of abortions, followed by PWM%, were the indicators with higher early detection, 79%, and 71%, respectively. Using the number of abortions, PWM%, and dead sows, the CUSUM model detected signs associated with PRRSV outbreaks up to eight weeks before the sow farm reported a change in PRRSV status using PF. Overall, the time to detect alarms was 4.4 weeks before the PF turned RT-PCR-positive. The FN and FP were 1.0% and 16.8%, respectively. The multivariate model with higher detection (100%) used the number of abortions and the number of dead sows. It was followed by the models using PWM% with neonatal losses (93%), dead sows and off-feed and dead sows and PWM% (detection of 89% for both), with an average early detection of 4.3 weeks (Table 1).

Conclusion

In conclusion, this study emphasizes the significance of integrating multiple data sources, including productivity and feed indicators, for the early detection sign potentially associated with PRRSV outbreaks. The findings underscore the limitations of relying solely on diagnostic methods, as PRRSV-associated signs were observed weeks before formal diagnosis in breeding herds using processing fluids. The study highlights the effectiveness of multivariate models, specifically MCUSUM, in early PRRSV detection. The combination of PWM% and Neonatal losses% in MCUSUM exhibited a 93% early detection rate, demonstrating the utility of considering multiple sow productivity indicators. In summary, analyzing productivity indicators can aid decision-making for producers and veterinarians, and combining different indicators in multivariate models proved to be highly sensitive and specific in early detection. Early investigation allows the decision-makers to intervene, plan disease control and elimination, and avoid spreading disease to other sites.

References

1. A. K. Baysinger, "The effect of PRRSV on reproductive parameters in swine herds," *Swine Health Prod.*, vol. 5, no. 5, 1997.

2. N. Nieuwenhuis, T. F. Duinhof, and A. van Nes, "Economic analysis of outbreaks of porcine reproductive and respiratory syndrome virus in nine sow herds," *Vet. Rec.*, vol. 170, no. 9, pp. 225–225, 2012, **doi: 10.1136/vr.100101**

3. A. Zhang, J. R. Young, S. Suon, K. Ashley, P. A. Windsor, and R. D. Bush, "Investigating the financial impact of porcine reproductive and respiratory syndrome on smallholder pig farmers in Cambodia," *Trop. Anim. Health Prod.*, vol. 49, no. 4, pp. 791–806, Apr. 2017, **doi: 10.1007/s11250-017-1264-1**

Table 1: Results from the CUSUM and MCUSUM models. Summary of results showing early detection, false negatives (%) and false positive (%) for the different productivity indicators.

MODEL	Production indicators	Early detection	Avg. early detection *	Range early detection (weeks)	FN%*	FP%*
CUSUM	Abort	79% (11/14)	-4.4	-8 to -1	1.0	16.8
	Dead sows	64% (9/14)	-4.5	-8 to -1	2.0	17.9
	PWM%	71% (10/14)	-4.4	-8 to -1	1.7	28.7
	Neonatal losses %	36% (5/14)	-4.5	-8 to -1	7.9	14.6
	Off-feed events	22% (2/9)	-4.5	-8 to -1	14.8	2.8
MCUSUM	Abort & dead sows	86 (12/14)	-4.3	-8 to -1	0.0	14.3
	PWM & neonatal losses	67 (6/9)	-3.8	-8 to -1	7.6	9.2
	Abort & off-feed events	93 (13/14)	-4.3	-8 to -1	0.5	9.0
	Dead sows & off-feed	57 (5/9)	-3.8	-8 to -1	0.8	5.4
	Dead sows & PWM	64 (9/14)	-4.3	-8 to -1	1.4	10.6
	Dead sows & neonatal losses	71 (10/14)	-4.2	-8 to -1	1.1	15.9
	Abort & PWM	71 (10/14)	-4.2	-8 to -1	2.7	9.9
	Abort & neonatal losses	79 (11/14)	-4.1	-8 to -1	1.2	12.5
	Off-feed & PWM	78 (7/9)	-4.0	-8 to -1	0.8	7.2
	Off-feed & neonatal losses	78 (7/9)	-4.3	-8 to -1	1.4	10.9

* Avg early detection = Average of early detection in weeks; FN% = False negatives (%); FP% = False positives (%)

4. D. J. Holtkamp, "Assessment of the economic impact of porcine reproductive and respiratory syndrome virus on United States pork producers," *J. Swine Health Prod.*, vol. 21, no. 2, pp. 72–84, Mar. 2013.

5. "Porcine Reproductive and Respiratory Syndrome (PRRS) | Iowa State University." Accessed: Nov. 14, 2023. [Online]. Available: **https://vetmed.iastate.edu/vdpam/FSVD/swine/index-diseases/porcine-reproductive**

6. K. Mertens, E. Decuypere, J. D. Baerdemaeker, and B. D. Ketelaere, "Statistical control charts as a support tool for the management of livestock production," *J. Agric. Sci.*, vol. 149, no. 3, pp. 369–384, Jun. 2011, **doi: 10.1017/S0021859610001164**

7. A. De Vries and J. K. Reneau, "Application of statistical process control charts to monitor changes in animal production systems1," *J. Anim. Sci.*, vol. 88, no. suppl_13, pp. E11–E24, Apr. 2010, **doi: 10.2527/jas.2009-2622**

8. C. Cornou, J. Vinther, and A. R. Kristensen, "Automatic detection of oestrus and health disorders using data from electronic sow feeders," *Livest. Sci.*, vol. 118, no. 3, pp. 262–271, Nov. 2008, **doi: 10.1016/j.livsci.2008.02.004**

9. G. S. Silva, M. Schwartz, R. B. Morrison, and D. C. L. Linhares, "Monitoring breeding herd production data to detect PRRSV outbreaks," *Prev. Vet. Med.*, vol. 148, pp. 89–93, Dec. 2017, **doi: 10.1016/j.prevetmed.2017.10.012**

10. W. C. Kayser, G. E. Carstens, K. S. Jackson, W. E. Pinchak, A. Banerjee, and Y. Fu, "Evaluation of statistical process control procedures to monitor feeding behavior patterns and detect onset of bovine respiratory disease in growing bulls," *J. Anim. Sci.*, vol. 97, no. 3, pp. 1158–1170, Mar. 2019, **doi: 10.1093/jas/sky486**

11. C. Merca *et al.*, "Veterinary syndromic surveillance using swine production data for farm health management and early disease detection," *Prev. Vet. Med.*, vol. 205, p. 105659, Aug. 2022, **doi: 10.1016/j.prevetmed.2022.105659**

12. R. Y. Murphy, T. Osaili, B. L. Beard, J. A. Marcy, and L. K. Duncan, "Application of statistical process control, sampling, and validation for producing Listeria monocytogenes-free chicken leg quarters processed in steam followed by impingement cooking," *Food Microbiol.*, vol. 22, no. 1, pp. 47–52, Jan. 2005, **doi: 10.1016/j.fm.2004.04.008**

13. W. A. Knauer, S. M. Godden, A. Dietrich, D. M. Hawkins, and R. E. James, "Evaluation of applying statistical process control techniques to daily average feeding behaviors to detect disease in automatically fed group-housed preweaned dairy calves," *J. Dairy Sci.*, vol. 101, no. 9, pp. 8135–8145, Sep. 2018, **doi: 10.3168/jds.2017-13947**

14. T. N. Madsen and A. R. Kristensen, "A model for monitoring the condition of young pigs by their drinking behaviour," *Comput. Electron. Agric.*, vol. 48, no. 2, pp. 138–154, Aug. 2005, **doi: 10.1016/j.compag.2005.02.014**

15. R. Crosier, "Multivariate Generalizations of Cumulative Sum Quality-Control Schemes," *TECHNOMETRICS*, vol. 30, no. 3, pp. 291–303, 1988.

16. C. Faverjon, L. P. Carmo, and J. Berezowski, "Multivariate syndromic surveillance for cattle diseases: Epidemic simulation and algorithm performance evaluation," *Prev. Vet. Med.*, vol. 172, p. 104778, Nov. 2019, **doi: 10.1016/j.prevetmed.2019.104778**

17. C. Rademacher, "Use of statistical process control in finishing records," *J. Swine Health Prod.*, vol. 12, no. 3, pp. 158–159, 2004.

18. D. H. Baum, "Statistical process control methods used to evaluate the serologic responses of pigs infected with three *Salmonella* serovars," *J. Swine Health Prod.*, vol. 13, no. 6, pp. 304–313, Nov. 2005.

19. M. Rogalewicz, "Some Notes on Multivariate Statistical Process Control," *Manag. Prod. Eng. Rev.*, 2012, **doi: 10.2478/v10270-012-0036-7**

https://doi.org/10.54846/am2024/9

Ear-vein blood swabs, oral swabs, and nasal swabs can be used with precision for PRRSV surveillance in weaning-age pigs

Onyekachukwu H. Osemeke; Giovani Trevisan; Gustavo Silva; Phillip C. Gauger; Daniel C. L. Linhares, VDPAM
College of Veterinary Medicine, Iowa State University, Iowa

Introduction

The porcine reproductive and respiratory syndrome virus (PRRSV) continues to challenge swine production globally.[1-4] Efficient PRRSV monitoring and surveillance (MOSS) is crucial for the early and accurate detection of PRRSV outbreaks and is pivotal to implementing a timely disease response program. MOSS focusing on the weaning-age pig population is the industry-accepted best strategy for investigating active PRRSV infections in breeding herds,[5] however, very few sample types have been characterized for surveilling this subpopulation. Some practitioners use swab samples to surveil weaning-age piglets for PRRSV; however, there needs to be a basis for ascertaining appropriate sample sizes and a context for interpreting laboratory results. The objective of this study was to compare the PRRSV reverse transcription-quantitative polymerase chain reaction (RT-qPCR) detection rates in serum samples to detection rates in oral swabs (OSw), nasal swabs (NSw) and ear-vein blood swabs (ESw) from naturally infected weaning age pigs under field conditions.

Methods

This study was designed to estimate the diagnostic performance (sensitivity and specificity) of the swab sample types compared to serum and to propose appropriate sample sizes for using these sample types for PRRSV MOSS. Three eligible PRRSV-positive herds in the Midwestern USA were selected for this study. Six hundred sixty-six pigs were sampled from 55 litters for serum, NSw, ESw, and OSw. All samples were tested for PRRSV RNA by RT-qPCR. Each swab sample type's sensitivity (Se) and specificity (Sp) were evaluated using serum as the reference. All statistical analyses were done on R statistical software.

Results

The Se and Sp were 83.3% and 99.7% for ESw, 83.3% and 98.8% for OSw, and 75.0% and 99.3% for NSw. The minimum number of piglets to be sampled to detect ≥ 1 PRRSV infected piglet at different prevalence scenarios are given in Table 1; for example, the minimum number of piglets to be sampled for serum, ESw, NSw, and OSw, to be ≥ 95% confident of detecting ≥ 1 infected piglet when PRRSV prevalence is ≥ 10% is 30, 36, 40, and 36, respectively.

Discussion and conclusion

Swine practitioners who use swab samples for surveillance now have contexts for interpreting laboratory data obtained from such investigations and a basis for determining appropriate sample sizes. While individual pig samples have their use in PRRSV surveillance, population-based sample types remain the most cost-efficient option for ascertaining the presence of PRRSV in breeding herds.

Table 1: Sample size estimates for serum (Srm), ear-vein blood swabs (ESw), nasal swabs (NSw), and oral swabs (OSw) assuming a ≥ 95% detection confidence, a perfect RT-qPCR test, and sampling without replacement

	Sample type			
Prevalence (%)	**Srm**	**ESw**	**NSw**	**OSw**
1	259	312	346	312
3	96	115	127	115
5	59	71	78	71
10	30	36	40	36
20	15	18	20	18

References

1. Balka, G. *et al.* Genetic diversity of PRRSV 1 in Central Eastern Europe in 1994-2014: Origin and evolution of the virus in the region. *Sci Rep* **8**, 7811 (2018).

2. Guo, Z., Chen, X. X., Li, R., Qiao, S. & Zhang, G. The prevalent status and genetic diversity of porcine reproductive and respiratory syndrome virus in China: A molecular epidemiological perspective. *Virol J* **15**, 1–14 (2018).

3. Sanhueza, J. M., Vilalta, C., Corzo, C. & Arruda, A. G. Factors affecting Porcine Reproductive and Respiratory Syndrome virus time-to-stability in breeding herds in the Midwestern United States. *Transbound Emerg Dis* **66**, 823–830 (2019).

4. Holtkamp, D. J. *et al. Assessment of the economic impact of porcine reproductive and respiratory syndrome virus on United States pork producers. Journal of Swine Health and Production* vol. 21 **http://lib.dr.iastate.edu/econ_las_pubs/50 (2013)**

5. Holtkamp, D. J. *et al.* Proposed modifications to porcine reproductive and respiratory syndrome virus herd classification. *J Swine Health Prod* 29, 261–270 (2021).

https://doi.org/10.54846/am2024/10

Environmental viability of *Mycoplasma hyopneumoniae*

Cassidy Cordon[1]; Albert Canturri[1,2]; Robert Evelsizer[3]; Maria Pieters[1,2,4]
[1]Department of Veterinary Population Medicine, College of Veterinary Medicine, University of Minnesota, St. Paul, Minnesota; [2]Veterinary Diagnostic Laboratory, University of Minnesota, St. Paul, Minnesota; [3] Elanco Animal Health Inc., Greenfield, Indiana; [4]Swine Disease Eradication Center, College of Veterinary Medicine, University of Minnesota

Introduction

Mycoplasma hyopneumoniae is the causative agent of enzootic pneumonia and one of the primary pathogens of the porcine respiratory disease complex, which contribute to economic production losses in the swine industry.[1] It has been proposed that due to *M hyopneumoniae* lacking a cell wall, it does not survive long in the environment, hence eliminating the bacterium via cleaning and disinfecting can be easily accomplished. However, Browne et al [2] performed a study *in vitro,* which resulted in *M hyopneumoniae* detected on surfaces for up to eight days, suggesting for the bacterium to survive outside the respiratory tract. Additionally, past studies by both Friis[5] and Goodwin[4] have shown the *in vitro* survival of *M hyopneumoniae.* Previous *M hyopneumoniae* studies have used culture or PCR methods to verify the presence of the bacterium on surfaces *in vitro,* however these methods have specific limitations. Isolation and culture of *M hyopneumoniae* is known to be difficult due to its fastidious requirements and slow growth, it is often overgrown by more fastidious bacteria such as *M hyorhinis.*[1] The PCR method, which targets DNA to detect *M hyopneumoniae,* while its highly sensitive, does not detect viability of the bacterium, thus detection does not differentiate between live and dead cells.[3] To combat these diagnostic limitations a new PCR method, reverse-transcription PCR (RT-PCR) was recently validated which targets messenger RNA (mRNA).[3] In contrast to DNA, mRNA is short-lived and their production is ceased after the cell dies, allowing for not only detection but viability of the bacterium to be determined.[3] Therefore, the objective of this study was to determine the viability of *M hyopneumoniae* in environmental samples over time by employing highly accurate diagnostic methods.

Methods

The study was performed at the Veterinary Isolation Facilities, of the University of Minnesota St. Paul Campus, after IACUC approval. The following materials: water (20 mL), feed (8g), dust (1.3g), plexiglass (20 cm^2), stainless steel (20 cm^2), cardboard (20 cm^2), and PVC (20 cm^2), were placed into an isolator, which was aerosolized over a 20-minute period, with 20 mL of lung homogenate containing $1x10^6$ CCU/mL of *M hyopneumoniae* strain 232, using a nebulizer. The isolators were hermetically closed, allowing the infective droplets to deposit on materials. Materials were sampled prior to and at one hour, one, two, three, four, and five days post-aerosolization of *M hyopneumoniae.* All samples were stored in RNALater and tested for genetic material via PCR.[6] Samples that tested positive for PCR were processed for additional viability testing via RT-PCR.[3] Results were expressed in Ct values, where ≤ 40 was considered a positive result. Results were analyzed using descriptive analysis.

Results and discussion

Mycoplasma hyopneumoniae was detected by PCR sporadically and inconsistently in all materials/surfaces starting at1 hour post-aerosolization. In addition, all materials (except stainless steel and PVC) had *M hyopneumoniae* detected through 4 days post-aerosolization. Ct values were high (> 30) in all samples, which indicated a relatively low bacterial load of *M hyopneumoniae* was recovered from materials. Results from this investigation were similar to the study performed by Browne et al[2], which showed less detection of *M hyopneumoniae* in stainless steel compared to dust and glass. However, in the Browne et al[2] study most detection on materials occurred at an incubation of 4°C. Additionally, the materials in this study that tested positive for PCR were tested for viability via RT- PCR. Viable *M hyopneumoniae* was not detected on any samples via RT-PCR. It is important to note that a potential limitation to the study was that some materials were placed closer to the nebulizer, therefore they may have received more infectious droplets compared to materials farther from the aerosolization point. Nevertheless, we anticipated all the materials would receive sufficient bacterial load due to the small-scale dimensions of the isolator in combination with the 20 minutes of aerosolization time, allowing for droplets to disperse across all surfaces. The outcome of this study sheds light on the type of materials and surfaces that favor the environmental persistence and detection of *M hyopneumoniae.* These results can directly influence cleaning and disinfection practices at the farm level, to support disease eradication efforts. Further studies to identify the possibility of indirect transmission via persistence of *M hyopneumoniae* on environmental surfaces would be beneficial.

References

1. Pieters, M. & Maes, D. (2019). Mycoplasmosis . In *Diseases of Swine* (11th ed.) Zimmerman et al. (editors). Wiley. (pp. 863-871).

2. Browne, C., Loeffler, A., Holt, H. R., Chang, Y. M., Lloyd, D. H., & Nevel, A. (2017). *Letters in Applied Microbiology, 64*(1), 2–7.

3. Canturri, A., Galina-Pantoja, L., Vonnahme, K., & Pieters, M. Distinguishing viable from non-viable *Mycoplasma hyopneumoniae* by PCR. *Proceedings of the International Pig Veterinary Society meeting.* (2022). Rio de Janerio, Brazil.

4. Goodwin, R.F. (1972). The Survival of *Mycoplasma suipneumoniae* in Liquid Medium, on Solid Medium, and in Pneumonic Tissue. In *Research in Veterinary Science* (Vol. 13, pp 203-204).

5. Friis, N.F. (1973). Resistance of porcine mycoplasmas to drying. *Acta Vet Scand* (14, 489-491).

6. Strait, E., Madsen, M., Minion, C., et al. (2008). Real-time PCR assays to address genetic diversity among strains of *Mycoplasma hyopneumoniae. Journal of Microbiology.* 46 (8): 2491-8.

https://doi.org/10.54846/am2024/11

Confirming day zero in a *Mycoplasma hyopneumoniae* herd closure

Amanda Sponheim[a,b]; Julio Alvarez[a,1,2]; Eduardo Fano[b,3]; Albert Rovira[a,c]; Emily McDowell[d]; Elise Toohill[e,4]; Laura Dalquist[f]; Maria Pieters[a,c,g]
[a]Department of Veterinary Population Medicine, College of Veterinary Medicine, University of Minnesota, St. Paul, Minnesota; [b]Boehringer Ingelheim Animal Health USA Inc., Duluth, Georgia; [c]Veterinary Diagnostic Laboratory, College of Veterinary Medicine, University of Minnesota, St. Paul, Minnesota; [d]Pipestone Veterinary Services, Pipestone, Minnesota, USA; [e]The Maschhoffs, Carlyle, Illinois; [f]Swine Vet Center, St. Peter, Minnesota; [g]Swine Disease Eradication Center, College of Veterinary Medicine, University of Minnesota, St. Paul, Minnesota
[1]Current affiliation: Centro de Vigilancia Sanitaria Veterinaria VISAVET, Universidad Complutense, Madrid, Spain;
[2]Current affiliation: Departamento de Sanidad Animal, Facultad de Veterinaria, Universidad Complutense, Madrid, Spain;
[3]Current affiliation: Pipestone Veterinary Services, Pipestone, Minnesota;
[4]Current affiliation: Carthage Veterinary Service, Ltd., Carthage, Illinois

Introduction

The welfare and economic impact of *Mycoplasma hyopneumoniae* infections on downstream production has led to the pursuit of control or eradication of the pathogen from swine populations globally.[1,2] Herd closure, medication, and vaccination is a common approach for *M hyopneumoniae* eradication in North America.[3] One of the most important components of a herd closure is the determination of day 0, which is based on the time point at which the last replacement gilts entering the herd were exposed to *M hyopneumoniae* and marks the beginning of the closure.[4] However, no specific diagnostic protocols are available for confirmation of successful exposure to define day 0. Therefore, the objective of this study was to develop sample collection approaches to confirm an entire population has been exposed to *M hyopneumoniae* using two common gilt exposure methods.[5]

Methods

Forty gilts, age 21 to 56 days, were ear-tagged for longitudinal sample collection at five commercial gilt developer units (GDUs) and exposed to *M hyopneumoniae* by natural contact or aerosolization. Study gilts originated from sources known to be negative to major swine pathogens, including *M hyopneumoniae,* and were sampled prior to exposure to confirm negative status, then every two weeks. Blood samples were collected for antibody detection, while laryngeal and deep tracheal secretions and pen based oral fluids were collected for PCR, and sampling continued until at least 85% of samples were positive by PCR. Sample collection schemes were developed to confirm population *M hyopneumoniae* infection following purposeful exposure. The Freecalc sample size for a finite population and specified cut-point number of positives algorithm function from the Surveillance package was used to perform the calculations in RStudio. A key modification was made to the Freecalc algorithm: non-infected animals were the target animal category of interest since the objective was to determine the sample size that would yield a high probability of detection of exposed herds in which the proportion of non-infected gilts was above a given threshold.

Results and discussion

Results indicated that detection of *M hyopneumoniae* varied based on sample type. Oral fluids showed the lowest detection in groups of gilts detected positive by other sample types. Detection by PCR in deep tracheal secretions was higher than in laryngeal secretions. Seroconversion to and PCR detection of *M hyopneumoniae* on oral fluids were delayed compared to PCR detection at the individual level. By two weeks post-exposure, at least 85% of study gilts in three GDUs exposed by aerosolization were PCR positive in deep tracheal secretions. Natural contact exposure resulted in 22.5% of study gilts becoming PCR positive by two weeks post-initial exposure, 61.5% at four weeks, and 100% at six weeks on deep tracheal secretions. Deep tracheal secretions required the lowest number of gilts to sample for the earliest detection compared to all other samples evaluated. As observed in one of the GDUs using aerosolization, demonstration of failure to expose gilts to *M hyopneumoniae* allowed for early intervention in the exposure protocol and delay of day 0, for accurate timing of the eradication protocol. Sampling guidelines proposed in this study can be used for verification of *M hyopneumoniae* infection of gilts following exposure to determine Day 0 of herd closure.

References

1. Dykhuis Haden, C., Painter, T., Fangman, T., Holtkamp, D., 2012. Assessing production parameters and economic impact of swine influenza, PRRS and *Mycoplasma hyopneumoniae* on finishing pigs in a large production system. *Proceedings of the American Association of Swine Veterinarians Meeting.* Orlando, FL., pp. 75-76.

2. Silva, G., Yeske, P., Morrison, R., Linhares, D., 2019. Benefit-cost analysis to estimate the payback time and the economic value of two *Mycoplasma hyopneumoniae* elimination methods in breeding herds. *Prev. Vet. Med.* 168, 95-102.

3. Holst, S., Yeske, P., Pieters, M., 2015. Elimination of *Mycoplasma hyopneumoniae* from breed-to-wean farms: A review of current protocols with emphasis on herd closure and medication. *J. Swine Health Prod.* 23, 321-330.

4. Yeske, P., Betlach, A., Holst S, Pieters M., 2020. Eradication of *Mycoplasma hyopneumoniae* from pig herds, in: Maes, D., Sibila, M., Pieters, M. (Eds.), *Mycoplasmas in Swine.* 1st ed. Acco Netherland, The Netherlands, pp. 229-245.

5. Sponheim, A., Alvarez, J., Fano, E., Rovira, A., McDowell, E., Toohill, E., Dalquist, L., Pieters, M. 2023. A diagnostic approach to confirm *Mycoplasma hyopneumoniae* "Day zero" for pathogen eradication. *Prev. Vet. Med.* 221, 106057. **https://doi.org/10.1016/j.prevetmed.2023.106057**

https://doi.org/10.54846/am2024/12

Probability of influenza A virus RNA detection at different pooling levels for commonly used sample types in breeding herds

D. Moraes[1]; O. H. Osemeke[1]; P. C. Gauger[1]; C. A. Moura[2]; G. Trevisan[1]; G. Silva[1]; D. C. L. Linhares[1]
[1]Department of Veterinary Diagnostic and Production Animal Medicine, Iowa State University, Ames, Iowa,
[2]Iowa Select Farms, Iowa Falls, Iowa

Introduction

Active surveillance of influenza A virus (IAV) in swine populations is essential for understanding the virus epidemiology under field conditions. Recently, family oral fluids (FOF) was described for IAV detection by reverse transcription real-time polymerase chain reaction (RT-rtPCR) with promising results.[1] Another recent study tested the effect of pooling udder wipes (UW) for IAV detection and showed that pooling 3 udder wipes was appropriate for IAV surveillance.[2] However, pooling swine samples for IAV surveillance without negatively impacting sensitivity would depend on the sample type and pool size. Therefore, this study aimed to compare the probability of IAV RT-rtPCR detection (PD) at different levels of pooling in FOF, UW, and nasal wipes (NW) obtained from an endemic swine breeding herd under field conditions.

Materials and methods

Forty-five IAV RT-rtPCR positive samples (15 FOF, 15 UW, and 15 NW) obtained from an IAV-positive herd from a previous study[1] were pooled at different levels (undiluted, 1/3, 1/5, and 1/10) with RT-rtPCR negative IAV samples obtained from herds that had laboratory evidence of being IAV negative for at least 3 months. All dilution levels for each sample type was tested in replicates of 6.[3] For analysis, samples were categorized into three groups based on the undiluted replicates' mean cycle threshold (Ct) value: Category A - Ct < 30; Category B - Ct between 30 - 34; Category C - Ct between 34 - 38. A probit regression model was used to estimate the PD for each pooling level for each sample type using the *brglm* package in R statistical software.[4] In addition, data from the Iowa State University Veterinary Diagnostic Laboratory from 2018 to 2022 were analyzed to understand the cycle threshold (Ct) value distribution by category (A, B, or C) for the sample types FOF, UW, and NW.

Results

For FOF and UW, the probability of IAV detection in Ct Categories A and B did not decrease when the dilution level increased to 1/10. In Category C for both sample types, the PD was observed to decrease as the dilution level increased, reaching a value of 18% at 1/10 dilution (Table 1). For NW, the PD remained constant despite dilution level for Category A, however it decreased with increasing dilution levels for Categories B and C (Table 1). Based on data from the Iowa State University Veterinary Diagnostic Laboratory from 2018 to 2022, 77% of all IAV RT-rtPCR positive oral fluid samples from breeding herds were from Categories A and B, while 23% were from Category C. For UW, Categories A and B contained 67% of the positive samples, with the remaining 33% in Category C. NW and nasal swabs were similar to oral fluid samples with 75% in Categories A and B and 25% in Category C.

Discussion and conclusion

This study assessed the pooling effect on the PD between different IAV surveillance sample types. These results suggest that about 70% to 80% of the time, the Ct values of all three sample types are within Categories A and B; practitioners should therefore consider pooling samples to improve coverage on a fixed testing budget, testing more samples with the same budget. Pooling population-based samples (ie, udder wipes, family oral fluids) allows cost reduction and potential gains on the probability of detection by increasing the coverage of pigs, pens, rooms, and/or sites sampled. Providing these comparative probabilities on IAV detection between different sampling approaches allows veterinarians and producers to make informed decisions on IAV monitoring practices in swine populations, especially in breeding farms with low prevalence of IAV (< 15%).

Table 1: IAV RNA probability of detection in family oral fluids, udder wipes, and nasal wipes with differing undiluted sample Ct values and dilution levels

Dilution level	Family oral fluids			Udder wipes			Nasal wipes		
	A Ct < 30	B Ct 30-34	C Ct 34-38	A Ct < 30	B Ct 30-34	C Ct 34-38	A Ct < 30	B Ct 30-34	C Ct 34-38
Undiluted	0.99[a]	0.99[a]	0.95[b]	0.99[a]	0.99[a]	0.99[c]	0.99[a]	0.99[b]	0.95[b]
1/3	0.99[a]	0.99[a]	0.83[b]	0.99[a]	0.99[a]	0.76[bc]	0.99[a]	0.99[b]	0.79[bc]
1/5	0.99[a]	0.99[a]	0.44[a]	0.99[a]	0.99[a]	0.57[b]	0.99[a]	0.73[ab]	0.50[ab]
1/10	0.99[a]	0.99[a]	0.18[a]	0.99[a]	0.99[a]	0.18[a]	0.99[a]	0.40[a]	0.21[a]

[abc] Least-square means within a column with different letters are statistically different ($P \leq .05$).

References

1. D. C. A. Moraes et al. Use of Family Oral Fluid to detect Influenza A Virus in the breeding herd. 54th AASV Annual Meeting 2023, United States. **https://doi.org/10.54846/am2023/125**

2. de Lara, Anne C., et al. "Effect of pooling udder skin wipes on the detection of influenza A virus in preweaning pigs." Journal of Veterinary Diagnostic Investigation 34.1 (2022): 133-135.

3. Osemeke, Onyekachukwu H., et al. "Effect of pooling family oral fluids on the probability of PRRSV RNA detection by RT-rtPCR." Preventive Veterinary Medicine 206 (2022): 105701.

4. Team, R.C., 2021. R: A language and environment for statistical computing. Vienna, Austria: R Foundation for Statistical Computing; 2020.

https://doi.org/10.54846/am2024/13

Comparison of fecal diagnostic tests for the detection of monophasic *Salmonella enterica* serovar I 4,[5],12:i:- and their association with clinical signs

N. Resende-De-Macedo[1], DVM, MS, PhD; B. Munguia- Ramirez[1], DVM, MS; B. Armenta Levya[1], DVM MS; N. C. Twu[1], DVM, MS; K. Sett[1]; R. Musco[1]; L. Gimenez-Lirola[1], MS, PhD; D. Baum[1], MS, DVM, PhD; A. Griffith[2]; R. Griffith[2], DVM, MS, PhD; R. Saltzman[3], DVM; K. Crawford[3], DVM, MS; K. Schwartz[1], DVM, MS; S. Bearson[4], PhD; F. Leite[5], DVM, MS, PhD
[1]Iowa State University Veterinary Diagnostic Laboratory, College of Veterinary Medicine, Ames, Iowa; [2]Department of Veterinary Microbiology and Preventative Medicine, College of Veterinary Medicine, Ames, Iowa; [3]VRI-AMVC, Ames, Iowa; [4]USDA-ARS, Ames, Iowa; [5]Boehringer Ingelheim Animal Health USA Inc., Duluth, Georgia

Introduction

Salmonella infection of swine can result in subclinical infection from a broad range of *Salmonella* serotypes that can potentially contaminate pork products and pose a threat to human health. *Salmonella* infection can also lead to clinical disease in pigs.[1] The most prevalent *Salmonella* serotype detected from diseased swine in the United States is monophasic *Salmonella enterica* serovar I 4,[5],12:i:-.[2,3] It is important to have an understanding of diagnostic tests available in order to properly diagnose clinical disease and evaluate shedding status. Feces are readily available samples, with testing options of qPCR as well as bacterial culture with or without *Salmonella*-specific enrichment. The purpose of this study was to compare the sensitivity of qPCR and direct plate culture and also to investigate their association with clinical signs.

Materials and methods

Forty pigs were randomly assigned to one of two treatment groups, negative control and *Salmonella* inoculated. Pigs were housed in separate rooms; each room contained 4 pens with 5 pigs each. The *Salmonella* inoculated group received an intranasal dose of 1 x 10^9 *Salmonella* I 4,[5],12:i:-. Fecal samples were collected on the following days post inoculation (dpi) -1, 2, 3, and 7. Samples were tested for the presence of *Salmonella* by **1)** enrichment culture following the method of Shippy et al., 2018 specific for the challenge strain,[4] **2)** routine VDL direct plate culture, using Brilliant Green agar with novobiocin and Hektoen Enteric agar, and **3)** routine VDL qPCR.[2] The enrichment culture for the challenge strain served as positive control to characterize shedding status. All pigs were monitored daily for clinical signs and the presence of diarrhea.

Results

All -1 dpi samples were negative for *Salmonella,* and all negative control group fecal samples remained negative by all methods throughout the study. All (100%, 20/20) inoculated pigs' fecal samples were enrichment culture positive for *Salmonella* on dpi 2, 3 and 7. There were 45% (9/20), 60% (12/20) and 65% (13/20) pigs' fecal samples that were direct plate culture positive for *Salmonella* on dpi 2, 3 and 7, respectively. qPCR detected 5% (1/20) positive fecal sample on dpi 2 and 15% (3/20) positive fecal samples on dpi 3. All fecal samples were qPCR-negative for *Salmonella* on dpi 7. Clinical signs peaked at 3 dpi with 50% (10/20) of pigs presenting diarrhea and was present in 30% (6/20) and 5% (1/20) of pigs in the inoculated group on dpi 2, and 7 respectively. No clinical signs were observed in the negative challenge control group or in any pig prior to challenge. When considering all post inoculation time points together, enrichment culture detected *Salmonella* in 100% (60/60) of samples, direct plate culture in 56% (34/60) of samples and qPCR in 6.7% (4/60) of samples (Table 1). Among these time points, diarrhea was present in 28% (17/60) of pigs at time of sampling. Of these samples, 100% (17/17) were *Salmonella* positive by enrichment culture, 82% (14/17) were positive by direct plate culture and 23% (4/17) were positive by qPCR (Table 1).

Discussion and conclusion

This study compared antemortem tests to detect the presence of *Salmonella*. Enrichment culture results demonstrated that all pigs shed *Salmonella* in all time points post inoculation. Direct plate culture detected much fewer *Salmonella*-positive samples, thereby inadequately reflecting the shedding status of a pig, but it did detect

Table 1: *Salmonella* detection results of different fecal tests among *Salmonella* inoculated pigs

	Diagnostic test		
Sample description	**Enrichment culture**	**Direct culture**	**qPCR**
Fecal samples from pigs with clinical signs at time of sampling	100% (17/17)[a]	82% (14/17)[a]	23.5% (4/17)[b]
All fecal samples	100% (60/60)[a]	56% (34/60)[b]	6.7% (4/60)[c]

Results represent number of positive samples/total number of samples. Different superscript letters within a row indicate statistical significance ($P < .05$).

Salmonella in most cases when diarrhea was present. qPCR results had very poor agreement with both shedding and clinical status. This study presents important considerations when interpreting and requesting routine fecal diagnostic tests for *Salmonella*.

References

1. Griffith RA, Carlson SA, Krull AC. Salmonellosis. *Diseases of Swine*. 2019.

2. Naberhaus SA et al., Emergence of *Salmonella* enterica serovar 4,[5],12:i:- as the primary serovar identified from swine clinical samples and development of a multiplex real-time PCR for improved *Salmonella* serovar-level identification. *J Vet Diagn Invest*. 2019 Nov;31(6):818-827.

3. Mainenti M. Salmonellosis in pigs: an update based on the ISU VDL diagnostic cases from 2012 to 2022. 2023 ISU James D. McKean Swine Disease Conference.

4. Shippy et al. Porcine Response to a Multidrug-Resistant Salmonella enterica serovar I 4,[5],12:i:- Outbreak Isolate. *Foodborne Pathog Dis*. 2018 May;15(5):253-261.

https://doi.org/10.54846/am2024/14

Characterization of hemolytic *E coli* and antimicrobial susceptibility from ISU-VDL submitted cases from 2010 to 2022

Rodrigo C. Paiva; Eric Burrough; Pablo Piñeyro; Marcelo Almeida
Department of Veterinary Diagnostic and Production Animal Medicine, Veterinary Diagnostic Laboratory, Iowa State University, Ames, Iowa

Introduction

Diarrhea due to enterotoxigenic *Escherichia coli* (ETEC) remains one of the main enteric challenges to US swine. The presence of different virulence factors has been associated with economic losses due increase mortality and treatment costs and decreased weight gain. Data from the Iowa State University Veterinary Diagnostic Laboratory (ISU-VDL) has shown a high prevalence of F4 (K88) and F18 isolates over the last twelve years. In addition, antibiotic susceptibility has changed during this period. Therefore, this study aimed to characterize changes over time in ETEC strains isolated at ISU-VDL based on fimbriae genotyping and antimicrobial susceptibility to provide comprehensive information to swine veterinarians.

Materials and methods

A database containing 3,240 *E coli* isolates recovered from ISU-VDL diagnostic cases associated with an ETEC disease diagnostic code (Dx-code - representing confirmation of disease by a pathologist based on individual case history, pathological examination, and laboratory testing) was analyzed. Laboratory testing included bacteriological culture, genotypification based on fimbriae gene detection by PCR (F4, F18, F41, F5, AIDA, EAEA), and antimicrobial susceptibility testing.

Results

Isolates with F18 and F4 fimbriae gene represented the most common adhesins; (F18: 2,211 [68%]; F4: 979 [30%]). In contrast, 2% of the isolates contained other fimbrial genes (EAEA, F5, F41, and AIDA). An increased number of cases associated with F18 fimbria has been observed since 2019 (Figure 1).

The proportion of isolates susceptible to enrofloxacin dramatically decreased for F4 (from 2014 to 2022) and F18 (from 2017 to 2022), respectively (Figures 2 and 3). There was no variation in the proportion of susceptibility to ceftiofur isolates. The proportion of isolates susceptible to florfenicol, gentamicin, neomycin, sulfadimethoxine, and trimethoprim-sulphamethoxazole decreased on F18 positive isolates since 2017 (Figure 3).

Figure 1: Occurrence of F4 and F18 *E coli* cases (2010-2022).

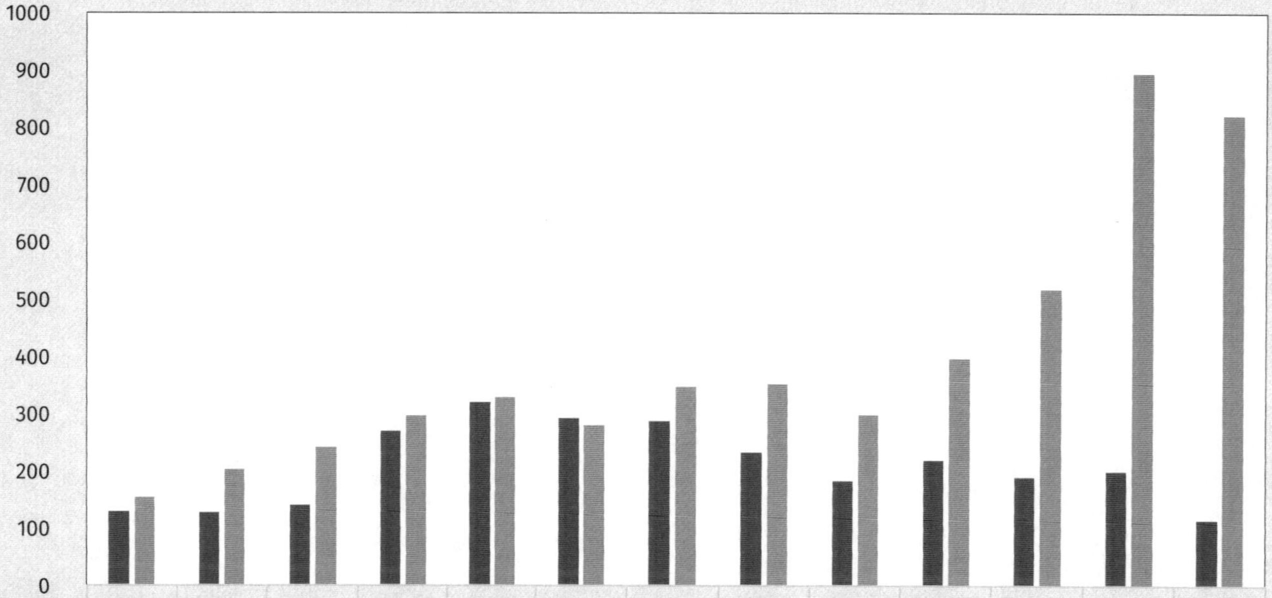

	2010	2011	2012	2013	2014	2015	2016	2017	2018	2019	2020	2021	2022
F4	128	126	140	269	319	292	287	232	182	218	190	199	115
F18	154	204	240	297	327	280	347	351	299	398	518	895	822

Figure 2: Antimicrobial susceptibility - *E coli* F4 (2010-2022).

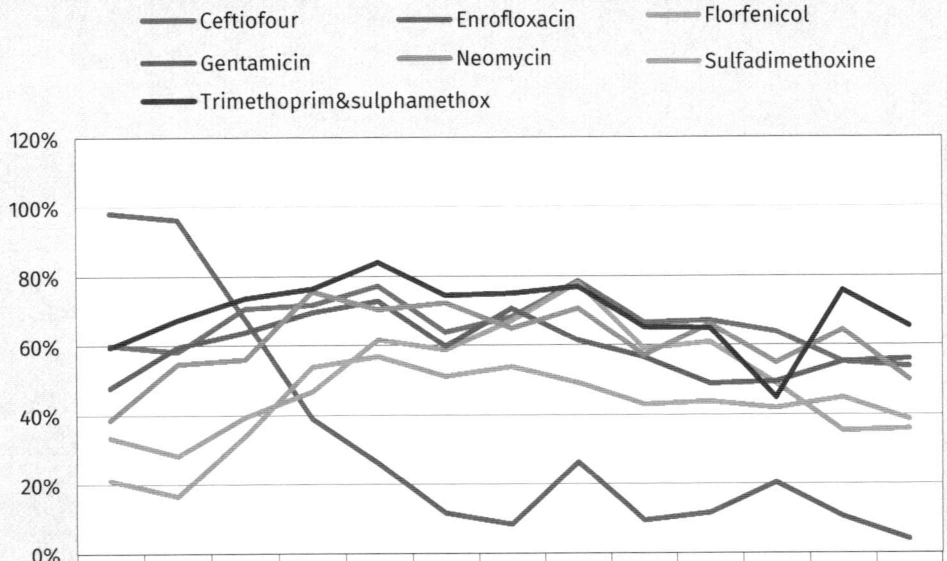

	2010	2011	2012	2013	2014	2015	2016	2017	2018	2019	2020	2021	2022
Ceftiofour	60%	58%	71%	72%	77%	64%	68%	78%	67%	67%	64%	55%	54%
Enrofloxacin	98%	96%	68%	39%	26%	12%	8%	26%	10%	12%	20%	11%	4%
Florfenicol	33%	28%	39%	47%	62%	59%	67%	78%	59%	61%	49%	35%	36%
Gentamicin	48%	60%	64%	70%	73%	60%	71%	61%	56%	49%	49%	55%	56%
Neomycin	38%	55%	56%	75%	70%	72%	65%	71%	57%	66%	55%	64%	50%
Sulfadimethoxine	21%	16%	34%	54%	57%	51%	54%	49%	43%	44%	42%	45%	38%
Trimethoprim&sulphamethox	60%	67%	74%	76%	84%	74%	75%	77%	65%	65%	45%	76%	65%

Figure 3: Antimicrobial susceptibility - *E coli* F18 (2010-2022).

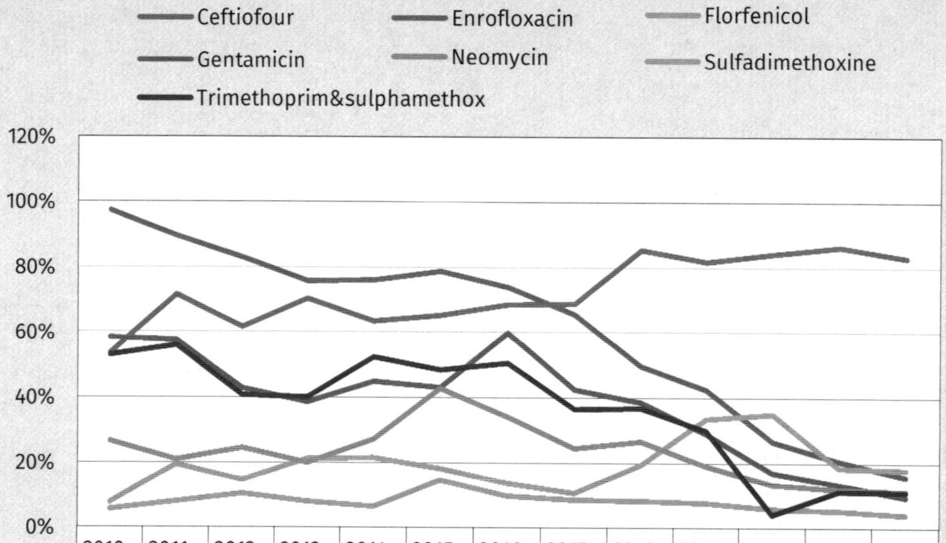

	2010	2011	2012	2013	2014	2015	2016	2017	2018	2019	2020	2021	2022
Ceftiofour	54%	71%	62%	70%	63%	65%	68%	69%	85%	82%	84%	86%	83%
Enrofloxacin	97%	89%	83%	76%	76%	79%	74%	65%	50%	43%	27%	20%	16%
Florfenicol	8%	19%	14%	21%	21%	18%	14%	11%	19%	33%	35%	18%	18%
Gentamicin	58%	57%	43%	39%	45%	43%	60%	42%	38%	29%	17%	13%	10%
Neomycin	26%	21%	25%	20%	27%	43%	34%	24%	27%	19%	13%	12%	11%
Sulfadimethoxine	6%	8%	10%	8%	7%	15%	10%	8%	8%	7%	6%	5%	4%
Trimethoprim&sulphamethox	53%	56%	41%	40%	52%	48%	50%	36%	37%	30%	4%	11%	11%

Discussion and conclusions

This study revealed important changes in the frequency of detection of different fimbrial genes and antimicrobial susceptibility patterns on *E coli* isolated from ISU-VDL cases. Identification of the genetic characteristics of *E coli* helps to understand how the pathogen has evolved to survive in the environment and in the host. Changes in antibiotic susceptibility pattern is a public health concern, and understanding these changes is important to help veterinarians in their decision-making on disease control.

References

1. Fairbrother JM and Nadeau É. *"Diseases of swine* (2019): 807-834.

2. Gyles CL and Fairbrother JM. *"Escherichia coli." Pathogenesis of bacterial infections in animals* 4 (2010): 267-308.

https://doi.org/10.54846/am2024/15

Paper sampling for passive environmental surveillance for swine pathogens

B. Armenta-Leyva, DVM, MS; B Munguía-Ramírez, DVM, MS; L. G. Giménez-Lirola BS, MS, PhD; J. J. Zimmerman, DVM, PhD
College of Veterinary Medicine, Iowa State University, Ames, Iowa

Introduction

Passive environmental surveillance (PES) detects pathogens circulating in the population of interest by sampling their environment. Unlike point-in-time environmental sampling, PES samplers remain in place over time, thereby increasing the probability of detection. In laboratory animal colonies, PES was shown to be as sensitive as sampling the animals themselves.[1]

Our long-term objective is to adapt PES to swine production. For this application, sampling devices would be placed in production facilities, collected at designated time periods, and tested for the pathogen(s) of interest. However, the first step in adapting PES to swine production is identifying the best material ("membrane") to use in PES samplers. In this pilot, we used PRRSV as the target.

Materials and methods

The performance in terms of target (PRRSV RNA) recovery of seven different candidate membranes was evaluated: (1) Whatman filter paper; (2) Flinders Technology Associates (FTA) cards; (3) polyester filter paper; (4) SmartSolve cardstock (soluble in water); (5) dry surface polyester; (6) tacky surface polyester; (7) starch foam sheet (soluble in water). In brief, membranes were placed inside a plastic bag and inoculated with 100 ul of four ten-fold dilutions (10^{-2} – 10^{-5}) of a PRRSV MLV (Ingelvac) that was rehydrated and diluted with PRRSV-free oral fluids. Inoculated membranes were stored overnight at 25°C in an environmental chamber. Thereafter, membranes were eluted with TE buffer and the eluate subjected to extraction and RT-qPCR. Total nucleic acid extraction was performed using the RealPCR*DNA/RNA Spin Column Kit (IDEXX Laboratories, Inc, Westbrook, Maine, USA) following the instructions of the manufacturer. All qPCR reactions were performed using the Magnetic Induction Cycler qPCR (Mic qPCR Cycler, Bio Molecular Systems, Australia) as directed by the manufacturer, using commercially available PCR reagents (RealPCR*PRRS Types 1-2 RNA Mix) for the detection of PRRSV (IDEXX).

Results and discussion

Target recovery was expressed in terms of percent recovery of the inoculum (Table 1). These results showed that PRRSV RNA can be recovered from various types of membranes, albeit more efficiently from some than others. Overall, these results are a promising start to our ultimate goal. Future experiments will focus on the optimization of target recovery from the membranes for swine pathogens and stability of pathogens on membranes by time and temperature.

Table 1: Summary of results by inoculum (PRRSV) dilution

PRRSV dilution	Candidate membranes* (% recovery relative to inoculum)						
	1	2	3	4	5	6	7
1×10^{-1}	21	30	24	5	1	24	10
1×10^{-2}	27	36	20	19	2	15	11
1×10^{-3}	24	28	26	23	3	13	8
1×10^{-4}	15	19	16	11	0	13	7

*1 = Whatman filter paper; 2 = Flinders Technology Associates (FTA) cards; 3 = polyester filter paper; 4 = SmartSolve cardstock (soluble in water); 5 = dry surface polyester; 6 = tacky surface polyester; 7 = starch foam sheet (soluble in water).

Reference

1. O'Connell KA, Tigyi GJ, Livingston RS, et al. Evaluation of in-cage filter paper as a replacement for sentinel mice in the detection of murine pathogens. *J Am Assoc Lab Anim Sci* 2021;60:243-250.

https://doi.org/10.54846/am2024/16

Evaluation of oral meloxicam to reduce signs of OCD associated lameness in developing boars

M. Hood[1], DVM; D. Reicks[1], DVM; P. Smith[2], RPh, Pharm D; B. Payne[2], DVM
[1]Reicks Veterinary Research and Consulting, [2]Veterinary Pharmaceutical Solutions

Introduction

Osteochondritis dissecans (OCD), with clinical signs such as lameness, dog-sitting and swollen, painful joints, is a developmental disorder of large, rapidly growing boars.[1] In recent years, this condition has become a growing concern in boars as it impacts production with reluctancy to mount a collection dummy and increases euthanasia rates. Meloxicam has been found to have a positive impact on clinical signs of lameness in sows.[2] The purpose of this study was to target the known time-period of stress for developing boars (ie, transport to and training in isolation) and implement a treatment protocol of a continuous oral anti-inflammatory drug (meloxicam) to reduce the occurrence of clinical OCD.

Materials and methods

Two replicates were conducted (Table 1). At isolation arrival (D0), boars were randomly divided into treatment (Tx) and control (Cx) groups by room. Tx boars were administered 90mg (6 mL) oral suspension drench of compounded 1.5% meloxicam (VPS) on D0 and soluble compounded 0.6% meloxicam (VPS) in the drinking water (targeting 0.1mg/kg/day) from D1-28. In Replicate 2, an additional sham treatment was utilized for Cx boars for both the drench and water medication to improve blinding. The same individual performed lameness scoring (0 - 5)[3] at four points in time: D0, D2, D28 and D75. Lameness scores were assigned in the following manner based on both gait and stance: 0 for boars with a normal stance and gait; 1 for boars with a normal stance and a stiff gait; 2 for boars with an uneven posture and a shortened gait with a hip swagger or head bob; 3 for boars with an uneven posture and reduced weight on an affected limb; 4 for boars unwilling to put weight on an affected limb; and 5 for a boar that cannot stand unaided. Treatments, semen quality and jump rate were assessed. Herd retainability to 18 months is an ongoing parameter.

Results

On D0 there was no statistical difference between the mean lameness scores. At D2 and D28, Tx boars had statistically lower mean lameness scores compared to Cx boars (Table 1). In addition, the Tx boars in Replicate 1 also had statistically lower mean lameness scores at D75. No safety concerns or side-effects were observed. There were no statistical differences in treatment rate or semen quality between the two groups. For both replicates in boars at 10 months of age, the percent removal for reasons associated with OCD are 17% for control and 11% for treatment boars. However, herd retention and production data through 18 months is being monitored.

Conclusion

Results indicate that use of oral meloxicam during the isolation period could be a tool to reduce OCD clinical lameness during and beyond the isolation period into the stud. A single dose of the meloxicam suspension improved lameness scores within a day and this was maintained with meloxicam soluble at the reduced rate. Herd longevity data analysis and further evaluation of this NSAID as a direct lameness treatment in boars and sows continues. No herd safety issues have been identified.

References

1. Carlson, C. The pathogenesis of osteochondrosis. U of MN. FeetFirst Sow Lameness Symposium II, 2010.

2. Pairis-Garcia, M et al. Behavioral evaluation of analgesic efficacy for pain mitigation in lame sows. *Anim Welfare* 2015. 24:93-99.

3. Mack, L et al. Group space allowance has little effect on sow health, productivity, or welfare in a free-access stall system. *J Anim Sci.* 2014 Jun;92(6):2554-67.

Table 1: Average lameness score (0 - 5) with 28-day treatment of oral meloxicam

		# Boars	Day 0 (pre-Tx)	Day 2	Day 28	Day 75
Replicate 1	Treatment	29	1.17	0.53	0.32	0.41
	Control	30	1.03	1.26	1.30	0.79
	P-value	–	0.54	< .01	< .01	< .05
Replicate 2	Treatment	24	0.71	0.63	0.79	1.00
	Control	24	0.79	1.17	1.38	0.83
	P-value	–	0.73	< .05	< .05	0.56

https://doi.org/10.54846/am2024/17

Repeat offenders: PRRSV clinical re-breaks from a whole genome perspective

J. P. Baker[1]; A. Rovira[2], DVM, PhD; K. VanderWaal[1], PhD
[1]Department of Veterinary Population Medicine, University of Minnesota, Saint Paul, Minnesota; [2]Veterinary Diagnostic Laboratory, University of Minnesota, Saint Paul, Minnesota

Introduction

Immunization-based control efforts on sow farms, such as gilt acclimation with vaccination or live virus inoculation, exist to ensure that herds are exposed to and have developed immunity against PRRSV. However, some farms experience repeated clinical PRRSV outbreaks caused by viruses with near-identical sequences; such "re-breaks" are incredibly frustrating for the practitioner and producer. The ability of a very closely related virus to induce a second clinical outbreak of disease suggests that the virus has changed, allowing it to reinfect a herd with preexisting "homologous" immunity. Thus, the objective of this study is to generate and characterize PRRSV whole genomes from viral isolates associated with clinical re-breaks of near identical PRRSV variants on single sow farms in order to examine patterns of change that could be related to viral immune escape in field settings.

Methods

A participating production system provided production data (weekly abortion counts) and diagnostic data (ORF5 sequences and weekly PCR diagnostics) for all sow farms from 2014 to the present. Vaccination data was also provided; the system uses only commercially available modified live vaccination (MLV). Weekly abortion counts were evaluated using statistical process control methods (EWMA) to identify spikes in abortion counts above the weighted average baseline. Processing fluid PCR CT values from samples concurrent with the abortion spike were investigated to confirm that they were PRRSV positive and that CT values were dropping, indicating that PRRSV likely contributed to the abortion spike. These two criteria (positive PCR diagnostics alongside abortion spikes) were used to define a "clinical break" of PRRS. A "clinical re-break" follows the same definition as above with additional criteria: **1)** the second clinical break must occur within 3-12 months of a previous break, and **2)** the ORF5 gene sequence obtained from the second clinical break must be ≥ 97% similar, at the nucleotide level, to the first clinical break. Upon fulfilling these criteria, viral isolates from these defined clinical breaks and re-breaks were utilized for whole genome sequencing.

Results

Suitable viral isolates were available for 13 clinical re-break events which satisfied the above criteria (see Figure 1 for example re-break case). These represent re-breaks on 12 farms, one of which experienced two, distinct re-break events caused by unrelated viral variants several years apart. The median number of days between re-breaks was 213.5 days (IQR = 159.8-230.8). A total of 26 full coverage (excluding 5' and 3' untranslated regions) whole genome sequences were generated from viral isolates obtained from samples submitted to the University of Minnesota Veterinary Diagnostic Laboratory. Out of the 13 re-break events, pairwise alignments of each re-break event revealed a single instance of a suspected recombination event between the primary and subsequent clinical breaks. Whole genome consensus sequences were then aligned to VR-2332 PRRSV Type 2 isolate (Genbank ID: NC_038291) to delineate ORFs. ORF1a/1b were the most variable, with median p-distances of 1.85% (IQR: 0.53-4.23%) and 2.25% (IQR: 0.45-8.28%), respectively. The most variable ORF coding for structural proteins (ORFs2-7) was ORF5, with a median p-distance of 1.60% (IQR: 0.675-1.98%).

Pairwise comparisons of ORFs 2 - 7 for each re-break event were performed to identify amino acid positions that were frequently changing between re-breaks. Any amino acid position that changed in ≥ 3 out of 13 re-break events was reported in Table 1. Given that some regions of ORF5 are known to be hypervariable, a Fisher's exact test was performed per amino acid site to determine if the proportion of re-break events with a substitution at that site was higher than expected. For each amino acid site, expected rates were defined based on the frequency with which the amino acid changed amongst ORF5 sequences with similar divergence times to the re-break pairs (in other words, the same amount of evolutionary time had passed). This was calculated from a time-scaled phylogenetic tree of all ORF5 sequences from within the system; pairwise comparisons (n = 164) were generated between closely related tips on the tree (nearest neighbor) with a branch length of ≤ 1.6 years (the median branch length between re-break pairs). GP5 amino acid positions 32, 33, 58, 94, 104, and 121 changed more frequently in re-break pairs than expected, as demonstrated in Table 1. Sites identified by previous studies are referenced in Table 1 and include studies demonstrating positive selection[1,2] and/or resistance to neutralization by homologous serum.[3,4]

Discussion

Re-break events caused by an endemic viral variant on a farm are not uncommon and do not appear to be primarily driven by recombination events between breaks; recombination was only observed in 1 of 13 re-break events. This suggests that smaller scale changes resulting from genetic drift or micro-evolution allow for closely related viral variants to evolve the capacity to re-establish within a population of which most members were previously exposed to a very similar virus. ORF5 amino acid substitutions identified in re-breaks in this study are consistent with previous phylogenetic studies, tending to occur within known hypervariable regions, which include positively selected amino acid positions.[1,2] For many of these sites, the accrual of genetic changes in re-break events was significantly higher than expected given background levels of variability. Additionally, many of these same sites have been shown, *in vitro* and *in vivo*, to generate increased resistance to neutralization by antisera generated against a parent or vaccine strain.[3-5]

Beyond ORF5, we incorporated analysis of ORFs 2-4, 6, and 7, assessing these for additional amino acid sites where substitutions frequently occurred between re-break events. Though we could not assess statistical significance due to a lack of comparative data for these genomic regions, some sites have been previously identified as sites undergoing positive selection, such as sites 42

Figure 1: Re-break case 13. **A.** Weekly abortion counts from late 2021 - early 2023. **B.** Weekly processing fluid CT values over time. **A:** Black dots indicate the number of abortions within the herd. Purple dashed line is upper control limit (EWMA analysis). Blue dashed line is moving average (6-week window). **B:** CT of 40 indicates a PRRSV-negative sample. Green stars indicated whole-herd vaccination events with lineage 5 MLV. Red stars indicate dates from which samples were used to generate viral isolates.

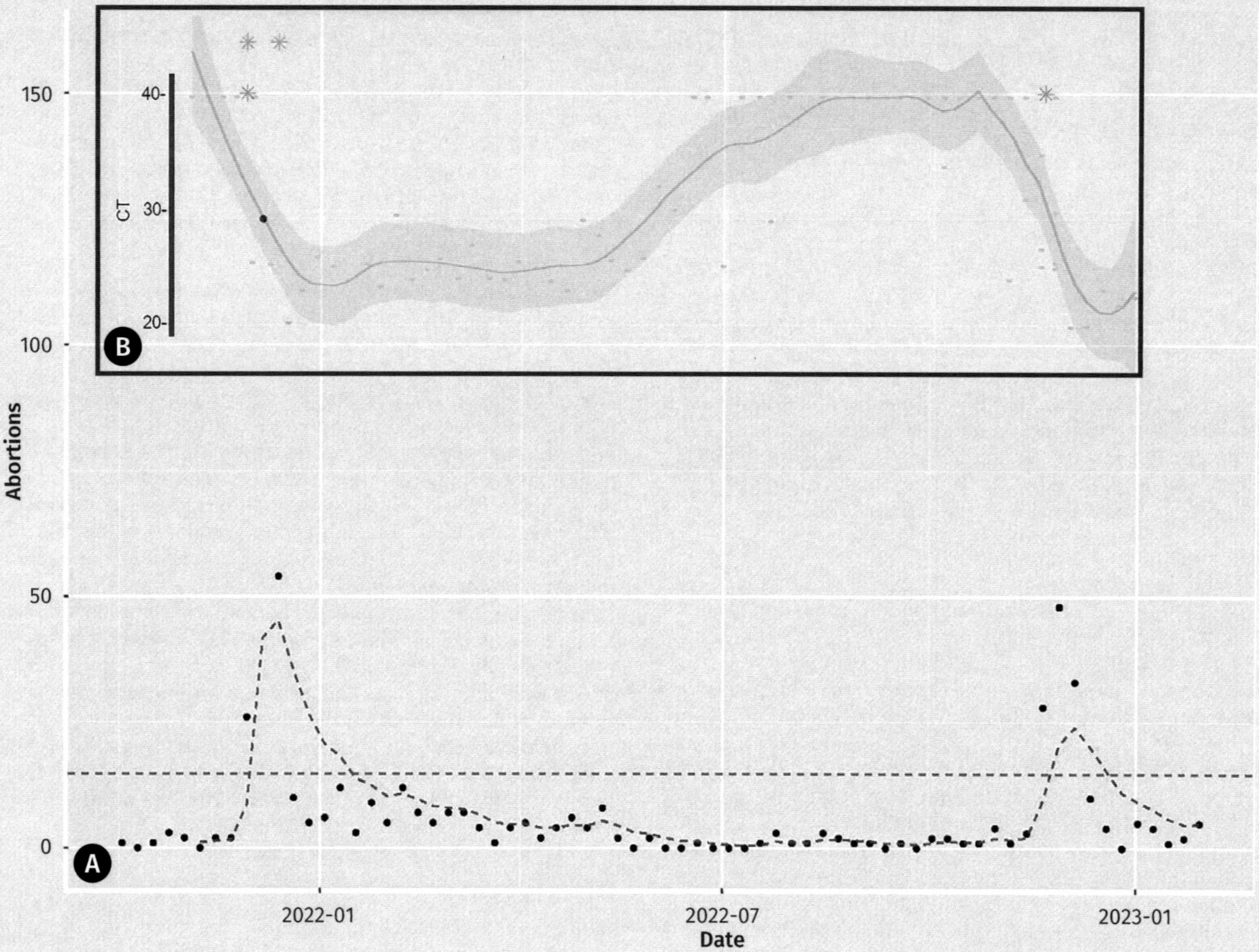

and 189 in GP2[1] (occurring in 30% of re-break pairs), or occur at notably high frequencies, such as sites 8 and 9 in the E protein (occurring in 38 - 46% of re-break pairs). Such sites warrant further research in experimental contexts.

Overall, our analysis has identified repeated clinical outbreaks of PRRSV attributable to closely related, yet distinct viral variants. The amino acid substitutions identified that accrue between a primary and subsequent outbreak are consistent with those that are observed to occur within hosts under heterologous immune pressure and have demonstrated, both *in vitro* and *in vivo*, to confer resistance to neutralization by homologous and heterologous anti-sera. To our knowledge, this is amongst the first evidence that sites associated with immune escape experimentally also occur in field clinical scenarios where immune escape is suspected. This is consistent with immune-driven evolutionary changes related to escape phenotypes that occur within hosts and demonstrates that these same genotypes can be identified associated with hypothesized immune escape events seen at the farm-population level. Improving our understanding of the antigenic diversity of PRRSV, as it occurs in the field, can help us better understand the persistence of the virus during herd closures, the failure of immunological interventions, and may illuminate methods for improvement of PRRSV control.

References

1. Guo J, Liu Z, Tong X, et al. Evolutionary dynamics of Type 2 Porcine Reproductive and Respiratory Syndrome Virus by whole-genome analysis. Viruses 2021, 13, 2469. **https://doi.org/ 10.3390/v13122469**

2. Paploski IAD, Corzo C, Rovira A, et al. Temporal Dynamics of Co-circulating Lineages of Porcine Reproductive and Respiratory Syndrome Virus. Front. Microbiol. 10:2486. **doi: 10.3389/fmicb.2019.02486**

3. Kim WI, Kim JJ, Cha SH, et al. Significance of genetic variation of PRRSV ORF5 in virus neutralization and molecular determinants corresponding to cross neutralization among PRRS viruses. Vet. Microbiol. Volume 162, Issue 1, 2013. **doi:10.1016/j.vetmic.2012.08.005**

4. Fan B, Liu X, Bai J, et al. The amino acid residues at 102 and 104 in GP5 of porcine reproductive and respiratory syndrome virus regulate viral neutralization susceptibility to the porcine serum neutralizing antibody. Virus Res. Volume 204. 2015. **doi: 10.1016/j.virusres.2015.04.015**

5. Evans AB, Loyd H, Dunkelberger JR, et al. Antigenic and Biological Characterization of ORF2–6 Variants at Early Times Following PRRSV Infection. Viruses 9, no. 5 (May 16, 2017). **https://doi.org/10.3390/V9050113**

Table 1: Prevalent sites of mutation between re-break pairs within structural glycoproteins. *P < 0.05; **P < 0.1; evidence of positive selection in references 1 - 2; evidence of experimental resistance to neutralization in references 3 - 5.

	AA position of mutation	Proportion of re-breaks (%)	Expected % (n = 164)	Citations for evidence of positive selection (PS) or experimental immune evasion (Ex)
GP5	58*	5/13 (38%)	17%	PS: 1-2 / Ex: 3
	33*	4/13 (31%)	7.8%	PS: 1-2 / Ex: 3
	104**	4/13 (31%)	14%	PS: 1-2 / Ex: 4
	32**	3/13 (23%)	7.8%	PS: 1 / Ex: 3, 5
	57	3/13 (23%)	18%	PS: 1-2 / Ex: 3, 5
	94**	2/13 (15%)	3.9%	–
	101**	2/13 (15%)	2.6%	–
	121**	2/13 (15%)	3.9%	PS: 2
GP2	42	4/13 (31%)	–	PS: 1
	86	4/13 (31%)	–	–
	189	4/13 (31%)	–	PS: 1
	193	4/13 (31%)	–	–
	10	3/13 (23%)	–	–
	11	3/13 (23%)	–	PS: 1
	20	3/13 (23%)	–	–
	88	3/13 (23%)	–	–
	191	3/13 (23%)	–	–
	197	3/13 (23%)	–	–
	193	3/13 (23%)	–	–
	240	3/13 (23%)	–	–
	255	3/13 (23%)	–	–
E	9	6/13 (46%)	–	–
	10	5/13 (38%)	–	–
	77	3/13 (23%)	–	–
GP3	151	4/13 (31%)	–	–
GP4	35	5/13 (38%)	–	PS: 1
	57	4/13 (31%)	–	–
	77	3/13 (23%)	–	–

https://doi.org/10.54846/am2024/18

Evaluation of the indirect transmission of *Mycoplasma hyopneumoniae*

Cassidy Cordon[1]; Albert Canturri[1,2]; Robert Evelsizer[3]; Maria Pieters[1,2,4]
[1]Department of Veterinary Population Medicine, College of Veterinary Medicine, University of Minnesota, St. Paul, Minnesota; [2]Veterinary Diagnostic Laboratory, University of Minnesota, St. Paul, Minnesota; [3] Elanco Animal Health Inc., Greenfield, Indiana; [4]Swine Disease Eradication Center, College of Veterinary Medicine, University of Minnesota

Introduction

Mycoplasma hyopneumoniae is an important agent in the swine industry as it causes economic losses due to being the causative agent of enzootic pneumonia, a chronic respiratory disease in swine.[1] The primary route of transmission for *M hyopneumoniae* is direct nose-to-nose contact.[2] *Mycoplasma hyopneumoniae* is not thought to survive long in the environment due to its lack of cell wall, and thus routes of indirect transmission have been poorly explored. However, previous research has shown that *M hyopneumoniae* could survive for one week at room temperature in lung tissue, which is comparable to infectious respiratory secretions.[3] Additionally, more recent studies have shown that *M hyopneumoniae* can survive for up to eight days at 4°C on materials such as dust.[4] Other species of mycoplasma (*Mycoplasma synoviae*) have been shown to infect poultry via indirect transmission of contaminated materials such as dust and food.[9] On the other hand, in a case study by Batista et al[5], when personnel followed basic biosecurity protocols and moved from a *M hyopneumoniae* infected to a naïve herd, no transmission from infected to susceptible pigs occurred. It is apparent that there are conflicting studies and a gap in knowledge in regard to the possibility of indirect transmission via *M hyopneumoniae*. Studies indicating the survival of the bacterium on surfaces has led to the hypothesis that *M hyopneumoniae* can remain viable outside the pig's respiratory system and in the environment, potentially posing a risk for indirect transmission to susceptible pigs. Therefore, the objective of this study were to assess the infection and disease development potential of a *M hyopneumoniae* experimentally and a naturally contaminated environment and to evaluate the indirect transmission capacity for susceptible pigs.

Methods

The study was performed under experimental conditions after approval by the IACUC of the University of Minnesota. A total of 12 pigs were housed at the Veterinary Isolation Facilities (St. Paul Campus) for this study. Pigs were randomly allocated to experimental groups, as shown in Table 1. Each experimental group consisted of two, three-week old gilts born to non-vaccinated dams from a *M hyopneumoniae* free herd. Nebulizers were used to aerosolize lung homogenate containing $1x10^6$ CCU/mL *M hyopneumoniae* strain 232 into an isolator. An isolator is a hermetically closed holding space where infective particles remain inside and passively deposit on environmental surfaces. Experimental group 1 was the positive control, in which pigs were placed into the isolator and directly aerosolized with *M hyopneumoniae*. Group 2 was the negative control in which the isolator was aerosolized with culture medium without *M hyopneumoniae*. Pigs in groups 3 - 5 were placed into the isolators one, three, and five days post-aerosolization, respectively. Group 6 consisted of pigs placed in an isolator naturally contaminated by pigs positive to *M hyopneumoniae*. Pigs in all groups remained in the isolator for 8 hours. Pigs were confirmed free of *M. hyopneumoniae* prior to exposure via PCR and ELISA.[6] At 7,14, 21, and 28-days post exposure (dpe) pigs were tested via tracheal secretion PCR. Additionally, on 11 dpe onward, clinical signs were assessed by cough scoring (observing pigs for 15 min in the pen).[7] At 28 dpe, serum was collected for ELISA testing and pigs were humanely euthanized via barbiturate overdose. Post-mortem, bronchial secretions were collected for PCR testing and lung lesions were scored macroscopically and histologically.[1,8] Results were expressed in Ct values, with a cut-off value of ≤ 40. Data was analyzed using descriptive analysis.

Table 1: Experimental study groups

Experimental Group	Exposure to *Mycoplasma hyopneumoniae*	Contamination of surfaces	Pigs placed into isolator
1 (positive control)	Direct	Experimental	During exposure
2 (negative control)	No	Experimental	During exposure
3	Indirect	Experimental	1 day post contamination
4	Indirect	Experimental	3 days post contamination
5	Indirect	Experimental	5 days post contamination
6	Indirect	Natural	Immediately post contamination

All pigs were placed into isolators, the positive and negative control groups were directly placed into the isolators and aerosolized with *M hyopneumoniae* or culture medium, respectively.

Results and discussion

The positive control group of pigs became infected with *M hyopneumoniae,* as confirmed by PCR, clinical signs, and lung lesions. Additionally, the negative control group of pigs remained negative for *M hyopneumoniae* throughout the duration of the study, showing the validity of the experimental model. Pigs indirectly exposed to an experimentally or naturally contaminated environment did not develop infection with *M hyopneumoniae,* regardless of being exposed to contaminated environments at different times. The results shown in this experimental model are supportive that *M hyopneumoniae* was not transmitted via indirect exposure of environmental surfaces. It is necessary to mention that this was a 28-day infection model which may have impacted the infection and disease development of the pigs due to *M hyopneumoniae* being slow to develop and colonize in the respiratory tract. A longer infection model may allow for further *M. hyopneumoniae* development and colonization to produce lesions and clinical signs. Results of this study indicated that in contaminated environments, *M hyopneumoniae* was not transmitted or did not cause infection via indirect exposure to susceptible swine. These results contribute to the body of knowledge related to current biosecurity and disinfection measures used to control *M hyopneumoniae*, as they indicate that it was unlikely for the bacterium to be transmitted to susceptible pigs via fomites. Moreover, the information generated in this study further supports disease eradication efforts.

References

1. Pieters, M. & Maes, D. (2019). Mycoplasmosis . In *Diseases of Swine* (11th ed.) Zimmerman et al. (editors). Wiley. (pp. 863-871).

2. Pieters, M., Betlach, A., Nathues, H., Maes, D. (2020). Epidemiology of *Mycoplasma hyopneumoniae* infections. In *Mycoplasmas in Swine*. CABI. (pp. 87-96).

3. Goodwin, R.F. (1972). The Survival of *Mycoplasma suipneumoniae* in Liquid Medium, on Solid Medium, and in Pneumonic Tissue. In *Research in Veterinary Science* (Vol. 13, pp 203-204).

4. Browne, C., Loeffler, A., Holt, H. R., Chang, Y. M., Lloyd, D. H., & Nevel, A. (2017). Low temperature and dust favour in vitro survival *of Mycoplasma hyopneumoniae*: time to revisit indirect transmission in pig housing. Letters in Applied Microbiology, 64(1), 2–7. **https://doi.org/10.1111/lam.126**

5. Batista, L., Pijoan, C., Ruiz, A., et al. (2004). Assessment of transmission of *Mycoplasma hyopneumoniae* by personnel. *J Swine Health Prod.* 12(2):75-77.

6. Strait, E., Madsen, M., Minion, C., et al. (2008). Real-time PCR assays to address genetic diversity among strains of *Mycoplasma hyopneumoniae. Journal of Microbiology.* 46 (8): 2491-8.

7. Nathues, H., Woeste, H., Doehring, S., et al. (2012). Detection of *Mycoplasma hyopneumoniae* in nasal swabs sampled from pig farmers. *Veterinary Record, 170*(24), 623. **https://doi.org/10.1136/vr.100817**

8. Woolley, L. K., Fell, S., Gonsalves, J. R., Walker, M. J., Djordjevic, S. P., Jenkins, C., & Eamens, G. J. (2012). Evaluation of clinical, histological and immunological changes and qPCR detection of Mycoplasma hyopneumoniae in tissues during the early stages of mycoplasmal pneumonia in pigs after experimental challenge with two field isolates. Veterinary Microbiology, 161, 186-195.

9. Marios, C., Picault, J., Kobisch, M., Kempf, I. (2005). Experimental evidence of indirect transmission of *Mycoplasma synoviae. Vet Res* 36, 759-769.

https://doi.org/10.54846/am2024/19

Comparison of 200 mg injectable iron dextran vs 400 mg in either a single timepoint or split-dosed into two timepoints on growth and hemoglobin in pigs from birth to end of finishing

F. A. Gomez Cruz[1], BS; S. Krantz[2], DVM; W. Lyons[3], DVM; J. L. Pierce[4], PhD
[1]Western College of Veterinary Medicine, University of Saskatchewan, Saskatoon, Saskatchewan; [2]Tosh Farms, Henry, Tennessee; [3]Pharmacosmos Inc., Dekalb, Illinois; [4]James Pierce Consulting, Nicholasville, Kentucky

Introduction

Piglets are born with low iron stores and thus require supplementation to promote development of hemoglobin and prevention of anemia. For decades, the industry standard dosage for parenteral iron in piglets has been 200 mg given intramuscularly (IM) between 1 and 3 days of age. Despite changes in genetics, nutrition, and wean age, the 200mg recommended dosage has remained unchanged. Recently published research has shown that an iron dosage higher than 200 mg has a better outcome on body weight and average daily gain,[1] in addition to better body temperature regulation and intestinal health in pigs.[2,3] Previous studies in increasing the dose of iron focused on giving 200 mg iron dextran to piglets at two different timepoints during the suckling period.[1-3] The objective of this study was to determine if a dosage of 400 mg of iron dextran split into two injections at a single timepoint is equivalent to dosing 200 mg at two different timepoints on the growth characteristics noted in previous studies.

Material and methods

The study was initiated on a 5,000 head commercial sow farm in Tennessee. The piglets were offspring of PIC C29™ sows bred to the PIC 337™ boar line. The study began by enrolling 590 piglets within the first 24 hours of birth. Enrollment consisted of selecting three gilts and three barrows per litter. The piglets selected within the litter were similar in size and weight. Piglets were blocked by sex, then randomly allocated into three treatment groups. A random list generator was used to allocate piglets to treatment. To ensure that the allocation was entirely randomized, the same investigator always chose the three gilts and three barrows from each litter. The same investigator would then select a pig from the cart without knowing the order in the list. An assistant would then assign the pig to a treatment group once it was removed from the cart. An important detail in this process was to ensure that all the pigs from the same sex were assigned before moving to the other sex. This prevented any mix-up in the randomization.

A common iron dextran product (Uniferon™ 200mg/mL) was used for all treatment groups. Uniferon™ is labeled for the treatment and prevention of iron deficiency anemia in baby pigs, thus it provides veterinarians flexibility in the dosing of the product. The treatment groups consisted of a single 200 mg Uniferon™ IM injection (control) given at 24 hours of age, a 400 mg Uniferon™ IM injection at a single timepoint given as two, 1mL injections on either side of the neck (single) at 24 hours of age, and two, 200 mg Uniferon™ IM injections six days apart (split) with the first being at 24 hours of age. Piglets were ear tagged with numbered and color-coded tags representing their respective treatment group. No cross-fostering of study pigs was allowed once enrollment was complete; however, non-study pigs were allowed to be fostered on or off the sows.

Hemoglobin was measured using a handheld meter (HemoCue 201™), and body weight was measured for all pigs at enrollment, 6 days of age, and at weaning. The pigs were weaned at an average of 22 days of age. At weaning, pigs were moved into a 4,800 head wean to finish site that was double stocked until pigs were approximately 9 weeks of age. All study pigs were housed in one barn on the site and were randomly distributed across all pens within that barn. Before removing half the pigs from the double stock around 9 weeks of age, pig weight and hemoglobin was measured. Three days prior to removing the first cut of top hogs from the site, all study pigs that had intact ear tags with legible numbers were weighed a final time and data was recorded. Data were analyzed as a randomized block design with three treatments using the linear model in R version 4.2.3.[4] Wean age was the covariate for weaning weight and Hb. Days on feed in the nursery were used covariates for end-of-nursery Hb and growth performance. Total days on feed were used as the covariate for the wean to finish growth performance data. Therefore, growth, and Hb results are reported as least-square means.

Results

At weaning, Hb in pigs that received both split and single injections was increased compared to control (12.8, 12.7 vs. 11.8 g/dL, $P < .05$) (Table 1). Weaning weight was improved in the split dose group compared to the control (15.4 vs 14.7 lbs. $P < .05$) (Table 2).

At the end of nursery, pigs receiving split injection were 2.3lbs heavier than the control (48.4 vs 46.1 lbs. $P < .05$) (Table 2). Concurrently, single injection pigs (400 mg or 2mL) were 1.9lbs (48lbs) heavier than the control ($P < .05$) (Table 2). Additionally, Hb values were higher for both single and split treatments compared to the control (9.96, 10.05 vs 9.43 g/dL, $P < .05$) (Table 1).

At the end of wean-to-finish period pigs that received 400 mg of iron in either the single or split dose were 14 lbs. heavier (266.5 vs 253.0, $P < .05$) (Table 3) than those that received the control (200 mg). Likewise, average daily gain was 0.10 lbs. per day more in pigs given 400 mg of iron, than those given 200 mg (1.89 vs 1.99 lb/day $P < .05$) (Table 3) regardless of dosing times.

Discussion

Previous studies showed the value of increasing the dose of iron above 200 mg. In production systems where pigs are picked up twice before they are 14 days of age, the addition of a second dose of 200 mg of iron on top of the farm's existing protocol can be

Table 1: Hemoglobin (Hb) least-square means at weaning and end of nursery

Treatment	LS means, g/dL	SE
Hb weaning		
Control	11.8[a]	0.098
Single	12.7[b]	0.095
Split	12.8[b]	0.104
Hb nursery		
Control	9.43[a]	0.145
Single	9.96[b]	0.123
Split	10.05[b]	0.139

Means with different superscripts differ (*P* < .05)

Table 2: Body weight (BW), least-square means at weaning and end of nursery

Treatment	LS means, lb	SE
BW, weaning		
Control	14.7[a]	0.232
Single	14.8[ab]	0.224
Split	15.4[b]	0.245
BW, nursery		
Control	46.1[a]	0.719
Single	48.0[ab]	0.612
Split	48.4[b]	0.688

Means with different superscripts differ (*P* < .05)

Table 3: Body weight (BW) and average daily gain (ADG) least-square means for wean to finish period

Treatment	LS means, lb	SE
BW, final		
Control	253[a]	2.37
Single	266[b]	2.31
Split	267[b]	2.63
ADG		
Control	1.89[a]	0.018
Single	1.98[b]	0.018
Split	1.99[b]	0.020

Means with different superscripts differ (*P* < .05)

seamless and add little cost in labor and product. However, many systems only pick up all pigs once for processing, therefore adding a second dose of iron at a second timepoint becomes cost and labor prohibitive. The results of this study prove that additional parenteral iron can be supplied to the pig in a cost-effective manner by doing it at a single time point with one injection for treatment and one for prevention of iron deficiency anemia. It is not recommended to exceed 1mL of iron dextran in a single injection site due to the viscosity of the product and the propensity for leakage at higher volumes. In production systems where pigs are only handled once prior to weaning, two, 1 mL injections should be given to avoid wastage and leakage.

References

1. Chevalier TB, Lyons W, Paczosa DB, Rentfrow GK, Lindemann MD. A second iron injection administered to piglets during lactation improves hemoglobin concentration, growth performance, and carcass characteristics at slaughter. *Journal of Animal Science*. 2023; 101, 1-8.
2. Pierce JL, Lyons W, Chevalier TB, Paczosa DB, Lindemann MD. Effects of an additional iron dextran injection administered to piglets on differential gene expression at weaning. Poster presented at: AASV Annual Conference; March, 2023; Denver, Colorado.
3. Johnson A, Dittrich B, Cole A, Prodell M, Lyons W, Heisel J, Fritz S, Jang YD. Effects of second iron injection before weaning on growth performance, hematological parameters, and fecal score of pigs fed nursery diets with different dietary iron levels under natural disease challenge. Poster presented at: Leman Swine Conference; September 2023; St. Paul, Minnesota.
4. R Core Team (2023). R: A Language and Environment for Statistical Computing. R Foundation for Statistical Computing, Vienna, Austria. https://www.R-project.org

https://doi.org/10.54846/am2024/20

Evaluation of sample types for the characterization of *Lawsonia intracellularis* shedding status

A. Vegter[1]; J. Morgan[2], DVM; G. Silva[1], DVM, MS, PhD; P. Gauger[1], DVM, MS, PhD; F. Leite[2], DVM, MS, PhD
[1]Iowa State University, Ames, Iowa; [2]Boehringer Ingelheim Animal Health USA Inc., Duluth, Georgia

Introduction

Lawsonia intracellularis is an obligate intracellular pathogen endemic to swine that causes porcine proliferative enteropathy (PPE) and is shed in feces.[1] Different sample types can be used to assess the shedding status of a herd and individual pig using *L intracellularis* specific PCR. These include oral fluids, pooled pen fecal samples, and the use of rectal swabs or feces. However, it is unclear how the sensitivity of these different sample types compare.[2-5] It is also unclear how accurately an oral fluid and pooled pen fecal sample reflect the shedding status of pigs in a pen. Therefore, this study aimed to evaluate differences among these sample types and assess the correlation between oral fluids, pooled pen fecal samples, and individual pig shedding status.

Table 1: Summary of the number of pens that were qPCR positive for *Lawsonia intracellularis* by oral fluid and pooled pen fecal samples

Sample type	Clinical herd	Subclinical herd
Oral fluid samples	100% (54/54)[a]	6.5% (4/62)[a]
Pooled pen fecal samples	85.2% (46/54)[b]	46.8% (29/62)[b]

Different superscript letters indicate statistical significance when comparing oral fluid and pooled pen fecal sample results within each herd ($P < .05$).

Methods

Two wean-to-finish farms with a recent history of *L intracellularis* positive diagnostics and no recent mass antibiotic treatment were identified. One farm had clinical signs of diarrhea with loose stools consistent with *L intracellularis* infection. The other farm was characterized as being sub-clinically affected as positive fecal PCR results were found without the presence of clinical signs. To investigate sample types, oral fluids and pooled pen fecal samples were collected from all pens at each farm. Pooled pen fecal samples were collected by selecting five fecal samples in a pen and placing them into a ziplock bag. A subset of nine pens from the subclinical farm and three pens from the clinically affected farm were randomly selected to have individual fecal swabs, and rectal fecal samples collected from every pig in the pen. Feces were collected digitally directly from the rectum for the fecal samples, while a sterile swab was used to directly swab the rectum and then placed into a StarSwab II transport container with liquid Stuart's media for the fecal swabs. Samples were then submitted to a veterinary diagnostic lab (VDL) for *L intracellularis* qPCR. All fecal samples, both from individual pigs as well as pooled pen fecal samples were processed utilizing a weight of 0.3 g in 1 mL of PBS, which is the maximum allowed in the MagMAX™ Total Nucleic Acid Isolation Kit (Thermo Fisher Scientific). All swabs were processed with the same extraction kit and subsequent qPCR, with the swabs being processed in 1mL of PBS. The Ct cutoff for a positive sample was 35 as determined by the diagnostic lab. Correlation coefficients and kappa test were used to compare and assess agreement between different sample types. Fisher's exact test was used to evaluate difference in the number of positive samples among different tests. An unequal variance t-test was used to evaluate the differences in Ct value for different sample types.

Results

In the subclinical herd, 6.5% (4/62) of oral fluids and 46.8% (29/62) of pooled pen fecal samples were qPCR positive (Table 1). In the clinical herd, 100% (54/54) of oral fluids and 85.2% (46/54) of pooled pen fecal samples were qPCR positive (Table 1). The level of positivity of pooled pen fecal samples and oral fluids were significantly different in both the clinical and subclinical herds ($P < .05$, Table 1). There was a weak correlation between oral fluids and pooled pen fecal sample results in the subclinical ($r = 0.208$) and the clinical ($r = 0.282$) herd. When comparing the average of all the oral fluid samples to the average of the pooled pen fecal samples for each farm, the oral fluids were 2.1 cycle threshold values (Ct) lower than the pooled pen fecal samples at the clinical farm ($P < .001$) and 0.7 Ct higher at the subclinical farm ($P < .002$). To assess the relationship between the pooled pen fecal samples and the individual animals in the corresponding pen, the average of the individual feces was compared to the pooled pen fecal. The clinical farm had a high correlation ($r = 0.963$) between the pooled pen fecal and the average Ct of the individual feces, while the subclinical farm had a weak correlation ($r = 0.197$) and minimal agreement ($k = 0.053$). Results from samples of the subset of pens with every animal sampled are summarized in Table 2.

Comparing individual pig fecal sampling methods, among both the clinical and subclinical herd, a greater number of fecal samples versus fecal swabs were positive (Table 3). In the clinical herd 47.6% (60/126) of fecal samples were positive while 36.5% (46/126) of the fecal swabs were qPCR positive ($P < .0969$, Table 3). In the subclinical herd, 24.8% (60/242) of the feces and 1.7% (4/242) of the fecal swabs were qPCR positive ($P < .0001$, Table 3). In the samples from the clinical herd there was a moderate agreement between the two sampling methods ($k = 0.582$). While in the subclinical herd, there was no agreement between fecal samples and fecal swabs ($k = 0.033$). When comparing the average Ct of the individual fecal swabs and feces, the fecal swab Ct was 2.4 Ct higher than the feces in the clinical herd. In the subclinical herd, the individual fecal swab average was 1.5 Ct higher than the average individual feces. There was no qPCR inhibition detected in any sample type.

Discussion and conclusions

In this study, clinical presentation and shedding levels were found to affect the sensitivity of the different sample types. The oral fluids samples did not have consistent results between the farms, with the clinical farm having significantly more positive results when compared to pooled pen fecal samples. In the subclinical herd, a pooled pen fecal sample led to significantly more positive samples than oral fluids. This suggests that pooled pen fecals may be a more suitable sample type for *L intracellularis*

Table 2: Summary of results for pens in which all individual pigs were sampled along with the collection of pen oral fluids and pooled pen fecal samples

Sample type	Clinical herd (3 pens)			Subclinical herd (9 pens)								
Oral fluid	26.1	26.7	31.5	>35	>35	>35	>35	>35	>35	>35	>35	>35
Pooled pen fecal	24.7	28.1	35	29.3	35	35	33.7	35	31.6	34.6	34.4	35
Fecal samples positive*	54.5%	56.4%	32.6%	6.7%	3.8%	25.9%	53.8%	37.0%	63.0%	0.0%	17.2%	14.8%
Average fecal sample Ct	31.2	31.4	33.6	34.8	35	34.4	34.0	34.5	34.0	35	34.7	34.8
Fecal swabs positive*	38.6%	56.4%	16.3%	0%	0%	3.7%	7.7%	0%	3.7%	0%	0%	0%
Average fecal swab Ct	32.9	32.5	33.6	-	-	35	35	-	34.9	-	-	-

* Results represent the percentage of positive samples.

Table 3: Results from the comparison of fecal sample and fecal swabs in clinical and subclinical herds

	Clinical herd	Subclinical herd
Feces	47.6% (60/126)[a]	24.8% (60/242)[a]
Fecal swab	36.5% (46/126)[b]	1.7% (4/242)[b]

Different superscript letters indicate statistical significance when comparing fecal sample and fecal swab results within each herd (*P* < .05).

monitoring than oral fluids. Regardless of the clinical status of the herd, fecal samples were found to be more sensitive than fecal swabs, and thus better suited for surveillance than fecal swabs. These findings highlight the importance of choosing the correct sample type and understanding their limitations when investigating *Lawsonia intracellularis* in pig populations.

References

1. Zimmerman, Jeffrey J., et al. "Proliferative Enteropathy." *Diseases of Swine,* Wiley-Blackwell, Hoboken, New Jersey, 2019, pp. 898–911.

2. Pedersen, Ken Steen et al. "Diagnostic performance of fecal quantitative real-time polymerase chain reaction for detection of Lawsonia intracellularis-associated proliferative enteropathy in nursery pigs." *Journal of veterinary diagnostic investigation : official publication of the American Association of Veterinary Laboratory Diagnosticians, Inc* vol. 25,3 (2013): 336-40. **doi:10.1177/1040638713480499**

3. Lindecrona, R H et al. "Application of a 5' nuclease assay for detection of Lawsonia intracellularis in fecal samples from pigs." *Journal of clinical microbiology* vol. 40,3 (2002): 984-7. **doi:10.1128/JCM.40.3.984-987.2002**

4. Nathues, H et al. "Quantification of Lawsonia intracellularis in porcine faeces by real-time PCR." *Journal of applied microbiology* vol. 107,6 (2009): 2009-16. **doi:10.1111/j.1365-2672.2009.04389.x**

5. Pieters, M., et al. "Determining if qPCR on oral fluids can reliably detect Lawsonia intracellularis in a commercial swine production environment." American Association of Swine Veterinarians Annual Meeting; 2015; Orlando, Florida.

https://doi.org/10.54846/am2024/21

Comparison of different sample types and assays for detecting *Cystoisospora suis* at the litter level

M. Wheeler[1], BS; E. Kettelkamp[2], DVM; J. Okones[3], DVM; N. Lauterbach [3], DVM; A. M. Betlach[2], DVM, PhD
[1]Iowa State University, College of Veterinary Medicine, Ames, Iowa; [2]Swine Vet Center, St. Peter, Minnesota; [3]Pharmgate Animal Health, Wilmington, North Carolia

Introduction

Coccidiosis caused by *Cystoisospora suis* (*C suis*), previously known as *Isospora suis,* is a common cause of diarrhea in piglets.[1] Clinical signs of coccidiosis include pasty, yellow diarrhea, high morbidity, and low mortality that is not responsive to antibiotic treatment, resulting in longer time to market due to lighter wean weights.[2] Diagnosis of coccidiosis is most reliable by identification of histologic lesions. Fecal samples are often collected for fecal flotation analysis; however, piglets excrete oocysts for a short time period causing variability in diagnostic sensitivity.[3] To aid in coccidiosis diagnosis, a *C suis* real-time polymerase chain reaction test (PCR) was validated by an accredited veterinary diagnostic lab for the detection of *C suis* genetic material. The objective of this study was to compare different sample types and diagnostic assays for the detection of *C suis* at the litter level under field conditions.

Materials and methods

A 2,500-head multiparous sow farm located in southwest Minnesota with a clinical history of a coccidia-like scour in piglets over ten days of age was enrolled in the study. Within 50 days of sampling, piglet coccidiosis was confirmed by histology and fecal flotation at an accredited diagnostic laboratory. A power calculation was performed to select 40 crates of piglets 10 to 18 days of age distributed across three farrowing rooms. The farrowing crates were allotted based on piglet size and clinical status and additionally blocked by parity group and room location. A fecal scoring system of 0 - 3 (0 - firm, 1 - formed, 2 - loose and pasty, and 3 - liquid) was used to allocate 20 crates with clinical scours and no scours, respectively. Fecal scores of 0 - 1 were considered non-clinical and 2 - 3 were clinical. From each litter, piglet feces (1 - 2g) from the crates, piglet rectal swabs, and environmental Swiffer® samples were collected. Five randomly selected piglets were rectally swabbed per litter. A Swiffer® Dry Sweeping Cloth wetted in 50 mL of phosphate buffer saline was used to swab the entire mat and both crate corners within a square foot of the sow, followed by liquid extraction into a sterile conical tube. Rectal swabs were pooled by litter and submitted with environmental samples for *C suis* PCR. Fecal samples were split at the lab for fecal flotation analysis and *C suis* PCR. The PCR cutoff value was a Ct-value of 42. The association between litter clinical status and parity and the probability of resulting positive for *C. suis* for each sample+diagnostic assay was evaluated using a multivariate logistic mixed model. Data analysis was performed using the 'glmer' package in R (version 4.2.3). Litter was considered the experimental unit (n = 40). Agreement of diagnostic result outcome (positive/negative) between the sample+diagnostic assay was evaluated via Cohen's kappa statistic using the 'psych' package in R. Statistical significance was deemed at a *P*-value < .05.

Results

A higher *C suis* positive rate was observed in fecal samples via PCR (38%, 15/40) compared to environmental samples via PCR (13%, 5/40), with fecal swabs via PCR (20%, 8/40) and feces via flotation (18%, 7/40) being intermediate. Non-clinical crates contributed 45% (18/40) to total positives compared to 42% (17/40) of clinical crates. The odds of having a positive result were 26 times greater for PCR with piglet fecal samples versus fecal float (OR: 26.79, *P* = .007). Agreement between the different sample+diagnostic assays ranged from no (Kappa = 0) to fair (Kappa = 0.21 - 0.40). Overall, there was no effect of clinical status or parity on the probability of resulting positive for *C suis* for any sample+diagnostic assay (*P* > .10).

Discussion

Cystoisospora suis continues to pose health challenges in the swine industry due to recent limited treatment availability and high environmental stability in farrowing environments. Based on results of this study, *C suis* piglet feces via PCR had greater odds of a positive result compared to feces via flotation, providing a new, alterative diagnostic method for veterinarians to detect *C suis* in the field. Limitations of this study include utilizing one sow farm for evaluation and the ability to consistently block by equivalent parity and clinical prevalence in one farrowing room. Further research should be conducted to validate the findings from this study and provide insight into how this sampling and diagnostic approach can improve the industry's knowledge on *C suis* detection and control.

References

1. Joachim A, Ruttkowski B, Sperling D, et al. Detection of *Cystoisospora* suis in faeces of suckling piglets – when and how? A comparison of methods. *Porcine Health Management.* 2018;4(20).

2. Aliaga-Leyton A, Friendship R, Dewey C, Todd C, Peregrine A, et al. *Isospora suis* infection and its association with postweaning performance on three southwestern Ontario swine farms. *J Swine Health Prod.* 2011;19(2):94-99.

3. Gong Q, Zhao W, Wang Y, Zong Y, Wang Q, Yang Ya, Yang Yi, Shi K, Li J, Leng X, Du R, Zhano Q, et al. Prevalence of coccidia in domestic pigs in China between 1980 and 2019: a systematic review and meta-analysis. *Parasit Vectors.* 2021;14(1):248.

https://doi.org/10.54846/am2024/22

Characterizing *E coli* isolates contributing to postweaning diarrhea in pigs in Ontario, Canada: A case-control study

Jenny Lin[1,2], BSc; Corinne Schut[1,2], BSc, MSc; John Morris Fairbrother[3], BSc, PhD; Brandon N Lillie[2], DVM, PhD, DACVP; Robert Friendship[1], DVM, MSc; Vahab Farzan[1,2], DVM, MSc, PhD

[1]Department of Population Medicine, Ontario Veterinary College, University of Guelph; [2]Department of Pathobiology, Ontario Veterinary College, University of Guelph; [3]Department of Pathology and Microbiology, Faculty of Veterinary Medicine, Université de Montréal

Introduction

Post-weaning diarrhea (PWD) poses an important challenge to the swine industry due to associated economic losses, welfare implications, and increased antimicrobial use.[1] Although PWD is a multifactorial disease, an important component is the causative agent, enterotoxigenic *Escherichia coli* (ETEC). *E coli* bacteria form part of the intestinal microbiome, however, some strains possess genes for virulence factors that contribute to the development of disease, such as fimbriae, adhesins, and enterotoxins.[2,3] In particular, *E coli* isolated from post-weaning diarrheic pigs are often enterotoxigenic *E coli* (ETEC), characterized by the expression of fimbriae and enterotoxins.[2,3] ETEC may be further virotyped, based on the combination of virulence factors.[3]

Fimbriae and afimbrial adhesins are virulence factors that allow bacteria to colonize the small intestine through interaction with receptors expressed on the small intestinal microvilli in susceptible animals, without morphological destruction of enterocytes.[2] Fimbriae designated F4 (previously K88) and F18 have been identified in most ETEC contributing to PWD, though some F4- and F18-negative virotypes exist.[4,5] Other fimbriae that are sometimes found on ETEC from PWD cases include F5 (K99), F6 (987P), and F41.[6] Afimbrial adhesins that have been identified include the adhesin involved in diffuse adherence (AIDA), intimin (EAE), and porcine attaching and effacing associated (Paa), though their significance in PWD is unknown.[7-10]

The hypersecretory diarrhea seen in PWD due to ETEC is the result of the bacterial secretion of one or more enterotoxins, such as heat-labile enterotoxin (LT) or heat-stable enterotoxins (Sta, STb, and EAST-1).[1] These enterotoxins induce excessive electrolyte and water secretion into the intestinal lumen while inhibiting their absorption, leading to diarrhea with subsequent dehydration and acidosis.[3]

Identification of the ETEC virotypes contributing to PWD can help inform the development of targeted vaccination programs. However, the prevalence of ETEC virotypes varies with geographic range and timeframe, highlighting the necessity of understanding and monitoring the regional prevalence of virotypes.

Objective

This project aimed to identify virulence factor genes present in *E coli* isolated from post-weaning pigs on a subset of swine farms in southern Ontario, Canada, and compare their distribution between healthy and sick pigs.

Methods

Sample collection

Fifteen swine farms with a current outbreak of PWD and a history of ETEC-associated disease located in southern Ontario, Canada were visited from August 2022 to August 2023 and fecal swabs were collected from 5 healthy pigs (controls) and 10 pigs with PWD (cases). PWD was diagnosed based on clinical signs of watery diarrhea, thin body condition, and an inflamed anus; the age matched controls were chosen from the same nursery batch based on the absence of PWD clinical signs.

E coli isolation

Fecal swabs were cultured on both Columbia blood agar (CBA) and MacConkey (MAC) plates. After incubation at 37°C for 24 hours, colonies with morphology and media differential characteristics suggestive of *E coli* were isolated on CBA for further testing.

Virotyping

The suspected *E coli* isolates were subjected to DNA extraction using InstaGene™ Matrix (Bio-Rad Laboratories (Canada) Ltd., Mississauga, ON, CA). Multi-plex PCR was used to determine the presence of virulence factor genes including those for F4, F18, F5, F6, F41, EAE, AIDA, Paa, STa, STb, Stx1, Stx2, LT, and EAST1. Virulence factor gene primers were separated into a total of 5 reactions: **1)** F4, F18, LT **2)** STa, Stx2 **3)** Stx2, STb **4)** EAE, Paa, F5, AIDA and **5)** F41, F6, EAST-1. In brief, each reaction consisted of 1 µL of extracted DNA template, 0.25 µM primers, 12.5 µL Taq FroggaMix (FroggaBio, Concord, Ontario, CA), and up to 25 µL of water. Thermocycling was performed at 30 cycles: **1)** Denaturing at 94°C for 30 seconds **2)** Annealing at 30°C for 30 seconds **3)** Extension at 72°C for 30 seconds. Multi-plex PCR products were analyzed with gel electrophoresis. Virulent *E coli* were defined as isolates that tested positive for at least one virulence factor gene.

Data analysis

A multi-level mixed effects logistic regression modelling method, with farm as a random effect, was used to compare virulent *E coli* shedding and virulence factors detected in case and control pigs.

Results

A total of 225 pigs (75 controls, 150 cases) across 15 farms were sampled, 122 (54.2%) of which tested positive for virulent *E coli*. The cases were more likely to shed virulent *E. coli* than controls ($P = .019$). Moreover, the isolation of virulent *E coli* was also related to season, with more pigs testing positive on farms visited in the spring ($P = .045$) and autumn ($P = .043$) compared to winter. A total of 197 virulent *E coli* isolates were identified. Hemolysis was noted in 188/197 (95.4%) virulent *E coli* isolates, including 52/55 (94.6%) control isolates and 133/142 (93.7%) case isolates.

The most frequently detected fimbriae genes were F4 (5/75 (6.7%) controls and 15/150 (10%) cases) and F18 (4/75 (5.3%) controls and 13/150 (8.7%) cases). At the farm level, 4 farms tested positive only for F4, 1 farm only for F18, 1 farm for F6/F18, 1 farm for F5/F18, 1 farm for F4/F6/F18, 1 farm for F4/F5/F6, whereas 6 farms had no isolated virulent *E coli* carrying fimbriae genes (Table 1). No statistically significant association was found between the detection of tested fimbriae genes and the health status of the pigs.

The most frequently detected enterotoxin genes were for EAST-1 (23/75 (30.7%) controls and 59/150 (39.3%) cases), LT (13/75 (17.3%) controls and 47/150 (31.3%) cases), and STb (13/75 (17.3%) controls and 45/150 (30.0%) cases) (Figure 1). At the farm level, 14 (93.3%) farms tested positive for EAST-1 but there was no statistically significant association between EAST-1 and health status of the pigs. In terms of LT and STb, 9 farms tested positive for LT/STb, 3 farms for LT, 1 farm for STb, while 2 farms had no isolated virulent *E. coli* carrying either LT or STb (Table 1). The odds of shedding *E coli* carrying the gene for LT ($P = .011$) and STb ($P = .010$) were 2.8 and 3.3 times greater in cases than controls, respectively Additionally, 11/60 (18.3%) LT positive pigs also tested positive for F18 and 10/60 (16.7%) tested positive for F4. 13/58 (22.4%) of STb positive pigs tested positive for F18 and 17/58 (29.3%) tested positive for F4. In contrast, the gene for EAE was detected in *E coli* isolated from 10/75 (13.3%) controls but only from 4/150 (2.7%) cases (Figure 1). At the farm level, 7/15 (46.7%) farms tested positive for EAE (Table 1). The odds of shedding *E coli* carrying the gene for EAE were 6.7 times greater in controls than cases ($P = .004$).

Discussion

The findings indicate that the limited success of current vaccination programs targeting F4/F18 may be in part due to the between-farm variation of *E coli* virotypes, with many enterotoxigenic strains not carrying genes for any tested fimbriae or adhesins. Virotyping of *E coli* isolated from individual farms with an outbreak of PWD may be worthwhile to gauge the potential effectiveness of vaccines. Though the presence of the genes for fimbriae did not differ significantly in *E coli* isolates recovered from controls and cases, this may be due to the limited fimbriae-positive isolates identified in this study. Additionally, LT and STb may be promising targets for future preventative programs given their frequency and association with disease. It is also possible that genetically resistant pigs shed ETEC but did not develop the clinical signs of PWD, which should be further investigated.

References

1. Nagy B, Fekete PZ. Enterotoxigenic *Escherichia coli* in veterinary medicine. *Int. J. Med. Microbiol.* 2005;295(6):443-454.
2. Abubakar RH, Madoroba E, Adenubi O, Morar D, Fasina, FO. Bacterial pathogens of pigs with particular reference to *Escherichia coli*: A systematic review and meta-analysis. *J. Vet. Med. Anim. Health.* 2017;9(7):159-185.
3. Dubreuil JD, Isaacson RE, Schifferli DM. Animal Enterotoxigenic *Escherichia coli*. *EcoSal Plus.* 2016; 7(1).
4. Frydendahl, K. Prevalence of serogroups and virulence genes in *Escherichia coli* associated with postweaning diarrhoea and edema disease in pigs and a comparison of diagnostic approaches. *Vet. Microbiol.* 2002; 85:169–182.
5. Luppi, A. Swine enteric colibacillosis: diagnosis, therapy and antimicrobial resistance. *Porc. Health Manag.*2017; 3(1):16 (2017).
6. Vereecke N, et al. Virotyping and genetic antimicrobial susceptibility testing of porcine ETEC/STEC strains and associated plasmid types. *Front. Microbiol. 2023;*14.

Table 1: Presence of virulence factor genes on 15 swine farms in Ontario, Canada

Virulence factors	Farm														
	1	2	3	4	5	6	7	8	9	10	11	12	13	14	15
F4	0	7	0	3	0	0	0	0	1	0	0	3	0	1	5
F5	0	0	0	0	0	0	0	0	0	0	0	0	1	0	1
F6	0	0	0	0	0	0	0	0	0	0	1	1	0	0	1
F18	0	0	0	0	0	0	0	4	0	0	2	5	6	0	0
F41	0	0	0	0	0	0	0	0	0	0	0	0	0	0	0
EAE	0	0	0	0	2	1	0	1	0	0	0	3	2	4	1
Paa	0	11	0	3	2	3	0	1	1	0	2	1	0	5	3
AIDA	0	0	0	0	0	2	0	0	0	0	1	4	4	1	7
LT	1	5	0	3	4	8	1	10	0	0	2	9	7	1	9
STa	0	7	0	0	0	0	0	1	0	0	0	2	6	1	11
STb	0	11	0	3	0	2	0	9	1	0	1	7	8	3	13
EAST-1	2	5	3	2	8	10	1	9	5	0	5	8	6	8	10
Stx1	0	0	0	0	0	0	0	0	0	0	0	0	0	0	0
Stx2	0	0	0	0	0	0	0	2	0	0	0	1	0	0	0

Figure 1: Frequency of virulence factor genes detected in *E coli* isolates recovered from case (n = 150) and control (n = 75) pigs.

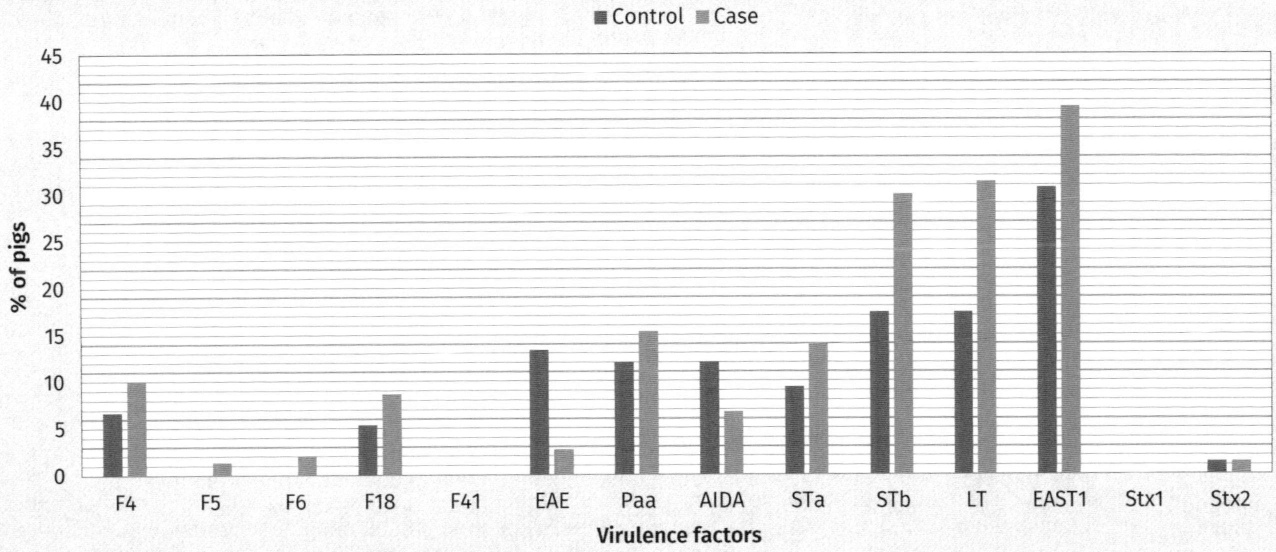

7. Batisson I, et al. Characterization of the Novel Factor Paa Involved in the Early Steps of the Adhesion Mechanism of Attaching and Effacing *Escherichia coli*. *Infect. Immun.* 2003;71(8):4516–4525.

8. Malik A, et al. Serotypes and intimin types of intestinal and faecal strains of eae+ *Escherichia coli* from weaned pigs. *Vet. Microbiol.*2006:114(1):82–93.

9. Ngeleka M, Pritchard J, Appleyard G, Middleton DM, Fairbrother JM. Isolation and Association of *Escherichia Coli* AIDA-I/STb, Rather than EAST1 Pathotype, with Diarrhea in Piglets and Antibiotic Sensitivity of Isolates. *J. Vet. Diagn. Invest.*2003;15(3), 242–252.

10. Sherlock O, Schembri MA, Reisner A, Klemm P. Novel Roles for the AIDA Adhesin from Diarrheagenic *Escherichia coli*: Cell Aggregation and Biofilm Formation. *J. Bacteriol.* 2004;186(23):8058–8065 (2004).

https://doi.org/10.54846/am2024/23

Comparison of meloxicam dosing protocols in periparturient sows and the effect on fever reduction and performance

E. Larsen[1], MS; A. Anderson-Reever[2], DVM, MS; B. Burton[2], DVM; B. Payne[3], DVM
[1]Lincoln Memorial University, Harrogate, Tennessee; [2]Suidae Health & Production, Algona, Iowa, [3]Veterinary Pharmaceutical Solutions (VPS), St. Peter, Minnesota

Introduction

The periparturient period is one of the most stressful events for sows and their piglets. Prolonged inflammation and fevers can result in negative implications for both the dam and the litter. Meloxicam is a nonsteroidal anti-inflammatory drug (NSAID) that has been shown to have analgesic, anti-inflammatory, anti-endotoxic, and antipyretic effects.[1] Internationally, meloxicam, in conjunction with antibiotics, has been approved for supportive therapy in the treatment of mastitis-metritis-agalactia syndrome and lameness.[2-5] In previous studies, a single meloxicam dosage has been shown to reduce rectal temperatures at 12- and 24-hours post-farrowing, increase litter wean weights, lower preweaning mortality rates and improve sow recovery post-farrowing.[6-9] Additional benefits of meloxicam dosing in sows include improved laying position of lactating sows, increases in colostrum quality and transfer of antibodies to piglets.[9-11] Although previous studies have indicated benefits to meloxicam administration, there is a lack of literature comparing dosing protocols. The objective of this study was to determine the effects of one versus two post farrowing doses of meloxicam on maternal fever reduction, piglet performance, and subsequent breeding metrics.

Materials and methods

In a 5,000-head commercial sow farm, 403 mixed parity sows, and their litters were randomly assigned to one of three treatment groups. Sows were enrolled in the study after the farm staff determined farrowing was complete. All breeding females were blocked by parity group (Parity Group 1 = gilts, PG2 = P2 - P3, and PG3 = P4+) to ensure equal distribution during randomization. The one-dose treatment group received a single treatment of compounded 1.5% meloxicam suspension (90mg/6mL, VPS, St. Peter, Minnesota) post-farrowing, while the two-dose treatment group received the same treatment post-farrowing as well as a second equal dose approximately 48 hours later. The control group did not receive any NSAID. Treatment was administered in the sow feeder instead of a direct oral drench. Sows that farrowed overnight were treated with meloxicam the following morning. Any sow that completed farrowing during the day was treated with meloxicam prior to the farm staff leaving for the day. Sow rectal temperatures were recorded upon farrowing completion and every 24 hours for 3 days post-farrowing. Fever was defined as a rectal temperature of ≥ 103°F and was the primary parameter of the study. Sows that had a fever over 104°F were additionally treated following the farm's standard treatment protocol with up to three days of ampicillin and flunixin. Litter records, treatment information, and subsequent retention and breeding data were recorded. Sows were excluded from the fever and lactation phase of the study if they died or were removed from the farrowing room prior to the completion of all rectal temperature measurements but remained in the second phase when measuring overall retention of the sows in the herd until the subsequent farrowing. Cross fostering of piglets was allowed within treatment groups but not tracked from litter to litter.

Results and discussion

Initial power calculations using a previous study for fever reduction, the primary parameter, indicated 210 sows per treatment group were necessary to detect statistical significance.[6] Due to time and labor constraints, this study was unable to achieve the necessary sample size for finding statistical differences for fever in the 6-week period of enrollment. Therefore, this study served as a pilot project, and metrics were analyzed for numeric trends only. The percentage of fevers immediately post-farrowing and before any NSAID treatment was numerically similar (16 - 20%) between treatment groups (Figure 1). Across the whole study, only subtle numeric trends were noted between treatment groups for fever reduction (5 - 7%). However, a higher fever percentage and greater (14 - 23% above PG2 and PG3 averages) measured response to meloxicam administration were observed in PG1 first-litter sows (Figure 2). At the 72-hour post-farrowing timepoint, fever percentage was lowest in the two-dose group, intermediate in the one-dose group, and highest in the control group (37.1%, 42.5%, 51.4%, respectively). While feverish control group sows post-farrow rectal temperatures increased during the first 24 hours, the one and two-dose Meloxicam groups fevers reduced by 0.47°F and 0.67°F, respectively.

Recorded preweaning mortality rate was lowest and pigs weaned per sow was highest in the two-dose meloxicam group. However, the raw data is not clear due to unaccounted for pigs and results. Likely, these missing pigs are due to bump weaning and recording in the records system, rather than truly missing or dead pigs.

In agreement with Payne,[12] subsequent reproduction after weaning showed sows receiving meloxicam had 7 - 9% more sows bred within 7 days post-weaning compared to the control group (Figure 3). Similarly, the average wean-to-first-service intervals increased from greatest to least number of meloxicam administrations with standard deviations decreasing (4.44 ± 1.46, 4.59 ± 1.89, 4.69 ± 2.56, respectively, Figure 4). The control group has the lowest retention rate (86.0%) from enrollment to one and a half months post-weaning compared to the one and two doses of meloxicam (both at 89.4%). Retention through subsequent farrowing will be reported fully once all sows farrow. These findings are directionally comparable to previously reported research all the way through the subsequent farrowing.[12]

This study was performed on a healthy sow herd. Today, peri-farrowing compounded meloxicam administration to sows is being used in healthy and challenged herds. The results of this study support the use of this selective-COX2 inhibiting NSAID in sows going through normal inflammation and stressor events

Figure 1: Percent of all sows with fever (rectal temperature ≥ 103°F).

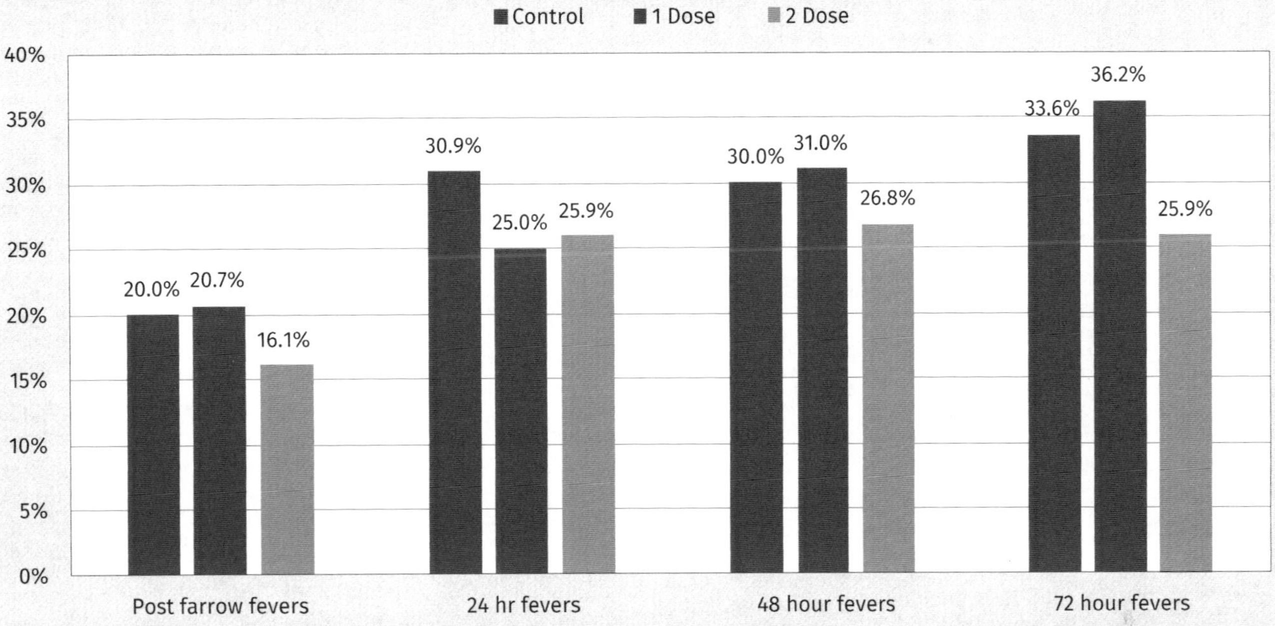

Figure 2: Percentage of parity group 1 gilts with fever (rectal temperature ≥ 103°F).

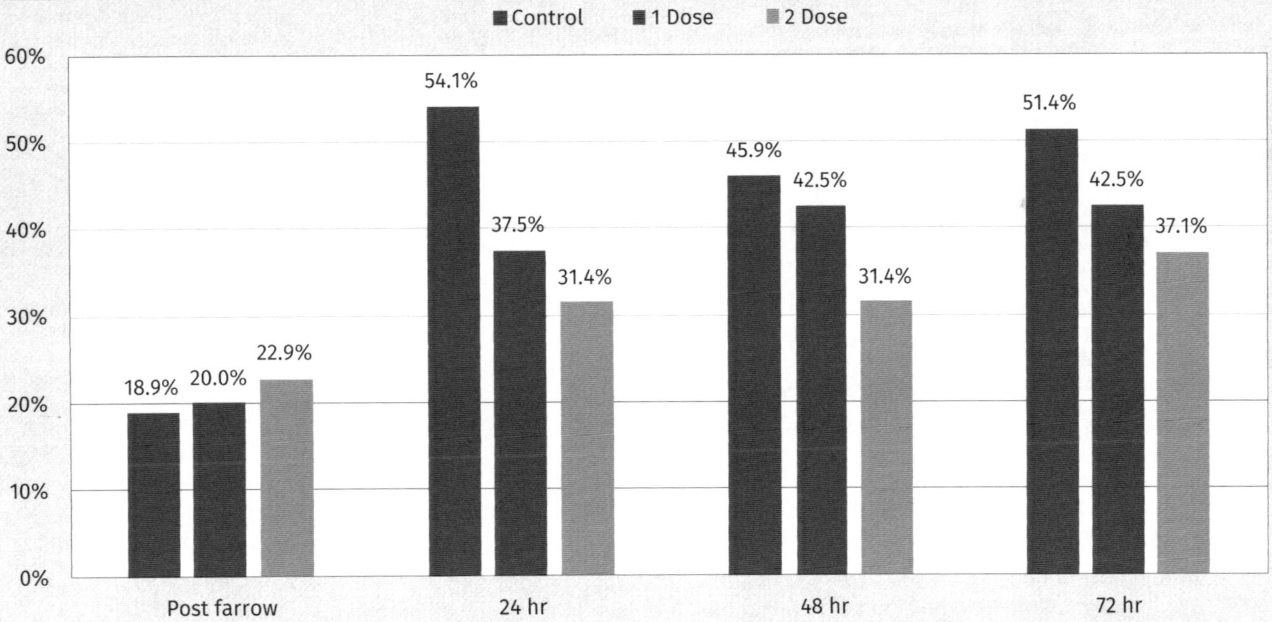

Figure 3: Percentage of all sows bred < 7 days post-weaning.

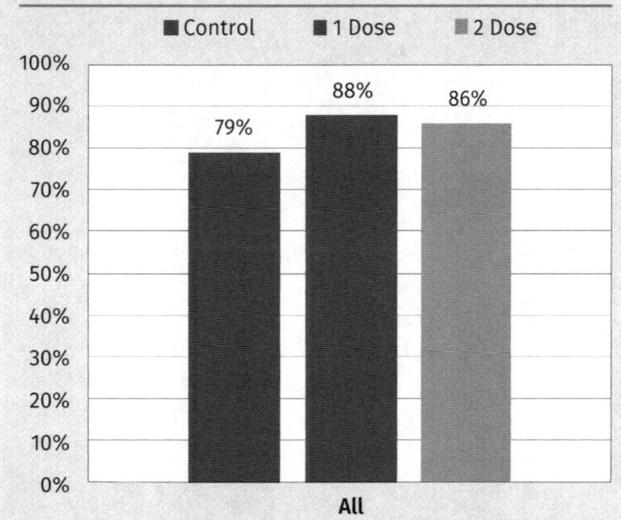

Figure 4: (Mean ± SD) Wean to first service interval, including all parities.

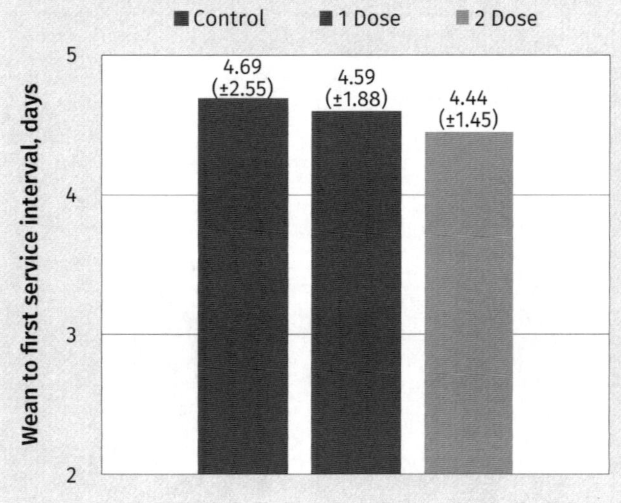

surrounding farrowing, uterine involution, and lactation. This study used a novel method of meloxicam administration (in the feeder vs oral drench) which required minimal labor. Through this method, meloxicam was easy to administer and was consumed readily by most sows. It is noted that the only true way to ensure the sow consumed the full intended dose of meloxicam is through oral drenching. Any sow that is not up and at the feed pan, should always have direct administration. As sow comfort and welfare are of the utmost importance to pork producers, the numeric trends identified in the study are supportive of previous studies and indicate it is likely worth exploring further benefits of a two-dose meloxicam protocol including timing (before and after farrowing).

References

1. Friton G, et al. Clinical and anti-inflammatory effects of treating endotoxin-challenged pigs with meloxicam. *Vet Rec*. 2006 Oct 21;159(17):552-7. **doi: 10.1136/vr.159.17.552**

2. Ubiergo A. Preliminary findings on the impact of meloxicam treatment on litter performance of sows with MMA syndrome. Proceedings 18th IPVS Congress, Hamburg, Germany, 2004. v2, 607.

3. Revilla E. Post-farrowing treatment of sows with meloxicam on the preweaning weight gain and mortality rate of the low birth weight piglets in subclinical MMA. Proceedings 19th IPVS Congress. Denmark. 2006.

4. Pairis-Garcia M, et al. Behavioral evaluation of analgesic efficacy for pain mitigation in lame sows. *Anim Welfare* 2015. 24:93-99.

5. Hood M, et al. Evaluation of oral Meloxicam to reduce symptoms of osteochondritis dissecans in developing boars. Allen D. Leman Swine Conference. 2023. p6.

6. Payne B, et al. Evaluation of meloxicam in sows post-farrowing on piglet performance. AASV. Colorado, 2023, pp209-10.

7. Finestra A, et al. Post-farrowing treatment of sows with meloxicam or tolfenamic acid on the preweaning mortality rate of the low birth weight piglets in subclinical MMA. Proceedings 19th IPVS Congress. Denmark, 2006, p477.

8. Keller F. Improved piglet survival after Metacam treatment of sows post farrowing. Proceedings 22nd IPVS. Korea, 2012, p249.

9. Mainau E, et al. Effects of meloxicam (Metacam) on post-farrowing sow behaviour and piglet performance. *Animal*. 2012, 6:3, pp494-501.

10. Koenders K, et al. Administering meloxicam to postpartum sows reduces within litter variation of IgG titres. Proceedings 20th IPVS. Mexico, 2014 p119.

11. Navarro E, et al. Oral meloxicam administered in sows at farrowing and its effects on piglet immunity transfer and growth., *Frontiers in Vet Science*. 2021, v8 art. 574250.

12. Payne B, et al. The impact of Meloxicam administered post-farrowing to healthy sows on their subsequent reproduction (phase 2). Allen D. Leman Swine Conference. 2023. p22.

https://doi.org/10.54846/am2024/24

Evaluation of non-steroidal anti-inflammatory water medications on PRRSV 1-4-4 L1C variant

Morgan Almeida[1]; Dave Ammann[2]; Gustavo Silva[3], DVM, PhD; Jeff Okones[4], DVM; Mark Wagner[2], DVM
[1]College of Veterinary Medicine, Iowa State University, Ames, Iowa; [2]Fairmont Veterinary Clinic, Fairmont, Minnesota;
[3]Veterinary Diagnostic and Production Animal Medicine department, Iowa State University, Ames, Iowa;
[4]Pharmgate Animal Health, Wilmington, North Carolina

Introduction

Though years of extensive research have targeted porcine reproductive and respiratory syndrome virus (PRRSV), it remains one of the largest sources of lost revenue in the swine industry at $664 million annually.[1] In nursery-age pigs, clinical signs can include, but are not limited to, fever, anorexia, depression, and a decreased average daily gain.[2] There is a need to find mitigation strategies and therapies that consistently aid in recovery, improve well-being, and decrease losses. This study aimed to determine if non-steroidal anti-inflammatory water medications could effectively reduce fever duration and improve gain in nursery-age pigs challenged with PRRSV 1-4-4 L1C variant.

Materials and methods

In this randomized controlled trial, seventy-four 21-day-old pigs were sourced from a PRRSV-naïve herd and tested negative for PRRSV RNA by PCR on serum prior to placement. Pigs arrived on day post-challenge (DPC) -5 to allow time for acclimation to the facility and feeding program; each animal received a 1 mL dose of ceftiofur. On DPC 0, all pigs were weighed and sorted into three weight groups: < 9 pounds, 10 - 11 pounds, or > 12 pounds (rounded for disparities). Once all animals were appropriately placed, a pig was taken from each group, randomly assigned a treatment group, tagged, reweighed, rectal temperature taken, given 1 mL of enrofloxacin, and challenge groups were inoculated with a 2 ml PRRSV 1-4-4 L1C variant (intramuscular). The inoculum consisted of 1 mL serum positive for PRRSV 1-4-4 L1C variant mixed with 200 ml of phosphate buffer solution. Each 2 mL IM injection of the final inoculum had a qPCR of 10^6 and a Ct of 16. Daily rectal temperatures were taken from DPC 0 and beyond (Agri-Pro Enterprises™). Baseline temperature was defined as 103.1°F, the average temperature of all pigs at DPC 0, and fever as ≥ 104.1°F. Pen average fevers began on DPC 4, and water medications were then administered to appropriate groups. The six treatment groups were alcohol-soluble acetylsalicylic acid (4 oz per gal stock), water-soluble acetylsalicylic acid (8 oz per gal stock), sodium salicylate (10 oz per gal stock), meloxicam (3 oz per gal stock), challenged non-treated group, and non-challenged non-treated group. Number of pigs per treatment group is found in Table 1. Water medications were mixed fresh daily, while daily water and feed intake were tracked. On DPC 15, the average temperature of the challenged non-treated group was below the 104.1°F fever baseline; water medications were stopped at this time, and a final weight was taken on each of the pigs.

Analysis was done using R to determine statistical significance. ANOVA was performed to assess differences in performance metrics between treatments, and the Tukey-Kramer method was used for post hoc comparison. To evaluate fever, a cut-off of [3] 104.1°F was used to categorize the pigs that presented fever or not per treatment. The Fisher exact test was run to assess the significance of the proportion between the number of pig days with fever between treatments.

Results

All challenged groups had lower weight gain compared to the non-challenged non-treated group. A statistical difference was not observed in weight gain among the challenged groups; however, there were numerical differences as noted in Table 1. Meloxicam statistically reduced the number of pig days with fever compared to the challenged non-treated and alcohol-soluble acetylsalicylic acid groups but had comparable results to water-soluble acetylsalicylic acid and sodium salicylate as noted in Table 1. No statistical difference regarding fever was observed in sodium salicylate, water-soluble acetylsalicylic acid, alcohol-soluble acetylsalicylic acid, or the challenged non-treated groups.

Table 1: Production information by treatment group

Treatments	Pens per treatment group	Pigs per treatment group	Weight gain (21 days)	Number of pigs days with fever
Alcohol-soluble acetylsalicylic acid	2	10	5.0[b]	76 (44.7%)[c]
Water-soluble acetylsalicylic acid	2	10	6.2[b]	65 (38.2%)[bc]
Challenged non-treated	4	22	6.8[b]	70 (37.0%)[c]
Sodium salicylate	2	10	7.9[b]	61 (35.9%)[bc]
Meloxicam	2	10	8.4[b]	53 (31.2%)[b]
Non-challenged non-treated	2	12	13.8[a]	21 (10.3%)[a]

[a,b,c] Within a column with different superscripts represent statistical differences among treatment groups ($P < .05$).

Discussion

Data suggested that pigs in groups experiencing fewer days with fever had a numerically greater weight gain. With nursery weight showing a strong correlation to finishing weight,[3] health must be prioritized at early post-weaning stages, and pharmacological interventions that act quickly and effectively to maintain intake and growth are essential; this was noted in the meloxicam and sodium salicylate groups, which had the greatest numerical weight gain. Clinical observation would suggest fewer partial or no-value pigs in the meloxicam and sodium salicylate groups. Cost-wise, each product's price per gallon of stock solution ranged between \$1.50 - \$6.00. Alcohol-soluble acetylsalicylic acid tended toward the lower end, whereas meloxicam tended to be near the higher end. Further research may be warranted, including a larger sample size and multiple manufacturers for each treatment group to validate weight gain and performance differences.

References

1. Holtkamp, D. J. & Kliebenstein, J. B. & Zimmerman, J. J. & Neumann, E. & Rotto, H. & Yoder, T. K. & Wang, C. & Yeske, P. & Mowrer, C. L. & Haley, C., (2012) "Economic Impact of Porcine Reproductive and Respiratory Syndrome Virus on U.S. Pork Producers", *Iowa State University Animal Industry Report* 9(1). **doi: 10.31274/ans_air-180814-28**

2. Zimmerman, J.J., Dee, S.A., Holtkamp, D.J., Murtaugh, M.P., Stadejek, T., Stevenson, G.W., Torremorell, M., Yang, H. and Zhang, J. (2019). Porcine Reproductive and Respiratory Syndrome Viruses (Porcine Arteriviruses). In Diseases of Swine (eds J.J. Zimmerman, L.A. Karriker, A. Ramirez, K.J. Schwartz, G.W. Stevenson and J. Zhang). **doi: 10.1002/9781119350927.ch41**

3. De Rodas, B. *Correlation of Weaning Numbers to Finishing Productivity. Summary of data from Purina Animal Nutrition Center.* August 14, 2015.

https://doi.org/10.54846/am2024/25

A comparison of ceftiofur crystalline free acid at birth to enrofloxacin at processing and the effects on umbilical defects and average daily gain

S. Lutz[1], BS; D. Amodie[2], MS; M. Jansen[2], DVM
[1]Auburn University, Auburn, Alabama; [2]Zoetis, Parsippany, New Jersey

Introduction

Neonatal pigs are exposed to a variety of bacterial pathogens early in life. This can lead to infection of the navel cord, joints, or even systemic infection.[1] Many swine practitioners implement antibiotic control programs in piglets at, or soon after, birth to improve overall health and reduce the likelihood for complications due to infection.[2] The objective of this study was to compare two antibiotic protocols, ceftiofur crystalline free acid (CCFA) at birth and enrofloxacin at processing, by evaluating umbilical defects and average daily gain (ADG).

Materials and methods

Piglets (n = 1250) from one commercial sow farm, born to sows of mixed parity, and greater than two pounds were randomly allocated within litter to three treatment groups in multiples of five (block) within 24 hours of birth (d0). Cultures of umbilical stump swabs were performed to confirm presence of susceptible bacterial pathogens. Treatment groups used in this study included a negative control group, T01 (n = 250) which received no antibiotic treatment; T02 (n = 500) received CCFA at birth (d0), and T03 (n = 500; current farm protocol) received enrofloxacin at processing (d3) (Table 1); both antibiotics were dosed based on label dose. On d0, weights, gender, and sow parity were recorded. T02 pigs received 0.15 mL CCFA (Excede for Swine®, Zoetis, Parsippany, New Jersey) IM in the neck at enrollment. T03 pigs received 0.11 mL enrofloxacin (Baytril®, Elanco, Greenfield, Indiana) IM in the neck on d3. All pigs were weighed prior to weaning and umbilical stumps scored based on a previously published method.[1] Navels were scored on a scale of 0 to 5 (Figure 1). A normal navel area with no signs of swelling or a hernia was scored as 0. A score of 1 indicated a scab but no redness or hernia. A score of 2 was a rosy red appearance of the umbilicus, but no hernia. A score of 3 indicated there was no hernia or scab, but the umbilicus was hard when palpated. A score of 4 was an umbilicus that was hard with pus when squeezed. A score of 5 indicated a hernia was present. Scores of 4 and 5 were classified as umbilical defects for the purpose of this study.

Pigs were observed again for umbilical defects 6 weeks post weaning and individual weights were recorded. At that time, palpable hernias or abscesses were classified as umbilical defects.

Average daily gain (ADG) was analyzed as a continuous variable with the Proc Mixed Procedure (SAS 9.4, Cary, North Carolina) while defects were analyzed as a binomial variable with the Proc Glimmix Procedure. Least squares means were calculated and compared by the two-sided t-test. All *P*-values were tested at the 5% alpha level.

Results and discussion

Results are shown in Table 2, Figure 2, and Figure 3. While both T02 and T03 had greater average daily gain (ADG) from birth to weaning, all treatments were similar when ADG was calculated in the nursery phase or from birth to end of nursery. Umbilical defects were similar among treatments in the nursery, but pigs administered CCFA at birth had reduced defects at weaning, and overall, compared to controls. Additionally, there were statistically fewer CCFA pigs that ever had an umbilical defect when compared to controls. Results suggest that CCFA at birth may have been more successful than enrofloxacin at processing in reducing the prevalence of umbilical defects; however, it is unknown if this advantage was due to treatment, timing, or both.

References

1. Reiman J, Folkerts C, Greiner L, et. al. Bacterial infection of the navel cord blood clot in neonatal piglets: prevalence and clinical outcomes. AASV Annual Meeting, 2010.

2. Reiman J, Greiner L, Lowe J, et. al Bacterial infection of the navel cord in neonatal piglets: A follow up on prevalence and clinical outcomes. AASV Annual Meeting, 2012.

Table 1: Treatment groups

Treatment group	T01: Control	T02: Proposed protocol	T03: Current protocol
Treatment	No antibiotic treatment	0.15 mL CCFA	0.11 mL enrofloxacin
Time of treatment	None	Day 0 at birth	Day 3 at processing
Number per group	250 pigs	500 pigs	500 pigs

CCFA = ceftiofur crystalline free acid

Figure 1: Umbilical defect scoring system.[1]

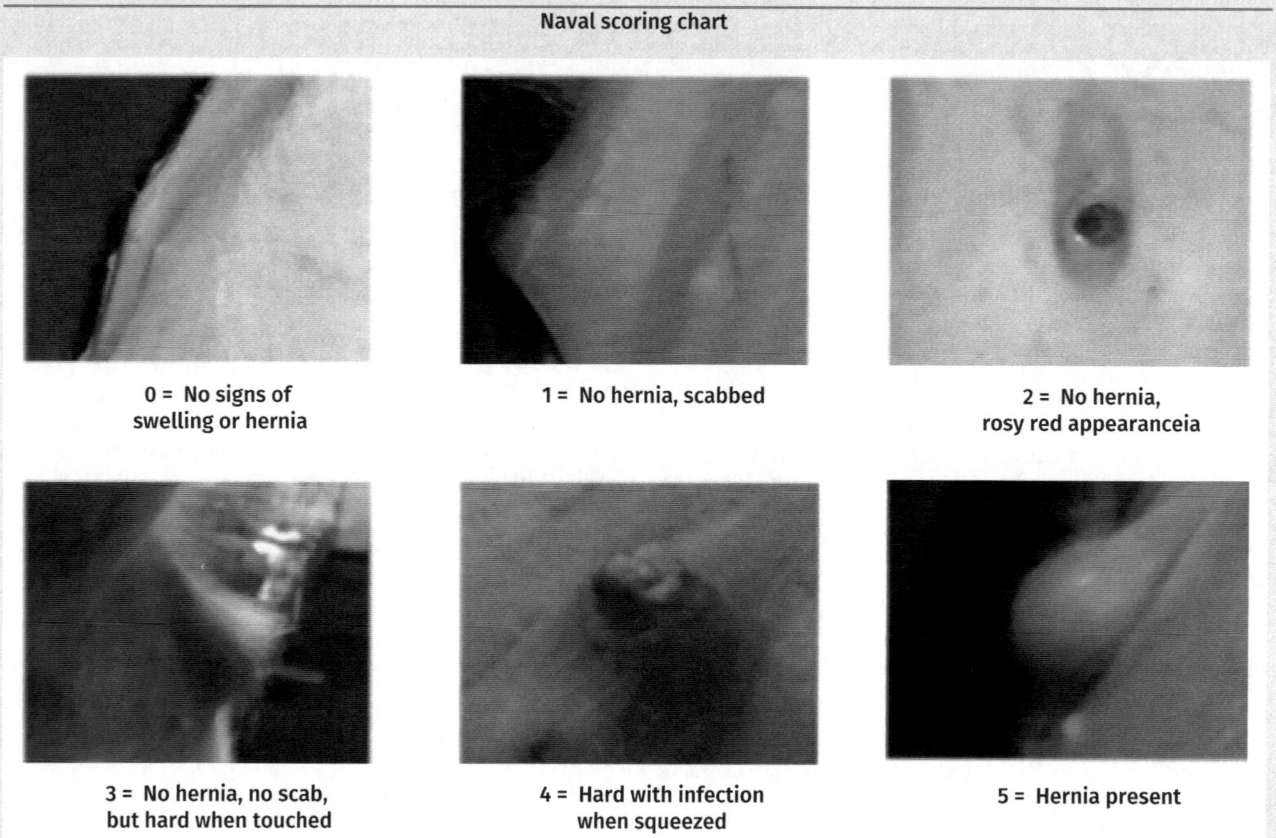

Table 2: Least squares means (± standard error of the mean, SEM) of % umbilical defects and ADG between treatment groups

Measure	T01: Control	T02: Proposed protocol	T03: Current protocol
Umbilical defects at weaning (%)	7.28 (0.02)^{a}	2.49 (0.01)^{b}	3.76 (0.01)^{ab}
Umbilical defects at nursery (%)	2.59 (0.01)^{a}	1.63 (0.01)^{a}	2.03(0.01)^{a}
Total umbilical defects (%)	8.68 (0.02)^{a}	3.87 (0.01)^{b}	5.32 (0.02)^{ab}
Birth weight at enrollment (lb)	3.356 (0.064)^{a}	3.371(0.062)^{a}	3.364 (0.062)^{a}
Birth to weaning ADG (lb/day)	0.406 (0.010)^{b}	0.425 (0.008)^{a}	0.425 (0.008)^{a}
Weaning to nursery ADG (lb/day)	0.769 (0.016)^{a}	0.788 (0.012)^{a}	0.800 (0.012)^{a}
Birth to nursery ADG (lb/day)	0.675 (0.013)^{a}	0.693 (0.011)^{a}	0.699 (0.11)^{a}

ab LSMeans ± SEM within a row with different superscripts differ, $P \leq .05$

Figure 2: Least squaresmeans, umbilical defects (%). Bars within a category, with different letter groupings, are significantly different.

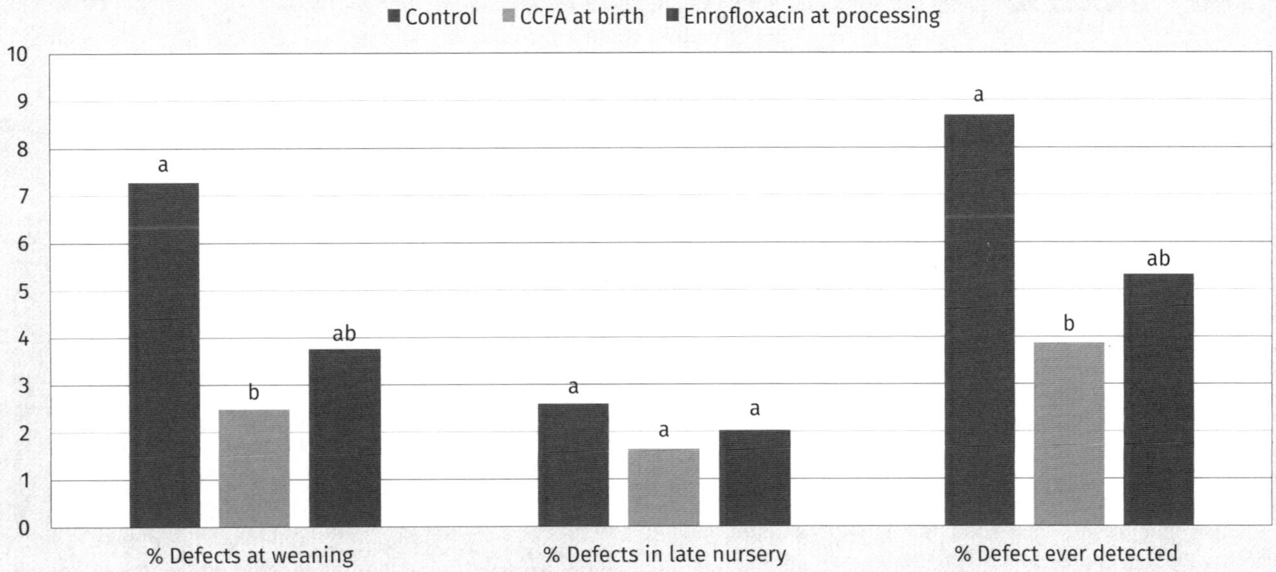

Figure 3: Least squaresmeans, average daily gain (lb/d). Bars within a category, with different letter groupings, are significantly different.

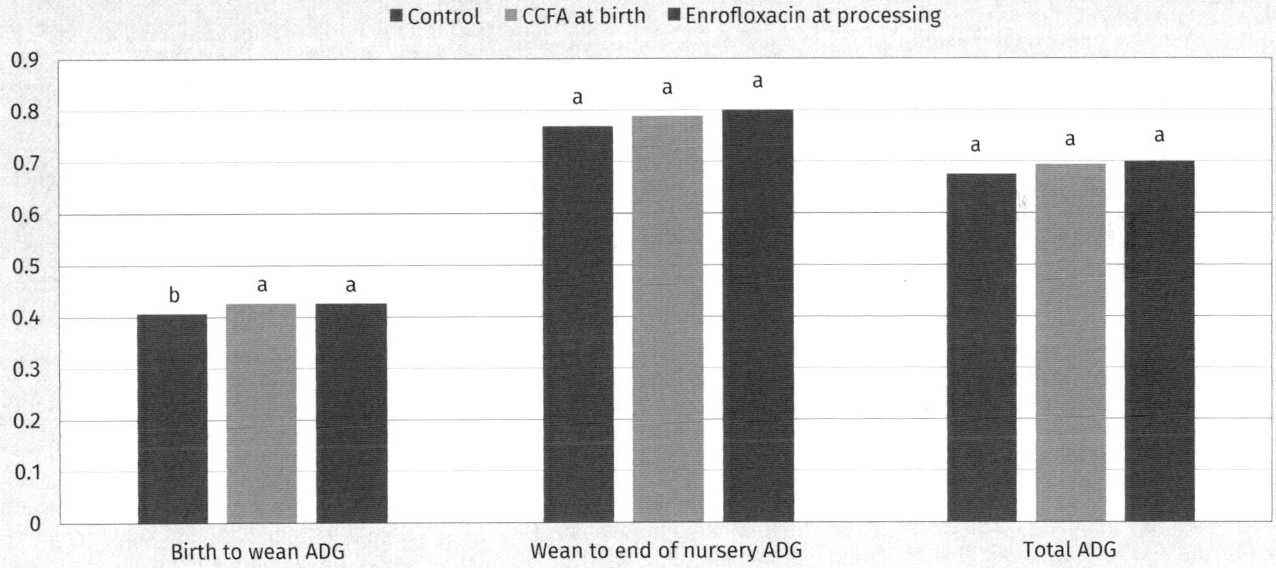

https://doi.org/10.54846/am2024/26

Evaluation of feed mill biosecurity using fluorescent powder detection

M.Wilhelm[1], BS; E. Kettelkamp[2], DVM; E. Spronk[2], DVM; A. M. Betlach[2], DVM, PhD
[1]Iowa State University, College of Veterinary Medicine, Ames, Iowa; [2]Swine Vet Center, St. Peter, Minnesota

Introduction

Feed delivery is a regular event on swine farms that interconnects swine facilities within a system and across a region. Feed and transport vehicles have been shown to serve as fomites for swine disease transmission.[1] Without any interventions, pathogens may be spread to swine farms or feed mills via foot or vehicular traffic post-contact with contaminated farms.[2] One factor that can help facilitate disease transmission is the cleanliness of truck cabs post-contamination. Previous research has shown that current truck cab cleaning methods do not fully disinfect cab interiors, emphasizing the need for biocontainment interventions to reduce the risk of disease transmission.[3] The objectives of this study were to demonstrate the role of feed delivery and vehicular movements on pathogen transmission within an integrated swine system using fluorescent powder (Glo-Germ™, GG) detection and to evaluate the potential impact of feed truck driver biosecurity measures on the likelihood of pathogen (ie, GG) spread during feed delivery events.

Materials and methods

An integrated swine system with an internal feed mill servicing commercial sow farms and wean-to-finish farms located in southwest Minnesota was selected for the study. The system had dedicated trucks delivering to different phases of production. While biosecurity procedures (eg, shoe cover usage) were encouraged by feed mill drivers, there were no set biosecurity standard practices outlined by the system.

A pilot study was conducted to assess the efficacy of five different mixtures/forms of Glo-Germ™ (ie, GG powder + GG gel, GG powder, GG gel, GG gel + piglet drying agent, GG powder + piglet drying agent) to determine which mixture would not only transfer from the driver's shoe into the truck, but also would be most readily detectable under Ultraviolet A light. The initial mixture of GG fluorescent powder and gel was derived from Ruston, et al.[4] Concurrently, GG was placed on various surfaces (grass, gravel, concrete, etc) and walked through 8 hours later to ensure that GG was detectable on various surfaces and within the truck cabs after being placed for an extended period of time (Figure 1). Based on pilot study results, GG powder and GG gel were utilized for the following study.

Figure 1: Transmission of Glo-Germ™ from point of contamination to the feed truck.

Feed mill drivers (n = 6) were observed delivering feed to a variety of swine facilities, in which 3 drivers serviced sow farms and 3 drivers serviced nursery, grow-finish sites. A mixture of GG fluorescent powder (9.66 g) and gel (3.33 g) was used to visualize contamination of swine material from the farm to the feed truck. Prior to the start of route, the GG mixture was placed on the ground below the feed bin lid handle, allowing the drivers to blindly step through it at the first site. Discrete cameras were placed at the first and last sites of feed driver's designed routes to observe the driver's biosecurity practices and confirm driver contamination via contact with the GG mixture. Ultraviolet A light was utilized to observe the transmission of GG from select sites along the delivery route, including the feed mill and truck cab. Additionally, environmental samples, using Swiffer® Dry Sweeping Cloths soaked in 50 mL of PBS, were collected from the truck cabs before and after the delivery route and from high traffic areas at the feed mill. Samples were collected and sent to the University of Minnesota Veterinary Diagnostic Laboratory to assess potential pathogen spread via the detection of Rotavirus genetic material by real-time PCR. At the end of the study, drivers completed a short survey to understand their current biosecurity practices relative to their cleaning and disinfection practices within and outside of their truck cabs.

Results

For drivers utilizing enhanced biosecurity practices (ie, disposable shoe covers or daily truck cab cleaning), GG was observed at 36% of downstream sites, compared to drivers utilizing fewer biosecurity practices, GG transfer was observed at 58% of the measured routes (Figure 2). Drivers that placed used shoe covers into trash cans inside the truck cab had improved biocontainment of

Figure 2: Comparison of Glo-Germ™ transmission between sow farm drivers and nursery, grow-finish (NGF) drivers.

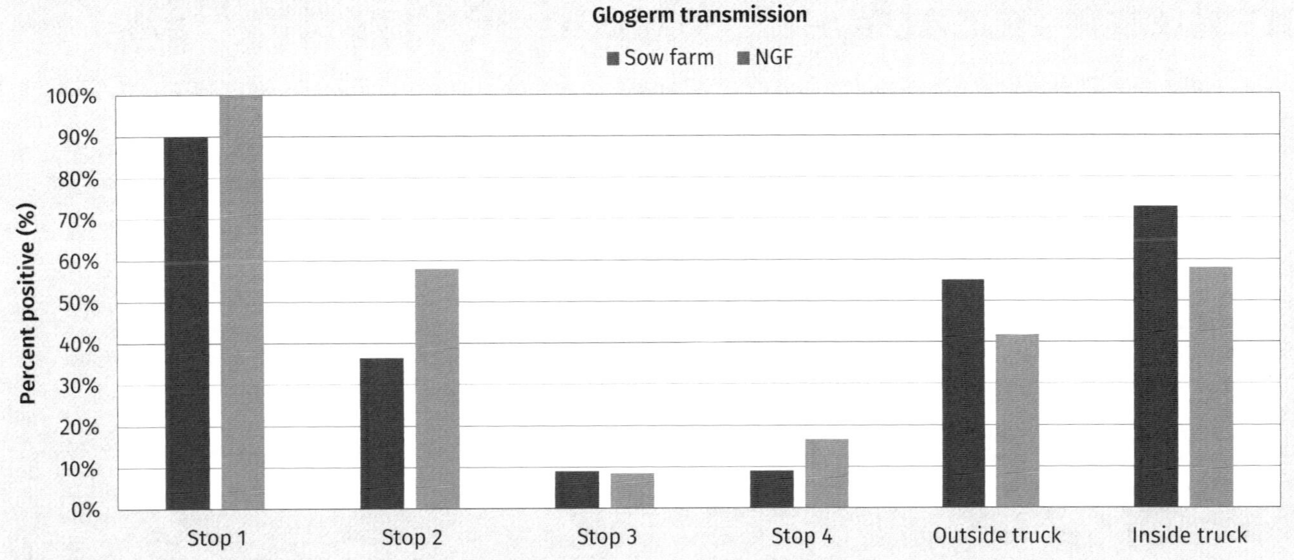

GG inside the truck cab compared to drivers that did not utilize shoe covers. Moreover, GG was observed 33% of the time in high traffic areas of the feed mill within 48 hours of observing driver routes. A lower relative viral load via Rotavirus Ct-values, was detected at the end of driver daily routes (36.01 average Ct value) compared to the beginning of the route (35.56 average Ct value), regardless of driver biosecurity practices.

Discussion

Overall, drivers that had increased biosecurity practices (ie, disposable shoe covers and daily truck cab cleaning) showed less GG transmission in the truck cabs and downstream sites than drivers implementing fewer biosecurity practices. One limitation of the study was that there was occasional overlap in delivery schedules. For instance, if GG was placed at one bin for one collection timepoint, at the next timepoint a different driver could deliver to the same bin and potentially contaminate themselves. As a result, there were multiple timepoints where it was unclear if GG was from the initial timepoint or the second. Another limitation was that drivers that maintained higher biosecurity practices observed lower Ct values in their truck cabs from the beginning of the study. Drivers were not required to clean the inside/outside truck cabs prior to the first sample collection. Consequently, these drivers could have started with a higher viral load than those drivers with lower biosecurity practices. These results elucidate the importance of implementing daily feed mill and delivery truck sanitation practices to enhance biocontainment and minimize the risk of infectious pathogen spread. These findings are consistent with previous research, further supporting the importance of feed mill biosecurity. Additional research is warranted to demonstrate the risk of potential pathogen contamination from feed delivery events into the respective barn on site.

References

1. Dee SA, et al. The risk of viral transmission in feed: What do we know, what do we do?. *Transboundary and Emerging Diseases.* 2020;67(6):2365-2371.

2. Galvis JA, et al. Modeling between-farm transmission dynamics of porcine epidemic diarrhea virus: characterizing the dominant transmission routes. *Prev Vet Med.* 2022;208:105759.

3. Elijah CG, et al. Understanding the role of feed manufacturing and delivery within a series of porcine deltacoronavirus investigations. *J Swine Health Prod.* 2022;30(1):17-23.

4. Ruston CR, et al. Evaluation of a staged loadout procedure for market swine to prevent transfer of pathogen contaminated particles from livestock trailers to the barn. *J Swine Health Prod.* 2021;29(5):234-243.

https://doi.org/10.54846/am2024/27

Interstate swine transportation network as a predictor of pathogen spread across the USA

Megan N. Thomas[1]; Phillip C. Gauger[1], DVM, PhD; Amy L. Baker[2], DVM, PhD; Tavis K. Anderson[2], PhD
[1]Department of Veterinary Microbiology and Preventive Medicine, College of Veterinary Medicine, Iowa State University, Ames, Iowa; [2]Virus and Prion Research Unit, National Animal Disease Center, USDA-ARS, Ames, Iowa

Introduction

In the face of a novel outbreak of an economically significant disease in domestic livestock, stopping the spread of the pathogen becomes a national issue. Mitigation efforts must be balanced with maintaining productivity and economic stability to preserve our country's safe food supply. Prior research conducted at a regional level demonstrated that swine pathogen spread was highly associated with pig movement.[1] However, it is difficult to estimate how a pathogen emerging in a given state will move across the country and, hence, which areas to target for restricted movement of swine in an effort to mitigate the spread of disease. Here, we used state-level data to create a national network of interstate swine movement and evaluated this network as an informative predictor of the spread of disease, using the porcine reproductive and respiratory syndrome virus (PRRSV) L1C 1-4-4 variant as a test case.

Methods

Transport data collection

We collected hog import records from 2020 - 2022 for 27 US states. We prioritized data collection from states with the largest hog inventory or significant interstate transport. The three years were chosen due to the relative ease of access to records following the adoption of electronic certificate of veterinary inspection (CVI) databases by the majority of states. We aggregated the data into an adjacency matrix describing movement of pigs between states: the matrix used all movements between states in 2021 (n = 23) or substitutes from 2020 (n = 2) or 2022 (n = 2) when necessary. Using this method, we were able to represent approximately 95% of annual interstate swine transport (55.4 million out of 58.3 million total head), as reported by the USDA Meat Animal Production, Disposition, and Income 2021 Summary.[2]

Network construction & statistics

The interstate transport adjacency matrix was used to construct a directed, weighted network of pig movement using the igraph R package (v1.5.01).[3] This package was used to calculate network centrality statistics including in- and out-strength (degree for a weighted graph), hub, and authority centrality. The network was visualized using the tidygraph R package (v1.2.3).[4]

Testing interstate transport as predictor of L1C 1-4-4 variant spread

All PRRSV L1C 1-4-4 variant sequences and metadata (n = 3,645) from 2020 to August 2023 were collected from a regional diagnostic lab. Metadata associated with the sequences included collection date, US state, sample type, and stage of production. The sequences were aligned and a maximum-likelihood tree was inferred using IQ-TREE (v2.2.6) following alignment using mafft (v7.490).[5,6] The dataset was representatively subsampled (n = 598) for computational feasibility using parnas (v0.1.4).[7] The resultant alignment was prepared for Bayesian phylogeographic analysis using Beauti (v1.10.4).[8] A generalized linear model was implemented in BEAST (v1.10.4) to represent the spatial diffusion of the PRRSV L1C 1-4-4 variant using US state as a discrete trait.[8] The graph weights from the interstate swine transport network were exported, log-transformed, and used as input as a predictor of spatial diffusion. The Bayesian inference implemented a GTR + Gamma substitution model, an uncorrelated relaxed clock, and GMRF Bayesian Skyride tree prior.

Results

Summary of interstate swine transport

Imports and exports are reflected by in- and out-strength centrality measures. Illinois, Missouri, Oklahoma, Nebraska, and South Dakota exported the most pigs in 2021 (Figure 1; Table 1). Iowa, Minnesota, Indiana, Kansas, and Missouri imported the most pigs in this same time frame. The number of pigs imported had a stronger association with the total hog census than the number of pigs exported. Six states for which we received import records (Kansas, Virginia, North Dakota, California, Georgia, and Idaho) ranked disproportionately high on imports relative to their total hog census (Table 1). These states have the potential to contribute more to disease spread than their census numbers alone would typically suggest. States for which import data was not yet received have inaccurate in-strength and authority centrality.

Authority centrality considers a state's significance based on the importance of states it is connected to (via hog import/export) and the number of pigs being imported into it. Authorities have relatively high incoming traffic, thus the three states with the highest authority were also the three states that imported the most hogs (Iowa, Minnesota, and Indiana; Table 1). These states are particularly susceptible to a pathogen being transmitted into their borders. Hub centrality measures the degree to which a state connects highly important locations in the network. Which states it connects to and the number of pigs moved between them are taken into consideration. Pathogens are likely to quickly spread from hubs to highly populated neighboring states. Illinois, Missouri, and Nebraska were recognized as the most important hubs in the network (Table 1). Four states (Illinois, Minnesota, Missouri, and Nebraska) were ranked in the top ten of both hubs and authorities.

Iowa, despite being the most hog dense state in the country, does not export a proportionate number of pigs and thus is not considered a significant hub. Thus, in this network, Iowa has the highest risk of disease entering the state, but a lower risk of disease leaving the state. The state of Virginia also had similar dynamics in this network. However, a caveat to these results is that we had not yet received North Carolina's import data at the time of analysis, and results could change as we incorporate these data.

Interstate transport predicts PRRSV variant spatial diffusion

The graph weights from the interstate swine transport network were tested for their predictive value in PRRSV L1C 1-4-4 variant spatial diffusion. The interstate transport network was correlated with the spatial dissemination of the 1-4-4 variant with a posterior inclusion probability of > 0.99 and a coefficient of 1.872 (95% HPD interval of [1.3448, 2.4255]), indicating the transport network was highly correlated with interstate spread of the 1-4-4 variant.

Discussion

These data demonstrate statistical support for interstate swine movement influencing the spread of a PRRSV variant and indicated a pattern that may be repeated with additional pathogens that should be investigated further. These results reinforce that an interstate swine transportation network has important implications for national preparedness for a foreign animal disease (FAD) outbreak that could likely spread with animal movement. Interrogating the network to focus control efforts will improve organization and efficiency. This network could also inform targeted vaccine design by identifying which antigens to include in autogenous vaccines based on the upstream location of the pathogen in the movement network. Understanding the role animal movement plays in pathogen spread will allow the identification of areas where targeted surveillance and intervention could be best directed in novel outbreaks and the control of endemic infections in swine.

In its current form, the network is useful for statically identifying states with two major types of risk: Risk of being a receiver and risk of being a donor of an emerging pathogen. States with high authority centrality, most notably Iowa, will have an extremely high risk of the pathogen entering their borders. States that are major hubs present a risk being the donor to key hog production states. These states could be some of the best targets for interventions focused on slowing the spread of a pathogen. Illinois, Minnesota, Missouri, and Nebraska could be considered extremely high priority states for disease mitigation because they rank in the top ten of both of these types of risks. Dynamically, these risks change based on where a pathogen enters the US pig population. In the future, the utility of the network could be improved by creating a simulation in which a pathogen can be detected in a certain state and the relative risk of spread to each state can be calculated from that specific scenario. Intervention efforts could also be modeled to determine the effect on both pork production and pathogen spread.

To improve the accuracy of this tool for FAD mitigation, hog import data for the remaining 23 US states must be received and aggregated into the network. Additionally, further communication is warranted in situations where import data (total number of hogs imported) provided by the state agency for our study does

Figure 1: Interstate swine transportation network. Black lines are weighted by the number of pigs moved from one state to another (arrowhead) annually. The size of the state circles (salmon) are scaled by $\log_2$ of the total pigs exported by that state annually.

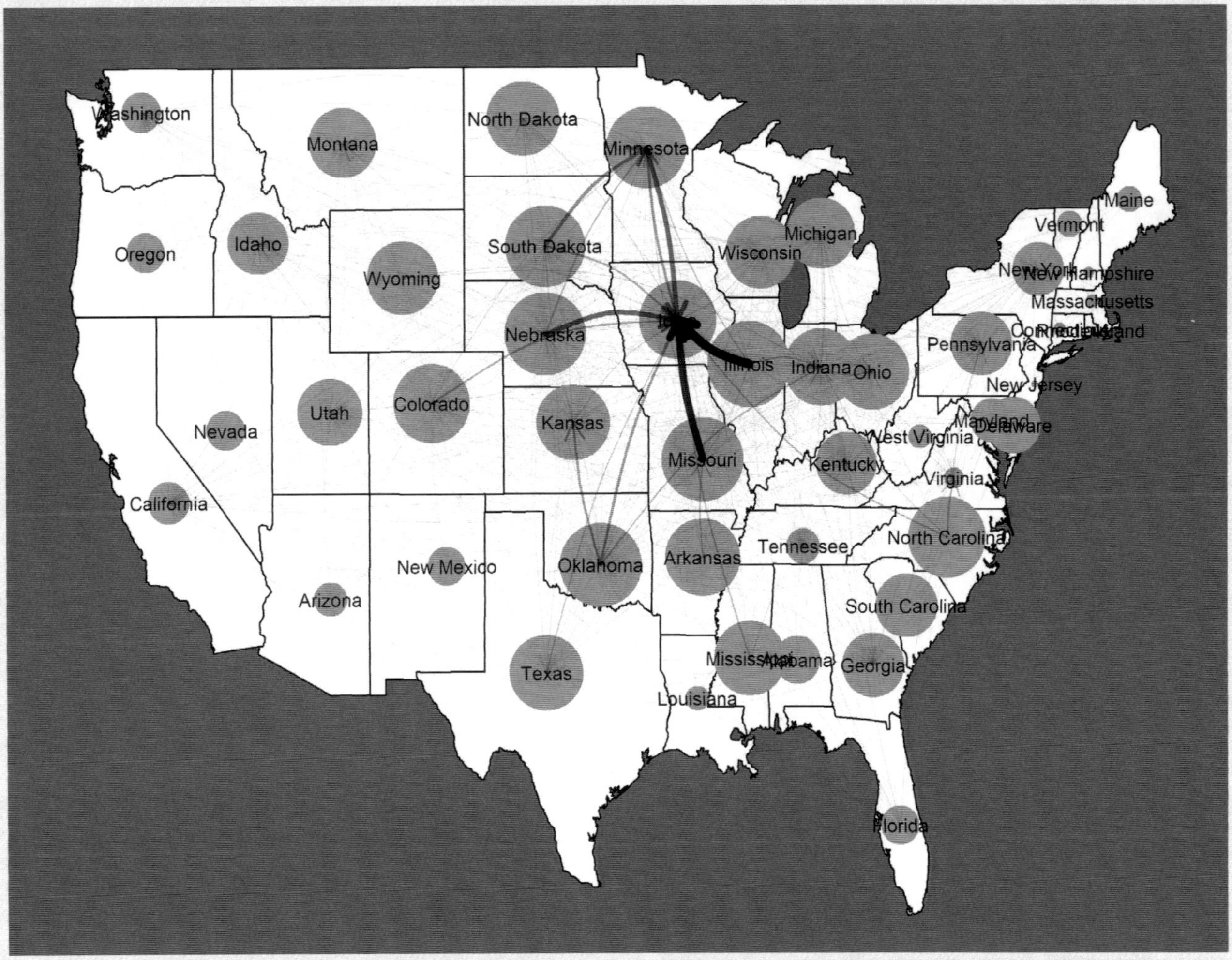

Table 1: Hog census, centrality rankings, and inclusion of import data for each US state. Yellow fill indicates states for which import data was not available at the time of network creation, thus in-strength and authority are not accurate and should be interpreted with caution. Green fill indicates states that were ranked in the top ten for both authority and hub centrality. Blue fill indicates states identified as having a disproportionate number of imports when compared to their hog census.

State	Hog census (USDA, 2021)	Census rank	In-strength rank	Auth. rank	Out-strength rank	Hub rank	Import data?
Iowa	24600000	1	1	1	9	16	Yes
Minnesota	9400000	2	2	2	6	4	Yes
North Carolina	8800000	3	28	29	7	8	No
Illinois	5450000	4	6	7	1	1	Yes
Indiana	4450000	5	3	3	10	12	Yes
Nebraska	3800000	6	9	10	4	3	Yes
Missouri	3750000	7	5	5	2	2	Yes
Ohio	2700000	8	7	6	12	18	Yes
Oklahoma	2080000	9	12	14	3	5	Yes
Kansas	2060000	10	4	4	14	11	Yes
South Dakota	2030000	11	11	8	5	6	Yes
Pennsylvania	1400000	12	10	11	24	25	Yes
Michigan	1310000	13	13	13	19	22	Yes
Texas	1080000	14	17	15	13	10	Yes
Utah	1000000	15	20	19	21	20	Yes
Colorado	620000	16	15	16	8	7	Yes
Kentucky	460000	17	22	17	25	23	Yes
Wisconsin	370000	18	16	12	18	14	Yes
Virginia	310000	19	8	9	43	41	Yes
Tennessee	260000	20	19	21	37	35	Yes
Montana	205000	21	23	22	20	19	Yes
South Carolina	188000	22	28	46	23	21	No
North Dakota	143000	23	18	18	16	13	Yes
Arizona	141000	24	28	28	38	39	No
Arkansas	126000	25	28	29	11	17	No
Mississippi	110000	26	27	24	15	9	Yes
California	99000	27	14	20	31	31	Yes
Wyoming	95000	28	28	29	17	15	No
New York	69000	29	21	25	28	29	Yes
Georgia	38000	30	24	23	22	24	Yes
Idaho	32000	31	25	26	26	30	Yes
Maryland	20500	32	28	33	30	32	No
Washington	17000	33	26	27	32	38	Yes
Florida	13000	34	28	29	36	33	No
Alabama	12000	35	28	33	29	26	No

Table 1 continued on next page

Table 1: Continued

State	Hog census (USDA, 2021)	Census rank	In-strength rank	Auth. rank	Out-strength rank	Hub rank	Import data?
Hawaii	10000	36	28	46	48		No
Oregon	9000	37	28	46	34	27	No
Massachusetts	9000	37	28	37	45	44	No
New Jersey	7500	39	28	42	47	47	No
Louisiana	6000	40	28	37	41	34	No
Delaware	5000	41	28	37	27	28	No
Maine	5000	41	28	33	40	37	No
Vermont	4000	43	28	37	39	43	No
Connecticut	4000	43	28	42	44	45	No
New Hampshire	3800	45	28	42	46	46	No
West Virginia	3000	46	28	33	42	40	No
Nevada	2500	47	28	37	33	42	No
Alaska	1900	48	28	46	48		No
Rhode Island	1500	49	28	42	48	48	No
New Mexico	1300	50	28	46	35	36	No

not match what is reported by the USDA NASS in the Meat Animal Production, Disposition, and Income 2021 Summary.[2] This highlights the challenge of gathering and verifying hog import data from 50 different state entities. This process is likely to improve with full adoption of electronic CVI management systems. However, it is our experience that agreements with individual companies or neighbor states may lead to some imports being tracked separately from the system, which makes it more difficult to ensure all pig movements are accounted for. Despite these challenges, the data we received represented 95% of reported hog imports in 2021 and this provides a strong basis for our proof of concept. Furthermore, the predictive value of the interstate swine transportation network should be tested for the spatial diffusion of other PRRSV variants and other pathogens (eg, influenza A virus) to test for repeatability of the correlation. Nonetheless, we developed a robust model of interstate pig movement, highlighting the roles different states played in the network, and showing the importance of interstate swine movement in the dissemination of the PRRSV 1-4-4 variant.

References

1. Makau D., Alkhamis M, Paploski I., et al. Integrating animal movements with phylogeography to model the spread of PRRSV in the USA, *Virus Evolution*. 2021;7(2). **doi:10.1093/ve/veab060**

2. USDA National Agricultural Statistics Service. Meat Animals Production, Disposition, and Income 2021 Summary. April 28, 2022. Accessed June 30, 2022. **https://usda.library.cornell.edu/concern/publications/02870v85d.**

3. Csardi, G., & Nepusz, T. The igraph software package for complex network research. *InterJournal, complex systems*. 2006;1695(5), 1-9.

4. Pedersen T. *tidygraph: A Tidy API for Graph Manipulation*. 2023. **https://tidygraph.data-imaginist.com**

5. Katoh, K., Misawa, K., Kuma, K. & Miyata, T. MAFFT: a novel method for rapid multiple sequence alignment based on fast Fourier transform. *Nucleic acids research*, 2002;30(14), 3059-3066.

6. Minh, B., Schmidt, H., Chernomor, O., et al. IQ-TREE 2: new models and efficient methods for phylogenetic inference in the genomic era. *Molecular biology and evolution*, 2020;37(5), 1530-1534.

7. Markin, A., Wagle, S., Grover, S., et al. PARNAS: objectively selecting the most representative taxa on a phylogeny. *Systematic Biology*, 2023; *72*(5), 1052-1063.

8. Suchard M., Lemey P., Baele G., et al. Bayesian phylogenetic and phylodynamic data integration using BEAST 1.10 *Virus Evolution* 2018;4. **doi:10.1093/ve/vey016**

https://doi.org/10.54846/am2024/28

Agreement between the detection of PCV2 via tongue tip samples, processing fluids, placental umbilical cord blood serum samples, and colostrum by PCR

Alexis Berte[1], BS; Nicolas Rippel[2], DVM; Keith Aljets[2], DVM; Kelly Greiner[3], DVM
[1]Iowa State University CVM; [2]Veterinary Medical Center, Williamsburg, Iowa; [3]Merck Animal Health

Introduction

Porcine circovirus type 2 (PCV2) is a single-stranded, non-enveloped DNA virus that can be found in nearly all swine herds. PCV2 can cause weight loss, slow growth, death, reproductive failure, porcine dermatitis and nephropathy syndrome, and postweaning multisystemic wasting syndrome.[1] In recent years, post-mortem tongue tip samples (TTS) have been used to detect porcine reproductive and respiratory syndrome virus (PRRSV) and influenza A virus of swine (IAV-S).[2] The objective of this study is to determine the agreement between the detection of PCV2 in TTS and validated methods including processing fluids (PF), placental umbilical cord blood serum samples (PUCS), and colostrum by polymerase chain reaction (PCR).

Materials and methods

A 7000 head sow farm with a previous history of detecting PCV2 via PCR in pooled PF was used for this study. The previous herd average PCR CT value for PCV2 was 33.7. If a litter contained at least one stillborn, it was enrolled into the trial. Four sample types were collected from 45 litters: colostrum, PUCS, TTS, and PF. Colostrum from 2 - 3 teats was collected within the first twelve hours of farrowing. PUCS were collected from 3 - 5 umbilical cords per litter and placed into a serum separator blood tube. The PUCS were centrifuged on farm, and the serum was collected. Using a disposable scalpel and rat tooth forceps, 2 - 3 centimeters of the stillborn tongue tip was removed. One to seven stillborn TTS from a litter were placed into a single tube with 0.5 mL of PBS. All boars within a litter were castrated using a disposable scalpel. At least 3 pairs of testicles per litter were collected and placed into a Ziploc bag. All sample types were stored in a 0°F freezer on the farm. TTS and PF were set out to thaw for 24 hours in a refrigerator and the resulting fluid was collected. An individual PCV2 PCR was performed on each of the samples. Samples with a CT less than 37 were considered positive. Cohen's Kappa statistics were run on the PCV2 PCR results to determine the positive and negative agreement between each of the four sample types.

Results and discussion

The level of agreement for PCV2 detection between sample types is listed in Table 1. TTS substantially agreed with PUCS and moderately agreed with colostrum for PCV2 in the herd studied, which had a relatively low screening prevalence of PCV2. The TTS were easy to collect and non-invasive. TTS can be used to test multiple diseases. Herds with a higher screening prevalence for PCV2 should be studied to validate how reliable TTS are for detecting PCV2 by PCR compared to other sample types when prevalence is higher. Since PCV2 is found in nearly all swine herds, surveillance testing is needed to assess the producer's herd stability to reduce costs associated with PCV2 clinical disease.

References

1. Segalés J, Sibila M. Revisiting Porcine Circovirus Disease Diagnostic Criteria in the Current *Porcine Circovirus 2* Epidemiological Context. Vet Sci. 2022 Mar 2;9(3):110. **doi: 10.3390/vetsci9030110.** PMID: 35324838; PMCID: PMC8953210.

2. Baliellas, Jordi. "Podcast: Jordi Baliellas on Using Tongue Tip Tissues for PRRS Monitoring." *Pig Progress*, Meet the Expert Podcast, 30 Aug. 2022, **www.pigprogress.net/health-nutrition/health/podcast-using-tongue-tip-tissues-for-prrs-monitoring/.**

Table 1: Kappa agreement between individual sample types

Individual sample types (PCV2)	Kappa	Agreement	Kappa agreement levels
TTS vs PUCS	0.79	Substantial	≤ 0 = No agreement
TTS vs colostrum	0.54	Moderate	0.1 – 0.2 = Slight
TTS vs PF	0.23	Fair	0.21 - 0.4 = Fair
PUCS vs colostrum	0.45	Moderate	0.41 - 0.6 = Moderate
PF vs PUCS	0	No agreement	0.61 - 0.8 = Substantial
PF vs colostrum	0	No agreement	0.81 - 1 = Almost perfect

TTS = tongue tip samples; PUCS = placental umbilical cord blood serum samples; PF = processing fluids

https://doi.org/10.54846/am2024/29

A pilot study evaluating the pharmacokinetics of tulathromycin and ketoprofen in the form of Draxxin KP® in swine

C. Vincent[1], BS; B. Fredrich[2], DVM; S. Krantz[3], DVM; H. Knych[4], DVM, PhD, DACVCP
[1]Mississippi State University College of Veterinary Medicine, Starkville Mississippi; [2]Zoetis, Parsippany, New Jersey; [3]Tosh Farms, Henry, Tennessee; [4]K.L. Maddy Equine Analytical Chemistry Laboratory, School of Veterinary Medicine, University of California- Davis, Davis, California

Introduction

Draxxin KP, indicated in cattle for the control of pyrexia and the treatment of bovine respiratory disease, is a long-acting single dose injectable that combines the antibiotic tulathromycin and the nonsteroidal anti-inflammatory (NSAID) ketoprofen. Ketoprofen, a non-selective COX inhibitor, inhibits the synthesis of prostaglandin E2 and prostaglandin F2 alpha.[1] Ketoprofen, as a mono-substance, has a relatively short half-life, but when combined with tulathromycin, such as in Draxxin KP, has been shown to provide sustained plasma concentrations and possesses an extended half-life in cattle.[2] Ketoprofen is currently approved for use in swine in Canada for the control and/or treatment of respiratory disease, mastitis- metritis- agalactia and as an antipyretic and analgesic agent.[1] Ketoprofen has been studied for analgesic use for castration and tail docking in Europe and is comparable to other NSAIDS that are currently used in the swine industry.[3] Draxxin KP is not approved for use in swine. The objective of this study was to determine the porcine pharmacokinetic profiles of Draxxin KP's active components in comparison to ketoprofen and tulathromycin when administered independently as mono-substances to swine.

Material and methods

Twenty-one PIC C29 gilts (8-9 weeks of age) were randomly selected and weighed (20.5 ± 5 kgs) on Day 0 of this study. Animals were blocked into groups of three animals each, and randomly assigned to one of three treatment groups yielding 7 animals per group. Treatment groups included: Draxxin® (tulathromycin 2.5 mg/kg, Zoetis) (T01); Ketofen® (ketoprofen 3 mg/kg, Zoetis) (T02); and Draxxin KP (ketoprofen 3 mg/kg, tulathromycin 2.5 mg/kg, Zoetis) (T03). All treatments were administered intramuscularly (IM) into the lateral aspect of the neck. The pigs enrolled in this study were porcine reproductive and respiratory syndrome virus negative, *Mycoplasma hyopneumoniae* positive, and, though not clinical at the time, had previously been exposed to Seneca Valley virus. Upon enrollment in the trial, the 21 pigs were housed together in the same room in a single pen. Ad libitum access to water and age-appropriate feed was granted. Selection criteria included no abnormalities found on physical exam and free from antibiotic administration for at least two weeks prior to trial commencement. To establish baseline concentrations, and confirm pig trial eligibility, on Day 0, 3 mL of blood was collected from jugular venipuncture for each pig into K_3-EDTA anticoagulation (EDTA) tubes and placed on ice. Preassigned treatments were then administered to each pig at Time 0 in blocked groups. Serial blood samples (3 mL) were collected into EDTA tubes at the following hour (h) timepoints post-administration of treatment (Time 0): 0.33, 0.67, 1, 1.5, 2, 3, 4, 6, 10, 24, 28, 32, 48, 53, and 72. All samples were immediately placed on ice post collection. Time point selection was optimized to capture ketoprofen plasma concentrations based on previously established PK studies.[3-5] Samples were centrifuged for 10 minutes and resulting plasma was stored in cryogenic tubes and frozen (4°C). Control blood samples (30 mL total) were collected from non-treated cohorts and centrifuged. The resulting plasma from the cohort pigs (20 mL total) was submitted in conjunction with trial samples to facilitate the validation of the laboratory assay alongside the treatment group plasma samples. Samples were shipped overnight on dry ice to the K.L. Maddy Equine Analytical Chemistry Lab (UC-Davis, California). Plasma drug concentrations were analyzed using liquid chromatography-tandem mass spectrometry as previously described.[5,6] A partial validation was performed for swine plasma. The limit of quantitation was 10 ng/mL and 1 ng/mL for ketoprofen and tulathromycin, respectively. The limit of detection was approximately 5 ng/mL and 0.5 ng/mL for ketoprofen and tulathromycin, respectively. Plasma pharmacokinetic parameters were determined for tulathromycin and ketoprofen when administered in combination (Draxxin KP) and as mono-substances using non-compartmental analysis and commercially available software (Certara Phoenix Winnonlin v 8.2, Printeton, New Jersey). Pig 2, from the Draxxin (mono-substance) treatment group, was eliminated from analysis due to abnormally low plasma concentrations. The area under the curve (AUC) was calculated using the log-linear trapezoidal rule and extrapolation to infinity using the last measured serum concentration divided by the terminal slope λ_z. The terminal rate constant (terminal slope; λ_z) was calculated by log-linear regression of the terminal portion of the concentration time curve. The terminal half-life ($t_{1/2}$) was calculated as $0.693/\lambda_z$.. C_{max} and T_{max} were determined directly from the plasma concentration time curve.

Results

No adverse reactions were noted in the treatment groups throughout the 72-h observation period. Pharmacokinetic parameters evaluated in this study are presented in Table 1. The $t_{1/2}$ and the T_{max} of ketoprofen as a mono-substance was comparable to values reported previously.[7] The C_{max} of the ketoprofen in the combined product was lower than what was observed in the mono-substance group but is comparable to C_{max} of ketoprofen in other studies.[3,7] The T_{max} and $t_{1/2}$ were longer for the combination product than Ketoprofen as a mono-substance. Tulathromycin in the combination product had a T_{max} and $t_{1/2}$ similar to the mono-substance group.

Discussion

Tulathromycin plasma concentrations and $t_{1/2}$ of the tulathromycin component of Draxxin KP were similar to those of Draxxin. The $t_{1/2}$ for ketoprofen as a mono-substance was prolonged when it was administered in combination with tulathromycin (Draxxin

Table 1: Pharmacokinetic parameters for treatment groups following IM injection of the assigned drug. Values are geometric mean and standard deviation half-life is harmonic mean and T_{max} is median and range. Parameters were generated using a non-compartmental analysis. [C_{max} is maximum plasma concentration. T_{max} is the time of maximal plasma concentration. $t_{1/2}$ is terminal half-life, AUC_{inf}, area under the serum concentration curve extrapolated to infinity.]

	Analyte	Tulathromycin	Tulathromycin + Ketoprofen
C_{max} (ng/mL)	Tulathromycin	610.6 (157.7)	1042.9 (983.0)
T_{max} (h) median (range)	Tulathromycin	0.33 (0.33-0.67)	0.33 (0.33-2.0)
Terminal half-life (h)	Tulathromycin	25.8 (4.6)	22.7 (3.9)
AUC_{inf} (h* ng /mL)	Tulathromycin	11484 (2183.6)	9700 (4196.0)
		Ketoprofen	**Tulathromycin + Ketoprofen**
C_{max} (ng/mL)	Ketoprofen	11670.1 (1761.1)	4579.5 (1019.1)
T_{max} (h) median (range)	Ketoprofen	0.67 (0.3-0.7)	1.00 (0.3-3.0)
Terminal half-life (h)	Ketoprofen	2.35 (0.5)	3.95 (0.66)
AUC_{inf} (h* ng /mL)	Ketoprofen	33797 (4469.5)	30747 (3751.7)

KP), compared to administration as a sole agent. The AUC_{inf} of ketoprofen as a combined product and a mono-substance are similar which shows the drug reaches the same plasma levels in swine. Clinical efficacy and the pharmacodynamics of Draxxin KP were not evaluated in this study.

References

1. The European agency for the evaluation of medicinal products Veterinary medicines evaluation unit. Ketoprofen Summary Report. 1996. Accessed on September 1st 2023. **https://www.ema.europa.eu/en/documents/mrl-report/ketoprofen-extension-pigs-summary-report-committee-veterinary-medicinal-products_en.pdf**

2. De Koster J, Boucher J, Tena J, Gehring R, Stegemann M. Co- formulation of ketoprofen with tulathromycin alters pharmacokinetic and pharmacodynamic profile of ketoprofen in cattle. *J Vet Pharm Ther.* 2021:69-92. **https://onlinelibrary.wiley.com/doi/epdf/10.1111/jvp.12999**

3. Nixon E, Almond G, Baynes R, Messenger K. Comparative plasma and interstitial fluid pharmacokinetics of meloxicam, flunixin, ketoprofen in neonatal piglets. *Front Vet Sci.* 2020;7. **https://doi.org/10.3389/fvets.2020.00082**

4. Fosse T, Toutain P, Spadavecchia C, Haga H, Horsberg T, Ranhaeim B. Ketoprofen in piglets: enantioselective pharmacokinetics, pharmacodynamic and PK/PD Modeling. *J Vet Pharm Ther.* 2012 **https://doi.org/10.1111/j.1365-2885.2010.01236.x**

5. Knych H, Arthur R, Steinmetz S, McKemie D. Pharmacokinetics of ketoprofen enantiomers following intravenous and oral administration to exercised Thoroughbred horses. *Vet J.* 2016:196-198. **https://doi.org/10.1016/j.tvjl.2015.09.018**

6. Barbaross A, Bardhi A, Gazzotti T, Mari G, Pagliuca G. A single LC-MS validated method of tulathromycin quantification in plasma, seminal plasma, and urine to be applied in pharmacokinetic study in bull. *Vet Med Int.* 2022. **https://doi.org/10.1002/dta.3270**

7. Raekallio M, Mustonen K, Heinonen M, Peltoniemi O, Sakkinen M, Peltoniemi S, Honkavaara J, Vainio O. Evaluation of bioequivalence after oral, intramuscular, and intravenous administration of racemic ketoprofen in pigs *Am J Vet Res.* 2008. **https://doi.org/10.2460/ajvr.69.1.108**

https://doi.org/10.54846/am2024/30

Vitamin A as a biomarker for livability in the nursery stage of swine production: Pilot project

Brianna McAleese[1]; Glen Almond[1]; James Mark Hammer[2]
[1]North Carolina State University College of Veterinary Medicine, Raleigh North Carolina; [2]Elanco, USA Greenfield, Indiana

Objective

To evaluate retinol serum levels as a biomarker for swine livability in the nursery stages of commercial swine production.

Animals

193 piglets (18 litters) from a variety of parous sows [parity 1 (2), parity 2 - 3 (2), parity 4 + (2)], located on three different farms with varying nursery production performance were chosen to be monitored throughout the course of this study.

Procedures

Initial blood samples were taken from selected piglets prior to weaning. Piglet mortality was recorded two weeks after pigs were weaned to different nursery farms. Final blood samples were collected, and mortalities recorded prior to/soon after piglets were moved to finishing farms. Both sets of blood samples were assessed for serum retinol levels by Elanco Animal Health using the BioAnalyt iCheck Fluoro system.

Results

Retinol serum levels of < 0.10 ppm, 0.10 - 0.15 ppm, 0.15 - 0.20 ppm, were associated with mortality rates of 40%, 10.22%, and 11.23%, respectively. There was no mortality associated with piglets that had retinol serum levels of 0.20 - 0.25 ppm. Collectively, pigs in this study had an initial pre-wean retinol serum level of 0.149 (SD: 0.023) and final pre-finishing retinol serum level of 0.151 (SD: 0.031).

Clinical relevance

Lower retinol serum levels in pre-weaned piglets were associated with higher mortality rates in the nursery stage of production, presumably due to the presence of inflammation decreasing retinol binding protein secretion from the liver. We demonstrated that vitamin A levels in the serum may have potential to determine inflammation status of a herd and predict mortality rates. Knowing the inflammation status of the herd can assist the practicing veterinarian in improving herd survivability rates with intervention strategies such as feed additives, management alterations or anti-inflammatories.

Introduction

Vitamin A is an essential vitamin that plays an important role in vision, reproduction, and embryonic development. Vitamin A is commonly obtained by adult pigs through their diet. It is obtained by nursing pigs through the lactating sow. It can also be obtained through injectable products. The National Research Council published in 2012 the daily vitamin A requirements for growing pigs, gestating sows, and lactating sows (Table 1). The information provided showed an indirect correlation between body weight and vitamin A requirement for growing pigs, with the highest requirements recommended for the smallest pigs. Common clinical signs of vitamin A deficiency include impaired vision, night blindness, decreased reproductive efficiency, and respiratory disorders. Offspring of deficient sows may have subsequent blindness, weakness, or malformations.[1]

Vitamin A has two available forms that can be obtained through the diet: provitamin A carotenoids and preformed vitamin A. Provitamin A carotenoids include both alpha-carotene and beta-carotene, which are mainly obtained through plant sources and must undergo gastric and pancreatic enzymatic breakdown into retinol to be absorbed.[2] Preformed vitamin A includes retinol and retinyl esters, which are mainly obtained through animal sources.[3] Vitamin A is a fat-soluble vitamin, and therefore, must be transported across the intestinal epithelium via micelles to be absorbed by enterocytes. This cascade of micelle formation is stimulated by presence of dietary fat, which triggers pancreatic release of lipase to break down the fats and create lipid components available for micelle formation. Micellar retinol can be efficiently absorbed by a specific retinol transporter protein STRA6, but micellar carotenoids cannot be absorbed as efficiently due to a lack of specific transporter, making retinol a more bioavailable source of vitamin A.[4] After transportation, the hepatocytes of the liver separate lipids from the retinoids to allow for either its storage or circulation. Hepatocytes transfer retinoids to hepatic stellate cells which are highly specialized cells that will store retinoids in the form of retinyl esters. Retinoids that are ready for circulation to peripheral tissues are bound to retinol binding protein 4 (RBP4) and secreted by the hepatocyte.[2] Once bound to retinol, RBP4 forms a complex with prealbumin in the plasma, also known as transthyretin, for travel through the bloodstream to peripheral tissues. The presence of prealbumin protects the RBP4 complex from degradation through glomerular filtration in the kidney.[5]

Table 1: Daily dietary vitamin A requirements of growing pigs allowed ad lib feed (90% dry matter), gestating sows, and lactating sows[1]

	Growing pigs						
Body weight (kg)	5 - 7	7 - 11	11 - 25	25 - 50	50 - 75	75 - 100	100 - 135
Vitamin A (IU/kg)	2200	2200	1750	1300	1300	1300	1300
			Gestation			**Lactation (21 days)**	
Vitamin A (IU/kg)			4000			2000	

In the case of liver disease, injury, or inflammation, hepatic retinol stores are lost due to the activation of hepatic stellate cells.[6] The loss of retinol storage results in a halt of RBP4 secretion from the liver, where it will instead accumulate due to lack of retinol available for transport.[7] During these events of infection or inflammation, the acute phase response begins hepatic synthesis of acute phase proteins, specifically C-reactive proteins (CRP) and Alpha-1-acid glycoprotein (AGP). This acute phase response inhibits RBP synthesis and quickly decreases RBP plasma concentrations due to its short half-life.[8]

In the case of zinc deficiency, hepatic synthesis of RBP4 is also decreased, leading to subsequent decreased levels of retinol serum concentrations.[7] In studies involving zinc deficiency affecting vitamin A levels, it was determined that zinc deficiencies do not result in decreased liver vitamin A concentrations, but instead decrease RBP4 serum concentration levels, therefore preventing the transportation of vitamin A to peripheral tissues.[9]

Retinol serum concentration levels are commonly used as a biomarker for vitamin A deficiency in humans when there is no inflammation present. Inflammation is accounted for by measuring the CRP and AGP markers in the serum.[7] To account for the effects of inflammation, there is a modified-relative dose response test that is not affected by these factors. Retinol is administered during the test to observe the reaction of RBP to determine if inflammation is present, or if the reaction is due to vitamin A deficiency. In this test, if RBP has accumulated due to deficiency, then within 5 hours of the dose there will be a large spike in retinol serum levels; if RBP has not accumulated due to inflammation, there will still be low retinol serum concentrations 5 hours after the dose.[10]

This study evaluated retinol serum concentrations with survivability in nursery stages of swine production to evaluate the use of vitamin A levels as a biomarker for inflammation.

Methods

This study was performed on commercial swine farms located in North Carolina. All animals were cared for under routine management practices. In our study, all three farrowing farms flowed to different nursery sites. Only one farm was tracked to the finishing site.

Initial blood samples were collected in pre-wean pigs on farms A-1, B-1, and C-1. As a routine management practice, all pigs from every farm were vaccinated for porcine circovirus-2, *Mycoplasma hyopneumoniae*, ileitis, and porcine reproductive and respiratory syndrome at weaning. Second samples were collected in two nursery farms and one finishing farm: A-2, B-3, and C-2. All samples were packaged and shipped appropriately to Iowa to be analyzed by Elanco Animal Health.

Phase 1: Farm selection

Farrowing farms were selected by the company veterinarian based on predicted nursery performance: Farm A-1 – average, Farm B-1 – low, and Farm C-1 – high. Nursery performance was ranked based on mortality, weight gain, and feed conversion impact cost; it was predicted based on the previous performance of these genetic lines. The genetic lines from farms A-1 and B-1 were sourced from Mississippi. The genetic lines from Farm C-1 were sourced from North Carolina.

Phase 2: Selection of specific parous sows, ear tag litters, & collection of first blood samples

Sows were chosen based on litter weaning date and parity. Six litters were chosen on each farm. Selected sow parities at Farms A-1 and B-1 were as follows: parity 1 (2), parity 2 - 3 (2), parity 4+ (2). The same was intended for Farm C-1, but due to available sows, chosen parities were as follows: parity 1 (2), parity 2 - 3 (3), parity 4+ (1).

All piglets from litters included in the study were individually ear tagged as blood samples were collected. Litter age and litter size at sampling varied between farms and litters (Table 2). At the first sample collection, the average litter age and litter size were 19.5 days and 10.7 piglets, respectively.

Phase 3: Track mortality in the nursery

Mortality was tracked at two weeks after being weaned to the nursery. Each farrowing farm was weaned to a different nursery farm: Farm A-1 → A-2, Farm B-1 → B-2, Farm C-1 → C-2. As checking for live pigs, pen numbers were recorded to be used at the final visit. Farm A-2 nursery was diagnosed with swine influenza virus after weaning and pigs were undergoing subsequent treatment with chlortetracycline followed by enrofloxacin.

Table 2: Distribution of sow parity, litter age, and litter size at first sample collection

Farrowing farm	Sow parity	Litter age (days)	Litter size at sampling (# of piglets)
Farm A-1	1	24	9
	1	24	12
	3	14	11
	3	15	10
	4	25	10
	5	25	12
Farm B-1	1	18	12
	1	19	7
	2	19	11
	2	22	9
	4	20	11
	5	17	13
Farm C-1	1	14	11
	1	22	13
	3	19	12
	3	19	12
	3	22	13
	4	14	5
Averages	–	19.5	10.7

Phase 4: Collect second blood samples and remove ear tags

The second round of samples were collected 45 days after initial sampling. The intention was to collect samples before pigs were moved to finishing farms, however, the B-2 nursery pigs had been moved to finishing farm B-3, three days before second sample collection. As second blood samples were collected, ear tags were removed from the pigs.

Results

The retinol serum level results were recorded in units of parts per million (ppm) and separated by farm (Table 3). Farm A-1 was observed to have an initial average retinol serum level of 0.148 ppm (SD: 0.023). Farm A-2 was observed to have a final average retinol serum level of 0.161 ppm (SD: 0.022). Farm A-1 was the only farm to increase average retinol serum levels, by 0.013 ppm, after being weaned and moved to the nursery. Farm B-1 was observed to have an initial average retinol serum level of 0.156 (SD: 0.020). Farm B-3 was observed to have a final average retinol serum level of 0.149 ppm (SD:0.035). Farm B-1 decreased average retinol serum levels by 0.007 ppm after being weaned, moved to the nursery, and moved to the finisher. Farm C-1 was observed to have an initial average retinol serum level of 0.144 ppm (SD: 0.026). Farm C-2 was observed to have a final average retinol serum level of 0.142 ppm (SD: 0.032). Farm C-1 decreased average retinol serum levels by 0.002 ppm after being weaned and moved to the nursery. Overall, the combined average of retinol serum levels between all farm samples increased from an initial 0.149 ppm (SD: 0.023) to a final 0.151 ppm (SD: 0.031).

The data was distributed into 5 different groups of retinol serum levels: < 0.10 ppm, 0.10 – 0.15 ppm, 0.15 – 0.20 ppm, 0.20 – 0.25 ppm, > 0.25 ppm (Figure 1). After grouping the data in this way, the mortality of each retinol serum level group was analyzed based on initial blood sample value (Table 4).

Table 3: Mean ± standard deviation on farrowing and nursery/finishing farms. Initial samples were taken in May 2023. Final samples were taken in June 2023.

Farrowing farm	Initial retinol serum levels (ppm)	Nursery/finishing farm	Final retinol serum levels (ppm)
Farm A-1	0.148 ± 0.023[ab]	Farm A-2	0.161 ± 0.022[a]
Farm B-1	0.156 ± 0.020[a]	Farm B-3	0.149 ± 0.035[ab]
Farm C-1	0.144 ± 0.026[b]	Farm C-2	0.142 ± 0.032[b]
Combined	0.149 ± 0.023	Combined	0.151 ± 0.031

[ab] Numbers with different superscripts differ at $P < .05$ (Tukey HSD Test)

Figure 1: Initial serum retinol concentrations distribution.

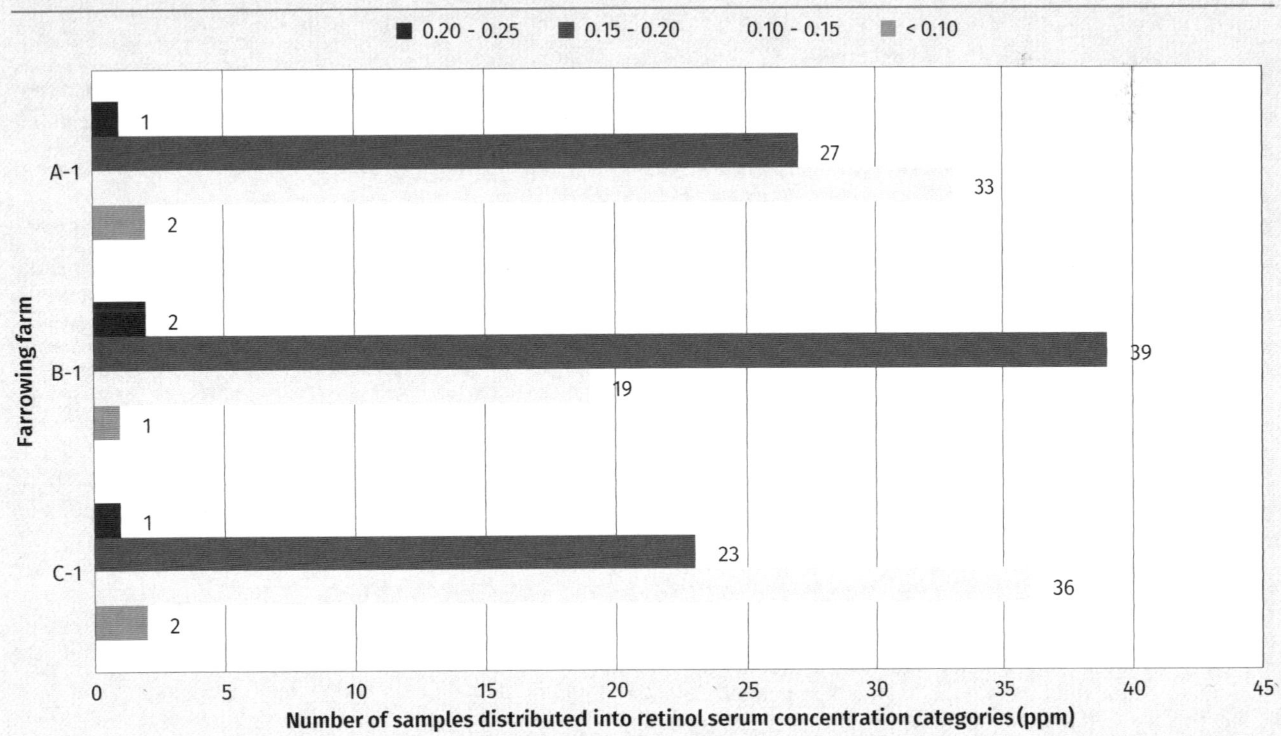

Table 4: Mortality distribution categorized by initial retinol serum concentrations (ppm)

	< 0.10	0.10 - 0.15	0.15 - 0.20	0.20 - 0.25	> 0.25	Total mortalities
Farm A-1	1	4	1	0	0	6
Farm B-1	0	1	4	0	0	5
Farm C-1	1	4	5	0	0	10
Total mortalities	2	9	10	0	0	21
Total pigs	5	88	89	4	0	
Mortality rate	40.0%	10.2%	11.2%	0.0%	0.0%	

Discussion

Summary

The purpose of this study was to observe the correlation between retinol serum levels and survivability in the nursery stages of swine production. According to our sample data, retinol serum levels of < 0.10 ppm, 0.10 - 0.15 ppm, 0.15 - 0.20 ppm, were associated with mortality rates of 40%, 10.22%, and 11.23%, respectively. There was no mortality associated with piglets that had retinol serum levels of 0.20 - 0.25 ppm.; however, few animals fell into this category. The results support the hypothesized association between lower retinol serum levels and higher mortality rates. In humans, retinol serum levels are commonly used as a biomarker for vitamin A deficiency, if able to rule out inflammation as the cause through measuring CRP and AGP markers in the serum.[7] However, since swine are fed a controlled and vitamin A sufficient diet, retinol serum levels should be directly correlated with inflammation since vitamin A deficiency should not be present.

Limitations

Limitations to this study include sample size, limited diagnostics completed, and no dietary analysis completed. The sample size only included an initial 193 pigs which is not enough to show significant data representative of a population. It is also important to note that inflammation from CRP and AGP markers were not measured in this study. No liver or dietary nutritional analyses were completed to ensure the sows of chosen litters were not deficient in vitamin A themselves.

Future directions

The future direction of this study should be to include a much larger sample size that can be considered representative of a population. Body weights should be measured along with retinol serum levels. It would also be helpful to measure the presence of inflammation biomarkers to come to a more definitive conclusion that vitamin A has a direct relationship with these markers. Finally, a dietary analysis should be conducted of sow diets. Gestating and lactating sows have much higher nutritional requirements than growing pigs, including when it comes to vitamin A intake. It may even be considered to monitor sow retinol serum levels throughout the period of 28-day lactation to see where there are peaks in inflammation and if that affects the serum levels of their nursing litters. Another point of interest would be to measure zinc serum levels, as deficiency can lead to decreased RBP4 production, and therefore decrease retinol serum levels.[9] This should be measured in the sows after day 90 of gestation, as zinc liver concentrations decrease rapidly at this stage.[11]

References

1. Cromwell, G. Nutritional requirements of pigs – management and nutrition. *Merck Veterinary Manual*. October 2022. Accessed November 7, 2023. **https://www.merckvetmanual.com/management-and-nutrition/nutrition-pigs/nutritional-requirements-of-pigs**

2. Blaner, WS, Li Y, Brun P, et al. Vitamin A absorption, storage and mobilization. *Subcell Biochem*. November 10, 2016. Accessed November 7, 2023. **doi: 10.1007/978-94-024-0945-1_4**

3. Kraemer K, Semba R, Eggersdorfer M, Schaumberg D. Introduction: the diverse and essential biological functions of vitamins. *Annals of Nutrition and Metabolism*. November 26, 2012. Accessed November 7, 2023. **doi: 10.1159/000343103**

4. Reboul E. Absorption of vitamin A and carotenoids by the enterocyte: focus on transport proteins. *Nutrients*. September 12, 2013. Accessed November 7, 2023. **doi: 10.3390/nu5093563**

5. Zabetian-Targhi F, Mahmoudi MJ, Rezaei N, Mahmoudi M. Retinol binding protein 4 in relation to diet, inflammation, immunity, and cardiovascular diseases. *Advances in Nutrition*. November 2015. Accessed November 7, 2023. **doi: 10.3945/an.115.008292**

6. Shirakami Y, Lee S, Clugston R, Blaner W. Hepatic metabolism of retinoids and disease associations. *Biochimica et Biophysica Acta*. January 2012. Accessed November 7, 2023. **doi: 10.1016/j.bbalip.2011.06.023**

7. Tanumihardjo S, Russell R, Stephensen C, et al. Biomarkers for nutritional development (BOND) – vitamin A review. *The Journal of Nutrition*. September 2016. Accessed November 7, 2023. **doi: 10.3945/jn.115.229708**

8. Rubin L, Ross A, Stephensen C, et al. Metabolic effects of inflammation on vitamin A and carotenoids in humans and animal models. *Advances in Nutrition*. March 15, 2017. Accessed November 7, 2023. **doi: 10.3945/an.116.014167**

9. Smith CJ, Brown ED, McDaniel EG, Chan W. Alterations in vitamin A metabolism during zinc deficiency and food and growth restriction. *The Journal of Nutrition*. April 1976. Accessed November 7, 2023. **doi: 10.1093/jn/106.4.569**

10. Suri D, Wirth J, Adu-Afarwuah S, et al. Inflammation adjustments to serum retinol and retinol-binding protein improve specificity but reduce sensitivity when estimating vitamin A deficiency compared with the modified relative dose-response test in Ghanaian children. *Current Developments in Nutrition*. July 15, 2021. Accessed November 7, 2023. **doi: 10.1093/cdn/nzab098**

11. Hostetler CE, Kincaid RL. Gestational changes in concentrations of selenium and zinc porcine fetus and the effects of maternal intake of selenium. *Biological Trace Element Research*. January 2004. Accessed November 7, 2023. **doi: 10.1385/BTER:97:1:57**

https://doi.org/10.54846/am2024/31

Practitioner perspectives on the use of McREBEL practices on sow farms

H. D. Dohlman[1]; C. J. LeFevre[2], DVM; J. Waddell[3], DVM, MBA; D. Linhares[4], DVM, MBA, PhD; R. Philips[3], DVM
[1]Iowa State University College of Veterinary Medicine, Ames, Iowa; [2]Carthage Veterinary Service, Carthage, Illinois; [3]Boehringer Ingelheim Animal Health USA Inc., Duluth, Georgia; [4]Veterinary Department of Production Animal Medicine, Iowa State University, Ames, Iowa

Introduction

In 2000, Dr. Monte McCaw[1] reported on the effectiveness of adopting a series of bio-management practices to decrease the spread of bacteria throughout a farrowing room, otherwise known as Management Changes to Reduce Exposure to Bacteria to Eliminate Losses, or MCREBEL. Since Dr. McCaw's report, McREBEL practices have been implemented in sow farms to assist in managing the infection of many swine pathogens. However, the original McREBEL practices have been modified to fit each farm's needs or protocols, leading to miscommunication and questioning throughout the swine industry of what practices are effective. Therefore, the objective of this study was to compare the knowledge and implementation of McREBEL practices across practicing veterinarians in the US.

Materials and methods

An anonymous, voluntary, web-based survey was utilized for the project. The survey was composed of 23 questions ranging from multiple choice (n = 12), true/false (n = 1), multiple answer (n = 5), ranking (n = 2), and open-answer (n = 3) question styles. The survey was split into five sections focusing on participant demographics, knowledge of McREBEL, current practices, timing of implementation, and opinions on the future of McREBEL. The survey was sent via email to 150 veterinarians comprising of 40 clinics and production companies. 71 practitioners with 1 to 50 years of experience completed the survey, resulting in a 47% response rate.

Results and discussion

Only one participant did not utilize McREBEL practices. The majority of participants considered the primary goal of McREBEL to decrease lateral PRRSV transmission between litters. All participants implemented McREBEL practices with PRRSV outbreaks along with other diseases. Over 50% of participants believe McREBEL protocols add 1 - 3% mortality, while 32% believe protocols add 3 - 5% mortality.

Participants were split on their opinion of whether or not McREBEL should be implemented when PRRSV prevalence is high. In a multiple-answer style question, participants were asked to select specific bio-management practices that were considered "original" or "modified" MCREBEL. For "original" MCREBEL, the top three choices across participants were: discontinuing rolling back litters after weaning, changing needles/blades between litter or piglets, and discontinuing fallback management. In regards to "modified" MCREBEL, the top three choices across participants were: cross-fostering allowed only within the first 24 hours of life, allowing fallback litter creation within 5 days of age, and cross-fostering allowed until 5 days of age. The majority of participants believed limiting cross-fostering was the most impactful practice on reducing viral spread, while containing testicles and tails during processing was the least impactful.

80% of participants believe McREBEL practices have unintended or negative consequences on their farms, with demotivation of workers being the most severe negative outcome. 58% of participants implement McREBEL once processing fluids begin to trend negative, which is the same selected timing of when most participants would move to a stricter or "traditional" McREBEL protocol. 58% of participants have a two-staged approach when implementing MCREBEL, with 60% implementing a "modified" McREBEL protocol first. 80% of practitioners end McREBEL at the completion of their disease elimination program.

The responses from this survey expose the variation of opinions and current practices of McREBEL protocols. This information can be put towards future research opportunities, as well as properly identifying Dr. McCaw's original McREBEL protocols versus modern bio-management plans in regards to disease elimination. It also highlights the importance, during vet-to-vet or vet-to-producer communications, to be specific about which practices they recommend implementing in herds undergoing disease management, as opposed to using general terms such as MCREBEL.

Reference

1. McCaw MB. Effect of reducing crossfostering at birth on piglet mortality and performance during an acute outbreak of porcine reproductive and respiratory syndrome. *Swine Health Prod.* 2000;8(1):15-21.

https://doi.org/10.54846/am2024/32

Patterns of feed disappearance under different health challenge scenarios utilizing Feed Allocation System (FAS) data

Mafalda Mil-Homens[1], DVM, MS; Edison Magalhaes[1], DVM, MS, PhD; Joel Stave[2], BS; Jordan Gebhardt[3], DVM, PhD; Tyler Holck[1], DVM, MS, MBA; Gustavo Silva[1], DVM, MS, PhD
[1] College of Veterinary Medicine, Iowa State University; [2] PrairieE Systems; [3] Kansas State University

Introduction

Diseases impact wean-to-finish performance and health, leading to economic losses and reduced productivity by hindering weight gain, decreasing feed conversion, reducing carcass weight, and elevating mortality rates. Monitoring key production indicators such as average daily weight gain, mortality, and feed conversion supports the swine producer, enabling a better understanding of patterns and significant deviations in the indicators during the presence and absence of disease.[1-3] This study used the Feed Allocation System (FAS), a PrairiE Systems Product, that producers and veterinarians leveraged to collect and analyze routinely collected feed consumption and mortality data on their farms. This FAS data can offer insights into the dynamic indicators under health challenges, allowing early interventions (eg, increased diagnostic testing, increased biosecurity), and a better understanding of what the impact on production would look like.

This preliminary study aims to assess patterns of feed consumption and the percentage of dead animals to better understand the impact of health challenges on pig performance.

Methods

Data from Prairie Systems© spanning 2020 to 2022 comprised 7 wean-to-finish lots from a swine production system in the Midwest US. The outcomes used for the analysis encompassed the daily cumulative count of dead animals per daily inventory and the feed consumption per head per 1000 pounds (FC) calculated by dividing the daily feed consumption per head by the weight (lbs.) and multiplying by 1000. The weight value was estimated based on the following formula:

$$\text{Weight (lbs.)} = \text{Cumulative daily average daily feed intake} + \left(\frac{\frac{\text{Average daily feed intake} \times \text{inventory}}{\text{Weighted average inventory}}}{\text{Weighted daily feed conversion}}\right)$$

The study subjects were grouped into cases and reference groups.

Reference groups

Reference groups were chosen based on low mortality, and PRRSV-negative nursery status to build a baseline that was the reference to compare with the patterns of the case groups. Three reference groups were tested:

1. Reference one, consisting of two lots originating from PRRSV-negative sow farms and negative for PRRSV through the nursery phase, averaging 1,048 animals per lot and, FC of 31.6 lbs., and finisher mortality rate of 3%;
2. Reference two, consisted of three lots originating from PRRSV-positive sow farms and negative PRRSV through the nursery phase, averaging 1,047 animals per lot, FC of 35.3 lbs., and finisher mortality rate of 2%;
3. Reference three, consisted of reference one and reference two together.

Cases

Two cases were considered for the analysis.

1. Case 1 consisted of a lot originating from a PRRSV-positive sow farm, had an initial inventory of 1,166, FC of 34.3 lbs., exhibited a mortality rate of 10%, and was considered PRRSV-positive on day 14 in the finishing phase.
2. Case 2 consisted of a lot originating from a PRRSV-negative sow farm, had an initial inventory of 1124, FC of 29.8 lbs., exhibited a mortality rate of 6%, and was considered PRRSV-positive on day 39 in the finishing phase.

Analysis

Two different methods were created to build the reference groups. A moving average and a weighted moving average, both with a 95% confidence interval. For the first method, a moving average of three days was calculated. For the second method, a weighted moving average with a period of three days, with weights of 0.2, 0.3 and 0.5 from the first to the latest day was calculated.

If the percentage of deaths from case 1 or case 2 had a value above the upper limit of the 95% confidence interval, then that would be considered a significant deviation from the reference, if the FC from case 1 or case 2 had a value below the lower limit of the 95% confidence interval, then that would be considered a deviation.

Results

The results showed that for both case 1 and case 2 the values of the percentage of deaths were above the reference groups before the PRRSV diagnostic became positive and that FC was below the reference group before the positive PRRSV diagnostic. An example of the patterns of the percentage of dead animals for case 1 can be observed in Figure 1, the pink line represents the percentage of dead animals compared to the reference group 2 in yellow. Furthermore, an example of the patterns of FC for case 1 can be observed in Figure 2, the pink line represents the FC compared to the reference group 2 in yellow. Concerning case 2, the percentage of dead animals can be observed in Figure 3, the pink line represents the percentage of dead animals compared to the reference group 2 in yellow, and the FC can be observed in Figure 4, the pink line represents the FC compared to the reference group 2 in yellow. A summary of results for cases 1 and 2 is shown in Tables 1 and 2, respectively.

Figure 1: Percentage of death patterns for case 1. The figure shows the pattern of the percentage of daily deaths (pink) for case 1 compared to reference 2 (yellow) using a moving average, with 95%CI for the reference (light blue). For case 1, the lot had a PRRSV positive diagnostic on day 14, represented in the figure by the vertical red dotted line and the label.

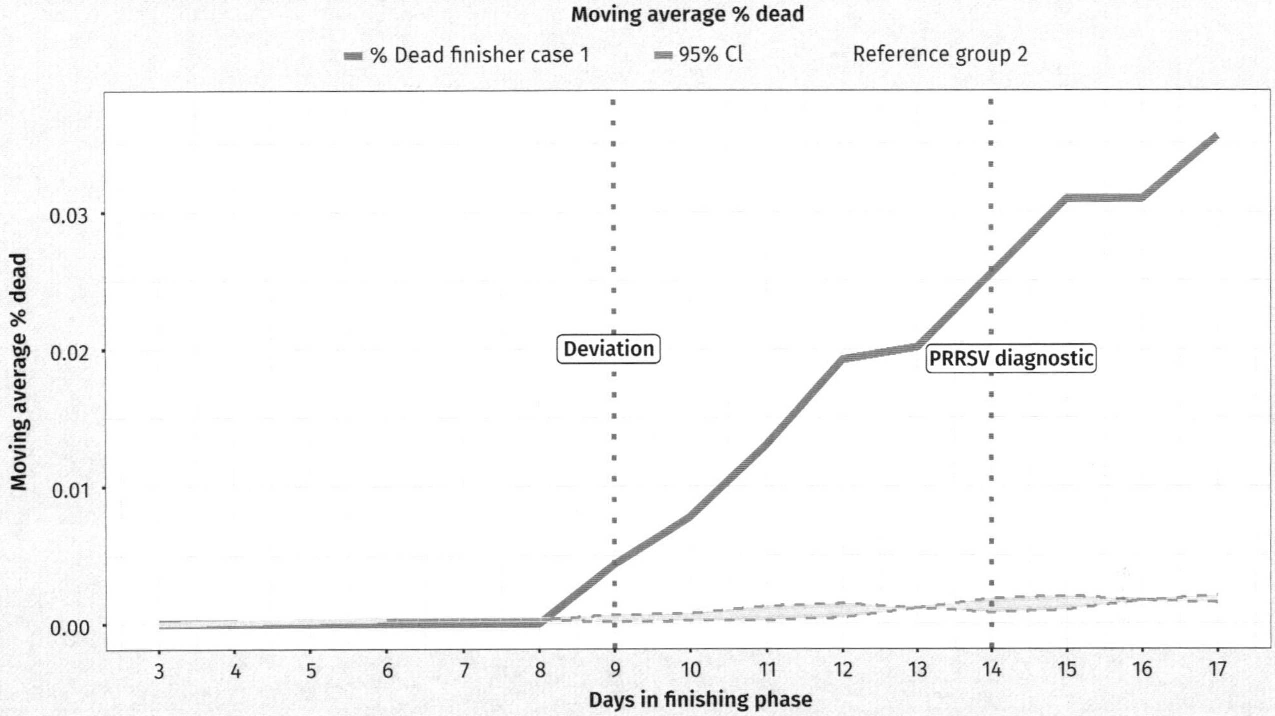

Figure 2: Feed consumption per 1000 pounds bodyweight patterns for case 1. The figure shows the pattern of feed consumption per 1000 pounds (pink) for case 1 compared to reference 2 (yellow) using a moving average with 95%CI for the reference (light blue). For case 1, the lot had a PRRSV positive diagnostic on day 14, represented in the figure by the vertical red dotted line and the label.

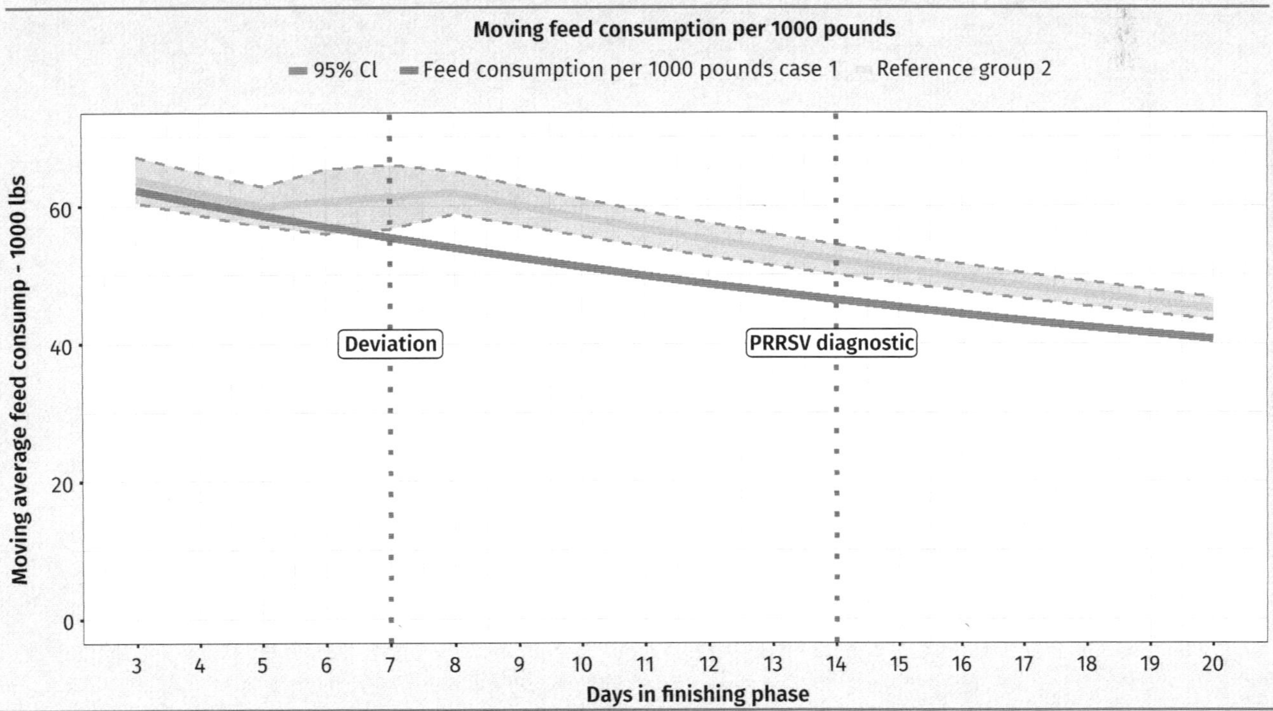

Figure 3: Percentage of death patterns for case 2. The figure shows the pattern of the percentage of daily deaths (pink) for case 2 compared to reference 2 (yellow) using a moving average, with 95%CI for the reference (light blue). For case 2, the lot had a PRRSV positive diagnostic on day 39, represented in the figure by the vertical red dotted line and the label.

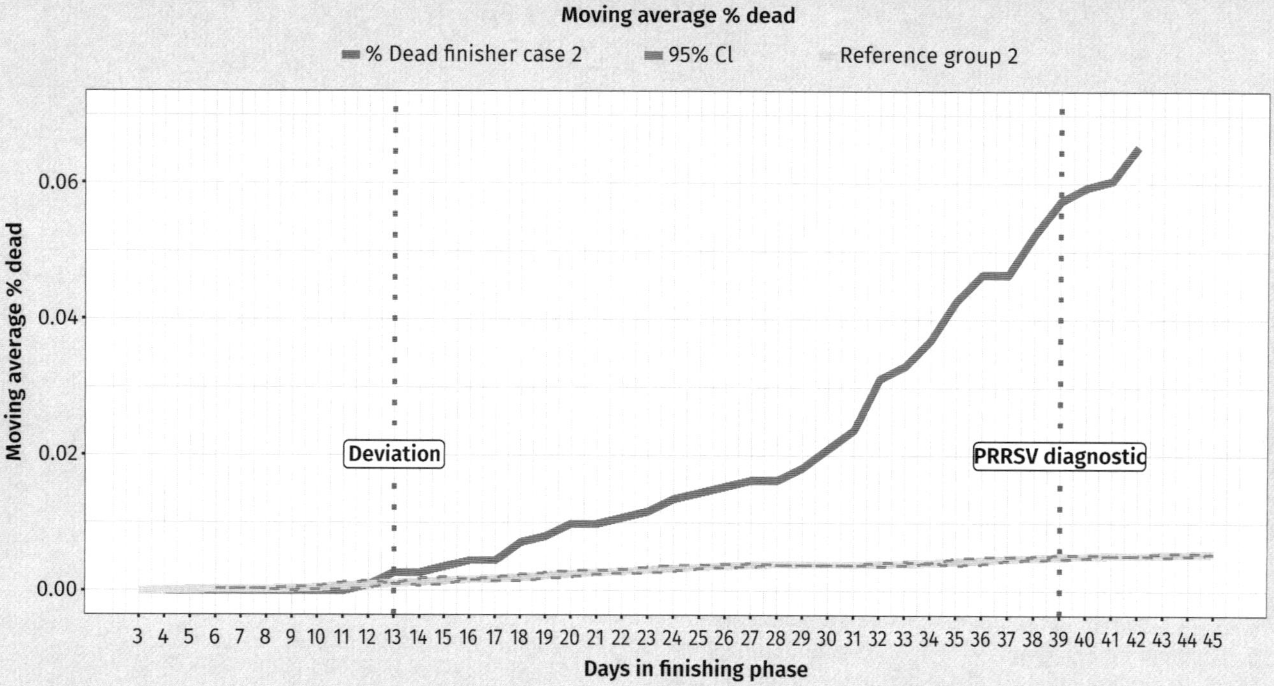

Figure 4: Feed consumption per 1000 pounds bodyweight patterns for case 2. The figure shows the pattern of feed consumption per 1000 pounds (pink) for case 2 compared to reference 2 (yellow) using a moving average with 95%CI for the reference (light blue). For case 2, the lot had a PRRSV positive diagnostic on day 39, represented in the figure by the vertical red dotted line and the label.

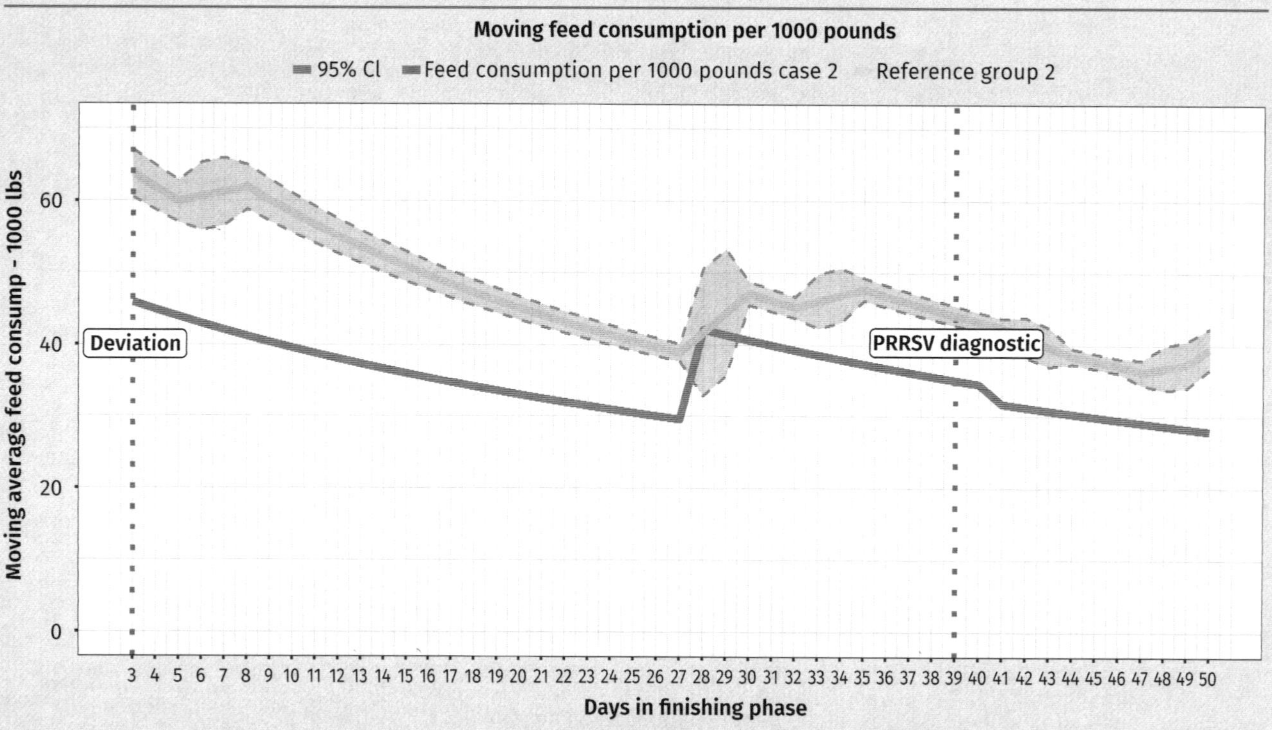

Table 1: Case 1 results. Table describing the results percentage of deaths and feed consumption for case 1. The red color represents the day when the model was above the upper limit of 95%CI for the percentage of deaths. The orange color represents the day when the model was below the lower limit of 95%CI for the feed consumption.

Case	Reference	Model	Indicator	Days in finishing phase 4	5	7	9	14
1	Sow farm PRRSV- finisher low mortality (reference 1)	Moving average	%Dead					PRRSV+
			Consumption					
		Weighted moving average	%Dead					
			Consumption					
	Sow farm PRRSV+ finisher low mortality (reference 2)	Moving average	%Dead					
			Consumption					
		Weighted moving average	%Dead					
			Consumption					
	Sow farm PRRSV+/ sow farm PRRSV- finisher low mortality (reference 3)	Moving average	%Dead					
			Consumption					
		Weighted moving average	%Dead					
			Consumption					

Table 2: Case 2 results. Table describing the results percentage of deaths and feed consumption for case 2. The red color represents the day when the model was above the upper limit of 95%CI for the percentage of deaths. The orange color represents the day when the model was below the lower limit of 95%CI for the feed consumption.

Case	Reference	Model	Indicator	Days in finishing phase 1	3	4	13	39
2	Sow farm PRRSV- finisher low mortality (reference 1)	Moving average	%Dead					PRRSV+
			Consumption					
		Weighted moving average	%Dead					
			Consumption					
	Sow farm PRRSV+ finisher low mortality (reference 2)	Moving average	%Dead					
			Consumption					
		Weighted moving average	%Dead					
			Consumption					
	Sow farm PRRSV+/ sow farm PRRSV- finisher low mortality (reference 3)	Moving average	%Dead					
			Consumption					
		Weighted moving average	%Dead					
			Consumption					

Feed consumption per 1000 lbs

For case 1, the methods had the earliest deviations occurred when the case was compared to reference 2. These methods had deviations on day 7 of the finishing phase (Table 1). FC had deviations on day 3 of the finishing phase for case 2 compared to all references (Table 2).

Percentage of dead animals

For case 1, the methods had the earliest deviations compared with reference 3 using the moving average. The deviations occurred on day 4 of the finishing phase (Table 1). For case 2, the methods had the earliest deviations compared with reference 3. These models had deviations on day 4 of the finishing phase (Table 2).

Discussion

The preliminary findings from this study suggest that indicators collected routinely at swine finishing sites offer valuable insights into the impact of diseases on production. Our analysis suggested that feed consumption and the number of deaths start having significant changes before the clinical diagnosis for PRRSV was made, this may be a valuable information for the swine producer since utilizing FAS to collect such data can be advantageous as it serves as a trigger for further investigation when significant deviations from the reference group occur, leading to implementation of measures at the farm, such as diagnostic testing, medication, and increased biosecurity in order to better manage the impact a disease has on the herd. Collecting FAS daily data can be helpful in developing an automated data collection and real-time surveillance system that promptly detects substantial deviations in the data.

References

1. C. Fourichon, "Application of ecopathological methods to the investigation of health problems on farms".

2. C. Fablet, N. Rose, B. Grasland, N. Robert, E. Lewandowski, and M. Gosselin, "Factors associated with the growing-finishing performances of swine herds: an exploratory study on serological and herd level indicators," *Porc. Health Manag.*, vol. 4, no. 1, p. 6, Mar. 2018, **doi: 10.1186/s40813-018-0082-9**

3. E. S. Magalhães *et al.*, "Whole-herd risk factors associated with wean-to-finish mortality under the conditions of a Midwestern USA swine production system," *Prev. Vet. Med.*, vol. 198, p. 105545, Jan. 2022, **doi: 10.1016/j.prevetmed.2021.105545**

https://doi.org/10.54846/am2024/33

Phytogenic Patch Plus (PPP) benefits nursery pig performance and health

M. J. Rincker[1], PhD; S. D. Carter[2], PhD; C. Wileman[2], MS; D. A. Nelson[1], PhD; M. D. Hemann[1], MS
[1]Distributors Processing, Inc., Porterville, California; [2]Department of Animal Science, Oklahoma State University, Stillwater, Oklahoma

Introduction

A multitude of factors contribute to weaning being a major stress period in a pig's life. The transition from milk to dry feed and the associated gastrointestinal tract development, exposure to new environmental health challenges, and commingling with new pen mates are all factors that can negatively impact nursery pig performance and health. It is estimated that improving the mortality rate by 1% (5% vs 6%) in wean-to-finish swine production has a positive impact on net income of $1.04 per head marketed (2,400 head group).[1] Thus, producers are constantly seeking viable tools to maximize swine health and performance and achieve the full value of a pig at market. The administration of treatments via the drinking water offers an effective method to optimize pig performance and aid in health challenge prevention or treatment. Phytogenic Patch Plus (PPP), commonly referred to as 'Triple P', is a liquid-blend of phytogenic extracts that is administered through the drinking water, and functions to maximize animal health and performance. The active plant compounds found in PPP have been reported to improve nutrient digestibility,[2] growth performance,[3] and odor control in swine.[4] Furthermore, these plant compounds are natural surfactants that can bind to membrane cholesterol of bacterial[5] and protozoal cells,[6] leading to cell lysis and death. Therefore, the objective of this experiment was to determine the effectiveness of PPP supplementation in the drinking water of nursery pigs on performance and health.

Methods

All experimental procedures were performed in accordance with the Oklahoma State University Institutional Animal Care and Use Protocol and were approved by this committee. The live animal research was completed at the Oklahoma State University Swine Research and Education Center (Stillwater, Oklahoma). One-hundred forty pigs (avg wt = 12.59 lb) were weaned on day 20 of age, blocked by weight, and randomly assigned to 1 of 14 nursery pens. There were 7 replicate pens per treatment with ten pigs per pen. Within the nursery, pens were randomly assigned to 1 of 2 water treatments: 1) Control (0 mL of PPP per L of stock solution), and 2) Phytogenic Patch Plus (16 mL of PPP per L of stock solution and administered continuously for 42 days). The water treatment was employed using a Dosatron water medicator set to a 1:128 ratio. A stock solution for each treatment was prepared daily to supply the desired treatment concentration.

Pigs were housed in a temperature-controlled nursery room and allowed ad libitum access to feed and water throughout the experiment. Pigs were fed standard, corn-soybean meal diets in four dietary phases (Phase 1: day 0 - 7, Phase 2: day 7 - 14, Phase 3: day 14 - 21, Phase 4: day 21 - 42), all meeting or exceeding NRC recommendations. No antibiotics or pharmacological levels of zinc or copper were used in these diets. Pigs and feeders were weighed weekly to determine ADG, ADFI, and G:F. Water meters were used to determine water disappearance, which was recorded at the same time every day.

On day 23 of the experiment, an acute LPS challenge was administered intramuscularly with lipopolysaccharide (LPS) from Escherichia coli serotype O55:B5 (lyophilized powder purified by phenol extraction; Sigma-Aldrich, St. Louis, Missouri). Pigs were weighed and then all pigs in each pen received an LPS dose of 10 ug LPS/kg body weight based on average pen weight. Within each pen, two pigs closest to the average weight of the pen were ear-tagged, weighed, rectal temperature recorded, and had blood collected prior to the LPS administration at hour 0. At hour 3, 6, 12, and 24 post-LPS injection, the ear-tagged pigs were again weighed, rectal temperature recorded using a digital thermometer, and blood collected via jugular venipuncture.

Data were analyzed as a randomized complete block design using analysis of variance procedures. Proc GLM, LS means and pdiff were used to determine the effects of PPP supplementation through the water by treatment. Proc Mixed was used to look at time effects of hour and week on body weight and rectal temperature. Pen served as the experimental unit.

Results

Daily water intake was similar between treatments throughout the entire study (Table 1).

Growth performance measures are presented in Figures 1 and 2. During Phase 1, a numeric increase of 16% was observed in ADG of nursery pigs supplemented with PPP vs the Control pigs (0.36 vs 0.31 lb), while G:F was also numerically improved by 15%. During the first three weeks of the study, pigs consuming PPP exhibited a 27% greater ($P = .06$) ADG and a 15% better ($P = .10$) G:F compared with Control pigs. During the LPS challenge, pigs consuming PPP exhibited a 5% increase in ADG ($P = .17$) compared with Control pigs. For the entire 42-d experiment, pigs consuming PPP had a 16% greater ($P < .05$) ADG, and a 15% improved G:F compared with the Control pigs.

Minimal treatment differences were observed on performance during the LPS-challenge period. However, pigs consuming PPP had a numerically lower mortality over the 42-day experiment (5.70 vs 1.40% for the Control and PPP treatments, respectively) (Table 1).

Discussion

While the administration of treatments via the drinking water supply can be an effective tool to use during health challenges, it is essential that these products are compatible with water programs, do not inhibit water system flow rates, and do not adversely affect palatability or water intake. Given that daily water intake was similar amongst both treatments, this would negate any concerns that the product may have on palatability or subsequent experimental results.

The supplementation of antibiotics and pharmacological levels of specific minerals (ie, Zn and Cu) are common in nursery diets. However, as these practices become more regulated, producers

Table 1: Effects of Phytogenic Patch Plus (PPP) on water intake, mortality and lipopolysaccharide (LPS)-challenge measures[1]

Parameter	Treatment[2] Control	PPP	SE
Water disappearance, gallons/pig/day			
Day 1-5	0.23	0.21	0.02
Day 6-7	0.28	0.29	0.02
Day 18-28	0.48	0.51	0.03
Day 29-42	0.71	0.73	0.06
Rectal temperature, °F	104.5	104.6	0.16
TNF-alpha, pg/mL	1,799.0	1,732.0	191.0
Day 0 - 42 mortality, %	5.70	1.40	0.29

[1] LPS-challenge was administered on day 23 of the study.

[2] Control consisted of 0 mL of Phytogenic Patch Plus (PPP) per L of stock solution; PPP consisted of 16 mL of PPP per L of stock solution and administered continuously for 42 days.

Figure 1: Growth performance measure, average daily gain. Control (0 mL of Phytogenic Patch Plus (PPP) per L of stock solution), PPP (16 mL of PPP per L of stock solution).

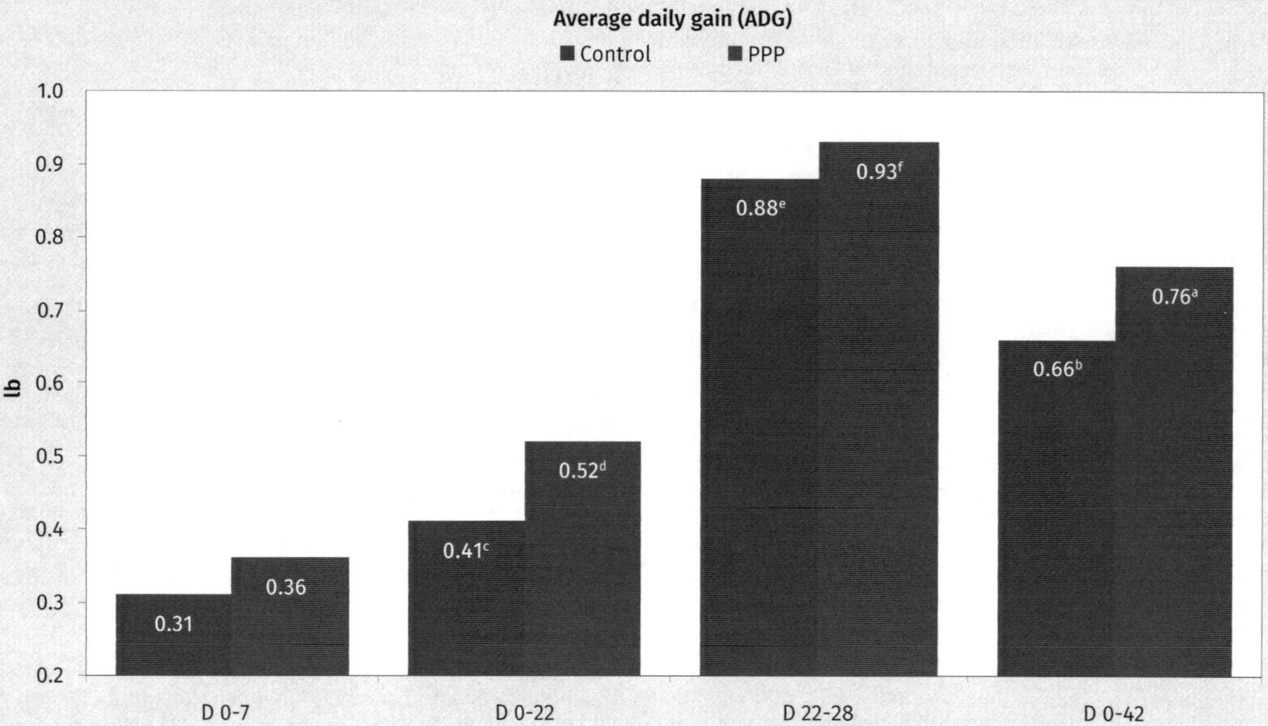

[a,b] Means differ $P < .05$

[c,d] Means differ $P = .06$

[e,f] Means differ $P = .17$

Figure 2: Growth performance measure gain:feed. Control (0 mL of (Phytogenic Patch Plus) PPP per L of stock solution), PPP (16 mL of PPP per L of stock solution).

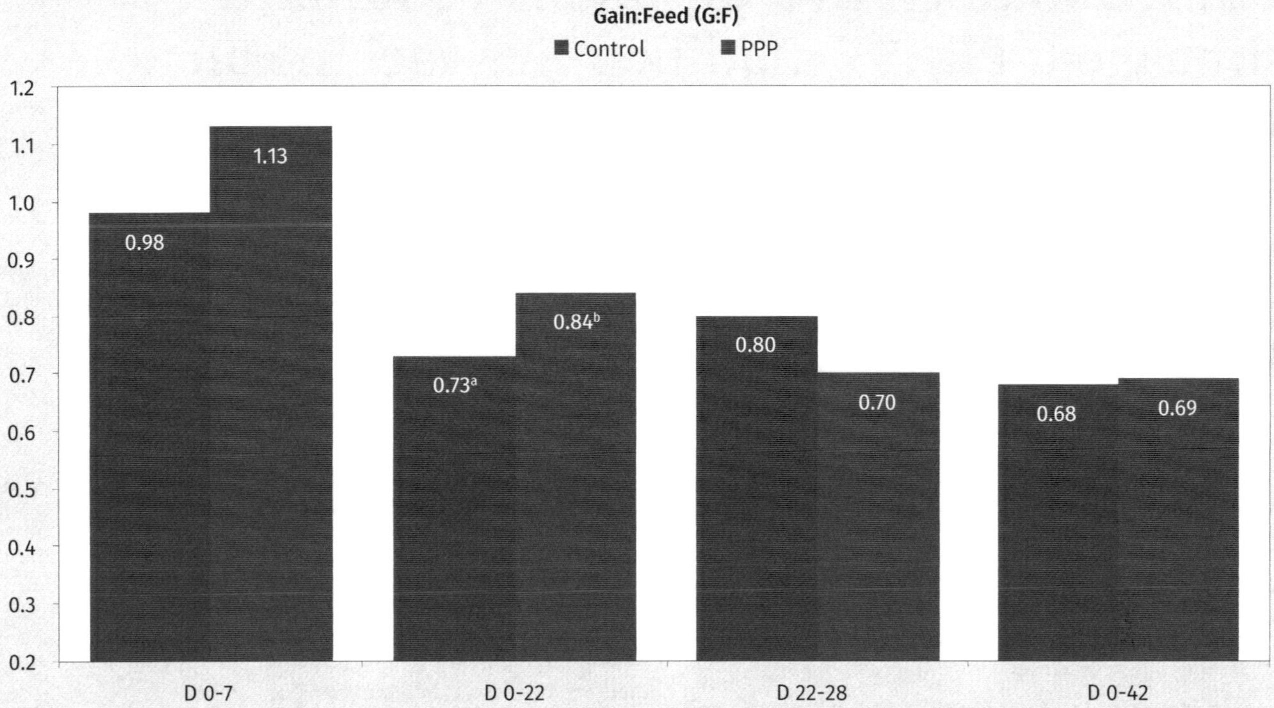

[a,b] Means differ P = .10

are looking for viable alternatives to help mitigate the typical stressors that the young pig experiences in the nursery phase. With the nursery diets used in the present study containing no antibiotics or pharmacological levels of minerals, the noted improvements in growth and efficiency from the current study support the use of PPP as a viable alternative in the early nursery period. The natural variation that is typical in this earliest nursery period likely contributed to a lack of statistical significance.

It is especially noteworthy that PPP significantly improved ADG during the LPS challenge. Although pigs consuming PPP did not exhibit an improvement in feed efficiency during the LPS challenge period, this may be due to the elevated average daily feed intake of the PPP-supplemented pigs (1.33 vs 1.13 lb./day; $P < .05$ for the Control and PPP respectively).

The ADG and mortality improvements observed in pigs consuming PPP during an LPS-challenge support a significant opportunity for the use of Phytogenic Patch Plus within the demanding environments of modern swine production systems.

References

1. Euken R and Schulz L. Assessing Economic Opportunity of Improving Mortality Rate in Wean-to-Finish Swine Production. Iowa State University Extension and Outreach Ag Decision Maker. 2021.

2. Johnson IT, Gee JM, Price K, Curl C, and Fenwick GR. Influence of saponins on gut permeability and active nutrient transport in vitro. *J. Nutr.* 1986. 116:2270-2277.

3. Mader TL and Brumm MC. Effect of feeding sarsaponin in cattle and swine diets. *J. Anim. Sci.* 1987. 65:9-15.

4. Hurst RJ, McFarlane JM, and Curtis SE. Sarsaponin reduces ammonia production by decomposing excreta in an in vitro system. J. Anim. Sci. 1990.

5. Walker RD, and Rincker MJ. Impact of Micro-Aid® on microbiota common to the intestinal flora of food and companion animals. 2012. Personal Communication.

6. Francis G, Kerem Z, Makkar HPS, and Becker K. The biological action of saponins in animal systems: a review. Br. J. Nutr. 2002. 88:587-605.

https://doi.org/10.54846/am2024/34

Evaluating the transition from a previously used PRRSV modified-live vaccine to PRRSGard®, a lineage 1, PRRSV modified-live vaccine on a commercial sow farm

B. Schantz
Diversified Management Company, Laurel, Nebraska

Introduction

Porcine reproductive and respiratory syndrome virus (PRRSV) can have a significant impact when introduced into a commercial sow farm, resulting in increased morbidity and mortality in sows, increased piglet loses, and reduced reproductive performance.[1] The use of modified-live (MLV) PRRSV vaccines to provide consistent immunity within sow herds has historically been used to reduce the impact of a lateral PRRSV introduction. PRRSGard® is a lineage 1, chimeric PRRSV MLV for the vaccination of healthy swine, three weeks of age or older, against PRRSV.[2] It has previously been shown to be safe in finishing animals, as well as efficacious in the face of PRRSV challenges.[3-5] Anecdotal reports also suggest that, in some cases, moving to a new PRRSV MLV after the introduction of a new, wild-type PRRSV strain on a sow farm can decrease the time needed to achieve PRRSV stability.

Following a lateral PRRSV introduction in November of 2022, a commercial sow farm decided to transition from a previously used PRRSV MLV to PRRSGard as part of their PRRSV elimination protocol. This project aimed to evaluate and describe that transition to PRRSGard on a commercial sow farm.

Materials and methods

A 5,500 head, commercial sow farm located in Nebraska (Farm A) broke with a new, wild-type PRRSV strain (RFLP 1-8-4) on 11/02/2022. The initial PRRSV elimination was initiated on 11/12/2022, when Farm A was exposed to PRRSV via live pig exposure throughout the sow farm and gilt developer unit, while animal entry into Farm A was stopped. Prior to this lateral PRRSV break, Farm A was routinely vaccinating all mature animals with a commercial PRRSV MLV.

Farm A was then vaccinated at three separate timepoints during the PRRSV closure with the commercial PRRSV MLV that was used on farm previously. Farm A was vaccinated on 12/1/2022, 01/03/2023, and 04/17/2023. From January through June of 2023, the farm continued to detect PRRSV in processing fluids, while the downstream performance of the pigs was considered to be poor.

In June 2023, additional mature gilts were brought into Farm A due to their extended closure. At this time, another round of PRRSV exposure was done in an effort to get Farm A back to PRRSV negative status. From July 2023 through August 2023, PRRSV continued to be detected in processing fluid samples. However, it was noted that the downstream performance of the pigs improved.

In September 2023, a decision was made to transition Farm A away from the previously used PRRSV MLV to PRRSGard, in an effort to further stabilize the herd for PRRSV and return to normal farm production. Two whole herd vaccinations were given to all mature animals on the farm. The first PRRSGard vaccination was administered on 09/27/2023 (0 weeks post-vaccination [wpv]). The second PRRSGard vaccination was administered on 11/01/2023 (5 wpv).

A diagnostic evaluation was done on Farm A before, during, and after the transition to PRRSGard. Processing fluids were collected weekly, starting eight weeks prior to the first vaccination (-8 to -1 wpv), between the first and second vaccination (0 to 4 wpv), and then eight weeks following the second vaccination (5 to 12 wpv). Serum was collected from due to wean pigs within gilt litters, as well as parity 2 and above (P2+) litters weekly, starting two weeks prior to the first vaccination (-2 to -1 wpv), between the first and second vaccination (0 to 4 wpv), and then for eight weeks following the second vaccination (5 to 12 wpv). Family oral fluids were collected from gilt litters and P2+ litters weekly, starting between the first and second vaccination (3 to 4 wpv), through eight weeks following the second vaccination (5 to 12 wpv).

Production data was collected from Farm A for a total of 20 weeks around the two whole herd vaccinations with PRRSGard, including conception rate, farrowing rate, pigs weaned per sow, and female mortality. This information was evaluated eight weeks prior to the first vaccination (-8 to -1 wpv), between the first and second vaccination (0 to 4 wpv), and then eight weeks after the second vaccination (5 to 12 wpv).

Results

At the time of submission, diagnostic information from Farm A has been collected through 6 wpv (Table 1). Complete diagnostic information, through 12 wpv, will be presented at the 2024 American Association of Swine Veterinarians meeting in Nashville, Tennessee.

Following the first whole herd vaccination, PRRSV was detected via PCR on processing fluid samples 2 wpv (average of 35.1 Ct on positive samples) and 6 wpv (33.0 Ct). The remaining five weeks of processing fluid samples post vaccination were negative for PRRSV on PCR.

Serum samples from due to wean pigs were negative for PRRSV from -2 to 3 wpv from both gilt and P2+ litters. Serum samples collected in due to wean pigs 4 wpv were positive for PRRSV in both gilt litters (23.9 Ct) and P2+ litters (35.1). Sequencing of the ORF5 region of the PRRSV positive sample with the lowest Ct revealed a RFLP 1-8-4 lineage 1H, which was 99.53% similar to the strain identified on Farm A in May 2023, and 96.35% similar to the strain identified in November 2022. Serum samples from due to wean pigs in gilt and P2+ litters were negative for PRRSV at 5 and 6 wpv.

Family oral fluids collected from gilt litters detected PRRSV via PCR at 4 wpv (33.4 Ct) and 6 wpv (33.6 Ct), while family oral fluids collected from P2+ litters detected PRRSV via PCR at 6 wpv (33.6 Ct).

Table 1: PRRSV testing results from Farm A prior to, during, and after PRRSGard transition

Timing to first PRRSGard vaccination	Processing fluids*	Wean pig serum (gilt litters)*	Wean pig serum (P2+ litters)	Family oral fluids (gilt litters)	Family oral fluids (P2+ litters)
-8 weeks	25.6				
-7 weeks	24.1				
-6 weeks	> 37				
-5 weeks	> 37				
-4 weeks	> 37				
-3 weeks	> 37				
-2 weeks	> 37	> 37	> 37		
-1 week	> 37	> 37	> 37		
0 week[1]	> 37	> 37	> 37		
1 week	> 37	> 37	> 37		
2 weeks	35.1	> 37	> 37		
3 weeks	> 37	> 37	> 37	> 37	> 37
4 weeks	> 37	23.9	35.1	33.4	> 37
5 weeks[2]	> 37	> 37	> 37	> 37	> 37
6 weeks	33.0	> 37	31.8	33.6	33.6

* Average Ct detected on PRRSV PCR for each sample type; > 37 indicates all submitted samples were negative for PRRSV

1 First whole herd PRRSGard vaccination

2 Second whole herd PRRSGard vaccination

At the time of submission, production information from Farm A has been collected through 5 wpv. During the whole herd vaccinations with PRRSGard, there were minimal reported disruptions at Farm A. Complete evaluation of production information, through 12 wpv, will be presented at the 2024 AASV meeting.

Discussion

Observations from Farm A indicate that PRRSGard, when administered to gilts and sows within a commercial setting, was safe and appeared to have minimal impact on production during the PRRSV elimination. The diagnostics collected, while not complete at the time of submission, indicate a low level of PRRSV on Farm A leading up to, during, and after vaccination with PRRSGard.

Veterinarians and producers will continue to evaluate the tools available to them to help control and mitigate the impact of PRRSV on sow farms. In this evaluation, PRRSGard, a lineage 1 PRRSV MLV, appears to be a viable option when considering vaccinating commercial sow herds with a PRRSV MLV.

References

1. Zimmerman, JJ., Karriker, LA., Ramirez, A., et. al. 2019. *Diseases of Swine, Eleventh edition.* Chapter 41, Porcine Reproductive and Respiratory Syndrome Viruses (Porcine Arteriviruses). P 685 - 708. Published 2019 by John Wiley & Sons, Inc.

2. PRRSGard label information. Accessed 15 Nov 2023. **https://pharmgate.com/usa/wp-content/uploads/sites/5/2023/11/PAH_PRRSGard_ProductProfile_1459-0623.pdf**

3. Chamba, F., Sui, J., Conarchy, B., et. al. *Experimental safety and efficacy of a unique MLV PRRSV vaccine: PRRSGard.* Proceedings 50th annual meeting of the American Association of Swine Veterinarians.

4. Chamba, F., Smith, C., Saltzman, R., et. al., *PRRSGard® was efficacious against a virulent PRRSV174 challenge.* Proceedings Allen D. Leman Swine Conference, 2020. Poster number 64.

5. Smith, C., Chamba, F., Pittman, J., et. al. *Evaluation of the response to PRRSGard® administration in weaned pigs.* Proceedings 51st annual meeting of the American Association of Swine Veterinarians.

https://doi.org/10.54846/am2024/35

Evaluation of Aivlosin® and tilmicosin to manage a natural porcine respiratory disease complex challenge in a commercial finishing setting

J. Madigan[1], DVM; N. Lauterbach[2], DVM; D. Fiene[2]
[1]Pillen Family Farms, Columbus, Nebraska, [2]Pharmgate Animal Health, Wilmington, North Carolina

Introduction

Managing respiratory disease in growing pigs is one of the critical challenges facing swine veterinarians today. Porcine respiratory disease complex (PRDC) is a multi-pathogen disease process that commonly includes viruses, such as porcine reproductive and respiratory syndrome virus (PRRSV), as well as bacterial pathogens such as *Mycoplasma hyopneumoniae* (MHP) and *Haemophilus (Glasserella) parasuis* (GPS).[1] As there are limited to no anti-viral medications available to swine veterinarians today, antibiotics are commonly used to reduce the impact of the bacterial pathogens commonly found in cases of PRDC. Aivlosin® (62.5% w/w Tylvalosin as Tylvalosin Tartrate; ECO Animal Health) is a second-generation macrolide antibiotic that is labeled for the control of swine respiratory disease (SRD) associated with *Bordetella bronchiseptica, Glasserella (Haemophilus) parasuis, Pasteurella multocida, Streptococcus suis,* and *Mycoplasma hyopneumoniae* in groups of swine intended for slaughter and female swine intended for breeding in buildings experiencing an outbreak of SRD.[2] The objective of this project was to evaluate two water soluble antibiotics, Aivlosin® and tilmicosin, in an effort to manage a natural PRDC challenge in a commercial finishing setting.

Materials and methods

A finishing site with eight finishing barns (Barns 1-8) in the Midwestern part of the United States, utilizing a single nursery site was identified for this evaluation. The pigs originated from multiple sow farms that were PRRSV, porcine epidemic diarrhea virus (PEDV) and MHP negative at the time of placement into the nursery. The pigs would have been moved from their nursery barns and placed in the selected finisher barns at roughly nine weeks of age. At placement, pigs were stocked at roughly the same density across all eight of the finishing barns. The week prior to moving from the nursery to the selected finisher barns, PRRSV was detected via PCR on oral fluid samples, which were confirmed positive on retest (lowest Ct of 27.8). After identification of PRRSV, and prior to movement to the finisher, all pigs received vaccination with a commercial PRRSV modified-live vaccine.

Previously, after diagnosing PRRSV, this system would have implemented a treatment of Pulmotil® AC (tilmicosin phosphate; Elanco Animal Health) at the labeled rate of 200 parts per million through the stock solution for five consecutive days,[3] starting at placement into the finisher.

For this evaluation, four of the barns received the previous protocol of Pulmotil® AC as outlined above (Barns 1, 3, 5, and 7). The additional four barns (Barns 2, 4, 6, and 8) received Aivlosin® at the labeled rate of 50 parts per million through the stock solution for 5 consecutive days.

Four of the eight barns (Barns 3, 4, 5, and 6) received an additional dose of their respective medication, seven days after the conclusion of the first round of medication. This resulted in Barns 3 and 5 receiving an additional five days of Pulmotil® AC, while Barns 4 and 6 received an additional 5 days of Aivlosin®. This decision was made by the attending veterinarian and was in response to the pig health.

Prior to administering medication, tissue samples were collected from the population at placement into the selected finishers. These samples were submitted to Iowa State University Veterinary Diagnostic Laboratory to confirm the presence of pathogens consistent with PRDC.

Production information is being collected from the time the pigs were placed into the selected finishers through marketing (22 weeks). Production information that is being evaluated includes percent mortality, number of treatments, percentage of cull animals at marketing, and truck weights at marketing from each barn.

Results

Through 19 weeks of the evaluation, the percentage of pigs that was lost via mortality across the two treatment groups shows a numeric difference in favor of the Aivlosin® treated barns. In the barns that received tilmicosin, the average percent mortality was 7.41%. In the barns that received Aivlosin®, the average percent mortality was 6.47%, an improvement in average percent mortality of 0.94%.

Additional production data between the eight barns will be presented at the 2024 American Association of Swine Veterinarians meeting in Nashville, Tennessee.

Tissue samples collected at placement into the finisher identified PRRSV in lung tissue via PCR (Ct 15.8). Histopathology indicated that a wild-type PRRSV was causing interstitial pneumonia. *Mycoplasma hyorhinis* was also identified on lung tissue via PCR (Ct 16.7). Additionally, GPS and *Bordetella bronchiseptica* were isolated from lung samples via culture. Influenza A testing was conducted on submitted lung samples, which returned negative results via PCR.

Discussion

This evaluation was conducted in a commercial setting with animals that had naturally been challenged with pathogens associated with PRDC, including PRRSV, *Bordetella bronchiseptica*, and GPS, as confirmed by diagnostics. Understanding the tools that are available for veterinarians to manage PRDC is critical to limiting the impact this disease complex has on growing pigs. In this evaluation, Aivlosin® was shown to be an effective antibiotic to manage the health of pigs in a commercial setting in the face of a natural PRDC challenge.

References

1. Zimmerman, JJ., Kariker, LA., Ramirez, A., et al. 2019. *Diseases of Swine, Eleventh edition.* Chapter 21, Respiratory system. P 396 & 397. Published 2019 by John Wiley & Sons, Inc.

2. Aivlosin label information. Pharmgate Animal Health website. Accessed 03 Nov 2023. **https://pharmgate.com/usa/wp-content/uploads/sites/5/2023/10/PAH_AivlosinWSG_Label_8.5x11_1123.pdf**

3. Pulmotil AC label information. Elanco Animal Health website. Accessed 14 Nov 2023. **https://www.elancolabels.com/us/pulmotil-ac**

https://doi.org/10.54846/am2024/36

Routine pen-side monitoring for PRRSV is possible using simple, hand-held device

Abraham Oommen, MSc, PhD; Govardhan Rathnaiah, DVM, PhD
MatMaCorp Inc. Lincoln, Nebraska

Introduction

Pig production has increased globally and is a major source of protein in many parts of the world.[1] Foreseeing this growth, ongoing surveillance to establish baseline levels of disease was proposed in the early 1980s as a means to monitor health and therefore continued production from healthy animals.[2,3] Today, oral fluids (OF) collected using in-pen chew ropes are monitored routinely for many disease-causing viruses including PRRSV.[4] Typically, viral nucleic acids detected in oral fluids is consistent with viremia in the upper respiratory tract or buccal tissues and this is also true for PRRSV.[5,6] Using a logistic regression model, it was shown that there is a relationship between the positive oral fluid testing results (RNA or antibody for PRRSV) and within pen prevalence.[7] In such testing, use of one rope per 25 pigs placed at regular intervals rather than randomly, resulted in the right proportion of pigs interacting with the ropes and thereby increasing the probability of detection.[8-11] Instead of pooling samples, collecting spatially oriented oral fluid samples from each pen and establishing their viral status with confidence was suggested as a viable economical option considering the fact the aggregated undiluted sample from each rope is a good representation of the group.[4] These studies confirmed that pathogens move from pig to pig and pen to pen in a spatially-dependent pattern of disease spread justifying routine pen-side monitoring as a good management practice for early detection and preventing economic loss. Easy to use pen-side monitoring systems are therefore necessary. Our goal was to evaluate the potential of pen-side monitoring of PRRSV using a hand-held real-time PCR system. Additionally, our goal was to determine if oral fluid samples collected from chew ropes in pens could be used without purifying the virus or the RNA from the virus.

Materials and methods

PRRS virus and samples

For determining sensitivity, specificity and limit of detection, laboratory cultured Type II PRRS virus strain FL12 (North American Strain) was obtained from Dr. Hiep Vu (University of Nebraska-Lincoln, Lincoln, Nebraska). PRRS virus was serially diluted in a negative oral fluid sample and an aliquot of 10^{-3}, 10^{-4} 10^{-5} and 10^{-6} dilutions were sent to the Veterinary Diagnostic Lab, (Neogen, Lincoln, Nebraska) to detect the PRRS virus using real-time PCR. They extracted the PRRSV RNA using MagMAX viral RNA isolation kit (ThermoFisher scientific) and detected PRRSV using VetMAX™ PRRSV EU & NA 2.0 Kit (ThermoFisher Scientific).

For validation of the method and assay developed in this study, clinical samples were obtained from The Pillen Family Farms (PFF, Columbus, Nebraska). PFF routinely screen pigs for PRRSV status in their farms by collecting OF samples. These samples are routinely tested at Iowa State University-Veterinary Diagnostic Laboratory (ISU-VDL). To provide samples for this study, they prepared two aliquots of the samples collected at their farm, one aliquot was sent to us, and we stored them at -80°C until use. Second aliquot was sent to ISU-VDL on ice, and they tested samples on arrival using well established procedures as described in Weiser et al.[12] Briefly, they extracted PRRSV RNA using MagMAX viral RNA isolation kit (ThermoFisher Scientific) and PRRSV was detected using RT-PCR assay kit EZ-PRRSV MPX 4.0 (Tetracore, Rockville, Maryland). Results were reported as Ct values for each sample.

Primers and probes

The probe and primers as previously described by Beuar et al[13] were used to design the assay for detection of type II PRRS (NA strain) with TaqMan probe- [FAM] CACCTATTCAATTAGGGCGAC-CG [BHQ1], forward primer- ATGTGTGGTGAATGGCACTG, and reverse primer- GCATGGTTCTCGCCAATTAAA targeted to the 3' UTR of the virus, which generates an amplicon of 98 bp in length. Pig *cox*1 gene was used as an internal control for sample preparation method and RT-qPCR reaction. For the pig *cox*1 gene, the TaqMan probe- [CY5] ACCCCCCGCAATGTCTCAATACCA [BHQ2], forward primer -ACAATTTTCTCCCTACACC, and reverse primer – TGATCAGACAAACAGGG were used. In both cases, TaqMan probes were synthesized by Eurofins USA and unlabeled primers were obtained from IDT Inc (Coralville, Iowa).

PRRS NA/Pig cox1 RT-qPCR assay

Two separate assays were designed to detect pig *cox*1 and PRRSV NA strain. A one-step RT-qPCR by combining reverse transcription and PCR into one reaction was also developed. For the reaction, approximately 100 µL of the sample to be run was added to lyophilized reaction pellets in a standard 0.2 mL PCR tube. In several cases, approximate sample volumes are given due to the fact that reactions can be done by moving samples with a syringe, in which case 2 drops are equivalent to about 100 µL volume on the average. The reaction was mixed well and placed in the MYRTA (MatMaCorp, Lincoln, Nebraska), a 3 well, hand-held thermocycler with real-time capabilities of detecting in two channels. One well of the device is larger than the other two and is used for sample tubes like the standard 1.5 mL Eppendorf tubes and the two smaller wells are compatible with 0.2 mL PCR tubes. For the MYRTA device, one channel is configured for FAM or similar dyes (with Excitation at 495 nm and Emission at 520 nm) and the second channel is configured Cy5.0 or similar dyes (with Excitation at 649 nm and Emission at 670 nm). RT-qPCR reactions were performed as follows: 42°C for 15 minutes, 94°C for 2 minutes followed by 40 cycles at 94°C for 10 seconds and 54°C for 60 seconds. The MYRTA device is run using pre-programmed incubation and cycling steps made available to the device using a programmable RFID tag. This tag is scanned on the device which has a NFC/RFID antenna embedded inside the front cover. The programming is done using a standard computer with a software application written for the purpose and transferred to the RFID tag using a RFID writer (HID 5022 CL).

Each Lyophilized Reaction Pellet used has 80 units of MMLV Reverse Transcriptase in a 50X MMLV RT Reaction Buffer (LGC/Biosearch Technologies), 7.5 Units of EconoTaq (LGC/Biosearch Technologies), 20 mM of each dNTPs, 1% Triton X-100, stabilizing agents and in the case of the PRRSV reaction; 30 µM each of PRRSV Forward and Reverse primers, 5 µM PRRSV TaqMan probe and in the case of the pig control reaction, 6 µM each of pig Cox1 forward and pig Cox1 reverse primers and 5 µM of pig Cox1 TaqMan probe. In the MYRTA device, the lyophilized reaction pellets are placed in standard 0.2 mL PCR tubes which are either yellow (for PRRSV detection) or pink in color (for pig cox1 detection). The tubes are placed within a plastic holder, where the left side (yellow tube) will detect FAM reactions and the right side (pink tube) will detect Cy5.0 reactions.

MYRTA device data access and analysis

Raw data generated by the MYRTA device used in this study is accessed from the device using a USB port at the back of the device. Fluorescence data is collected using appropriate filters and through a photodiode and then processed through an analog to digital converter and finally stored as CSV files inside the device. Each run will include test name, kit used, date of run as well as run time, Ct value calculations for data collected from each channel as well as raw fluorescence data collected by photodiodes in each channel.

Processing of oral fluids for direct detection (detection without RNA isolation)

Samples collected from the pen and provided in 15 mL screw cap tubes were mixed well and an aliquot was moved to a 1.5 mL standard Eppendorf tube. This is summarized in Figure 1. As shown in Step 1 this tube was left standing at room temperature for 5 to 10 minutes to clarify the sample. Using a transfer pipette, 100 µL of clarified OF was transferred to a tube with 100 µL of SDT Lysis Buffer (MatMaCorp, Lincoln, Nebraska). This is shown in Step 2 of Figure 1. This tube was placed in the large well of the MYRTA device and incubated at 95°C for 2.0 minutes, as shown in Step 3. As shown in Step 4, using a 1 mL disposable syringe (Becton-Dickinson BD 309659), 800 µL of a dilution buffer was added to sample and mixed well by moving sample in and out of the syringe 3 to 4 times. After mixing, about 400 µL of sample was drawn into the syringe, a filter (Tisch, SF18106) was attached to the syringe and by pressing the plunger down gently, 2 drops of sample were dispensed into each of the two PCR tubes that are attached to the tube holder. This is shown in Step 5 of Figure 1. The reaction is mixed well after layering with 2 drops of mineral oil and the tube holder with tubes are placed in the MYRTA device (Step 6) and cycled as described above.

Results

Oral fluid samples are physiologically and biochemically complex with extraneously introduced matter that substantially impedes the ability to directly detect any nucleic acid molecule, let alone RNA which is known to be unstable even in very stable and clean environments. In this study, direct detection means the use of an oral fluid (OF) sample in a nucleic acid amplification reaction (like PCR) with the intent of detecting the nucleic acid without the isolation of viral particles or the isolation of and purification of RNA from the sample.

Figure 1: Steps showing the processing of one oral fluid sample to demonstrate the simplicity of the method. Step 1 shows a 1.5 mL tube with with 1 mL of OF sample. Step 3 incubation is for 2 minutes at 95°C. For Steps 4 and 5, any standard 1 mL syringe can be used.

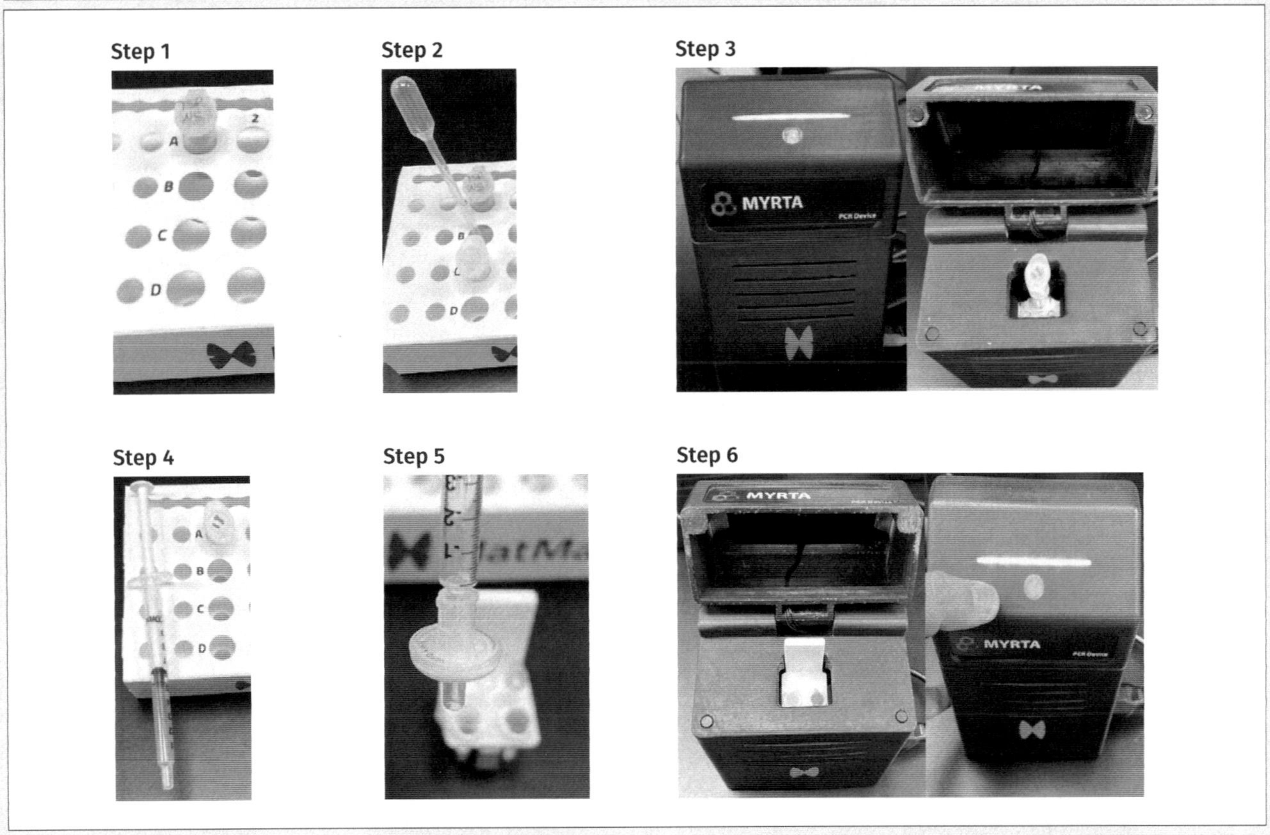

Raw data accessed from each run on the MYRTA device can be used to assess the quality of the run. An example run from an oral fluid sample that tested positive for PRRSV is shown in Figure 2. A negative sample for the PRRSV is shown in Figure 3. In this case, the pig cox1 gene is positive in both samples suggesting that the reactions are working as expected. The negative sample shows the PRRSV signal to be flatlined (in blue) and the system reports the test as not detected (ND). If there is a signal, then the Ct value is calculated and reported in the downloadable data.

Limit of detection using spiked negative samples

Laboratory cultured Type II PRRS virus strain FL12 was used to create a set of dilutions. Equal volumes of the dilution was mixed with an equal volume of SDT Lysis Buffer and 100 µL of the sample was used in subsequent real time PCR as described above for direct detection of samples. Each dilution was done in triplicate and the results are shown in Table 1. As shown in Table 1, the negative samples (without any spiked virus) does not show presence of the virus which was reported in all 3 replicates as not detected (ND). The 10^{-6} dilution was also called by the device as not detected (ND) which was detected as Ct 36.36 when the same sample was checked using purified RNA. This is shown in Table 2. Additionally, the $10^{-5.5}$ dilution gave a Ct of 35 in 2 out of 3 replicates suggesting that direct detection (without RNA purification) can determine samples with 35 Ct and below with good confidence. The 10^{-5} dilution of the same sample (Table 2) shows that without RNA purification, the sensitivity drops by almost 1 Ct point (Table 1) and this can be expected due to the fact that direct detection is done in crude samples with no RNA isolation or purification.

Validation of the method using clinical samples collected at a pig farm

To validate the direct detection of PRRSV in oral fluid without RNA isolation, we used the samples collected from different locations at Pillen family farms. In total, 46 samples collected at various sites from Pillen family farms were evaluated using the new method. These samples were also tested earlier at ISU-VDL by another well-established method and their results were compared with the results obtained with the MYRTA device. This is summarized in Table 3 for the Positive Samples and Table 4 for the Negative Samples. It is important to point out that a) the validation samples from the field were stored at -80°C for close to 2 years and aliquots were removed at least two times for analysis before the evaluations with the MYRTA device and b) the determination of Ct values at ISU-VDL were done using samples shipped overnight on ice and never frozen. Considering the nature and the age of the samples tested, all 36 samples showed host DNA (pig Cox1) suggesting that the OF samples were mostly intact except for the virus particles. In the positive samples tested, samples showed a higher Ct value (compared to the ISU-VDL data) when done on the MYRTA and this is consistent with samples stored over a long period of time and subject to freeze thaw cycles. Even then, only 1 sample that had Ct 31 or lower (ISU data) showed not detected (ND) suggesting that the MatMaCorp assay is very specific for virus RNA and a certain minimum level may be needed for optimal detection. On the other hand, the results seen here in conjunction with the limit of detection study (Table 1) indicate that sensitivity up to 35 Ct may be achievable on fresh samples. What is even more encouraging is the fact that negative samples (Table 4) showed 100% concordance with previous data indicating lack of false positives.

Figure 2: Test result from PRRSV positive sample. The Y axis shows data as arbitrary fluorescence units (AFU). The calculated Ct value for internal positive control Cox1, was Ct 25 and for PRRSV was Ct 31.

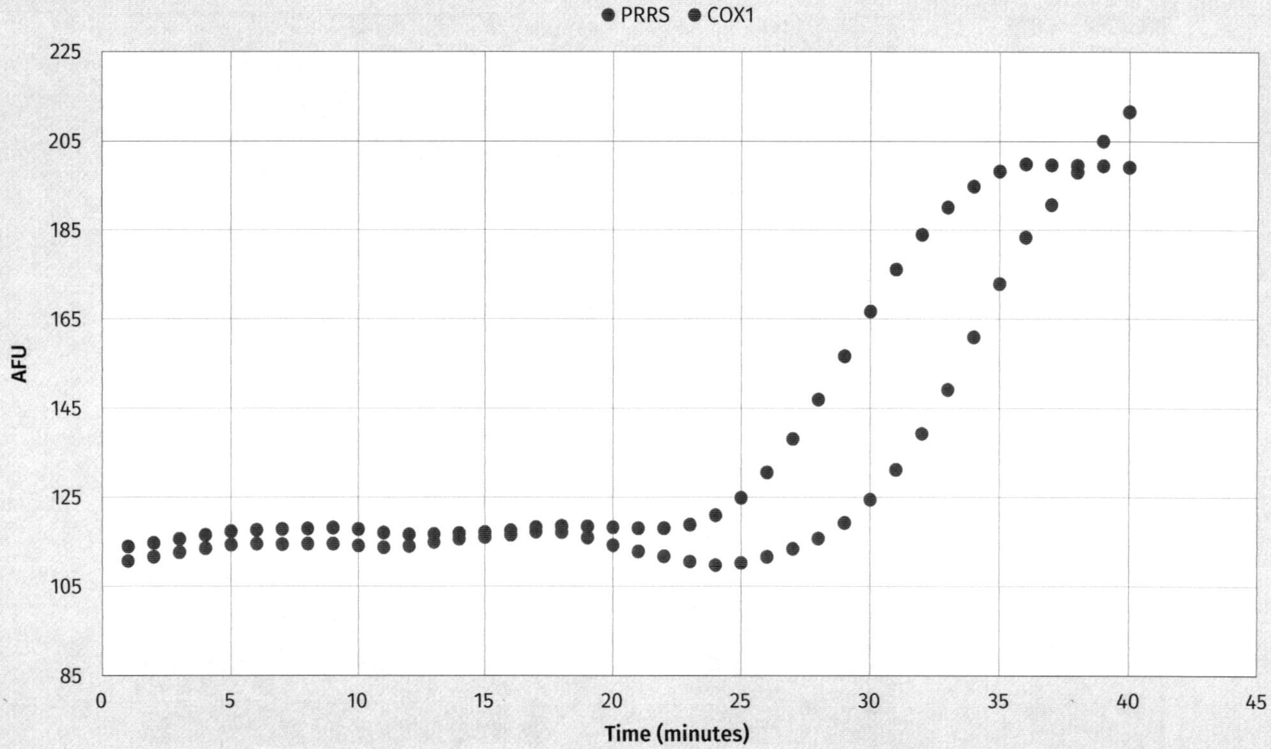

Figure 3: Test result from PRRSV negative sample. The Y axis shows data as arbitrary fluorescence units (AFU). The calculated Ct value for internal positive control Cox1, was Ct 28 and PRRSV was reported as ND (not detected).

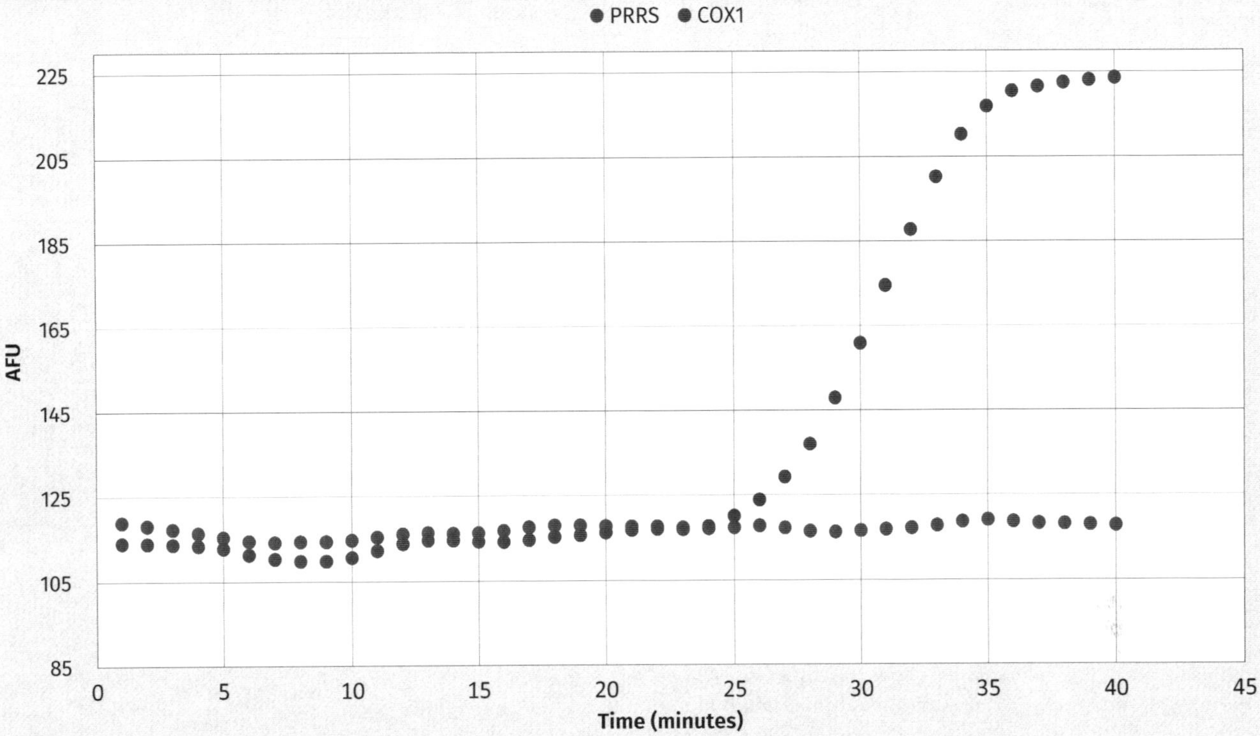

Table 1: Direct detection (no RNA isolation) of virus on MYRTA

PRRSV virus dilution	Rep 1	Rep 2	Rep 3	Avg Ct
10^{-3} dilution of stock PRRSV in negative oral fluid	29	30	29	29.33
10^{-4} dilution of stock PRRSV in negative oral fluid	32	33	32	32.33
10^{-5} dilution of stock PRRSV in negative oral fluid	33	33	35	33.67
$10^{-5.5}$ dilution of stock PRRSV in negative oral fluid	35	35	ND	35
10^{-6} dilution of stock PRRSV in negative oral fluid	ND	ND	ND	ND
Negative oral fluid	ND	ND	ND	ND

ND = not detected

Discussion

In a recent summary of suggested guidelines in PRRSV surveillance, Heano-Diaz et al[4] put forward the notion that testing for surveillance is different from testing for diagnostic purposes. They further elaborated on this concept by saying that specificity is important in surveillance more than sensitivity and that increased sample size and frequency of testing can adequately compensate for the decreased sensitivity. What we have developed is a surveillance tool for the detection of PRRSV that is almost as sensitive as a laboratory test but done with crude sample lysate and without RNA purification. The possibility of detecting fresh unfrozen field samples with Ct values between Ct 34 and Ct 35 will enable confident surveillance at pen side as long as false positives are not that common. Our ability to detect the internal control separately from the virus and within 40 cycles create stringent conditions under which false positive calls should be very limited or extremely rare. This requires further field validation with fresh samples.

Furthermore, the ease of performing the test onsite rather than sending it out to a centralized lab saves time and money. While confirmatory tests can be done by a centralized lab, onsite herd management decisions can be made rapidly preventing spread of disease and reducing economic loss. This simple method involving direct detection can also be applied to development of methods for detection of other RNA and DNA viruses that infect farm animals.

Table 2: Virus detection after RNA Isolation

PRRS virus concentration	Reported Ct Value
10^{-3} dilution of stock PRRSV in negative oral fluid	25.74
10^{-4} dilution of stock PRRSV in negative oral fluid	29.38
10^{-5} dilution of stock PRRSV in negative oral fluid	32.42
10^{-6} dilution of stock PRRSV in negative oral fluid	36.36
Negative oral fluid	ND

ND = not detected

References

1. *Global Pig Industry and Trends,* October 13, 2021. Accessed November 8, 2023. **https://www.feedandadditive.com/global-pig-industry-and-trends/**

2. Schwabe C. The current epidemiological revolution in veterinary medicine. Part I. *Prev Vet Med.* 1982;1:5–15.

3. Schwabe C. The current epidemiological revolution in vet medicine. Part II. *Prev Vet Med.* 1993;1:3–16.

4. Henao-Diaz A, Gimenez-Lirola L, Baum DH, Zimmerman J. Guidelines for oral fluid-based surveillance of viral pathogens in swine Porcine Health Management, 2020;6:28 **https://doi.org/10.1186/s40813-020-00168-w**

5. Prickett KW, Simer R, Yoon K, Zimmerman J. Oral-fluid samples for surveillance of commercial growing pigs for porcine reproductive and respiratory syndrome virus and porcine circovirus type 2 infections. *J Swine Health Prod.* 2008;16:86–91.

6. Bjustrom-Kraft J, Woodard K, Giménez-Lirola L, Rotolo M, Wang C, Sun Y, et al. Porcine epidemic diarrhea virus (PEDV) detection and antibody response in commercial growing pigs. *BMC Vet Res.* 2016;12:1–8.

7. Olsen C, Wang C, Christopher-Hennings J, Doolittle K, Harmon KM, Abate S, et al. Probability of detecting porcine reproductive and respiratory syndrome virus infection using pen-based swine oral fluid specimens as a function of within-pen prevalence. *J Vet Diagnostic Investig.* 2013;25:328–35.

8. White D, Rotolo M, Olsen C, Wang C, Prickett J, Kittawornrat A, et al. Recommendations for pen-based oral-fluid collection in growing pigs. *J Swine Health Prod.* 2014;22:138–41.

9. Holmes A, Kittawornrat A, Goodell C, Panyasing Y, Hoff S, Subramanya K, et al. Behavioral aspects of oral fluid sample collection. In: Ellis R, editor. The 95th Annual Meeting of the CRWAD. *Chicago: CRWAD;* 2014. p. 138.

10. Seddon YM, Guy JH, Edwards SA. Optimising oral fluid collection from groups of pigs: effect of housing system and provision of ropes. *Vet J.* 2012;193:180–4.

11. Rotolo ML, Sun Y, Wang C, Gimenez-Lirola L, Baum DH, Gauger PC, et al. Sampling guidelines for oral fluid-based surveys of group-housed animals. *Vet Microbiol.* 2017;209:20–9.

12. Weiser AC, Poonsuk K, Bade SA, Gauger PC, Rotolo M, Harmon K, Gonzalez WM, Wang C, Main R, Zimmerman JJ. Effects of sample handling on the detection of porcine reproductive and respiratory syndrome virus in oral fluids by reverse-transcription real-time PCR. *J Vet Diagnostic Investig.* 2018 30(6):807-12.

13. Beura LK, Subramaniam S, Vu HL, Kwon B, Pattnaik AK, Osorio FA. Identification of amino acid residues important for anti-IFN activity of porcine reproductive and respiratory syndrome virus non-structural protein 1. *Virology.* 2012; 433(2):431-9.

Table 3: Validation using frozen positive samples

Sample ID	ISU PRRSV Ct	MYRTA PRRSV Ct	MYRTA Pig cox1
	Positive samples		
1	26.4	31	25
2	25.5	31	25
3	26.1	30	26
4	26	30	26
5	26.4	35	27
6	26	32	26
7	29.5	33	26
8	30.9	34	27
9	30.9	32	28
10	30.9	33	28
11	33.5	ND	24
12	32	34	27
13	35	ND	29
14	30.3	33	28
15	32.3	34	25
16	30.7	34	27
17	35.7	ND	25
18	33	ND	28
19	32	ND	32
20	33.1	34	26
21	33.1	ND	26
22	33.4	ND	28
23	33.6	ND	27
24	33.3	35	29
25	31.6	33	28
26	35.6	ND	28
27	27.1	31	26
28	28.5	34	31
29	30.5	35	28
30	33.2	35	29
31	34.2	ND	25
32	28.2	32	26
33	31.1	33	25
34	34	ND	28
35	31	ND	28
36	32.8	ND	31

ND = not detected. Data generated at ISU-VDL was with samples that were not frozen and shipped overnight on ice. Data generated from samples were shared with MatMaCorp by PFF.

Table 4: Validation using frozen negative samples

Sample ID	ISU PRRSV Ct	MYRTA PRRSV Ct	MYRTA Pig cox1
Negative samples			
1	> 37	ND	26
2	> 37	ND	25
3	> 37	ND	27
4	> 37	ND	25
5	> 37	ND	25
6	> 37	ND	26
7	> 37	ND	30
8	> 37	ND	28
9	> 37	ND	27
10	> 37	ND	24

ND = not detected Data generated at ISU-VDL was with samples that were not frozen and shipped overnight on ice. Data generated from samples were shared with MatMaCorp by PFF.

https://doi.org/10.54846/am2024/37

Assessing the changes in antimicrobial use in grow-finish pigs after the introduction of PRRSV in a naïve farrow-to-finish system

I. Machado, DVM, MS; T. Petznick, DVM; A. P. Poeta Silva, DVM, MS, PhD; L. Karriker, DVM, MS;
D. C. L. Linhares, DVM, MBA, PhD; G. S. Silva, DVM, MS, PhD
Iowa State University, Ames, Iowa

Introduction

Porcine reproductive and respiratory syndrome (PRRS) is a viral swine disease with immunosuppressive effects on swine that can exacerbate the clinical impacts of bacterial infections, such as *Streptococcus suis* and *Glaesserella parasuis*. As a result, herd veterinarians need to employ additional antimicrobial usage (AMU) to provide care to the affected animals, resulting in an increase in treatment costs.[1-6] Moreover, the responsible use of antimicrobials is important for both animal and human health and requires constant improvement of antimicrobial stewardship (AMS).[7] Briefly, the five core principles of AMS defined by AVMA are: (#1) Commit to stewardship, (#2) Advocate for a system of care to prevent common diseases, (#3) Select and use antimicrobial drugs judiciously, (#4) Evaluate antimicrobial drug use practices, and (#5) Educate and build expertise. Therefore, this study aimed to assess the changes in AMU in a grow-finish pig population following a PRRS virus (PRRSV) outbreak in the sow farm, which was previously naïve to the virus.

Methods

Daily data on the number of animals transferred into or out of the growing lots, mortality, and AMU through water (oral) and injectable routes were obtained from nursery and finishing phases from a farrow-to-finish genetic multiplier system. A descriptive analysis of the active substances, antimicrobial class, and route of administration was performed. The frequency of use for each antimicrobial was calculated as the ratio of the total number of medicated animals to the total number of animals for the overall usage in the system.

For statistical analysis purposes, the growing pig population was categorized into three PRRSV statuses, determined by the PRRSV circulation within the sow farm: naïve, positive epidemic, and positive endemic. Considering the lot placement week, the evaluated data for the nursery phase was 17 weeks for the naïve status, 15 weeks for the positive epidemic status, and 17 weeks for the positive endemic status. In the finishing phase, the reported lot placement data was 14 weeks for the naïve status, 12 weeks for the positive epidemic status, and 13 weeks for the positive endemic status.

The pig treatments per animal days at risk (PTDR) metric was calculated per route (water and injectable) by dividing the total number of treatment days in the lot, ie, days in which animals were medicated, by the number of animal days at risk. A negative binomial regression model was applied to analyze the overall PTDR (water and injectable) and injectable PTDR, and the estimated marginal means for each PRRSV status of each phase were assessed. Moreover, a pairwise comparison using the Tukey test was performed to identify statistical differences between each PRRSV status group. For water PTDR, Kruskal-Wallis and Dunn tests were performed to assess statistical differences between PRRSV status groups. For all statistical analyses, a significance level using a *P*-value < .05 was considered.

Results

In this study, a total of 166 lots (104,270 animals) were evaluated for AMU for the nursery phase and 39 lots (30,793 animals) for the finishing phase. Among injectable antimicrobials used in the nursery phase, ampicillin was the most used, with 49% of total injections, followed by lincomycin (31%) and enrofloxacin (20%). For the oral antimicrobials provided through the water route, neomycin was the most used, with 60% of total use, followed by penicillin (31%), sulfamethoxazole (6%), and tylvalosin (3%). For the injectable antibiotics in the finishing phase, lincomycin was the most used, with 72% of total injections, followed by enrofloxacin (28%). Regarding oral antimicrobials, neomycin was the most used, with 97% of total use, followed by tylvalosin (3%).

In the nursery phase, there was a statistical difference between the PRRSV statuses for overall PTDR. The naïve status had 739 PTDR, the positive epidemic status had 2803 PTDR, and the positive endemic status had 1860 PTDR. In summary, there was a 380% and 251% AMU increase in the positive epidemic and endemic status compared to the naïve status. Regarding the pairwise comparison between the groups, all PRRSV status groups were statistically different from each other. Similarly, in the injectable PTDR, there was a statistical difference between the PRRSV statuses. The naïve status had 432 PTDR, the positive epidemic status had 949 PTDR, and the positive endemic status had 627 PTDR. In brief, there was an AMU increase of 219% and 145% in the positive epidemic and endemic status compared to the naïve status. All PRRSV status groups were statistically different from each other (Table 1).

In the finishing phase, there was no statistical difference between the PRRSV statuses for overall PTDR. The naïve status had 5306 PTDR, the positive epidemic status had 5332 PTDR, and the positive endemic status had 5109 PTDR. On the other hand, in the injectable PTDR, there was a statistical difference between the PRRSV statuses. The naïve status had 489 PTDR, the positive epidemic status had 1344 PTDR, and the positive endemic status had 1114 PTDR. In brief, there was an AMU increase of 274% and 228% in the positive epidemic and endemic status compared to the naïve status. The naïve status group was statistically different from the positive epidemic and endemic group (Table 1).

Discussion

The results of this retrospective study revealed an association between PRRSV status and AMU in the growing phase from a farrow-to-finish genetic multiplier system. Additionally, this study provided useful information that supports three of the AMS's five core principles as defined by the AVMA: (#2) Advocate for a system of care to prevent common diseases, (#4) Evaluate antimicrobial drug use practices, and (#5) Educate and build expertise. The PRRSV outbreak in the source herd was associated with an

Table 1: Comparison between PRRSV statuses regarding changes in antimicrobial usage in grow-finish pigs

Phase	Characteristic	PRRSV status	PTDR EMM[1]	95% CI[2]
Nursery	Overall PTDR[3] Water & injectables	Naïve	739[A]	599; 912
		Positive epidemic	2803[C]	2160; 3638
		Positive endemic	1860[B]	1505; 2299
	PTDR[3] All injectables	Naïve	432[A]	404; 463
		Positive epidemic	949[C]	872; 1032
		Positive endemic	627[B]	585; 671
Finishing	Overall PTDR[3] Water & injectables	Naïve	5306[A]	4821; 5840
		Positive epidemic	5332[A]	4878; 5828
		Positive endemic	5109[A]	4605; 5669
	PTDR[3] All injectables	Naïve	489[A]	368; 650
		Positive epidemic	1344[B]	1032; 1750
		Positive endemic	1114[B]	819; 1517

[1] PTDR EMM = pig treatment per animal days at risk estimated marginal means
[2] 95% CI = confidence interval
[3] PTDR = pig treatments per animal days at risk
ABC = different letter indicates statistical differences (*P*-value < .05) between status lots.

AMU increase of up to 380% in positive epidemic compared to naïve status and 251% in the positive endemic status compared to naïve status in both water and injectable route treatments in the grow-finish phase. It is worth informing that the site caretakers followed a standard operation protocol to identify sick animals and properly medicate them, and it remained the same across the studied period. These results highlighted the importance of advocating for a system of care to prevent common diseases, such as the PRRSV, calling for research and initiatives that integrate PRRSV infection control and responsible antimicrobial stewardship in swine production systems.

References

1. Galina, L., Pijoan, C., Sitjar, M., Christianson, W. T., Rossow, K., & Collins, J. E. (1994). Interaction between Streptococcus suis serotype 2 and porcine reproductive and respiratory syndrome virus in specific pathogen-free piglets. *The Veterinary Record*, 134(3), 60-64. **doi.org/10.1136/vr.134.3.60**

2. Dee, S., Guzman, J. E., Hanson, D., Garbes, N., Morrison, R., Amodie, D., & Galina Pantoja, L. (2018). A randomized controlled trial to evaluate performance of pigs raised in antibiotic-free or conventional production systems following challenge with porcine reproductive and respiratory syndrome virus. *PLoS One*, 13(12), e0208430. **doi.org/10.1371/journal.pone.0208430**

3. Done, S. H., & Paton, D. J. (1995). Porcine reproductive and respiratory syndrome: clinical disease, pathology and immunosuppression. *The Veterinary record*, 136(2), 32-35. **doi.org/10.1136/vr.136.2.32**

4. Halbur, P., Thanawongnuwech, R., Brown, G., Kinyon, J., Roth, J., Thacker, E., & Thacker, B. (2000). Efficacy of antimicrobial treatments and vaccination regimens for control of porcine reproductive and respiratory syndrome virus and Streptococcus suis coinfection of nursery pigs. *Journal of clinical microbiology*, 38(3), 1156-1160. **doi.org/10.1128/jcm.38.3.1156-1160.2000**

5. Thanawongnuwech, R., Brown, G. B., Halbur, P. G., Roth, J. A., Royer, R. L., & Thacker, B. J. (2000). Pathogenesis of porcine reproductive and respiratory syndrome virus-induced increase in susceptibility to Streptococcus suis infection. *Veterinary pathology*, 37(2), 143-152. **doi.org/10.1354/vp.37-2-143**

6. Dee, S., Guzman, J. E., Hanson, D., Garbes, N., Morrison, R., Amodie, D., & Galina Pantoja, L. (2018). A randomized controlled trial to evaluate performance of pigs raised in antibiotic-free or conventional production systems following challenge with porcine reproductive and respiratory syndrome virus. *PLoS One*, 13(12), e0208430. **doi.org/10.1371/journal.pone.0208430**

7. American Veterinary Medical Association: AAFP/AAHA antimicrobial stewardship guidelines. 2022. Available at: **https://www.avma.org/resources-tools/avma-policies/aafpaaha-antimicrobial-stewardship-guidelines**

https://doi.org/10.54846/am2024/38

Nutritional approach to mitigation of clinical *Streptococcus suis* in the nursery: Learnings and considerations for application

B. N. Smith, DVM, PhD; C. González-Vega, MS, PhD; M. McGhee, MS, PhD; M. Oelschlager, MS, PhD; K. Price, MS, PhD
Cargill Animal Nutrition, Minneapolis, Minnesota

Introduction

One of the most notable production challenges facing the North American pork industry is pathogenic disease, which contributes significantly to decreased growth performance and increased financial losses due to mortality. One such pathogen is *Streptococcus suis* (*S suis*), which occurs commonly in the nursery period both as a primary and secondary systemic bacterial infection. Due to changes in virulence and success of traditional medical management of *S suis*,[1-3] there has been increased interest in the development of alternative strategies to help support affected pigs during the nursery period, such as nutritional modulation of the inflammatory response. Normal metabolic processes result in some contributions to total body inflammation and impacts on immune function.[4] While the immune system has compensatory mechanisms to modulate both pro-inflammatory and pro-oxidative processes to minimize damage to host tissues, these compensatory mechanisms can become overwhelmed in scenarios such as pathogenic infections.[5] From a nutritional perspective, we can approach reducing total inflammatory burden through the restriction of essential nutrients, such as amino acids or energy, to decrease metabolic contributions to inflammation, or through providing bioactive compounds that demonstrate anti-inflammatory or anti-oxidative activity to support compensatory mechanisms. In attempt to develop a feeding program to nutritionally support *S suis*-infected nursery pigs, the former approach has been evaluated and summarized over a series of eight internal nursery trials.

Methods

Eight trials (trials A-H) summarized in this report included a control treatment consisting of a standard nursery formulation meeting or exceeding NRC dietary recommendations and a nursery formulation where standardized ileal digestible lysine (SID lysine) was restricted by approximately 20% relative to standard formulation levels. In all trials evaluated, this restriction was applied to the first two nursery dietary phases only (approximately 21 days), and all pigs received a common third and final nursery dietary phase. Trials were conducted at either a nursery or wean-to-finish research facility supplied by one of two commercial production systems in the Midwest. The source populations for pigs utilized in these trials were negative for porcine reproductive and respiratory syndrome virus (PRRSV) and *Mycoplamsa hyopneumoniae*. Primary *S suis* infection was confirmed by bacterial culture on non-respiratory tissue samples collected from untreated, acutely affected individuals within our test populations. Primary clinicals signs of *S suis*-related disease observed across both test populations in untreated, acutely affected individuals were polyarthritis and meningitis. Note that for trials F-H there were health status changes that occurred for both pig sources supplying their respective research facilities: introduction of a F18 enterotoxigenic *Escherichia coli* (ETEC) (trials F and G) and a decrease in severity of *S suis* clinical presentation (trial H). Standard antimicrobial interventions as prescribed by the herd veterinarian for each pig source (in-feed, water, and injectable treatments) were permitted across all studies, though antimicrobial selection and application varied between the two trial facilities. Pig and feed weights were recorded by pen at the beginning and end of each phase for the calculation of growth performance metrics (final body weight [BW], average daily gain [ADG], average daily feed intake [ADFI], feed conversion ratio [FCR]). Mortality and removal from study events (all trials) and individual treatments (excluding trial C) were recorded by pen. Data were analyzed using the lme4 package of R 4.1.2, and the statistical models included the fixed effect of treatment, as well as the random effects of barn (if applicable) and location within barn. Mortality and removals were analyzed as generalized linear mixed models with a binomial distribution. Note that the data presented in Table 1 represent the analyzed means of the respective treatments from each study; no further statistical analyses for comparisons of these treatments across studies have been performed.

Results

In the context of a primary *S suis* challenge, restriction of SID lysine applied in the first two dietary phases resulted in reduction of growth performance (Table 1, trials A - E) and lower average final BW at the end of the nursery period when compared to the control treatment (50.00 lbs. vs 50.44 lbs., respectively), which is expected when limiting an essential nutrient for growth. However, the restriction of SID Lysine also decreased the average probability of mortality by 0.95 percentage points relative to the control (1.75% vs 2.70%, respectively) over the entire nursery period. Restriction of SID lysine also decreased the average number of individual antimicrobial treatments by approximately 14% compared to the control diets (66.5 vs 85.8 treatments, respectively). While these results look promising, changes to the health status of our test populations (as described in the Methods section) in trials F-H resulted in differences in the clinical presentation and appreciated contributions of nursery morbidity and mortality by *S suis*. When these trials are included in the comparison, growth performance was more negatively impacted and the positive impact on probability of mortality was minimized (Table 1, trials F - H).

Discussion

Comparison of these eight trials suggests that under primary *S suis* disease, restriction of dietary SID lysine early in the nursery period may improve survivability of affected pigs. Lysine's role as the first limiting amino acid in most swine diets means that it has considerable impact on growth rate[7] and associated metabolic demand for growth processes. There are studies that demonstrate a period of feed restriction in the early nursery period (as a percent reduction of expected *ad libitum* intake) is beneficial for health, but this manner of restriction results in restriction of all nutrients, and health outcomes of these studies were largely focused

Table 1: Summary of effects of restricted SID lysine[1] in first two dietary phases compared to a standard nursery formulation on closeout nursery growth performance and health status of pigs facing naturally established *Streptococcus suis* infection

	Trial	n/trt	Final BW (lbs)	ADG (lbs/d)	ADFI (lbs/d)	FCR	Probability of mortality (ppt)	Injections (%)
Primary *S suis* disease	A	1,320	0.02	0.01	0.02	0.01	-0.07	0.00
	B	546	0.90	0.03	0.05	0.02	-1.38	-45.58
	C	690	-1.10	-0.03	-0.02	0.03	-0.85	Not recorded
	D	440	-1.10	-0.02	-0.03	0.01	-1.32	0.00
	E	588	-1.10	-0.03	-0.03	0.02	-1.26	-12.66
Non-primary *S suis* disease	F	264	-1.60	-0.06	-0.04	0.03	1.74	17.97
	G	264	-2.33	-0.07	-0.04	0.03	1.47	23.17
	H	280	-2.10	-0.07	-0.06	0.01	0.36	0.00
	AVG trials A-E		-0.44	-0.01	0.00	0.02	-0.95	-14.56
	AVG all trials		-1.03	-0.03	-0.02	0.02	-0.12	-2.44

Abbreviations: *S suis*, *Streptococcus suis*; trt, treatment; BW, body weight; ADG, average daily gain; ADFI, average daily feed intake; FCR, feed conversion ratio; ppt, percentage point

[1] SID lysine was restricted by approximately 20% relative to standard formulation levels for the first two dietary phases of the nursery only (approximately 21 days)

on indicators of enteric disease.[5,6] If we consider the selection for rapid lean growth in pigs and the interaction of hormones that support nutrient utilization on immune system function,[4] using lysine alone as a restrictive approach to reduce growth rate could be a viable means to support immune system function without applying restriction to other essential nutrients. That said, the connection of selection for rapid, lean growth and its effect on sensitivity to disease remains somewhat unclear and requires ongoing investigation. There is also evidence that lysine restriction may lead to increased susceptibility to certain infectious diseases due to its requirement for protein synthesis, so the positive impact observed in these studies may not translate to other pathogens.[8] This may partially explain why changes in health status observed in latter trials, either introduction of a different pathogen or decreased disease prevalence, resulted in decreased benefits from lysine restriction. In conclusion, a thorough understanding of the current health status and the primary pathogens causing clinical disease is necessary to increase likelihood of success when applying nutritional strategies to support nursery pigs facing pathogenic disease.

References

1. Neila-Ibanez, C., Brogaard, L., Pailler-Garcia, L., Martinez, J., Segalas, J., Segura, M., Heegard, P.M., Aragon, V. Piglet innate immune response to *Streptococcus suis* colonization is modulated by the virulence of the strain. *Vet. Res.* 2021;52(145): 1-15. **doi: 10.1186/s13567-021-01013-w**

2. Segura, M., Aragon, V., Brockmeier, S. L., Gebhart, C., de Greeff, A., Kerdsin, A., O'Dea, M. A., Okura, M., Salery, M., Schultz, C., Valentin-Weigard, P., Weinert, L. A., Wells, J. M., Gottschalk, M. Update on *Streptococcus suis* Research and Prevention in the Era of Antimicrobial Resistance: 4th International Workshop on *S. suis* [Conference Report]. *Pathogens.* 2020;9(374): 1-39. **doi: 10.3390/pathogens9050374**

3. Segura, M., Calzas, C., Grenier, D., Gottschalk, M. Initial steps of the pathogenesis of the infection caused by *Streptococcus suis*: fighting against nonspecific defenses. *FEBS.* 2016;590: 3772-3799. **doi: 10.1002/1873-3468.12364**

4. Prunier, A., Heinonen, M., Quesnel, H. High physiological demands in intensively raised pigs: impact on health and welfare. *Animal,* 2010;4(6): 886-898. **doi: 10.1017/S175173111000008X**

5. Yang, S., Lian, G. ROS and diseases: role in metabolism and energy supply. *Molecular and Cellular Biochemistry.* 2020;467: 1-12. **doi: 10.1007/s11010-019-03667-9**

6. Floc'h, N. L., Knudsen, C., Gidenne, T., Montagne, L., Merlot, L., Zemb, O. Impact of feed restriction on health, digestion, and faecal microbiota of growing pigs housed in good or poor hygiene condition. *Animal.* 2014;8(10): 1632-1642. **doi: 10.1017/S1751731114001608**

7. Menegat, M. B., Dritz, S. S., Tokach, M. D., Woodworth, J. C., DeRouchey, J. M., Goodband, R. D. A review of compensatory growth following lysine restriction in grow-finish pigs. *Transl. Anim. Sci.* 2020;4: 531-547. **doi: 10.1019/tas/txaa014**

8. Liao, S. F., Wang, T., Regmi, N. Lysine nutrition in swine and the related monogastric animals: muscle protein biosynthesis and beyond. *SpringerPlus.* 2015;4(147): 1-12. **doi: 10.1186/s40064-015-0927-5f**

https://doi.org/10.54846/am2024/39

Sperm quality and fertility of antibiotic-free preserved boar semen stored in Androstar Premium® extender at 5°C

R. Grossfeld[1], PhD; J. Quackenbush[2]; C. Simmet[1], DVM; K. Simmet[1], PhD
[1]Minitüb GmbH, Tiefenbach, Germany; [2]Minitube USA Inc., Verona, Wisconsin

Introduction

Adding antibiotics to semen extenders is a standard procedure in artificial insemination (AI) of livestock to avoid bacterial growth in semen doses during the preservation period. The necessity for antimicrobial control results from the unavoidable presence of bacteria in raw semen naturally stemming from the male and, to a smaller extent, from the environment. With adaptation of the World Health Organization's one-health concept, researchers and the animal breeding industry are seeking strategies to reduce or replace antibiotics in semen extenders.

Hypothermic preservation of boar sperm at +5°C allows to omit the addition of antibiotics to boar semen extenders as bacteria will remain in a static state at this temperature.[1,2] Therefore, bacteria that are possibly present in diluted boar semen do not proliferate and will not harm the sperm cells.[2]

The main challenge of hypothermic storage without antibiotics is the prevention of chilling injuries to spermatozoa and bacterial growth during the cooling of the semen doses to 5°C. Androstar Premium® boar semen extender allows hypothermic and antibiotic-free storage of boar semen by protecting the integrity of the sperm cells during cooling and by containing antimicrobially active substances.[1]

Storage of boar semen without antibiotics at 5°C in practice

The cooling of the diluted boar semen to +5°C requires a controlled cooling curve over several hours to allow an adaption of the spermatozoa to lower temperatures. However, cold damage to boar spermatozoa is an important concern during storage at such low temperatures.[3]

For further development and incorporation of the hypothermic storage concept into AI practice, determining a cooling-rate frame that concurrently is tolerated by spermatozoa and inhibits bacterial growth is required. Therefore, several cooling regimes of diluted boar sperm were tested for their suitability in hypothermic storage, as demonstrated in Figure 1.[3]

Split sample ejaculates from different boars (n = 8), diluted in Androstar Premium® were cooled to +5°C according to cooling curves A-E. Sperm analyses for total motility were performed after 24, 72 and 144 h of storage at +5°C. All semen samples were evaluated with the AndroVision CASA system for motility after 30 minutes of incubation at 38°C.

Figure 2 shows the results of the motility analysis after different cooling curves at the respective time points of analysis.

In samples submitted to cooling rate E, the percentages of motile spermatozoa were lower ($P < .01$) after 24, 72 and 144 h storage compared to samples cooled at the other cooling rates.

Hypothermic storage is also the only sustainable tool to combat multi-drug resistant strains of bacteria, as the low temperature reliably inhibits the growth of *Serratia marcescens* and *Klebsiella oxytoca*. In a study from Maaßen at al[4] trial semen samples were spiked with ~10^2 CFU/mL of these bacteria species. Storage at +5°C for 144h inhibited the growth of both bacterial species and maintained the sperm quality, whereas bacterial counts increased to more than 10^{10} CFU/mL in the +17°C samples used as positive controls. This was accompanied by an increase in the sperm agglutination and the loss of motility and membrane integrity.

In-vivo fertility after hypothermic storage of boar semen

Several trials have been conducted in the past to prove the suitability of antibiotic-free conservation of diluted boar semen at 5°C.[5,6] In the study of Waberski et al[5] pooled semen from 23 boars (3 boars per pool) were split into two groups. The semen samples in one group were cooled to and stored at 5°C while the samples in the other group were stored at 17°C as a control. The samples were subsequently transported to a sow farm and used for AI. The insemination trial resulted in similar field fertility with 14.1 ± 0.2 vs 14.5 ± 0.2 total born piglets after AI with semen stored at 17°C and 5°C, respectively (Table 1).

Figure 1: Samples cooled by Curve A, B, C, D and E had an initial cooling rate of 0.01, 0.06, 0.09, 0.14 and 0.65°C min^{-1}, respectively, until reaching 25°C. Then, samples were cooled to 10°C at the rates of 0.03, 0.03, 0.05, 0.06 and 0.31°C min^{-1}, respectively. Finally, samples were cooled to 5°C at a rate of 0.01°C min^{-1} (Curves A, B, C and D) and at 0.02°C min^{-1} (Curve E) (adapted from Paschoal et al[3]).

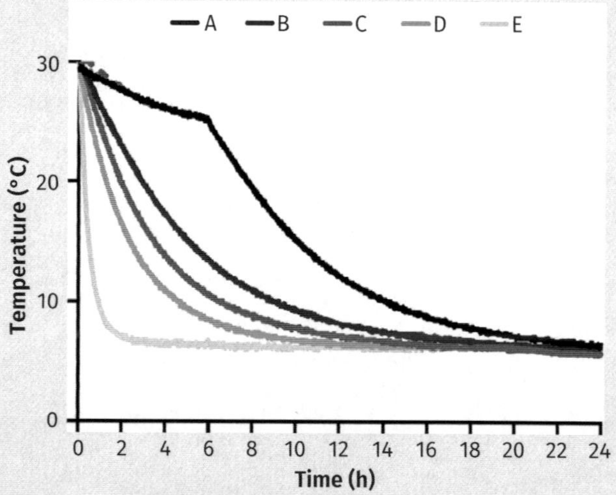

Figure 2: Total sperm motility of sperm samples cooled according to different cooling regimes after 24, 72 and 144h of storage at +5°C (modified from Paschoal et al[3]).

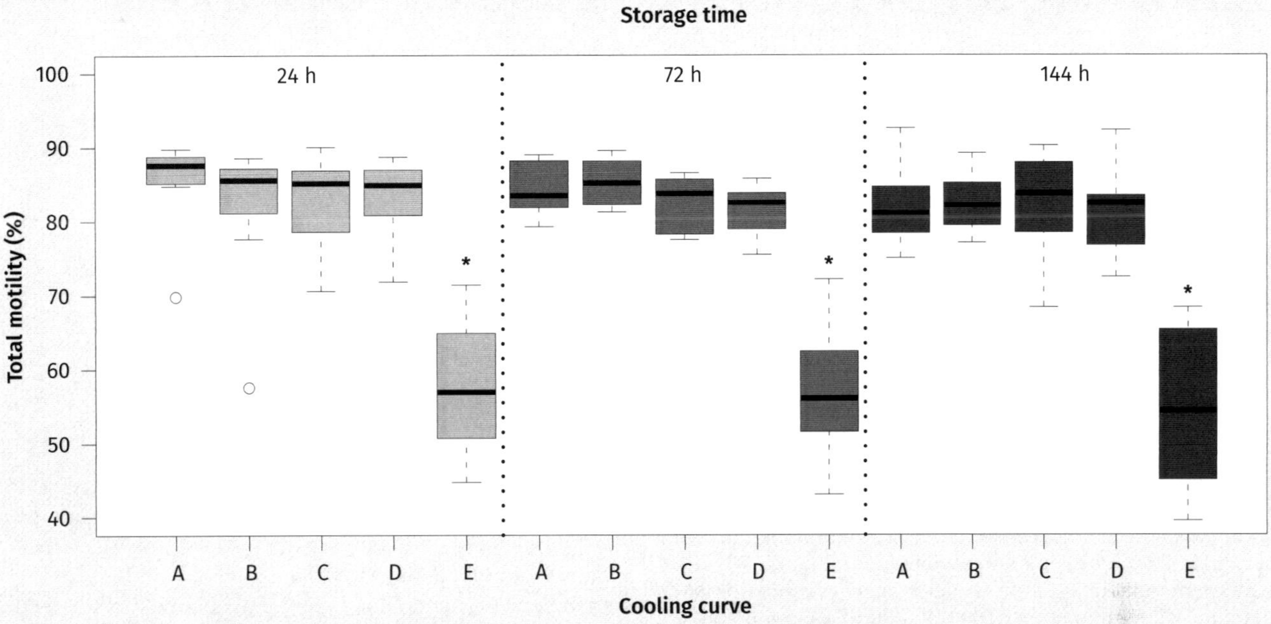

Table 1: Field fertility of boar semen samples stored with either one of the following temperatures and extenders:
17°C: Semen stored in Androstar® Premium with antibiotics (w/AB; 0.25 g/L gentamicin sulphate)
5°C: Semen stored in Androstar® Premium without antibiotics (w/o AB).
There were no differences in farrowing rate and number of total born piglets ($P > 0.05$) (adapted from Waberski et al[5]).

Semen storage	Sows, n (gilts, n)	Farrowing rate (%)	Total born piglets (n)
17°C with AB	406 (71)	93.1	14.1 ± 0.2
5°C without AB	411 (87)	92.0	14.5 ± 0.2

Conclusion

Hypothermic storage of boar semen allows to omit the use of antibiotics in diluted boar semen. Several cooling curves are suitable for cooling boar sperm to +5°C. Therefore, different cooling regimes in practice are possible making the use of hypothermic storage of boar sperm usable in different environments of AI centers. In addition, equal fertility results after hypothermic storage, compared to regular AI processes, pave the way for eliminating the use of antibiotics in boar sperm production.

References

1. Luther, A. M., Nguyen, T. Q., Verspohl, J. & Waberski, D. Antimicrobially active semen extenders allow the reduction of antibiotic use in pig insemination. *Antibiotics* 10, (2021).

2. Menezes, T. de A. *et al.* Antibiotic-free extended boar semen preserved under low temperature maintains acceptable in-vitro sperm quality and reduces bacterial load. *Theriogenology* 149, 131–138 (2020).

3. Paschoal, A. F. L. *et al.* Determination of a cooling-rate frame for antibiotic-free preservation of boar semen at 5°C. *PLoS One* 15, 1–14 (2020).

4. Maaßen, I. K., Luther, A.-M., Verspohl, J. & Waberski, D. Storage of Extended Boar Semen at 5 °C Inhibits Growth of Multi-Drug Resistant Serratia marcescens and Klebsiella oxytoca while Maintaining High Sperm Quality. *Antibiotics* 12, 857 (2023).

5. Waberski, D. *et al.* Sperm function in vitro and fertility after antibiotic-free, hypothermic storage of liquid preserved boar semen. *Sci Rep* 9, 1–10 (2019).

6. Reckinger, F., Riesenbeck, A., Waberski, D. & Waßmuth, R. Hypothermic preservation of boar semen at 5°C: A field test. *Reproduction in Domestic Animals* 58, 27–28 (2023).

https://doi.org/10.54846/am2024/40

Novel RT-qPCR workflow for simultaneous identification of rotavirus species A, B and C

R. Tebbs; A. Balogh; B. Qiu; P. Broska; K. Kice; D. Meza; R. Conrad; Y. Zein-Phillipson
Thermo Fisher Scientific, 2150 Woodward Street, Austin, Texas

Introduction

Rotaviruses are one of the most prevalent etiological agents causing severe diarrhea in young pigs. Although rarely fatal, disease severity can increase with concurrent infection or environmental stressors, which can further delay weight gain and time-to-market. Adult pigs become resistant to clinical disease as their immune system matures. Of the 9 known rotavirus species, rotavirus A, B, and C (RVA, RVB, and RVC) are the most important species associated with diarrhea in piglets.

Rotaviruses (*Reoviridae*) are double-stranded RNA (dsRNA) viruses with linear genomes comprising 11 gene segments. RNA viruses depend on RNA-dependent RNA polymerase (RdRp) to replicate their genome during infection. The poor fidelity of RdRps results in numerous mutations between members of the same RNA virus species.[1] Molecular detection of rotavirus must account for this complexity. Here, we describe a multiplex reverse transcriptase, real-time PCR (RT-qPCR) design that can differentiate RVA, RVB, RVC, and an external positive control in a single reaction. The VetMAX™ Rotavirus A/B/C Primer/Probe Mix, showed excellent ability to selectively amplify rotavirus species A, B, and C and showed no indications of amplifying non-rotavirus targets *in silico* and *in vitro*. When compared to a commonly used published workflow against a large panel of rotavirus-infected porcine samples, this new design demonstrated a lower average Cq and identified a larger number of rotavirus positives samples. The VetMAX™ Rotavirus workflow successfully identified rotavirus in porcine feces, intestine, and environmental samples both individually and in pools of 5.

Materials and methods

VetMAX™ Rotavirus A/B/C reagents

The VetMAX™ Rotavirus A/B/C Reagents uses TaqMan™ probes for real-time PCR detection. The Primer/Probe Mix assigns different fluorescent dyes to each rotavirus target and the internal positive control to differentiate between the four targets (Figure 1). The fluorescent dyes are ABY™ (RVA), FAM™ (RVB), VIC™ (RVC), and Cy5® (internal positive control).

VetMAX™ Rotavirus Primer/Probe design

The VetMAX Rotavirus Primer/Probe Mix targets two separate RVA gene targets and two separate RVB gene targets to increase coverage of known rotavirus sequences. The primers and probes were selected based on a large panel of RVA, RVB, and RVC

Figure 1: Amplification of RVA, RVB, RVC, and the internal positive control (IPC) in a single reaction mix.

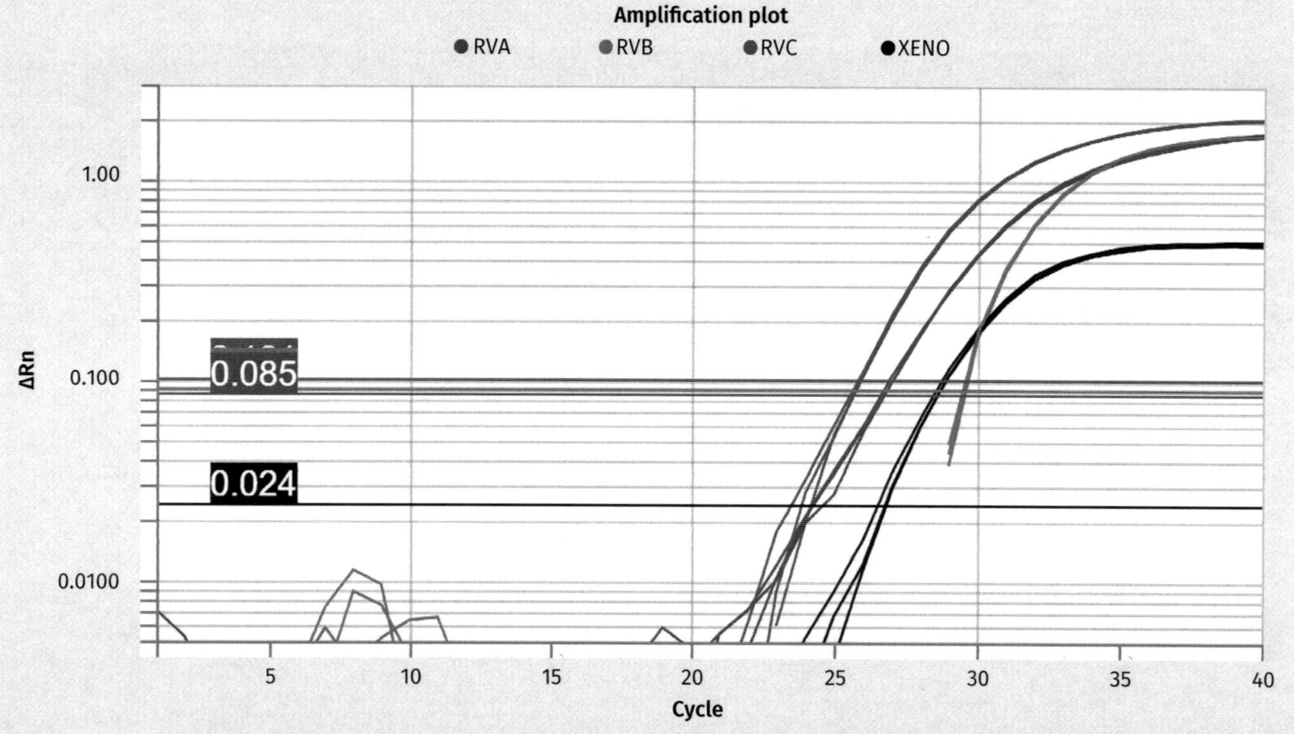

sequences obtained from NCBI, collaborators, and samples sequenced internally. A proprietary software was used to determine percent coverage of the Primer/Probe Mix to the rotavirus sequences. *In silico* the coverage was > 96% for porcine rotavirus A, B, and C (Table 1).

Samples

A total of 339 porcine samples infected with rotavirus were obtained from Iowa State University's Veterinary Diagnostic Lab (ISU-VDL), the University of Minnesota Veterinary Diagnostic Laboratory (UMN-VDL), and the Carthage Veterinary Service. Sample descriptions included feces (208), fecal swabs (51), rectal swabs (5), feedback (20), fluid (2), homogenate (13), oral fluid (3), processing fluid (1), tissue composite (2), intestinal tissue (25), environmental samples (5), and other (4).

Nucleic acid extraction

Total nucleic acid was extracted from porcine samples using the MagMAX™ CORE Nucleic Acid Purification Kit automated on the KingFisher™ Flex Purification System following the Complex Workflow as described in the MagMAX CORE User Guide. Xeno™ Control RNA was added to the MagMAX CORE Lysis Solution as recommended in the User Guide to gauge extraction efficiency and inhibitor presence. The extracted nucleic acid was used immediately in RT-qPCR testing and then stored long term at -20°C.

Table 1: Percent coverage of known rotavirus sequences by the VetMAX Rotavirus Primer/Probe Mix

Sequences	RVA	RVB	RVC
Number analyzed	290	197	162
Number detected	288	194	156
Percent detected	99.3%	98.5%	96.3%

RT-qPCR

All extracted nucleic acid samples and controls were heat-denatured to allow for efficient RNA transcription prior to PCR amplification. Samples and controls were heat-denatured at 95°C for 3 minutes, allowed to cool and briefly centrifuged prior to adding to the RT-qPCR reaction mix. Three controls were included in each RT-qPCR run, namely **(1)** "No Template Control" (NTC) consisting of nuclease-free water, **(2)** "Negative Extraction Control" (NEC) prepared during nucleic acid extraction with 1X PBS, and (3) "Positive Control" consisting of the RVA/RVB/RVC/Xeno RNA Control. The RT-qPCR reactions were run on either the 7500 Fast Real-Time PCR System; the QuantStudio 5 Real-Time PCR System, 96-well, 0.1 mL; or the QuantStudio 5 Real-Time PCR System, 96-well, 0.2 mL. The real-time PCR instruments were run using the thermal profiles described in Tables 2 and 3.

Data analysis

Design and Analysis software version 2.7.0 was used to analyze run files from both instruments. The run threshold for each dye channel was set based on the ΔRn of the rotavirus positive control using Control-Based Threshold (CBT). 10% CBT was set by multiplying the ΔRn at cycle 40 for each dye in the control sample by 0.1. The baseline was set using a Manual Baseline setting of 3-15. However, porcine samples with high concentrations of rotavirus having Cq values under 15 gave better results using Auto Baseline. When testing porcine field samples for presence of rotavirus we found that Auto Baseline worked best for RVA and RVC, but Manual Baseline worked best for RVB.

Data interpretation

Results from the controls (positive control, NTC, and NEC) were examined to determine if the PCR run was valid (see Table 4). Rotavirus samples with Cq values < 37 was considered positive, between 37 and 39.9 was considered suspect positive, and ≥ 40 was considered negative.

Table 2: Fast run mode for the thermal cycling conditions

Stage	Description	Step	Reps	Temp	Time
Reverse transcription	RNA → cDNA	1	1	50°C	5 minutes
Enzyme activation	Activate Taq Polymerase	2	1	95°C	10 minutes
PCR	Template denaturation	3a	40	95°C	3 seconds
	Template annealing/extension	3b		60°C	30 seconds

Table 3: Standard run mode for the thermal cycling conditions

Stage	Description	Step	Reps	Temp	Time
Reverse transcription	RNA → cDNA	1	1	48°C	10 minutes
Enzyme activation	Activate Taq Polymerase	2	1	95°C	2 minutes
PCR	Template denaturation	3a	40	95°C	15 seconds
	Template annealing/extension	3b		60°C	45 seconds

Table 4: Valid runs gave results from the RT-qPCR controls within the listed parameters

Stage	RVA Cq value	RVB Cq value	RVC Cq value	Xeno Cq value
VetMAX™ Rotavirus A/B/C RNA	24 - 28	27 - 31	24 - 28	24 - 28
NTC (water)	Undetected	Undetected	Undetected	Undetected
NEC (PBS or water extraction)	Undetected	Undetected	Undetected	26 - 34

Results

Analytical sensitivity

Two hundred and eighty-four porcine samples were obtained from State VDL's that were identified to be infected with RVA, RVB and/or RVC. These samples were extracted and tested with both the VetMAX rotavirus design and a published method[2] using the same Master Mix on the same Quant Studio 5 Real-Time PCR instrument. The VetMAX Rotavirus workflow was run using Fast PCR cycling conditions and the published method was run under the PCR cycling conditions recommended by the authors. The results showed that the VetMAX design identified 4 RVA, 7 RVB, and 1 RVC targets in the samples not identified by the published method (Table 5). However, the VetMAX workflow did miss 1 RVA and 1 RVB target that was identified by the published method. The Cq values for each target missed by one of the two Primer/Probe mixes is reported in Table 6. The average difference in Cq values between the two primer/probe mixes for samples identified by both methods were compared and the VetMAX workflow demonstrated an average Cq that was 1.9 less for RVA, 2.5 less for RVB, and 0.2 higher for RVC compared to the published method.

Analytical specificity

The VetMAX Rotavirus Primer/Probe Mix demonstrated no detection of non-rotavirus sequences within GeneBank indicating high analytical specificity for rotavirus *in silico*. In vitro studies with the Rotavirus Primer/Probe Mix showed no false positive detection with a panel of porcine samples infected with non-rotavirus pathogens associated with porcine enteric disease (Table 7).

Level of detection (LoD)

The VetMAX Rotavirus Primer/Probe Mix LoD was determined using dilutions of the rotavirus RNA control templates. Each rotavirus RNA template was tested separately with 20 replicates at each concentration. The Limit of Detection was defined as the lowest concentration to give a minimum of 19 out of 20 replicates testing positive (95% of replicates test positive). For this study a Cq value below 40 was considered positive. The analytical sensitivity for RVA, RVB, and RVC was 25 copies, 40 copies, and 10 copies, respectively, as shown on Table 8.

Sample pooling

The rotavirus Primer/Probe Mix was used on sample pools consisting of one rotavirus-positive porcine sample mixed with 4 negative porcine samples in equal volumes prior to nucleic acid extraction. Each rotavirus-positive sample was tested with and without pooling (Figure 2). The results show amplification of rotavirus-positive samples with and without pooling; however, the pooled sample give a higher Cq value due to dilution. A rotavirus-positive sample in a pool of 5 shows an average Cq increase of 3.0 compared to the same sample run individually.

Discussion

Designing molecular workflows for rotavirus is challenging due to high strain diversity and lack of conserved genomic targets. Furthermore, GenBank contains rotavirus sequences that are often poorly annotated, and the number of sequences differ greatly between gene segments. The VetMAX Rotavirus Primer/Probe Mix was developed from a database of rotavirus sequences obtained from collaborations and from whole genome sequencing done within Thermo Fisher as part of this project in addition to porcine rotavirus sequences from GenBank. Amplification of RVA and RVB was increased by combining two independent designs for these two species. Amplification of all three species of rotavirus was further expanded by addition of degenerate primers and probes to compensate for the high degree of genetic variability associated with rotavirus. The VetMAX Rotavirus Primer/Probe Mix is a complex RT-qPCR design based on the number of primers and probes needed for optimal coverage, but the complexity of the Primer/Probe mix showed no negative impact on analytical specificity, level of detection, or ease of use in these studies.

Acknowledgments

We would like to thank Dr. Phil Gauger and Laura Bradner from Iowa State University, the Minnesota Veterinary Diagnostic Laboratory, and Nicloe Eddy from Carthage Veterinary Service for supplying porcine samples. We would like to thank Dr. Jianqiang Zhang from Iowa State University, and Dr. Stephanie Rossow and the Minnesota Veterinary Diagnostic Lab for providing rotavirus sequences.

References

1. Castro, C., Arnold, J.J., and Cameron, C.E. (2005) Incorporation fidelity of the viral RNA-dependent RNA polymerase: a kinetic, thermodynamic, and structural perspective. *Virus Res*, 107:141. **https://doi.org/10.1016/j.virusres.2004.11.004**

2. Marthaler, D., N. Homwong, K. Rossow, M. Culhane, S. Goyal, J. Collins, J. Matthijnssens, and M. Ciarlet (2014). Rapid detection and high occurrence of porcine rotavirus A, B, and C by RT-qPCR in diagnostic samples. *J. Virol. Methods* 209:30-34. **https://doi.org/10.1016/j.jviromet.2014.08.018**

Table 5: Number of samples not detected in vitro out of 285 samples by the VetMAX Rotavirus A/B/C multiplex versus a published rotavirus A/C & B method

Method	RVA	RVB	RVC
VetMAX Rotavirus A/B/C	1	1	0
Published rotavirus A/C & B	4	7	1

Table 6: Cq values for samples missed by either the VetMAX Rotavirus A/B/C or the published A/C & B method

Method showing missed call	RVA		RVB		RVC	
	VetMAX	Published	VetMAX	Published	VetMAX	Published
VetMAX Rotavirus A/B/C	UND	36.7	UND	32.2	–	–
Published rotavirus A/B & C	8.7	UND	27.3	UND	13.5	UND
Published rotavirus A/B & C	15.5	UND	33.9	UND	–	–
Published rotavirus A/B & C	16.1	UND	35.1	UND	–	–
Published rotavirus A/B & C	19.3	UND	36.2	UND	–	–
Published rotavirus A/B & C	–	–	36.7	UND	–	–
Published rotavirus A/B & C	–	–	36.8	UND	–	–
Published rotavirus A/B & C	–	–	36.9	UND	–	–

Table 7: Exclusion panel consisting of non-rotavirus enteric pathogens used to determine the analytical specificity of the Primer/Probe Mix

ID	Agent positive for	Sample type	Univ16S*	RVA	RVB	RVC	Xeno
1	Porcine sapovirus (Cq = 15.7)	Fecal swab	13.2	Und	Und	Und	29.2
2	Porcine sapovirus (Cq = 32.3)	Fecal swab	10.4	Und	Und	Und	29.9
3	Porcine sapovirus (Cq = 13.4)	Fecal swab	5.2	Und	Und	Und	30.9
4	Porcine sapovirus (Cq = 20.3)	Fecal swab	5.2	Und	Und	Und	31.2
5	Porcine sapovirus (Cq = 28.6)	Fecal swab	4.5	Und	Und	Und	31.0
6	Lawsonia (Cq = 16.2)	Feces	11.2	Und	Und	Und	30.0
7	Lawsonia (Cq = 16.2)	Feces	5.0	Und	Und	Und	30.5
8	Lawsonia (Cq = 16.4)	Feces	9.5	Und	Und	Und	30.1
9	Brachyspira hyodysenteriae	Media	6.8	Und	Und	Und	29.7
10	Salmonella choleraesuis	Media	< 3	Und	Und	Und	29.9
11	Salmonella typhimurium	Media	< 3	Und	Und	Und	29.2
12	Clostridium perfringens	Media	4.5	Und	Und	Und	29.3
13	Clostridium difficile	Media	4.4	Und	Und	Und	29.5
14	PEDV positive	Environmental	4.9	Und	Und	Und	30.1
15	PDCoV positive	Environmental	5.0	Und	Und	Und	29.9
16	TGEV positive	Environmental	5.0	Und	Und	Und	29.7

* Univ16S is a universal real-time PCR assay that detects bacterial DNA based on the 16S gene sequence and was included in this study to demonstrate the presence of microbial nucleic acid in the sample. The bracketed Cq values are from the donating institution where available.

Table 8: Level of detection for RVA, RVB and RVC

Species	Copies/rxn	No. positive	Cq range	LoD
RVA	10	12	37.6 - 39.7	25
	25	20	35.4 - 37.7	
	50	20	34.1 - 35.8	
	100	19	32.7 - 33.8	
RVB	5	0	UND	40
	20	18	38.0 - 39.7	
	40	20	36.3 - 38.4	
	80	20	36.1 - 37.0	
RVC	5	13	37.1 - 40.0	10
	10	20	35.7 - 38.1	
	20	20	35.2 - 36.8	
	40	20	34.2 - 35.3	

Figure 2: Amplification of RVA, RVB, RVC, and the internal positive control (IPC) in a single reaction mix.

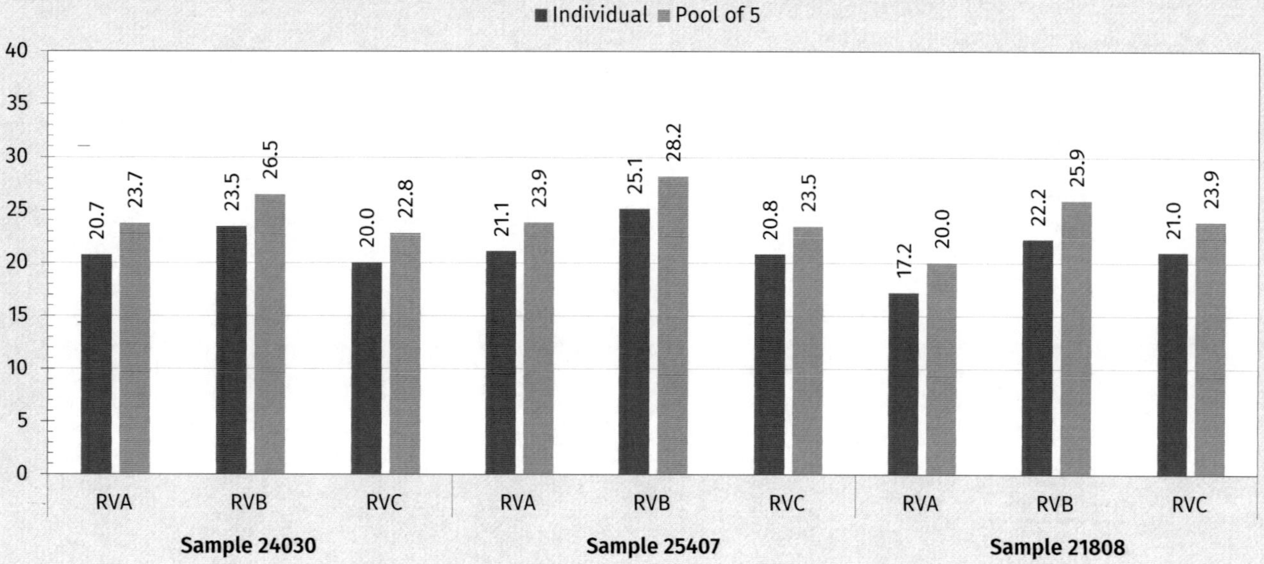

https://doi.org/10.54846/am2024/41

Evaluation of the efficacy of Enterisol® Ileitis administered orally by gel and drinking water in a wean-to-finish *Lawsonia intracellularis* challenge model

Joel Nerem[1], DVM; Amanda Sponheim[2], DVM; Daniel Hanson[3]; Erin Little[3], MS; Justin Rustvold[2], Roy Edler[3], MS; Fernando Leite[2], DVM, MS, PhD
[1]Pipestone Veterinary Services, Pipestone, Minnesota; [2]Boehringer Ingelheim Animal Health USA Inc., Duluth, Georgia; [3]Pipestone Applied Research, Pipestone, Minnesota

Introduction

Labor continues to be a challenge in the swine industry. Traditionally, oral live vaccines have been administered through the drinking water and by oral drench. Oral drenching allows for vaccination in the sow farm; however, it requires the handling of each pig individually. Here we present gel administration, which consists of utilizing a gel as a delivery vehicle for the vaccine, similar to water, taking advantage of the oral route of administration to allow for vaccine administration without the need to individually handle each pig. This has the advantages of requiring less labor and facilitating management of the med-free window. The latter is due to pigs being vaccinated prior to weaning, thus avoiding potential concurrent administration of nursery antibiotics. The objective of this study was to evaluate the efficacy of the oral administration of Enterisol® Ileitis by gel and to compare this method of administration to that of the drinking water.

Methods

A randomized, controlled trial was conducted with four different treatment groups: 1) Non-vaccinated, *Lawsonia* challenged control; 2) Enterisol® Ileitis FF vaccinated by gel and *Lawsonia* challenged; 3) Enterisol® Ileitis FF vaccinated by drinking water and *Lawsonia* challenged; 4) Enterisol® Ileitis FF and Enterisol Salmonella T/C® vaccinated by gel and *Lawsonia* challenged. A total of 45 litters and 540 pigs were randomly assigned to each treatment, after blocking for sow parity and balancing for weight and gender between each treatment. Gel administration was performed by adding 120 mL of Underline Gel containing 12 full doses of Enterisol® Ileitis FF vaccine to each litter of treatment groups 2 and 4. Treatment group 4 also received a full dose of Enterisol Salmonella T/C® by gel. This administration took into consideration an average of 12 pigs per litter and the consumption of 10 mL of gel per pig. The gel vaccine mixture was applied to the farrowing crate mat. Treatment group 3 received Enterisol® Ileitis FF in the drinking water at 6 weeks of age.

Experimental *Lawsonia* challenge was conducted with a seeder pig model. A group of non-vaccinated pigs designated to be "seeders" were challenged with a dose of 10^{10} *L intracellularis* organisms/pig. When pigs reached 12 weeks of age (study day 69), all pigs in the trial were individually weighed, and all four treatment groups were mixed per pen with the addition of three seeders per pen. To confirm that seeder pigs induced the transmission and challenge of *Lawsonia*, 32 pigs per treatment group were randomly selected for serial blood and fecal collection. These samples were tested for *Lawsonia* by ELISA and PCR, respectively. Mortality was recorded and pigs were removed from the study in case of welfare concerns. The primary parameters of this study were average daily weight gain (ADG) and mortality from the time of commingling treatments and seeders at 12 weeks of age to when pigs were first sent to market at 20 weeks of age.

Results

The non-vaccinated and challenged treatment group 1 had the lowest ADG from challenge to final weight at 1.82lbs. This was significantly ($P < .001$) lower than the ADG observed in vaccinated treatment groups 2, 3 and 4 which were of 2.05, 2.02, and 2.03lbs, respectively (Table 1). No significant difference was observed between the gel and water administration treatment groups. The percentage of removal and mortality in treatment group 1 was 7.45%, being significantly ($P < .01$) higher than that observed in treatment groups 2 (4.31%), 3 (4.11%) and 4 (3.17%) (Table 1). Again, no significant differences were observed between the vaccinated treatment groups. Evaluation of diagnostic test results revealed that all treatment groups had an increase in the number of PCR positive fecal samples (Table 2) as well as ELISA positive samples over time (Table 3), confirming that all treatment groups were challenged with *Lawsonia*.

Discussion and conclusion

Oral vaccination through the drinking water has been employed with Enterisol® Ileitis for more than twenty years in the United States swine industry.[1] The oral route of administration allows for the advantage of not necessitating individual pig handling, which saves labor and minimizes stress to both the one performing the vaccination and the pig. This study compared the efficacy of Enterisol® Ileitis when administered by the drinking water to a novel method of oral administration, which allows pigs to consume the vaccine with a gel prior to weaning. Ileitis, caused by *L intracellularis*, is a disease known to cause growth performance loss as well as mortality in grow finish pigs.[2] When comparing the efficacy of water and gel administration, both were significantly superior to non-vaccinated controls in growth performance and in preventing mortality and removals. The level of efficacy of these two oral administration methods was also numerically similar and not statistically different from one another, further demonstrating their efficacy. Additionally, the use of Enterisol Salmonella T/C® did not impair the efficacy of Enterisol® Ileitis, and the group which received both of these vaccines by gel was the group with the lowest numerical mortality values. These results suggest that the oral administration of Enterisol® Ileitis by gel is efficacious, representing a novel administration option to be used in the field.

Table 1: Average daily weight gain (ADG), mortality and removals among different treatment groups from the time of challenge to study end point

Treatment Group	ADG	Mortality	Removal	Removal and mortality
1	1.82 ± 0.02[b]	5.93 ± 0.82[b]	1.53 ± 0.43	7.45 ± 0.92[b]
2	2.05 ± 0.02[a]	3.69 ± 0.85[ab]	0.62 ± 0.44	4.31 ± 0.95[a]
3	2.02 ± 0.02[a]	2.88 ± 0.85[a]	1.22 ± 0.44	4.11 ± 0.95[a]
4	2.03 ± 0.02[a]	2.57 ± 0.84[a]	0.60 ± 0.44	3.17 ± 0.93[a]

Different superscript letters indicate statistical significance ($P < .05$); values represent least square means and standard error.

Table 2: Percent of fecal samples positive for the presence of *Lawsonia* among different treatment groups at the time of challenge (day 69) and following timepoints

	Study day						
Treatment group	Day 69	Day 84	Day 99	Day 112	Day 123	Day 140	Day 148
1	3.5%	73.3%	75.9%	35.7%	17.9%	0%	0%
2	3.5%	40.0%	73.3%	16.7%	3.3%	0%	0%
3	6.7%	22.6%	33.3%	3.3%	10.7%	0%	0%
4	3.2%	58.1%	50.0%	6.3%	12.5%	0%	0%

Table 3: Percent of serum samples positive for the presence of antibodies against *Lawsonia* among different treatment groups at the time of challenge (day 69) and following timepoints

	Study day			
Treatment group	Day 69	Day 84	Day 123	Day 148
1	0.0%	6.7%	82.1%	71.4%
2	3.3%	6.7%	53.3%	36.7%
3	3.2%	6.5%	28.6%	18.5%
4	0.0%	9.4%	62.5%	41.9%

References

1. Leite et al., Enterisol Ileitis, what we have learned in twenty years controlling *Lawsonia intracellularis*. 2021. 52nd AASV, p. 157-160.
2. Vannucci et al., Proliferative Enteropathy in Diseases of Swine. 2019.

https://doi.org/10.54846/am2024/42

Field experience with the oral administration of Enterisol® Ileitis by gel

M. McDowell[1], DVM; A. Sponheim[2], DVM; T. Bates[2]; F. Leite[2], DVM, MS, PhD
[1]Pipestone Veterinary Services, Pipestone, Minnesota; [2]Boehringer Ingelheim Animal Health Inc., Duluth, Georgia

Introduction

A recent study evaluated the oral administration of Enterisol® Ileitis by gel applied on farrowing crate heat mats pre-weaning.[1] The vaccination strategy used gel as a delivery vehicle for the vaccine, similar to water, allowing pigs to be administered vaccine without the need to be individually handled. Furthermore, the approach allowed for pre-weaning vaccination to avoid potential concurrent administration of nursery antibiotics with Enterisol® Ileitis. This proof-of-concept study indicated that the oral administration of Enterisol® Ileitis by gel was efficacious, representing a novel oral administration option to be used in the field. Here we report findings from the monitoring of four wean to finish cohorts which were vaccinated with Enterisol® Ileitis by gel, discussing clinical impression and diagnostic data, which further suggests the efficacy of this vaccination strategy.

Methods

Four cohorts, each comprised of a group of wean to finish pigs housed at separate sites among three different producers with concerns of clinical ileitis in finishing pigs, were enrolled in this case study. Sow farm caretakers vaccinated pigs four days prior to weaning with Enterisol® Ileitis administered by gel on the farrowing crate heat mat as described previously.[1] At 10 - 12 weeks of age, pigs were randomly selected and tagged for longitudinal serum sample collection every four weeks. Pools of five fecal samples per pen were collected from the same pens every two weeks. Fresh and fixed tissue sets were collected in the event of clinical concerns or loose stools. Blood samples were spun to obtain serum, which was tested using the SVANOVIR® *L intracellularis*/Ileitis-Ab-X ELISA (Svanova, Sweden), while pooled pen based fecal samples were processed for DNA extraction, then *Lawsonia intracellularis* PCR at the Iowa State University Veterinary Diagnostic Laboratory. At each collection time point, the entire barn was evaluated for clinical signs and the number of pens observed with diarrhea were recorded. In instances of mortality, necropsy was performed, and tissues were submitted for diagnostic workup, inclusive of IHC for *L intracellularis*.

Results and discussion

Diagnostic and clinical results are summarized by producer in Table 1.

Lawsonia intracellularis was detected by PCR and ELISA within each cohort, demonstrating that these vaccinated pigs were exposed. The sites were chosen based on concerns of the occurrence of clinical ileitis prior to the implementation of this vaccination strategy. Clinical impressions combined with diagnostic results in these cases demonstrate that pigs vaccinated with Enterisol® Ileitis by gel were protected against disease.

Clinical impressions of diarrhea were found in cohort 2 (producer 2) and in cohort 3 (producer 3). At the time point where loose stools were observed in cohort 2, only one loose stool was found in each of two pens, which yielded high Ct values (33.4 and 33.5), indicating that *Lawsonia* was likely not the cause.[2,3] For cohort 3, looseness was attributed to water quality by the attending veterinarian. At the two time points where clinical signs were observed in this cohort, only one of the five pooled fecal samples were PCR positive (Ct values 32 and 26.7), further indicating that *Lawsonia* was likely not the primary cause of looseness.

Mortality events were recorded in cohorts 3 and 4. Mortality in cohort 1 was attributed to gastric ulcers at 14 weeks of age. Among these pigs, diarrhea was not observed and *Lawsonia* was only found in tissue of one pig. At the same site, there were several mortalities at 17 weeks of age which were also due to ulcers. PRRSV and influenza were identified in these cases, and laboratory diagnosis of intestinal tissues did not reveal *Lawsonia*. The clinical event that led to tissue collection at cohort 3 was also ascribed to gastric ulcers.

In this case study of four wean to finish groups of pigs, no events of mortality or diarrhea were attributed to *L intracellularis*. Although these sites had prior concerns of ileitis, clinical impressions together with diagnostic results showed that pigs vaccinated with Enterisol® Ileitis by gel were exposed to *L intracellularis* and protected against disease.

References

1. Nerem J, Sponheim A, Hanson D, Little E, Rustvold R, Edler R, Leite F. Evaluation of the oral administration of *Lawsonia intracellularis* vaccine by gel. Proceedings of the Allen D. Leman Swine Conference. 2023; 94.

2. Burrough ER, Rotolo ML, Gauger PC, Madson DM, Schwartz KJ. Correlation of *Lawsonia intracellularis* semi-quantitative fecal polymerase chain reaction assay results with the presence of histologic lesions of proliferative enteropathy and positive immunohistochemical staining. *J Swine Health Prod*. 2015;23(4):204-207.

3. Helm ET, Burrough ER, Leite FL, Gabler NK. *Lawsonia intracellularis* infected enterocytes lack sucrase-isomaltase which contributes to reduced pig digestive capacity. *Vet Res*. 2021;52:90. **https://doi.org/10.1186/s13567-021-00958-2**

Table 1: Summary of diagnostic and clinical results by the producer

Sow farm	Producer/ cohort	Diagnostic test[1,2] and clinical impression[3]	Diagnostic and clinical data # suspect or positive/# tested (%)						
			Approximate age (weeks)						
			10	12	14	16	18	20	22
1	1/1	PCR	0/5 (0)	0/5 (0)	5/5 (100)	5/5 (100)	2/5 (40)		
		ELISA		3/30 (3.3)		13/30 (43.3)		28/29 (96.7)	
		Clinical	0/10 (0)	0/10 (0)	0/10 (0)	0/10 (0)	0/10 (0)		
	2/2	PCR		1/5 (20)	2/10 (20)	2/10 (20)	2/10 (20)	0/10 (0)	0/10 (0)
		ELISA			21/60 (35)		38/59 (64.4)		43/60 (72)
		Clinical		0/24 (0)	0/24 (0)	2/24 (8.3)	0/24 (0)	0/24 (0)	0/24 (0)
2	3/3	PCR		1/5 (20)	1/5 (20)	1/5 (20)	0/5 (0)	0/5 (0)	0/5 (0)
		ELISA			3/15 (20)		9/15 (60)		1/15 (6.7)
		Clinical		0/10 (0)	5/10 (50)	10/10 (100)	0/10 (0)	0/10 (0)	0/10 (0)
	3/4	PCR		0/5 (0)	3/5 (60)	0/5 (0)	2/5 (40)	1/5 (20)	1/5 (20)
		ELISA			5/15 (33.3)		6/15 (40)		7/15 (47)
		Clinical		0/16 (0)	0/16 (0)	0/16 (0)	0/16 (0)	0/16 (0)	0/16 (0)

[1] *Lawsonia intracellularis* PCR. Samples with Ct value < 35 considered positive.

[2] *L intracellularis* ELISA. Samples with percent inhibition of ≥ 20% considered suspect, ≥ 30% considered positive.

[3] Number of pens observed with loose feces.

https://doi.org/10.54846/am2024/43

Assessing detection and interaction of *Lawsonia intracellularis* and PCV2 in low and high-performance wean-to-finish flows in different PRRSV detection scenarios

G. A. Cezar[1], DVM, MS; F. Leite[2], DVM, MS, PhD; E. Fano[2], DVM, MS, PhD; R. Philips[2], DVM; J. Waddell[2], DVM; K. Dion[1], DVM; E. Magalhães[1], DVM, MS; G. Trevisan[1], DVM, MBA, PhD; G. Silva[1], DVM, MS, PhD; D. C. L. Linhares[1], DVM, MS, PhD
[1]Department of Veterinary Diagnostic and Productive Animal Medicine, Iowa State University, Ames, Iowa; [2]Boehringer Ingelheim Animal Health USA Inc., Duluth, Georgia

Introduction

Multiple pathogens can affect the respiratory system and have been described in the swine respiratory disease complex, demonstrating the importance of disease co-infections. However, mixed infections among respiratory and enteric pathogens remain poorly described and investigated. Porcine reproductive and respiratory syndrome virus (PRRSV) causes systemic infection and can affect the gut microbiome, which could favor the colonization of enteric pathogens.[1,2] Opriessnig et al[3] demonstrated that PRRSV co-infection with porcine circovirus type 2 (PCV2) influences the expression of porcine circovirus-associated disease. This study aimed to assess *Lawsonia intracellularis*, PRRSV, and PCV2, pathogens that cause both systemic as well as enteric infection, to describe their detection dynamics in wean-to-finish groups, and investigate their association with performance indicators.

Methods

Seventy-six groups of growing pigs, from PRRSV-stable (n = 49) or PRRSV-negative sow farms (n = 27), from a producer, were tested over time, from weaning to market. Spatially distributed oral fluid samples were collected every two weeks for 16 collecting points between 3 - 33 weeks of age. Thirty-six groups were selected and categorized as Low and High performance based on mortality and average daily gain (ADG) using the 1st and 4th quartiles of data, and the presence of disease diagnosis assigned by a diagnostician. Samples were tested by quantitative PCR for PRRSV RNA, and *Lawsonia* and PCV2 DNA. All pig groups were vaccinated for *Lawsonia* between 5 - 9 weeks of age and for PCV2 at weaning age (3 weeks). Pigs were not vaccinated for PRRSV and were categorized into groups according to the time of the first PRRSV detection. These categories were: nursery (3 - 9 weeks of age); early finish (11 - 13 weeks of age) and late finish (14 weeks onward). The difference between the positivity of oral fluids for *Lawsonia* and PCV2 at each sampling age, cumulative mortality, and ADG among the High/Low performance and PRRSV detection groups was assessed using linear regression models (R v.4.2.3).

Results

The study included 46,446 growing pigs and 4,000 samples, which were characterized into 19 High and 17 Low-performance groups. Statistical differences were found in mortality (Low-performance, $\bar{x}$ = 12.1%; High-performance, $\bar{x}$ = 6.9%) and ADG (Low-performance, $\bar{x}$ = 1.6 lb/day; High-performance, $\bar{x}$ = 1.8 lb/day) among performance groups. Two of the 36 groups did not have PRRSV detection. Table 1 demonstrates the results of mortality and ADG of *Lawsonia* and PCV2-positive groups among the PRRSV detection categories. Lower detection of *Lawsonia* was observed in the groups of High performance regardless of the PRRSV detection group (Figures 1 a,b,c). When PRRSV was first detected in the early finish phase, a lower detection of *Lawsonia* at week ages 23 - 33 was observed in the High-performance groups (Figure 1b). Statistical difference in the oral fluid positivity rate was found on the week of age 25 (High $\bar{x}$ = 26.4%; Low $\bar{x}$ = 62.5%). In the comparison among PRRSV detection groups, the early finish PRRSV detection with Low performance had higher *Lawsonia* detection

Table 1: Average mortality and daily weight gain from High and Low-performance groups among the different PRRSV detection categories

Pathogen	PRRSV first detection	Mortality		ADG (lbs)		Statistical difference in oral fluid positivity
		Low	High	Low	High	
Lawsonia	Nursery (n = 4)	14.9%^{a}	7.8%^{a}	1.59^a	1.58^a	No
	Early finish (n = 14)	13.60 %^{a}	7.36%^{b}	1.58^a	1.64^a	Yes
	Late finish (n = 13)	13.8%^{a}	5.3%^{b}	1.53^a	1.63^b	Yes
PCV2	Nursery (n = 4)	14.9%^{a}	7.8%^{a}	1.59^a	1.58^a	No
	Early finish (n = 14)	13.60 %^{a}	7.36%^{b}	1.58^a	1.63^a	Yes
	Late finish (n = 14)	13.2%^{a}	6.2%^{b}	1.54^a	1.63^a	Yes

^{n} = number of groups
^{ab} Different letters indicate a significant statistical difference between groups within the PRRS first detection categories by pathogen.

Figure 1: The average percentage of positive oral fluids detection of *Lawsonia intracellularis* and porcine circovirus type 2 among PRRSV detection groups. The blue line represents High-performance groups, and the red line represents Low-performance groups. **(A)** PRRSV 1st detection in nursery pigs; **(B)** PRRSV 1st detection in early finish pigs; and **(C)** PRRSV 1st detection in late finish pigs.

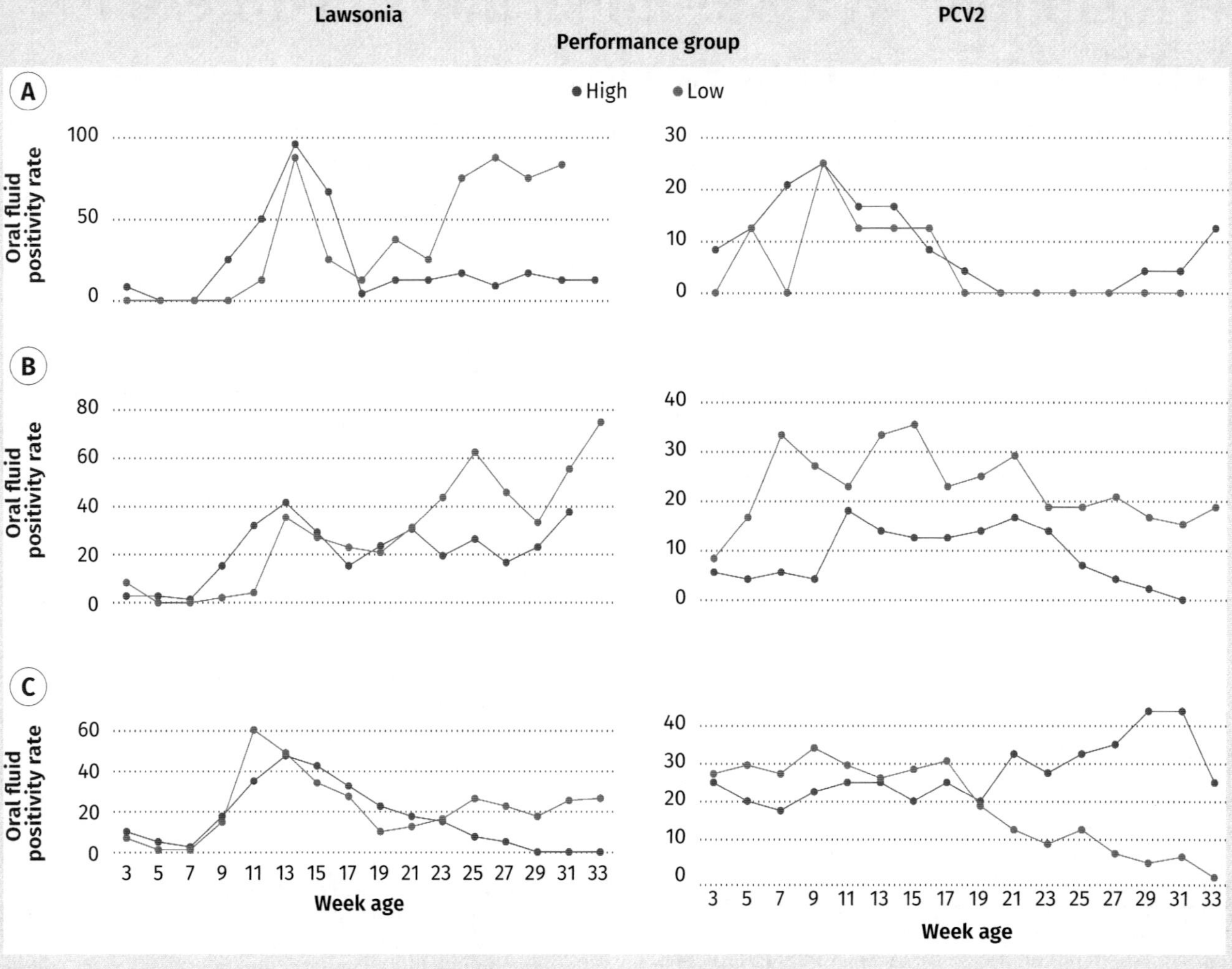

than the late finish groups with Low and High performance on weeks of age 23-27. Regarding PCV2, the early finish PRRSV detection had higher PCV2 detection comparing the performance groups (Figure 1c), with a statistical difference at the week of age 15 (High $\bar{x}$ = 12.5%; Low $\bar{x}$ = 35.42%). On the other hand, when PRRSV was first detected in the late finish, the High group had higher detection of PCV2 from weeks 23-33 (Figure 1c). Statistical difference in the oral fluid positivity rate was found at the age of 27 (High $\bar{x}$ = 35%; Low $\bar{x}$ = 6.2%) and 29 (High $\bar{x}$ = 32.2%; Low $\bar{x}$ = 1.3%).

Discussion

Under the conditions of this study, groups of pigs with greater indices of ADG and lower mortality had lower levels of *Lawsonia* detection. PCV2 had similar overall detection among both High and Low-performance groups. These findings occurred in a scenario where 34 of 36 groups became PRRSV-positive at different time points. When PRRSV detection occurred in the early finish phase, an increase in the detection of both *Lawsonia* and PCV2 was observed in the Low-performance groups, demonstrating a possible association between PRRSV and these pathogens. Unexpectedly, groups of High performance had the first PRRSV detection in late finish and had higher detection of PCV2 in the weeks of age 23 onward compared with Low-performance groups, demonstrating that the increased detection of PCV2 without expression of PCVAD might not affect pig performance in PCV2 vaccinated pigs. The interaction among these pathogens merits further investigation to understand the performance impact of different co-infections and detection dynamic scenarios from agents affecting multiple physiological systems.

References

1. Zhao, J., Wan, S., Sun, N., et al. Damage to intestinal barrier integrity in piglets caused by porcine reproductive and respiratory syndrome virus infection. Vet. Res. 52, 93 (2021).

2. Argüello H, Rodríguez-Gómez IM, Sánchez-Carvajal JM, Pallares FJ, Díaz I, Cabrera-Rubio R, Crispie F, Cotter PD, Mateu E, Martín-Valls G, Carrasco L, Gómez-Laguna J. Porcine reproductive and respiratory syndrome virus impacts on gut microbiome in a strain virulence-dependent fashion. Microb Biotechnol. 15(3), 1007-1016 (2022).

3. Opriessnig, T, Halbur, PG. Concurrent infections are important for the expression of porcine circovirus-associated disease. Virus Research. 164, 20-32 (2012).

https://doi.org/10.54846/am2024/44

Sequivity® IAV-S NA: A newly licensed commercial vaccine for the protection against influenza in pigs

Pravina Kitikoon[1]; Susan Knetter[1]; Mark Mogler[1]; Chandra Morgan[1]; Allison Hoehn[1]; Supraja Puttamreddy[1]; Faris Jirjis[1]; Erin Strait[2]; Ruud Segers[3]
[1]Merck Animal Health, Rahway, New Jersey; [2]Previously with Merck Animal Health, Rahway, New Jersey; [3]MSD Animal Health, The Netherlands

Introduction

Vaccination is currently the most effective method in the swine industry to control the economic burden caused by influenza A virus in swine (IAV-S). The virus has two major envelope glycoproteins, hemagglutinin (HA) and neuraminidase (NA) that are used to further characterize the virus into subtypes and clades.[1] The HA protein binds to sialic acid receptors on the target cells and serves as the main target for viral neutralizing antibodies detected via a hemagglutinin inhibition (HI) test.[2] The viral NA protein removes the sialic acid receptors enabling the release of newly produced virus to spread in the host.[3] Antibodies to NA thus have a functional effect by limiting virus spread and enhancing virus clearance, and can be detected by a neuraminidase inhibition (NI) test.[4,5] Up until now the viral HA protein has been the primary focus of IAV vaccines for inducing protective antibodies. Commercially available whole inactivated virus (WIV) vaccines are formulated based on standardized levels of multiple HA proteins targeted to protect against circulating viruses with homologous or antigenically similar HA.[2] Autogenous WIV vaccine and non-replicating HA RNA Particle vaccine (Sequivity Prescription HA) produced to match circulating virus in the swine herds have been widely used in the US. These IAV-S vaccines have been used primarily in gilts and sows to not only reduce disease and virus circulation in the herd, but also to provide IAV-specific maternally derived antibodies (MDA) to their piglets. However, IAV-S still remains a major issue in nursery and growing pigs, as IAV-S contributes to the porcine respiratory disease complex problem and these two age groups have the highest IAV detection rate.[6,7] The more sustainable way to protect pigs into the nursery and growing phase is by establishing active IAV-S immunity by vaccination. The HA-specific MDA is known to mask clinical signs in suckling pigs but it will not prevent IAV infection and transmission.[8,9] Further, it will interfere with the immune response if later pigs receive a similar HA-based vaccination.[9-11] Thus a different vaccine concept is needed to provide protection against IAV in the pig population.

This novel Sequivity® IAV-S NA (Sequivity NA) vaccine is the first to target only the NA protein of IAV-S. The NA protein is a non-competing antibody target to HA-specific MDA (if present) that could still stimulate the pig's immune response. This provides a new approach for controlling IAV in the pig population by allowing for a higher chance of successful vaccination when used in pigs born from sows vaccinated with a HA-based vaccine (ie, Sequivity HA vaccine). The objectives of these studies were to evaluate the immunogenicity and efficacy of Sequivity NA against disease associated with different strains of IAV-S in IAV-negative pigs housed in a controlled environment, and to investigate the safety profile in field conditions.

Materials and methods

Efficacy studies

Vaccine

The adjuvanted frozen Sequivity® IAV-S NA (Sequivity NA) vaccine was produced based on RNA Particle (RNA-P) technology. The vaccine was formulated based on four NA proteins representing the major NA clades of U.S. IAV-S ($N1_{classic}$, $N1_{pandemic}$, $N2_{1998b}$ and $N2_{2002a}$). At point of use the frozen vaccine was thawed and mixed with Microsol Diluvac Forte®, an oil-in-water adjuvant containing mineral oil and Vitamin E. The vaccines were administered intramuscularly (IM) in a 1 mL dose to pigs at three days of age (DOA) and boostered three weeks later. The placebo vaccine was formulated in the same manner as the vaccine with a green fluorescent protein RNA-P replacing NA RNA-Ps. After each vaccination, systemic reactions were observed daily while local injection site reactions were observed and measured every other day for 15 days following each vaccination.

Study design

The Sequivity NA vaccine efficacy including the onset of immunity (OOI) and duration of immunity (DOI) was tested against infection with five different strains of the H1N1, H1N2 and H3N2 subtypes.[12] A total of 308 IAV-S seronegative pigs were enrolled into five separate studies conducted in biosafety level 2 containment facilities. In each study, whole litters of pigs were obtained and litter stratified into two treatment groups, 'vaccine' or 'placebo' and further randomized into subgroups, 'necropsy' or 'shedding'. Pigs in four of the five studies were challenged intratracheally at 8 weeks of age (WOA) to evaluate OOI; and pigs in the fifth study were challenged at 15 WOA to assess the DOI. The challenge viruses (Table 1) were obtained from the National Veterinary Service Laboratory, Ames, IA as part of the USDA Swine Influenza Virus surveillance program to monitor IAV-S genetic evolution. Study 1 was the DOI trail where pigs were challenged with a 2019 H1N1 virus at 15 WOA or 85 days post second vaccination (DPV2). Studies 2 - 5 were the OOI trials where pigs were challenged with different IAV-S isolates at 8 WOA or approximately 36 DPV2. In all studies, pigs were humanely euthanized and necropsied at five days post challenge (DPC) to evaluate macroscopic lung lesions and obtain bronchoalveolar lavage fluids (BALF) to evaluate virus titer in the lower airways via virus titration assay. Serum samples were collected prior to each vaccination and weekly until 35 DPV2 for OOI studies. For the DOI study, two additional serum samples were collected at 56 and 84 DPV2. The NI antibody titers were determined according to a previously published enzyme-linked lectin assay with modifications.[12] Nasal swabs were collected daily from pigs in the 'shedding' subgroups post-challenge to assess duration and magnitude of nasal shedding. The nasal shedding titers and virus titers in BALF were assessed via virus isolation and titration and reported as $\log_{10}$ $TCID_{50}$ per mL.

Table 1: Influenza A virus isolates used for pig challenge

Study (n)	Strain name	NA clade	% Amino acid identity to Sequivity NA vaccine[1]
Study 1 (64)	A/swine/IN/A02429505/2019 (H1N1)	$N1_{classic}$	96.97 % (heterologous strain, homologous clade)
Study 2 (64)	A/swine/MN/A01483170/2014 (H1N1)	$N1_{pandemic}$	100.00 % (homologous strain and clade)
Study 3 (52)	A/swine/OK/A01409770/2014 (H1N2)	$N2_{1998b}$	93.84 % (heterologous strain, homologous subclade)
Study 4 (64)	A/swine/IL/A01475495/2014 (H1N2)	$N2_{2002a}$	100.00 % (homologous strain and subclade)
Study 5 (64)	A/swine/MO/A01840724/2015 (H3N2)	$N2_{2002b}$	91.32 % (heterologous strain and subclade)

[1] Sequivity NA vaccine consists of $N1_{classic}$, $N1_{pandemic}$, $N2_{1998b}$, and $N2_{2002a}$ antigen fractions

Field safety study

A field safety study was performed in three separate geographical locations, each at a sow herd to follow safety post 3 DOA vaccination and at a nursery to follow safety post booster vaccination. Sites of observations were as follows: Site 1 was in Minnesota for both sow herd and nursery; Site 2 was in South Dakota for sow herd and Minnesota for nursery; and Site 3 was in Illinois for both sow herd and nursery. A total of 748 healthy pigs received the initial vaccination (at 3 - 5 DOA) and 715 pigs remained to receive the second vaccination (at 18 - 23 DOA). Pigs received the vaccine in the order of handling, alternating between two separate Sequivity NA vaccine serials. Both serials were formulated with the maximum antigen titer in order to assess safety. Approximately 240 - 261 pigs per farm were enrolled in the study with about half of the pigs vaccinated per serial. Medications were recorded and pigs that died during the study were examined post-mortem to determine the cause of death. The study investigator observed the animals at 1 hour after each vaccination and at least once daily for 21 days to observe for general health and local or systemic reactions suspected to be related to vaccine administration. In addition, the injection site was palpated on days 3, 4, and 5 post vaccination. Any visible injection site reactions were followed daily until resolution.

Results

Efficacy studies

The Sequivity NA vaccine induced robust serum NI antibodies to all four vaccine fractions (Figure 1). After one vaccination, over 85.1% (132/155) of the vaccinated pigs had positive titers to both N1 vaccine fractions. At 7 DPV2, 99.35% (154/155) of the vaccinated pigs had high geometric mean titers (≥ 966) to all four vaccine fractions. The titers remained high and were positive up until 84 DPV2. The immune response to vaccination was protective against all five IAV-S strains, as the macroscopic lung lesions in vaccinated pigs were significantly reduced compared to placebo pigs (Figure 2). Sequivity NA vaccine was shown to reduce over 85% lung lesions when pigs were infected with homologous NA clade and subclade to the vaccine antigen. Protection extended against a heterologous NA subclade as pigs infected with an H3N2 virus with a heterologous N2 subclade ($N2_{2002b}$) had 78.4% less pneumonia compared to their placebo group. The duration and magnitude of nasal shedding were significantly reduced by 2 to 4 days post-challenge and 10- to 2,570-fold, respectively in vaccinated pigs infected with a homologous subclade. Lastly, virus in the lungs of vaccinated pigs was either completely eliminated or significantly reduced ($P < .0001$) compared to the placebo group (Table 2). The duration of immunity demonstrated against a 2019 H1N1 ($N1_{classic}$) virus was at least 12 weeks post second vaccination.

Safety

In the efficacy studies (Studies 1 - 5), no pigs had any systemic reactions after their first or second vaccinations. After the first vaccination, none of the pigs developed any local injection site reactions. Nine percent of pigs (14 of 155) had local injections site reactions post second vaccination, all of which were ≤ 1.5 cm. All reactions resolved within the 15-day observation period.

In the field safety study, only three of the 748 pigs had a brief anaphylactic-type reaction following either the first or second vaccination. All three pigs recovered within 10 minutes without any intervention. Local injection site swellings were observed in 4.1% of enrolled pigs post first vaccination and 12.5% after the second vaccination. The swellings were primarily small, < 1.5 cm in diameter, and no swellings were larger than 10 cm in diameter. The mean duration of injection site swellings for all three sites post first and second vaccinations were ≤ 2.1 days and ≤ 9.6 days, respectively. None of the injection site swellings progressed to abscesses; and pigs did not show any sign of discomfort during the post vaccination period.

Discussions

These studies demonstrated safety, immunogenicity, and efficacy of the novel Sequivity NA vaccine for use in pigs as young as three days of age. Vaccination induced high levels of NI antibody response to all four vaccine antigens. The efficacy studies demonstrated OOI occurred as early as five weeks and lasted at least 12 weeks post second vaccination (DOI). Sequivity NA significantly reduced pneumonia from infection with five strains, including strains which were homologous to the vaccine, heterologous within the same NA clade or subclade, or heterologous in a different subclade (Study 5). The vaccine reduced the duration and magnitude of nasal shedding and reduced virus in the lower airway or completely eliminated the virus in the case of H1N1 infection. In the OOI and DOI trials, no systemic reactions were observed and only limited numbers of vaccinated animals had small injection site swellings after the second vaccination. When formulated at a dose 10X higher than standard production serials, the vaccine was proven safe with moderate transient local injection site reactions particularly after second vaccination. These local injection swellings completely resolved within an average of 10 days.

In summary, this IAV vaccine for pigs is the first of its kind to be commercially available based solely on the viral neuraminidase protein. The two dose regimen fits with current swine production handling schedules of piglets (processing and weaning) to reduce animal stress and human labor. The robust and long-lasting NI antibody response can protect pigs into nursery and growing

Figure 1: Neuraminidase inhibition (NI) titer to N1classic **(A)**, N1pandemic **(B)**, N21998b **(C)** and N22002a **(D)** vaccine antigen from vaccinated pigs prior to first and second vaccinations (DPV1 and DPV2, respectively) and weekly post second vaccination. A titer of ≤ 20 is negative. NI titers from placebo pigs were negative at all testing timepoints (data not shown).

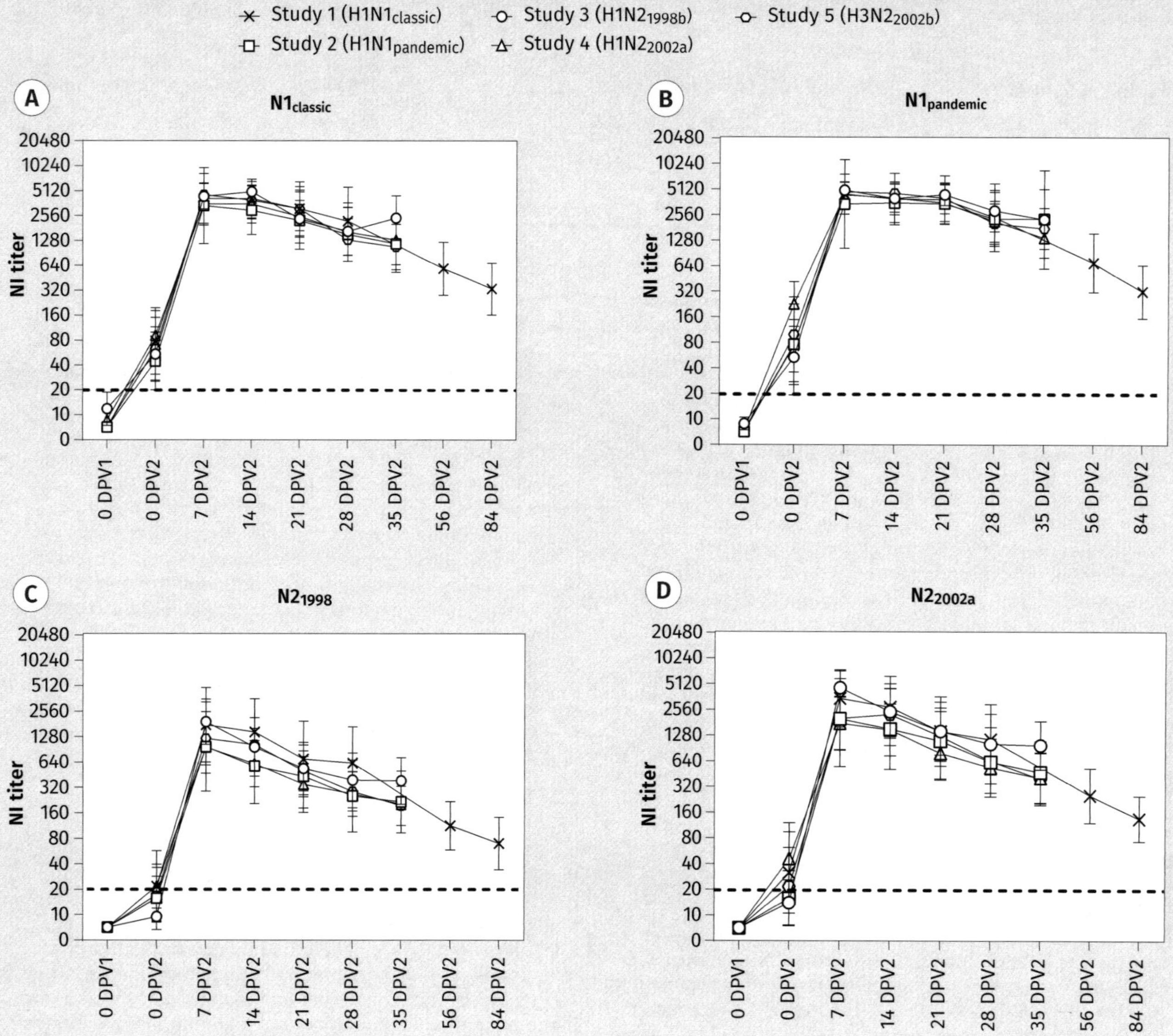

phases vulnerable to the effects of IAV-S infection. Furthermore, a strategic use of Sequivity NA in pigs in conjunction with HA-targeted vaccines in sows can allow for whole herd control of IAV-S. Reduction of IAV-S circulation in swine herds can decrease the chance of virus reassortment and zoonotic potential to support the One Health approach as well.

References

1. Liang Y. Pathogenicity and virulence of influenza. *Virulence* 2023;14:2223057.

2. Cox RJ. Correlates of protection to influenza virus, where do we go from here? *Hum Vaccin Immunother* 2013;9:405-408.

3. McAuley JL, Gilbertson BP, Trifkovic S, et al. Influenza Virus Neuraminidase Structure and Functions. *Front Microbiol* 2019;10:39.

4. Kaplan BS, Vincent AL. Detection and Titration of Influenza A Virus Neuraminidase Inhibiting (NAI) Antibodies Using an Enzyme-Linked Lectin Assay (ELLA). *Methods Mol Biol* 2020;2123:335-344.

5. Sandbulte MR, Gao J, Straight TM, et al. A miniaturized assay for influenza neuraminidase-inhibiting antibodies utilizing reverse genetics-derived antigens. *Influenza Other Respir Viruses* 2009;3:233-240.

6. Gebhardt JT, Tokach MD, Dritz SS, et al. Postweaning mortality in commercial swine production II: review of infectious contributing factors. *Transl Anim Sci* 2020;4:485-506.

7. USDA Influenza A Virus in Swine Surveillance. *Surveillance Summary for Second Quarter Fiscal Year 2023: January 1 to March 31, 2023* 21.

8. Cador C, Herve S, Andraud M, et al. Maternally-derived antibodies do not prevent transmission of swine influenza A virus between pigs. *Vet Res* 2016;47:86.

Figure 2: Percent lung lesions associated with IAV-S infections at five days post challenge in the vaccinated and placebo pigs from five separate studies. *Indicates significant difference of the % lung lesion compared between the vaccinated group to their corresponding placebo group ($P < .05$).

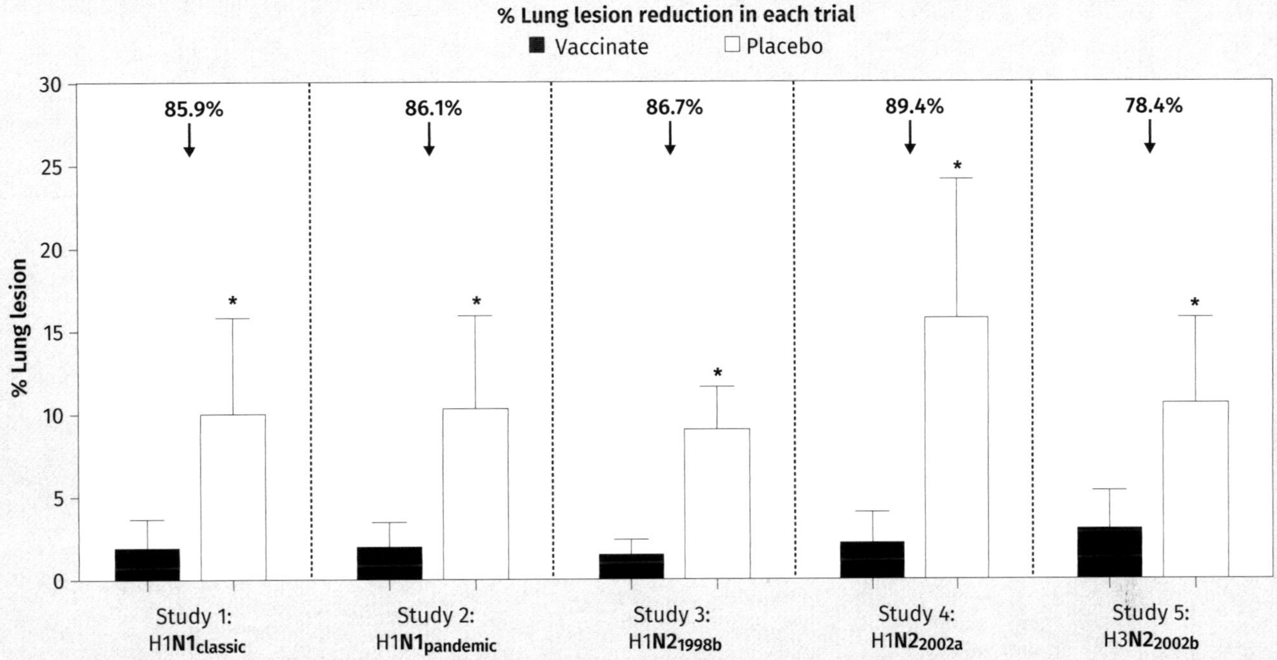

Table 2: Virus titers in lungs (BALF) collected from vaccinated and placebo pigs at five days post challenge and reduction of duration (days) and magnitude (titer fold reduction) of nasal shedding in vaccinated pigs compared to placebo pigs. *Indicates significant difference of the virus titer compared between the vaccinated group to their corresponding placebo group ($P < 0.05$).

Study	Treatment	Virus in lungs		Nasal shedding in vaccinates compared to placebos	
		% Positive (# of total)	Titer (log_{10} $TCID_{50}$/mL)	Duration reduction	Magnitude reduction
Study 1: $H1N1_{classic}$	Vaccinate	0.0 (0/20)	0.50*	2 days	18- to 25-fold
	Placebo	44.4 (8/18)	1.44		
Study 2: $H1N1_{pandemic}$	Vaccinate	0.0 (0/20)	0.50*	4 days	115- to 2,570-fold
	Placebo	85.0 (17/20)	3.24		
Study 3: $H1N2_{1998b}$	Vaccinate	13.3 (2/15)	0.84*	3 days	63- to 741-fold
	Placebo	100.0 (16/16)	4.29		
Study 4: $H1N2_{2002a}$	Vaccinate	45.0 (9/20)	1.62*	No	10- to 40-fold
	Placebo	100.0 (20/20)	3.96		
Study 5: $H3N2_{2002b}$	Vaccinate	5.0 (1/20)	0.53*	No	No
	Placebo	100.0 (20/20)	3.69		

9. Loeffen WL, Heinen PP, Bianchi AT, et al. Effect of maternally derived antibodies on the clinical signs and immune response in pigs after primary and secondary infection with an influenza H1N1 virus. *Vet Immunol Immunopathol* 2003;92:23-35.

10. Kitikoon P, Nilubol D, Erickson BJ, et al. The immune response and maternal antibody interference to a heterologous H1N1 swine influenza virus infection following vaccination. *Vet Immunol Immunopathol* 2006;112:117-128.

11. Rose N, Herve S, Eveno E, et al. Dynamics of influenza A virus infections in permanently infected pig farms: evidence of recurrent infections, circulation of several swine influenza viruses and reassortment events. *Vet Res* 2013;44:72.

12. Kitikoon P, Knetter SM, Mogler MA, et al. Quadrivalent neuraminidase RNA particle vaccine protects pigs against homologous and heterologous strains of swine influenza virus infection. *Vaccine* 2023;41:6941-6951.

https://doi.org/10.54846/am2024/45

Use of a hemagglutinin (HA) and neuraminidase (NA) vaccine program to stabilize a sow farm for influenza

Evan Koep[1], DVM; Marisa Rotolo[2], DVM, PhD; Alison Buetel[3]; Perry Harms[3], DVM, MS; Mario Zavala[3]; Joel Nerem[1], DVM; Channing Sebo[4], DVM; Brett O'Brien[4], VMD, MS

[1]Pipestone Veterinary Services, Pipestone, Minnesota; [2]Formerly with Pig Improvement Company, Hendersonville, Tennessee; [3]Pig Improvement Company, Hendersonville, Tennessee; [4]Merck Animal Health, Lenexa, Kansas

Introduction

Influenza A virus in swine (IAV-S) is a prevalent and costly respiratory pathogen for the United States swine industry.[1] The greatest impact on production performance typically occurs in the nursery period with the most common source of infection being the weaned pig. Thus, the cornerstone of control in growing pigs is sourcing IAV-S negative pigs from the sow farm. Whole herd vaccination programs in tandem with different management tools (biosecurity, McRebel, all-in-all-out farrowing rooms, improved sanitation practices, etc) have been used to successfully stabilize sow farms for IAV-S (elimination of active IAV-S circulation).[2,3] However, replacement gilts continue to serve as a potential source of IAV-S re-introduction due to the fact that it is difficult to effectively immunize them with traditional hemagglutinin-based (HA) vaccines in the face of long lived maternally derived antibodies (MDA). Merck Animal Health has a newly approved neuraminidase-based (NA) influenza vaccine, Sequivity® IAV-S NA (Sequivity NA), that is labeled for use in pigs 3 days of age (DOA) or older against influenza suptypes H1N1, H1N2, and H3N2. This new vaccine provides an opportunity to address the MDA interference issue by allowing for an IAV-S vaccination program that includes the use of Sequivity IAV-S HA prescription vaccine (Sequivity prescription HA) in sows while administering the new neuraminidase-based vaccine to their offspring. The purpose of this study was to evaluate **1)** the impact of whole herd vaccination with Sequivity prescription HA on IAV-S virus circulation in a sow herd and **2)** the impact of vaccinating suckling piglets destined to be replacement gilts with Sequivity NA.

Materials and methods

A multiplication sow farm (1,700 PIC Line 3 sows) that was endemically infected with IAV-S was used in this study. This sow farm produces its own replacement gilts which are reared at an off-site nursery and finisher complex from 3 to 20 weeks of age (WOA). Diagnostic monitoring performed before the stabilization program began confirmed the presence of one pandemic H1N1 virus on farm. The farm had no prior history of IAV-S vaccine use. Active IAV-S circulation was found at a low prevalence in due-to-wean pigs, nursery and finisher pigs up to 12 WOA. Clinical signs of IAV-S were characterized as mild (bilateral serous nasal discharge with minimal to no cough).

The vaccine program used was a double mass vaccination of all animals on farm staggered 3 weeks apart (initial mass vaccination occurred on study week 0). Pigs that were > 12 WOA were vaccinated with a Sequivity IAV-S prescription vaccine that was built using an HA gene that was 100% homologous to the circulating wild-type pandemic H1N1 strain isolated from the farm. All pigs in the flow ranging from 3 DOA to 12 WOA were vaccinated with Sequivity IAV-S NA vaccine. Following the initial mass vaccination, piglets continued to be vaccinated with Sequivity IAV-S NA at 3 DOA and again at 3 WOA. Gilts and sows were vaccinated 3 weeks prior to farrowing with Sequivity prescription HA starting at study week 6. All mature gilts, sows, and boars maintained a quarterly mass vaccination program of Sequivity prescription HA vaccination.

Additional procedures that were put in place to help reduce IAV-S circulation within the sow farm included: strict McRebel during study weeks 3 - 9, emptying all pigs from the on-site nursery at week 3, and enhanced biosecurity procedures (unidirectional employee traffic flow, changing gloves between every litter, etc).

The IAV-S circulation status of the farrowing house was monitored by collection of udder wipes (UW) from due-to-wean piglets. Two 4" x 4" sterile cotton gauzes were saturated in 20 mLs DMEM which were used to wipe down all teats from randomly selected litters. Thirty UW were collected 1 week before the study began, study weeks 0, 1, 2, 6, 11, 13, 15, 18 and then monthly thereafter. More intensive UW monitoring (60 UW) occurred at critical times following the second mass vaccination with Sequivity prescription HA on study weeks 3, 7, 8, and 9. Oral fluid (OF) monitoring was performed in the gilt development unit (GDU) and on-site nursery rooms throughout the study. All UW and OF were tested individually by IAV-S PCR at Iowa State University Veterinary Diagnostic Laboratory.

A subset of replacement gilts were tagged and monitored for post-vaccination serological response at 3 different phases of the study: Group A was enrolled at study week 0 when wild-type IAV-S was circulating (baseline status), Group B was enrolled at study week 6 which was shortly after mass vaccinations were complete, and Group C was enrolled at study week 15 when IAV-S circulation could no longer be detected in the herd (post-stabilization). Table 1 outlines the schematic of the sampling performed. These replacement gilts were bled at 3 DOA and at 3, 6, 9, 12, 15, and 18 WOA and were tested by Nucleoprotein (NP) ELISA, Hemagglutination Inhibition (HI) assay, and Neuraminidase inhibition (NI) enzyme-linked lectin assay. NP ELISA titers ≤ 0.5 were categorized as positive and this assay was used in tandem with oral fluid (OF) monitoring to evaluate the presence of IAV-S exposure since neither Sequivity prescription HA nor Sequivity NA vaccines contain IAV-S nucleoprotein. HI assays were run on sera using the pandemic strain > 99% homologous to the farm strain. NI assays were run using the N1-pandemic protein contained within the Sequivity NA vaccine which is approximately 96% homologous to the farm strain as well as NI assays to a N2-2002a protein (specific to the Sequivity NA vaccine). Oral fluids were obtained from each pen housing the subset of tagged replacement gilts weekly from 4-10 WOA and then biweekly until 18 WOA and these were tested by IAV-S PCR. It is important to note that all replacement gilts were raised in an off-site nursery and finisher unit until roughly

Table 1: Replacement gilt diagnostic monitoring program

Study phase	Gilt group	Enrolled (study week)	N	Diagnostic sampling schedule
Baseline	A	0	30	**Serum:** 3 DOA, 3, 6, 9, 12, 15, 18 WOA **OF:** 4 - 10 WOA weekly, 12 - 18 WOA biweekly
Mass vaccination	B	6	40	**Serum:** 3 DOA, 3, 6, 9, 12, 15, 18 WOA **OF:** 4 - 10 WOA weekly, 12 - 18 WOA biweekly
IAV-S stable	C	15	40	**Serum:** 3 DOA, 3, 6, 9, 12, 15, 18 WOA **OF:** 4 - 10 WOA weekly, 12 - 18 WOA biweekly

OF = oral fluid

20 weeks of age. This off-site nursery and finisher unit also had its own IAV-S elimination program in place that coincided with the sow farm to ensure groups B and C remained IAV-S negative. Group A gilts which were housed at the off-site nursery and finisher unit were vaccinated with Sequivity prescription HA at 12 and 15 WOA as part of the mass vaccination program in place to help stabilize this site prior to IAV-S negative groups being weaned there.

Descriptive statistics were used to summarize diagnostic results obtained over the course of the study. Serological titers were log transformed to calculate least square means and then back transformed. A positive response rate to the vaccine was defined by more than a 2 fold titer increase in NI titer post-vaccination. The results shared herein summarize the study through week 18. More results will be shared at the annual AASV meeting.

Results

Over the course of the study the sow farm remained in high health with no notable heath changes occurring.

All udder wipes were 100% IAV-S PCR negative starting at study week 3 (week of second mass vaccination) and remained negative as of writing of this paper (study week 20). All pigs that were retained in the flow starting on study week 9 when the sow farm was declared IAV-S stable, have remained 100% negative for IAV-S on nursery oral fluids. The onsite GDU which housed replacement gilts > 20 WOA has remained IAV-S PCR negative on oral fluids over the course of the entire evaluation.

Group A IAV-S serology results are summarized in Table 2. Mean NI pandemic titers for Group A demonstrate a nearly 7 fold increase in titers by 6 WOA (1097) with a peak mean titer at 9 WOA of 1638. It is interesting to note that while Group A pigs had maternally derived antibodies to the farm's pandemic virus present as indicated by 63% of piglets at 3 DOA being NI positive, 20% HI positive and 17% NP ELISA positive there was 100% responder rate in piglets vaccinated with Sequivity NA. The NP ELISA and HI titers in Group A piglets remained negative after initial maternal antibodies waned which demonstrates minimal wild-type IAV-S exposure post-weaning. Likewise, oral fluid results confirmed low positive IAV-S PCR (CT > 32) up to 8 WOA and were thereafter negative on PCR. There was a slight increase in Group A HI titers at 15 WOA but this was expected as this sampling timepoint was three weeks following Sequivity prescription HA mass vaccination in the off-site finisher unit where they were housed.

The preliminary serological results for Group B (see Table 3) showed 100% serological response to Sequivity NA vaccination as demonstrated by a nearly 8 fold increase in vaccine specific NI titers (N2-2002a) which had a peak mean titer of 1412 three weeks after their second NA vaccination. It is important to note that even after a single dose of Sequivity NA vaccine at 3 DOA that vaccine-specific NI titers showed an increase (mean titer = 286) by 3 WOA.

All oral fluid results for Group B were IAV-S PCR negative. Likewise, all NP ELISA results were negative and HI titers continued to decline post-weaning which also confirms the lack of wild-type IAV-S exposure on farm post-stabilization.

Results for Group C are not available as of the writing of this paper.

Discussion

The use of a combination Sequivity prescription HA and Sequivity IAV-S NA whole herd vaccine program coupled with on farm biosecurity was able to effectively control IAV-S circulation on the sow farm by 9 weeks after the start of the study. Serology assays demonstrate the DIVA capabilities of the Sequivity IAV-S NA vaccine used in growing pigs as well as Sequivity prescription HA vaccine in sows to differentiate vaccine versus wild-type virus exposure. This attribute will help veterinarians better evaluate the effectiveness of their IAV-S vaccination programs.

Table 2: Group A replacement gilt post-vaccination IAV-S serology

Parameter	3 DOA	3 WOA	6 WOA	9 WOA	12 WOA	15 WOA
Mean NI (pandemic) titer	30	31	1097	1638	352	98
% NI (pandemic) positive	63	82	100	100	100	100
Mean NP ELISA S/P	0.716	1.606	0.827	0.770	0.948	1.015
% NP ELISA positive	16.7	0	0	4	0	0
Mean HI (pandemic) titer	< 20	< 20	< 20	< 20	< 20	< 20
% HI (pandemic) positive	20	0	0	8	10	44

Table 3: Group B replacement gilt post-vaccination IAV-S serology

Parameter	3 DOA	3 WOA	6 WOA	9 WOA
Mean NI (pandemic) titer	42	26	204	73
% NI (pandemic) positive	85	76	100	94
Mean NI (vaccine) titer	< 20	286	1412	330
% NI (vaccine) positive	0	91	100	100
Mean NP ELISA S/P	0.889	0.808	0.911	0.691
% NP ELISA positive	10	6	0	0
Mean HI (pandemic) titer	180	33	< 20	< 20
% HI (pandemic) positive	95	56	47	9

References

1. Haden, Cara, et al. Assessing production parameters and economic impact of swine influenza, PRRS, and *Mycoplasma hyopneumoniae* on finishing pigs in large production system. *AASV meeting proceedings*. 2012; 75-76.

2. Thompson, R. et al. Elimination of two consecutive swine influenza subtypes in a large breed to wean herd". *AASV meeting proceedings*. 2014; 221-223.

3. Sebo, Channing, et al. Influenza A vaccination using the SEQUIVITY™ Technology. *AASV meeting proceedings*. 2019; 242-244.

https://doi.org/10.54846/am2024/46

Vaccine compliance monitoring program for growing pigs vaccinated with Circumvent® CML

Kurt Kuecker, DVM; Kate Dion, DVM; Jon Tangen, DVM
The Hanor Company, Enid, Oklahoma

Introduction

Vaccination of swine is a critical component of raising healthy pigs. Vaccination programs are typically designed to help prevent or mitigate the impact of diseases known to affect pig populations. Three of the most economically important pathogens facing pork producers include porcine circovirus type 2, *Mycoplasma hyopneumoniae*, and *Lawsonia intracellularis*.[1] Vaccination compliance monitoring can be used to ensure pigs are getting vaccinated properly.[2,3] Typically, vaccines are administered using a multi-dose program and often before pigs are 10 weeks of age. Multiple factors can affect the immune response elicited following vaccination including vaccine formulation, proper timing of vaccination, health status, level of maternally derived antibody (MDA) of the pig at the time of vaccination, the number of vaccine doses administered and vaccine administration compliance.

Proper vaccine administration is essential for optimal efficacy. Typical errors can be centered around the following: Proper vaccine storage and handling, using the appropriate equipment (proper needle size, clean and functioning syringes that deliver the correct dose volume, using the proper equipment to restrain and identify pigs receiving vaccine), ensuring the correct vaccination administration process (creating a minimal stress environment, checking for good ventilation and barn temperatures, correct barn, pigs of the correct age, using approved restraint methods and injection techniques, experience level of the vaccinator). Avoiding these errors will help to ensure the desired dose of vaccine is administered to all pigs in the group at the proper time. Thus, utilizing vaccine compliance monitoring via post-vaccination serology in US swine herds is a method to help signal when vaccine auditing is necessary.

Methods

A commercial sow herd in Illinois was used in this study. Litters with the oldest pigs (that were the next due to wean) and representing three parity groups were identified. Average weaned pig age for the farm was 21 days of age at the time of the study. Five sows each from parities 1, 2 - 3 and ≥ 4 (P1, P2-3, and P3+, respectively) were enrolled, tagged with a unique number, and vaccinated. Pigs in this study were vaccinated by farm employees with a 1 mL dose of Circumvent® CML at both one day prior to weaning (approximately 3 weeks of age (WOA) and subsequently at 6 WOA. Serum was collected from all study pigs immediately prior to the initial Circumvent CML vaccination (3 WOA), at the time of second Circumvent CML vaccination (6 WOA) and three weeks later (9 WOA).

All serum was tested by the following assays at the Kansas State Veterinary Diagnostic Laboratory (KSVDL): porcine circovirus type 2 serotype B indirect florescent antibody (PCV2b IFA), PCV2 real-time polymerase chain reaction (PCR) (pooled in groups of 5 by KSVDL lab staff) and *Mycoplasma hyopnuemoniae* (Mhp) IDEXX ELISA (*Mhp* Ab Test, IDEXX Laboratories, Inc). Serum from the same animals were pooled at each test date and each pool was comprised of animals from the same parity group. All serum was also tested by *Lawsonia intracellularis* (LI) ELISA testing (SVANOVIR® L. intracellularis / Ileitis-Ab, INDICAL Sweden AB) at the Merck Animal Health laboratory in Lawrence, Kansas. Descriptive statistics was used to summarize the results. Percent positives, arithmetic means for Mhp ELISA S/P ratios and LI ELISA percent inhibition (PI) and IFA geometric means (geomean) were calculated per assay. Cutoff values for each assay are described in Table 1.

Results

All samples were PCV2 PCR negative at each test date.

The 3 WOA test measured pre-vaccination maternally derived antibody (MDA) levels. The 6 WOA test measured a combination of maternal antibody decay and the initial vaccination immune response and the 9 WOA test measured the immune response 3 weeks after the completion of the vaccination protocol.

The PCV2b IFA titers at 3 WOA (MDA) were low-to-moderate for pigs from all parity groups with 93% of pigs having an IFA titer of 1:320 or lower (2 pigs had a titer of 1:640). The 3 WOA IFA geomean for all parties was 1:88 (range < 1:40 - 640). At 6 WOA, the IFA geomean decreased to 1:47 (range < 1:40 - 320) with 90% of pigs having titers ≤ 1:40. At 9 WOA, the PCV2b IFA geomean had increased to 1:1437 (range 1:40 - 1:10240) and 77% of pigs had a titer greater than 1:640. (Table 2 and Figure 1)

Figure 2 shows the change in Mhp S/P values for each pig. The average Mhp ELISA S/P was 0.756 at 3 WOA (range: 0.166 - 1.801). At 3 WOA, average S/P values were higher in pigs from parity 1 sows vs parity 2 - 3 sows and parity 3+ (average: 1.001, 0.68 and 0.547, respectively). At 6 WOA, Mhp ELISA S/P values decreased in 29/30 pigs (96.6%) compared to 3 WOA values (6 WOA average S/P: 0.405, range: 0.059 - 1.350). At 9 WOA, the average Mhp ELISA S/P value for all pigs at 9 WOA increased to 1.332 (range: 0.208 - 1.807), with 25/30 pigs (83.3%) having positive S/P values.

The Mhp ELISA S/P ratios in 5/30 pigs (16.7%) decreased from 3 WOA to 9 WOA. Of those five animals, the 9 WOA Mhp ELISA S/P ratios for 3 pigs remained in the positive range (S/P = 0.564, 0.752 and 0.987), 1 pig's S/P was in the suspect range (0.366) and 1 pig's s/p in the negative range (0.208). Figure 2 graphs the changes in S/P for these pigs (1, 2, 12, 16 and 25) and demonstrates the typical pattern of MDA interference.

Figure 3 shows the change in LI ELISA percent inhibition (PI) for each pig. The 3 WOA mean LI ELISA PI was 17.195 (range: -15.667 to 61.271). The number of pigs in the positive, suspect, and negative range at 3 WOA was 7 (23%), 5 (17%) and 18 (60%), respectively. At 6 WOA, the average LI ELISA PI rose to 31.458 (range: 6.566 to 77.826). The percentage of PI directional change was nearly equal; 16/30 (53%) pigs had higher PI values and 14/30 (47%) pigs had lower PI values. At 9 WOA, the percentage of pigs with positive LI ELISA PI values increased to 27/30 (90%) with 2 pigs in the suspect range (PI values = 21.282 and 23.418) and 1 pig with a negative PI value of 17.583. The average PI for the 27 positive pigs was 67.150 (range: 31.388 - 93.592).

Table 1: Test result interpretation

Assay result interpretation	Lawsonia ELISA PI	Mhp ELISA S/P	PCV2b IFA titer
Positive	≥ 30	≥ 0.4	> 1:40
Suspect	20-30	0.3-0.4	1:40
Negative	< 20	< 0.3	< 1:40

Table 2: PCV2b IFA data summary

Pig age	3WOA	6WOA	9WOA
Number tested	30	30	30
Count ≤ 1:40	13	27	1
Percent ≤ 1:40	43.3%	90%	3.3%
Count 1:80 - 1:320	15	3	6
Percent 1:80 - 1:320	50%	10%	20%
Count ≥ 1:640	2	0	23
Percent ≥ 1:640	7%	0.0%	76.7%
Lowest IFA titer	< 1:40	< 1:40	< 1:40
Highest IFA titer	1:640	1:320	1:10240
IFA Geometric mean, all pigs	1:88	1:47	1:1437
IFA Geometric mean, P1 group	1:70	1:65	1:2941
IFA Geometric mean, P2 - 3 group	1:70	1:40	1:557
IFA Geometric mean, P3+ group	1:139	1:40	1:1810

Figure 1: PCV2b IFA titers.

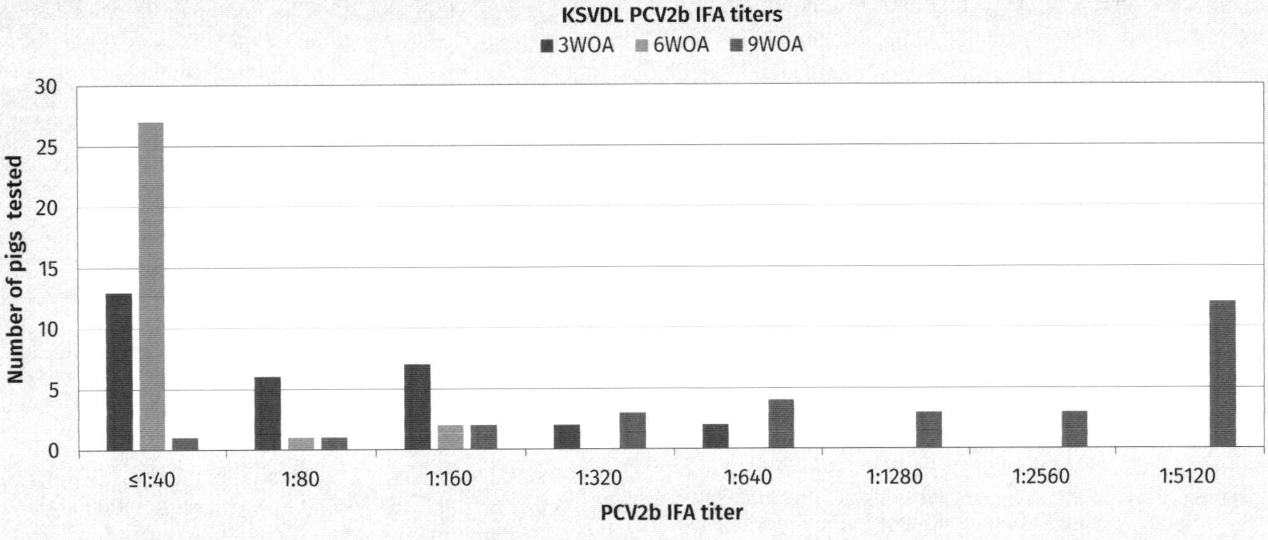

Figure 2: IDEXX Mhp ELISA S/P results.

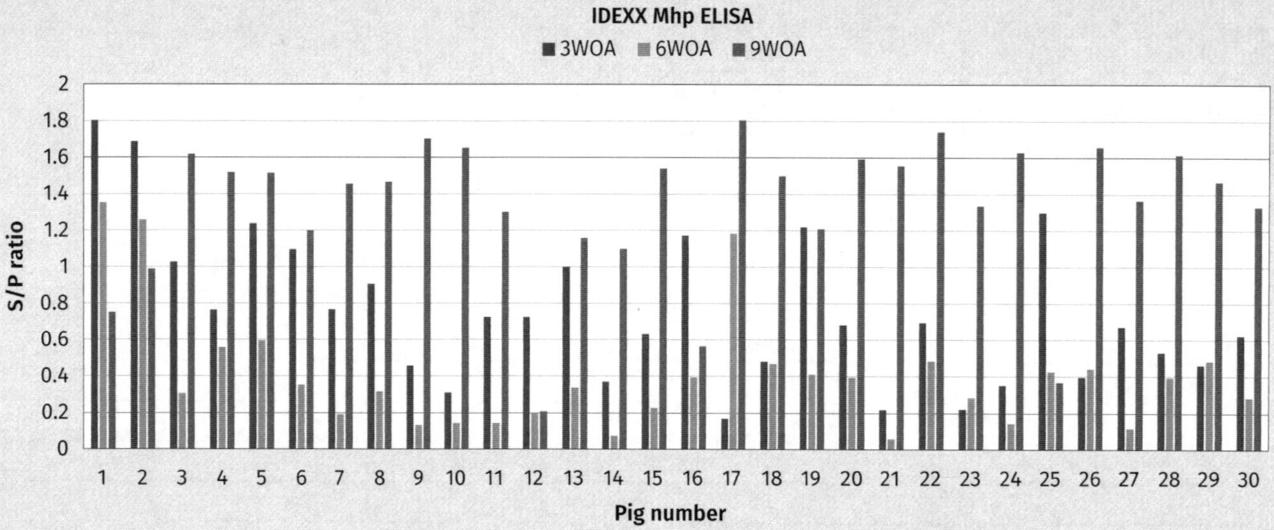

Figure 3: SVANOVIR Lawsonia ELISA percent inihibition results.

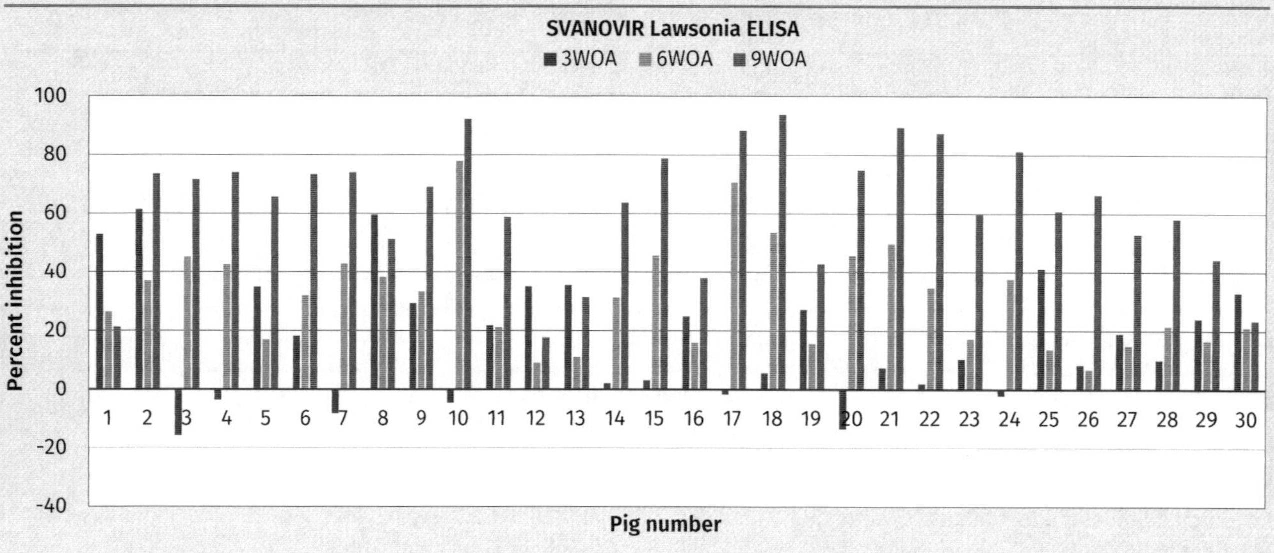

Discussion

Vaccination of pigs post-weaning is a critical component of the overall health program that helps ensure optimal production performance through the growing period. Inadequate response to vaccination may result in higher-than-expected disease burden which often leads to higher morbidity and mortality from otherwise preventable diseases during the post-wean growth phase. The goal of compliance monitoring is to objectively verify that a group of pigs was properly vaccinated. When groups of pigs receive vaccines that elicit an expected serologic profile, the resulting antibody response can be compared to the predicted acceptable population post-vaccination response. The acceptable vaccine response rates can often be provided with the help of the vaccine manufacturer. This type of monitoring can help identify when something is awry with your vaccination program.

Compliance testing can be used to assess the seroconversion of a group of pigs and/or to evaluate workers that administer vaccines (the vaccine crew). There are multiple reasons for poor seroconversion in a group of pigs, including post-vaccination illness, high MDA levels at the time of vaccination and improper vaccine administration. Verifying post-vaccination antibody levels will give assurance that the vaccine was administered at the correct time and using proper methods.

Compliance blood testing protocols are usually implemented at 3 - 5 weeks after the final vaccination to allow enough time for post-vaccination seroconversion. This timing can be dependent on the type of serological assay being used and may differ per antigen being detected. Expected level of seroconversion will vary for reasons that include: Vaccine type and dosing/timing protocol, number of days between vaccination and testing, health of pigs when vaccinated, level of maternal antibody at time of vaccination, and proper vaccine administration.

Since all samples tested negative for PCV2 PCR, the PCV2 IFA titer rise was attributed to vaccine response, not PCV2 post-infection immune response. The overall ending seroconversion data for PCV2 IFA titer, Mhp ELISA S/P and LI ELISA PI showed robust increases in over 83.3%, 83.3% and 90% of pigs, respectively, therefore the vaccination of this group of pigs was deemed successful.

We as a company have determined that if the percentage of pigs in the tested group have a negative Mhp ELISA S/P > 30%, it warrants a re-vaccination and further investigation into the cause of the poor serum antibody levels. The re-vaccination is done with 1 mL dose of Circumvent® CML as quickly as able on the farm. The vaccination handling and administration processes are also audited at the sow farm and nursery levels as an issue at either location could have resulted in the poor serum antibody level. Subsequent groups of pigs within that same flow are then more intensely monitored via serum collection and testing.

Circumvent CML is a novel 3-way vaccine that combines the antigen-adjuvant components of Porcilis ileitis and Circumvent PCV M G2 vaccines. The immune response to each of the monovalent vaccines has been established in field surveys[2,3] and controlled laboratory studies.[4] These data demonstrate the ending antibody responses to CML are similar to predecessor vaccines and demonstrate the post-vaccination antibody response to CML can be used as part of a comprehensive compliance monitoring program.

References

1. Neumann E, Ramirez A, Schwartz K, ed. Swine Disease Manual 5th Ed. 2020

2. Girard T, Thomas P. Evaluation of diagnostic tests for Circumvent PCV-M compliance monitoring. AASV. 2016; 250

3. Pittman J, Thacker B, Johnson J. Vaccination implementation and compliance in a large production system. Proc 21st IPVS. 2010: 217

4. Thacker B, Armbruster G, O'Brien B. Efficacy of two dose Porcilis™ Ileitis regimen against experimental *Lawsonia intracellularis* challenge. AASV. 2021; 138

https://doi.org/10.54846/am2024/47

From farm to bottle – checkpoints in the selection and production of protein-based prescription vaccines to meet the needs of evolving disease pressures

Alan J. Young, PhD
Medgene Labs, Brookings, South Dakota

Introduction

Protein-based prescription-platform vaccines represent a new approach to targeting disease control needs in the swine industry. In concept, the approach takes the sequences of field isolates to generate a library of ready-made vaccine targets that can be formulated into veterinarian-prescribed formulations to meet individual, regional, or national needs. In practice, the technology to produce vaccine targets relevant to current disease challenges that successfully generate the desired immune response involves advanced sequence monitoring, artificial-intelligence-assisted predictive algorithms for vaccine design, and both computer-assisted and bench assay methods to ensure the effective delivery of appropriate structure and amount for immunogenic proteins. We will report on the proprietary methods used by Medgene to monitor field variation in current disease pressures, in-house methods to ensure a diverse and complete library of proteins necessary to meet the needs of swine veterinarians, as well as proprietary tools we have developed to ensure complete control over the process used to generate final vaccines.

What are platform-based vaccines?

The goal of any vaccine is to convince the immune system that it has already seen an infectious agent, and therefore will have a robust and protective immune response upon later encounters with that agent. The key to vaccination is to induce the correct protective antibodies and T cells to identify and eliminate the virulent infectious agent without the associated damage induced by actual infection. Traditional commercial and autogenous vaccines have used either weakened versions of the live agent itself, or alternatively killed whole viruses or bacteria to attempt to generate immune response against the target agent (Figure 1). This approach relies upon the assumption that, while the vaccine contains all the proteins found in the native organism, the immune system will provide a strong neutralizing response against at least the ones necessary for protection. In practice, that requires either an extended (> 5 year) development and licensing period to make fully licensed commercial products that can be "on the shelf" for producers, or a relatively shorter but undefined development period to isolate, grow, and produce autogenous vaccines that can only be used in a restricted way. Over the past several decades our knowledge has significantly advanced of both how viruses and bacteria cause disease, as well as how the immune system specifically protects against those diseases. We can now use technology to specifically produce or isolate the target proteins that are critical to generate protection, and in fact the specific regions of those proteins that must be in the vaccine to protect against disease (Figure 2). Using current USDA guidelines, these vaccines can either be produced with identity to farm-specific isolates similar to autogenous vaccines, or alternatively a library of vaccine antigens can be pre-developed for off-the-shelf usage by producers.[1] These approaches are particularly suited to platform-based approaches, where a conserved production system is used to generate target proteins at will, by keeping all components conserved

Figure 1: Outline of the method of production of traditional commercial or autogenous vaccines. The infectious agent is isolated, grown, and either weakened or killed agent is bottled for use as vaccine.

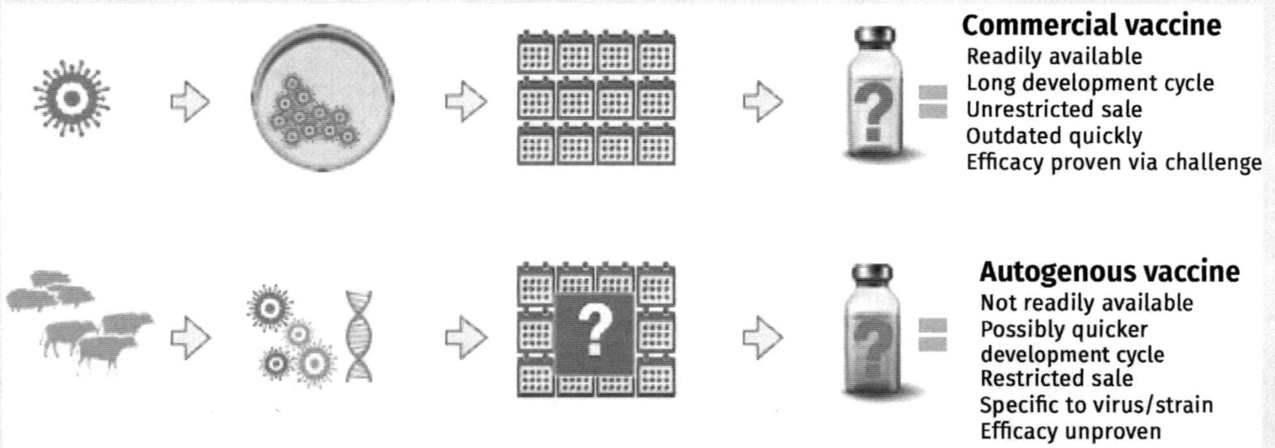

Figure 2: Outline of the method of production of platform-based vaccines, including platform-prescription. The nucleic acid code for the known target protein from the infectious agent is electronically transmitted and used as a "plug-and-play" cassette to generate target proteins to stimulate a neutralizing immune response.

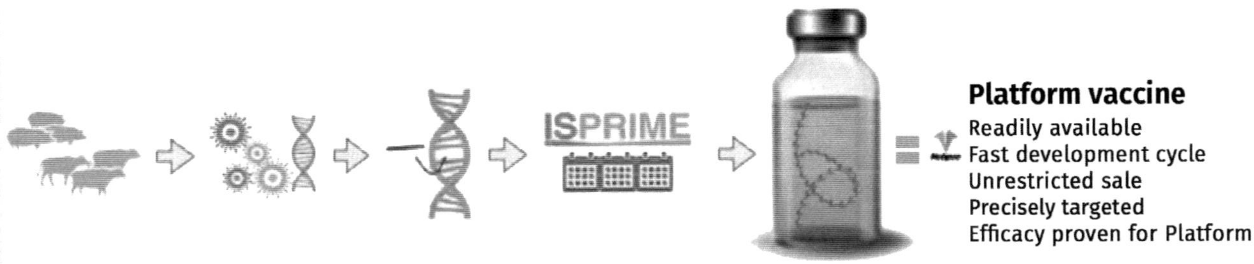

while specifically changing only a specific target of interest to generate new vaccines. For example, one yearly influenza vaccine in use for human medicine uses a baculovirus-based platform system to generate new target hemagglutinin proteins relevant for that year's dominant strains of influenza.[2] Platform vaccines, therefore, share the advantages of traditional commercial vaccines allowing "off-the-shelf" solutions for many agents, while also facilitating the "custom design" characteristics of autogenous vaccines. Current USDA regulations set out the guidelines for this approach in their description of "Prescription-Platform Vaccines".[3,4]

DNA, RNA, or protein?

There are multiple ways to produce a platform-based vaccine. The key factor is to provoke an immune response against the target protein (the immune system mostly responds to proteins in bacteria and viruses for protection). The key event for any of these methods is to get the immune system exposed to the target protein in a form that promotes the development of strong and specific antibody and cellular responses. The difference between platforms is how that protein is generated and provided to the immune system. As a short refresher, living cells produce protein in a conserved way - DNA found in the genes provides a library of "codes" that provide a template for the production of every protein in the cell. The DNA itself is converted into RNA as a messenger that takes the code from the genes to the protein production apparatus and ultimately assembles individual amino acids into the final protein itself. During virus infection, the virus itself delivers either DNA or RNA to the host cell that ultimately produces the necessary proteins to recreate the virus that eventually destroy the cell, liberating the new virus that can then infect other cells. It is these proteins, found on the "outside" of viruses, that form the targets necessary to provoking a strong and protective immune response. DNA and RNA vaccines create the nucleic acids that code for the target protein (for example, hemagglutinin from influenza) which is then inserted into host cells that, in turn, produce the target protein for recognition by the immune system.[5] In contrast, protein-based vaccines, such as those produced in baculovirus systems such as those used by Medgene, use a cell-culture system to produce and quantify the target protein which is then inactivated to kill any live components of the production system prior to formulating the final vaccine.[6] This allows Medgene to directly control and verify all aspects of the target production process, including not only the required dose but also the quality of the final protein delivered to the immune system. The remainder of this paper will address the detailed processes used by Medgene to ensure control over the consistency, purity, and quality of the final vaccine the promotes protective immune responses.

What are protective antibodies?

Ultimately, the goal of any vaccine is to protect against the negative effects associated with infection. As already mentioned, all viruses and bacteria contain many genes and proteins, only a few of which can be used as targets to limit infection and disease. This is similar to cutting a new key at the hardware store as a substitute for the original. In this case, the key is the pathogen protein regions that stimulate protective responses, and the lock is the neutralizing antibodies. Our goal is to produce a duplicate of that key in our vaccines. To this end, the specific shape of the key blank is irrelevant - it can be the same as the original, but it doesn't have to look the same. The critical aspect is the shape of the notches and ridges - that must match, or it will not properly engage the lock (or induce the correct antibody) (Figure 3). Similarly, if we target proteins on the surface of a virus that are uninvolved in infection, or even target regions of the important proteins that are outside the functional region, we aren't providing the appropriate template for the immune system to provide a protective immune response. Clearly, there are certain characteristics of target proteins that make them appropriate targets - they must be on the outside of the infectious organism, exposed to the immune system and likely to be involved in establishing interactions with the host. These protein structures tend to be conserved among virus families, allowing extension of data from human to animal vaccines and vice-versa. Furthermore, we have many years of research data on viruses and bacteria that provides direct information on the appropriate targets for providing protective vaccines. What Medgene's ISPrime system implements is integrating that knowledge into generation of protective vaccines.

What is ISPrime?

The platform-prescription guidelines were developed to meet the needs of animal agriculture through the generation of responsive vaccines to emerging needs. While the technology to produce platform vaccines has been available for many years, the new guidelines permit rapid implementation of new vaccine formulations to address the ebb and flow of disease in field production. At a practical level, Medgene recognized early that through appropriate science and surveillance, we could generate not only a rapidly responsive process, but actually proactively generate

Figure 3: While traditional protein homology plots are obtained by comparing the full sequence for an organism or protein to the desired target, what is critical is matching only the region necessary to stimulate protective antibodies. The "framework" region, like the handle of a key, is not important, as long as the "functional" region, like the blade, matches the neutralizing antibody.

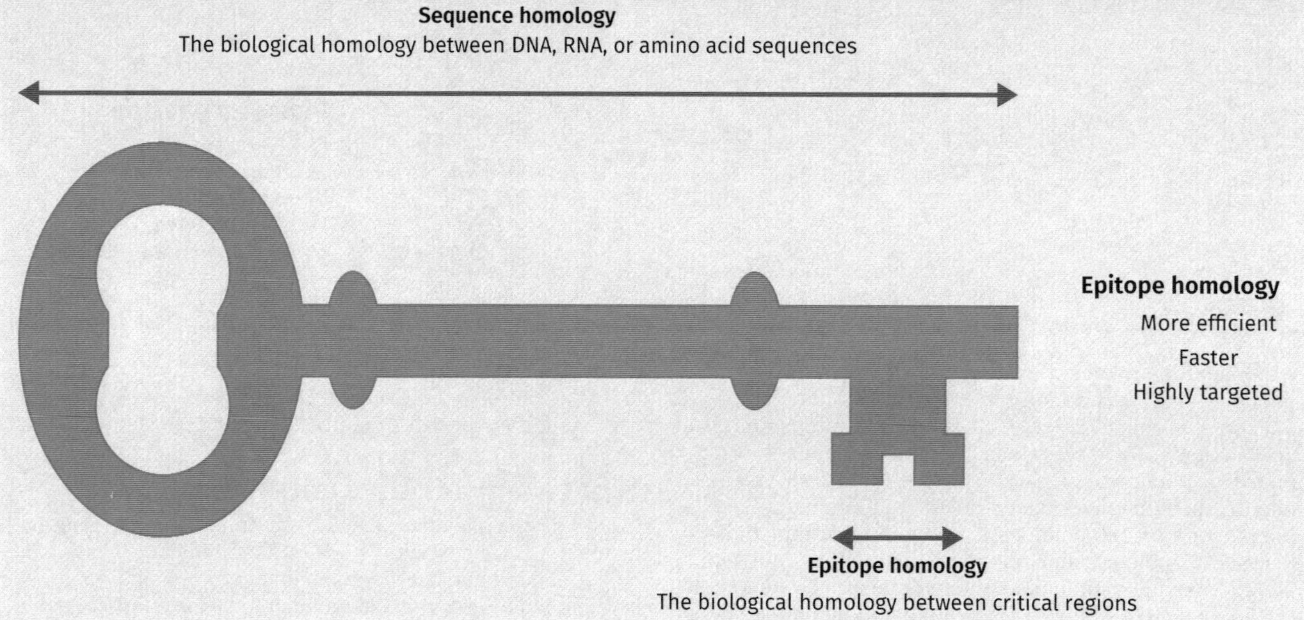

a library of banked vaccine targets that could provide national, as well as regional solutions for the animal health industry. To this end, Medgene developed the ISPrime system, which encompasses a complete process to identify the need and targets for newly emerging diseases, produce Master Seeds and antigen for those disease targets, and formulate solutions for the animal health industry as final vaccine formulations suited to the needs of individual veterinarians (Figure 4). Once fielded, we must also continue to work with veterinarians to monitor needs for adjustment to the vaccine formulations and continually update our system accordingly. To be effective, our system involves continual surveillance and updating of our antigen library. A key aspect of the system is our commitment to ensure that the library provides complete coverage of industry needs in a rapidly responsive system, rather than adopting an autogenous approach of reactive (and slower) production of antigen sequences based upon individual outbreaks.

Where do we get our field data?

The starting point for a robust and inclusive vaccine antigen library is strong field surveillance of prevalence and field diversity. While a primary source of this information comes to us from relationships with our producers, we also use publicly available databases to examine emerging diseases and monitor diversity of our viral targets. Under existing legislation, the USDA currently provides funding to monitor field sequences for influenza. In addition, a number of other databases exist which provide updates to field sequences of other pathogens both within the United States and worldwide. These sequences are entered into our databases to monitor for the emergence of new agents, sequences, and strains. While all this information is valuable, a primary question lies in what sequences represent a need for new vaccines, or which new sequences can efficiently be covered by our pre-existing antigen library.

What is epitope homology?

As mentioned previously, the key to a successful vaccine is targeting not only the correct proteins within the infectious agent to provide protection, but actually targeting the specific regions within those proteins. The region we need to duplicate within our antigen is the specific site of the viral or bacterial protein that interacts with the host, that when blocked eliminates or adversely affects the ability of the agent to cause infection. Returning to our key analogy above, everybody knows that if you need a new key cut, the part that matters is the blade, NOT the handle. The key blank that is selected at the store does not particularly matter, only the notches and ridges on the blade. At any hardware store, this analysis is done by computer, where the computer uses methods programmed into it to analyze the key and indicate an appropriate blank that can then be turned into a duplicate key. Medgene has used research developed by researchers over the past four decades to perform a similar analysis on target proteins, identifying critical regions for the immune response within those proteins, and then focusing on those regions to ensure precision homology to our vaccine targets.[7-11] Those changes that affect the specific epitope regions are considered for inclusion in our antigen library, while pre-existing constructs can be used to immediately address industry needs in a matter of days. Represented by a "probability profile" across the sequence of the protein, this provides a visual readout of the similarity or differences between the critical regions of native viral proteins and our construct library (Figure 5). This process allows us to create a comprehensive selection of target antigens that provide broad coverage across known agents circulating in the field.

Figure 4: The ISPrime system is used to describe all the science-based approaches that go into generation of Medgene's platform-based prescription vaccines. These involve inputs from both field diagnostics as well as public surveillance databases, computer-assisted scientific methods to analyze those sequences and predict appropriate vaccine antigens, and finally additional systems to optimize and control the quantity and quality of target antigen produced in the vaccine. Finally, feedback from the field continues to optimize and refine our library of vaccine antigens to have as much "on the shelf" as possible to meet the needs of producers.

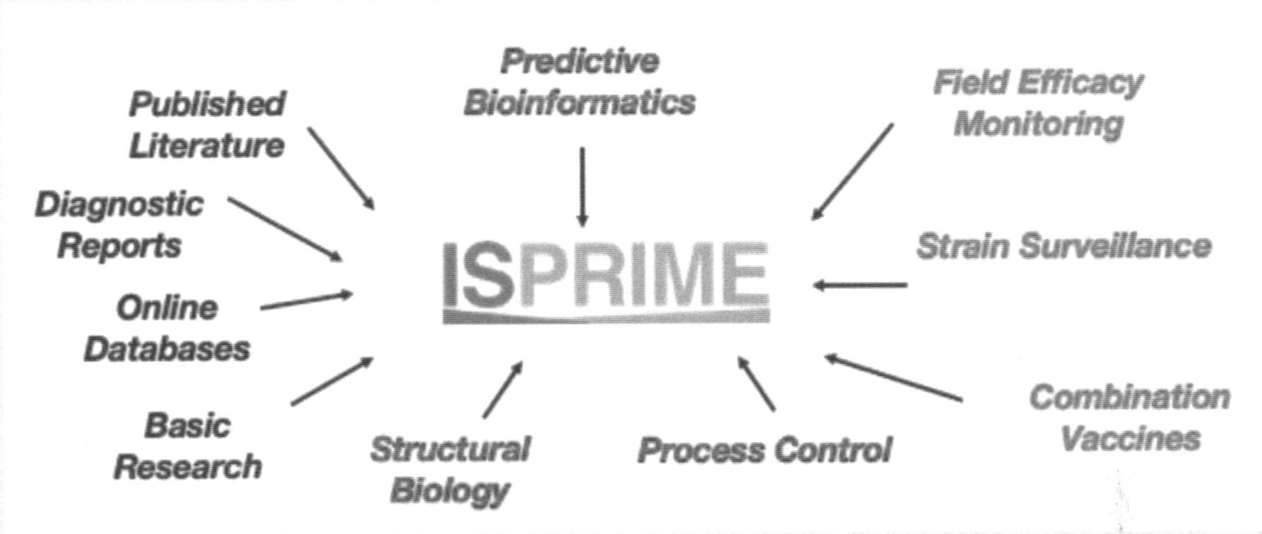

Figure 5: As an example, our epitope homology plots illustrate similarities and differences between constructs. Briefly, the red line represents a "probability cutoff" that any given region of the protein is an exposed, target epitope, with those regions of the protein above the line being likely epitopes. The pattern for each protein will be representative of the "immunological reactivity" of that protein. The provided plot shows a comparison between two amino acid sequences illustrating two circled regions where there are predicted differences within two regions of the proteins. These two proteins would NOT be similar regardless of overall sequence homology (which in this case is greater than 95%) and a new vaccine antigen would need to be produced.

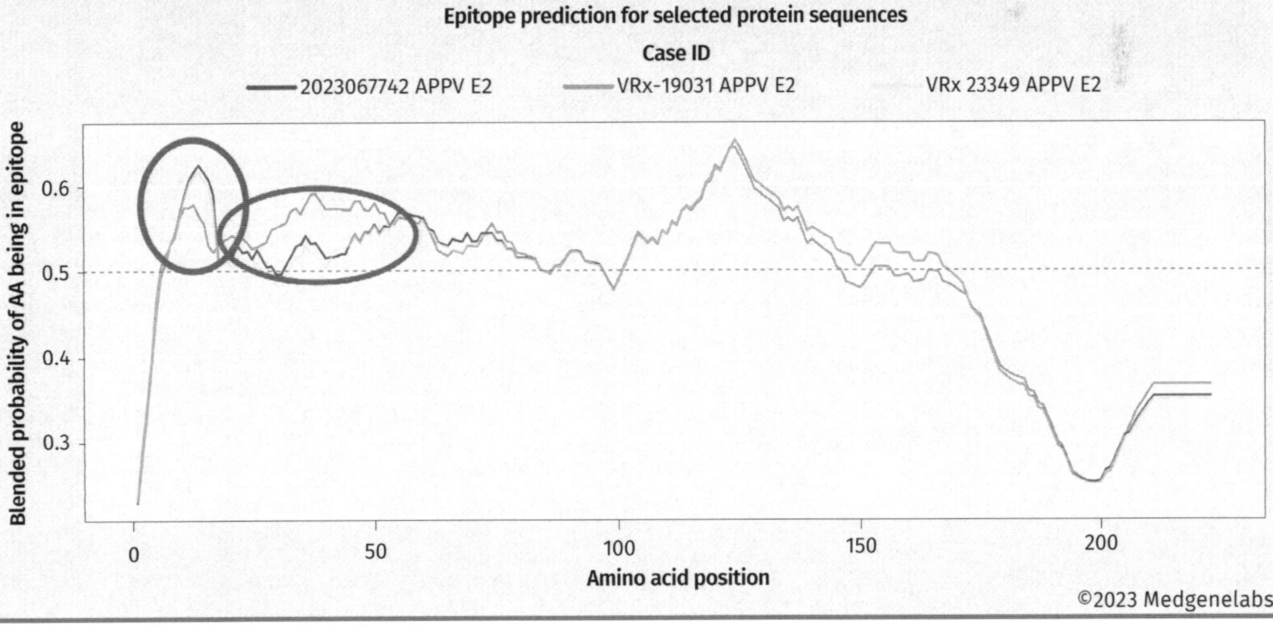

How do we control our processes?

The key component to generating protective antibody responses to the sequences is in final control over the dose and structure of the target protein presented to the immune system. Our in vitro systems permit that control, and the use of basic science principles allows us to predict how proteins will fold and express those key regions, or epitopes, that generate protection. Upon receipt of a new sequence from the field, we do a computer-assisted comparison of the amino acid sequence to our existing library to determine if there is need for a new antigen, or if existing stock is appropriate. Our ISPrime staff make that determination, and communicate with the client. If an existing antigen is sufficient, that antigen is formulated into a new vaccine (unless already available as finished product) and available within a matter of days or weeks. If a new vaccine formulation is needed, we generate a new antigen construct in a proprietary process that takes up to several months to release. For all our finished antigens, we analyze the quality of the antigen, and using a proprietary process formulate precise protein levels for each antigen into the final vaccine.

Where do we go next?

The processes developed and in place at Medgene are designed to provide uniform vaccines with science-based methods to generate target antigens that will perform in the field. One advantage of prescription-platform vaccine technology is our ability to continually refine and update our construct library and vaccine formulations based on industry needs in the field. To accomplish this, the processes outlined above do not stop at vaccine shipment, but continue to provide feedback to refine and update vaccines to meet current needs. Together, the ISPrime system can provide either off-the-shelf vaccines from our construct library, or personalized vaccine formulations as required by industry.

References

1. Erdman, M. M., Clough, N. E. and Hauer, P. J., Review of updated regulations and product license categories for veterinary vaccines in the United States. *J Am Vet Med Assoc* 2020. 257: 1142-1147.

2. Cox, M. M., Izikson, R., Post, P. and Dunkle, L., Safety, efficacy, and immunogenicity of Flublok in the prevention of seasonal influenza in adults. *Ther Adv Vaccines* 2015. 3: 97-108.

3. APHIS, U.-. Veterinary Services Memorandum No. 800.213. In Biologics, U. C. f. V. (Ed.). *USDA Animal and Plant Health Inspection Service,* Washington, DC 2015.

4. USDA-APHIS, Veterinary Services Memorandum No. 800.214. *In Service,* U. A. a. P. I. (Ed.). USDA-APHIS, Washington, DC 2018.

5. Poria, R., Kala, D., Nagraik, R., Dhir, Y., Dhir, S., Singh, B., Kaushik, N. K., Noorani, M. S., Kaushal, A. and Gupta, S., Vaccine development: Current trends and technologies. *Life Sci* 2024. 336: 122331.

6. Hong, Q., Liu, J., Wei, Y. and Wei, X., Application of Baculovirus Expression Vector System (BEVS) in Vaccine Development. *Vaccines (Basel)* 2023. 11.

7. He, L. and Zhu, J., Computational tools for epitope vaccine design and evaluation. *Curr Opin Virol* 2015. 11: 103-112.

8. Guarra, F. and Colombo, G., Computational Methods in Immunology and Vaccinology: Design and Development of Antibodies and Immunogens. *J Chem Theory Comput* 2023. 19: 5315-5333.

9. Zeng, X., Bai, G., Sun, C. and Ma, B., Recent Progress in Antibody Epitope Prediction. *Antibodies (Basel)* 2023. 12.

10. Desta, I. T., Kotelnikov, S., Jones, G., Ghani, U., Abyzov, M., Kholodov, Y., Standley, D. M., Beglov, D., Vajda, S. and Kozakov, D., The ClusPro AbEMap web server for the prediction of antibody epitopes. *Nat Protoc 2023.* 18: 1814-1840.

11. Desta, I. T., Kotelnikov, S., Jones, G., Ghani, U., Abyzov, M., Kholodov, Y., Standley, D. M., Sabitova, M., Beglov, D., Vajda, S. and Kozakov, D., Mapping of antibody epitopes based on docking and homology modeling. *Proteins 2023.* 91: 171-182.

https://doi.org/10.54846/am2024/48

The effects of fungal and bacterial infections and their toxin metabolites on pig health and performance and the impacts of a *Bacillus*-based mitigant in the presence of complex disease syndromes

A. D. Woodward[1], PhD; J. D. Spencer[1], PhD; E. Galbraith[2], MS; B. Arnold[1], BS
[1]United Animal Health Inc., Sheridan, Indiana; [2]Microbial Discovery Group, Oak Creek, Wisconsin

Introduction

Pig exposure to a combination of insults, such as viral, fungal and bacterial pathogens and their toxin metabolites, as well as environmental stressors, can lead to heightened inflammation, immunosuppression, and compromised gut integrity.[1-3] Synergies from multiple insults can contribute to elevated health and performance challenges and result in significant economic losses, amounting to over $200 billion in direct and indirect costs over the last decade.[4,5] Managing threats of complex disease syndromes requires a comprehensive approach including focused management strategies and effective tools that reduce growth losses, protect the gut barrier function, and promote healthy gut microbiota.

While many products are available to handle a single challenge, few have the capability to assist over a range of challenge scenarios that affect modern pig production. M-Mobilize® (United Animal Health, Sheridan, Indiana), a patented, specialized hydrated sodium-calcium aluminosilicate with novel *Bacillus subtilis* strains, can serve as a tool to ameliorate the effects of many of these co-challenges, including mycotoxins, fungal pathogens, bacterial pathogens, and common stressors in the production environment, thus reducing the severity of complex disease syndromes caused by viruses and opportunistic pathogens. The combination of specialized ingredients has been shown to improve performance, reduce pathogenic bacteria, and enhance immune function in nursery pigs and sows. The following four scenarios evaluate the effects of different insults to performance and health and demonstrate the effectiveness of M-Mobilize® to ameliorate losses during challenges.

M-Mobilize® use during mycotoxin challenges

Methods

A total of twenty controlled, in vivo research trials have been performed utilizing a low mycotoxin control, a high deoxynivalenol (DON, vomitoxin; average 3 ppm added DON) control, and a high DON + 1.5 lbs./ton M-Mobilize® treatment in nursery pigs fed treatment diets for 21 d. In each trial, animals were placed in pens, divided into replicates, and randomly assigned a treatment. Treatment diets were formulated to meet or exceed common industry recommendations, and animals were fed their treatment diets for the duration of each experiment. Pen weights and feed intakes were recorded weekly and utilized to calculate average daily gain (ADG) and average daily feed intake (ADFI); these metrics were used to calculate feed efficiency. Individual trial data were then analyzed using the Mixed Procedure of SAS with the main effect of treatment and the random effect of replicate. Finally, the complete 20-trial data set was subjected to a meta-analysis (n = 6,465, $P < .10$) using the Mixed Procedure of SAS to determine overall efficacy of M-Mobilize® to recover growth loss due to a DON challenge.

Results and discussion

Across all trials, the effects of mycotoxins on performance were apparent, as 14 of the 20 trials showed statistically significant response ($P < .10$). Overall, nursery pigs fed high DON diets decreased ($P < .001$) ADG and ADFI by 12.7% and 11.9%, respectively, compared to pigs fed a diet with low mycotoxins. Inclusion of M-Mobilize® in high DON diets resulted in recovery of 53.0% of lost gain (Figure 1) and 20.7% of lost feed intake, leading to a 56.5% recovery of feed efficiency, compared to pigs fed high DON diets with no mitigation effort.

These results demonstrate the effectiveness of M-Mobilize® to consistently restore lost performance due to high DON concentrations in the diet.

M-Mobilize® use during fungal challenges

Methods

Sows utilized in an industry scale, controlled research trial had feed that went out of condition quickly and generated mold. Feed was fed for approximately 7 days during the breeding period, with one-half of Sow Group 1 (n = 111, SG1) receiving non-moldy feed and one-half of Sow Group 2 (n = 110, SG2) receiving visibly moldy feed. Measurements of reproductive performance were recorded to determine the effect of fungus on sows.

In a concurrent trial, M-Mobilize® (1.5 lbs./ton) was provided as an additional treatment in SG2 (n = 110). Similar measurements of reproductive performance were recorded to determine the efficacy of M-Mobilize® to improve performance in the face of a fungal challenge.

All data were checked for normality, and any data point ± 3 standard deviations from the mean was removed from the data set. Data were then analyzed using the Mixed Procedure of SAS with the main effect of treatment and the random effect of replicate. Statistical significance was set at $P < .05$, with trends towards significance showing at $P < .15$.

Figure 1: Recovery (% ADG) of lost growth with M-Mobilize® during controlled DON contamination for 21 days in nursery pigs. Individual trial (red bars) and meta-analysis average (blue line) growth recovery shown.

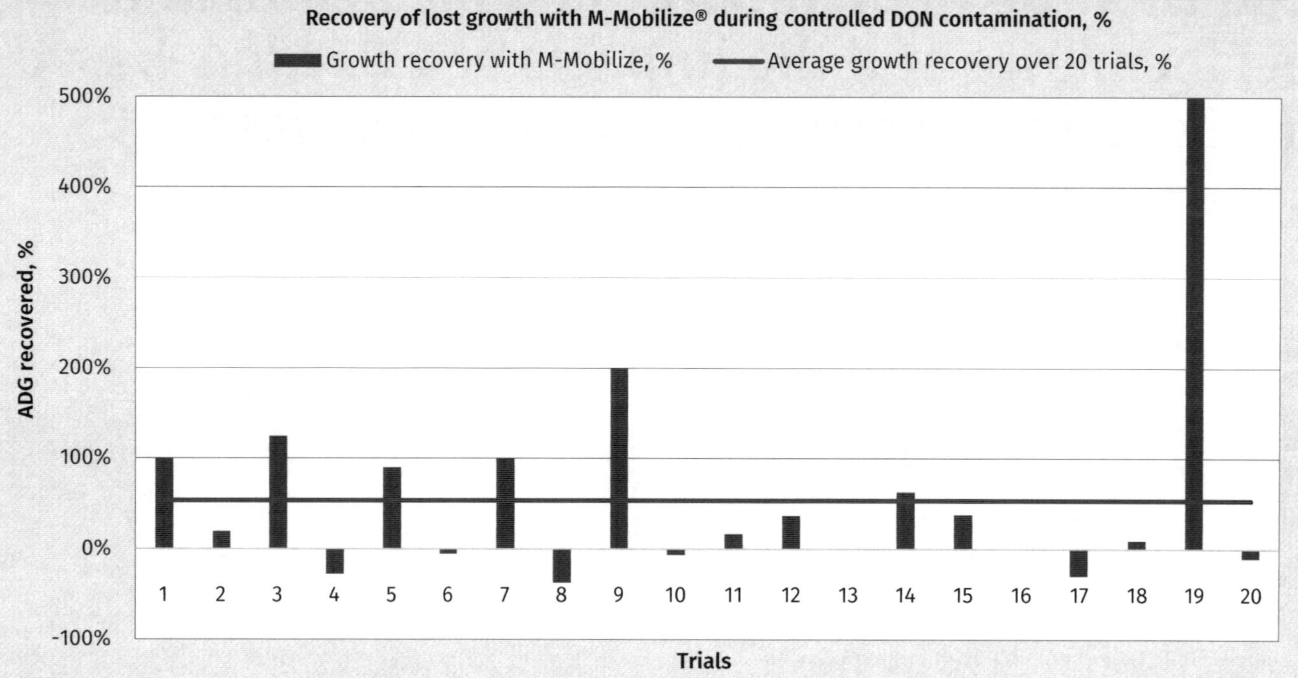

Results and discussion

Sows fed visibly moldy feed had a tendency for reduced ADFI ($P = .127$, Table 1), resulting in decreased pregnancy positive ($P = .040$) and weaned sow utilization ($P = .040$) compared to SG1. Moldy feed also resulted in a 5.5% increase in sows that were never in heat (data not analyzed) in SG2 vs SG1.

Sows in SG2 receiving M-Mobilize® in mold contaminated diets showed improvements in reproductive performance compared to SG2 with no mitigant (Table 2). Sows fed M-Mobilize® had a tendency toward increased number in heat ($P = .121$), with numerical increases in pregnancy positive and weaned sow utilization ($P > .200$) compared to SG2 with no mitigant.

Results show that M-Mobilize® benefits sow reproduction when moldy feed is present.

M-Mobilize® use during bacterial challenges

Methods

To determine the effects of M-Mobilize® during bacterial challenges, nursery pigs were fed standard commercial diets with or without M-Mobilize® (1.5 lbs./ton) immediately post-weaning in a controlled enterotoxigenic *Escherichia coli* (ETEC) challenge trial. Pigs were administered an oral gavage of F18+ ETEC (10^{10} cfu/pig) of swine origin on study days 18 and 19.

Rectal swabs (n = 16) were taken at both peak infection (3 days post-gavage) and recovery (19 days post-gavage). Genomic DNA was extracted and tested using a panel of qPCR assays specific for virulence and marker genes commonly associated with enteric disease in swine.

Gene quantities were analyzed utilizing the Mixed Procedure of SAS, and statistical ordination procedures were used to determine the strength of correlation between pathogen genes and treatment.

Results and discussion

During peak challenge, pigs receiving M-Mobilize® had reduced quantities of ETEC adhesion and toxin genes compared to untreated pigs. *E coli* heat-labile toxin (*LT*) and shiga toxin (*Stx2*) genes were decreased ($P < .05$), and *F18* and *F41* fimbrial adhesin genes tended to be decreased ($P < .10$) in pigs receiving M-Mobilize® compared to the untreated control.

During recovery, M-Mobilize® pigs continued to show fewer *E coli* detections and lower quantities of virulence genes compared to the untreated control. Pigs fed M-Mobilize® had decreased *eaeA* ($P = .02$), an *E coli* gene associated with intestinal attachment and effacement, and a tendency toward decreased *E coli* heat-stable toxin genes (*STb*, $P = .06$) compared to the untreated control.

In both time points, ordination graphs depicted strong correlation between enteric pathogen genes and the untreated control, compared to very little correlation with M-Mobilize®.

Results of this trial demonstrate that M-Mobilize® may reduce the quantity of adhesion and toxin genes associated with F18+ ETEC, thus benefiting the nursery pig during times of enteric pathogen challenge.

Table 1: Effect of visibly moldy feed on reproductive performance of sows in a commercially run sow facility

	Sow group 1 – no moldy feed	Sow group 2 – moldy feed	*P*-value
# of sows allotted	111	110	
Farrowed, n	107	102	.488
Farrowed, %	96.40	92.73	
Weaned, n	106	99	.280
Weaned, %	99.07	97.06	
Heat check, n	100	94	.846
Heat check, %	94.34	94.95	
Bred, n	100	94	
Wean-to-estrus, days	4.16	4.25	.215
OvuGel doses, n	1.05	1.09	.378
Preg positive, n	94	80	.040
Preg positive, %	94.00	85.11	
Weaned sow utilization, %	88.68	80.81	.040

Table 2: M-Mobilize® effect on reproductive performance in sows consuming visibly moldy feed in a commercially run sow facility

	No mitigant	M-Mobilize®	*P*-value
# of sows allotted	110	110	
Farrowed, n	102	107	.306
Farrowed, %	92.73	97.27	
Weaned, n	99	107	.973
Weaned, %	97.06	100.00	
Heat check, n	94	106	.121
Heat check, %	94.95	99.07	
Bred, n	94	106	
Wean-to-estrus, days	4.25	4.30	.574
OvuGel doses, n	1.09	1.07	.719
Preg Positive, n	80	92	.766
Preg positive, %	85.11	87.07	
Weaned sow utilization, %	80.81	85.98	.320

M-Mobilize® use during induced environmental stress

Methods

A simulated environmental stress trial was utilized to determine the effects of a feed and water deprivation, which reduces gut integrity and replicates malabsorptive scenarios experienced in production, and M-Mobilize® use on performance. Nursery pigs (n = 168) were weaned at 21 days of age and allotted to pens based on initial body weight in a completely randomized block design of 14 replicates of 4 pigs/pen. Pens were assigned one of three treatments: Control, no removal of feed or water (CTRL), feed and water stress challenge (FWS), and FWS + M-Mobilize® at 1.5 lbs./ton (FWS+M). Feed and water stress challenge consisted of removal of feed and water for 24-hr at weaning, followed by water removal again on day 7 post-weaning.

Pen body weights and feed intakes were recorded at weaning, d 6, and d 13 to calculate body weight, average daily gain, average daily feed intake, and feed conversion ratio. On d 8, blood was collected from 2 pigs/pen, pooled, and serum separated to analyze the effects of stress on serum acute phase proteins.

All data were checked for normality, and any data point ± 3 standard deviations from the mean was removed from the data set.

Data were then analyzed using the Mixed Procedure of SAS with the main effect of treatment and random effect of replicate. Statistical significance was set at $P < .05$.

Results and discussion

There was no difference in ADG, ADFI, or feed efficiency between CTRL and FWS pigs from d 0 - 6 and 0 - 13 (Table 3). However, from d 0 – 6, FWS+M pigs had increased ADG ($P = .030$) by 57.1% and 37.5% compared to CTRL and FWS, respectively, with no differences in ADFI or F/G. After both challenge periods, FWS+M pigs had increased ADG ($P = .028$) by 24.0% compared to both CTRL and FWS; F/G was improved ($P = .025$) 15.0% and 12.8% with FWS+M compared to CTRL and FWS, respectively.

Although not different, feed and water deprivation numerically reduced haptoglobin, C-reactive protein, and major acute phase protein in serum in FWS pigs compared to CTRL. Pigs on FWS+M treatment had numerically increased serum haptoglobin and major acute phase protein concentrations compared to both CTRL and FWS, while C-reactive protein was intermediate between CTRL and FWS.

Results demonstrate that feeding M-Mobilize® to nursery pigs can prevent performance lag and immunocompromise due to environmental challenges related to weaning stress.

Conclusion

Stressors are additive in their negative impact on pig performance and health. Results from multiple in-vivo studies show M-Mobilize®, a *Bacillus*-based feed additive, delivers benefits in the presence of multiple insults, including mycotoxins, fungal and bacterial pathogens, and environmental stressors. The negative effects on performance due to mycotoxicosis or fungal contamination can be ameliorated with M-Mobilize® inclusion during the challenge period. Additionally, M-Mobilize® can reduce the presence of toxin and adherence genes during an ETEC challenge. Finally, when M-Mobilize® is presented during weaning stress, both performance and immune function are improved. Taken together, these data suggest that the negative impacts of stacked stressors on pig health and performance may be mitigated by use of M-Mobilize® in diets.

Table 3: Growth performance and serum protein concentrations of nursery pigs subjected to a feed and water stress challenge and fed M-Mobilize®

	Control	Feed and water stress	FWS + M-Mobilize®	SEM	*P*-value
Start wt., lbs.	13.35	13.37	13.18	0.62	.449
Day 0 - 6					
End wt., lbs.	14.21	14.32	14.47	0.60	.562
ADG, lbs.	0.14^{b}	0.16^{b}	0.22^{a}	0.02	.030
ADFI, lbs.	0.28	0.29	0.30	0.02	.907
F/G, lb./lb.	2.08	1.87	1.42	0.64	.611
Day 0 - 13					
End wt., lbs.	16.56^{y}	16.68^{y}	17.23^{x}	0.69	.068
ADG, lbs.	0.35	0.34	0.39	0.02	.315
ADFI, lbs.	0.47	0.43	0.48	0.02	.102
F/G, lb./lb.	1.38	1.33	1.27	0.07	.684
Day 0 - 13					
ADG, lbs.	0.25^{b}	0.25^{b}	0.31^{a}	0.02	.028
ADFI, lbs.	0.37	0.36	0.39	0.02	.346
F/G, lb./lb.	1.53^{a}	1.49^{a}	1.30^{b}	0.06	.025
Serum parameters[1]					
Haptoglobin, mg/mL	462	443	478	16	.382
CRP, ng/mL	11250	7892	9634	1366	.320
MAPP, ng/mL	855	689	890	105	.340
SAA, ng/mL	114	89	93	34	.898

[ab or xy] Means within a row with uncommon superscripts differ, $P < .05$ or $P < .10$, respectively.

[1] CRP = C-reactive protein; MAPP = major acute phase protein; SAA = serum amyloid A.

References

1. Pierron A et al. Impact of mycotoxin on immune response and consequences for pig health. *Anim Nutr.* 2016;2:63.

2. Gao Y et al. The compromised intestinal barrier induced by mycotoxins. *Toxins.* 2020;12(10): 619.

3. Horn N et al. Impact of acute water and feed deprivation events on growth performance, intestinal characteristics, and serum stress markers in weaned pigs. *J Anim Sci.* 2014;92(10):4407.

4. Mahoney et al. Post-weaning feed and water outage event with concurrent mycotoxin exposure additively impacts gut morphology and performance of nursery pigs. *Proceedings of the 54th AASV Annual Meeting.* 2023;193.

5. Barratt AS et al. Framework for estimating indirect costs in animal health using time series analysis. *Front. Vet. Sci.* 2019;6:190.

https://doi.org/10.54846/am2024/49

The utilization of an integrated technology platform in a commercial research setting captures individual pig data for improved management strategies

G. Krahn, PhD; J. Spencer, PhD; C. Eden; M. Johnston; B. Altman
United Animal Health Inc., Sheridan, Indiana

Introduction

Historically, management decisions in swine production have often been based on metrics that are largely representative of averages within populations. Averages fail to reveal the large variation that commonly exist within large populations, as well as factors influencing this performance variation, resulting in greater opportunity costs across pork production systems. The use of modern technology that integrates radio-frequency identification (RFID) of individual animals, such as LeeO, as well as computerized feeding systems, provides a deeper understanding of factors influencing individual animals. This generates insights that can lead to the implementation of management strategies and commercial solutions for improved systemwide efficiency. The use of a modern, integrated technology platform in a large-scale commercial research setting has enabled greater understanding of factors influencing sow productivity, and the importance of lactation performance on systemwide throughput. Individual pig data has also allowed greater ability to develop management strategies for optimal performance in various subpopulations within commercial settings.

Materials and methods

Since 2018, United Animal Health has utilized the LeeO individual animal identification system, a precision livestock farming system that uses RFID technology, to track and record individual pig data, including litter information, weights, mortalities and other on farm activities. To meet the objectives of understanding **(1)** sow daily lactation feed intake (DLFI) curves, **(2)** the impact of lactation feed intake on progeny, and **(3)** nutritional needs of small and large growing pigs, individual pig data was collected from June 2018 - November 2023 at United Animal Health's Research Farms. All pigs received RFID tags, and LeeO was used to collect individual pig data. Data was then analyzed to examine potential factors impacting sow and progeny performance.

To understand and predict sow DLFI curves, farrowing rooms at the Donald E. Orr Swine Farm were equipped with the Big Dutchman feed system, which recorded individual daily lactation feed intake on 4,720 lactation events (total days = 81,390). Daily intake data was used to create DLFI curves and calculate sow lactation average daily feed intake (ADFI). Mixed models were used to predict each sow's DLFI based on day of lactation, parity and season. Variation from the predicted DLFI values were used to categorize low feed intake events and determine their impact on lactation performance.

Since LeeO provides the ability to trace individual pig data back to the sow and littermates, we evaluated how sow lactation feed intake impacts downstream progeny performance. Individual pig weights were recorded in LeeO at birth, weaning, end of nursery, and end of finishing. Data were utilized to determine the impact of lactation ADFI on wean weight, end nursery weight, and mortality using mixed models in R statistical software. Additionally, during a separate mixed model analysis, individual pigs were sorted by birth weight or wean weight to evaluate the effect of body weight on mortality and downstream end of finishing weight.

Finally, to determine the nutritional requirements of small versus large growing pigs, individual nursery exit weights were used to sort pigs (N = 2,394) into bottom and top 50% weight groups in a grow-finish trial. Four dietary treatments were fed: **1)** Standard Diet, **2)** Standard Diet with added fat **3)** Increased Dig Lys:ME and **4)** Decreased Dig Lys:ME. Bottom 50% pigs were fed treatments 1, 2 & 3 while top 50% pigs were fed treatments 1, 2 & 4. Data were analyzed using mixed models in R statistical software with the main effect of treatment.

For all analysis, statistical significance was set at $P < .05$, with trends towards significance showing at $P < .10$.

Results and discussion

Daily lactation feed intake and impact on progeny performance

Data analysis revealed that daily lactation feed intake is the lowest in the first few days right after farrowing, but gradually increases until peak DLFI intake occurs in mid to late lactation (Figure 1). Throughout the entire lactation period, parity 1 sows have the lowest intake while older parity sows (parity 3+) have the highest intake. Sows with 2 or more low feed intake events had the lowest lactation ADFI ($P < .05$), reduced sow lactation weight change ($P < .05$), and the litters they nursed were the lightest at weaning ($P < .05$) and had the highest pre-wean mortality ($P < .05$, Table 1). Due to variation in genetics, management, nutrition, health, etc, the average values in Figure 1 will not match every farm in the industry, but the same general intake pattern can be applied to every farm, especially as it relates to reducing low intake events.

Litters nursed by higher ADFI sows had reduced pre-wean mortality ($P < .05$) and percentage of light weight pigs ($P < .05$), thus creating more pigs weaned ($P < .05$) that were heavier weight at weaning ($P < .05$, Table 2). The effects of lactation ADFI continued post-weaning as increased sow lactation ADFI resulted in a higher percentage of pigs alive by the end of the nursery, and pigs were heavier in weight ($P < .05$). This collective advantage in number of pigs and weight resulted in litters with more total pounds at weaning and at the end of nursery when lactation ADFI was increased. Managing sows during lactation continues to be an essential component of a sow's productivity, while optimizing litter and downstream production. Therefore, an emphasis should be placed on management practices that help maintain feed intake during lactation.

Figure 1: Daily lactation feed intake by day of lactation and parity.

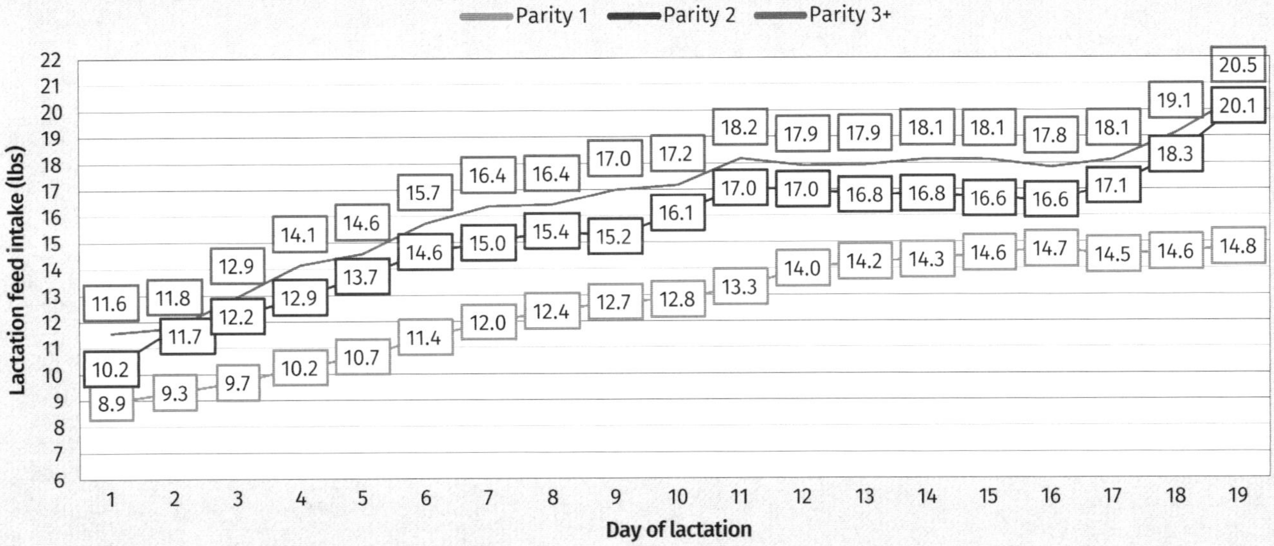

Table 1: Total number of low feed intake events and the impact on lactation performance

Number of deviations	Number of sows (% total)	Lactation ADFI (lbs.)	Sow lactation weight change (%)[1]	Average wean weight (lbs.)[2]	Pre-wean mortality[3]
0	2,308 (51%)	15.64[a]	1.4%[a]	13.94[a]	7.9%[a]
1	1,179 (26%)	14.64[b]	-0.7%[b]	13.80[b]	8.0%[a]
2+	1,013 (23%)	13.27[c]	-3.4%[c]	13.45[c]	8.7%[b]
All	4,500	14.84	-0.2%	13.79	8.1%

[1] Lactation weight change = sow weight at weaning - post-farrow weight.

[2] Adjusted for litter size, sow parity, sex and wean age (21-days).

[3] Mortality from cross-fostering (24h after birth) to weaning

[abc] Columns with different superscript are statistically different ($P < .05$)

ADFI = average daily feed intake

Impact of body weight on subsequent performance and nutritional requirements of small versus large pigs

Wean weight is strongly associated with downstream growth performance and livability (Table 3). Lower weight pigs at weaning remained lighter at the end of the end nursery phase ($P < .01$) and at the end of finishing ($P < .01$) and had higher nursery mortality ($P < .01$), whereas these metrics were improved with increased wean weight.

To reduce group variation, improve the performance of bottom end light weight pigs and reduced feed cost in the top end heavy weight pigs, different feeding techniques can be used. In the bottom 50% of the population, feeding higher levels of lysine increased market weight ($P < .05$) and reduced the number of days until market compared to both standard diet and added fat (Figure 2). In the top 50% pigs, reducing lysine did not decrease the end weight ($P > .10$), but did reduce the feed cost per pig ($P < .01$), compared to standard diet or added fat.

Summary

Individualized pig data generates insights that can lead to the implementation of management strategies and commercial solutions for improved systemwide efficiency. Analysis of data captured at United Animal Health's DEO Research Farm further confirms the importance of strategies to drive sow lactation feed intake, resulting in improved sow reproductive fitness and downstream progeny performance. Focused management and nutritional strategies towards sub-populations can improve system performance and maximize the potential of the entire population.

Table 2: Impact of lactation ADFI on weaning and nursery performance and total throughput

Lactation ADFI (lbs.)	Number of pigs (% total)	Wean weight[1]	Percent pigs ≤ 8 lbs.	Number weaned	Pre-wean mortality[2]	Total pounds weaned[1]	Percent alive end nursery	Average nursery weight[3]	Total pounds end nursery[3]
10	4,858 (4%)	13.87[g]	6.4%[g]	12.96[e]	8.6[c]	165.9[f]	87.8%[c]	46.9[f]	586.1[d]
11	9,411 (8%)	13.09[g]	5.8%[fg]	12.99[e]	8.5[c]	169.9[e]	87.7%[c]	47.7[f]	595.3[d]
12	13,128 (11%)	13.28[f]	5.2%[efg]	13.17[cd]	7.5[bc]	174.0[d]	89.1%[bc]	48.5[e]	615.9[c]
13	16,593 (14%)	13.44[ef]	4.8%[def]	13.20[c]	7.0[ab]	176.6[c]	89.4%[bc]	48.8[de]	619.3[c]
14	18,876 (16%)	13.58[de]	4.6%[cde]	13.16[d]	7.6[bc]	178.3[c]	88.9%[bc]	49.2[cd]	623.8[c]
15	18,760 (16%)	13.67[cd]	4.1%[bcd]	13.31[b]	6.8[ab]	181.2[b]	89.5%[ab]	49.5[bc]	632.3[c]
16	16,724 (14%)	13.75[bc]	3.8%[abc]	13.31[b]	6.8[ab]	182.8[b]	89.6%[ab]	49.9[b]	638.8[bc]
17	12,567 (11%)	13.91[ab]	3.2%[ab]	13.32[b]	6.7[ab]	185.4[a]	89.9%[ab]	50.6[a]	649.4[ab]
18	8,237 (7%)	13.99[a]	2.9%[a]	13.45[a]	6.2[a]	187.1[a]	90.4%[a]	50.8[a]	658.6[a]
All	119,154	13.56	4.6%	13.24	7.2%	179.6	89.3%	49.2	627.1

[1] Adjusted for litter size, sow parity, sex and wean age (21-days).
[2] Mortality from cross-fostering (24h after birth) to weaning
[3] Adjusted for sow parity, sex and wean age and days post-weaning (42 days).
Columns with different superscript letters are statistically different ($P < .05$).
ADFI = average daily feed intake.

Figure 2: Feeding strategies for bottom and top 50% pigs.

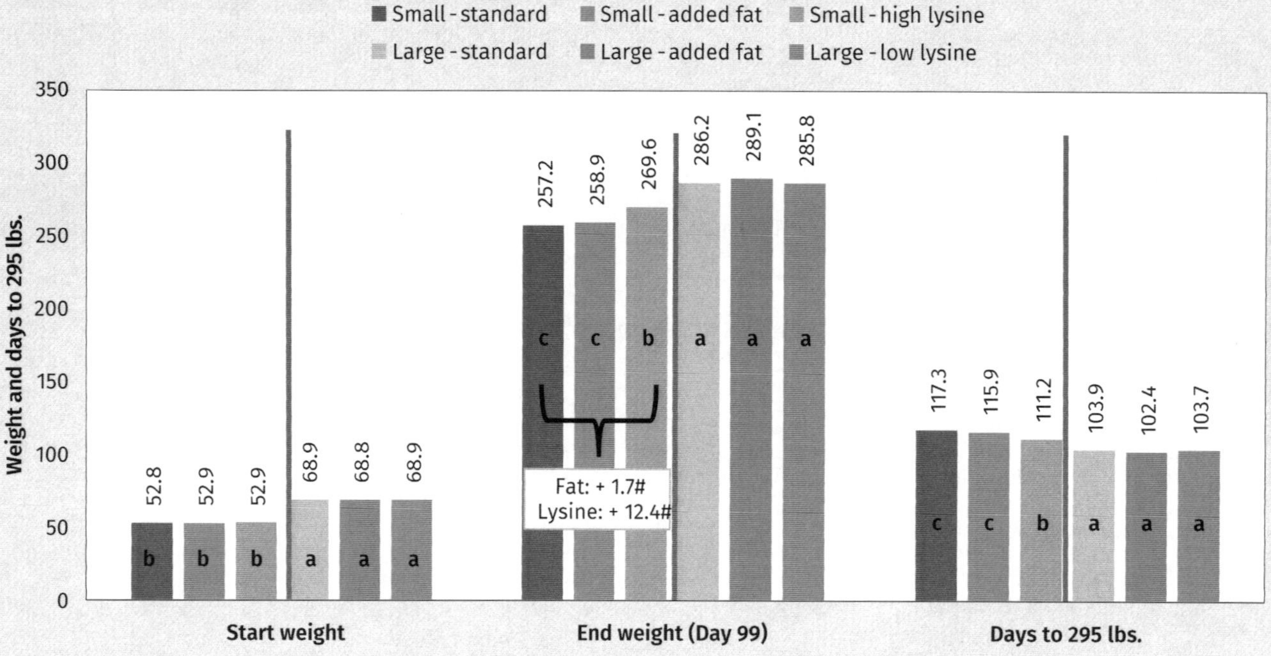

Table 3: Effect of wean weight on nursery weight, nursery mortality and finishing weight

Wean weight	Number of pigs (% of total)	Nursery weight[1]	Nursery mortality[2]	Finishing weight[3]
5.0 - 7.0	1,787 (2%)	32.2^{m}	$28.6\%^{i}$	231.6^{k}
7.0 - 8.0	2,406 (2%)	35.7^{l}	$9.9\%^{h}$	241.9^{j}
8.0 - 9.0	4,086 (3%)	37.8^{k}	$7.1\%^{g}$	246.1^{j}
9.0 - 10.0	6,441 (6%)	40.6^{j}	$5.0\%^{f}$	254.0^{i}
10.0 - 11.0	9,484 (8%)	42.9^{i}	$4.3\%^{ef}$	257.9^{h}
11.0 - 12.0	12,236 (10%)	45.1^{h}	$3.7\%^{e}$	262.4^{g}
12.0 - 13.0	15,322 (13%)	47.4^{g}	$2.9\%^{d}$	267.2^{f}
13.0 - 14.0	16,925 (14%)	49.5^{f}	$2.5\%^{cd}$	272.3^{e}
14.0 - 15.0	15,791 (14%)	51.6^{e}	$2.3\%^{bc}$	277.4^{d}
15.0 - 16.0	13,144 (11%)	53.4^{d}	$2.0\%^{ab}$	280.3^{c}
16.0 - 17.0	9,313 (8%)	55.4^{c}	$1.7\%^{a}$	286.1^{b}
17.0 - 18.0	5,512 (5%)	57.1^{b}	$1.7\%^{a}$	288.2^{b}
≥ 18.0	4,514 (4%)	59.6^{a}	$1.4\%^{a}$	296.0^{a}
All	116,961	48.7	3.4%	274.2

* Data collected at United Animal Health Research Farms from June 2018-November 2023.

1 Adjusted for sow parity, sex, wean age and age post-weaning (42 days). N = 116,961 pigs.

2 Pigs rejected at nursery placement or died during the nursery period.

3 Adjusted for sow parity, wean age, sex, and age post-weaning (21-weeks). N = 15,375 pigs.

Columns with different superscript letters are statistically different ($P < .05$).

https://doi.org/10.54846/am2024/50

Evaluation of Endovac-Porci® on diarrhea of purebred boars

[1]D. Reicks, DVM; [2]Curtis Stutheit
[1]Reicks Veterinary Research & Consulting (RVRC), Saint Peter, Minnesota; [2]ENDOVAC® Animal Health, Columbia, Missouri

Introduction

Looseness has been a challenge at some boar studs in purebred boars. Primarily, the looseness is present when total dissolved solids in the drinking water is high. Previous internal studies by RVRC to address this have not been successful. This included three studies with versions of chlorine dioxide treatment, egg antibodies in feed and water (including a specific blend specifically targeting certain bacteria following extensive fecal culturing), drinker type (traditional nipple, Bekalo, Arato, and trough), and antibiotics. The only treatment showing a significant benefit was Tiamulin with Tetracycline in the water for 5 days. However, relapses typically occur within several months requiring repeat medication. Endovac-Porci is a novel vaccine with a core antigen common to all gram-negative bacteria that is without negative active endotoxins and thus may provide broad spectrum protection. Additionally, the "IMMUNE Plus" component of Endovac-Porci is said to boost the immune response as an immunostimulant.

Materials and methods

Boars were randomized upon arrival and half given 1ml of Endovac-Porci followed by another 1 mL dose 2 weeks post arrival. Pure large white and pure duroc boars were utilized for the study numbering 28 treatment and 31 control boars. Boars were monitored daily for watery diarrhea. No treatments were given for diarrhea during the study. The percentage of boars with more than 1 diarrhea day were also evaluated to discern chronic versus acute one-off episodes. Percent normal sperm in both groups and total number 3 billion doses evaluated for 12 weeks (2 sperm production cycles) after the second dose.

Results

The number of diarrhea days in the control group were 159 vs 29 in the Endovac-Porci group resulting in an average number of diarrhea days of 5.1 in the control and 1.0 in the Endovac-Porci group. The percentage of boars with more than 1 diarrhea day was 32.3% in the control group and 10.7% in the Endovac-Porci group. The quantity of normal morphological sperm was 89.7 billion vs. 87.5 billion in favor of the Endovac-Porci group (23.2 and 22.0 doses respectively). Chronic boars were defined as having more than 1 diarrhea day and numbered 10 boars in the control group and 3 in the Endovac-Porci group. Trash rate was significantly higher for chronic boars in the control vs. Endovac-Porci group where there was no difference in normal stool boars between the groups, suggesting there is significant stress on boars that affects sperm production when having chronic diarrhea (Table 1).

Conclusion

Endovac-Porci significantly reduced the number of diarrhea days and number of boars with chronic diarrhea which resulted in more useable doses.

Table 1: Results

	# boars	Ave diarrhea days	% with chronic diarrhea	# doses	Trash rate
Endovac-Porci	28	1.1	10.7	23.2	5.8
Control	31	5.1	32.3	22.0	7.4
P-value	–	0.013	0.022	0.01	0.24

https://doi.org/10.54846/am2024/51

The use of a qualified, approved porcine cell line for the manufacture of an African swine fever virus emergency vaccine stockpile

G. Calzada, DVM, MSc, PhD; H. Bessoff, DVM
Aptimmune Biologics, Inc., 1234 Hanley Industrial Court, St. Louis, Missouri

Introduction

In recent years, African swine fever virus (ASFV) has developed into the major negative factor affecting the Chinese and Southeast Asian pork industry resulting in catastrophic financial and supply issues worldwide. ASFV has affected more than 50 countries in Africa, Europe, Asia, and the Pacific regions, with serious effects on food security.[1] The confirmation of the presence of ASFV in Haiti and the Dominican Republic[2] has made the risk of ASFV entering the United States (US) even more explicit than previously contemplated. The potential spread of ASFV in the US would cause tremendous economic damage to the American pork industry resulting in a loss of billions of dollars to pork producers.[3] Hence, it is important that innovative approaches are created to address existing challenges for ASFV vaccine development and production in the event of an emergency. A recent review of the status of global ASFV vaccine development covering all types of vaccine development efforts identified the recombinant live attenuated vaccines (LAV) developed by gene knockout technology as the most optimistic prospects as short/medium-term vaccine candidates in the future,[4] At present in the US there are several gene-deleted ASFV LAV candidates constructed from the highly virulent ASFV Georgia 2007 (ASFV-G). Among the viruses are ASFV-G-Δ9GL/ΔUK, ASFV-G-ΔMGF and ASFV-G-ΔI177L. Although it was demonstrated that these engineered viruses elicit protective immunity against challenge with the virulent ASFV-G,[5] they only replicate in porcine macrophages, and the required adaptation to grow in heterologous species cell lines results in modification of the viral genome that could affect their vaccine efficacy.[6] A major limitation for commercial ASFV vaccine production has been the lack of a porcine cell line capable of supporting the replication of ASFV vaccine candidates. Efficient ASFV replication is observed in primary porcine macrophages and differentiated porcine monocytes. However, due to lack of specified pathogen free pig colonies, lot-to-lot variation in harvested cells, risk of spread of other diseases, and labor-intensive collection, this cell type cannot be used as a reliable platform for safe and efficient vaccine production. Alternatively, some African green monkey derived cell lines such as Vero and COS-7 cells have been used to adapt ASFV strains as experimental vaccines.[6,7] However, adaptation of ASFV to replicate in non-porcine cell lines leads to viral genomic changes that are associated with lower viral yields, limited replication in vivo and/or failure to elicit a protective immune response. Reportedly, another cell line capable of supporting the growth of an ASFV vaccine candidate is available,[8] however, to our knowledge, it is not at the stage of commercial development since it is uncertain whether it will meet the strict USDA Center for Veterinary Biologics (CVB) requirements for the creation of a master cell stock (MCS). Thus, the selection of a validated cell line that will serve as the substrate to produce the vaccine at commercial scale is very important. Aptimmune Biologics, Inc is in a unique position to contribute as it has developed a USDA CVB approved MCS of the porcine macrophage cell line ZMAC-4 which is readily permissive to ASFV replication. A recent publication demonstrated that the ZMAC-4 cell line supports the growth of several ASFV field isolates, including the highly virulent ASFV-G without the need for adaptation.[9] This study also revealed that the replication of the wild-type ASFV-G in ZMAC-4 cells occurred with similar kinetics and yielded similar titers as in primary porcine bone marrow cells. Based on these observations, the ZMAC-4 cell line provides a sensible choice for ASFV commercial vaccine production.

Methods

Cells and viruses: ZMAC-4 cells were grown as previously described.[9] The ASFV gene-deleted vaccine candidates ASFV-G-ΔMGF and ASFV-G-D9GL/DUK were licensed by Aptimmune from the USDA. Growth kinetics and virus titrations were assessed in ZMAC-4 cells or primary porcine macrophages. The presence of virus was determined by hemadsorption (HAD),[9] and virus HAD_{50} titers (HAD_{50}/mL) were calculated by using the Reed and Muench method.

Sequencing: Briefly, DNA from ASFV-G-Δ9GL/ΔUK ZMAC-4 passages 5 and 10 was extracted using a modified HIRT method and quantified with a Qubit fluorometer using Thermo Fisher high sensitivity DNA assay. Libraries were prepared using Illumina's Nextera XT DNA library preparation kit and sequenced on either an Illumina Nextseq 500 or Nextseq 550. Sequencer 5.4.6 and GSNAP were used to generate p5 and p10 consensus sequences by aligning Illumina data to the reference sequence ASFV isolate Georgia 2007/1 (GenBank: FR682468.2). Passages 5 and 10 alignments were analyzed in TABLET for discrepancies against the reference sequence.

Results

We compared the primary porcine macrophage and ZMAC-4 infection susceptibility to different field isolates of ASFV from genotypes I, II, IV and X. The isolate samples were obtained from the USDA APHIS Foreign Animal Disease Diagnostic Laboratory ASFV collection. The results (Table 1) show that the ASFV tested isolates grew in ZMAC-4 cells and primary macrophages to similar levels. These results confirm previous studies[9] demonstrating that the ZMAC cell line is an excellent substrate for ASFV isolation.

To explore the use of ZMAC-4 cells for ASFV vaccine development, we performed a set of experiments to test the permissiveness of the cells to ASFV gene-deleted vaccine candidates ASFV-G-ΔMGF and ASFV-G-D9GL/DUK. These deletion mutant viruses can grow efficiently in primary porcine macrophages.[5] Hence, since the ZMAC-4 features are very similar to porcine macrophages, we investigated if gene-deleted candidates could grow on ZMAC-4 cells. First, ZMAC-4 cells were inoculated at a multiplicity of infection (MOI) of 0.05 with either ASFV-G-ΔMGF or the parental ASFV-G, for a total of ten 1 mL-vials for each virus. The vials were incubated at 37°C and two samples of each series were collected at 0, 24-, 48-, 72-, and 96-hours post-infection (hpi) and frozen for posterior viral quantitation by HAD. Our results demonstrated

Table 1: Titers of ASFV strains representing genotypes I, II, V and X grown on ZMAC-4 cells and primary porcine macrophages (PPM)

ASFV Strain	Titer (log10 HAD_{50}/mL) ZMAC-4	PPM
Georgia 2007	8.50	8.17
LEE, Kenya 1965	7.50	7.50
Haiti 1969	7.50	7.50
Lisbon 60	7.25	7.50
Brazil '78	7.50	7.50
Dominican Republic I	7.50	7.50
9/Zaire II 3/7/89	7.50	7.50
Tengani 1962 Malawi	6.75	6.00

that the growth kinetics of ASFV-G-ΔMGF were almost identical to the parental virus with titers increasing progressively until they reached the highest titers at approximately 72 hpi (Figure 1), confirming the high permissiveness of the ZMAC-4 cells to ASFV infection. We wanted to verify that the efficient replication of the gene-deleted vaccine candidate ASFV-G-ΔMGF on ZMAC-4 cells was not an isolated event. Subsequently, we repeated the experiment but this time with the deletion mutant ASFV-G-Δ9GL/ΔUK at different MOIs and as expected, this vaccine strain was also capable of growing in the cells without need for adaptation (Figure 2). Afterwards, blind serial passages of the ASFV-G-Δ9GL/ΔUK were performed on ZMAC-4 cells and after 4 - 5 days the viral fluid was harvested for titration. Through all passages the ASFV-G-Δ9GL/ΔUK vaccine virus readily replicated in the cells with titers higher than 10^7 HAD_{50}/mL (Figure 3) indicating that these vaccine viruses don't require adaptation to grow in the cells, and it is likely that there will be no significant genetic changes in the virus genome when using ZMAC-4 cells as the substrate to grow a leading ASFV vaccine candidate. Whole genome sequencing was performed to demonstrate if continuous passaging of the ASFV-G-Δ9GL/ΔUK gene-deleted virus on ZMAC-4 cells caused significant mutations in the virus genome. The analysis showed that there were no sequence differences between the ASFV-Δ9GL/ΔUK ZMAC-4 passage 5 and 10, meaning the virus remained stable in between passages. Only two large deletions corresponding to the ASFV 9GL and DP96R genes in both passages were identified.

Discussion

The DHS S&T Master Question List (MQL) under the medical veterinary countermeasures section[10] mentions the need to identify a lead vaccine candidate that can be used to produce 5 to 10 million emergency vaccine doses for stockpiling the National Animal Vaccine Veterinary Countermeasures Bank (NAVVCB). There are several promising modified live virus candidate vaccines which reported protection against same strain ASFV challenges with up to 100% protection.[11] However, a lead vaccine candidate requires a validated cell line to produce the necessary ASFV vaccine doses for the NAVVCB. In agreement with previous results,[9] we have demonstrated that the USDA CBV-approved cell line ZMAC-4 cell line can propagate wild type ASFV and ASF vaccine candidate strains. The ZMAC-4 cell line offers a high-yield stable vaccine manufacturing process that in our estimation can easily provide the required number of doses of a leading live attenuated ASFV vaccine candidate in a short period of time and without changes in the virus genome.

References

1. ASF Situation update - African swine fever (ASF) - FAO emergency prevention system for animal health. Accessed 2021, August 19. **http://www.fao.org/ag/againfo/programmes/en/empres/ASF/situation_update.html**

2. Jean-Pierre RP, Hagerman AD and Rich KM (2022) An analysis of African Swine Fever consequences on rural economies and smallholder swine producers in Haiti. Front. *Vet. Sci.* 9:960344.

3. Carriquiry, M., Elobeid, A., Swenson, D., Hayes, D. (2020) Impacts of African Swine Fever in Iowa and the United States. *Working Paper* 20-WP 600. Center for Agricultural and Rural Development, Iowa State University. **https://www.card.iastate.edu/products/publications/pdf/20wp600.pdf**

4. Wu, K., Liu, J., Wang, L., Fan, S., Li, Z., Li, Y., Yi, L., Ding, H., Zhao, M., Chen, J. (2020) Current State of Global African Swine Fever Vaccine Development under the Prevalence and Transmission of ASF in China. *Vaccines*, 8:531.

Figure 1: ASFV-G-ΔMGF growth kinetics on ZMAC-4 cells.

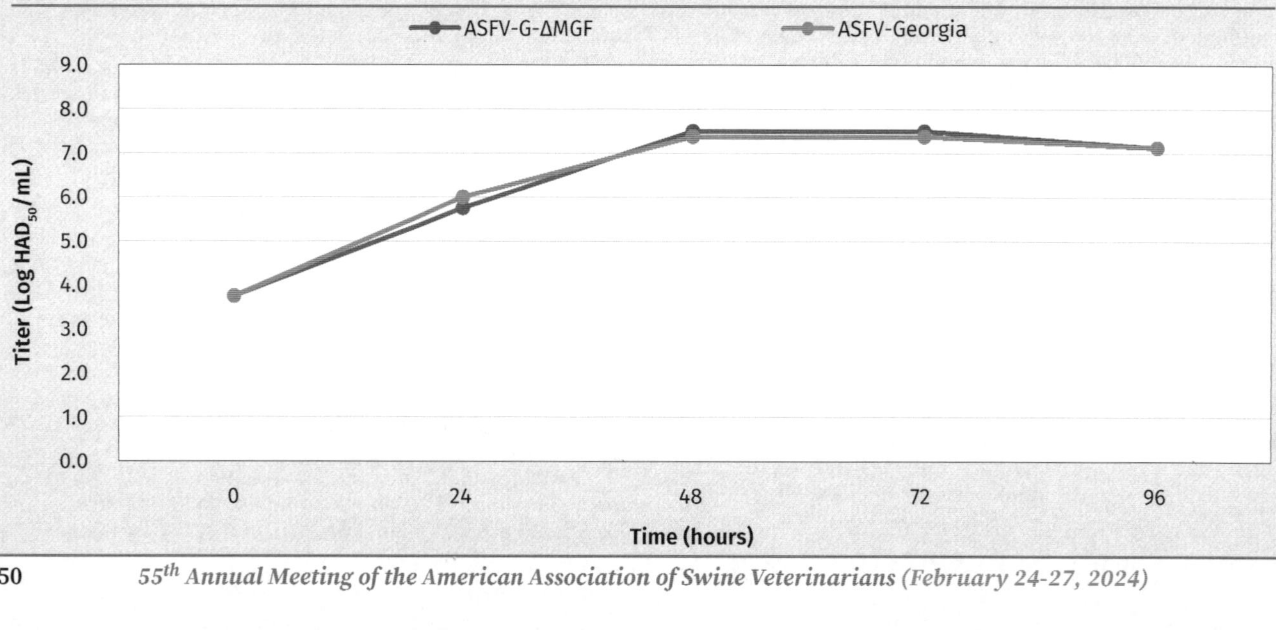

Figure 2: ASFV-G-Δ9GL/ΔUK growth kinetics on ZMAC-4 cells.

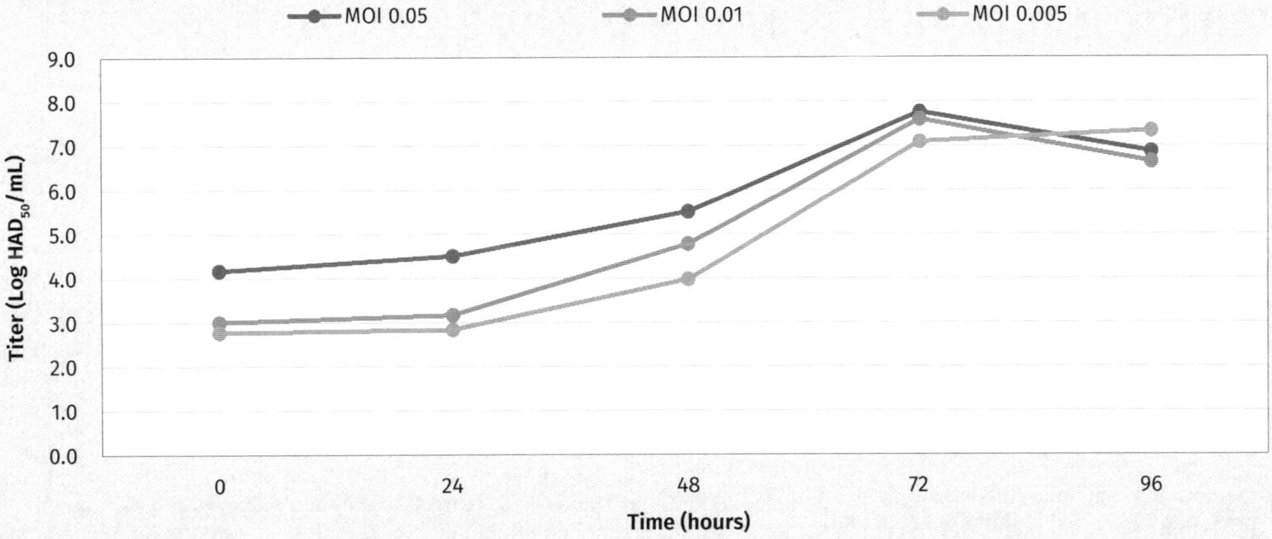

Figure 3: Titers of serial passages of ASFV-G-Δ9GL/ΔUK.

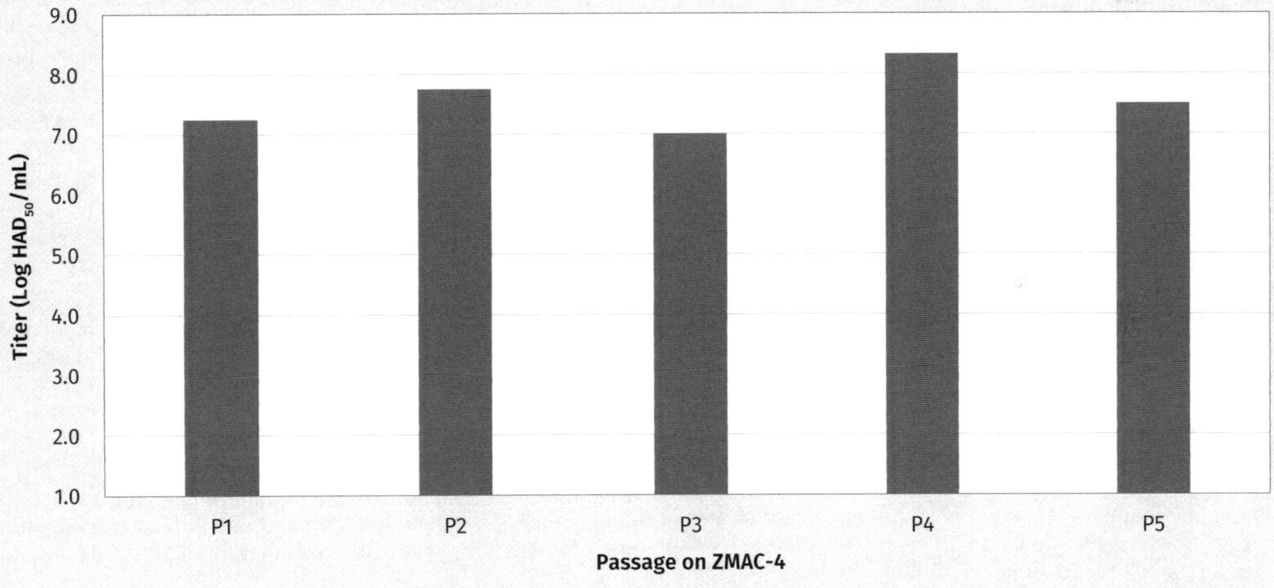

5. Gladue, D.P., Borca, M.V. Recombinant ASF Live Attenuated Virus Strains as Experimental Vaccine Candidates. *Viruses 2022*, 14, 878.

6. Krug, P. W., Holinka, L. G., O'Donnell, V., Reese, B., Sanford, B., Fernandez-Sainz, I., Gladue, D. P., Arzt, J., Rodriguez, L., Risatti, G. R., & Borca, M. V. (2015). The progressive adaptation of a Georgian isolate of African swine fever virus to Vero cells leads to a gradual attenuation of virulence in swine corresponding to major modifications of the viral genome. *J. Virol.*, 89(4), 2324–2332.

7. Sánchez, E.G., Riera, E., Nogal, M., Gallardo, C. Fernández, P., Bello-Morales, R., López-Guerrero J.A., Chitko-McKown, C.G., Richt, J., Revilla, Y. (2017) Phenotyping and susceptibility of established porcine cells lines to African Swine Fever Virus infection and viral production. *Sci Rep.*;7(1):10369.

8. Borca, M.V., Rai, A., Ramirez-Medina, E., Silva, E., Velazquez-Salinas, L., Vuono, E., Pruitt, S., Espinoza, N., Gladue, D.P. (2021). A cell culture-adapted vaccine virus against the current African swine fever virus pandemic strain. *J Virol* 95:e00123-21.

9. Portugal, R., Goatley, L.C., Husmann, R.L., Zuckermann, F.A., Dixon, L.K. (2020) A porcine macrophage cell line that supports high levels of replication of OURT88/3, an attenuated strain of African swine fever virus. *Emerging Microbes and Infections*. 9(1):1245-1253.

10. DHS Science and Technology. Master Question List for African Swine Fever Virus (ASFV). March 09, 2023. **https://www.dhs.gov/sites/default/files/2023-07/2023_06_07_st_asfv_mql.pdf**

11. Brake D. A. (2022). African Swine Fever Modified Live Vaccine Candidates: Transitioning from Discovery to Product Development through Harmonized Standards and Guidelines. *Viruses*, 14(12), 2619.

https://doi.org/10.54846/am2024/52

Evaluation of a multi-strain probiotic on post-weaning mortality and weight gain in pigs with natural enteric challenge[1]

A. Anderson Reever[1], MS, DVM; B. Burton[1], DVM; K. Bretey[2], DVM; B. Spannagel[2], MVS
[1]Suidae Health and Production, Algona, Iowa; [2]Kemin Industries Inc., Des Moines, Iowa

Introduction

Post-weaning diarrhea is a prevalent, significant, and costly health concern for swine producers. *Escherichia coli* (*E coli*) and other pathogenic organisms opportunistically flourish in the gastrointestinal tract of piglets, especially following stress events in the nursery and grow-finish stages. ENTEROSURE™ is a multi-strain product containing *Bacillus licheniformis* 12G-3 (G3), *Bacillus amyloliquifaciens* FxA (FxA) and *Bacillus subtilis* PB6 (PB6). *In vitro* studies have shown that ENTEROSURE™ can inhibit the growth of *E. coli* strains found in commercial production.[2] Thus, this study was conducted to evaluate the in vivo effects of ENTEROSURE™ on mortality and average daily gain when administered to nursery pigs with natural enteric challenges in a commercial research facility.

Experimental design

The study was conducted in a Midwest swine nursery site. The flow had a history of enteric disease challenges (including F18 *E coli*). A total of 2,016 female and male pigs were randomly assigned to the control or ENTEROSURE™ (1 lb/ton of feed) group for the 35-day nursery study. Pigs were randomly allotted into two rooms with a total of 36 pens (n = 18) and 2 pull pens per treatment group were left available. In accordance with the farm's typical sorting methods, pigs were also visually categorized by size (ie, small, medium, large, and extra-large). Average weights of small, medium, large, and extra-large pigs were 11.8 lbs, 13.8 lbs, 15.3 lbs, and 17.7 lbs, respectively. Equivalent numbers of each size of pigs were allotted to each treatment group. There were 4 placement events where pigs were tagged, weighed, and randomly allotted into pens over the course of 10 days.

Diagnostic sampling

Oral fluid samples (a treatment composite and a control composite) were collected bi-weekly and submitted to Iowa State University Veterinary Diagnostic Laboratory (ISU VDL) for real-time reverse transcriptase polymerase chain reaction (RT-qPCR) analysis for porcine reproductive and respiratory virus (PRRSV), porcine epidemic diarrhea virus (PEDV), and influenza A virus-swine (IAV-S). Five sets of intestinal tissue samples from mortalities exhibiting clinical diarrhea (3 from the ENTEROSURE™ group and 2 from the control group) were submitted to ISU VDL for histopathology and culture and sensitivity throughout the trial.

Nutrition information

The control diet was a typical Midwest corn/soybean meal diet with no probiotics included. The ENTEROSURE™ diet was the same as the control diet plus the inclusion of ENTEROSURE™ (1 lb/ton). A composite feed sample, including a minimum addition of 2 oz from each appropriate pen, was collected from the control and ENTEROSURE™ groups each week of the trial. The feed sample testing showed that the colony forming units per gram (CFU/g) in the finished feed met product inclusion specifications for ENTEROSURE™.

Statistical analysis

All data was analyzed by the One-Way ANOVA procedure using the JMP statistics software (Cary, North Carolina). Pigs were used as the experimental unit due to individual weight measurement. Treatment was considered as a fixed effect and pens and rooms were considered random effects. In addition, size of pig was also included as a fixed effect. Large (L) and extra-large (XL) pig groups were combined for analysis. LSMEANS procedure was used to compare the differences between the individual treatments. Differences were considered significant if the *P*-values were ≤ .05 and considered a statistical trend if the *P*-values were ≥ .05 and ≤ .1.

Results

Diagnostic results

The *E coli* strain isolated from the trial site showed significant resistance to antibiotics (Table 1) and was genotyped as F18 *E coli* (hemolytic, smooth/mucoid) (Table 2) by ISU VDL.

PEDV was not detected in any of the oral fluid samples during the trial (data not shown). As depicted in Table 3, PRRSV and IAV-S were detected in the oral samples at multiple time points.

Growth performance

ENTEROSURE™ significantly improved average daily gain (ADG) per pig by 0.02 lb (Figure 1) and 0.89 lb in weight gain per pig over 35 days (Figure 2).

ENTEROSURE™ significantly improved ADG in the medium pigs by 0.04 lb compared to control pigs ($P < .05$), and a numerical difference was seen in the small pigs ($P = .07$; Figure 3). The same ENTEROSURE™ effect was seen in medium pigs on weight gain per pig with the improvement being statistical, and the small pigs showed a trend ($P = .08$; Figure 4).

No differences were observed in mortality between the two treatments (Table 4). ENTEROSURE™ did not reduce the percentage of pigs placed in the pull pen compared to the control group (Table 4).

Conclusions

There was moderate enteric colibacillosis (*E coli*) disease as indicated by clinical signs (diarrhea and ill-thrift leading to a 7.71% pull rate), elevated mortality levels (6.71%) and confirmatory laboratory tissue diagnostics. PRRSV and influenza virus were also present, but at relatively low levels as indicated by laboratory testing of oral fluids. ENTEROSURE™ did not decrease mortality. However, ENTEROSURE™ significantly improved ADG by 0.02 lb/pig and weight gain by 0.89 lb/pig compared to control pigs. Specifically, ENTEROSURE™ showed significant differences in ADG and weight gain improvements in the medium-sized pigs; similar difference was seen with small-sized pigs but only a statistical trend was observed. Additional trials with higher levels of disease pressure are warranted to confirm the results of this study and to determine if additional benefits in weight gain and mortality can be detected.

Table 1: Food animal susceptibility testing – hemolytic *E coli* (smooth/mucoid)

Antimicrobial	*Int/MIC
Ampicillin	R / > 16.0000
Ceftiofur	R / > 8.0000
Clindamycin	R / > 16.0000
Danofloxacin	NI / > 1.0000
Enrofloxacin	R / > 2.0000
Florfenicol	R / > 8.0000
Gamithromycin	NI / > 8.0000
Gentamicin	R / > 16.0000
Neomycin	R / > 32.0000
Penicillin	R / > 8.0000
Sulfadimethoxine	R / > 256.0000
Spectinomycin	R / > 64.0000
Tetracycline	R / > 8.0000
Tiamulin	R / > 32.0000
Tildipirosin	NI / 4.0000
Tilmicosin	R / > 16.0000
Trimethoprim/ Sulphamethoxazole	R / > 2.0000
Tulathromycin	NI / 64.0000
Tylosin (Tartrate/Base)	NI / > 32.0000
Positive growth control	OK / 3554.333

* Int/MIC = Interpretation/Minimum Inhibitory Concentration;

R = Resistant, NI = no interpretation available based on antimicrobial, organism, species, and tissue combination: MIC levels are given in mcg/mL.

Table 2: *E coli* genotyping

*******Escherichia coli* **genotyping**	
Gene	**Result**
EAST1 (toxin)	Negative
LT (toxin)	Positive
STa (toxin)	Positive
STb (toxin)	Positive
Stx1 (toxin)	Negative
Stx2 (toxin)	Positive
Stx2e (toxin)	Positive
F18 (pilus)	Positive
F41 (pilus)	Negative
K88 (pilus)	Negative
K99 (pilus)	Negative
987P (pilus)	Negative
AIDA (adhesin)	Negative
EAEA (adhesin)	Negative
PAA (adhesin)	Positive

Organism: *Escherichia coli* hemolytic

Table 3: PRRSV and IAV-S testing via RT-qPCR

Test	Control CT value	Treatment Ct value
Entry PRRSV	30.1	31.7
Entry IAV-S	Negative	Negative
2-week PRRSV	32.3	32.0
2-week IAV-S	28.0	38.3
4-week PRRSV	31.0	30.0
4-week IAV-S	30.9	25.9
Ending PRRSV	28.5	31.3
Ending IAV-S	32.4	34.5

PRRSV CT values ≥ 36 considered negative

IAV-S CT values ≥ 40 considered negative

Figure 1: Average daily gain (ADG) per pig in control and ENTEROSURE groups in natural enteric challenge conditions.

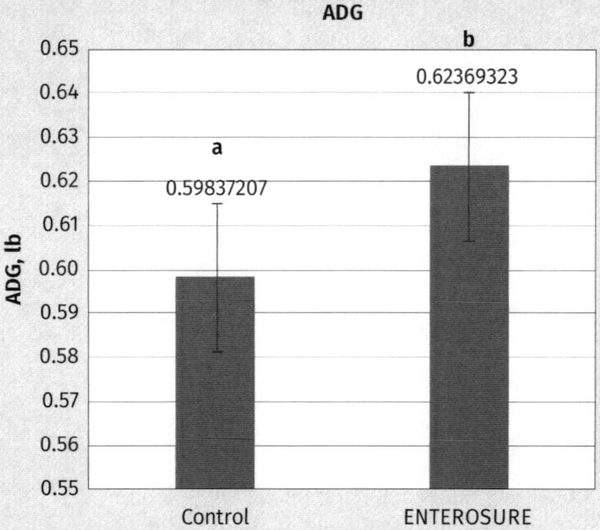

[a,b] Superscripts indicate significant differences between treatments, $P < .05$

Figure 2: Weight gain/pig in control and ENTEROSURE groups in natural enteric challenge conditions.

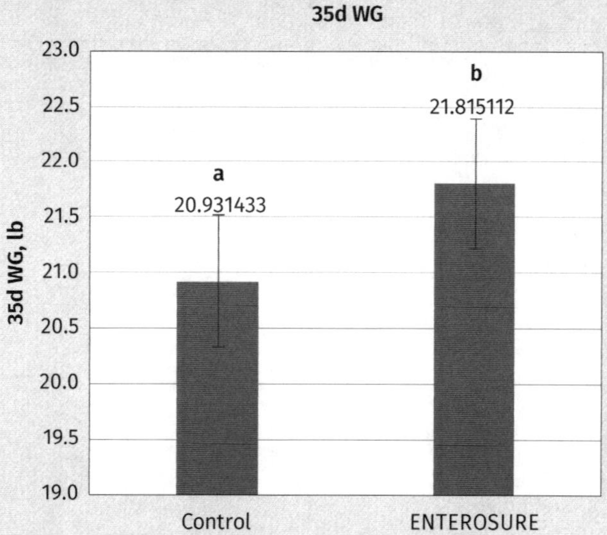

[a,b] Superscripts indicate significant differences between treatments, $P < .05$.

Figure 3: Average daily gain per pig by initial body size category for control and ENTEROSURE treatment groups in natural enteric challenge conditions.

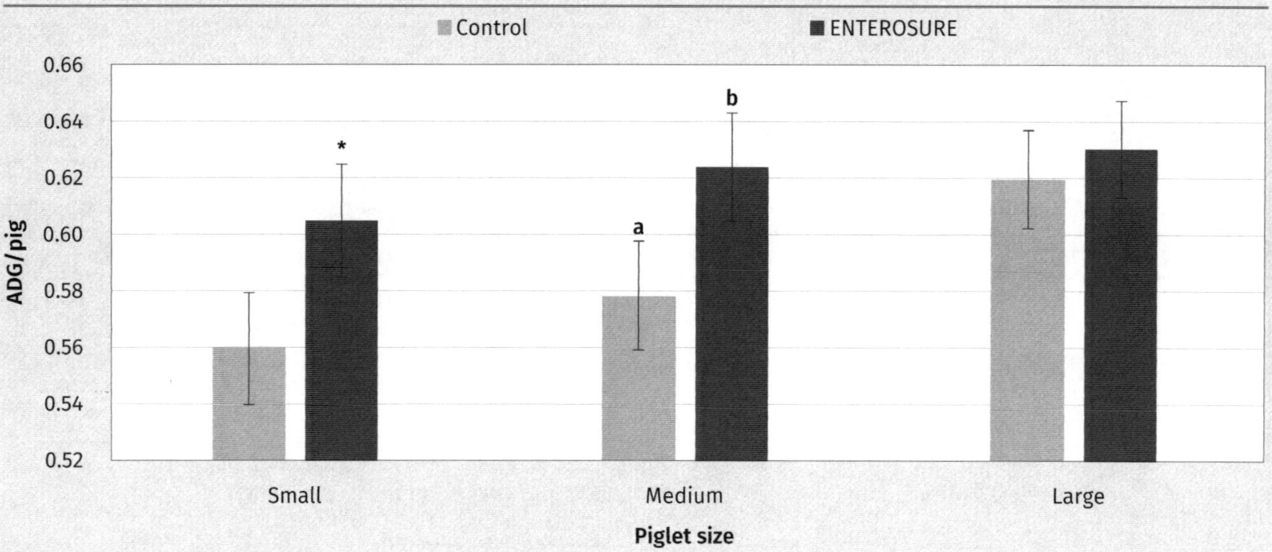

[a,b] Superscripts indicate significant differences between treatments, $P < 0.05$.

[*] Indicates trend of $P = .07$.

Piglets size: Small (avg wt/pig = 11.8 lb), medium (avg wt/pig = 13.8 lb), large (avg wt/pig = 16.3 lb).

Figure 4: Weight gain per pig in control and ENTEROSURE groups in natural enteric challenge conditions.

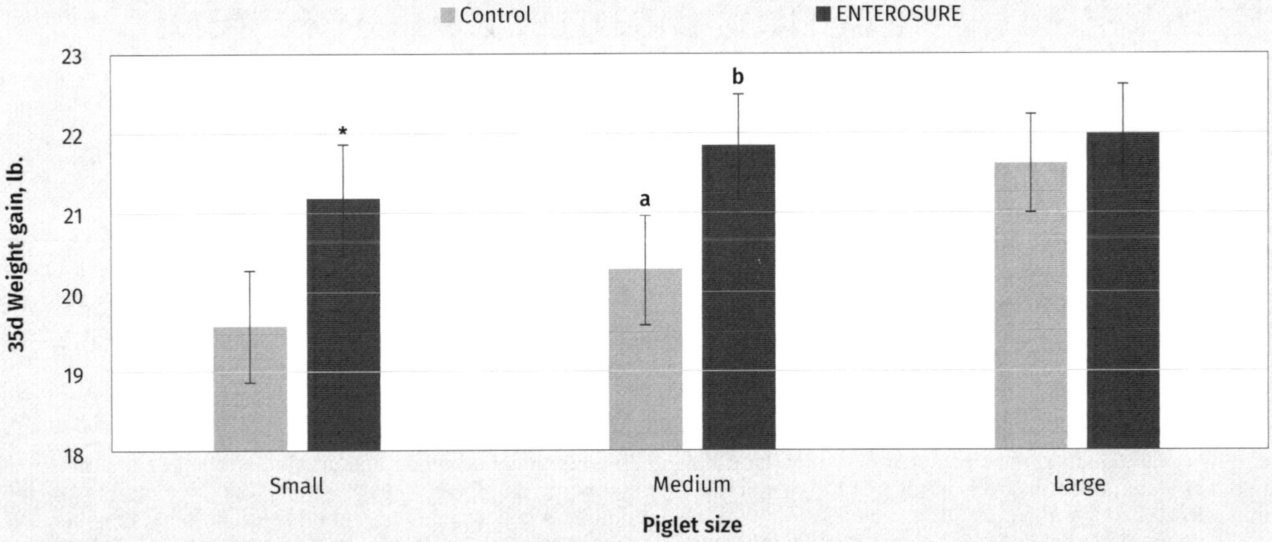

[a,b] Superscripts indicate significant differences between treatments, $P < .05$.

[*] Indicates trend of $P = 0.08$.

Piglets size: Small (Avg. wt/pig=11.8 lb), Medium (Avg. wt/pig=13.8 lb), Large (Avg. wt/pig=16.3 lb).

Table 4: Mortality % and pull % by treatment group and combined

	Mortality %	Pull %
Control	6.80%	7.76%
ENTEROSURE	6.62%	7.67%
Combined	6.71%	7.71%

References

1. Kemin Internal Document: TD-23-9253
2. Kemin Internal Document: TD-23-9244

https://doi.org/10.54846/am2024/53

Evaluation of protective efficacy of six commercial PRRSV-2 MLV vaccines against the emergent 1-4-4 L1C variant (L1C.5) virus in weaned pigs under experimental conditions

Gaurav Rawal, DVM, PhD; Marcelo Almeida, DVM, PhD; Phillip Gauger, DVM, PhD; Fangshu Ye, PhD;
Chris Rademacher, DVM; Ethan Aljets, BS; Loni Schumacher, DVM, PhD; Benjamin Hollis; Derald Holtkamp, DVM, MS;
Joseph Thomas, DVM, MS; Mallory Wilhelm, BS; Sydney Brosnahan; Julie Fisher; Jinhui Zhu, BS;
Alyona Michael, DVM, PhD;Patrick Halbur, DVM, PhD; Jianqiang Zhang, MD, PhD
Department of Veterinary Diagnostic and Production Animal Medicine,
College of Veterinary Medicine, Iowa State University, Ames, Iowa

Introduction

Porcine reproductive and respiratory syndrome virus (PRRSV) continues to evolve, and new highly virulent PRRSV strains periodically emerge.[1-3] Since October 2020, a new PRRSV 1-4-4 L1C variant has emerged in the United States and has been associated with high mortality and morbidity of pigs.[4,5] According to the recently refined PRRSV-2 ORF5-based genetic classification system, the PRRSV 1-4-4 L1C variant is referred to as L1C.5 virus.[6] Our group has conducted an experimental study in four-week-old pigs and compared the virulence and transmissibility of PRRSV 1-4-4 L1C.5, 1-4-4 L1C.1, 1-4-4 L1A, 1-4-4 L1H, and a previously described highly virulent 1-7-4 L1A isolates.[7] The experimental pig study confirmed that the emergent L1C.5 virus is highly virulent. Currently, six commercial PRRSV-2 modified live virus (MLV) vaccines are available in the USA to combat PRRS: Ingelvac PRRS MLV and Ingelvac PRRS ATP from Boehringer Ingelheim, Fostera PRRS from Zoetis, Prime Pac PRRS RR from Merck, Prevacent PRRS from Elanco, and PRRSGard from Pharmgate. However, the protective efficacy of these vaccines against the PRRSV L1C.5 strain has not been thoroughly evaluated and compared. The objective of this study was to compare the protective efficacy of these six PRRSV MLV vaccines against the PRRSV L1C.5 strain in a weaned pig model under experimental conditions.

Materials and methods

The experimental design is shown in Table 1. Seventy PRRSV naïve pigs at 3–4 weeks of age were blocked by weight and then randomly divided into seven groups with ten pigs per group, one group per room. Microchips were intramuscularly implanted at the base of the left ear on the scutiform cartilage of each pig. At 0 day post vaccination (DPV), Group 1 (G1) received virus-negative medium as a negative control and G2 through G7 were intramuscularly vaccinated with Vaccine A (Prime Pac PRRS), Vaccine B (Ingelvac PRRS ATP), Vaccine C (Fostera PRRS), Vaccine D (Ingelvac PRRS MLV), Vaccine E (Prevacent PRRS), and Vaccine F (PRRSGard), respectively, using the doses labeled by the vaccine manufacturers. At 35 DPV, all pigs were challenged with PRRSV 1-4-4 L1C.5 isolate USA/MN/01775GA/2021 grown in ZMAC cells via intramuscular (2mL/pig) and intranasal (2mL/nostril) inoculation at a dose of 10^6 TCID50/pig. All surviving pigs were euthanized and necropsied at 49 DPV (14 days post challenge [DPC]).

Pigs were monitored daily for clinical signs including mortality, lethargy, and anorexia. Microchip temperature was taken from individual pigs once daily from 0 - 14 DPV, once weekly during 14 - 35 DPV, and once daily during 35 - 49 DPV. Pigs were weighed at 0, 7, 14, 21, 28, 35, and 49 DPV. Individual serum samples were

Table 1: Experimental design of the PRRSV vaccination and challenge pig study

Group (number of pigs)	Vaccine	Day 0 (vaccination) 0 DPV	Day 35 (challenge) 35 DPV 0 DPC	Day 49 (necropsy) 49 DPV 14 DPC	Sample collection
G1 (N = 10)	Negative control (virus-negative medium)	IM; 2 mL	L1C.5	Surviving pigs	Individual serum, samples collected at 0, 2, 4, 7, 10, 14, 21, 28, 35, 37, 39, 41, and 49 DPV; lung and brain tissues collected at necropsy
G2 (N = 10)	A: Prime Pac PRRS (Merck Animal Health)	IM; 1 mL/dose	L1C.5	Surviving pigs	
G3 (N = 10)	B: Ingelvac PRRS ATP (Boehringer Ingelheim)	IM; 2 mL/dose	L1C.5	Surviving pigs	
G4 (N = 10)	C: Fostera PRRS (Zoetis)	IM; 2 mL/dose	L1C.5	Surviving pigs	
G5 (N = 10)	D: Ingelvac PRRS MLV (Boehringer Ingelheim)	IM; 2 mL/dose	L1C.5	Surviving pigs	
G6 (N = 10)	E: Prevacent PRRS (Elanco)	IM; 1 mL/dose	L1C.5	Surviving pigs	
G7 (N = 10)	F: PRRSGard (Pharmgate)	IM; 1 mL/dose	L1C.5	Surviving pigs	

collected at 0, 2, 4, 7, 10, 14, 21, 28, 35, 37, 39, 41, and 49 DPV. At necropsy, gross lung lesions were scored in a blinded fashion and fresh and formalin-fixed lung and brain tissues were collected.

Serum, lung, and brain samples were tested by three quantitative PRRSV real-time reverse transcription-polymerase chain reaction (RT-PCR) assays: 1) a commercial PRRSV screening PCR (VetMAX™ PRRSV NA&EU v2, Thermo Fisher Scientific); 2) each vaccine-specific PCR as described previously[8,9]; and 3) L1C.5 challenge virus-specific PCR. Formalin-fixed lung and brain tissues were examined in a blinded fashion for histopathological lesions and PRRSV immunohistochemistry (IHC) staining. Serum samples collected at 0, 7, 10, 14, 21, 28, 35, 41, and 49 DPV were tested by a commercial PRRS X3 ELISA antibody kit, and these serum samples were also tested for fluorescence focus neutralization (FFN) antibody titers against their homologous vaccine viruses and the L1C.5 challenge virus.

Temperatures, anorexia, and lethargy scores, virus loads in serum samples determined by PCR, PRRS X3 ELISA antibody titers, and FFN antibody titers were analyzed using a linear mixed model. The average daily weight gain (ADG), PRRSV RNA levels in tissues (eg, lung and brain), gross lung lesion scores, microscopic lesion scores, and IHC scores were analyzed using a one-way ANOVA test. For all analyses, SAS was used and a *P*-value ≤ .05 was considered significant.

Results

It is noteworthy that one pig in G3 was dead after being bled at 10 DPV and, therefore, only nine pigs in G3 were used for analysis. During 0 - 35 DPV before challenge with the PRRSV L1C.5 isolate, pigs in G1 - G7 groups did not develop fever, anorexia, or lethargy. The average weights among G1–G7 groups were similar at each of the time points (0, 7, 14, 21, 28, and 35 DPV).

After challenging pigs with the PRRSV L1C.5 isolate at 35 DPV through the end of the study at 49 DPV (14 DPC), the patterns of fever development were different among groups. In G1 group, the first fever peak developed at 1 DPC, then body temperature tended to drop at 2 - 3 DPC but started to increase again at 4 DPC reaching the second peak level during 6 - 10 DPC, followed by temperature decrease to < 40.5°C for most pigs after 12 DPC. Pigs in G2 group developed and maintained fever (~41.5°C) during 1 - 8 DPC, had body temperature of ~41°C during 9 - 10 DPC, and then decreased to < 40.5°C for most pigs at 13 - 14 DPC. In G3 and G5 groups, over 50% of pigs developed and maintained a fever (40.5 - 41°C) during 1 - 2 to 9 - 10 DPC followed by temperature decrease. In G4 group, most pigs had a fever (40.5 - 41°C) during 3 - 8 DPC and all pigs had no fever (< 40.5°C) at 11 DPC and thereafter. Pigs in G6 group had a sharp fever peak (~41.5°C) during 1 - 4 DPC followed by the gradual decrease of temperature with all pigs having no fever (< 40.5°C) at 11 DPC and thereafter. Pigs in G7 group had a fever (> 40.5°C) during 1 - 13 DPC.

After challenge with the PRRSV L1C.5 isolate, pigs in G1 and G2 groups overall had significantly higher anorexia and lethargy scores than pigs in G7, G4, G3, G5, and G6 groups (Table 2). After L1C.5 virus challenge, the mortality of pigs in G1–G7 groups was 5/10, 5/10, 1/9, 1/10, 1/10, 0/10, and 2/10, respectively. When ADG was compared, the G2 pigs had the numerically lowest ADG, although it was not significantly different from that in G1, G3, and G7 groups. Pigs in G6 had the numerically highest ADG, although it was not significantly different from that in G3, G4, G5, and G7 pigs. But, pigs in G4 through G6 had significantly higher ADG compared to G1 and G2 pigs (Table 2).

After challenge with L1C.5 virus, G1 pigs had significantly higher viremia levels compared to pigs in G2 - G6; in addition, pigs in G4 - G6 had significantly lower viremia levels than pigs in G2, G3, and G7 (Table 2). G1 pigs had significantly higher percentage of lung tissues with gross lesions compared to pigs in G2 - G7, but gross lung lesions in G2 - G7 pigs did not have a clear-cut trend of differences. Regarding the microscopic lung lesions and PRRSV IHC scores in lung tissues, G6, G4, and G3 pigs had significantly lower levels compared to pigs in G1 and G2. Pigs in G6 had the lowest level of PRRSV RNA in lung tissues as examined by the challenge virus-specific PCR. When PRRSV IHC scores in brain tissues were compared, G6 and G3 pigs had significantly lower scores than G2 pigs. Pigs in G6 had the lowest level of PRRSV RNA in brain tissues as examined by the challenge virus-specific PCR.

In G1 group, all pigs (10/10) were PRRSV ELISA antibody negative from 0 DPV through 35 DPV, 7/10 pigs developed detectable ELISA antibody at 41 DPV (6 DPC), and all pigs surviving through 48–49 DPV (13–14 DPC) were PRRSV ELISA antibody positive. At 0 and 7 DPV, all pigs in G2 - G7 groups were PRRSV ELISA antibody negative. In G2 - G7 groups, 50%, 44.4%, 60%, 40%, 80%, and 30% of pigs developed detectable ELISA antibody at 10 DPV, 50%, 100%, 80%, 70%, 100%, and 60% of pigs developed detectable ELISA antibody at 14 DPV, respectively, and almost all pigs were ELISA antibody positive at 21 DPV and thereafter. Pigs in G6, G5, G3, and G7 had significantly higher ELISA antibody S/P ratios than pigs in G4 and G2. Throughout the study (0 - 49 DPV), no FFN antibody titers against any vaccine virus and the challenge L1C.5 virus were detected in G1 pigs. Pigs in G2 - G7 remained FFN antibody negative against the respective homologous vaccine virus from 0–14 DPV; they started to develop low levels of FFN antibody against the homologous vaccine virus at 21 DPV and the antibody titers continued to increase through 49 DPV. G6 pigs had the highest FFN antibody titers against the homologous vaccine virus compared to other groups at 35 DPV before virus challenge. G2 - G7 groups remained FFN antibody negative against the L1C.5 challenge virus during 0 - 28 DPV. At 35 DPV, very low levels of FFN antibody against the L1C.5 virus were developed in G6 group. Pigs in G2 through G7 all had increasing FFN antibody titers against the L1C.5 challenge virus at 41 DPV (6 DPC) and 49 DPV (14 DPC).

Discussion and conclusion

This study provides a head-to-head comparison of six commercial PRRSV-2 MLV vaccines for their protection against the emergent PRRSV L1C.5 strain in weaned pigs. Under the conditions of this study, Vaccines A through F showed varying degrees of protection against the L1C.5 virus challenge. The parameters such as mortality, anorexia and lethargy scores, body temperature, ADG, viremia level, gross lung lesions, histopathological lung lesions, viral loads in lung and brain tissues, and PRRSV IHC scores in lung and brain tissues were evaluated. Veterinarians can choose the vaccine(s), based on their assessments of these parameters, to protect against the PRRSV L1C.5 strain. Although vaccines A through F developed neutralizing antibodies against the homologous vaccine virus at 35 DPV before virus challenge, the neutralizing antibody titers against the L1C.5 strain were low at 35 DPV prior to virus challenge, suggesting that other immune mechanisms are contributing to protective efficacy, too. Cytokines and cell-mediated immune responses inducted by these vaccines remain to be evaluated and compared.

Table 2: Summary of PRRSV vaccination and challenge pig study (35 - 49 days post vaccination corresponding to 0 - 14 days post challenge)

Group	G1; unvaccinated /L1C.5	G2; Vaccine A / L1C.5	G3; Vaccine B / L1C.5	G4; Vaccine C / L1C.5	G5; Vaccine D / L1C.5	G6; Vaccine E / L1C.5	G7; Vaccine F / L1C.5	The trend (the order of high to low)
Overall anorexia and lethargy score comparison	a (more severe)	a (more severe)	b	b	b	b	b	G1, G2 > G7, G4, G3, G5, G6
Mortality*	5 out of 10	5 out of 10	1 out of 9	1 out of 10	1 out of 10	0 out of 10	2 out of 10	
Mean ADG (lb)	0.1376[a]	-0.0279[a]	0.5857[ab]	0.6874[b]	0.6810[b]	0.7714[b]	0.3386[ab]	
Challenge virus viremia level	a	c	c	b (lower)	b (lower)	b (lower)	ac	G1, G7 > G2, G3 > G4, G5, G6
Gross lung lesion (% of lung with lesions)	a (more severe)	b	cd	c	ce	be	bde	G1 > G2, G6, G7, G5, G3, G4
Lung microscopic lesions	a (more severe)	a (more severe)	b	b	ab	b	ab	G1, G2, G5, G7, G3, G4, G6
Lung PRRSV IHC scores	a (higher)	ac	b	b	bc	b	bc	G1, G2, G7, G5, G4, G3, G6
Viral load in lung by L1C.5 PCR	a (highest)	ab	b	b	ab	c (lowest)	b	G1, G2, G5, G7, G4, G3 > G6
Brain PRRSV IHC scores	ab	a	b (lower)	ab	ab	b (lower)	ab	G2, G1, G7, G5, G4, G3, G6
Viral load in brain by L1C.5 PCR	a	a	b	a	b	c (lowest)	a	G1, G2, G7, G4 > G3, G5 > G6

Notes: * The pigs that naturally died or were euthanized due to severe body conditions were counted as mortality. Different letters (a, b, c, d) indicate statistical differences.

Acknowledgments

We would like to acknowledge Iowa Pork Producers Association, American Association of Swine Veterinarians Foundation, and Iowa Livestock Health Advisory Council for funding this study. We thank numerous students and the Iowa State University Laboratory Animal Resources staff for helping with sample collection and animal care.

References

1. Zimmerman J, Dee S, Holtkamp DJ, Murtaugh M, Stadejek T, Stevenson G, Torremorell M, Yang H, Zhang J. 2019. Porcine reproductive and respiratory syndrome viruses (porcine arteriviruses). In *Diseases of Swine*. 11th edition. Zimmerman JJ, Karriker LA, Ramirez A, Schwartz KJ, Stevenson GW, and Zhang J, editors. Hoboken: Wiley-Blackwell. 685-708.

2. van Geelen AGM, Anderson TK, Lager KM, Das PB, Otis NJ, Montiel NA, Miller LC, Kulshreshtha V, Buckley AC, Brockmeier SL, et al. 2018. Porcine reproductive and respiratory disease virus: Evolution and recombination yields distinct ORF5 RFLP 1-7-4 viruses with individual pathogenicity. *Virology*. 513:168-179.

3. Wang X, Marthaler D, Rovira A, Rossow S, Murtaugh MP. 2015. Emergence of a virulent porcine reproductive and respiratory syndrome virus in vaccinated herds in the United States. *Virus Res*. 210:34-41.

4. Trevisan G, Li G, Moura CAA, Coleman K, Thomas P, Zhang J, Gauger P, Zeller M, Linhares D. 2021. Complete Coding Genome Sequence of a Novel Porcine Reproductive and Respiratory Syndrome Virus 2 Restriction Fragment Length Polymorphism 1-4-4 Lineage 1C Variant Identified in Iowa, USA. *Microbiol Resour Announc*. 10:e0044821.

5. Kikuti M, Paploski IAD, Pamornchainavakul N, Picasso-Risso C, Schwartz M, Yeske P, Leuwerke B, Bruner L, Murray D, Roggow BD, et al. 2021. Emergence of a new lineage 1C variant of porcine reproductive and respiratory syndrome virus 2 in the United States. *Front Vet Sci*. 8:752938.

6. Yim-im W, Anderson T K, Paploski I, VanderWaal K, Gauger P, Kreuger K, Shi M, Main R, Zhang J. 2023. Refining PRRSV-2 genetic classification based on global ORF5 sequences and investigation of their geographic distributions and temporal changes. *Microbiol Spectr. e0291623*.

7. Rawal G, Almeida M, Gauger P, Zimmerman J, Ye F, Rademacher C, Armenta-Leyva B, Munguia-Ramirez B, Tarasiuk G, Schumacher L, Aljets E, Thomas J, Zhu J-H, Trexel J, Zhang J. 2023. In vivo and in vitro characterization of the recently emergent PRRSV 1-4-4 L1C variant (L1C.5) in comparison with other PRRSV-2 lineage 1 isolates. *Viruses*. 15:2233.

8. Rawal G, Yim-Im W, Chamba F, Smith C, Okones J, Francisco C, Zhang J. 2022. Development and validation of a reverse transcription real-time PCR assay for specific detection of PRRSGard vaccine-like virus. *Transbound Emerg Dis*. 69:1212-1226.

9. Rawal G, Krueger K, Yim-im W, Li G, Gauger P, Almeida M, Aljets E, Zhang J. 2023. Development, evaluation, and clinical application of PRRSV-2 vaccine-like real-time RT-PCR assays. *Viruses*. 15:2240.

https://doi.org/10.54846/am2024/54

Lab and real-world variation in an inflammatory biomarker, using and interpreting the results

S. Weiland, MS; J. Hansen, MS, PhD
Elanco Animal Health, Greenfield, Indiana

Introduction

Vitamin A plays essential roles in vision, epithelial surface maintenance, the immune system, reproduction, and embryonic growth and development.[1] In fact, vitamin A was labeled as an anti-infective, as early as 1928.[2] As an essential vitamin in swine, vitamin A must be obtained through the diet. Shortly after consumption, vitamin A is transported to the liver for storage as retinyl palmitate. Transport of vitamin A from the liver to target tissues almost exclusively occurs in the form of retinol bound to retinol binding protein (RBP) and transthyretin complex.

The acute phase response of immune system activation alters synthesis of many proteins, including a significant decrease in production of both RBP and transthyretin.[3] This inverse relationship between proinflammatory response and RBP production serves as the basis for utilizing serum retinol as a biomarker of inflammation. Elanco Animal Health has previously characterized the reliability of this biomarker in the field, as well as the connection between biomarker levels and expected growth performance.[4,5] However, the correct application of this biomarker requires an understanding of variation observed in the field, a reliable and accurate analysis of the sample, and application to expected performance outcomes. Our goal was to describe the biomarker variation observed among various farms, as well as variation observed when sending duplicate samples to various labs, and ultimately the implications of these factors on interpreting biomarker results.

Methods

Our field results summary includes 3,907 samples from 54 barns across the United States. On each farm, all piglets from 6 sows were bled just prior to weaning. Selected litters were balanced across sow parity (2 parity 0, 2 parity 1 -2, 2 parity 3+). Blood samples were spun to separate out serum, and serum samples analyzed by Elanco Animal Health using our internal analytical system or by the veterinary diagnostic lab at Iowa State University (ISU).

Additionally, serum samples for the lab ring test were collected from 10 grow finish pigs in a contract research facility. Samples were analyzed using the Elanco method, then homogenized to create five pooled samples representing the widest range of retinol levels. Each sample was then split and sent to each lab as duplicates. Samples were randomly coded to blind labs to duplicate samples. The following veterinary diagnostic labs were included in the test: ISU, Michigan State University, and Kansas State University, along with the Elanco internal assay and a World Health Organization reference lab. Labs were instructed to analyze samples for serum retinol levels using their protocols for analysis. Results among laboratories were compared using Microsoft Excel and JMP. Coefficients of variation were calculated for duplicate samples using the average results among labs as the mean.

Results

When averaging biomarker levels across each farm, 39% of farms fell below the Elanco research threshold where the largest impacts on performance and mortality were observed, and all barn averages fell below the Puls 1994 retinol threshold, still used by ISU and other veterinary diagnostic labs (Figure 1).[6] Compared to farm averages, individual pig data results in a larger bell curve with 47% of samples falling below the Elanco research threshold, and 99% falling below the Puls threshold (Figure 1). These differences highlight the importance of looking at individual data distribution for a farm, rather than the farm average.

Among the five labs included in the ring test, variation in results for the same sample ranged from 0 to 0.28 ppm and coefficient of variation ranged from 0% to 18%. A pattern was observed as labs consistently reported high or low values compared to other labs included over the wide range of samples tested (Figure 2).

We applied these lab differences to the individual pig biomarker levels to create a simulation representing the differences that could be expected when sending field samples to various labs. The results demonstrate how, depending on lab used, the same samples could generate results ranging from extremely low to adequate biomarker levels (Figure 3).

Discussion

Of the 3,907 samples analyzed to-date, the overwhelming majority fall much lower than the expected range of 0.400 - 0.500 ppm previously reported by Puls and still being used as a reference range by many labs.[6] Our data more closely resemble serum retinol ranges recently reported by Greiner et al.[7] This recent data would indicate our industry expectations for serum retinol levels likely need to be adjusted. Also, the range in serum retinol levels observed by Greiner et al[7] and over the 54 barns included in this paper indicates the inflammatory stress among and within production systems is highly variable.

Our simulation of individual pig results sent to various labs demonstrates that an extreme range between labs would be found analyzing the same samples. These differences among labs are large enough to completely change interpretation of results. The consistently high or low levels reported from various labs would indicate the differences observed are likely due to differences in protocol, technique, or standard, rather than random variation. Regardless of the reason behind these differences, these data emphasize the importance of using a consistent lab for serum retinol analysis. Likewise, these differences also emphasize the importance of understanding levels at which performance and mortality are impacted for any testing system used to properly interpret and utilize the inflammatory biomarker.

Figure 1: Individual and farm average inflammatory biomarker results among 54 barns sampled (3,907 samples).

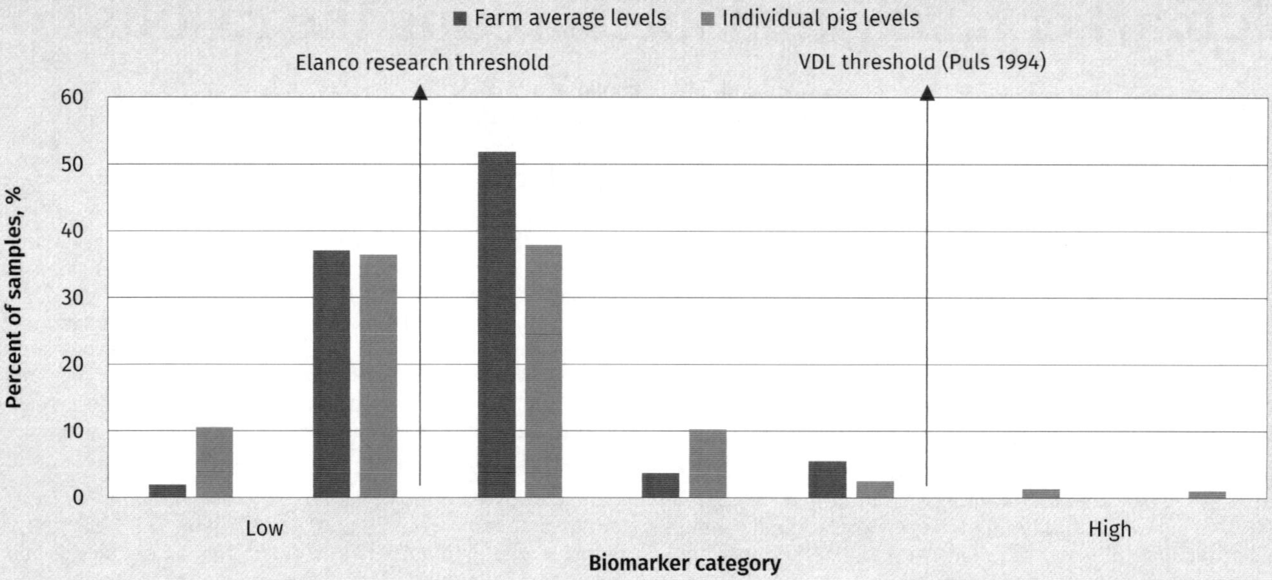

Figure 2: Lab ring test results by laboratory.

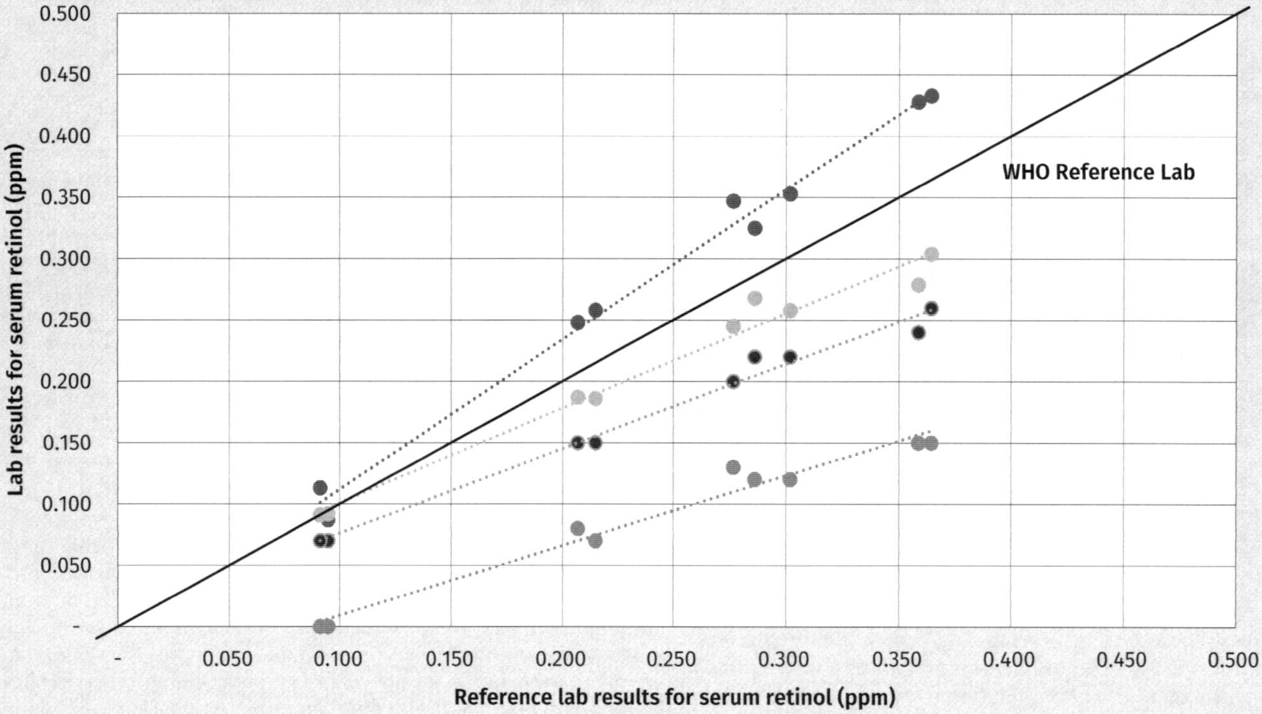

Figure 3: Simulation of individual pig results if sent to various labs, based on lab ring test results.

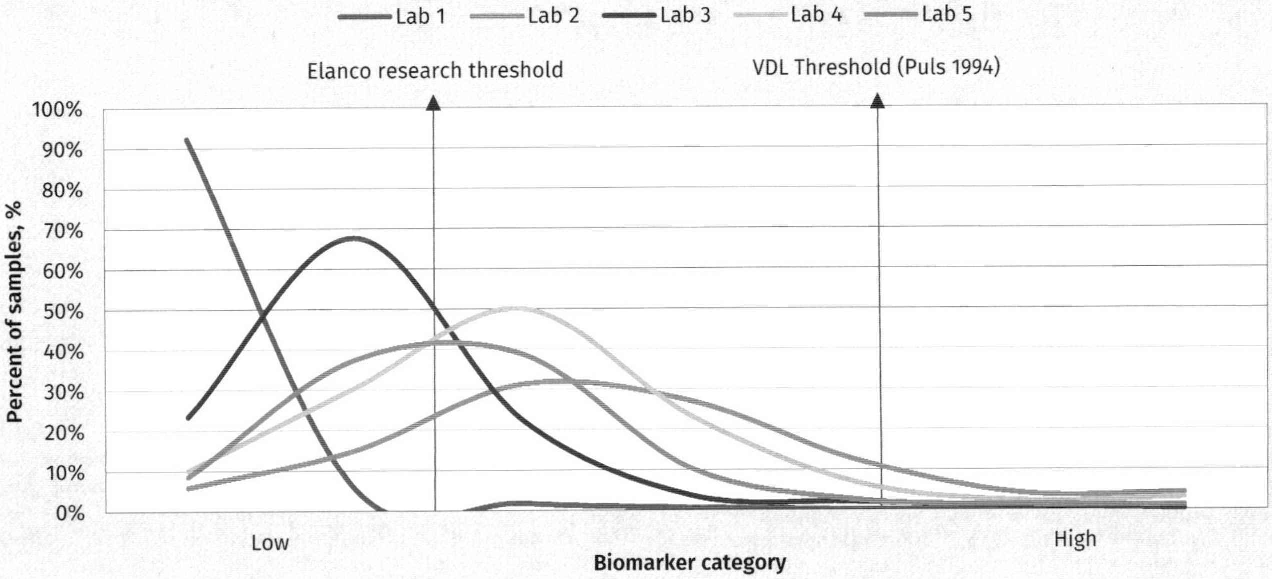

References

1. Tanumihardjo SA, Russell RM, Stephensen CB, et al. Biomarkers of nutrition for development (BOND)—vitamin A review. *ASN*. 2016 (Supplement: Biomarkers of nutrition for development (BOND) expert panel reviews, part 4): 1816S-1848S. **doi:10.3945/jn.115.229708**

2. Green HN and E Mellanby. Vitamin A as an anti-infective agent. *Br Med J*. 1928; 2: 691-696.

3. Rubin LP, Ross AC, Stephensen CB, et al. Metabolic effects of inflammation on vitamin A and carotenoids in humans and animal models. *Adv Nutr* 2017;8:197–212. **doi:10.3945/an.116.014167**

4. Chang P, and R Song. Vitamin A potentially may be used as indicator of health status in pigs. Feedstuffs 2020: 18-19.

5. Petznick T, Chang P, and R. Song. Improved performance feeding NutriQuest Baseline in sows and related impact on inflammation biomarker levels. *AASV Annual Meeting* 2023; 95-99. **doi:10.54846/am2023/39**

6. Puls, R. Vitamin Levels in Animal Health. *Sherpa International* 1994: 2nd ed.

7. Greiner L, Elefson S, Radke S, et al. A survey of vitamin and trace mineral ranges for diagnostic lab reporting from conventionally raised swine. *J Swine Health Prod*. 2022;30(5):282-291. **https://doi.org/10.54846/jshap/1286**

https://doi.org/10.54846/am2024/55

Translating influenza A virus sequencing: Improving vaccine selection

Melissa Hensch[1], DVM; Marie Culhane[2], DVM, PhD
[1]Phibro Animal Health Corporation, Teaneck, New Jersey; [2]College of Veterinary Medicine, University of Minnesota, St. Paul, Minnesota

Introduction

The North American swine population has been plagued by influenza A virus (IAV) for decades, experiencing IAV-associated morbidity, mortality, and economic losses. What makes IAV a constant challenge is its propensity for genetic changes and reassortment,[2] making vaccine selection against this moving target difficult, and cross-protection against multiple variants seemingly unattainable. Dogma asserts that vaccine protection against influenza is optimized when the vaccine strain matches the infecting, ie, farm, strain.[10,12] To determine if the strains match, sequence analyses are usually focused on the hemagglutinin (HA) gene and protein, comparing the nucleotide and amino acid identity and, for H3 IAV, focusing on six amino acids that are known to be key to antigenicity.[1] Genetic analyses can also be expanded to include comparisons of additional putative key amino acids of H1 IAV[9] and the neuraminidase (NA) gene and protein sequences[5] to more thoroughly inform vaccine strain selection. Several case studies from current production systems have demonstrated the utilization and value of this additional information.

HA sequencing

Today's HA sequence analyses include subtyping, clade determination, and calculating nucleotide and amino acid identity.[16] Comparing key amino acids that have been identified as highly important to IAV antigenicity against which the pig's immune response is targeted is also a necessary step for vaccine selection. In addition to the key amino acids identified for H3 IAVs of humans, equines, and swine,[6] published research suggests that there may be key amino acids for H1 IAVs that should also be investigated.[7,9] However, due to the substantial diversity of H1 IAV in swine,[3,17] there is variation in which amino acids are key for each H1 clade. Nevertheless, examinations of amino acids at epitope sites of H1s provide a deeper look into strain comparison. Such deep dives may help investigate potential vaccine failure when the sequence identity is relatively high (> 95%) yet variable immune response and disease prevention are observed. When exploring multiple vaccine options with highly similar sequences, expanded amino acid comparisons may help determine which of the available strains is the best candidate for autogenous vaccine production (Tables 1, 2).

NA sequencing

Neuraminidase (NA) is important in generating an immune response from the pig to IAV; specifically, the clade of NA may be imparting a greater contribution to cross-protection than previously thought.[5] Including NA in addition to HA sequence analyses, NA sequencing allows the comparison of different influenza viruses of both of the surface antigens. Current NA analyses are usually limited to comparisons of clades as it has been shown that not only are there multiple N1 and N2-types circulating in US pigs (Figures 1 and 2),[16] but there are also antigenic differences between different clades of NA. However, comparisons of percent NA identities are rarely done since very little is known about what percent identity is important for cross-protection and cross-reaction between strains with different NAs. While the last five years of research has provided more information on NA than ever before,[11,14,15] it is still not fully understood if a vaccine that contains multiple virus strains must have the NA and HA derived from the same candidate virus to generate a cross-protective immune response. For example, it is not clearly understood if a bivalent vaccine containing an H1N2 with alpha H1 and N2.1998B and an H3N2 with 3.2010.1 H3 and N2.2002B would provide pigs on farm sufficient cross-protection against a natural infection with an H1N2 with alpha H1 and N2.2002B. However, research in mice has shown that whole particle pandemic H1N1 vaccine-induced anti-N1 antibodies that cross-reacted with an avian H5N1 virus,[4] suggesting that antibodies against NA help overcome mismatching of vaccine and infecting strain (Figures 1, 2).

Table 1: Example of HA (H3) analysis results. Clade classification, percent identities for both nucleotides and amino acids, and key amino acid sites are provided. Results suggest that although there is high identity for both HA nucleotides and amino acids between the farm strain and vaccine strain, there are differences at 3 of the 7 key amino acid sites.[1] More specifically, the vaccine strain has a histidine (H) at position 159 but the farm strain has a phenylalanine (F) (H159F) and the other differences are E189K; and Y193S. The differences of H159F and E189K involve charge and polarity properties which may impart antigenic changes in the HA protein. Similarly, E189K involves a difference in hydrophobicity, another physiochemical property change[8] that may alter the presentation of the antigen to the immune system.

Strain	US clade	Global clade	% Nucleotide identity to vaccine strain	% Nucleotide identity to farm strain	% Amino acid identity to vaccine strain	% Amino acid identity to farm strain	Key amino acids at site						
							145	155	156	158	159	189	193
Vaccine	Human-like H3	3.2010.1	100	97.59	100	97.17	N	T	H	N	H	E	Y
Farm	Human-like H3	3.2010.1	97.59	100	97.17	100	N	T	H	N	F	K	S

Table 2: Example of HA (H1) analysis results. Clade classification, percent identities for both nucleotides and amino acids, and key amino acid sites are provided. Results suggest that although there is high nucleotide identity yet < 95% identity for amino acids between the farm strain and vaccine strain. However, there are only differences in 2 of the 23 amino acids at the epitope sites.[9] This suggests that differences in amino acid identity are occurring outside of the possible epitope sites. Nevertheless, the vaccine strain has an isoleucine (I) at position 174 but the farm strain has a valine (V) (I174V) and the other difference is lysine (L) at position 272 but the farm strain has glutamic acid (E); K272E. The difference of I174V is only a change in size of the amino acid chain but K272E involves a difference in hydrophobicity, a physiochemical property change[8] that may alter the presentation of the antigen to the immune system.

Strain	US clade	Global clade	% Nucleotide identity to vaccine strain	% Nucleotide identity to farm strain	% Amino acid identity to vaccine strain	% Amino acid identity to farm strain	Amino acids differences at putative key amino acid site		
							174	272	Note: at sites 69, 119, 121, 127, 129, 133, 140, 152, 167, 174, 185, 188, 208, 214, 215, 221, 255, 258, 269, 272, 288, 307, and 309 the strains had identical amino acids
Vaccine	Delta 2	1B.2.1	100	95.82	100	93.48	I	K	
Farm	Delta 2	1B.2.1	95.82	100	93.48	100	V	E	

Figure 1: Heatmap of H1 and NA combinations, 11/2/2022 to 11/1/2023.

Count of H1 and NA combinations

	N2.1998	N2.1998B	N2.2002A	N2.2002B	N1.classical	N1.pandemic
alpha	1	5		8	44	8
gamma					85	
pdmH1	1			1		20
delta1A	1			2		
delta1B			1	14		
delta2		75		1		

Figure 2: Heatmap of H3 and NA combinations, 11/2/2022 to 11/1/2023.

Count of H3 and NA combinations

	N2.1998	N2.2002A	N2.2002B	N2.2016	HA human-to-swine 2022
2010.1		4	65	1	
2010.2				2	
cluster_IVA		2	25		
cluster_IVB			4		
HA-human-to-swine 2022	1		2		3

Use of sequence analysis for on-farm vaccine decisions

The following case studies are provided to further demonstrate how more in-depth sequence analyses can be used to improve vaccine selection and control influenza in pig farms.

Case study #1

A customer provided several IAV strains for sequence analysis. The IAV strains were obtained from a severe influenza outbreak in growing pigs. Pigs were 13 weeks of age, located in a finisher in the Midwest-United States. Clinical signs included cough and increased mortality. Pigs had been vaccinated with a commercial porcine circovirus type 2 (PCV2) and mycoplasma vaccine at weaning and were porcine reproductive and respiratory syndrome (PRRS) negative. Pigs were vaccinated with a killed autogenous, pentavalent IAV vaccine at 9 weeks of age. Diagnostic tests of diseased lungs were PCR negative for PRRS and PCV2 but positive for IAV and an H1 of alpha clade 1A. positive sequence for H1, Alpha-clade 1A.1.1.3 This farm strain's HA sequence was 99.06% identical to the HA sequence of the vaccine strain, so cross-protection was expected, yet vaccine performance was perceived to be non-optimal. Visits to the farm did not reveal any environmental stressors or coinfections and it was believed, but not verified, that vaccine compliance was as directed. Upon further H1 key amino acid analysis, 51 of 52 amino acids at H1 epitope sites were matched, further supporting cross-protection by HA analyses alone. However, the NA sequence from the farm strain was of N1-classical clade which was distinct (ie, heterosubtypic) from the N1-pandemic clade in the vaccine strain. This N1 mismatch is believed to have caused the vaccine failure.

Case study #2

A killed autogenous vaccine containing an H1 strain of Alpha clade 1A.1.1.3 was given to pigs in a farm that had previously been infected with an Alpha H1N2. The autogenous vaccine strain also contained an Alpha H1 clade 1A.1.1.3, thus the vaccine strain and farm strain were of the same HA clade. The farm veterinarian perceived the results as a vaccine failure. On further HA analysis of amino acids, only 42 of 52 at epitope sites were a match, demonstrating a degree of variability that could explain reduced cross-protection. Also, NA-analysis revealed that the vaccine strain contained an N1-classical while the field virus was an N2-2002B. These new comparison techniques highlight the power of the additional analysis in determining the potential efficacy of autogenous vaccines.

Case study #3

A fully integrated swine system in the US has conducted a comprehensive IAV surveillance and vaccination program for several years. Although the IAV program is robust and successful in many ways, there is still room for improvement. To that end, the system has conducted more in-depth analyses of the NA subtypes detected in their farms and ensured that the most commonly detected NA subtypes identified, ie, N2 2002B, N2 1998B, and N1 classical, are included in the strains that comprise their multivalent autogenous vaccines. While data is still being collected, initial analyses of the N1 subtypes alone have revealed that two distinct clades are circulating between 2020 and 2023 – pandemic N1 and classical N1. Furthermore, within the classical N1 clade, there is a dominant clade (lower portion of the tree in Figure 3)

Figure 3: Neighbor-joining tree of 122 N1 sequences. The tree was constructed with Geneious Tree Builder (Version 2023.2) using a Jukes-Cantor genetic distance model and a Neighbor-Joining tree build method. There were 47 farm strains, 74 reference strains, and 1 vaccine strain included and the tree was out grouped to reference strain classicalSwine|CY028173|A/swine/Iowa/2/1987|H1N1|. At the top of the tree is a classical N1 clade with reference strains from the 1990s. The middle of the tree shows the pandemic (pdm) N1 clade of the 2000s. The bottom portion of the tree shows two subclades of classical N1s from the 2000s. The dominant clade, ie, the clade containing the most (28/47) sequences from the farm contains the majority of the sequences from 2020-2023, the current vaccine strain (*), and a potential vaccine candidate (C). The vaccine candidate groups closely with 2022-2023 farm strain N1s.

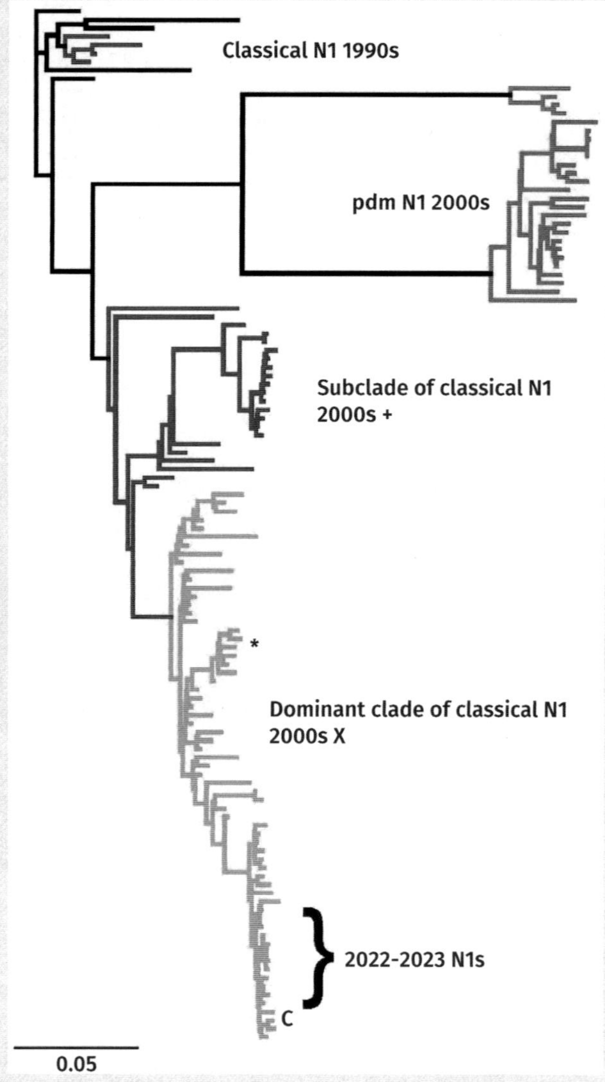

and a subclade (middle portion of the tree in Figure 3). Between the N1 clades, the average identity is 89%, suggesting that cross-protection between the N1s in the dominant clade and subclade of classical N1, despite being both in the classical N1 clade, might be less than optimal. Furthermore, the current H1N1 vaccine strain contains an N1 classical that, although it falls within the dominant clade in the bottom portion of the tree (as indicated by the asterisks), is on average 95.8% identical to the farm strains within the dominant clade. An H1N1 farm strain with an alpha clade 1A.1.1 and N1 classical from August 2022, indicated by a C in Figure 3, might be a better vaccine candidate, as it shares > 99% amino acid identity with 28 other farm strains (bracketed) in the dominant classical N1 clade. When vaccine candidates are chosen for the system's annual vaccine update, careful attention must be paid not only to H1 but also to N1 (Figure 3).

Conclusions

There have been several advances in swine influenza control in the 21st century and much of the knowledge gained can be attributed to genetic analyses of IAV that have taken us from knowing the type, to discerning the subtype, to classifying the clades, and to characterizing the amino acids of the antigenic proteins. The swine veterinarians and pork producers of the United States have contributed to the knowledge base through their willingness to participate in influenza A virus surveillance programs.[13] The influenza researchers have used the surveillance stream to conduct relevant research both in the laboratory and in farms. The vaccine manufacturers and allied industries have collaborated with veterinarians, producers, and researchers to help ensure that safe vaccines and more protective vaccines are available and implemented appropriately. This is not only for the benefit of the pig, but human health also benefits due to control and reduction of IAV that are shared between humans and pigs. The next few decades may lead us to elimination of influenza if and when vaccines are properly selected and immunization results in broad cross-protection, but only if we continue to accumulate data and share the knowledge.

References

1. Abente, Eugenio J., et al. "The Molecular Determinants of Antibody Recognition and Antigenic Drift in the H3 Hemagglutinin of Swine Influenza A Virus." *Journal of Virology*, vol. 90, no. 18, Sept. 2016, pp. 8266–80, **https://doi.org/10.1128/JVI.01002-16**

2. Anderson, Tavis K., et al. "Characterization of Co-Circulating Swine Influenza A Viruses in North America and the Identification of a Novel H1 Genetic Clade with Antigenic Significance." *Virus Research*, vol. 201, Apr. 2015, pp. 24–31, **https://doi.org/10.1016/j.virusres.2015.02.009**

3. Chang, Jennifer, et al. "OctoFLU: Automated Classification for the Evolutionary Origin of Influenza A Virus Gene Sequences Detected in U.S. Swine." *Microbiology Resource Announcements*, vol. 8, no. 32, Aug. 2019, pp. e00673-19, **https://doi.org/10.1128/MRA.00673-19**

4. Handabile, Chimuka, et al. "Inactivated Whole Virus Particle Influenza Vaccine Induces Anti-Neuraminidase Antibodies That May Contribute to Cross-Protection against Heterologous Virus Infection." *Vaccines*, vol. 10, no. 5, May 2022, p. 804, **https://doi.org/10.3390/vaccines10050804**

5. Kaplan, Bryan S., et al. "Evolution and Antigenic Advancement of N2 Neuraminidase of Swine Influenza A Viruses Circulating in the United States Following Two Separate Introductions from Human Seasonal Viruses." *Journal of Virology*, vol. 95, no. 20, Sept. 2021, p. e0063221, **https://doi.org/10.1128/JVI.00632-21**

6. Lewis, Nicola S., et al. "Substitutions near the Hemagglutinin Receptor-Binding Site Determine the Antigenic Evolution of Influenza A H3N2 Viruses in U.S. Swine." *Journal of Virology*, vol. 88, no. 9, May 2014, pp. 4752–63, **https://doi.org/10.1128/JVI.03805-13**

7. Lorusso, Alessio, et al. "Genetic and Antigenic Characterization of H1 Influenza Viruses from United States Swine from 2008." *The Journal of General Virology*, vol. 92, no. Pt 4, Apr. 2011, pp. 919–30, **https://doi.org/10.1099/vir.0.027557-0**

8. O., Rodolfo, et al. "Decoding the Building Blocks of Life from the Perspective of Quantum Information." *Advances in Quantum Mechanics*, edited by Paul Bracken, InTech, 2013, **https://doi.org/10.5772/55160**

9. Rajao, Daniela S., et al. "Antigenic and Genetic Evolution of Contemporary Swine H1 Influenza Viruses in the United States." *Virology*, vol. 518, May 2018, pp. 45–54, **https://doi.org/10.1016/j.virol.2018.02.006**

10. Rajão, Daniela S., and Daniel R. Pérez. "Universal Vaccines and Vaccine Platforms to Protect against Influenza Viruses in Humans and Agriculture." *Frontiers in Microbiology*, vol. 9, 2018, p. 123, **https://doi.org/10.3389/fmicb.2018.00123**

11. Sandbulte, Matthew R., Phillip C. Gauger, et al. "Neuraminidase Inhibiting Antibody Responses in Pigs Differ between Influenza A Virus N2 Lineages and by Vaccine Type." *Vaccine*, vol. 34, no. 33, July 2016, pp. 3773–79, **https://doi.org/10.1016/j.vaccine.2016.06.001**

12. Sandbulte, Matthew R., Anna R. Spickler, et al. "Optimal Use of Vaccines for Control of Influenza A Virus in Swine." *Vaccines*, vol. 3, no. 1, Jan. 2015, pp. 22–73, **https://doi.org/10.3390/vaccines3010022**

13. Vincent, A., et al. "Review of Influenza A Virus in Swine Worldwide: A Call for Increased Surveillance and Research." *Zoonoses and Public Health*, vol. 61, no. 1, Feb. 2014, pp. 4–17, **https://doi.org/10.1111/zph.12049**

14. Walz, Lisa, et al. "Neuraminidase-Inhibiting Antibody Titers Correlate with Protection from Heterologous Influenza Virus Strains of the Same Neuraminidase Subtype." *Journal of Virology*, vol. 92, no. 17, Sept. 2018, pp. e01006-18, **https://doi.org/10.1128/JVI.01006-18**

15. Wang, Yang, et al. "Antigenic Characterization of Influenza and SARS-CoV-2 Viruses." *Analytical and Bioanalytical Chemistry*, vol. 414, no. 9, Apr. 2022, pp. 2841–81, **https://doi.org/10.1007/s00216-021-03806-6**

16. Zeller, Michael A., Tavis K. Anderson, et al. "ISU FLUture: A Veterinary Diagnostic Laboratory Web-Based Platform to Monitor the Temporal Genetic Patterns of Influenza A Virus in Swine." *BMC Bioinformatics*, vol. 19, no. 1, Nov. 2018, p. 397, **https://doi.org/10.1186/s12859-018-2408-7**

17. Zeller, Michael A., Anugrah Saxena, et al. "Use of the ISU FLUture Multisequence Identity Tool for Rapid Interpretation of Swine Influenza A Virus Sequences in the United States." *Journal of Veterinary Diagnostic Investigation: Official Publication of the American Association of Veterinary Laboratory Diagnosticians, Inc*, vol. 34, no. 5, Sept. 2022, pp. 874–78, **https://doi.org/10.1177/10406387221111128**

https://doi.org/10.54846/am2024/56

Dollars in the bucket (stock): What's getting into the pigs?

P. Smith, RPh, Pharm D; B. Shields, PhD; B. Payne, DVM
Veterinary Pharmaceutical Solutions (VPS), St. Peter, Minnesota

Introduction

Managing losses

Iowa State University Extension and Outreach's estimate of returns for farrow to finish operations had an average of $22.87/head loss (R: $49.47/head loss - $9.39/head profit) in the first nine months of 2023.[1] Producers need to conserve and prioritize expenditures in such an unprofitable market. However, pigs may still get infected from disease and need to be treated. During an outbreak of disease in a large population of pigs veterinarians often prescribe water medication due to the ability to quickly treat the entire group for multiple days. It is desirable that treatment controls disease in the shortest number of days, treating clinical signs, preventing additional pigs from becoming clinically ill, and preventing lingering morbidity in the group to minimize the cost of therapy and subsequent production costs of clinical signs maximize the return for the producer.

Medication handling and mishandling's role

Effectiveness of the medication to achieve the desired result are impacted by the pathogen, the pig and the effective application of the medication. We focus on the medication aspect in this paper as there are several variables that may impact the potency (activity) of the antibiotic from the time of opening the package to absorption of the antibiotic. Once a diagnosis has been made, an antibiotic is selected which the bacteria is sensitive to its mode of action. Next, an appropriate rate of the dosing must be made to achieve therapeutic levels. This proper medicine and science can be negated by improper handling of the product. Failure at any point may lead to failure of therapy, increased clinical signs, higher mortality, increase in chronically ill pigs and poor production performance. Additionally, the veterinarian may prescribe treatment for a longer period or for multiple rounds. Responsible use of antibiotics in the swine industry is often objectively measured by units of medication per pig (ie, mg/pig). Longer or more frequent treatments will increase the mg/pig metric. In this paper, we will limit our scope on selection and common use to beta-lactam antibiotics delivered through the drinking water of pigs.

Mode of action for beta-lactams

The class of beta-lactams includes penicillins, cephalosporins, carbapenems, monobactams and beta-lactamase inhibitors which all contain the beta-lactam ring. Penicillin-binding proteins (PBPs) are enzymes integral to peptido-glycan cross-linking, a necessary step in cell wall assembly for gram-negative and gram-positive bacteria. Beta-lactams covalently bond to the PBPs, thus interrupting the assembly of peptidoglycan cell walls.

Beta-lactam selection

There are no FDA-approved, commercially labeled, orally administered beta-lactams available for swine today in the United States. Internationally, there are several registered amoxicillins available for swine. In the United States (US), penicillin G potassium (benzylpenicillin, R-Pen®, Huvepharma, Peachtree, Georgia) is labeled for the treatment of erysipelas in turkeys. Its use in swine is considered extra-label drug use (ELDU) and is permitted for veterinary prescription under AMDUCA (Animal Medicinal Drug Use Clarification Act) for swine. Both pediatric amoxicillin and compounded amoxicillin soluble are available in the US and permitted for prescription from a veterinarian under AMDUCA and 21 CFR 530.13 when there is no medically appropriate drug higher on the decision-tree.[2,3] The decision-tree of antibiotic selection guides the veterinarian to antibiotics with FDA-approved labels for the pathogen and disease of concern in the species of interest as the first go-to product. If efficacy of an FDA-labeled drug is not satisfactory, ELDU is allowed when failure to treat with an ELDU-drug "may result in the animal's suffering or death". Specifically, AVMA (American Veterinary Medical Association) calls out enhancing production is not an appropriate reason to choose ELDU.[2] Compounding is allowed if an ELDU use product is not available or if the ELDU-drug use is not satisfactory to the therapeutic outcome desired. When compounding drugs, compounders and food-animal veterinarians must adhere to specific Federal and state laws, such as never compounding from bulk active pharmaceutical ingredients (APIs) for food animals and a withdrawal period must be established as there is a zero-tolerance for residues with compounded drugs. State laws must be followed from both the state of the compounder and the location of the pigs in which the compounded drug is prescribed and used. It is recommended that veterinarians utilize pharmacists with expertise in livestock compounding to educate themselves on local and Federal regulations.

We will use *Streptococcus suis* (*S suis*) as an example of a target pathogen and the disease sequelae caused by *S suis*. There are labeled water soluble antibiotics available with indications: Florcon® (Aspen Veterinary Resources, Ltd, Loveland Colorado) and Aivlosin® (Pharmgate Animal Health, Wilmington, North Carolina). Concurrent use or back-to-back use of macrolides (feed and water) are contraindicated. This veterinarian and his colleagues have observed adverse reactions including those listed on an injectable form of the active ingredient, perianal inflammation and diarrhea.[4] These undesirable reactions causing additional clinical signs in pigs and lack of efficacy in the field situations are examples of criteria necessary to move to the next step of the AMDUCA flow-chart, ELDU selection. In this case, water soluble penicillin G potassium can be evaluated for its efficacy and any reactions in the field since it is an FDA-approved drug, and it only requires changing species from turkeys to swine for ELDU. Its label indicates that new preparations of the stock solutions should occur every 12 hours.[5] Likely this is due to the instability of penicillin G potassium in water causing a large decrease in potency once the powder comes in contact with stock solution water.[6] Lower potency in water decreases the availability of active ingredients for absorption into the body which may already have low bioavailability due to the acidity in the stomach environment.[6] Human pediatric amoxicillin is available in 7.5g of active ingredient per bottle. Due to its anticipated use in humans, its formulation makes it a colloidal suspension for administration versus a true solution; therefore, it is not able to achieve therapeutic dosages in large pig barns via water delivery without falling out of solution within the stock bucket.[7] Supply and availability of this product has been inconsistent in the last several years.[8] Veterinarians that experience lack of effective therapy with ELDU drugs should evaluate all the tools they have been licensed to use, including compounding drugs to reduce suffering and mortality in their patients. If the commercially registered drugs have not satisfied the necessary outcomes, compounded amoxicillin is available to veterinarians to prescribe. Livestock veterinarians should only prescribe compounded amoxicillin that is compounded from FDA-approved finished form dosages vs bulk API, which would be illegal in livestock.

Veterinarians are also interested in choosing nutrients or medications that are miscible with other products in the water system to concurrently administer to maximize therapeutic outcomes. Examples are the concurrent administration of a nonsteroidal anti-inflammatory drug (NSAID) or vitamin/electrolytes preparations with an antibiotic. This is common practice to prescribe either to replace electrolytes or to reduce fevers and inflammation while treating the underlying primary or secondary bacterial pathogens. Due to inherent properties, later discussed, limitations and choice of antibiotic may be a part of the larger therapeutic prescription by a veterinarian.

Resistance patterns of bacteria

Resistance to antibiotics is a global concern in both animals and humans. Resistance can be from bacterial production of beta-lactamases, decreased penetration to the target site, changes to membrane-associated proteins associated with penicillin binding and active pumping out of material from the periplasmic space of gram-negative bacteria.[9]

University of Minnesota's Chagas st al reported high variation of *S suis* resistance for nine years across antibiotics ranging from 89.67% of samples resistant to oxytetracycline, down to 0.51% for ampicillin, with other beta-lactams at 7.01% for penicillin and 2.21% for ceftiofur.[10] Note that ampicillin is considered the surrogate for amoxicillin in sensitivity testing. With a 14-fold variation amongst beta-lactams reported, it is not surprising that some clinical outcomes may vary between beta-lactams in the field. Similar findings were observed on Iowa State University Veterinary Diagnostic and Production Animal Medicine's Swine Bacterial Susceptibility Profiles from 2003-2021.[11]

Potency and degradation over time

Veterinarians' and producers' purchase decisions often are based on the overall value of antibiotics. One method to evaluate value is the cost per unit of active ingredient (mg or g) on the package. What is often overlooked is the actual value of the active ingredient over time. A unit of active ingredient is more valuable for treating pigs if it maintains activity for a longer period compared to a similar unit of active ingredient whose activity is degraded or depleted over the period in which treatment occurs. High performance liquid chromatography (HPLC) using validated methods to compare against USP standards (US Pharmacopeia) is a best practice to measure potency over time using area under the curve (AUC). Additionally, further degradation can be visualized via HPLC and interpreted from distortions in the curve. As an example, amoxicillin that is not properly formulated and compounded will lead to degradation byproducts such as amoxicillin penicilloic acid, amoxicillin penilloic acid I and II, diketopiperazine amoxicillin and 4-hydroxyphenylglyl amoxicillin (4-HPA).[12] A study conducted high-performance liquid chromatography (HPLC) of two amoxicillin solutions (VPS Amoxicillin Red and a competitor) compared potency 24-hours after they were prepared per instructions. There was a notable difference between the two products' stability of the amoxicillin molecule and findings of degradation products in the competitive product (Figure 1). These degradation products do not have the intended antibiotic activity from the parent amoxicillin. The same is true about penicillin G potassium degradation over time. In an experiment to emphasize the breakdown over time, the stock solution was tested at time 0, 20 hours and 50 hours, well beyond the labeled instructions of only 12 hours. The HPLC chart shows the decreasing potency and increase in degradation products (penicilloaldehyde, penillic acid and penilloic acid) over this period (Figure 2).

Minimum inhibitory concentrations (MIC) are used to quantify in vitro antibiotic concentrations needed to prevent bacterial growth over time, often 24 hours. Interpretation of MIC data must be made knowing that it may not 100% correlate in vivo. MIC testing also assumes there is no degradation of the antibiotic over time. However, this may not be true in the field if the antibiotic consumed is degraded or loses potency before consumption. FDA approved antibiotics guarantee a level of potency when handled per package instructions until the expiration date. Compounded drugs are similar and must adhere to standards for potency until the beyond use date (BUD) which is shorter than the expiration date of the FDA-approved final dosage forms which they are compounded from and based on testing and USP regulations. Pharmaceutical companies should know the impact on potency during normal use of products in the field.

Beta-lactams are heat, pH and light sensitive, experience water-based degradation, and are highly susceptible to chlorine and oxidative stresses.[13-16] Some products handle external factors better than others based on how the active ingredient is formulated with other ingredients. Hawkins reported on the differences of human oral amoxicillin compared to the compounded amoxicillin (VPS) noting that the human oral amoxicillin resulted in underdosing of pigs, was not fully soluble and was less consistent at dosing over a given period.[7]

Discussion

Water soluble delivery of antibiotics is often selected to treat large populations of pigs of various sizes to provide systemic antibiotics for multiple days, without the safety risks and labor costs associated with injecting all pigs with needles. Antibiotics through the drinking water have additional benefits above feed grade antibiotics in that they may be started immediately instead of waiting for the feed bins, lines and feeders to be empty as well as feed consumption decreases before water consumption thus potentially limiting access to the antibiotics sooner when administered via feed.

Figure 1: Zoomed in HPLC graph of samples collected from stock solutions of Amoxicillin Red (VPS, solid line) and another amoxicillin (competitor, dashed line) after 24 hours of preparation per compounded package instructions. Penilloic acid and 4-HPA are degradation products of amoxicillin that are not active.

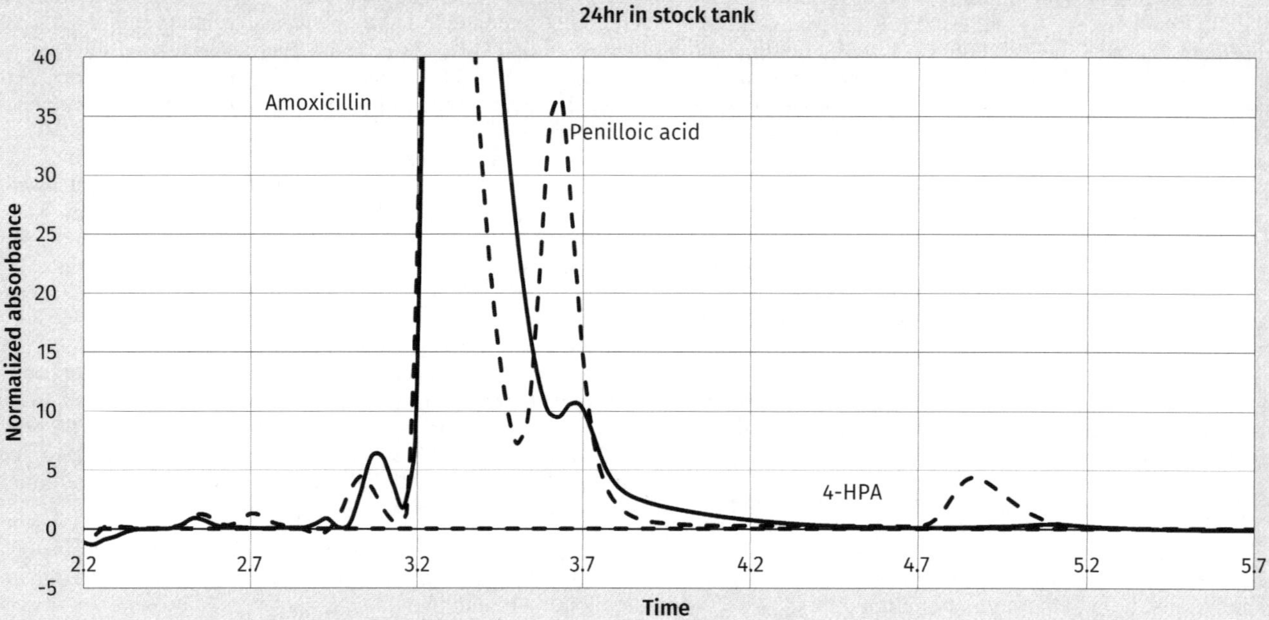

*Peaks have been normalized and are not to be used for comparing relative potency.

Figure 2: HPLC graph of samples collected from stock solution of penicillin G potassium at time 0, 20 hours and 50 hours after preparation per package instructions. Penicilloaldehyde, penillic acid and penilloic acid are degradation products of penicillin G that are not active.

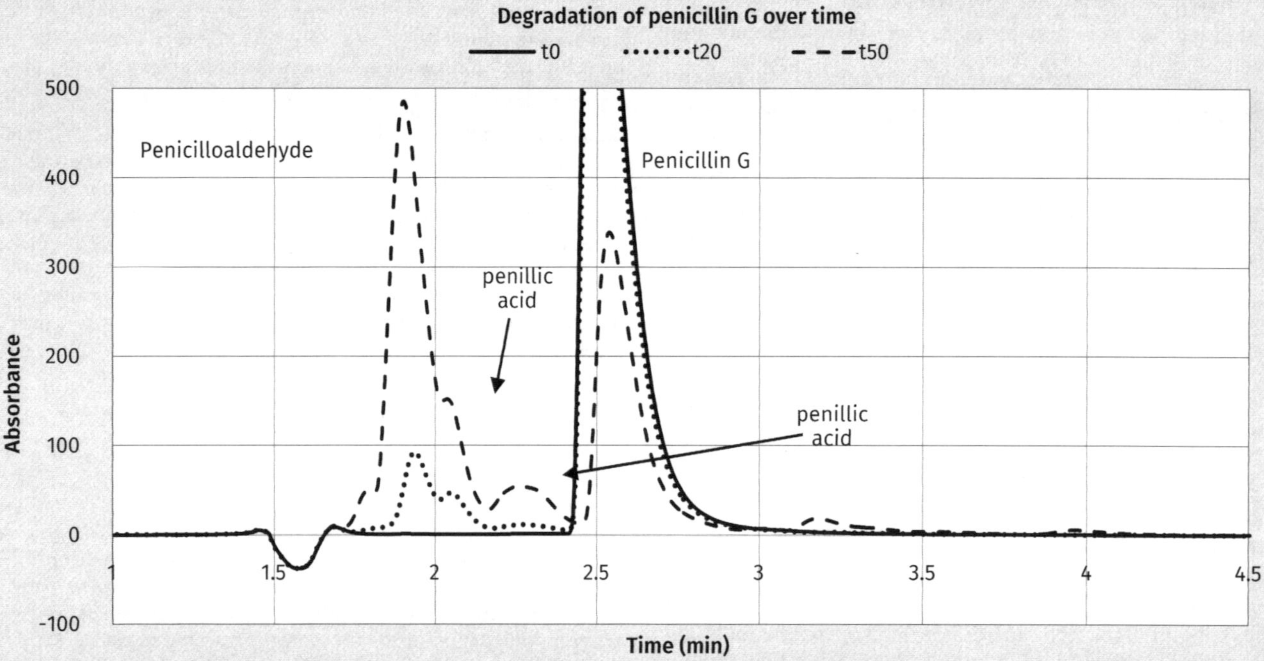

*Labeled instructions indicate fresh solution should be made every 12 hours.
The longer times are used to emphasize degradation and decrease in potency over time.

Veterinarians must follow the AMDUCA, ELDU and compounding regulations while prescribing. Additionally, it is essential to understand how the ELDU and compounded drugs react in certain environments, whether they remain stable and potent during the stock solution lifespan and or whether they are unstable or degrade in a water environment. While coadministration of drugs is common to achieve necessary therapeutic outcomes, one should not do so unless they know the miscibility and cross-reactivity of the drugs. Pharmacists are underutilized resources for livestock veterinarians and should be consulted. Our clients spend their money on veterinary expertise and the drugs we prescribe. We should maximize the value in the stock solution bucket by emphasizing proper handling of the medications that allow the most milligrams administered to reach the site of action rather than breaking down before they arrive.

References

1. Schulz L. Iowa State University Extension and Outreach. Estimated Livestock Returns: Swine. ISU. Accessed November 12, 2023. **http://www2.econ.iastate.edu/estimated-returns/Farrow%20to%20Finish%20PDF.pdf**

2. American Veterinary Medical Association. Animal Medicinal Drug Use Clarification Act (AMDUCA). AVMA. Accessed November 12, 2023. **https://www.avma.org/resources-tools/animal-health-and-welfare/animal-health/amduca**

3. Code of Federal Regulations: Title 21. Ch I. Subch E. Part 530. US National Archives. Accessed November 12, 2023. **https://www.ecfr.gov/current/title-21/chapter-I/subchapter-E/part-530#530.13**

4. Merck Animal Health. S is for Swine: Nuflor-S (florfenicol). Merck AH. Accessed November 12, 2023. **https://www.merck-animal-health-usa.com/wp-content/uploads/sites/54/2022/06/NUFLOR-S-florfenicol-Detailer.pdf**

5. Huvepharma. R-Pen (penicillin G potassium). Huvepharma. Accessed November 12, 2023. **https://www.huvepharma.us/wp-content/uploads/2021/03/R-Pen-0.500-BU-Film-P08-9102BF-Rev-02-2021-FDA-Approved-04142021.pdf**

6. Nielsen P, et al. Bioavailability of penicillin V after oral administration to and fasted pigs. *J. Vet. Pharmacol. Therapy*. 17.160.162.1994. **https://doi.org/10.1111/j.1365-2885.1994.tb00228.x**

7. Hawkins P, et al. Amoxicillin – Let's get the dose right! *Proc AASV*. 2009; 223-30.

8. American Academy of Pediatrics. Amoxicillin Shortage: antibiotic options for common pediatric conditions. AAP. Accessed November 12, 2023. **https://www.aap.org/en/pages/amoxicillin-shortage-antibiotic-options-for-common-pediatric-conditions/#:~:text=There%20is%20a%20national%20shortage,been%20impacted%20by%20the%20shortage**

9. Pandey N, Cascella M. Beta-Lactam Antibiotics. [Updated 2023 Jun 4]. In: StatPearls [Internet]. Treasure Island (FL): StatPearls Publishing; 2023 Jan. **https://www.ncbi.nlm.nih.gov/books/NBK545311/**

10. Chagas S, et al. Antimicrobial resistance of *Streptococcus suis* isolated from diseased pigs in the USA – a retrospective study from 2014 to 2022. *Proc Leman Conference*. St Paul, MN. 2023:37-8.

11. Iowa State University Veterinary Diagnostic and Production Animal Medicine. Swine bacterial susceptibility profiles. IA St Univ. Porcine 2003-2021. Accessed November 12, 2023. **https://vetmed.iastate.edu/vdpam/research/disease-topics/bacterial-susceptibility-profiles**

12. Nägele E, et al. Structure elucidation of degradation products of the antibiotic amoxicillin with ion trap MS^n and accurate mass determination by ESI TOF, *J of the Amer Soc for Mass Spectrometry*, 2005;16:10:1670-1676, **https://doi.org/10.1016/j.jasms.2005.06.002**

13. Binson G, et al. Overcoming stability challenges during continuous intravenous administration of high-dose amoxicillin using portable elastomeric pumps. *PLoS One*. 2019; 14(8):e0221391.

14. Strobel M, Soderland P. Soluble forms of amoxicillin and treatment of animals. US Patent #7244720.

15. Li D, et al. Determination of penicillin G and its degradation products in a penicillin production wastewater treatment plant and the receiving river. *Water Research*. 2008;1-2(42):307-317.

16. He X, et al. Degradation kinetics and mechanism of β-lactam antibiotics by the activation of H2O2 and Na2S2O8 under UV-254nm irradiation. *Journal of Hazardous Materials*. 2014;279:375-383.

https://doi.org/10.54846/am2024/57

A *Streptococcus suis* targeted microbial solution in gestation and lactation diets affected the microbiome and *S suis* populations in sow herds and their piglets

S. Hernandez[1]; S. Burmeister[1]; J. Rehberger[1]; L. Leonard[1]; D. Murray[2], DVM; E. Davis[1], PhD; A. Smith[1], PhD
[1]Arm and Hammer Animal and Food Production, Waukesha, Wisconsin ; [2]New Fashion Pork, Jackson, Minnesota

Abstract

S suis resides asymptomatically in the gastrointestinal tract, reproductive tract and respiratory tract of pigs but the microbe's ability to cross the dermal and mucosal layers can cause infections in other areas of the lymphatic system resulting in arthritis and sepsis as well as crossing the blood-brain barrier causing meningitis and other pathologies. The purpose of this study was to determine the level of *S suis* in sow herds before and after the implementation of a *S suis* targeted microbial solution (CERTILLUS) to evaluate its efficacy against *S suis* populations. Three farms were selected to determine the levels and prevalence of *S suis* by plate counts and using 16S rDNA gene primers specific for lactic acid bacteria (LAB). Three timepoints were identified for this study: 1) A baseline sampling; 2) Sows from one farm were placed on CERTILLUS targeted microbial solution and sampled after 3 months (Farm C only); 3) Sows returned to a basal diet and sampled after 3 months. Each sampling resulted in 45 samples per farm for 135 total samples per timepoint; these samples consisted of rectal and vaginal swabs of sows and rectal swabs from suckling piglets. Samples were processed to determine colony forming units (CFU) per swab on Columbia CNA agar to isolate *Streptococcus* and samples were stored to compare microbial communities. A maximum of five isolates per swab were selected and grown in standard conditions and isolates were subjected to a multispecies PCR to determine presence of *S suis*. There were detectable levels of *S suis* at all farms and timepoints. *S suis* levels and prevalence at Farm A increased from sampling 1 to sampling 2 (2.1 log CFU to 3.2 log CFU; 14.7% to 23.8% of samples); however, levels and prevalence at Farm B stayed stable (3.2 log CFU to 3.2 log CFU; 24.4% to 23.1% of samples). *S suis* levels at Farm C significantly decreased ($P < .01$) after CERTILLUS implementation across all swab types: Rectal swabs: 3.4 log CFU to 2.2 log CFU, vaginal swabs: 4.0 log CFU to 1.8 log CFU, piglet swabs: 3.4 log CFU to 1.3 log CFU. Post-treatment counts increased to 2.9 log CFU for rectal and 3.5 log CFU for piglet samples; levels stayed constant in vaginal samples. Prevalence of *S suis* significantly decreased ($P < .01$) at Farm C, Rectal swabs: 53.3% to 26.7%, vaginal swabs: 67% to 20% and piglet swabs: 53.3% to 6.7%. Post treatment, prevalence increased to 46% in rectal and 62% in piglets; prevalence stayed constant in vaginal samples. Amplicon sequencing using LAB specific primers on Farm C, indicated a significant difference in the Bray Curtis beta diversity between baseline and treated sows for rectal ($P < .001$) and vaginal samples ($P < .004$). In vaginal samples, the targeted microbial solution decreased the relative abundance of *S suis* from 18% to 2% ($P < .01$). Our results indicate that sows placed on a CERTILLUS targeted microbial solution had reduced prevalence and lower levels of *S suis* which would lower the chances of sows transferring potentially pathogenic *S suis* to their offspring.

Introduction

Streptococcus suis is an important pathogen in streptococcal diseases in pigs. This pathogen can cause systemic infection and causes many morbidity and mortality issues especially in nursery piglets worldwide and this can cause economic losses for the swine industry.[2] It is believed that early colonization of *S suis* happens at or around the time of birth but infection of *S suis* usually affects nursery piglets around four weeks after weaning.[4] *S suis* resides asymptomatically in the gastrointestinal tract, reproductive tract and respiratory tract of pigs but the microbe's ability to cross the dermal and mucosal layers causes infections in other areas of the lymphatic system resulting in arthritis and sepsis as well as crossing the blood-brain barrier causing meningitis and other pathologies.[3] *S suis* is considered a zoonotic pathogen meaning that it can cause disease in humans as well as in pigs and is considered a common cause of meningitis in some Asian countries that are exposed to or consume uncooked pork products.[4] There are several different methods for combatting *S suis* infection in swine, however, antimicrobial usage is becoming less effective with the emergence of multidrug-resistant strains of *S suis*. Vaccines are widely used but many vaccines are only targeted to certain serotypes of *S suis* and therefore do not encompass all serotypes that can cause infection. *Bacillus* species have become the most desirable form of probiotics due to their ability to produce spores which allow them to be stored for long periods of time, endure high heat from processing and offer better stability in acidic environments. Some *Bacillus* species can inhibit pathogens, have antioxidant abilities and have immunomodulatory effects.[5] The purpose of this study was to determine the level of *S suis* in sow herds before and after the implementation of a *S suis* targeted microbial solution (CERTILLUS) to evaluate its efficacy against *S suis* populations.

Methods

Streptococcus suis identification

Three farms were selected to determine the levels and prevalence of *S suis* and observe the overall LAB community. Three timepoints were selected for this field trial; 1) A sampling was completed for baseline levels at each farm (December 2021), 2) Sows from one farm were placed on a *S suis* targeted microbial solution (CERTILLUS, June 2022) and 3) Sows returned to an untreated diet (October 2022). Each sampling resulted in 45 samples per farm for a total of 135 samples per timepoint. Rectal and vaginal swabs were collected from the same lactating sow source and a rectal swab was obtained from a neonatal piglet in her litter. Swab samples were processed to determine colony forming units (CFU) per swab on Columbia CNA agar and samples were stored for further evaluation of total LAB population. A maximum of five isolates were selected from each sample and grown for 48 hours at 37°C in Todd Hewitt Broth (THB), selected isolates were subjected to a multispecies PCR to determine the presence of *S suis*, *S parasuis*[1] and *S equi* subsp *zooepidemicus*. Isolates were subjected to random amplified polymorphic DNA (RAPD) PCR to determine genetic diversity. Seven different Arm & Hammer™ *Bacillus* strains were utilized to screen 165 *S suis* isolates and a CERTILLUS targeted microbial solution was designed using the two strains (Arm & Hammer *Bacillus* strains 839 and 1999) that showed the best efficacy against the *S suis* populations.

Lactic acid bacteria (LAB) amplicon sequencing

The swab samples from timepoints one and two were selected from Farm C for analysis of the LAB community through amplicon sequencing. Total genomic DNA was isolated from each swab samples using the Qiagen DNeasy PowerSoil HTP 96 Kit. DNA was then sent to the University of Illinois Urbana-Champaign for library preparation by the Functional Genomics Unit at the Roy J. Carver Biotechnology Center. Library preparation was carried out using a Fluidigm 48 Access Array IFC. Master mix for amplification was created using Roche High Fidelity Fast Start Kit and 20x Access Array loading reagent according to Fluidigm protocols, and Fluidigm Illumina linkers with unique barcodes were added to each well. Samples were amplified with the forward NLAB2F primer (5'-GGCGGCGTGCCTAATACATGCAAGT-3') and the reverse WLAB1R primer (5' TCGCTTTACGCCCAATAAATCCGGA-3') to amplify the lactic acid bacterial population.[6] Following library prep the samples underwent sequencing on an Illumina NovaSeq 6000 by the DNA Services laboratory at the Roy J. Carver Center.

Sequencing results were analyzed using QIIME2 (version 2023.2). Single end, forward reads were imported into QIIME2, and DADA2 was used to correct errors in raw reads and generate error-corrected ASVs. Rare ASVs that were present in less than 3% of samples were removed through filtering. Taxonomy was assigned to ASVs using a Naïve Bayes Silva classifier made from the Silva 138 SSURef NR99 full-length taxonomy that was trained using the NLAB2F and WLAB1R primers. Filtering was done to remove ASVs that were assigned as Eukarya, Archaea, and any ASV not assigned at the phylum level. Diversity analyses were done at a depth of 12000, which was determined to be a sufficient sequencing depth using an alpha rarefaction curve. Bray-Curtis beta diversity was analyzed using a PERMANOVA with a pairwise comparison. Bray-Curtis distance matrices were used to create Principal Coordinate Analysis (PCoA) plots using RStudio. Differential abundance analysis was completed using GraphPad (version 9.5.1) to compare the average percentages of *S suis* detected in each group of swabs using a Mann-Whitney test.

Results

Over the course of this study, *S suis* isolates were collected from the three different timepoints (December 2021, June 2022 and October 2022) and a total of 165 isolates were screened against a panel of seven different *Bacillus* strains. After the baseline sampling in December against 52 different *S suis* isolates, a *Bacillus* formulation of strains 839 and 1999 were used as a targeted microbial solution. This formulation was consistent for inhibition across the other two timepoints, even after the product was removed from the system.

The percentage of all samples containing at least one *S suis* isolate (prevalence) was greatest at the Baseline sampling on Farm C compared to all other timepoints ($P < .05$). The percentage of samples positive for *S suis* at Baseline for Farm C was 60.5% and decreased to 20% after the implementation of CERTILLUS, after CERTILLUS was removed *S suis* prevalence increased to 43.2% ($P < .01$). The percent of samples determined to have *S suis* did not differ across the sampling timepoints at Farms A and B where the CERTILLUS product was not implemented (Figure 1).

The percentage of isolates (proportion) determined to be *S suis* was greater at the Baseline sampling at Farm C across sample types, with the largest number of isolates detected in the vaginal samples. Once CERTILLUS was implemented, the percentage of *S suis* detected decreased across all sample types (sow vaginal and rectal and piglet samples). After CERTILLUS was removed from the diet, the percent of *S suis* detected in the samples increased across all sample types (Figure 2).

The *S suis* targeted microbial solution decreased the relative abundance of *S suis* from 18% to 2% in vaginal samples after implementation ($P < .01$; Figure 3). The relative abundance of *S suis* stayed stable in both the sow rectal samples (2%) and the piglet samples (1%).

Amplicon sequencing using LAB specific primers on Farm C indicated a difference in the Bray-Curtis beta diversity between baseline and treated sows for vaginal and rectal samples ($P < .08$ and $P < .001$ respectively; Figure 4 and Figure 5). There was no difference in the Baseline and treated samples for the piglet rectal samples received ($P < .317$).

Discussion

In this study, three farms were monitored for the presence of *S suis* in sows and suckling piglets. *S suis* isolates from the Baseline sampling were successfully inhibited by one or more Arm & Hammer proprietary *Bacillus* strains and two strains were used to formulate the *Bacillus*-based product (CERTILLUS). Prevalence of *S suis* was highest at Baseline on Farm C, which was the farm chosen to include the *S suis* targeted microbial solution for this trial. After the product was implemented, the levels and prevalence of *S suis* decreased significantly in all sow samples that were tested. This trend extended to the neonatal piglets after CERTILLUS was implemented into the sow diet. The number of *S suis* isolates detected in the sampling decreased across all sample types. At the Baseline sampling, the largest number of isolates detected was in the sow vaginal samples at 45.3% which decreased to 8% during the CERTILLUS treatment period. When product was removed from the farm, the levels and prevalence increased across all sample types. When comparing the total microbial populations from samples tested at Baseline and the CERTILLUS treatment period on Farm C, there was a decrease in the relative abundance of *S suis* in the vaginal samples from 18% to 2%. The vaginal canal is regarded as one of the modes of inoculation of *S suis* into the systems of young piglets and may explain the decrease in the piglet rectal samples. However, in this study we did not swab the nasal cavity or the trachea of the young piglets so there still could have been potentially pathogenic isolates found there. According to the total microbial data analyzed at Farm C, there was an increase in the Shannon alpha diversity in the vaginal and rectal swabs taken from sows while the piglet samples stayed the same. When comparing the Bray-Curtis beta diversity, there was a shift in the populations recovered from both the vaginal and rectal samples from sows, there was no shift in populations in the piglet populations in this trial. In this study, sows placed on a targeted microbial solution had reduced prevalence of *S suis* which lowered the transfer of potentially pathogenic *S suis* to her offspring.

Figure 1: *S suis* prevalence (%) by farm in all sample types (ab $P < .05$); Farm C shows the difference in prevalence before and after the implementation of CERTILLUS. Farm C was on CERTILLUS for 3 months prior to the sampling in June 2022.

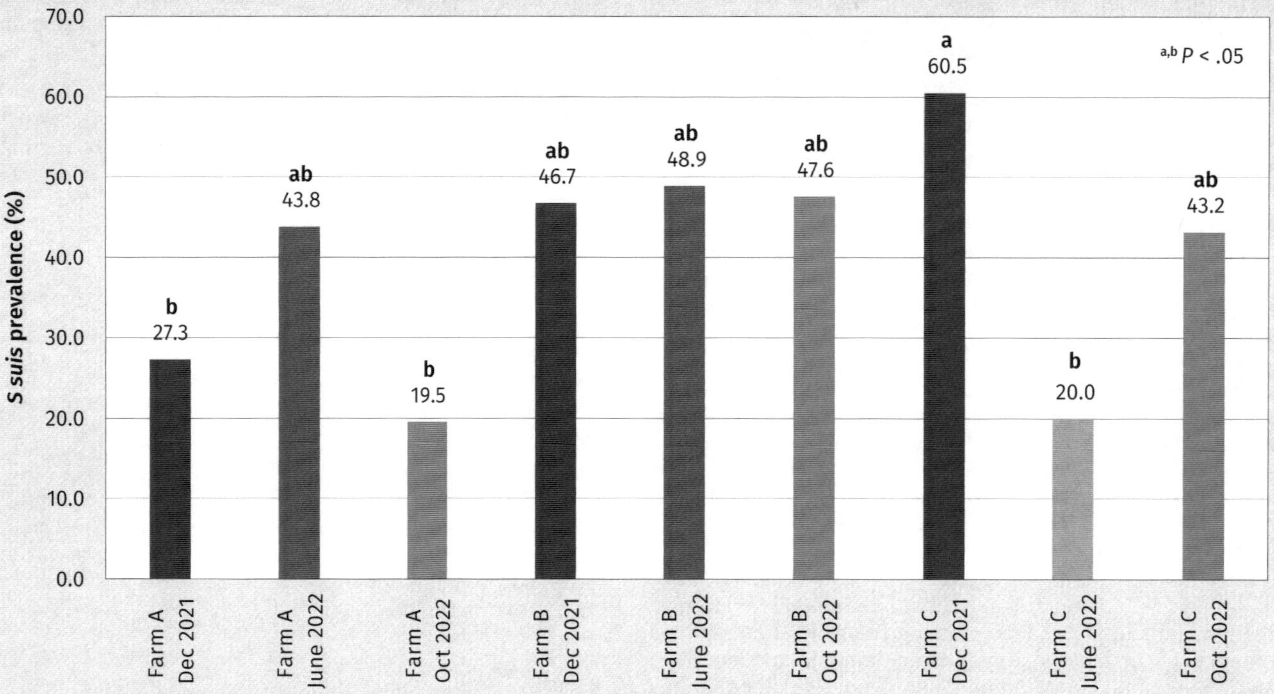

Figure 2: Proportion of isolates grown on selective agar that were *S suis* positive from each sample type on Farm C before and after CERTILLUS implementation.

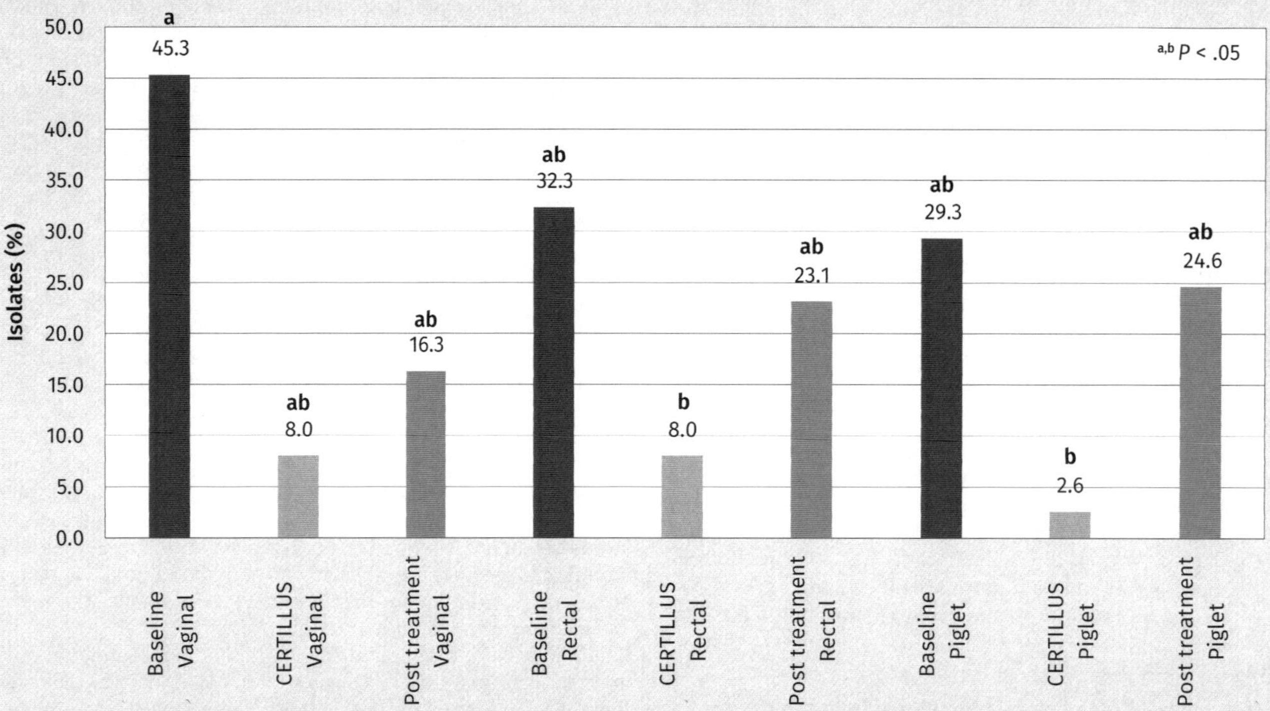

Figure 3: Relative abundance of bacterial species present in samples before and after implementation of CERTILLUS at Farm C using 16S rDNA gene primers specific for lactic acid bacteria.

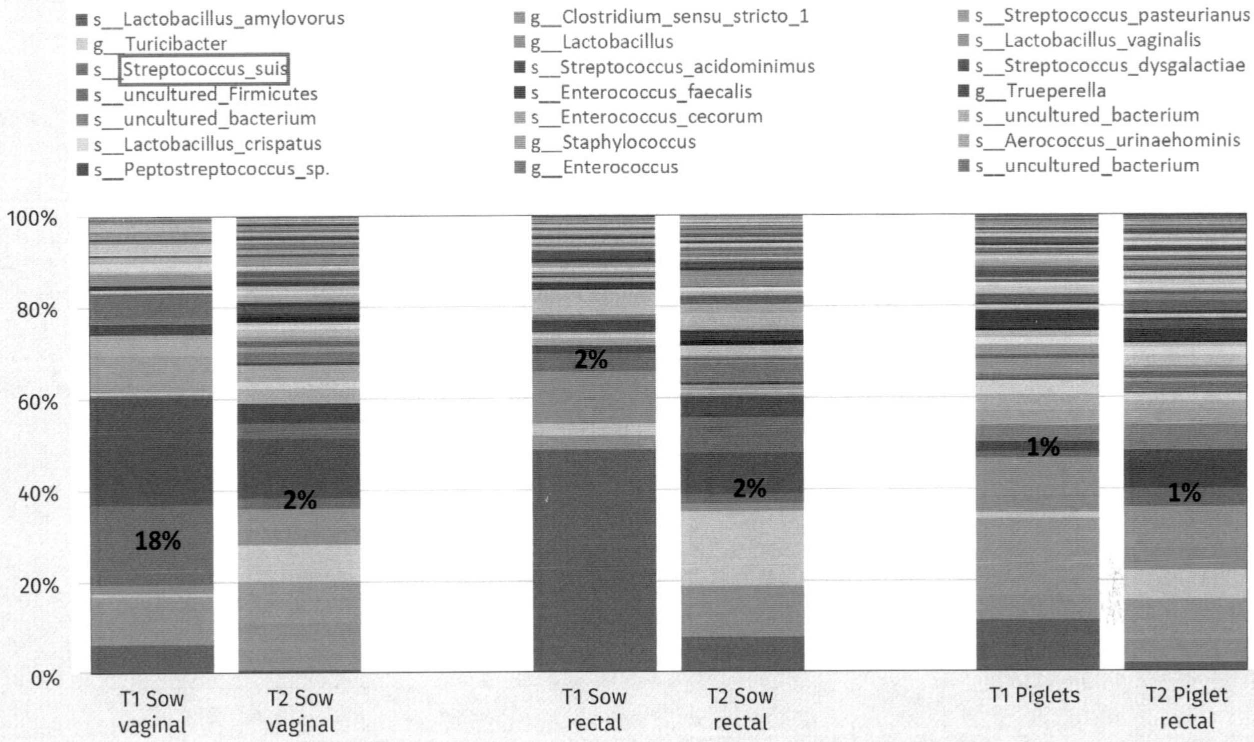

Continued work looking at the genetic similarities of *S suis* detected in the respiratory tract of nursery piglets and those found within the vaginal tract of sows would allow for assumptions to be made for the early inoculation of *S suis*.

References

1. Okwumabua O, O'Connor M, Shull E. 2003. A polymerase chain reaction (PCR) assay specific for *Streptococcus suis* based on the gene encoding the glutamate dehydrogenase. FEMS Microbiology Letters 218:79-84.

2. Arenas J., Zomer A., Harders-Westerveen J., Bootsma J., De Jonge M., Stockhofe-Zurwieden N., Smith H.E., De Greeff A. 2020. Identification of conditionally essential genes for *Streptococcus suis* infection in pigs. Virulence. Vol 11, No. 1: 446-464.

3. Rui L., Weiyi L., Yu M., Hong Z., Jiao Y., Zhe M., Hongjie F. 2018. The serine/threonine protein kinase of *Streptococcus suis* serotype 2 affects the ability of the pathogen to penetrate the blood-brain barrier. Cellular Microbiology. **DOI:10.1111/cmi.12862**

4. Segura, M. 2020. Streptococcus suis Research: Progress and Challenges. Pathogens. 9(9), 707. **https://doi.org/10.3390/pathogens9090707**

5. Elshaghabee F.M.F., Rokana N., Gulhane R. D., Sharma C., Panwar H. 2017. *Bacillus* As Potential Probiotics: Status, Concerns, and Future Perspectives. Front. Microbiol. Vol. 8, 2017. **https://doi.org/10.3389/fmicb.2017.01490**

6. Bokulich, N. A. & Mills, D. A. Differentiation of mixed lactic acid bacteria communities in beverage fermentations using targeted terminal restriction fragment length polymorphism. *Food Microbiology* **31**, 126–132 (2012).

Figure 4: Amplicon sequencing using 16S rDNA gene primers specific for lactic acid bacteria at Farm C indicated a difference in the Bray-Curtis beta diversity between baseline and treated sows for vaginal samples ($P < .008$).

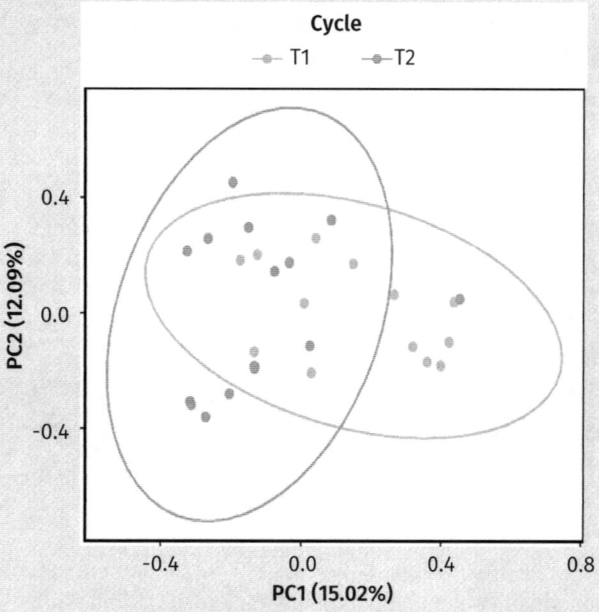

Sample size	Permutations	pseudo -F	p-value	q-value
29	999	2.1	0.008	0.008

Figure 5: Amplicon sequencing using 16S rDNA gene primers specific for lactic acid bacteria at Farm C indicated a difference in the Bray-Curtis beta diversity between baseline and treated sows for rectal samples ($P < .001$).

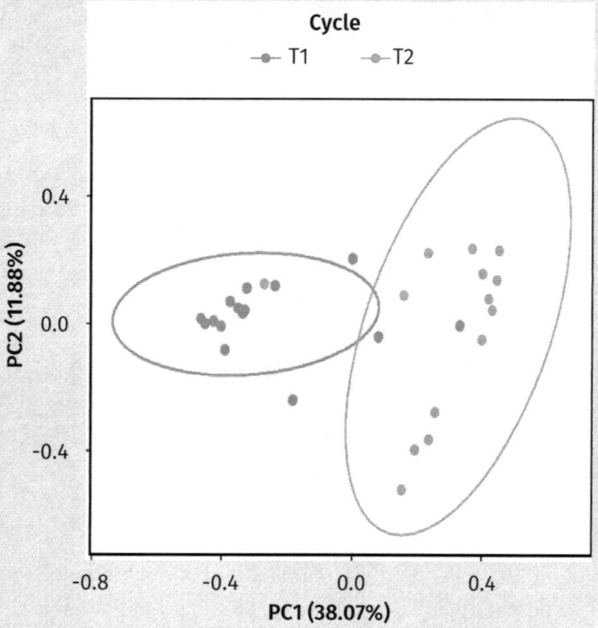

Sample size	Permutations	pseudo-F	p-value	q-value
29	999	9.2	0.001	0.001

https://doi.org/10.54846/am2024/58

How to choose a disinfectant

Gisele Ravagnani, DVM, MSc, PhD
Technical Marketing Manager at Lanxess

Introduction

Biosecurity is a set of measures that aim to prevent the introduction of diseases or pathogens into a herd and, in parallel, control the spread of pathogens that are already present. Disinfection is an important measure that, when used correctly, reduces the amount of pathogens present on a given surface. Disinfection is the process of eliminating many or all pathogenic microorganisms, other than bacterial spores, on inanimate objects.

With so many disinfectants on the market, what characteristics should I evaluate before buying one? Since not all disinfectants work the same, the selection of appropriate disinfectants for use in animal facilities requires consideration of multiple factors, including:

1. **Efficacy:** Spectrum of activity, performance under field conditions, including factors such as hard water, organic load, temperature variations, stability in use, low contact time;
2. **Application:** Wide application method possibilities (eg, vehicles, hard surfaces, boot baths, equipment) to ensure maximum coverage and pathogen elimination;
3. **Safety:** Choose a product that is safe for humans, animals and the environment;
4. **Cost in-use:** Optimize cost-benefit without compromising performance.

Chemical disinfectants are often the first line of defense against these pathogens; however, guidelines for their selection and usage often are defined less strictly. Because disinfectant use is critical in preventing the spread of pathogens causing diseases, it is an essential component of any animal facility management. Chemical disinfectants can be further categorized into three classes according to their method of action—denaturants, reactants, and oxidants.

Denaturants, such as quaternary ammonium compounds, phenolics, and alcohols, act by disrupting protein and lipid structures, making these products particularly effective against lipophilic enveloped viruses.

Disinfectants that are reactants, including aldehydes (formaldehyde and glutaraldehyde) and ethylene oxide, form and break covalent bonds, altering DNA, RNA, and protein structure and synthesis.

Oxidant disinfectants are the largest group and include halogens—hypochlorites, chlorine dioxide, and iodine—peroxides, and peroxymonosulfates. These disinfectants oxidize proteins, enzymes, and amino acids, making their spectrum of activity relatively broad. Oxidants are inexpensive and fast-acting and are often considered to be mycobacteriocidal, sporocidal, and fungicidal; they also have the ability to kill nonenveloped viruses.

Efficacy

When choosing a product to best address your needs, you should keep in mind what you are disinfecting against, what is the pathogen of concern? Is it a virus, bacteria, fungi or a combination of one or more? For example, some viruses tend to weaken the immune system of pigs, which could lead to a secondary bacterial infection, which can further complicate matters.

A disinfectant must be able to eliminate the most resistant pathogens under conditions that vary among facilities, depending on the biosecurity level, water hardness, temperature and organic load.

A study performed by the University of Minnesota tested eight commercially available disinfectants, bleach and 3% sodium hydroxide for efficacy against PCV2. Only four of the products tested delivered the required 5 $\log_{10}$ reduction in PCV2 but exposure times ranged from 10 minutes to 12 hours.

Environmental factors such as temperature can also play a role in the ability of a products performance or speed of kill. Some disinfectant chemistries, like glutaraldehyde, are negatively impacted by cold temperatures and require higher concentrations or longer contact time in order to be effective. It is critical to select a disinfectant that will not be impacted by the environmental conditions in which it will be used.

Application

Disinfectants must only be used for the item or area specified on the label. Additionally, the required application method and contact time should be provided by the manufacturer. Application methods may involve spraying, fogging or misting, wiping, immersing, or mop-on methods. Appropriate safety measures (eg, personal protective equipment) should be used during all disinfection processes; including the application process.

Safety

It is very important to check the product label to ensure that the disinfectant under evaluation is Environmental Protection Agency (EPA) registered to kill as many of the pathogens as possible.

Disinfectants are regulated in the US by the EPA and efficacy claims must be reviewed and approved before a manufacturer can make claims in marketing.

Cost in-use

Cost in-use should be calculated based on the price of the product divided by the dilution factor. And always remember to take into consideration all the points mentioned above. Try to answer the following questions: Will this disinfectant be effective against the pathogen (s) I am experiencing a problem with?

Will the product be effective under farm conditions?

Is this product safe for operators, the environment and animals?

Ideally, the disinfectant you select should have the: ability to kill a broad spectrum of pathogens, can be used safely under real world conditions and be cost effective when comparing products in-use.

A comprehensive biosecurity program consists of vaccination, medication and hygiene. Hygiene includes cleaning and disinfection practices, which are often overlooked and not given the same level of importance as vaccines or antibiotics. Preventing disease is always preferable and usually more cost effective than treating disease. Cleaning and disinfection is a simple concept to understand. Reduce the organic contaminate level present by chemical and physical means, which will then allow the disinfectant to kill pathogens.

When you provide animals a clean and healthy environment, you allow them to grow and thrive at their maximum potential. The return on investment will be measured in healthier animals with greater productivity, less therapeutic costs and lower mortality.

References

1. Campagna MV.; Faure-Kumar E.; Treger JA., et al. Factors in the selection of surface disinfectants for use in a laboratory animal setting. Journal of the American Association for Laboratory Animal Science, 2016, 55 (2): 175-188.

2. Rutala WA, Weber DJ. Infection Control and Hospital Epidemiology, 2014, 35 (7): 855-865.

3. Kim, HB, Lyoo, KS, Joo, HS. Efficacy of different disinfectants in vitro against porcine circovirus type 2. The Veterinary Record, 2009, 164: 599-600.

4. Jang Y, Lee J, So B, et al. Evaluation of changes induced by temperature, contact time, and surface in the efficacies of disinfectants against avian influenza virus. Poultry Science, 2014, 93: 70-76.

https://doi.org/10.54846/am2024/59

Effect of monoglyceride and organic acid supplements on growth performance in enterotoxigenic *E coli*-infected nursery pigs

Ying Chen; Mark Whitney; Ross Wolfenden
Animal Nutrition, Eastman Chemical Company, Kingsport, Tennessee

Introduction

Enterotoxigenic *E coli* (ETEC) has been continually recognized as a major issue in nursery pigs and has led to increased mortality and production loss in the US. Organic acids and monoglycerides of short-chain and medium-chain fatty acids have been reported to effectively prevent the proliferation of pathogenic bacteria like ETEC. When used in combination, there is a synergistic effect observed in the in-vitro microbiology tests. In this study, we investigated the effects of supplementing a combination of organic acid and monoglyceride additives in feed with and without added antibiotics and pharmacological levels of ZnO on growth performance of nursery pigs infected with pathogenic ETEC F18.

Methods

A total of 1,008 pigs (~21 d of age, 5.8 kg of initial BW) from a commercial sow farm were assigned to 4 treatments (12 replicates, n = 21 pigs/pen, mixed sex) following 2×2 factorial experiment design and fed for 42 d in 4 phases (D0 - 8, D9 - 21, D22 - 35, and D36 - 42, respectively). Treatments were (1) Positive control (PC), basal diet with in-feed antibiotics in phase 1 - 3, and 3000 ppm of Zn in phase 1, 2000 ppm of Zn in phase 2 and 125 ppm of Zn in phase 3 and 4 (ie, pharmacological levels); (2) PC supplemented with the combination of Acitra™ G20C (formic acid-based acid blend on inert carrier) and Entero-Nova™ 410C (blend of monoglycerides of short-chain and medium-chain fatty acids on inert carrier), Acitra G20C was added at 3 kg/MT in phase 1 diet and 1.5 kg/MT in phase 2 diet, while Entero-Nova 410C was added at 3 kg/MT in phase 1 and 2 diets, and 1 kg/MT in phase 3 and 4 diets; (3) Negative control (NC), basal diet without in-feed antibiotics and pharmacological levels of ZnO; (4) NC supplemented with the combination of Acitra G20C and Entero-Nova 410C (same supplementation rate as Treatment 2). Diets contained nutrients meeting or exceeding NRC requirements. Within 24 hours of arrival to the nursery barn, *E coli* vaccine was applied to all pigs via water. At the event of mortality, dead pigs were submitted for necropsy and confirmed ETEC F18 infection. Growth performance was measured for each phase. Data were analyzed using JMP software with pen serving as the experimental unit and differences considered significant at $P < .05$.

Results

Throughout the trial, pigs fed diets without in-feed antibiotics and pharmacological levels of ZnO had lower ($P < .01$) BW, ADG and ADFI compared with pigs fed diets added with antibiotics and pharmacological levels of ZnO. During phase 2 of the trial, pigs fed diets supplemented with the combination of Acitra and Entero-Nova additives had higher ($P < .05$) ADG and ADFI and higher ($P < .05$) BW on Day 22 of trial than the pigs in control groups without additive supplements. In phase 3 and 4, when Acitra G20C was removed and Entero-Nova 410C was reduced to1 kg/MT, there was no significant improvement on growth performance.

Conclusions

A holistic approach is needed when addressing ETEC issues. Nutritional interventions using feed additives can effectively enhance intestinal and overall health status of nursery pigs, subsequently alleviating the negative impacts on growth performance in nursery pigs caused by ETEC infection. Our results suggested the combination of Acitra G20C and Entero-Nova 410C improved growth performance of nursery pigs infected with ETEC F18 when supplemented at tested dosage in phase 1 and 2 diets than control groups without the additives.

https://doi.org/10.54846/am2024/60

Results from a diagnostic survey of biological measurements in healthy, sick, and lame growing pigs across the US swine industry for consideration in determining bone mineralization

B. Williams[1], PhD; M. Tokach[2], PhD; J. Woodworth[2], PhD; J. DeRouchey[2], PhD; R. Goodband[2], PhD; J. Gebhardt[3], DVM, PhD; J. Bergstrom[4], PhD; S. Hough[4], DVM; M. Rahe[5], DVM; P. Sitthicharoenchai[5], DVM; C. Siepker[6], DVM

[1]Seaboard Foods, Guymon, Oklahoma; [2]Department of Animal Science and Industry, College of Agriculture, Kansas State University, Manhattan, Kansas; [3]Department of Diagnostic Medicine/Pathobiology Animal Medicine, College of Veterinary Medicine, Kansas State University, Manhattan, Kansas; [4]dsm-firmenich, Animal Nutrition & Health; [5]Department of Population Health and Pathobiology, College of Veterinary Medicine, North Carolina State University, Raleigh, North Carolina; [6]Veterinary Diagnostic and Production Medicine, College of Veterinary Medicine Iowa State University, Ames, Iowa

Introduction

Due to improper bone mineralization in pigs, producers in the US swine industry experience economic losses and animal welfare concerns. Diagnosticians rely on bone mineralization as a diagnostic tool for identifying cases of lameness in pigs. However, because numerous factors can contribute to lameness, identifying the underlying cause can be difficult. There are different diagnostic assessments to measure bone mineralization, including bone ash, density, and histopathological examination; however, there is limited data comparing these different measurements across pigs with different health status. Recent work evaluating bone mineralization has compared the difference between bone ash procedures available for use. There are differences in percentage bone ash when comparing bones that had the lipid extracted (defatted) vs. those without the lipid extraction (non-defatted) before ashing of the bone.[1] The authors observed a reduction in standard deviation when the bone is defatted before ashing, and this response was more apparent in finishing pigs compared to nursery pigs due to the increased fat accumulation in bones as pigs grow. Also, there are differences in percentage bone ash and bone density among fibulas, metacarpals, 2nd ribs, and 10th ribs in nursery pigs and differences between analytical methods.[2] Similarly, defatting bones prior to ashing reduces the amount of variation between the bones compared to not extracting the lipid.[1]

Currently, reference values used by diagnosticians for bone mineralization and serum vitamin D have been adapted from previous work.[3,4] Typically, the pigs submitted to diagnostic laboratories are sick or lame and their recent dietary intakes are unknown but likely reduced and highly variable, possibly causing their nutritional status or bone mineralization to be lower than that of healthier pigs in the general population. Thus, interpretation of these results may potentially lead to misdiagnosis relative to the entire population. Recently, work evaluated different bones and various analytical measurements between healthy and unhealthy pigs.[5] The authors observed that healthy pigs have increased serum Ca, P, and vitamin D compared to unhealthy pigs. They also observed no statistical difference between the health statuses for non-defatted bone ash; however, healthy pigs had increased defatted bone ash compared to the unhealthy pigs. Therefore, our objective in the current study was to evaluate the effect of different pig types (healthy, lame, or sick) from pigs in commercial production sites across the Midwestern US on the assessment of bone mineralization. Furthermore, this study evaluated the effect of different bones and analytical methods on the assessment of bone mineralization in growing pigs.

Methods

A total of 192 pigs from 64 commercial production sites across 14 different production systems in Minnesota and Iowa were used in the diagnostic survey. All samples were collected from December 2021 to February 2022. At each site, the staff veterinarian selected 3 pigs to be euthanized and used in the diagnostic survey: **1)** a clinically normal pig, **2)** a pig with locomotive issues and evidence of clinical lameness, and **3)** a pig from within a hospital pen with an assumed recent low feed intake (sick pig). Pigs ranged in age from nursery to market weight, with the three pigs sampled from each site representing the same age and phase of production.

A blood sample was taken from the jugular vein of each pig prior to euthanasia. Blood samples (10 mL) were centrifuged (3000 × *g*, 15 min) within 2 h after collection and serum was stored at -20°C until analysis. Serum samples were split into three, 2 mL aliquots and analyzed for Ca and P (Iowa State University Veterinary Diagnostic Lab) and $25(OH)D_3$ (Heartland Assays, Ames, Iowa). A 10 mL urine sample was collected shortly after euthanasia via cystocentesis using a 20 gauge × 2.5 cm needle and a 12 mL syringe, then frozen at -20°C until analyzed for Ca, P, and creatinine (Iowa State University Veterinary Diagnostic Lab, Ames Iowa).

The metacarpal, 2nd rib, 10th rib, and fibulas were analyzed for bone density, bone-breaking strength, bone ash (de-fatted and non-defatted), and bone Ca and P. The leftover extraneous soft tissues and cartilage were removed from the bones prior to assessment. Archimedes principle was used for bone density. Briefly a dry bone weight was collected, then bones were submerged in ultra-purified water under a vacuum with 1.06 kg per cubic centimeter for a minimum of 4 h. Bones were then weighed while suspended in a vessel of ultra-purified water, and the weight used to calculate bone density. Bone breaking strength was determined with an Instron (Instron 5569, NV Lab, Norwood, Massachusetts). Briefly each bone was held by two supports spaced 30 mm apart and were broken by a wedge lowered on the center of the bone at a speed of 100 mm per min and a maximum pressure of 5,000 kg. The force was measured by a pressure-sensitive cell, and peak of maximum force was recorded. For bone ash, both the defatted and the non-defatted methods were used.[6] For defatted bone ash, all bones were placed in Soxhlet extractors containing petroleum ether for 7 d as a means of removing water and fat. Bones were then dried at 105°C for 7 d and then ashed in a muffle furnace at 600°C for 24 h. For non-defatted bone ash, all bones were dried at 105°C for 7 d and then ashed in a muffle furnace at 600°C for 24 h. These methods were used to determine total bone ash weight and percentage ash relative to dried bone weight. For bone Ca and P, a portion of the residual defatted bone ash (.05 g) from the metacarpal and 10th ribs were placed in a tube with nitric acid and incubated at 60°C for 6 h. The solution was then diluted using a 20x dilution of ultra-purified water and the Ca (AOAC 985.01, 2006) and P (AOAC 985.01, 2006) quantity were measured by Inductively Coupled Plasma Optical Emission Spectrometer (ICP-OES). Concentrations of Ca and P were calculated as a percentage of the total bone ash.

For the statistical analysis, bone data were analyzed by fitting a linear mixed model using the lmer function from the lme4 package in R (version 3.5.1 (2018-07-02), R Foundation for Statistical Computing, Vienna, Austria), pig type, bone, and the associated interaction included as fixed effects. System and pig type within system were included as random effects to account for systems having multiple sites sampled and multiple bones being collecred from each pig type with each site. Correlation coefficients (r) between dietary levels of Ca and P and individual bone analysis were determined using the lm function in R with dietary levels being the predictor variables and individual bone analysis being the response. Results were considered significant at $P \leq .05$ and marginally significant at $.05 < P \leq .10$.

Results

Healthy pigs had greater serum 25(OH)D_3 and serum Ca levels compared to sick pigs, with lame pigs intermediate, but significantly different than the healthy and sick pigs ($P < .05$; Table 1). Healthy pigs had greater serum P compared to sick and lame pigs ($P < .05$) with no difference between sick and lame pigs ($P > 0.10$).

There was no significant difference for urine Ca between the different pig types ($P > .10$). Sick pigs were excreting greater concentrations ($P < .05$) of P and creatinine in urine compared to healthy pigs, with lame pigs intermediate. There was no difference observed between the pig types for Ca:creatinine and P:creatinine ratios ($P > .10$). There was no difference in serum 25(OH)D_3, between pigs sampled from curtain barns (21.7 ng/mL) compared to pigs sampled from tunnel barns (19.8 ng/mL; $P > .10$).

For the main effect of pig type, there was no difference between pig types for defatted bone ash percentage ($P = .122$; Table 2). Healthy and lame pigs had increased defatted bone ash weight compared to sick pigs ($P < .05$). There was a significant difference between pig types for percentage non-defatted bone ash, non-defatted bone ash weight, bone density, and breaking strength ($P < .05$). For non-defatted bone ash, sick pigs had greater percentage bone ash compared to healthy pigs, with lame pigs intermediate; however, healthy and lame pigs had greater bone ash weight compared to sick pigs. Healthy and lame pigs had greater bone density compared to sick pigs. Healthy and lame pigs required greater force to break bones compared to sick pigs ($P < .05$). Healthy and lame pigs had increased grams of Ca and P in defatted bone ash compared to sick pigs ($P < .05$).

For the main effect of bone, there were significant differences for defatted and non-defatted bone ash weight, defatted and non-defatted percentage bone ash, bone density, and bone breaking strength between bones ($P < .05$). For defatted bone ash weight, metacarpals and 10th ribs had greater bone ash weight compared to the fibula and 2nd rib ($P < .05$). Fibulas had the greater defatted percentage bone ash compared to the 2nd and 10th ribs, with the metacarpals having the lowest defatted percentage bone ash ($P < .05$). Metacarpals and 10th ribs had greater non-defatted bone ash weight compared to fibulas and 2nd ribs ($P < .05$), with the 2nd ribs being the lowest. For non-defatted percentage bone ash, 10th ribs had the greatest percentage bone ash, metacarpals the lowest, with 2nd ribs and fibulas intermediate ($P < .05$). Fibulas had the

Table 1: The effect of pig type on serum and urine analysis

	Pig type							
	Healthy		Lame		Sick			
Item	**Mean**	**Range**	**Mean**	**Range**	**Mean**	**Range**	**SEM**	**P =**
Serum analysis[1]								
25OHD_3, ng/mL	25.2[a]	5.6 - 78.9	21.0[b]	4.6 - 64.0	15.5[c]	2.8 - 71.8	1.88	< .001
Ca, mg/dL	10.4[a]	6.9 - 14.4	9.7[b]	6.7 - 13.4	9.2[c]	6.6 - 14.5	0.16	< .001
P, mg/dL	10.2[a]	6.4 - 18.0	9.4[b]	5.5 - 15.8	8.8[b]	4.5 - 14.7	0.26	< .001
Urine analysis[2]								
Ca, mg/dL	10.9	1.0 - 35.6	11.1	1.0 - 41.0	9.5	1.0 - 35.6	1.88	.740
P, mg/dL	50.9[b]	5.5 - 142.9	75.2[ab]	5.5 - 903.0	96.9[a]	5.5 - 286.0	11.76	.003
Creatinine, mg/dL	103.1[b]	9.1 - 255.9	135.1[ab]	3.2 - 378.3	151.3[a]	16.4 - 381.1	14.08	.022
Calcium: creatinine	0.13	---	0.19	---	0.13	---	0.048	.456
Phosphorus: creatinine	0.63	---	0.71	---	0.94	---	0.151	.303

[abc] Means within a row with different superscripts differ ($P < 0.05$).

1 Serum Ca and P were measured at Iowa state Veterinary Diagnostic Lab (Ames, Iowa). The vitamin D serum analysis was conducted at Heartland Assays (Ames, Iowa).

2 Urine was collected from the bladder of each pig and analyzed for Ca, P, and creatinine at Iowa State Veterinary Diagnostic Lab (Ames, Iowa).

Table 2: The effect of pig type on bone analysis[1]

	Pig type							
	Healthy		Lame		Sick			
Item	**Mean**	**Range**	**Mean**	**Range**	**Mean**	**Range**	**SEM**	***P* =**
Defatted bone ash[2]								
Bone ash, g	4.02[a]	0.37 - 11.54	3.55[a]	0.20 - 14.12	2.71[b]	0.14 - 10.86	0.337	.001
Bone ash, %	62.4	51.0 - 67.9	62.3	49.3 - 69.1	61.8	47.3 - 69.8	0.29	.122
Non-defatted bone ash[3]								
Bone ash, g	4.23[a]	0.31 - 13.16	3.73[a]	0.21 - 13.59	2.92[b]	0.16 - 12.18	0.354	.001
Bone ash, %	51.7[b]	38.1 - 62.3	52.3[ab]	34.8 - 65.8	53.1[a]	36.2 - 65.1	0.53	.037
Bone density, g/mL[4]	1.33[a]	1.16 - 1.56	1.33[a]	1.16 - 1.52	1.31[b]	1.14 - 1.44	0.007	.001
Breaking strength, kg	69.1[a]	3.1 - 328.3	61.1[a]	4.5 - 280.1	44.9[b]	0.9 - 196.8	5.61	.001
Defatted bone ash content[5]								
Ca, g	2.47[a]	0.30 - 6.62	2.22[a]	0.21 - 7.41	1.62[b]	0.18 - 4.62	0.214	.001
P, g	1.18[a]	0.14 - 3.18	1.05[a]	0.10 - 3.47	0.76[b]	0.08 - 2.17	0.100	.001
Ca, %	48.0	29.9 - 66.8	47.4	33.1 - 65.3	45.9	31.3 - 62.1	0.648	.064
P, %	23.0[a]	14.3 - 33.6	22.7[ab]	15.6 - 31.2	21.9[b]	15.2 - 30.3	0.354	.050

[abc] Means within a row with different superscripts differ ($P < .05$).

1 The mean of all 4 bones measured are combined for each pig type.

2 All bones were cleaned of tissue and then placed in Soxhlet extractors containing petroleum ether for 7 d as a means of removing water and fat. Bones were then dried at 105°C for 7 d, and then ashed in a muffle furnace at 600°C for 24 h.

3 All bones were cleaned of tissue and then dried at 105°C for 7 d and then ashed in a muffle furnace at 600°C for 24 h.

4 Bone density was measured on each bone based on Archimedes principle.

5 After bone ash was completed, the ash from the metacarpal and 10th ribs were digested and analyzed for Ca and P using ICP by the KSU Soils lab (Manhattan, Kansas).

greatest bone density, with metacarpals having the lowest bone density, with 2nd and 10th ribs intermediate, with 10th ribs having greater bone density than 2nd ribs ($P < .05$). Metacarpals had the greatest bone breaking strength ($P < .05$), with the 2nd ribs being the lowest, and fibulas and 10th ribs intermediate, with 10th ribs having greater bone breaking strength than fibulas. There were no differences between defatted bone ash of metacarpals and 10th ribs for grams and percent Ca and P ($P > 0.10$).

Average serum 25(OH)D_3 was 25.2 ng/mL for healthy pigs, 20.9 ng/mL for lame pigs, and 15.3 ng/mL for sick pigs. The average non-defatted bone ash was 55.4% for healthy pigs, 56.6% for lame pigs, and 57.2% for sick pigs. The average defatted bone ash was 62.3% for healthy pigs, 62.5% for lame pigs, and 62.7% for sick pigs. For serum 25(OH)D_3, there is variation across the different pig types across all systems.

Discussion

The measurement of bone mineralization is routinely conducted alongside histopathologic evaluation in diagnostic cases to assess metabolic bone disease in pigs. The most common measurement to assess bone mineralization is bone density and percentage bone ash. Reference values adapted by diagnosticians have been adopted from research generated in the last 50 years.[3,4,7] Recent work has evaluated different analytical methods to assess bone mineralization in different ages of pigs.[2,5] Currently, there is limited research comparing the differences between bone mineralization and serum analysis across pigs of different health statuses. This diagnostic survey is one of the first large-scale studies to compare different bones and analytical measurements between healthy, lame, and sick pigs. The goal of the diagnostic survey is to compare the results from different analytical measurement across systems, health statuses, and diet diet formulation strategies in pigs raises in the Midwestern US.

Percentage bone ash has been the preferred measurement to determine bone mineralization in growing pigs.[8-10] The two main procedures used to measure percentage bone ash are the defatted and non-defatted techniques.[1] The major difference between the two methods is with defatting the bones, the lipid and water is extracted from the bone using ether in a Soxhlet device which circulates petroleum ether around the submerged bones for 7 d, while the non-defatted technique bypasses this step and leaves the lipid inside the bone during the ashing process. Previous research has shown the bone of a mature animal is comprised of 10% fat.[11] By extracting the lipid from the bone, the variation across all bones measured is reduced, and this variation is larger in finishing pigs compared to nursery pigs[1,5]. In finishing pigs, the defatted percentage bone ash is 14 to 15 percentage units greater than the same bone measured using the non-defatted bone ash method[5]; whereas in nursery pigs, the defatted percentage bone ash is 11 to 12 percentage units greater than non-defatted bones[2]. Similar results were observed between nursery and finishing bones using both methods for bone ash.[1]

When investigating bone metabolic disease cases in pigs, it is common practice to measure serum vitamin D due to its critical role in the absorption and retention of Ca and P.[12] Unlike measuring bone mineralization, this procedure is non-invasive and can provide important insights into the pig's metabolic status. Vitamin D can be obtained by pigs through dietary supplementation or synthesized in the skin from 7-dehydrocholesterol upon exposure to sunlight. Upon exposure to ultraviolet-B (UV-B), 7-dehydrocholesterol is converted to vitamin D_3.[13] Vitamin D_3 must then undergo a hydroxylation step in the liver to be converted to the primarily inactive form, 25(OH)D_3. The 25(OH)D_3 enters circulation and can be stored and later hydroxylated in the kidney to the active metabolite, 1,25-$(OH)_2D_3$, as needed. To assess vitamin D status in pigs, diagnosticians primarily measure serum levels of 25(OH)D_3, which can be increased through greater dietary supplementation with vitamin D_3 or 25(OH)D_3.[4,14,15] Several factors can affect serum vitamin D levels in pigs, including the dietary form of vitamin D and supplementation levels, growth stage, and sunlight exposure.

Along with serum vitamin D levels, Ca and P can also be analyzed in the serum to determine circulating levels. Due to efficient homeostatic mechanisms of serum Ca and P, it is difficult to detect differences due to dietary changes.[16] Feeding deficient levels of P to pigs for an extended period can lead to low serum P. Recent research observed reduced serum P in nursery pigs fed 57% of their P requirement estimate for 28-d.[2] In the current study, lame and sick pigs had lower serum P than healthy pigs, with the healthy pigs having 16% greater levels of P in serum compared to sick pigs. The reduction in serum P for sick pigs is speculated to be due to a feed intake reduction compared to the healthy pigs. Serum P ranged from 4.5 to 18.0 mg/dL across the different pig types.

In summary, this diagnostic survey demonstrates differences between healthy, lame, and sick pigs for serum Ca, P, and vitamin D, as well as differences across the various bones and analytical measurements to assess bone mineralization. By using the defatted bone ash method, the amount of variation across bones and different pig types was reduced, but no differences were observed between healthy, lame, and sick pigs. For non-defatted bone ash, there is more variation, with sick pigs having greater percentage bone ash compared to healthy pigs because of lipid mobilization from the bones of sick pigs. When comparing all swine production systems included in this diagnostic survey, we observed considerable variation across different measurements, with serum vitamin D having a wide range between the different pig types. Although serum vitamin D had the highest mean serum concentration for healthy pigs, there were healthy pigs below the reference range threshold for clinically normal serum vitamin D analysis. This data can be used to help guide future diagnostic investigations with baseline levels of different methodologies to determine bone mineralization in pigs.

References

1. Wensley, M. R, C. M. Vier, J. T. Gebhardt, M. D. Tokach, J. C. Woodworth, R. D. Goodband, and J. M. DeRouchey. 2020. Technical note: assessment of two methods for estimating bone ash in pigs. *J. Anim. Sci.* 98:1-8. **doi:10.1093/jas/skaa251**

2. Williams, H. R., T. E. Chin, J. T. Gebhardt, M. D. Tokach, J. C. Woodworth, J. M. DeRouchey, R. D. Goodband, J. R. Bergstrom, M. C. Rahe, C. L. Siepker, P. Sitthicharoenchai, and S. M. Ensley. 2023a. The effect of bone and analytical methods on the assessment of bone mineralization response to dietary phosphorus, phytase, and vitamin D in nursery pigs. *In progress.*

3. Field, R. A., M. L. Riley, F. C. Mello, M. H. Corbridge, and A. W. Kotula. 1974. Bone composition in cattle, pigs, sheep, and poultry. *J. Anim. Sci.* 39:3. **doi:10.2527/jas1974.393493x**

4. Arnold, J., D. M. Madson, S. M. Ensley, J. P. Goff, C. Sparks, G. W. Stevenson, T. Crenshaw, C. Wang, and R. L. Horst. 2015. Survey of serum vitamin D status across stages of swine production and evaluation of supplemental bulk vitamin D premixes used in swine diets. *J. Swine Health Prod.* 23:28-34.

5. Williams, H. R., J. T. Gebhardt, M. D. Tokach, J. C. Woodworth, R. D. Goodband, J. M. DeRouchey, J. R. Bergstrom, C. W. Hastad, Z. B. Post, M. C. Rahe, C. L. Siepker, P. Sitthiccharoenchai, and S. M. Ensley. 2023b. The effect of bone and analytical method on the assessment of bone mineralization in response to dietary phosphorus, phytase, and vitamin D in finishing pigs. *In progress.*

6. Williams, H. R., M. D. Tokach., J. C. Woodworth., J. M. DeRouchey, R. D. Goodband, and J. T. Gebhardt. 2023c. The comparison between analytical methods to measure bone mineralization of different bones and age of pigs. *In progress.*

7. Hagemoser, W. A., J. P. Goff, T. P. Sanderson, and J. S. Haynes. 2000. Osteopenic disease in growing pigs: diagnostic methods using serum and urine calcium and phosphorus values, parathormone assay, and bone analysis. *J. Vet. Diagn. Invest.* 12:525-534. **doi:10.1177/104063870001200606**

8. Vier, C. M., S. S. Dritz, F. Wu, M. D. Tokach, J. M. DeRouchey, R. D. Goodband, M. A. D. Gonçalves, U. A. D. Orlando, and J. C. Woodworth. 2019a. Effects of standardized total tract digestible phosphorus on growth performance of 11-to 23-kg pigs fed diets with or without phytase. *J. Anim. Sci.* 97:4032-4040. **doi:10.1093/jas/skz255**

9. Vier, C. M., S. S. Dritz, F. Wu, M. D. Tokach, J. M. DeRouchey, R. D. Goodband, M. A. D. Gonçalves, U. A. D Orlando, K. Chitakasempornkul, and J. C. Woodworth. 2019b. Standardized total tract digestible phosphorus requirement of 24- to 130-kg pigs. *J. Anim. Sci.* 97(10):4023-4031. **doi:10.1093/jas/skz256**

10. Lee, S. A., L. V. Lagos, M. R. Bedford, and H. H. Stein. 2021. Quantities of ash, Ca, and P in metacarpals, metatarsals, and tibia are better correlated with total body bone ash in growing pigs than ash, Ca, and P in other bones. *J. Anim. Sci.* 6:1-6. **doi:10.1093/jas/skab149**

11. Lian, J. B., G. S. Stein, E. Canalis, P. G. Robey, and A. L. Boskey. 1999. Bone formation: osteoblast lineage cells, growth factors, matrix proteins, and the mineralization process. In: Favus, M. J., editor. Primer on the Metabolic Bone Diseases and Disorders of mineral Metabolism, 4th ed., Philadelphia, PA: Lippincott Williams & Wilkins, chap. 3.

12. O'Doherty, J. V., D. A. Gahan, C. O'Shea, J. J. Callan, and K. M. Pierce. 2010. Effects of phytase and 25-hydroxyvitamin D_3 inclusions on the performance, mineral balance and bone parameters of grower-finisher pigs fed low-phosphorus diets. *Animal.* 4:1634-1640. **doi:10.1017/S1751731110000807**

13. Crenshaw, T. D. 2001. Calcium, phosphorus, vitamin D, and vitamin K in swine nutrition. In: A. Lewis and L. L. Southern, editors, *Swine Nutrition.* 2nd ed. CRC Press, Boca Raton, FL. p. 187-212.

14. Burild, A., C. Lauridsen, N. Faqir, H. M. Sommer, and J. Jakobsen. 2016. Vitamin D_3 and 25-hydroxyvitamin D_3 in pork and their relationship to vitamin D status in pigs. *J. Nutr. Sci.* 5:1-6. **doi:10.1017/jns.2015.28**

15. Flohr, J. R., J. C. Woodworth, J. R. Bergstrom, M. D. Tokach, S. S. Dritz, R. D. Goodband, and J. M. DeRouchey. 2016. Evaluating the impact of maternal vitamin D supplementation on sow performance: II. Subsequent growth performance and carcass characteristics of growing pigs. *J. Anim. Sci.* 94:4643-4653. **doi:10.2527/jas2016-0410**

16. Koch, M. E. and D. C. Mahan. 1985. Biological characteristics for assessing low phosphorus intake in growing swine. *J. Anim. Sci.* 60:699-708. **doi: 102527/jas1985.603699x**

https://doi.org/10.54846/am2024/61

Recent research to support optimum levels of dietary vitamins, including 25-OH-D_3 from HyD®, for improved performance of nursery pigs

J. Bergstrom, PhD; S. Hough, DVM
dsm-firmenich, Animal Nutrition & Health, Plainsboro, New Jersey

Introduction

Historically, vitamins have been the least researched category of essential nutrients in swine nutrition, largely because they make up such a small portion of the diets and are difficult to study. Nevertheless, there has been significantly more research on vitamins in the last 10 years and the interest in vitamin nutrition remains high.

While the swine industry continues to make remarkable progress in improving the efficiency and sustainability of producing healthy and nutritious pork, producers are continually challenging themselves to improve further. The incentives and sense of need for further reduction in the use of antibiotics for food animals has increased, with greater emphasis being placed on the importance of sustaining or improving swine health through management, genetics, and nutrition.

The term "precision nutrition" sometimes enters swine nutrition discussions nowadays, but this concept for improving efficiency and sustainability is made more challenging by continuous genetic progress, changes in housing and management schemes, and some lag in understanding any associated changes in nutritional requirements for optimizing swine productivity, health, and welfare. For example, while the average and potential growth rates, F/G and litter size have continuously been improving, producers have become more concerned about increasing mortality and culling rates.

During the 1960s, one predominant factor found to contribute to losses associated with vitamin E-Se deficiency in sows and young pigs on many swine farms was the movement of sows and pigs to complete confinement.[1] Later, Mutetikka and Mahan[2] demonstrated that sows housed in confinement benefitted from dietary supplementation with vitamin E and Se, and that supplementation was necessary to achieve a status like that of sows receiving these nutrients from access to fresh pasture forage. Even though sow and nursery diets are commonly fortified to exceed the NRC[3] requirement estimates for vitamin E nowadays, it is not uncommon to find nursery pigs with a low vitamin E status after weaning, or cases of Mulberry Heart Disease in rapidly growing young pigs. The current basis for many vitamin E recommendations is previous research with less prolific gilts/sows and weaning ages ≥ 21 days. Larger litters can result in pigs being weaned with a lower vitamin E status, since the amount of colostrum and milk available for each suckling pig may be reduced, limiting the ingestion and accumulation of essential vitamin E transferred from the sows. Piglets weaned at a young age and a lower vitamin E status are more prone to becoming deficient, since pigs typically experience a rapid decline in serum and tissue levels of vitamin E after an abrupt weaning.

Most domestic pigs are now born and raised indoors without exposure to the UVB rays of sunlight (to allow for synthesis of vitamin D_3 via skin), and sows naturally provide very low levels of vitamin D in colostrum and milk. Therefore, it is not surprising that the vitamin D status of many young pigs is found to be low at weaning. Although low plasma levels of 25-OH-D_3 are often found in pigs suspected of having "metabolic bone disease" or vitamin D deficiency symptoms, serum levels below the currently suggested minimum (of 18 ng/mL of 25-OH-D_3) for 2- to 14-week-old pigs can also be found in pigs without any obvious compromises in their growth or health.[4] Nevertheless, an increased vitamin D status reduces the likelihood of vitamin D deficiency and can improve immunological functions and optimize bone development and health. However, even with common industry levels of vitamin D_3 fortification at 7- to 14-times the NRC[3] requirement estimate for nursery diets (220 IU vitamin D_3/kg), the vitamin D status of nursery pigs is frequently considered to be low.

Existing swine NRC[3] requirement estimates for many of the vitamins are largely outdated. Consequently, swine nutritionists generally recognize the importance of supplemental vitamins and provide levels that are well above the minimal recommendations of the NRC.[3,5,6] Nevertheless, there is considerable variation among nutritionists in the levels of supplementation used, particularly for the fat-soluble vitamins. Also, there is increasing use of alternative sources for some vitamins and minerals, such as the more biologically relevant 25-OH-D_3 as a source of vitamin D, but the levels and application are also variable. Over the last several years, dsm-firmenich (previously DSM) Animal Nutrition & Health has been collaborating with industry partners to perform research that improves and supports our understanding of the optimum levels of vitamin supplementation for nursery pigs. The results from this research can help nutritionists to supplement vitamins more precisely to optimize performance, nutritional status, and health of nursery pigs.

Commercial experiment #1

During the winter of 2020-2021 in southern Minnesota, a group of pigs in a commercial research nursery was used to evaluate several dietary micronutrient strategies during an expected, severe, natural PRRS challenge. A total of 1,056 pigs were weaned (21 d of age, 6.49 kg BW) and used to compare the effects of 4 supplementation strategies on the growth performance and health from d 0 to 42 post-weaning. The 4 supplementation strategies consisted of the following: 1) Control (C), "industry-type" diet with 10,410 IU vitamin A/kg, 550 IU vitamin D_3 + 44 µg 25-OH-D_3/kg from HyD®, 70 IU vitamin E/kg, and 250 mg vitamin C/kg; 2) C+ESe, C with vitamin E increased to 195 IU/kg and added 0.15 mg Se/kg from an organic source; 3) C+AEC, C with increased vitamin A at 22,420 IU/kg, 25-OH-D_3 at 55 µg/kg, vitamin E at 295 IU/kg, and vitamin C at 298 mg/kg; and 4) C+Wadd, C diet and a proprietary drinking water additive from d 0 to 21. The levels of vitamins in the feed for each strategy were maintained throughout the 42-d experiment. There were 12 pens of 22 pigs allotted to each supplementation strategy in a RCBD based on initial BW. Pens were weighed and feed consumption for each was recorded weekly, and ADG, ADFI, and F/G were calculated for each week, diet phase, and the overall period. All treatments of individual pigs with injectable treatments were recorded during the study, as well as all mortalities + culls. Blood samples were collected from 2 pigs per pen on d 1, 21 and 42 of the experiment to assess serum retinol, α-tocopherol, and 25-OH-D_3 status.

At the end of the study, ADG and ending BW were numerically greatest (+11% and +4%, respectively) for pigs that received C+AEC or C+Wadd (Table 1). Pigs fed C+ESe had growth performance like those fed C, but pigs that received C+AEC or C+Wadd had the best F/G ($P < .01$, -6%) and lowest percentage of mortalities + culls ($P < .06$; C = 27.4%, C+ESe = 30.5%, C+AEC = 21.4%, C+Wadd = 22.5%). Collectively, these differences resulted in pigs receiving C+AEC or C+Wadd having numerically greater Income-Over-Feed-Cost/pig and revenue/pen at d 42 post-weaning. There were no differences between any of the experimental treatments for serum retinol, α-tocopherol, or 25-OH-D_3 within each timepoint sampled, but serum retinol and 25-OH-D_3 were greater on d 21 and 42 when compared to d 1, while serum α-tocopherol was greatest on d 1 and substantially lower on d 21 before increasing by d 42. These results demonstrate that higher, optimum dietary levels of vitamins A, D, E, and C can improve performance and reduce mortality when pigs experience a severe PRRS challenge, but solely increasing vitamin E along with added organic Se was not beneficial under the conditions of this experiment. Interestingly, there were not any differences amongst treatments in the serum vitamin levels tested, but all the diets contained substantial levels of 25-OH-D_3 (HyD®) and much greater levels of vitamin A, E, and C than suggested by the NRC.[3]

Commercial experiment #2

Last year at this meeting, we presented findings from a research collaboration with Cooperative Research Farms that demonstrated improvements in nursery pig performance from feeding 3,500 IU vitamin D/kg diet from d 0 to 42 post-weaning when compared to 2,000 IU vitamin D/kg.[7] The improvements in ADG and ADFI from feeding 3,500 IU total vitamin D/kg using a combination of 1,500 IU vitamin D_3 with 50 µg 25-OH-D_3 (= 2,000 IU from HyD®) occurred regardless of whether vitamin E was supplemented at 70 IU/kg or 140 IU/kg. Naturally, serum levels of vitamin E on d 42 post-weaning were greater in pigs that received 140 IU vs 70 IU/kg diet, but supplementation for vitamin D using 50 µg 25-OH-D_3/kg diet resulted in a further increase in serum vitamin E status at d 42 post-weaning. These novel findings stimulated interest in conducting further research. Therefore, another nursery study to evaluate 50 µg/kg of added vitamin D from 25-OH-D_3 (HyD®) with either 70 or 140 IU vitamin E/kg diet was completed during the summer of 2023.

A total of 1,440 pigs (PIC, Camborough × line 800) were weaned (20 d of age, 5.5 kg BW) and used to compare the effects of 3 dietary fortification strategies for vitamin D and vitamin E on growth performance from d 0 to 42 post-weaning. The 3 vitamin fortification strategies were fed over 3 dietary phases; d 0 to 7 (P1), d 7 to 21 (P2), and d 21 to 42 (P3); and consisted of 1) Control (C), "industry-type" diet that met or exceeded NRC[3] requirement estimates for each diet phase with 1,500 IU vitamin D_3 and 70 IU vitamin E/kg diet; 2) HyD, C with 50 µg (or 2,000 IU) 25-OH-D_3 added per kg diet; and 3) HyD+E, HyD with 140 IU vitamin E/kg diet. There were 16 pens of 28 - 32 pigs allotted to each fortification strategy in a RCBD based on initial BW. Pens were weighed and feed consumption for each was recorded weekly, and ADG, ADFI, and F/G were calculated for each week, diet phase, and the overall period.

During P1, ADFI was highest ($P < .05$) for pigs fed HyD, and pigs fed HyD or HyD+E had greater ($P < .05$) ADG compared to C (Table 2). Pig fed HyD were heavier ($P < .05$) on d 7 compared to pigs fed C, with pigs fed HyD+E intermediate, but there were no differences in F/G during P1 (Table 1). During P2, ADFI was highest ($P < .05$) for pigs fed HyD, and pigs fed HyD or HyD+E had greater ($P < .05$) ADG compared to C. Pig fed HyD or HyD+E were heavier ($P < .05$) on d 21 compared to pigs fed C, but there were no differences in

Table 1: Commercial experiment #1 - pig growth and economic performance[1]

Dietary vitamin strategy:	"C"	"C+ESe"	"C+AEC"	"C+Wadd"		
Vitamin A, IU/kg:	10,410	10,410	22,420	Diet "C" + Proprietary Water Additive, d 0-21		
Vitamin D3, IU/kg:	550	550	550			
25-OH-D3 (HyD®), µg/kg:	44	44	55			
Vitamin E, IU/kg	70	195	295			
Vitamin C, mg/kg:	250	250	298			
Overall (d 0 to 42)					**SEM**	***P*-value**
ADG, lb	0.53	0.53	0.59	0.59	0.025	.139
ADFI, lb	0.82	0.84	0.88	0.86	0.028	.414
F/G	1.56^{ab}	1.60^{a}	1.50^{ab}	1.47^{b}	0.029	.007
Day 42 BW, lb	41.1	41.9	42.6	42.8	0.82	.373
Pigs/pen that received injectable, %	61.4	62.5	54.2	48.5	5.05	.184
Dead + Culled, %	27.4	30.5	21.4	22.5	3.02	.056
Revenue/pen, $	477.12	460.80	530.76	524.58	26.98	.152
Income-over-feed-cost/pig, $	20.90	20.93	22.00	21.81	0.62	.392

1 Growth and feed data collected and analyzed using Imperial weights. Within a row, means without a common superscript are different ($P < .05$). Least square means based on 12 pen-observations per dietary treatment. Each pen began with 22 pigs per pen.

Table 2: Commercial experiment #2 - pig growth and economic performance[1]

Dietary vitamin strategy:	"C"	"HyD"	"HyD+E"		
Vitamin D3, IU/kg:	1,500	1,500	1,500		
25-OH-D3 (HyD®), IU equiv./kg:	0	2,000	2,000		
Vitamin E, IU/kg:	70	70	140		
Phase 1 (d 0 to 7)				**SEM**	***P*-value**
Day 0 BW, kg	5.54	5.55	5.54	0.02	.84
ADG, g	103[a]	117[b]	110[b]	4	< .01
ADFI, g	130[a]	137[b]	132[a]	2	< .01
F/G	1.27	1.18	1.23	0.04	.09
Day 7 BW, kg	6.27[a]	6.37[b]	6.31[ab]	0.03	.02
Phase 2 (d 7 to 21)					
ADG, g	401[a]	419[b]	412[b]	5	< .01
ADFI, g	482[a]	502[b]	490[a]	6	< .01
F/G	1.20	1.20	1.19	0.01	.94
Day 21 BW, kg	11.91[a]	12.22[b]	12.13[b]	0.10	< .01
Phase 3 (d 21 to 42)					
ADG, g	591	589	598	8	.49
ADFI, g	896	892	909	12	.35
F/G	1.52	1.51	1.52	0.02	.85
Overall (d 0 to 42)					
ADG, g	445	453	453	5	.16
ADFI, g	628	636	637	7	.32
F/G	1.41	1.40	1.41	0.01	.72
Day 42 BW, kg	24.33	24.64	24.70	0.22	.21
Initial inventory, n	480	480	480		
Removals/morbidity, n	4	1	8	–	–
Mortalities, n	9	3	6	–	–
Ending inventory, n	467	476	466	–	–
Avg. feed cost, CA$/tonne	702.9[a]	707.3[b]	707.3[b]	0.85	< .01
Feed cost/pig, CA$	18.64	18.93	19.07	0.20	.11
Income-over-feed-cost/pig, CA$	30.03	30.35	30.33	0.29	.44

[1] Growth and feed data collected and analyzed using Metric system. Within a row, means without a common superscript are different ($P < .05$). Least square means based on 16 pen-observations per dietary treatment. Each pen began with 28, 29, 31 or 32 pigs per pen, depending on the block.

F/G during P2. There were no differences in growth performance during P3 or overall (d 0 to 42). Nevertheless, in conclusion, even for pigs of good health, performance differences during P1 and P2 resulted in pigs fed HyD or HyD+E having numerically greater BW (+0.34 kg) and Income-Over-Feed-Cost (+CA $0.31/pig) on d 42 when compared to pigs fed C.

Summary

The dietary vitamin requirements for optimizing pig performance are generally greater than the requirement estimates suggested by the NRC.[3] Swine nutritionists recognize the importance of supplemental vitamins and generally provide levels that are well above the minimal recommendations of the NRC.[3,5] Because there is limited research to understand the precise vitamin requirements for optimizing nursery pigs' performance, there is considerable variation in the levels of vitamins supplemented in commercial feeds.[6] Past research to establish nutrient requirements has most often been conducted in archetypal conditions, such as university farms, which may not adequately characterize the requirements for optimal performance of modern genotypes on commercial farms. Our recent collaborations to improve the understanding of vitamin requirements for nursery pigs are beneficial. The latest vitamin research demonstrates that nursery pigs experiencing severe PRRSv can benefit from a more optimum dietary vitamin strategy (OVN®).[8] Also, the performance and nutritional status of nursery pigs in commercial farms can be improved by feeding 3,500 IU vitamin D/kg diet; especially using the combination of 1,500 IU vitamin D_3 with the maximum of 2,000 IU (50 µg)/kg diet of the more biologically relevant 25-OH-D_3 from HyD®; when compared to lower levels of vitamin D supplementation.

References

1. Mahan, D. C., Are We Still Having Vitamin E and Selenium Deficiencies in Pigs? 2007. Are We Still Having Vitamin E and Selenium Deficie - Swine **illinois.edu**

2. Mutetikka, D. B., and D. C. Mahan. Effect of Pasture, Confinement, and Diet Fortification with Vitamin E and Selenium on Reproducing Gilts and Their Progeny. *Journal of Animal Science*, 1993, 71:3211-32-18.

3. National Research Council (NRC). *Nutrient Requirements of Swine*, 11th Revised Edition. 2012.

4. Arnold, J., et al. Survey of serum vitamin D status across stages of swine production and evaluation of supplemental bulk vitamin D premixes used in swine diets. *Journal of Swine Health and Production*, 2015, 23(1):28-34.

5. Faccin, J. E. G., et al. Industry survey of added vitamins and trace minerals in U.S. swine diets. *Translational Animal Science*, 2023, 7, txad035. **https://doi.org/10.1093/tas/txad035**

6. Hinson, R. B., et al. Reduced vitamin supplementation with fat-soluble vitamins A, D, and E added at National Research Council requirements may not be adequate for optimal sow and progeny performance. *Journal of Swine Health and Production*, 2022, 30(2):79-94. **https://doi.org/10.54846/jshap/1259**

7. Bergstrom, J., S. Hough, and K. Herkelman. The effects of different levels of supplemental vitamin D, 25-OH-D_3, and vitamin E on nursery pig performance and vitamin status. *Proceedings of the 54th AASV Annual Meeting*, 2023, pp. 134-138.

8. Optimum Vitamin Nutrition (OVN). Swine, 2022. **https://www.dsm.com/anh/products-and-services/products/ovn.html#swine**

https://doi.org/10.54846/am2024/62

AgriSafe Total Farmer Health®: Engaging the One Health and Total Worker Health® models in addressing the wellness of producers and those who are devoted to the health of their livestock

L. Emanuel[1], BSN, RN; C. Halverson[2], BSN, RN, COHN-S
[1]Community Health Director, AgriSafe Network, North Bend, Nebraska; [2]Total Farmer Health Coach, AgriSafe Network, Covington, Louisiana

Background

AgriSafe was born out of the need to advocate for agricultural workers, their families and the professionals who care for them and their livestock.

In 2016, Total Farmer Health® (Figure 1) was created and aligns closely with the CDC One Health initiative. AgriSafe is an affiliate of CDC Total Worker Health®. As such, the AgriSafe Total Farmer Health® model drives curriculum development and outreach methods to engage health care professionals and producers. Work exposures on any given day are many and varied. The traditional occupational health approach does not match the worker's health and safety needs. Grounded on evidence-based data/information, programs and resources have been developed and continuously updated to address those needs.

Results

AgriSafe' s holistic approach recognizes the interwoven relationship of physical, mental, and emotional components impacting total wellness. Training and educational modules address respiratory health, ergonomics, women's health, safety, and health issues impacting individuals along the age continuum.

Figure 1: AgriSafe Total Farmer Health Model®.

This presentation will focus on noise-induced hearing loss (NIHL) and its heavy impact on mental wellness. NIHL is common in agricultural work. In fact, agriculture is listed in the top three occupations that cause hearing loss.[1] NIHL is permanent and can lead to a diminished quality of life. Studies have shown it can contribute to isolationism, depression, and anxiety. Other consequences include social withdrawal and cognitive decline especially for the elderly, and for those living in rural areas whereas resources and treatment are difficult to access.[2]

Pig squeals average approximately 100dB or more and over time may contribute to permanent hearing loss. Noise is considered dangerous at 85 decibels and above. OSHA compliant industries mandate hearing protection for personnel when exposed to 90 decibels or greater as an 8-hour Time Weighted Average (TWA) sound level. Noise work exposures decibels coupled with TWA sound levels are predictors in considering a Standard Threshold Shift in audiograms.[3]

Swine veterinarians and veterinarian technicians' exposure to dangerous noise levels set them up to negotiate a lifetime hearing loss disability. Allow AgriSafe to be your occupational health resource when designing a hearing conservation program.

Additionally, the work-related stress and demands of veterinarians leading to fatigue, burnout, and depression contribute to the overall health of those caring for livestock. High physical workloads, particularly in large animal/farm practice increase the vulnerability to poor mental health and physical health outcomes. Coping strategies for reducing work-related stress will look different.[4] Moreover, AgriSafe realizes your resilience and work-life balance is key to achieving a healthy mind and body. Programs are available specifically for this unique population of professionals and AgriSafe will connect you with a list of resources.

Discussion

If you come into contact with sounds that exceed the 85 dB level, it is recommended that you wear hearing protection. Hearing protection comes in many forms and AgriSafe will provide a quick guide to finding the right protection to fit the wearer.[5] Come by the AgriSafe table to learn more about a hearing conservation program designed to address your needs.

Good hearing health begins with a baseline hearing exam. Consider taking an exam while here at the conference. Vanderbilt University is offering hearing screenings on-site during this conference. Check the times and take advantage of this wonderful opportunity!

This AASV conference provides opportunities for discussion on hearing health and mental health - because they are related. Mental wellness matters directly impact the health of your practice, your support system and networks, and personal life.

References

1. National Institute on Deafness and Other Communication Disorders. (2020, April 24). It's a Noisy Planet. USDA Department of Health and Human Services, National Institutes of Health. Accessed November 9,2023. **https://www.noisyplanet.nidcd.nih.gov/have-you-heard/not-always-quiet-farm**

2. N. Hernandez, DPT, Hearing Loss Statistics 2023: More Common Than You Might Think. National Council on Aging. 2023, Accessed November 15, 2023. **https://www.ncoa.org/adviser/hearing-aids/hearing-loss-statistics**

3. Occupational Safety and Health Administration. (2008, December 12). 1910.95-Occupational noise exposure. U.S. Department of Labor, Occupational Safety and Health Standards. Accessed November 15, 2023. **https://www.osha.gov/laws-regs/regulations/standardnumber/1910/1910.95**

4. Pohl R, Botscharow J, Böckelmann I, Thielmann B. Stress and strain among veterinarians: a scoping review. Ir Vet J. 2022. June 21, 2022. Accessed November 15, 2023. **doi: 10.1186/s13620-022-00220-x**

5. Central States Center of Agricultural Safety and Health. Protecting Farmer's Hearing. University of Nebraska Medical Center. Accessed on November 09, 2023. **https://www.unmc.edu/publichealth/cscash/_documents/_resources/resources-protecting-farmers-hearing**

https://doi.org/10.54846/am2024/63

A veterinarian's influence on worker safety

Bryn Jensson

Unlocking the power of veterinary influence: A seminar on safety advocacy

Join us for an insightful seminar that sheds light on the pivotal role veterinarians play in shaping and enhancing safety measures across various production types. As a veterinarian, your influence extends far beyond the well-being of animals; it reaches the very core of ensuring the safety of employees, workers, producers, and youth.

The veterinarian's sphere of impact

In the diverse environments where veterinarians engage with people, from farms and production facilities to youth events and educational institutions. Understand the unique position veterinarians hold in disseminating crucial information and implementing safety protocols.

Training for safety excellence

Learn about success stories showcasing how veterinarians have effectively trained individuals in new processes, procedures, and safety measures with safety at the forefront. Learn about how you can integrate safety information in with your daily tasks to make a huge impact on their interactions with this diverse audience.

Advocacy for a safer workplace

Examine the ways in which veterinarians can actively advocate for and contribute to creating safer working environments. Discuss strategies to collaborate with industry partners, safety professionals, and other stakeholders to implement and enforce safety standards.

Building strong safety cultures

It is vital that we continue to gain insights into fostering a culture of safety throughout agriculture through effective communication and leadership. Understand the psychological aspects of safety and how veterinarians can positively influence the mindset and behavior of those they interact with. Join us in recognizing the profound impact veterinarians can have on the safety landscape. Together, let's create a safer and healthier future for all.

https://doi.org/10.54846/am2024/64

Flies play a role in the spread of antimicrobial resistance (AMR): Literature review

Gene Spellman, BS
Central Life Sciences, Schaumburg Illinois

Introduction

Flies have been an overlooked disease vector in swine production. Fly control is now recommended as an important plank of swine biosecurity protocols and disease risk reduction.[1-3] Flies can carry and transmit important swine disease viruses such as porcine reproductive and respiratory syndrome virus (PRRSV),[1] porcine epidemic diarrhea virus (PEDV),[2] Senecavirus A (SVA)[3] and now African swine fever virus (ASFV).[4] Flies can carry infectious bacteria,[5] and antibiotic-resistant bacteria.[6] Flies can travel up to four miles.[7] Fly control in swine till now has been labor-intensive, sporadically effective, and non-coordinated, using multiple products with harsh chemicals. ClariFly 267 (diflubenzuron) insect growth regulator larvicide was developed and approved for swine as an in-feed fly control product. Earlier research showed ClariFly 267 in-feed larvicide achieved a 99% reduction in housefly populations in swine finishing facilities.[8] This fly control option offers a labor-free, convenient tactic for producers considering fly control strategies. With this convenient effective control option, veterinarians can now analyze fly control as a tool to possibly reduce AMR in swine production.

Flies and AMR review summary

When researching the topic of flies and AMR, the amount of data is overwhelming. We believe the following summary of original peer-reviewed research papers and review articles summarizes the importance of insects vectoring several pathogens in livestock systems. There is substantial scientific literature to support the notion that house flies and other insects transport pathogenic and antimicrobial resistant (AMR) microbes on the insect's surface, insect excreta, and internally in the alimentary canal (reviewed in Khamesipour et al).[5] This report is not represented to be all-inclusive of the available literature.

Extended spectrum beta-lactamase (ESBL)-producing *Escherichia coli* bacteria were found that were characterized as identical in both flies (house flies and blow flies) and manure from poultry farms.[11] *Escherichia coli* O157:H7 can infect house fly adults and infected flies may vector these pathogenic bacteria through feeding and excreta.[12,13] House flies have been found to contaminate animal feed stuff with fecal coliforms in less than 6 hours.[16,17] Ahmad et al[9] demonstrated *Escherichia coli* O157:H7 could be vectored by house flies to colonize digestive tracts of cattle.

House flies and German cockroaches have been found containing multiple antibiotic resistant *Enterococcus*, which, are identical to those found in manure from a confined swine operation.[10] Piglets and house flies have been found with similar patterns of antibiotic resistance genes indicating there is likely frequent transfer of bacteria between pigs and flies.[18] Infections of *Enterococcus faecalis* OG1RF:pMV158 in house flies were not only vectored mechanically by the flies, but bacteria proliferated within the house fly alimentary tract.[14]

Within the house fly, Shiga Toxin genes have been shown to be horizontally transferred within *Escherichia coli* strains.[19] Several *Enterococcus* species were evaluated for impact on developing larval house flies (egg to adult) more flies were produced with non-pathogenic species, yet known virulent bacteria species were not different from their non-virulent counterparts.[16] This indicates that virulence factors to humans and animals does not impact house flies.

Bacteria are interacting within insects, often being digested within house flies, yet may be sharing antimicrobial resistant genes, and maintaining infections through house fly metamorphosis from larvae to adults (reviewed in Nayduch and Burrus).[15] House flies have been implicated in both moving AMR bacteria on farms and in dissemination of plasmids with AMR genes between farms.[21]

As vectors of AMR microbes, insects provide transportation for microbes both within and between animal production facilities and human habitations (reviewed by Zurek and Ghosh[6]), insect control should be an integral component of AMR management plans for all livestock producers.

For typical swine production systems, the most important insects to have a management plan in place for include filth flies such as house flies and cockroaches such as German cockroaches. Insect population management should be incorporated into every livestock management plan due to insect's ability to vector microbes to animals, food and feed stuffs, and humans as well as the capacity insects have to generate mixing of antimicrobial genetics of microbes within the insect's alimentary canal (AMR mechanisms reviewed in Schwarz et al[20]).

References

1. Otake S, Dee SA et al. Transmission of porcine reproductive and respiratory syndrome virus by houseflies, (Musca domestica Linnaeus). *Vet Rec* 2003. 152: 73-76.

2. Allison G, et al. PEDV positive bioassay reveals houseflies (Musca domestica) can transmit infectious PEDV to pigs. *Proceedings 50th AASV 2019.*

3. Allison G, et al. PEDV, PRRSV, and Seneca Valley virus PCR positive flies (Musca domestica). What now? *Proceedings 49th AASV 2018.* 394-396.

4. Olesen, A, et al. Infection of pigs with African swine fever virus via ingestion of stable flies (Stomoxys calcitrans) *Transboundary Emerging Diseases 2018,* 65: 1152-1157.

5. Khamesipour F, et al. A systematic review of human pathogens carried by the housefly (Musca domestica L.). *BMC Public Health 2018,* 18:1049.

6. Zurek L, et al. Insects Represent a Link between Food Animal Farms and the Urban Environment for Antibiotic Resistance Traits. *Appl Environ Microbiol. 2014 Jun;* 80(12): 3562–3567.

7. Nazni WA, Luke H et al. Determination of the flight range and dispersal of the house fly, Musca domestica (L.) using mark release recapture technique. *Trop Biomed 2005 Jun;*22(1):53-61.

8. Spellman, G, et al. New labor-free fly control ClariFly 267 (diflubenzuron) achieves 99% reduction in Housefly (*Musca domestica)* populations in swine facilities. *Proceedings 51st AASV 2020.* 175-176.

9. Ahmad, A., Nagaraja, T.G. and Zurek, L., 2007. Transmission of Escherichia coli O157: H7 to cattle by house flies. *Preventive veterinary medicine, 80*(1), pp.74-81.

10. Ahmad, A., Ghosh, A., Schal, C., & Zurek, L. (2011). Insects in confined swine operations carry a large antibiotic resistant and potentially virulent enterococcal community. *BMC microbiology, 11*(1), 23.

11. Blaak, H., Hamidjaja, R. A., van Hoek, A. H., de Heer, L., de Roda Husman, A. M., & Schets, F. M. (2014). Detection of extended-spectrum beta-lactamase (ESBL)-producing Escherichia coli on flies at poultry farms. *Appl. Environ. Microbiol., 80*(1), 239-246.

12. Buma, R., Sanada, H., Maeda, T., Kamei, M., & Kourai, H. (1999). Isolation and characterization of pathogenic bacteria, including Escherichia coli O157: H7, from flies collected at a dairy farm field. *Medical entomology and zoology, 50*(4), 313-321.

13. Buma, R., Maeda, T., Marutaka, Y., Kamei, M., Nagamune, H., & KOURAI, H. (2004). Vectorial capacity of larvae, pupae and adult of housefly (Musca domestica) for Escherichia coli O157: H7 and the possibility of transmission from source to human. *Medical Entomology and Zoology, 55*(2), 95-106.

14. Doud, C. W., & Zurek, L. (2012). Enterococcus faecalis OG1RF: pMV158 survives and proliferates in the house fly digestive tract. *Journal of medical entomology, 49*(1), 150-155.

15. Nayduch, D., & Burrus, R. G. (2017). Flourishing in filth: house fly–microbe interactions across life history. *Annals of the Entomological Society of America, 110*(1), 6-18.

16. Ghosh, A., Akhtar, M., Holderman, C., & Zurek, L. (2014). Significance and survival of Enterococci during the house fly development. *Journal of medical entomology, 51*(1), 63-67.

17. Ghosh, A., & Zurek, L. (2015). Fresh steam-flaked corn in cattle feedlots is an important site for fecal coliform contamination by house flies. *Journal of food protection, 78*(3), 567-572.

18. Literak, I., Dolejska, M., Rybarikova, J., Cizek, A., Strejckova, P., Vyskocilova, M., ... & Klimes, J. (2009). Highly variable patterns of antimicrobial resistance in commensal Escherichia coli isolates from pigs, sympatric rodents, and flies. *Microbial Drug Resistance, 15*(3), 229-237.

19. Petridis, M., Bagdasarian, M., Waldor, M. K., & Walker, E. (2014). Horizontal transfer of Shiga toxin and antibiotic resistance genes among Escherichia coli strains in house fly (Diptera: Muscidae) gut. *Journal of medical entomology, 43*(2), 288-295.

20. Schwarz, S., Loeffler, A., & Kadlec, K. (2017). Bacterial resistance to antimicrobial agents and its impact on veterinary and human medicine. *Advances in Veterinary Dermatology, 8*, 95-110.

21. Usui, M., Shirakawa, T., Fukuda, A., & Tamura, Y. (2015). The role of flies in disseminating plasmids with antimicrobial-resistance genes between farms. *Microbial Drug Resistance, 21*(5), 562-569.

https://doi.org/10.54846/am2024/65

A discussion of piglet production losses associated with *Mycoplasma hyopneumoniae* (Mhp) elimination programs in low vs profitable market environments

D. A. Baumert, DVM
Zoetis, Parsippany, New Jersey

Introduction

Mycoplasma hyopneumoniae (Mhp) is the bacterial agent resulting in clinical mycoplasmal pneumonia or enzootic pneumonia in swine populations. Clinical disease due to Mhp is worldwide in distribution and common in North America. Economic losses related to Mhp are associated with decreased feed efficiency, reduced average daily gain, and increased medication cost.[1] Management protocols to eliminate Mhp from swine breeding herds and subsequently to produce Mhp-negative flows of commercial pigs have been used to eliminate the performance-related cost of the disease. In addition to the cost of vaccines and antimicrobials that are typically used in Mhp elimination programs, Mhp elimination programs may also have production-related costs that should be considered during the planning phases of a disease elimination effort.

Disease elimination programs

Routine production practices have developed over time in order to maximize production outputs from the various segments of swine production systems. Disease elimination programs, by nature, must modify routine production practices in order to change (eliminate) the presence of an undesirable disease agent(s) currently present within a production system. The degree of change or modification to be applied to routine production practices will vary based on the specific disease and specific farm system characteristics.

Mhp-specific elimination programs

An Mhp elimination program will typically involve a herd closure period lasting approximately 8 months. This herd closure period, which is not a normal production practice, is an important component of an overall plan to eliminate the active shedding of Mhp. The elimination of Mhp shedding is necessary in order to move the breeding herd population to an Mhp-negative status. During this herd closure period, gilts will not be allowed to enter the breeding herd to replace sows lost to culling or to mortality. These gilt 'non-entries' will allow the development of 'holes' in the breeding herd inventory resulting in missed breeding targets, missed piglets farrowed, and missed pigs weaned over time.

Extent of herd closure and missed gilt entries

Routine maintenance of breeding herd inventory generally involves 50% annual replacement rates in response to routine sow culling and mortality.[2] A cessation of voluntary sow culling can reduce by half the typical 4% monthly replacement rate needs at the sow farm. However, the portion of the 4% monthly replacement rate attributed to sow mortality and involuntary culls (1.5-2% per month) is more difficult to manage in a significant, practical manner.[3] This mortality-related sow attrition will result in 'inventory holes' or inventory reductions equivalent to approximately 2% of each group of due-to-breed animals; the extent of these holes will increase by an additional 2% each month that the gilts are not allowed to enter the herd. These holes in breeding herd inventory result in missed breeding targets and subsequently become missed piglet production and missed piglets for transfer or sale. See Figure 1.

Production impact

Sow herd attrition due to mortality and unavoidable removal of culls results in a decrease in breeding herd inventory by approximately 2% per month. This 2% loss will be cumulative month-over-month for as long as gilt replacements are not entering the herd. While preliminary overstocking of gilts may manage the mortality-related inventory losses during the first four months of a breeding herd closure period, overstocked gilts will not typically be available during the second 4 months of an 8-month herd closure period. This 2% loss of breeding herd inventory will result in a 2% deficiency in achieving breeding herd targets during the first month that replacement gilts are not available. Moreover, this 2% deficiency in meeting breeding targets will continue to increase cumulatively for each additional month that replacement gilts are not available.

Mathematical impact

A mathematical description of the total monthly missed breeding herd targets from month 5 through month 9 of a typical 8-month herd closure period relative to annual pig production can be stated as: $(2\% + 4\% + 6\% + 8\% + 10\%)/12 = 2.5\%$. Using this 2.5% loss of annual breeding herd targets as an estimate for the annual piglet production loss, a herd that had expected to produce 25 PSY (pigs/sow/year) would produce only 24.4 PSY. At a time when a weaned pig may be valued at $50/head,[4] this loss of 0.6 PSY is equivalent to a loss of $30/sow ($50/pig x 0.6 PSY) during the year that the Mhp elimination program and the associated herd closure is occurring. See Figure 2.

Impact of low-profit environment

In the current (2023) low-profit environment, the effect of a Mhp elimination program-related loss of 0.6 PSY is very different than the effect of a 0.6 PSY loss in a typical, profitable environment. When the sale or transfer value assigned to a weaned pigs is $15/head, rather than $50/head, the value of the 0.6 annual PSY loss subsequent to Mhp elimination program-related missed breeding targets is significantly reduced. A loss of 0.6 PSY in a $15/pig down-market environment is only a $9 loss per sow rather than the $30 loss per sow that would be realized in a typical $50/weaned pig market environment. See Figure 3.

Figure 1: Missed breeding targets general impact on production.

Figure 2: Missed breeding targets calculated impact on production.

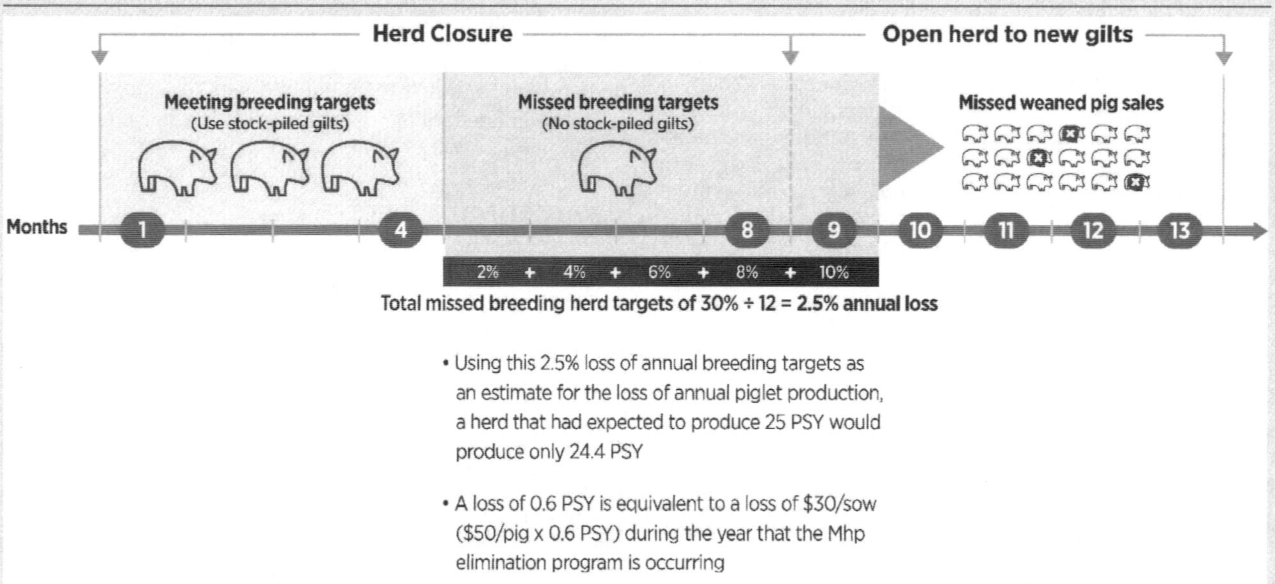

Figure 3: Missed breeding targets calculated impact in a low-profit environment.

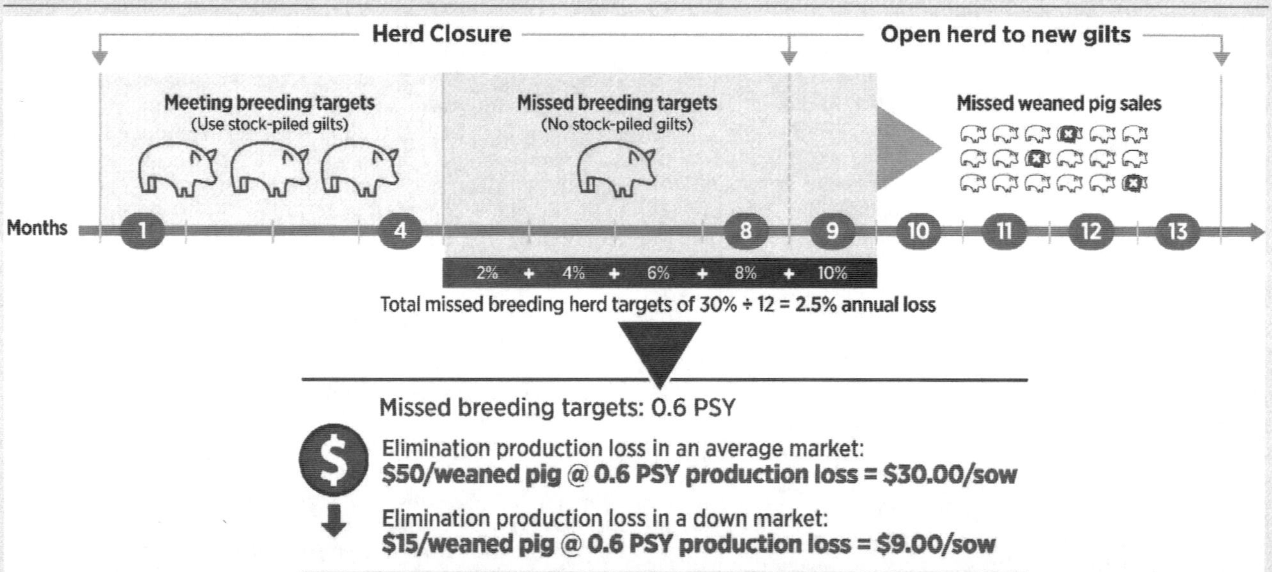

Discussion

One management option for reducing the impact of herd closure on missed breeding targets, and the subsequent loss of weaned pig production, is the use of an off-site breeding project. An off-site breeding project allows for the receipt and breeding of replacement gilts that would have been destined for delivery to the sow herd site. Gilts that are received, bred, and held (at the offsite facility) can assist the sow farm in meeting its breeding targets. However, an off-site breeding project can be a management challenge as it can require the acquisition of an additional facility and the staffing of that facility. A detailed discussion of the operation of an off-site breeding project is beyond the scope of this current discussion.

The key message to emphasize is that, although it may not seem intuitive initially, a down market environment is a financially beneficial time to consider an Mhp elimination program as the economic impact of the production losses, related to missed breeding targets, are significantly less. Producers should consider this as an opportunity, rather than waiting for the overall market economy to improve.

When producers or production systems are ready to put an Mhp elimination program in place, having considered the economic and production impacts, Zoetis representatives and technical services veterinarians are here to support them with general guidelines and recommendations as part of Mhp Guardian*, Zoetis' four-step Mhp elimination program to help producers and vets build personalized, customized plans.

**Note: The program provides general guidelines. Zoetis cannot guarantee their applicability or accuracy in regard to individual circumstances. Zoetis shall not be liable for any loss arising from the use of the materials, including any loss resulting from any errors in the content contained herein, or from any actions taken in reliance thereon.*

References

1. Holst S, Yeske P, Pieters M. Elimination of Mycoplasma hyopneumoniae from breed-to-wean farms: A review of current protocols with emphasis on herd closure and medication. *J Swine Health Prod*. 2015;23(6):321-330

2. Moroni JL. How to achieve the optimal replacement rate of the sow herd. September 3, 2023. Accessed October 20, 2023. **https://topigsnorsvin.us/news-us1/how-to-achieve-the-optimal-replacement-rate-of-the-sow-herd/**

3. Paiva R, Moura C, Thomas P, et al. Risk factors associated with sow mortality in breeding herds under one production system in the Midwestern United States. February 24, 2023. Accessed October 20, 2023. **https://pubmed.ncbi.nlm.nih.gov/36867926/**

4. National Direct Delivered Feeder Pig Report (Quarterly); USDA Livestock, Poultry, and Grain Market News Division; USDA Agricultural Marketing Services. Accessed November 3, 2023. **https://usda.library.cornell.edu/concern/publications/2801pg37r?locale=en&page=2#release-items**

https://doi.org/10.54846/am2024/66

A diagnostic comparison of breeding herd mass vaccination versus rolling herd vaccination for porcine circovirus type 2 (PCV2)

A. Farkas[1], DVM; T. Painter[2], DVM; M. Jansen[2], DVM
[1]Carthage Veterinary Service, Carthage, Illinois; [2]Zoetis, Parsippany, New Jersey

Introduction

While the industry continues to lack a clear definition of sow herd PCV2 stability, one of the tools available to address instability is the use of vaccine in sows. However, a definitive optimal protocol has not yet been identified as the "gold standard." Previous publications have shared results from using "rolling" (or ongoing) vaccination,[1,2] but this has not been compared to the use of whole herd vaccination. The objective of this project was to evaluate differences between the use of whole herd and rolling vaccination against PCV2 in sow herds.

Materials and methods

This project was conducted in a 2800-head sow farm located in western Illinois.

Phase I—Mass vaccination

The project began in March 2019 when concern around PCV2 stability in a few sow herds was raised and clinical PCV2-associated disease (PCVAD) was seen in downstream pigs. At that time, whole herd vaccination was administered and a diagnostic plan outlined to understand the impact of the vaccination on virus detection (via polymerase chain reaction, or PCR) in serum of downstream pigs and antibody levels (through immunofluorescent assay, or IFA) in piglets at weaning (Table 1). Samples were collected on a quarterly basis for one year after the whole herd vaccination.

Phase II—Rolling vaccination

The second phase began after observation of a decline in herd immunity as measured by the antibodies present in piglets at weaning around six months post-whole herd vaccination. A decision was made to target a more frequent booster vaccination. As such, a rolling vaccination was implemented in the lactation phase (day 10). This was put in place starting in May 2021. PCV2 PCR of processing fluids were then monitored on a weekly basis with Ct values charted in an individuals control chart (I Chart, Minitab® Statistical Software v20.1.1, State College, Pennsylvania). Ct values of greater than or equal to 26 were considered "in control" as previously published.[1]

Table 1: Quarterly diagnostic sample protocol

Population sampled	Age (weeks)	Test
Piglets	3	PCV2 IFA
Downstream pigs	10	PCV2 PCR
	15	
	30	

Phase III—Final diagnostic collection

After approximately 2.5 years of rolling vaccination, a final round of diagnostics was collected to mirror the "quarterly" diagnostics in phase I.

Results

PCR testing

The results from PCR testing are shown in Figures 1 and 2. Serum PCRs show that only a small number of downstream pigs sampled were positive in the post-mass vaccination quarterly samplings, and none of the serum samples from downstream pigs were positive in the samples collected after implementation of rolling vaccination. The mean weekly processing fluids Ct values increased from 32.77 before rolling vaccination to 35.26 after rolling vaccination. Additionally, only one sample fell below the cutoff of 26 after the implementation of rolling vaccination, while previously, after only mass vaccination, several samples fell at or below the 26 Ct threshold previously established.

Antibody testing

Immunofluorescent assay values from wean pigs are shown in Figure 3. At 3 months post-mass vaccination, nearly all pigs sampled were antibody positive at weaning. Some pigs had titers as high as 5120. The percent of positive pigs decreased by 6 months post-mass vaccination, and by 9 months post-mass vaccination, essentially all pigs were negative at weaning. However, at 12 months post-mass vaccination, antibody levels began increasing again despite the fact that no additional vaccination had been

Figure 1: Individual value plot of serum PCRs in downstream pigs by age and by weeks post-mass vaccination.

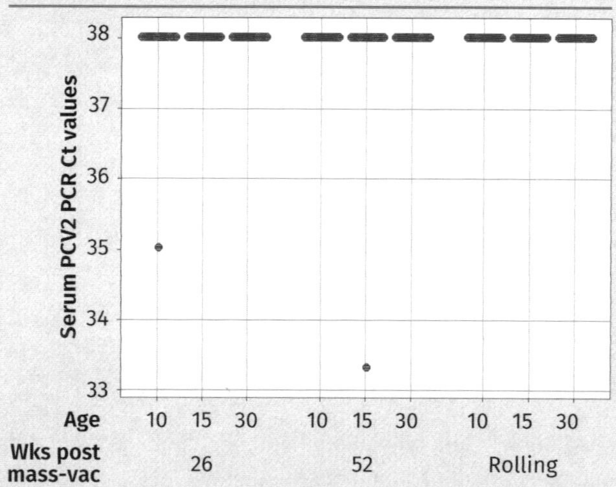

Figure 2: Individuals chart (I Chart) of processing fluids PCV2 PCR Ct values by week before and after rolling vaccination implementation.

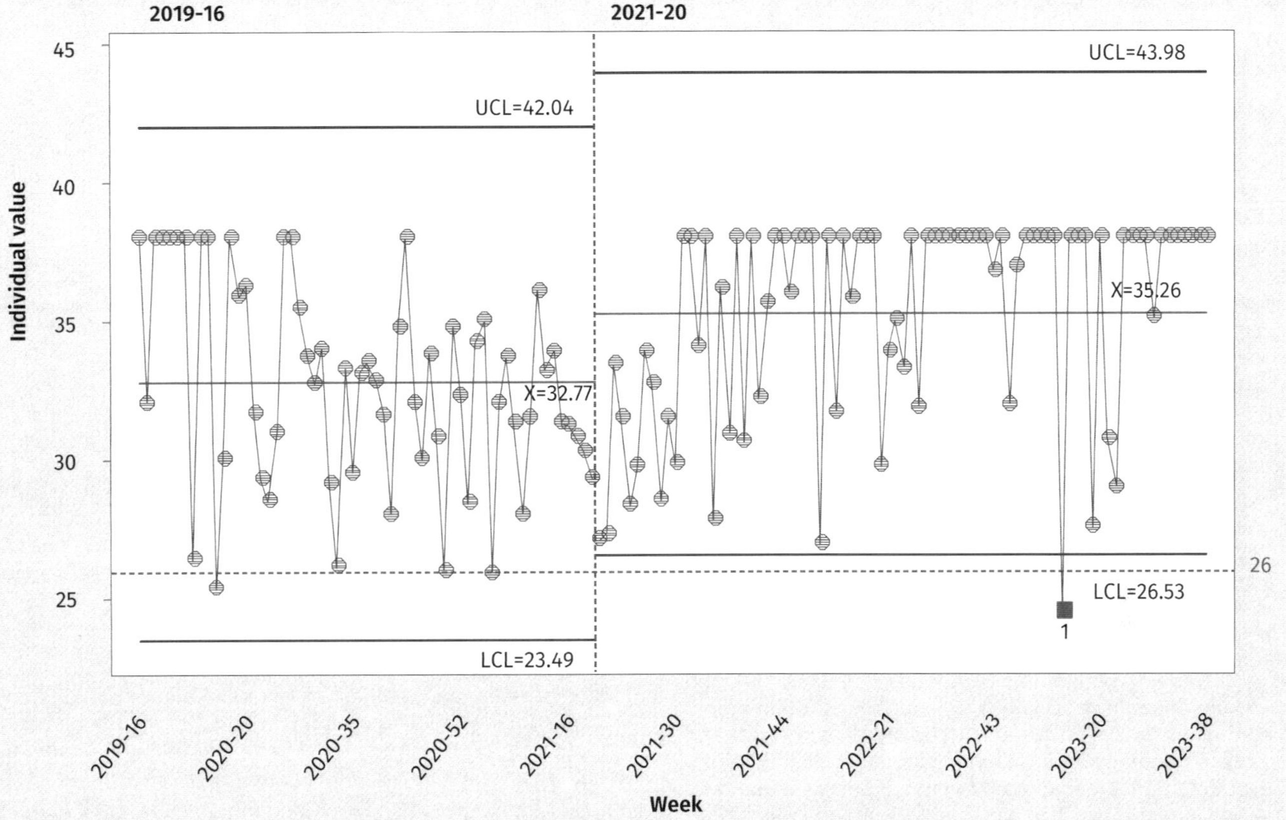

Figure 3: Interval plot displaying 95% confidence interval for the mean PCV2 IFA of wean-age pigs by months post-whole herd vaccination and after rolling vaccination.

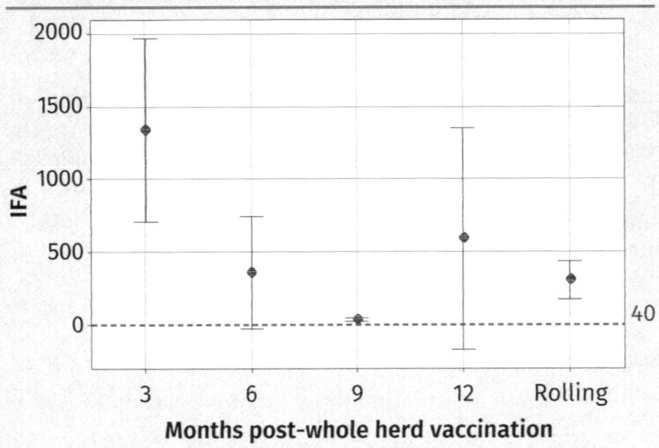

Individual standard deviations are used to calculate the intervals.

administered to the herd. This may indicate viral activity in the herd. After the implementation of rolling vaccination, several pigs were antibody positive at weaning, but at moderate levels of ≤ 1280.

Discussion

This comparison between whole herd vaccination and rolling vaccination has shown that using rolling PCV2 vaccination in sows results in lower virus detection in processing fluids and lower downstream pig viremia. Additionally, antibodies are present in piglets at weaning, indicating passive immunity is being provided at that time. The ultimate goal at the onset of this study was to reduce occurrence of PCVAD in downstream pigs, and by reducing viral activity and increase herd immunity, rolling vaccination appears to be the best protocol to utilize in the sow herd. More studies should be conducted in other herds to understand if these results are repeatable.

References

1. Farkas A, Eggers K, Guzman JE, Jansen M, Galina L. Rolling PCV2 Vaccination in Sows: Experiences with Lactation Vaccination and Processing Fluids Surveillance. Allen D. Leman Conference, 2022.

2. Vennekotter A, Copeland S, Jablonski E, Guzman JE. Rolling PCV2 Vaccination in Sows: Experiences with Mid-Gestation Vaccination and Processing Fluids Surveillance. Allen D. Leman Conference, 2022.

https://doi.org/10.54846/am2024/67

Use of PrimeStore® MTM for stabilizing oral fluid samples and getting higher yields of nucleid acids, especially for PRRSV surveys

Tilman Kuehn[1], Dr med vet, MBA; Jerry Torrison[2], DVM, PhD; Katharina Zimmer[1], MSc
[1]aCare Lab, Leipzig, Germany; [2]Longhorn Vaccines and Diagnostics, Bethesda, Maryland

Introduction

Oral fluid samples have become successfully established in the porcine health management of pig farms worldwide. They are accepted as a common diagnostic sample in group-housed nursery, grow-finish and adult swine. An essential prerequisite for successful use is first of all good strategic placement in the barn and careful sample management. The increasing importance of cross-species pathogen monitoring continues to pose a major logistical challenge. One of the main problems is that with a high degree of environmental and fecal contamination there is also a high enzyme activity in the samples, which can have a considerable influence on the quality of the results. And besides, many pathogens require special preservation in the sample in order to be detectable in significant amounts even after potentially days of transport to the diagnostic site and under sometimes extreme temperature conditions. Commonly used transport media such as saline solution or native sampling are only suitable to a limited extent here. Degeneration of the pathogen DNA or RNA up to the point of complete disappearance falsifies the result of the actual situation on the farm and also reduces the possibilities of further specification, eg, for subtyping or sequencing. With the PrimeStore® MTM, a stabilisation medium is available that can achieve exactly this stability in the sample, with the added benefit of simplified handling, as the samples are non-infectious and therefore easier to ship.

Questions

In our studies, the preservation medium PrimeStore® Molecular Transport Medium (MTM) from Longhorn Vaccines and Diagnostics (Bethesda, MD) was tested in comparison to native samples with regard to its pathogen stabilization. For this purpose, oral fluid samples were taken from domestic pig herds in Germany with expected positive results especially for PRRSV and IAV-S, and were stored native in classic transport tubes and also in the PrimeStore® MTM. In particular, we have addressed the following questions in studies:

1. Does the use of PrimeStore® MTM for porcine oral fluid sample placed immediately into this have a positive impact on outcome quality for PCR results, especially for RNA-viruses, and especially for the yield on an equivalent PRRSV-PCR result compared to the undiluted neat sample?
2. Can a porcine oral fluid sample placed immediately into PrimeStore® MTM be kept up to five days at ambient temperature yield an equivalent or better PRRSV-PCR result compared to the undiluted neat sample stored at a freezer or refrigerator?
3. Do samples preserved with PrimeStore® MTM have improved nucleic acid quality with regard to further investigations, eg, sequencing or vaccine strain testing?

Methods

Longhorn's PrimeStore® MTM is a molecular medium that provides the user with near actual pathogen content at the time of specimen collection by preserving and stabilizing DNA and RNA. According to the manufacturer, the medium is suitable for the storage and stabilization of a variety of different sample matrices and body fluids (oral fluids, untreated nasal lavage specimens, nasopharyngeal, oral/pharyngeal swabs, sputum) and retains its effectiveness even under extreme temperature conditions. This involves reliable stabilization, preservation and inactivation of pathogens such as viruses and gram-positive/gram-negative bacteria for safe transport and subsequent diagnostic processing. Samples can be frozen for long term storage and should not be affected by multiple freeze-thaw cycles. For oral fluids we used it in a ratio of 3:1/MTM:sample.

For the sample taking of the oral fluids we used the IDEXX Oral Fluid Collection Kit. For sample processing and nucleic acid extraction, we were provided with Indicals`s IndiMag 48/48s instrument and the IndiMag Pathogen IM48 cartridges for automated extraction of the nucleic acids (viral RNA and DNA, and bacterial DNA) from a broad range of animal sample types.

Within the scope of this studies, Indical provided specific test kits for the respective pathogens to be tested. These were the Virotype PRRSV RT-PCR Kit and the Virotype Influenza A RT-PCR Kit. The Virotype PRRSV RT-PCR Kit allows the detection of PRRSV 1 and PRRSV 2 in samples from swine and is also capable to detect the new highly pathogenic strains. The Virotype Influenza A RT-PCR Kit enables the identification of 16 HA subtypes of the influenza A virus (IAV-S) and is validated for multiple sample types. Both kits thus provide highly reliable results in the interest of this studies.

We also attempted to characterize the yield and quality of the RNA obtained. First by using the NanoDrop Spectrophotometer from Thermo Fisher Scientific, which is suitable for a simple, rapid, accurate quantification and qualification of DNA, RNA, and proteins. However, the system cannot distinguish between degraded or intact RNA. There are therefore other ways to assess the quality (also called integrity) of the isolated RNA. A quick and inexpensive method is electrophoresis by running the RNA sample in an agarose gel and observing the resulting ribosomal RNA (rRNA) bands. If the bands are clear, bright and of the expected size compared to the molecular standard ladder, the RNA is of good quality. The absence of these bands, together with the presence of a streak, indicates that the RNA sample is degraded.

Last but not least, the results of the Sanger sequencing were compared. This is usually carried out for PRRSV in the ORF5 and ORF7 open reading frames. From native samples, a Ct value < 30 is a prerequisite and even this is not always promising. The number of sequenced nucleotides and the query cover value are relevant for a good interpretation. Query coverage is the percentage of sequence aligned to a sequence in a genebank. Low query coverage results from short sequences and does not allow a reliable comparison with the sequences of the database.

Results

The qRT-PCR results are showing a lower Ct value for both PRRSV and IAV-S in the samples placed in PrimeStore® MTM compared to the native samples. Especially for PRRSV, the median Ct value of the PrimeStore® MTM samples shows a significant reduction of 2 to 3 (sometimes more) Ct-values compared to the native samples (Figure 1). This effect increases with longer storage time. The effect was similar for IAV-S, although not quite as clear. We were able to show that even oral fluid samples stored in the refrigerator lose several Ct values within a few days and that this does not happen in comparison to PrimeStore®MTM. In the evaluation of different storage conditions, a higher yield of RNA and lower Ct values could mostly be demonstrated.

This also applies to the quality, measured with NanoDrop or on the gel.

There was also a very impressive effect during sequencing. Samples preserved by PrimeStore® MTM had significantly better query covers and the proportion of sequencable samples was higher. This made better interpretations possible.

Figure 1: Ct value for PRRSV in the samples placed in PrimeStore® MTM compared to the native samples.

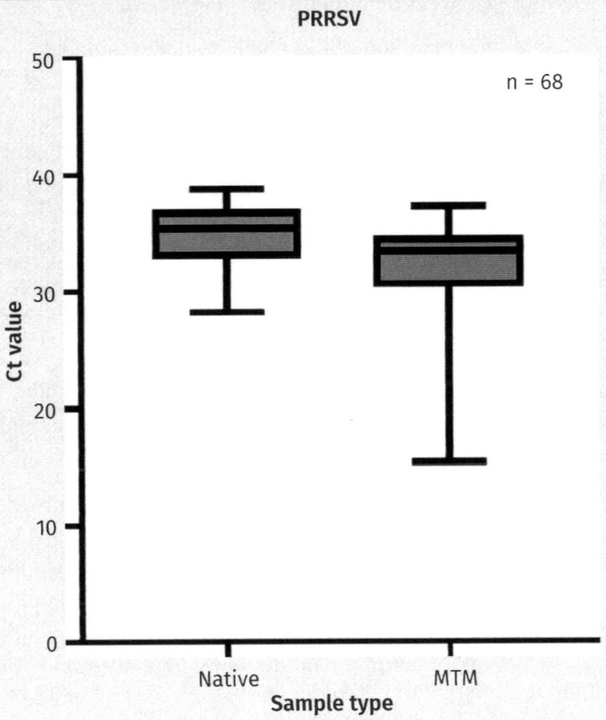

Discussion

These results are implying a higher to significantly higher viral RNA content and better nucleic acid quality in the PrimeStore® MTM preserved samples and thus a higher pathogen preservation by the PrimeStore® MTM compared to the native samples.

Accordingly, the major advantage of using PrimeStore® MTM for RNA-stabilizing sample transport is evident, as the initial pathogen content is almost maintained compared to the native sample. PrimeStore® MTM thus enables reliable diagnosis and the subsequent initiation of accurate measures. This means that samples that were previously not suitable for analysis due to a high Ct value or poor RNA quality are now available for further analysis. In addition, the proportion of samples that can be used for further testing for subtypes or field/vaccine strain differentiation will increase significantly.

https://doi.org/10.54846/am2024/68

Effect of feeding EnduraSow™ on performance and nasal microbiome of sows and nursery pigs infected with porcine reproductive and respiratory syndrome virus

Huyen Tran[1], PhD; Aileen Joy Mercado[1], PhD; Stacie Crowder[2], PhD; Brenda de Rodas[1], PhD
[1] Purina Animal Nutrition, Gray Summit, Missouri; [2]PMI Additives, Arden Hills, Minnesota

Introduction

Porcine reproductive and respiratory syndrome virus (PRRSV) is one of the most devastating endemic diseases in swine, which is responsible for approximately $664 million annual losses to the US pork production.[1] Data from our previous study[2] indicated that feeding EnduraSow™ (ES), a patent-pending blend of activated medium chain fatty acids, organic acids, a probiotic cocktail, and other compounds, to PRRSV-positive sows improved the reproductive performance of sows, and reduced mortality of nursery pigs while supporting nursery growth and feed conversion. Therefore, the objective of this trial was to evaluate the impacts of EnduraSow™ on the performance and nasal microbiome of gestating and lactating sows and nursery pigs infected with PRRSV.

Materials and methods

Two PRRSV-positive commercial sow farms (5,000 sows each; PRRSV strain 1-3-2) were used to evaluate the impact of ES fed to sows. Farm 1 was fed a commercial gestation and lactation diet without ES. Farm 2 was fed the same gestation and lactation diet with 0.375% ES. Reproductive performance and lactation feed intake data were recorded weekly for 6 months for calculation of pre-weaning mortality, non-productive sow days, days to estrus, and daily lactation feed intake. Blood samples were collected from a subset of parity 2 and 3 sows (n = 15/farm) at days 30 and 90 of gestation and day 14 of lactation, as well as in nursery pigs (n = 15/farm) at day 21 post-weaning for PRRSV PCR viremia testing. Nursery pigs were housed by sow farm origin, and nursery performance was recorded for mortality rate, average daily gain, average daily feed intake, and feed:gain ratio calculation. For the evaluation of feeding ES on microbiome, nasal swabs were collected from the sows and nursery pigs at the same time as blood samples were collected and subjected to 16S rRNA sequencing (V3-V4 region, Ilumina).

Results and discussion

Results from this experiment showed that sows fed ES had greater reproductive performance compared to the control sows, as evidenced by reductions in the pre-weaning mortality in piglets by 15.9%, non-productive sow days by 5.2 days, and time interval from weaning to first estrus by 1.1 day. In addition, sows fed ES had greater daily lactation feed intake by 0.87 lb relative to the control.

The nasal microbiome of sows was dominated by Proteobacteria, Actinobacteria, Firmicutes, and Bacteroidetes, which accounted for 98.4% and 99.7% of relative abundance in gestation and lactation, respectively. Results from Linear discriminant analysis Effect Size show an increase in Firmicutes (16.7 vs 6.2%; $P < .001$) in gestating sows and a reduction in Proteobacteria, Bacteroidetes, and Fusobacteria in lactating sows fed ES ($P < .01$). In addition, both gestating and lactating sows fed ES had greater Firmicutes:Bacteroidetes ratio (by 19X and 32X, respectively), which in turn may confer potential benefits in terms of mitigating the severity of PRRSV infection and improving performance outcomes.[3,4]

Nursery pigs from sows fed ES had a significant reduction in mortality rate (5.9% vs 17.8%; $P < .001$) compared to pigs from sows fed the control diet. In addition, ES nursery pigs had a greater average daily gain (+19%; $P = .01$), resulting in a 2.3 lb increase in final nursery body weight, and tended to have greater feed conversion (+4.8%; $P = .07$). Regarding immune response, feeding ES to sows improved immunity in their progenies as indicated by a higher cycle threshold (Ct) in nursery pigs from sows fed ES compared to the control (CT = 27.2 vs. 23.8; $P = .03$). Furthermore, results from processing fluid testing in piglets indicate that sows fed ES produced negative piglets at 6 months after feeding ES, while control sows did not produce PRRSV-negative piglets at any time during the trial.

Results from 16S rRNA sequencing indicated that feeding ES to sows significantly shifted nasal microbial beta diversity (weighted and unweighted unifrac, $P < .01$) of nursery pigs. Nursery pigs from PRRSV-positive sows fed ES had a balanced nasal microbiome, where ES decreased *Mycoplasma hyorhinis* by 54% and *Staphylococcus* sp by 29%, and increased *Lactobacillus reuteri* by 1.7X and *Lactobacillus johnsonii* by 2X, relative to the control.

Conclusions

Data from this study indicate that feeding EnduraSow™ to PRRSV-positive sows during gestation and lactation stabilized the sows sooner than non-supplemented sows and provided benefits toward balancing the nasal microbiome of sows and their progenies. Nursery pigs from sows fed EnduraSow™ had greater health and performance outcomes and lower nursery mortality.

References

1. Holtkamp DJ, Kliebenstein JB, Neumann EJ, et al. 2013. Assessment of the economic impact of porcine reproductive and respiratory syndrome virus on United States pork producers. *J Swine Health Prod.* 21(2):72-84.

2. Huyen Tran, Aileen Joy Mercado, Stacie Crowder, and Brenda de Rodas. 2023. Effect of feeding EnduraSow™ feed additive on performance of sows and nursery pigs infected with porcine reproductive and respiratory syndrome virus. Allen D. Leman Swine Conference research abstract page 121 (September 2023, St. Paul, Minnesota).

3. Constance, L. A., Thissen, J. B., Jaing, C. J., McLoughlin, K. S., Rowland, R. R., Serão, N. V., Cino-Ozuna, A. G., and Niederwerder, M. C. 2021. Gut microbiome associations with outcome following co-infection with porcine reproductive and respiratory syndrome virus (PRRSV) and porcine circovirus type 2 (PCV2) in pigs immunized with a PRRS modified live virus vaccine. *Veterinary Microbiology, 254,* 109018. **https://doi.org/10.1016/j.vetmic.2021.109018**

4. Argüello H, Rodríguez-Gómez IM, Sánchez-Carvajal JM, Pallares FJ, Díaz I, Cabrera-Rubio R, Crispie F, Cotter PD, Mateu E, Martín-Valls G, Carrasco L, Gómez-Laguna J. 2022. Porcine reproductive and respiratory syndrome virus impacts on gut microbiome in a strain virulence-dependent fashion. *Microb Biotechnol.* 5(3):1007-1016. **doi: 10.1111/1751-7915.13757**

https://doi.org/10.54846/am2024/69

Pathogen antibody surveillance of a sow farm, 2015 – 2023

David H. Baum[1], BS, MS, DVM, PhD; Katie Woodard[1], DVM, MS; Brennan Sorkin[2], BS, MPH; Jeffrey Zimmerman[1], DVM, PhD; Luis Gimenez-Lirola[1], BSBC, MS, PhD; Gustavo Silva[1], DVM, MS, PhD; Maria Jose Clavijo[3], DVM, PhD; Siyu Xiao[1], DVM; Alexandra Henao-Diaz[1], DVM, MS, PhD; Suzy Block[1], BS; Kent Doolittle[4], BS; Silvia Zimmerman[4], DVM, MS; Marisa Rotolo[3], DVM, MS, PhD; Perry Harms[3], DVM, MS

[1] Department of Veterinary Diagnostic and Production Animal Medicine, Iowa State University, Ames, Iowa; [2]Southern New Hampshire University, Manchester New Hampshire; [3]PIC, Nashville, Tennessee; [4]IDEXX, Westbrook, Maine

Introduction

Average and range ($\bar{x}$ R) charts are widely in many industries or production settings to monitor the stability of processes over time.[1] Simple to use, they detect obvious changes in a process as values that are "out-of-bounds" (outside three-sigma from the grand average) or subtle changes in a process as non-random patterns relative to the historic average. Thus, these charts use production data collected over time to judge the most recent metrics. In swine medicine $\bar{x}$ R charts are useful for characterizing enzyme-linked immunosorbent assay (ELISA) quantitative (sample/positive, S/P) values obtained from monthly anti-pathogen antibody surveillance results. The $\bar{x}$ chart is the average ($\bar{x}$)quantitative ELISA result (eg, S/P) of the herd's monthly sample test results. The R chart is the range (R) of S/P values for each set of samples. The $\bar{x}$ R charts upper and lower limits (established using historical data) identify when the current results exceed normal variation. Particular patterns of the samples' averages and ranges also identify special cause variation. Thus, these charts can be used to objectively characterize any population's pathogen status. For instance, "This farm's pathogen status is 'Not Detected' based on its in-control $\bar{x}$ S/P values that have averaged between 0.XXX (which will likely be at or below the kit manufacturer's S/P cutoff value) and 0.XXY (the lower limit around the mean S/P). When samples have been collected at regular and frequent intervals the $\bar{x}$ values detect post-infection antibody production and distinguish those from non-specific ELISA reactions. Post-infection antibody production will appear as an increase in S/P values over time, associated with shed and spread of the pathogen within the population. Non-specific reactors can be identified in the range chart. These ranges appear as an out-of-bounds S/P range with a frequency equal to or less than that expected for the assay. Efficient surveillance and management techniques are used to prevent outbreaks of pathogens, reduce mortality rates, and optimize production efficiency in pig farms. This paper describes the use of $\bar{x}$ charts of an anonymous anti-pathogen ELISA results for surveillance of a sow farm. The emphasis of this paper is on the analysis of ELISA surveillance data, not any particular pathogen.

Materials and methods

The farm submitted surveillance samples at ~5-week intervals from January 2015 through May 2023 although the number of samples collected and submitted varied over the course of the surveillance period. The farm became infected with the pathogen of interest, was depopulated, and repopulated with pathogen-negative females in January 2017. Quantitative ELISA results were obtained using a commercially available test performed at Iowa State University Veterinary Diagnostic Laboratory. Surveillance test results were sorted chronologically and put into a pivot table in an Excel spreadsheet. The pivot table included the date the sample was received, the pig ID number, and the quantitative result of the ELISA test. Thereafter, S/P $\bar{x}$ charts were created and interpreted according to Wheeler and Chambers.[1] Signals of instability in the process were: **1)** averages outside more than 3 standard deviations from the historic average or range; **2)** 2 of 3 consecutive averages or ranges outside 2 standard deviations, but on the same side of their respective averages; **3)** 4 of 5 consecutive averages or ranges outside one standard deviation, but on the same side of their respective averages; or **4)** 8 consecutive averages or ranges on either side of the average or average range. Thus, antibody surveillance stability was defined as the absence signals indicating instability.

Results

Averages and limits for the $\bar{x}$ and R charts were first calculated globally (Figure 1). The presence of out-of-bounds values or patterns of values justified recalculating their limits (Figure 2). Infection by the pathogen was detected in the latter part of the first 13 $\bar{x}$ chart (Figure 2) averages and ranges. This was followed by additional signals of instability, eg, increased averages and ranges and values higher than the positive/suspect cutoff value for the ELISA (0.400). These persisted for 16 sample collections then dropped precipitously on the 30th set of samples starting in January of 2016. The process was stable and low through August 2020 with just one Range (R) out of 40 consecutive samples' R (1/40 = 0.025) greater than the positive-suspect cut-off for the ELISA, but results from November 2020 to April 2023 (33 sample sets) are characterized by uncommon variation. Among the 33 sample sets, 5 have out-of-bounds values and 7 have out-of-bounds R values. Three of these R's had S/Ps > 0.400. The proportion of out-of-bounds R's (3/33 = 0.091) for the period November 2020 to April 2023 was higher than observed for (January 2017 to August 2020), but does not match the pattern of antibody seen in late 2015 – 2016.

Discussion

$\bar{x}$ charts are a robust method for objectively characterizing antibody test results' variation and a pathogen's behavior within a population, such as a sow farm. The $\bar{x}$ chart is the monthly average of ELISA S/P values plus their respective out-of-bounds (3 sigma) limits whereby recent ELISA results are interpreted in the context of past pathogen antibody results. Interpretation is based on the type of variation that characterizes the results' $\bar{x}$ and R. Common variation occurs when results are within the out-of-bounds values and contain no pattern in their distribution.[1] Special cause variation is characterized by the presence of out-of-bound values and/or patterns in their distribution. In September 2015 the $\bar{x}$ showed anti-pathogen antibody production in the sow farm. The R chart detected early potential signals of infection in mid-2015. An anamnestic response appeared in the $\bar{x}$ R charts

Figure 1: Average S/P (cl), range (cl), and limits (ucl and lcl of X-bar) are calculated globally. The pathogen status has varied over the course of surveillance.

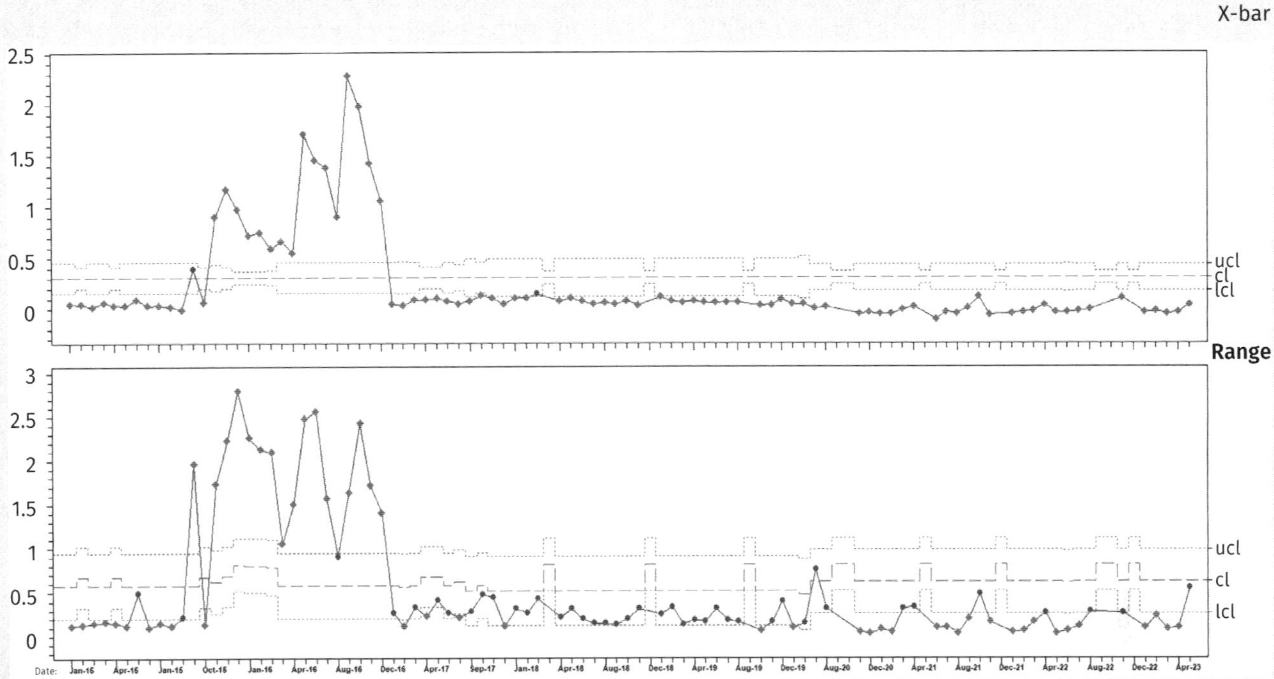

X-bar: cl: 0.3064047 ucl: 0.4369267 lcl: 0.1758829 **Subgroup size: 86**

Range: cl: 0.6194553 ucl: 0.9820209 lcl: 0.2568893

Figure 2: The cl's, ucl's, and lcl's calculated from Figure 1 data but differentiated by the location of data patterns and out-of-bounds X-bar and/or ranges.

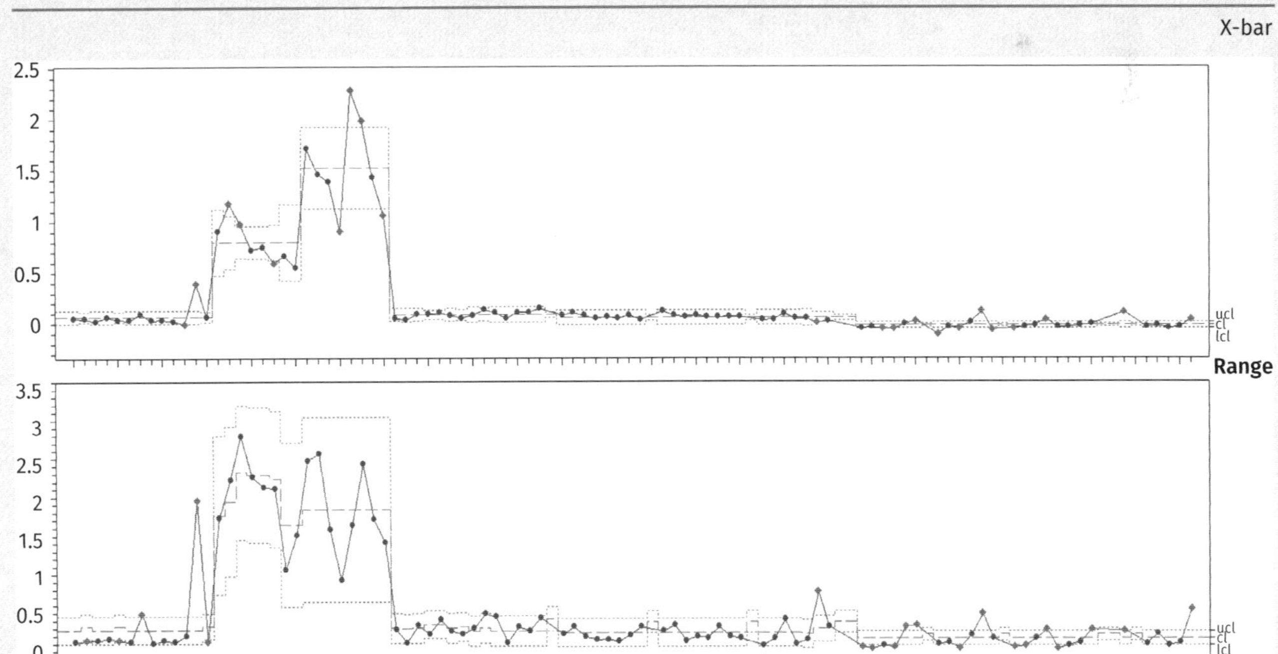

X-bar: cl: -0.02197688 ucl: 0.005742133 lcl: -0.04969588 **Subgroup size: 86**

Range: cl: 0.1596259 ucl: 0.2530546 lcl: 0.06619721

between March – December 2016. From January 2017 through August 2020 the $\bar{x}$ R chart was characterized by common cause variation. The average S/P remained well below the cut-off for the ELISA (0.400). One R value exceeded the upper control limit of the R chart in March of 2020. Thus, we concluded the herd was not infected by the pathogen on the basis of no evidence of antibody production by the pathogen. Four years monitoring showed variation in $\bar{x}$ R and instances where S/P averages and ranges were out-of-bounds. By assessing the frequency of S/P values outside the expected bounds, it is possible to differentiate non-specific reactions from genuine infection. For instance, if the diagnostic specificity of the ELISA is 0.97, one out-of-bounds range would be expected to occur in every 37 sets of samples. Therefore, we concluded that the period January 2017 through April 2023 showed instances of "special cause" and not infection. Thus, $\bar{x}$ R charts enabled objective evaluation of the sow farm's pathogen status over time and assisted in evidence-based decision-making concerning herd infection status and management practices. Using historical test results and this simple-yet-effective analysis, more accurate test interpretation can be achieved and costly management decisions, eg, herd closures, may be avoided.

Reference

1. Wheeler, D.J.; Chambers, D.S. Shewhart's Control Charts. In *Understanding Statistical Process Control*, Second ed.; SPC Press, Inc: Knoxville, Tennessee, 1992; pp. 37 - 54.

https://doi.org/10.54846/am2024/70

Scoring of swine lung lesion images: A comparison between an artificial intelligence algorithm and human expert evaluators

R. Valeris-Chacin[1]; B. Garcia-Morante[2]; M. Sibila[2]; A. Canturri[3,5]; I. Ballarà Rodriguez[4]; I. Bernal Orozco[4]; R. Jordà Casadevall[4]; P. Muñoz[4]; M. Pieters[5]

[1]Department of Veterinary Pathobiology, College of Veterinary Medicine and Biomedical Sciences, Texas A&M University, Texas; [2]Unitat Mixta d'Investigació IRTA-UAB en Sanitat Animal, Centre de Recerca en Sanitat Animal (CReSA), Campus de la Universitat Autònoma de Barcelona (UAB), Bellaterra, Catalonia, Spain; [3]Veterinary Diagnostic Laboratory, College of Veterinary Medicine, University of Minnesota, Minnesota; [4] Swine Business Unit, Laboratorios HIPRA, S.A., Amer, Girona, Spain; [5]Department of Veterinary Population Medicine, College of Veterinary Medicine, University of Minnesota, Minnesota

Introduction

Cranioventral pulmonary consolidation (CVPC)[1] is a common macroscopic morphologic pneumonic pattern observed in slaughter inspections and is often associated with *Mycoplasma* (*M*) *hyopneumoniae* infection.[1-5] Current CVPC scoring methods are considered subjective and not fully reliable.[6-10] In addition, lung lesion scoring can be a time-consuming and expensive activity, requiring the physical location of the evaluator at the slaughterhouse. Thus, there is a dire need to implement simple, fast, valid, and easily standardizable lung lesions scoring methods that allow for reliable and consistent data output.

Recent progress in the field of artificial intelligence (AI) has allowed the development of high-performance algorithms introduced to the veterinary image analysis domain.[11,12] In pigs, computer vision systems (CVS) based on machine learning algorithms have been shown to properly identify and score pneumonia and pleurisy from digital images captured at slaughterhouses.[13,14] However, the evaluation of the performance of CVS algorithms targeting swine lung lesions is an underdeveloped area. Therefore, we aimed to compare the pneumonic scores provided by an Artificial Intelligence (AI)-based scoring system with standard CVPC scores assigned by five human expert evaluators from pig lung images obtained at the slaughterhouse. Since lung lesion scoring was expected to vary both within and between evaluators, the intra- and inter-expert evaluator variability were assessed and compared to the repeatability of the AI DIAGNOS algorithm.

Methods

For this study, a total of 1,050 dorsal view images of swine lungs were taken at a slaughterhouse in northeast Spain. An artificial intelligence algorithm (AI DIAGNOS) was used to recognize and quantify CVPC on the images using the Madec scoring system.[15] The scoring performed using the AI-based method was then compared to the scoring performed by five human expert evaluators. Total Madec score, Madec score per lung lobe, and percentage of affected lung area (%LA)[16] were used as outcomes for the evaluation of AI DIAGNOS. Balanced accuracy, adjusted balanced accuracy, macro ROC area under the curve, and intra-class correlation (ICC) were calculated to estimate the accuracy of AI DIAGNOS scoring, the inter-expert evaluator variability, and the variability within expert evaluators and within AI DIAGNOS algorithm, as appropriate.

Results

The lung scores provided by the expert evaluators were, in general, significantly higher than those from AI DIAGNOS algorithm (P-value < .005), with the exception of the right cardiac lobe. AI DIAGNOS algorithm showed low accuracy in the multiclass classification setting. Conversely, in the binary classification setting, AI DIAGNOS algorithm achieved moderate accuracy in all lung lobes except for the diaphragmatic. The median of the difference in total Madec score between the AI DIAGNOS algorithm and the expert evaluators was 0 (Figure 1). Similarly, the difference in %LA between the AI DIAGNOS algorithm and the expert evaluators had a median of 0% (Figure 2).

A moderate inter-expert evaluator variability was observed in this study with high heterogeneity of the ICC values depending on the lung lobe (range = 0.29 - 0.6). Overall, the intra-expert evaluator variability was low and similar among the different outcomes and lung lobes. Additionally, the observed ICC values suggest that the intra-expert evaluator variability differed among expert evaluators. By contrast, AI DIAGNOS algorithm scoring was identical per lobe per image, obtaining an ICC of 1 in all outcomes.

Discussion

Information on lung lesions from routine official meat inspections is a priority demand of stakeholders in the pig industry.[17] The system, however, is flawed and lacks sensitivity due to the amount of variation attributed to the official meat inspectors.[6-10] Two other CVS in pigs for detecting and scoring respiratory lesions caused by infectious agents have been developed and tested.[13,14] The systems were developed to automatically score pleurisy associated to *Actinobacillus pleuropneumoniae*,[14] and to detect and size CVPC.[13] Several factors could have negatively affected the performance of AI DIAGNOS algorithm: the lower number of lung lobes scoring 2, 3 and 4 compared with those scoring 0 and 1 that were used to train the AI algorithm, the presence of artifacts (eg, blood inspiration, atelectasis due to mechanical compression, damaged or folded lung lobes) or scarring due to resolved CVPC, and the inter-expert evaluator variability. It is important to highlight that all these factors are present during routine meat inspections.

Pathological findings for which there are different levels of gradation to choose from, as pneumonia, typically exhibit larger variation among the meat inspectors.[10] Even though the evaluators in our study were experts in assessing CVPC from pig lungs, a moderate to high inter-evaluator variability was observed depending

Figure 1: Distribution of the difference in total Madec scores between AI DIAGNOS and expert evaluators. The expert evaluators' Madec scores were subtracted from the corresponding AI DIAGNOS scores to calculate the differences. The 25% percentile is -2, the median is 0, and the 75% percentile is 1.

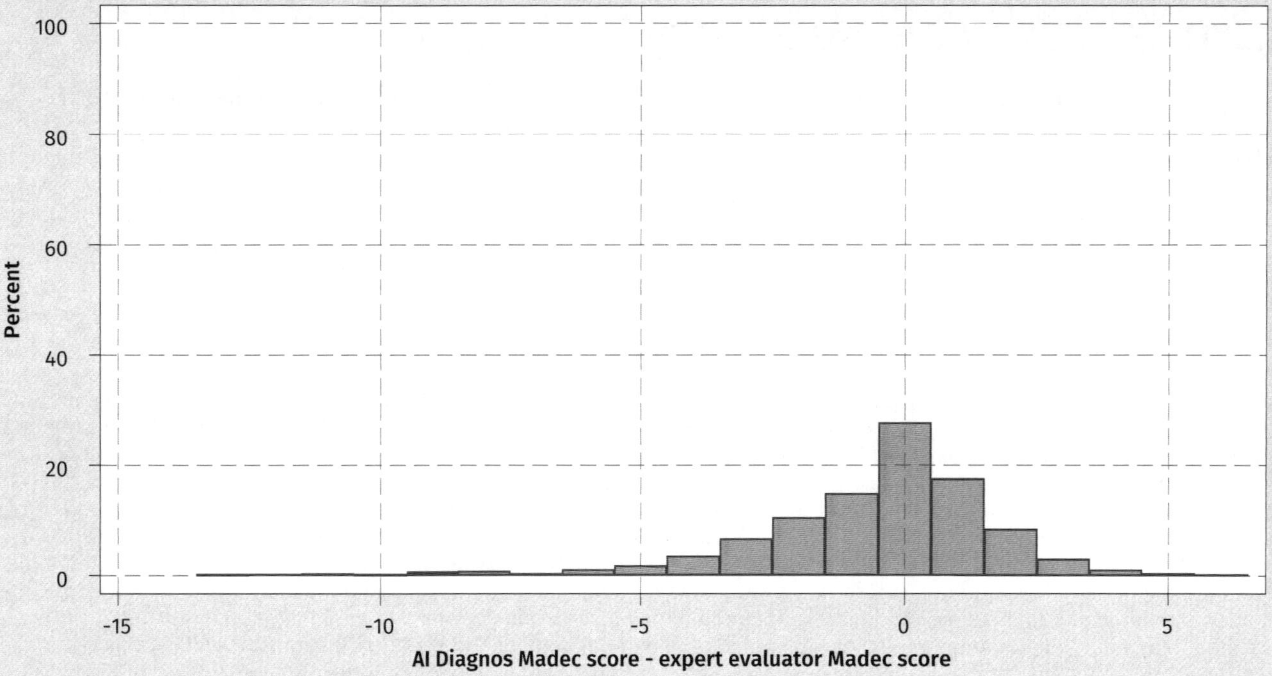

Figure 2: Distribution of the difference in the percentage of affected lung area between AI DIAGNOS and expert evaluators. The percentage of affected lung area from the experts' evaluation was subtracted from the corresponding AI DIAGNOS value to calculate the difference. The 25% percentile is -4.25%, the median is 0%, and the 75% percentile is 2.75%.

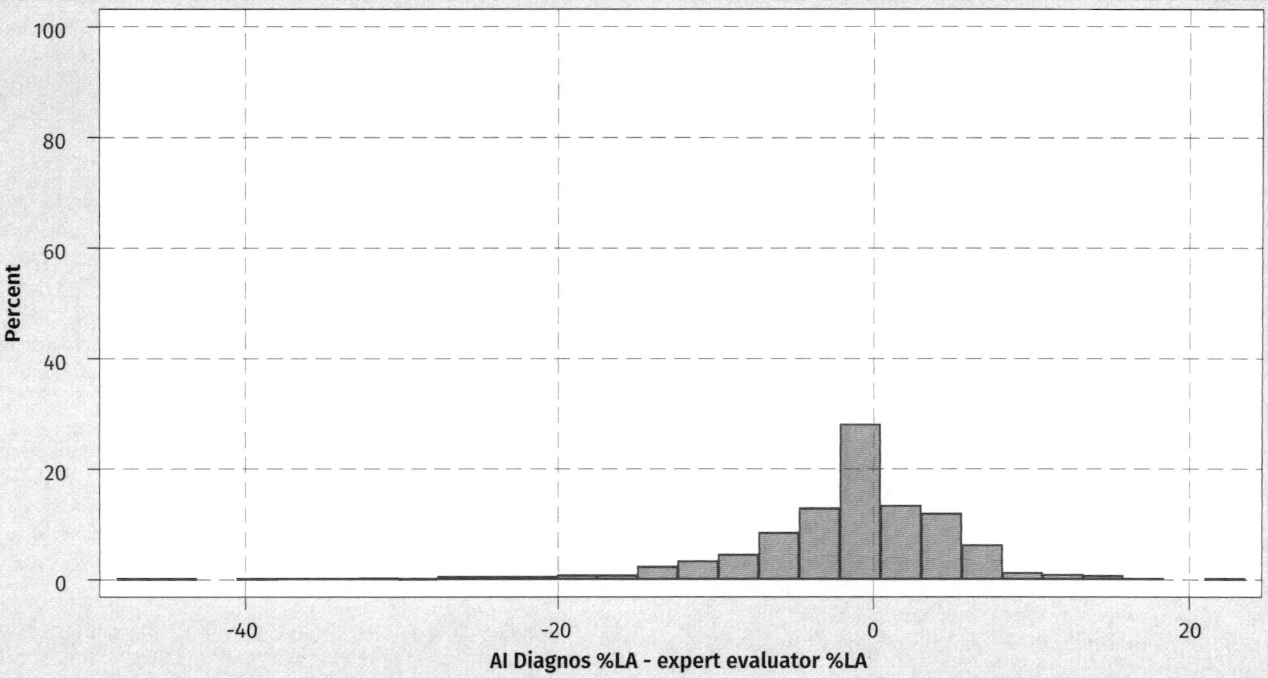

on the lung lobe (ie, ICC ranging 0.29 - 0.6), and this agrees with results reported in other studies.[7-10] The lack of a standardized meat inspection process is also one of the main causes of the low intra-observer reliability sometimes observed.[8,10,18] However, the overall intra-expert reliability estimates were very high in this study, although they slightly differed among expert evaluators (ie, ICC ranging 0.89 - 0.99). Hence, the results indicate that evaluators had overall a high level of consistency within their own scoring process of CVPC, which confirms their expertise. Remarkably, the AI DIAGNOS algorithm scoring was identical per lobe per image, showing perfect consistency. This capability has the potential of effectively reduce the inherent variability that exists between and within human evaluators, addressing a crucial issue in current scoring methods.

The use of AI methods has revolutionized many different aspects of health sciences, especially by enhancing our capabilities to extract quantitative information from digital images. To date, few CVS have been developed to monitor abattoir lesions in pigs.[13,14,19,20] The CVS employed herein was conceived as a fully automatic method for systematic detection, but also scoring of CVPC. The algorithm AI DIAGNOS delivers the Madec score, a well-known and broadly used scoring method in pigs at slaughter.[15] It also calculates results in reference to %LA,[16] thus providing additional value by reporting the total weighted percentage of affected lung.

Conclusion

Under the conditions of this study, AI DIAGNOS discriminated between healthy and diseased lung lobes with moderate accuracy. Therefore, AI DIAGNOS could be a good alternative to the detection of CVPC during slaughter inspections, especially considering its perfect consistency in the scoring compared with the human evaluators. However, the algorithm needs further improvements whenever the goal is to correctly predict the degree of pneumonia. Altogether, AI DIAGNOS has the potential to automatically score *M hyopneumoniae*-associated lung lesions and facilitate processing of data, which may aid solving problems linked to conventional CVPC scoring methods.

References

1. Fraile, L., Alegre, A., López-Jiménez, R., Nofrarías, M., & Segalés, J. (2010). Risk factors associated with pleuritis and cranio-ventral pulmonary consolidation in slaughter-aged pigs. *The Veterinary Journal, 184*(3), 326–333. **https://doi.org/10.1016/j.tvjl.2009.03.029**

2. Fablet, C., Marois-Créhan, C., Simon, G., Grasland, B., Jestin, A., Kobisch, M., Madec, F., & Rose, N. (2012). Infectious agents associated with respiratory diseases in 125 farrow-to-finish pig herds: A cross-sectional study. *Veterinary Microbiology, 157*(1–2), 152–163. **https://doi.org/10.1016/j.vetmic.2011.12.015**

3. Merialdi, G., Dottori, M., Bonilauri, P., Luppi, A., Gozio, S., Pozzi, P., Spaggiari, B., & Martelli, P. (2012). Survey of pleuritis and pulmonary lesions in pigs at abattoir with a focus on the extent of the condition and herd risk factors. *The Veterinary Journal, 193*(1), 234–239. **https://doi.org/10.1016/j.tvjl.2011.11.009**

4. Meyns, T., Van Steelant, J., Rolly, E., Dewulf, J., Haesebrouck, F., & Maes, D. (2011). A cross-sectional study of risk factors associated with pulmonary lesions in pigs at slaughter. *The Veterinary Journal, 187*(3), 388–392. **https://doi.org/10.1016/j.tvjl.2009.12.027**

5. Paz-Sánchez, Y., Herráez, P., Quesada-Canales, Ó., Poveda, C. G., Díaz-Delgado, J., Quintana-Montesdeoca, M. del P., Plamenova Stefanova, E., & Andrada, M. (2021). Assessment of Lung Disease in Finishing Pigs at Slaughter: Pulmonary Lesions and Implications on Productivity Parameters. *Animals, 11*(12), 3604. **https://doi.org/10.3390/ani11123604**

6. Bonde, M., Toft, N., Thomsen, P. T., & Sørensen, J. T. (2010). Evaluation of sensitivity and specificity of routine meat inspection of Danish slaughter pigs using Latent Class Analysis. *Preventive Veterinary Medicine, 94*(3–4), 165–169. **https://doi.org/10.1016/j.prevetmed.2010.01.009**

7. Eckhardt, P., Fuchs, K., Kornberger, B., & Köfer, J. (2009). A study of the reliability of findings obtained during meat inspection of slaughter pigs. *Wiener Tierarztliche Monatsschrift, 96*(5–6), 145–153.

8. Enøe, C., Christensen, G., Andersen, S., & Willeberg, P. (2003). The need for built-in validation of surveillance data so that changes in diagnostic performance of post-mortem meat inspection can be detected. *Preventive Veterinary Medicine, 57*(3), 117–125. **https://doi.org/10.1016/S0167-5877(02)00229-5**

9. Hoischen-Taubner, S., Blaha, T., Werner, C., & Sundrum, A. (2011). Repeatability of anatomical-pathological findings at the abattoir for characteristics of animal health. *Archiv Fur Lebensmittelhygiene, 62*(3), 82–87.

10. Schleicher, C., Scheriau, S., Kopacka, I., Wanda, S., Hofrichter, J., & Köfer, J. (2013). Analysis of the variation in meat inspection of pigs using variance partitioning. *Preventive Veterinary Medicine, 111*(3–4), 278–285. **https://doi.org/10.1016/j.prevetmed.2013.05.018**

11. Hennessey, E., DiFazio, M., Hennessey, R., & Cassel, N. (2022). Artificial intelligence in veterinary diagnostic imaging: A literature review. *Veterinary Radiology & Ultrasound, 63*(S1), 851–870. **https://doi.org/10.1111/vru.13163**

12. Sandberg, M., Ghidini, S., Alban, L., Capobianco Dondona, A., Blagojevic, B., Bouwknegt, M., Lipman, L., Seidelin Dam, J., Nastasijevic, I., & Antic, D. (2023). Applications of computer vision systems for meat safety assurance in abattoirs: A systematic review. *Food Control, 150*, 109768. **https://doi.org/10.1016/j.foodcont.2023.109768**

13. Bonicelli, L., Trachtman, A. R., Rosamilia, A., Liuzzo, G., Hattab, J., Mira Alcaraz, E., Del Negro, E., Vincenzi, S., Capobianco Dondona, A., Calderara, S., & Marruchella, G. (2021). Training Convolutional Neural Networks to Score Pneumonia in Slaughtered Pigs. *Animals, 11*(11), 3290. **https://doi.org/10.3390/ani11113290**

14. Trachtman, A. R., Bergamini, L., Palazzi, A., Porrello, A., Capobianco Dondona, A., Del Negro, E., Paolini, A., Vignola, G., Calderara, S., & Marruchella, G. (2020). Scoring pleurisy in slaughtered pigs using convolutional neural networks. *Veterinary Research, 51*(1), 51. **https://doi.org/10.1186/s13567-020-00775-z**

15. Madec, F., & Kobisch, M. (1982). Bilan lésionnel des poumons de porcs charcutiers à l'abattoir. *Journées de La Recherche Porcine, 14*(8), 405–412.

16. Christensen, G., Sorensen, V., & Mousing, J. (1999). Diseases of the respiratory system. In Straw B, D'Allaire S, Mengeling W, & Taylor DJ (Eds.), *Diseases of Swine* (pp. 913–940). lowa University Press.

17. van Staaveren, N., Doyle, B., Hanlon, A., & Boyle, L. (2019). Multi-Stakeholder Focus Groups on Potential for Meat Inspection Data to Inform Management of Pig Health and Welfare on Farm. *Agriculture, 9*(2), 40. **https://doi.org/10.3390/agriculture9020040**

18. Steinmann, T., Blaha, T., & Meemken, D. (2014). A simplified evaluation system of surface-related lung lesions of pigs for official meat inspection under industrial slaughter conditions in Germany. *BMC Veterinary Research, 10*(1), 98. **https://doi.org/10.1186/1746-6148-10-98**

19. Brünger, J., Dippel, S., Koch, R., & Veit, C. (2019). 'Tailception': using neural networks for assessing tail lesions on pictures of pig carcasses. *Animal, 13*(5), 1030–1036. **https://doi.org/10.1017/S1751731118003038**

20. McKenna, S., Amaral, T., & Kyriazakis, I. (2020). Automated Classification for Visual-Only Postmortem Inspection of Porcine Pathology. *IEEE Transactions on Automation Science and Engineering, 17*(2), 1005–1016. **https://doi.org/10.1109/TASE.2019.2960106**

https://doi.org/10.54846/am2024/71

Revolutionizing disease prevention through plant produced, orally delivered vaccines

T. Raines, PhD; R. Sibbel, DVM; J. Escobar PhD; J. Filbey PhD
Mazen Animal Health, Inc

Introduction

Mazen Animal Health Inc has built a novel vaccine platform using plants as a biofactory to produce recombinant vaccines and therapeutics. This unique approach enables oral delivery of vaccines in an edible form with distinct advantages over today's injectables, topicals and water-based vaccines. Our cross-species platform has the flexibility to be used in swine, poultry, livestock, aquaculture, pets, and wildlife, and has the potential for vaccine solutions for a wide variety of diseases. The Mazen Oralogics™ platform has been proven through our first vaccine, an orally delivered, maize based vaccine for sows that has shown efficacy over multiple trials with the ability to provide lactogenic immunity to newborn piglets against porcine epidemic diarrhea virus (PEDV), a devastating disease with much needed solutions.

Methods

Constructs containing the PEDV S1 subunit antigen were designed for plant transformation. The constructs were used to transform the gene encoding the S1 antigen into the maize genome resulting in expression driven by seed specific promoters to target protein accumulation in the kernel. The leading maize line with the highest antigen expression was advanced through development and scale-up for animal trials.

A randomized, masked, placebo-controlled pivotal efficacy study was designed to evaluate a novel, maize-produced vaccine, orally delivered to sows during gestation to protect their piglets from a PEDV challenge at three- to five- days-of-age. A total of 28 naïve, PEDV antibody negative sows were randomized to two groups: Vaccine (n = 20) and Control (n = 8). Treatments were administered, as a top dress on normal rations, at approximately 9-weeks, 5-weeks and 2-weeks prior to farrowing, with three consecutive days of Vaccine or Control Product feeding at each time point. All piglets were challenged on the same day with virulent PEDV. Animals were monitored for 11 days for clinical disease and mortality.

Results

All sows developed serum neutralizing antibodies against PEDV after challenge, confirming that all litters were exposed. The PEDV challenge resulted in death of 60% of piglets born to control sows, with a death rate per litter ranging from 45% to 100% (Table 1). Therefore, the challenge was virulent enough to evaluate vaccine efficacy. The observed survivability in test treated litters ranged from 0.25 to 0.83 while the observed survivability for control treated litters was 0.00 to 0.55. The vaccine demonstrated a statistically significant, clinically relevant increased survivability in litters farrowed from Vaccine treated sows (Group 1) compared to Control treated sows (Group 2) with an estimated increase in survivability of 0.68 (90% CCI 0.19 to 1.17). This was a statistically significant effect with a one-sided *P*-value of .0131.

The pivotal efficacy study successfully demonstrated the ability to orally vaccinate gestating sows with a subunit vaccine to provide protection to their newborn litters in a PEDV challenge.

Conclusions

To date, the power of the intestinal mucosal immune system has been underutilized in controlled animal vaccination approaches. The intestinal mucosa is where many pathogens and antigens are first perceived, triggering a robust innate immune response often referred to as the first line of defense. By taking advantage of oral delivery of antigens through feed, we have proven it is possible to illicit robust immunity that is also able to passively protect offspring through milk. This approach will not only offer solutions to diseases that are not well controlled by systemic injectable vaccines but will offer numerous advantages to the producers as well. Among these advantages are reduced labor needs along with increased worker safety, biosecurity and animal welfare. Additionally, Mazen's vaccines are room temperature stable, eliminating the cost of temperature-controlled supply chain and storage. Mazen has proven this novel approach of oral vaccination through multiple successful sow trials providing lactogenic immunity to newborn piglets challenged with PEDV. The Mazen Oralogics™ technology has the opportunity to provide much needed solutions to control diseases more effectively and efficiently than current options.

Table 1: Survivability of piglets from Group 1 (vaccinated sows) and Group 2 (unvaccinated sows) after PEDV challenge

Study Population	Estimate	Standard Error	Lower 90% confidence interval	Upper 90% confidence interval	One-sided *P*-value
Group 1 vs Group 2	0.68	0.285	0.19	1.17	0.0131

https://doi.org/10.54846/am2024/72

Increasing survival in the wean-to-finish pig

B. McNeil[1], MS; A. Cross[1], PhD, MBA; C. Mainquist-Whigham[2], DVM, MS; J. Sonderman[1], PhD; T. Rathje[1], PhD, MBA
[1]DNA Swine Genetics, Columbus, Nebraska; [2]Pillen Family Farms, Columbus, Nebraska

Introduction

Across the industry producers are seeing an increased prevalence of mortality regardless of genetic line. From 2017 to 2021, the industry experienced an increase in birth to harvest mortality of 2.04%, including 1.3% prior to weaning and 0.74% post weaning.[1] While health plays a significant role in the productivity of a system, improving the performance of the wean-to-finish pig requires a multi-factorial approach. Strategies can include daily management practices at the sow farm such as minimizing fall behind litters or system changes such as improving wean age. Setting the pig up for success in the nursery by creating the right environment and getting pigs up multiple times per day by mat feeding. Lastly, genetic selection can improve survivability and performance. Overall, producers should look at a systems-based approach to make improvements in wean-to-finish survival.

Sow farm strategies

As terminal sires have been selected for lean growth and thus lateness of maturity, weaned pigs respond to the weaning and nursery transition process differently today than they did 10 - 15 years ago. Setting a system up for success in the wean-to-finish flow may require interventions at the sow farm level. The creation of nurse sows or the process of bringing in additional females into a farrowing room to spread the number of pigs over more teats is a common process in the industry. This process is thought to decrease pre-wean mortality. However, creating nurse sows can have detrimental effects on the wean-to-finish performance of the market pig. To further investigate this, a trial with commercial pigs (DNA L241 females x DNA L600 sires) compared a 0% nurse sow (NoNurse) treatment to a 7% nurse sow (Nurse) treatment. Sows (n = 618), representing parities 1 through 6, were randomly assigned to farrowing rooms by farrow date. Within treatments pigs were identified as a fall-behind, a littermate to a fall-behind, or from a litter with no fall-behinds at day 3 - 5 of age. Fall-behind pigs in the Nurse treatment were placed on a nurse sow, pigs in the NoNurse treatment were left in their original litters. Pre-wean and nursery mortality were not significantly different between treatments ($P > .10$), however finishing mortality was significant with the NoNurse treatment at 2.45% and the Nurse treatment at 3.92% ($P < .10$; Table 1).

Evaluating wean age is another option for producers to consider when looking at ways to improve mortality. Across multiple trials, wean age has had a significant impact on survivability.

Table 1: Mortality rate of pigs from a farrowing room with and without nurse sows

	NoNurse	Nurse	*P*-value
Pre-wean mortality (%)	12.75	12.27	.484
Nursery mortality (%)	2.34	2.64	.388
Finish mortality (%)	2.45	3.92	.058

In a research barn trial, 2400 commercial pigs (DNA L241 x DNA L600) with wean ages between 20-26 days, a 0.45% mortality improvement per additional day of wean age was observed ($P < 0.01$). This result is dependent on a farm's current wean weight and mortality. For the DNA L600's mortality by weaning weight curve, increasing wean age is more beneficial for lighter pigs than heavier pigs. For example, increasing wean age by 2 days is worth 0.5% decrease in nursery mortality for pigs that weighed 5 kg, but only 0.25% decrease in nursery mortality for pigs that weighed 7 kg. In another study following commercial pigs from a 5600 heads sow farm, each additional day of wean age was worth 176 g of wean weight and 0.12% nursery mortality.

Nursery strategies

Preparing the receiving site correctly can often be an overlooked part of the survival equation. A poor receiving location can set the pig back and introduce pathogens that will increase mortality and morbidity within groups. Internal Pillen Family Farms data from four newly built 15,000 head nursery barns shows that with each turn, mortality increased (Figure 1) potentially due to biofilm build-up. To reduce the mortality, a detergent is needed in the wash process to break down the biofilm. By utilizing a detergent, mortality was reduced 33% across the Pillen Family Farms nursery system.

The first week post weaning can be the most critical time to the success of a wean-to-finish pig. Offspring from later maturing sire lines can be slower to start in the nursery, therefore, getting pigs up and encouraging them to find feed and water can decrease hard starting pigs. A trial where pigs were mat fed 6 times per day showed a 0.42 kg improvement in weights at day 14 ($P < 0.5$).

Genetic strategies

Genetic companies can make selection decisions to improve mortality. DNA Genetics has tracked over 200,000 pedigreed commercial pigs from nucleus sires. Within the same sire line, sires can have a large variation in total mortality of their offspring with sires ranging from 12 - 28% mortality, specifically in wean-to-finish mortalities can range from 2 - 11%. By accurately measuring these differences and incorporating them into a breeding program, significant improvements in survivability can be made across the industry.

Conclusion

When benchmarking a production systems mortality to the industry there are many factors to consider based on what makes the production system unique. Regardless of those differences, it is important for producers to evaluate the steps they are taking to minimize wean-to-finish mortality. Some steps may need to be implemented at the sow farm and others at the nursery, but overall a system approach is needed when tackling large scale issues like increasing wean-to-finish survival.

Figure 1: Nursery mortality by turns in four newly built 15000 head barns.

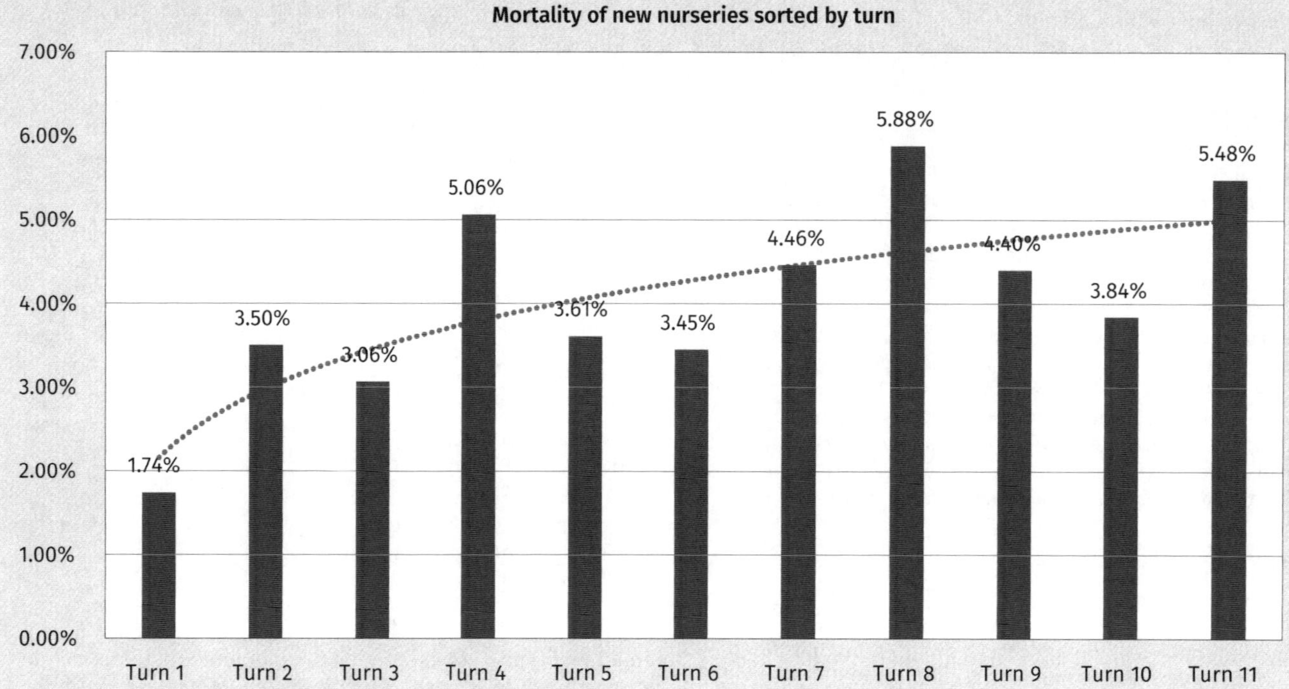

Reference

1. MetaFarms. (2021). Production Analysis Summary for U.S. Pork Industry: 2017-2021. **https://porkcheckoff.org/wp-content/uploads/2022/05/MetaFarms-2021-Production-Analysis-for-NPB.pdf**

https://doi.org/10.54846/am2024/73

Addressing important market issues in 2024

Andrew Kleis, AIF®
Insight Wealth Group, West Des Moines, Iowa

2024 is going to be a critical year for the global economy and global markets. The battle against inflation is ebbing without the long-feared recession. The battle ahead is about maintaining a healthy economy while easing the restrictive interest rate policies that helped defeat inflation. Can it be done without causing economic strife?

Jerome Powell and the Federal Reserve (Fed) made it clear in their December 2023 Federal Open Market Committee (FOMC) meeting that interest rate hikes are all but done. Their latest projections of three interest rate cuts in 2024 gave the market momentum going into the end of the year. It was Powell's comments about a "soft landing", however, which were even more optimistic.

"I have always felt, since the beginning, that there was a possibility, because of the unusual situation, that the economy could cool off in a way that enabled inflation to come down without the kind of large job losses that have often been associated with high inflation and tightening cycles. So far, that's what we're seeing. That's what many forecasters on and off the committee are seeing".

That type of optimistic sentiment is not one we hear frequently from the podium at the Federal Reserve. And the market loved it. They loved it so much, in fact, that they assumed it meant an even more positive outlook than the Fed was promising. According to CME Group, the market was pricing in six interest rate cuts in 2024, with the first beginning in March (Figure 1).

As with all things in the world, the reality is somewhere in between the Fed's optimism and the market's rose-colored glasses. But don't be surprised if the Fed starting a little later than March, or doing less than six cuts, causes some volatility as we get to the end of the first quarter.

More important than interest rate policy, however, is the strength of the economy. And there we find additional reason to be hopeful.

First, there is substantial liquidity in the economy today that has not been there in the past. Combining an economy flush with cash with potentially falling interest rates is a recipe for success in public markets (Figure 2).

And that liquidity in the system is in the right hands: with consumers. There has been much talk about the assets available to consumers. It is falling – but it's falling from otherworldly highs. On an inflation adjusted average, consumers typically have approximately $1 trillion in the bank. Today that number is more than $4 trillion! (Figure 3).

More importantly, that additional income is going to the right places today. We know you've heard all the panicked talk about "rising credit card debt" and "rising loan delinquencies". The people sounding the alarm bell aren't wrong – these figures are going up. But from where? And to where? Right now credit card delinquencies have risen off its artificial, pandemic influenced lows. It has risen right back up to...pre-pandemic levels (Figure 4).

The same has happened to the percent of household income required to pay for debt service each month. It's actually lower than we saw pre-pandemic and lower than we saw earlier this century. Simply put, personal balance sheets are in a surprisingly good place (Figure 5).

Figure 1: CME Fedwatch tool – meeting probabilities.

CME FEDWATCH TOOL - MEETING PROBABILITIES

MEETING DATE	325-350	350-375	375-400	400-425	425-450	450-475	475-500	500-525	525-550
1/31/2024				0.0%	0.0%	0.0%	0.0%	10.3%	89.7%
3/20/2024	0.0%	0.0%	0.0%	0.0%	0.0%	0.0%	6.8%	62.7%	30.5%
5/1/2024	0.0%	0.0%	0.0%	0.0%	0.0%	5.6%	52.9%	36.1%	5.3%
6/12/2024	0.0%	0.0%	0.0%	0.0%	5.1%	48.6%	37.7%	8.2%	0.5%
7/31/2024	0.0%	0.0%	0.0%	4.3%	41.9%	39.3%	12.7%	1.7%	0.1%
9/18/2024	0.0%	0.0%	3.8%	37.4%	39.6%	15.9%	3.0%	0.3%	0.0%
11/7/2024	0.0%	2.5%	25.9%	38.9%	24.0%	7.4%	1.2%	0.1%	0.0%
12/18/2024	2.0%	21.3%	36.3%	27.0%	10.7%	2.4%	0.3%	0.0%	0.0%

Figure 2: M2 money supply.

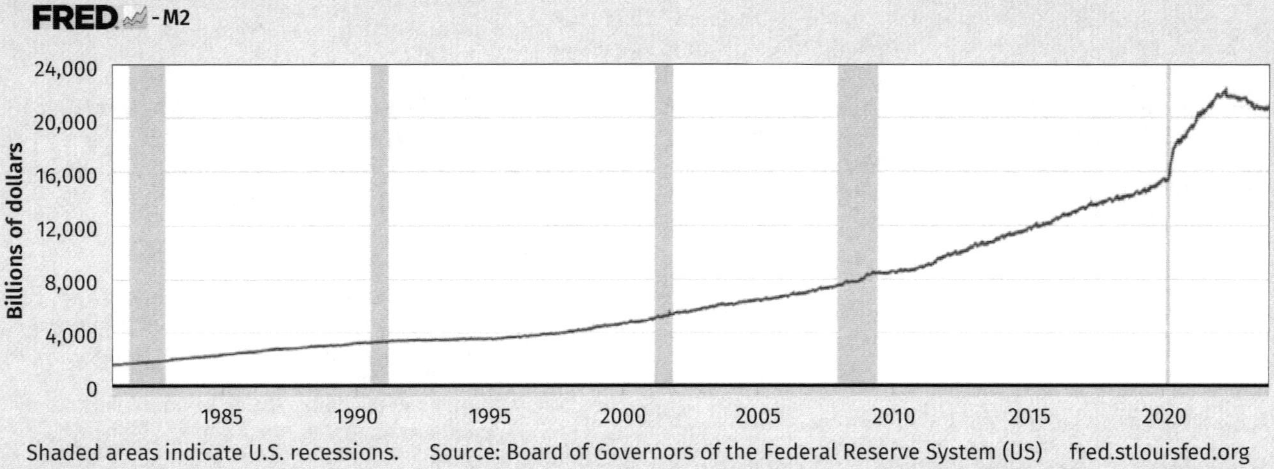

Figure 3: Households; checkable deposits and currency; asset, level.

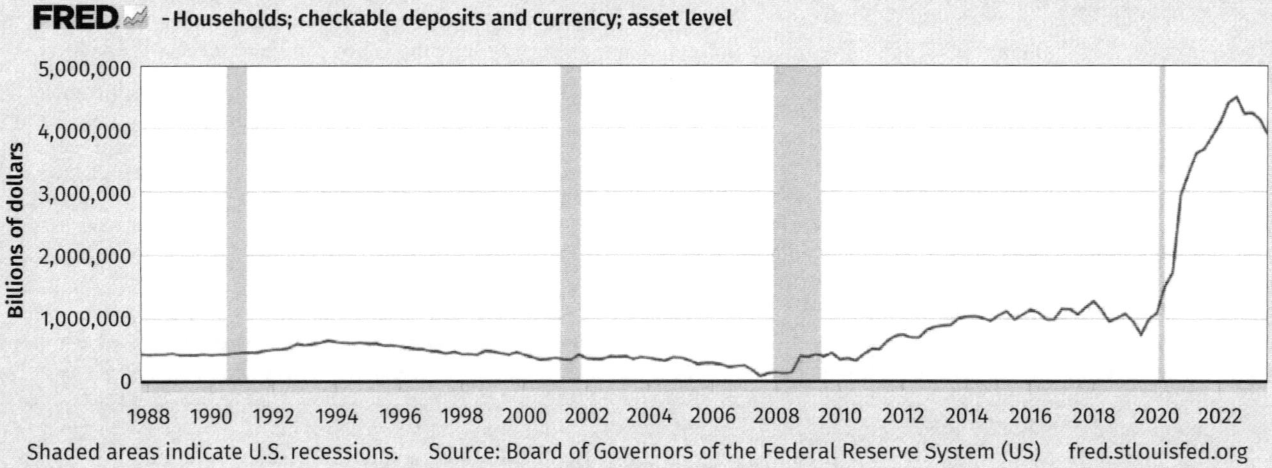

Figure 4: Delinquency rate on credit card loans, all commercial banks.

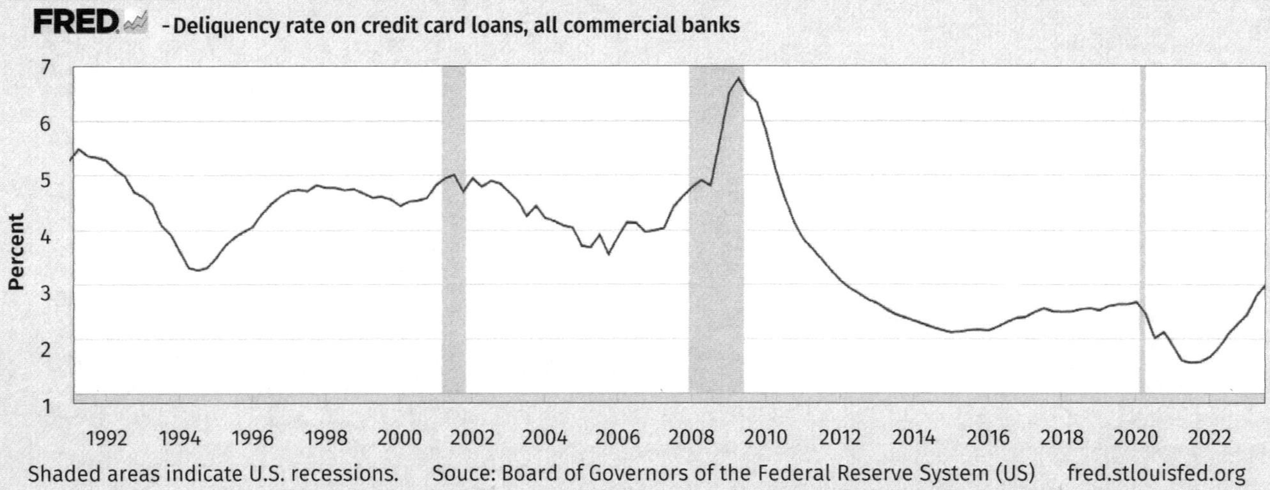

Figure 5: Household debt service payments as a percent of disposable personal income.

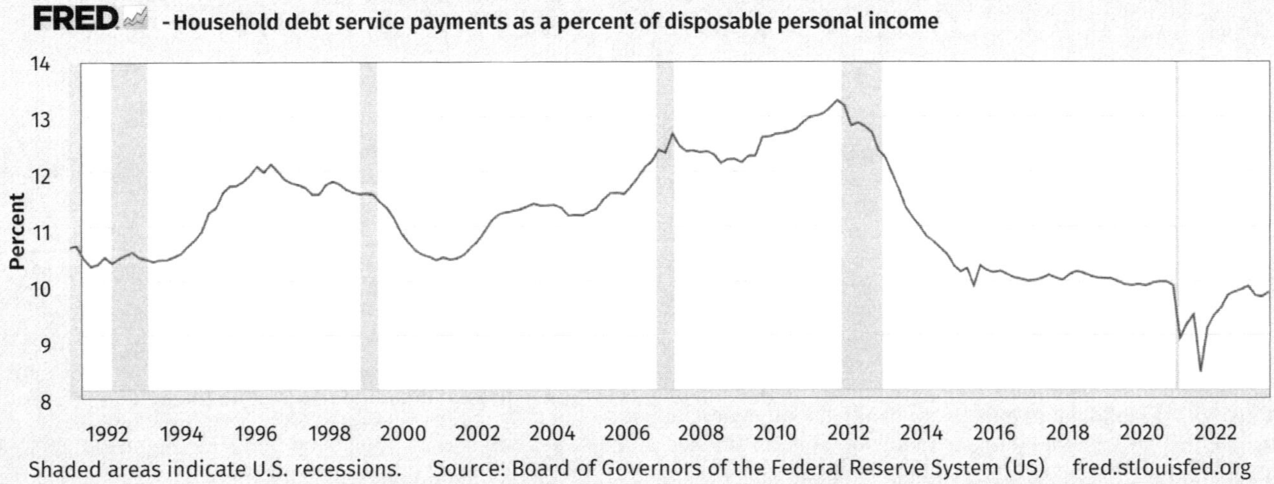

So where does this leave us? A year in which interest rates will fall while the consumer is in solid shape. That's not a recipe for disaster.

It's not to say drama can't or won't happen. It will. Between the time this is written and the AASV meeting there is the potential for another government shutdown. We are facing a dramatic Presidential election. And international affairs are as drama filled as we've seen in some time.

But the base case leaves room for optimism today. In our oral presentation we'll be looking forward to updating that base case and addressing some timely topics such as the impact of AI on public markets, how the election will impact the economy, and the implications of rising debt on currencies and the market. We look forward to expanding on those items with you in Nashville.

https://doi.org/10.54846/am2024/74

Effect of a unique essential oil blend on mortality of nursery pigs during a known F18 *E coli* challenge

N. Rippel[2], DVM; C. Scharlau[1], MS; R. Fent, PhD; W. Johnson[1]
[1]Ralco Nutrition, Marshall, Minnesota; [2]Veterinary Medical Center, Williamsburg, Iowa

Introduction

Mortality and morbidity during the nursery phase attributed to *Escherichia coli* is a common issue industry wide. Previous research in our lab has identified unique and specific properties of many phytonutrients. With this knowledge, several essential oils and proprietary essential oil blends were identified as candidates for this study. These candidates were then screened in a lab against 27 field isolates to test for their efficacy in inhibiting *E coli* growth by the minimal inhibitory concentration (MIC) method. Upon comparing the abilities of the individual essential oils and the essential oil blends to inhibit growth of *E coli*, the most advantageous blend was selected to be evaluated *in vivo*. For this study, a nursery site was chosen due to its persistent and recurring *E coli* challenge even with the use of an *E coli* vaccine. An isolate from this nursery site was evaluated in the lab using the MIC method which confirmed that the selected essential oil blend was efficacious in inhibiting *E coli* growth at a low concentration. Therefore, this essential oil blend was implemented in two studies at this nursery site to evaluate its effects on mortality during an F18 *E coli* challenge.

Methods

Two studies were conducted at a commercial nursery barn in eastern Iowa to evaluate the impact of a unique essential oil blend (EO) on piglet mortality during an F18 *E coli* challenge. Pigs were weaned at 19 days of age into an all-in/all- out nursery. Pigs evaluated in this study were PRRSV positive when they were weaned from the batch farrowing sow farm. All pigs received an oral *E coli* vaccine 3 days prior to weaning.

Study 1

In an eight-week study, 4,117 weaned pigs were used to evaluate two treatments. Pigs were split among 8 rooms and rooms were assigned to one of two treatments with 4 rooms placed on each treatment. The treatments evaluated were: 1) Control or 2) EO. Pigs designated to the EO treatment were administered the unique essential oil blend via water delivery at a rate of 64 oz per 5 gallons of stock solution and diluted at a ratio of 1:128 by water medicator beginning 24-hours after arrival at the nursery site and continuing for 7 days. Number of mortalities per day were recorded for each room. Weekly totals were reported. Mortality numbers were analyzed using the Chi Squared test in the FREQ procedure in SAS.

Study 2

In a subsequent 8-week study, 5,017 pigs were used to evaluate four treatments. Pigs were split among 8 rooms and rooms were assigned to one of four treatments with 2 rooms placed on each treatment. The treatments evaluated were: 1) Control, 2) EO at a rate of 64 oz per 5 gallons stock solution for 7 days, 3) EO at a rate of 64 oz per 5 gallons stock solution for 14 days, or 4) EO at a rate of 64 oz per 5 gallons stock solution for 7 days followed by EO at a rate of 32 oz per 5 gallons stock solution for 7 days. All EO treatments were delivered via water and stock solutions were diluted at a ratio of 1:128 by water medicator beginning on the day of placement. Number of mortalities per day were recorded for each room. Weekly totals were reported. Mortality numbers were analyzed using the Chi Squared test in the FREQ procedure in SAS.

Results

Study 1 results are presented in Figures 1 and 2. In Study 1, mortalities decreased ($P < .05$) with the use of EO in weeks 2 (3.3% Control v 0.9% EO), 3 (3.4% Control v 2.1% EO), and 7 (0.5% Control v 0.1% EO). Cumulatively, EO administration through the water resulted in decreased ($P < .05$) mortality each week from week 2 through week 8 of the nursery period. At the end of week 8 the cumulative mortality was 12.8% for the control and 8.6% for the group administered EO.

Study 2 results are presented in Figures 3 and 4. In Study 2, mortality was decreased ($P < .05$) with the use of EO in week 2. With the decrease in mortality observed during week 2 for all EO treatments, cumulative mortality was decreased ($P < .05$) for pigs administered EO beginning in week 2 and continuing throughout the study. At the end of week 8, the cumulative mortality was 4.0% for the 14-d EO administration, 6.3% for the step down EO administration, 6.9% for the 7-d EO administration and 8.9% for the control group.

Discussion

As had been the case in recent groups, nursery pigs at this site again experienced *E coli* challenge during these studies. In Study 1, the use of EO for the first 7 days was able to significantly reduce mortality particularly during the early weeks in the nursery. While there was a slight increase in mortalities after the treatment was stopped, the EO usage delayed and reduced these mortalities compared to the spike in mortalities experienced by the control group in week 2. This observation led to the treatment structure evaluated in Study 2, where EO administration time was increased to 2 weeks, both at a full and a half dose. While the 7-day EO treatment again reduced mortalities particularly during the early nursery period, the 14-day EO treatment at the consistent rate further reduced mortalities. These studies indicate the unique essential oil blend delivered through the water during the early nursery period significantly reduced mortalities compared to a control during an active *E coli* challenge, when both groups of pigs were previously administered an *E coli* vaccine.

Figure 1: Effect of a unique essential oil (EO) blend on weekly nursery mortality during an *E coli* challenge (Study 1).

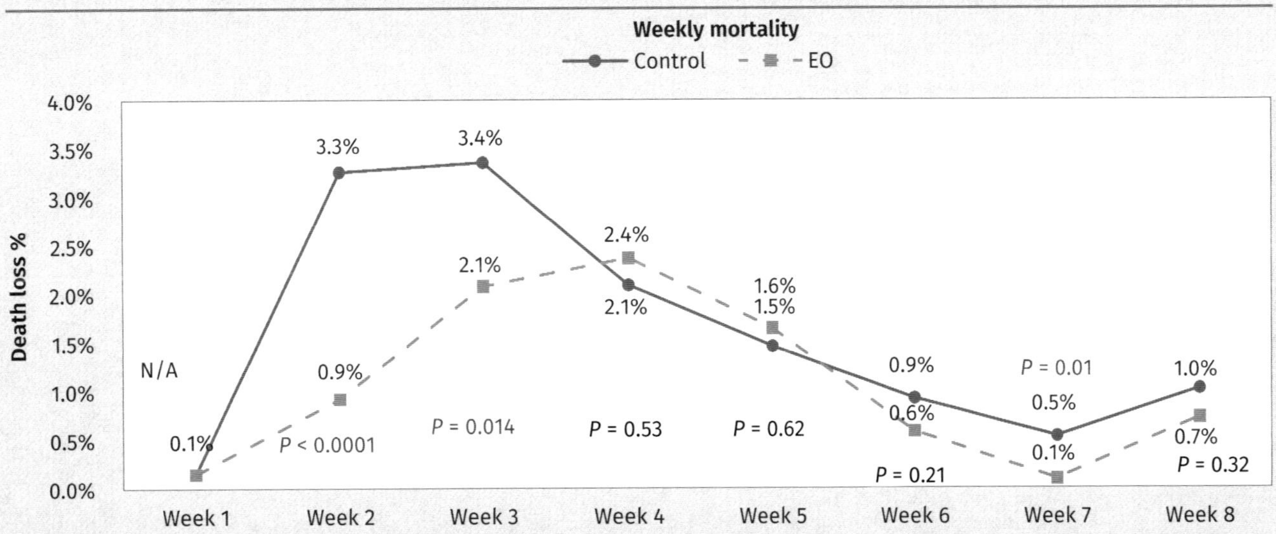

Figure 2: Effect of a unique essential oil (EO) blend on cumulative nursery mortality during an *E coli* challenge (Study 1).

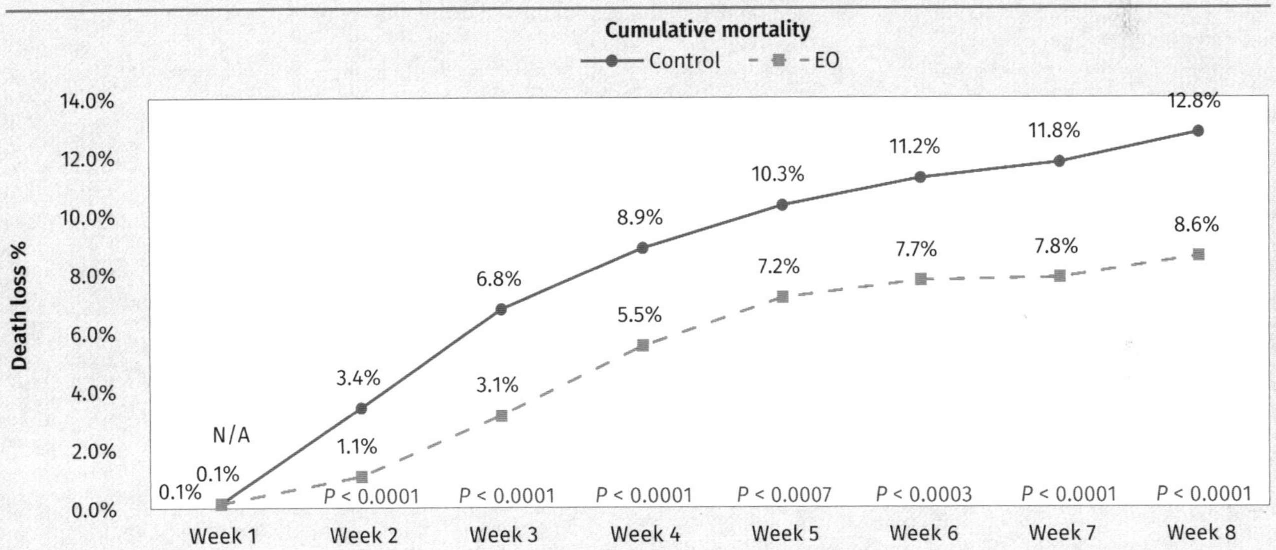

Figure 3: Effect of administration duration and level of a unique essential oil (EO) blend on weekly nursery mortality during an *E coli* Challenge (Study 2).

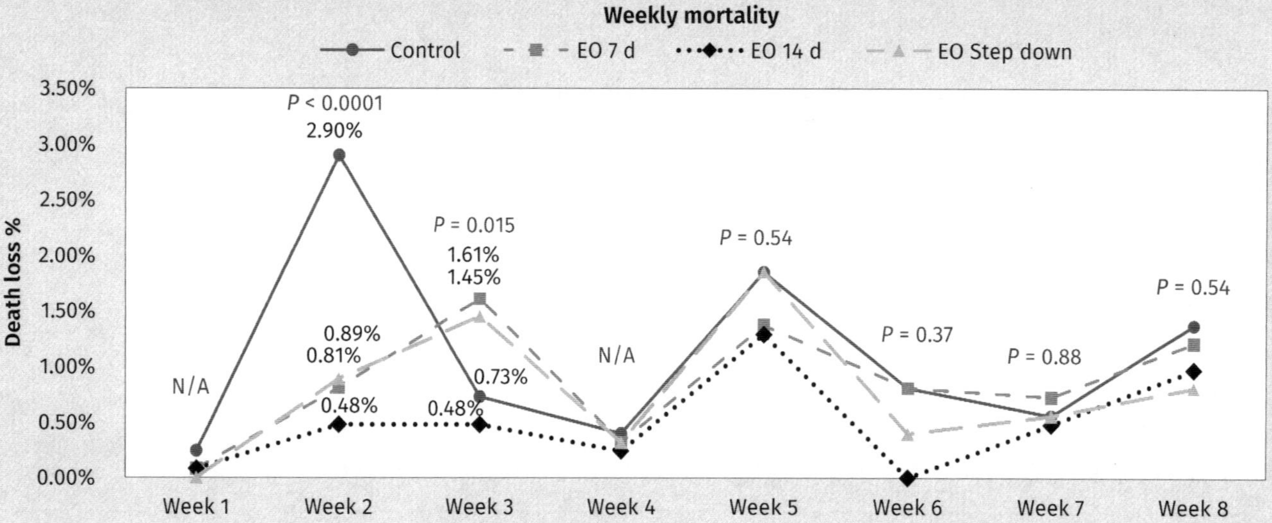

Figure 4: Effect of administration duration and level of a unique essential oil (EO) blend on cumulative nursery mortality durnig an *E coli* challenge (Study 2).

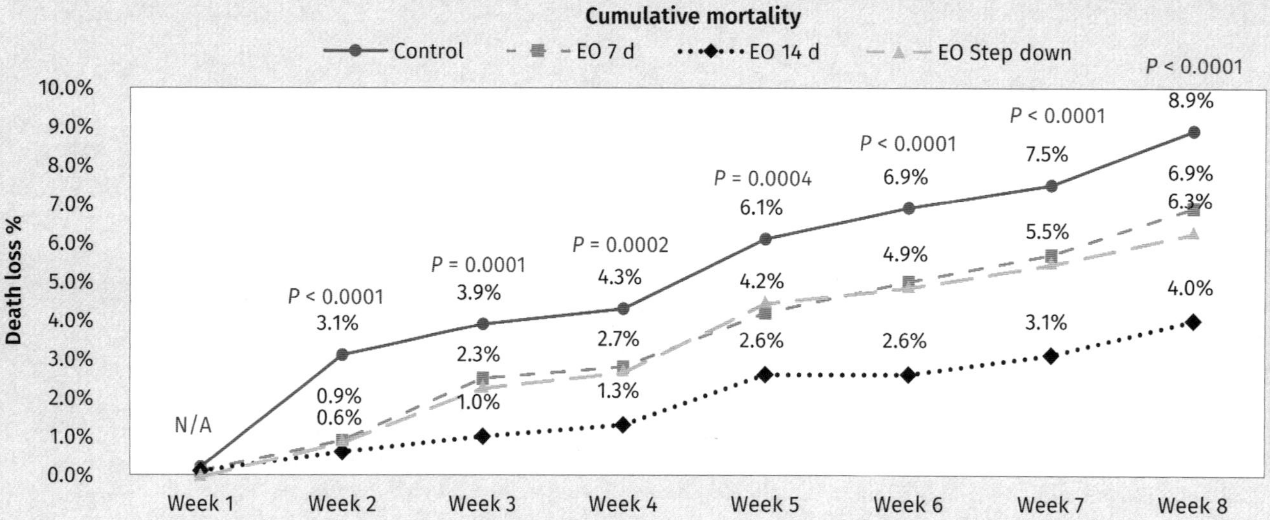

https://doi.org/10.54846/am2024/75

Soybean meal: Tactical option for swine respiratory disease mitigation: DVM perspective

Paul Yeske; Aaron Gaines; Jennifer Garret; R. Dean Boyd

Introduction

Soybean meal (SBM) is a superb amino acid and energy source for pig diets. Soybeans also contain an abundance of functional molecules (FM) in the non-nutritive fraction,[1] as well as bioactive peptides that arise during digestion of major SBM proteins.[2] There is considerable information on their benefits in relation to human health, but research with food animals is in the early stages of development.[3]

The disease-mitigating effects of SBM in pigs was reported in 2010 by researchers at the Hanor Company.[4,5] Swine respiratory disease infection (SRD) was unexpectedly encountered during a study with growing pigs that involved low SBM (typical) and high SBM diets. Pigs fed low SBM diets exhibited a profound loss in growth and feed conversion efficiency (FCE), which is common during periods of high immune stress. However, pigs fed diets with higher SBM content (38 - 50%) grew as though no active infection was present. These results were confirmed by another group who conducted a similar study design, almost simultaneous with ours, and who also encountered SRD pathogens unexpectedly (Gene Gourley, M.S., personal communication).

Results from the Hanor study exposed the hidden cost of SRD and illustrated the effectiveness of SBM in mitigating impaired growth. The observation that SBM can mitigate the effects of SRD is strengthened by results from a commercial study where the ability of pigs to thrive improved by increasing dietary SBM content.[6] SRD imposes significant hidden costs in FCE loss, increased medical treatment and potentially fewer pigs that thrive in late nursery through early growth. These costs are hidden because they are 'folded' into barn 'closeout' data, but they can add $3 - 6/ pig to cost (range $2 - 9).

SRD complex is prevalent in many North American systems during January - April, when barns are closed for cold weather. The most problematic swine respiratory diseases include *Mycoplasma hyopneumoniae*, porcine reproductive and respiratory syndrome (PRRSV), influenza A virus in swine, *Streptococcus suis*, *Hemophilus parasuis*, and *Actinobacillus suis pleuropneumoniae*. Influenza is also a problem in the fall for some systems, but the economic cost of winter-spring SRD tends to be more persistent.

The Hanor study

The original experiment was designed to determine whether SBM level was a factor in meeting the dietary lysine requirement for pigs fed the metabolic modifier, ractopamine hydrochloride (RAC). In other words, did the level of dietary lysine required to maximize growth and FCE depend on the amount of SBM used to meet the lysine requirement? The simultaneous investigation of response to dietary lysine and SBM level was a fortunate design feature when SRD infection was unexpectedly encountered. Given this circumstance, the question evolved beyond how amino acid specifications were met to whether the nutraceutical components of SBM could mitigate respiratory disease impaired growth. In the end, the high SBM (H-SBM) regimen profoundly improved the level of growth and FCE achieved when compared to low SBM (L-SBM) fed pigs.

Study design

Dietary treatments were arranged as a 4 x 2 factorial; having 4 standardized ileal digestible (SID) lysine levels and 2 SBM levels (L-SBM, H-SBM). Pigs in all 8 diet subclasses were fed RAC (5 ppm) with whole-body growth rate (ADG), FCE and carcass gain being the key measures. A total of 420 castrate pigs (PIC terminal genetics) with an average body weight of 217 lbs were used. Data were collected each week for 21 d, after which all pigs were harvested. The commercial research facility that Hanor used was an academic equivalent setting, since pig density (room, pen) was low and environmental stressors normally encountered in commercial sites were reduced. The latter was accurate until we encountered an active infection of SRD complex of pathogens. The experimental framework is shown in Table 1.

Diet formula design

Diets were formulated to provide 0.65, 0.75, 0.85 or 0.95% SID lysine for both SBM regimens. The level of SBM and crude protein (CP) for each diet subclass is shown in Table 2. Diet composition is shown in Table 3. Four diets were manufactured, and summit blended at the research farm to create the 8 experimental diets using Feed Logic equipment. Details of dietary nutrient content are available in the paper by Boyd and Gaines.[7] Nutrient level was confirmed by analysis.

Unexpected infection with multiple respiratory pathogens

Pigs unexpectedly became infected with multiple respiratory pathogens (SRD complex) that triggered systemic inflammation. SRD causes growth and FCE to plummet compared to what is typical for unaffected pigs. Upon learning of the infection, all pigs received medication by water and clinical diagnostics and physical inspection of pigs was initiated under the direction of Tara Donovan, DVM (The Hanor Co). Serology revealed that pigs were PRRSV and porcine circovirus (PCV2) positive. There were PCV2 lesions and *Streptococcus suis* infection was also confirmed. The inflammatory nature of these viruses has been proven by the presence of circulating pro-inflammatory cytokines.[8]

The study continued as planned because SRD infection is a production reality. In North America, SRD complex is prevalent in many systems when barns are closed due to cold weather. In addition, SRD is often a problem for weaned pigs that are derived from PRRSV active sow farms. Taken together, SRD infection can span 6 - 7 months of the year.

SBM mitigated respiratory disease-impaired growth

The overall effect of SBM level on whole-body and carcass growth and FCE is shown in Table 4. Since the response to dietary lysine level was similar for both SBM levels (SBM x lysine level, $P > .29$, all criteria), the average response to SBM level is shown for the 21-d study. The growth suppressing effects of SRD is illustrated

Table 1: Factorial arrangement of dietary lysine and soybean meal (SBM) level diets

	SID lysine,[1,2] %				Total pens and pigs	
Diet SBM level	0.65	0.75	0.85	0.95	EU pens	Pigs[3]
Low	6 (52)	6 (53)	6 (52)	6 (53)	24	210
High	6 (52)	6 (53)	6 (52)	6 (53)	24	210

[1] Standardized ileal digestible, SID.

[2] Experimental unit pens of pigs shown prior to parenthesis. Total pigs per treatment subclass is shown in parenthesis.

[3] Average initial whole-body weight for treatments, 216 ± 1.5 lbs.

Table 2: Level of dietary soybean meal (SBM) and crude protein (CP) content

		SID lysine,[1] %			
Diet SBM level	**Unit**	0.65	0.75	0.85	0.95
Low	lbs/ton	273	316	362	406
High	lbs/ton	374	462	553	637
Low	CP, %	16.0	16.0	16.0	16.0
High	CP, %	16.0	17.4	18.7	20.1

[1] Standardized ileal digestible, SID.

Table 3: Ingredient composition of summit blend diets[1]

Diet[2]	**1**	**4**	**5**	**8**
SID Lysine, %	**0.65**	**0.95**	**0.65**	**0.95**
Ingredient, lbs/ton	**Low SBM**		**High SBM**	
Corn ground, 8.5%	1521	1474	1482	1239
Soybean meal, 47.5%	273	406	374	637
Corn gluten, 62.0%	109	0	39	0
L-lysine.HCl	2.5	7.0	0	0
L-threonine	0	3.0	0	0
DL-methionine	0	1.5	0	0
Choice white grease	46	61	58	80
Mono-calcium Phos.	21.5	20.9	21.0	19.6
Limestone	16.7	15.8	16.0	14.2
Salt	8.0	8.0	8.0	8.0
VTM premix	2.0	2.0	2.0	2.0
Ractopamine, 9 g/lb	0.5	0.5	0.5	0.5
Total	**2000**	**2000**	**2000**	**2000**

[1] Standardized ileal digestible, SID.

[2] Diets 1, 4 were manufactured and blended by Feed Logic equipment to derive diets 2, 3. Diets 5, 8 were likewise blended to derive diets 6, 7.

Table 4: Main effect of SBM level on whole-body and carcass growth in pigs fed the metabolic modifier, ractopamine for 21 d [1,2]

Item	Unit	Low SBM	High SBM	SEM	Probability, less than	% advantage
No. pens	–	24	24	–	–	–
No. pigs	–	210	210	–	–	–
Initial body weight	lbs/pig	216.9	216.5	1.6	0.88	–
Whole-body ADG[3]	lbs/d	1.99	2.18	0.1	0.04	9.5
Whole-body FCE[3]	ratio	2.97	2.74	0.1	0.05	- 7.7
Initial carcass est [4]	lbs/pig	160.9	160.9	–	–	–
Carcass yield[5]	%	74.6	74.7	0.50	ND	–
Carcass weight	lbs/pig	192.4	196.2	1.4	0.05	2.0
Carcass ADG[3]	lbs/d	1.33	1.50	0.1	0.09	12.8
Carcass FCE[3]	ratio	4.35	3.93	0.1	0.05	- 9.7

[1] Dietary SBM levels are shown in Table 2.
[2] Least squares means calculated using whole-body initial weight as covariate.
[3] ADG = average daily gain; FCE = feed conversion efficiency ratio, FCE, calculated as feed/gain.
[4] Initial carcass weight estimated by multiplying initial live body weight x 0.745
[5] Calculated as carcass weight (head on) at plant/live weight at the research site (< 12 h to loading)

by comparing the response of L-SBM and H-SBM diets; the latter resulted in profound improvements in whole-body (ADG, +9.5%; FCE, -7.7%) and carcass growth (ADG, +12.8%; FCE, -9.7%).

The net result of H-SBM diets, during the infection, was improved carcass weight gain (3.8 lbs/pig) and significant feed savings. If H-SBM fed pigs were harvested on a weight constant basis to L-SBM pigs (31.5 lbs carcass gain/pig), they would have required 13.2 lbs less feed per pig.

This was the first study to show that SBM can reduce high immune stress-impaired growth. Until this report,[9] efforts to reduce disease impaired growth by nutritional means had been unremarkable. We propose that SBM is a nutritional and 'prescriptive' ingredient when SRD is encountered in commercial practice; prescriptive, meaning that SBM is a credible, tactical option for respiratory disease mitigation in pigs (term first suggested by Dr. Lisa Weaver, personal communication).

Growth response was dependent on SBM level

Although the shape of FCE response curves was similar for both SBM regimens, dietary SBM level caused the curves to separate into 2 planes of genetic expression (Figure 1). The same pattern was observed for growth rate (Figure 2). This suggests an important principle in relation to genetic expression when confronted with disease. SBM content influenced the level of growth permitted, with growth depending next on amino acid intake. To illustrate, when higher lysine levels (amino acids) were fed to improve performance by apparently less healthy pigs (L-SBM), FCE and ADG became worse not better (Figures 1, 2). Stated alternatively, attempting to overcome health constrained growth is counterproductive.

In the hierarchy of growth regulation, amino acid intake is important but secondary to health status. Physiologically, the body responds to systemic immune stress by prioritizing nutrients for life-sustaining processes overgrowth (eg, IGF-1 declines.[9,10] Ultimately, amino acid intake provides the building blocks for muscle growth that health status permits.

Basis for disease mitigating effects of SBM

The biological basis for the mitigating effects of SBM on SRD impaired growth is unclear, but it is unlikely to be nutritional. SBM level, not nutrient content, is the variable in play and the difference in SBM functional molecules (FM) intake is markedly different between SBM levels. These components are most likely involved in mitigating mechanisms. Functional molecules in SBM are remarkably diverse and abundant. The isoflavone (ISF) fraction of FM was used as a 'marker' to track relative FM differences between SBM levels over the lysine response curve. The disparity in ISF content between dietary cohorts increased as lysine level increased (Figure 3).

Possible mechanisms for disease mitigation by functional molecules

We are indebted to human medical research for our understanding of disease mitigating mechanisms of soybeans. A variety of disease-uncoupling actions are exerted simultaneously by an equally diverse set of FM. Some mechanisms improve or balance the animal's survival response to infection (no. 2, 3, 4 below), while other mechanisms operate to reduce pathogen concentration (no. 1). Disease mitigation mechanisms include:[2]

1. Anti-viral
2. Oxidation stress reduction
3. Anti-inflammatory
4. Improve function of existing immune cells
5. Preserve intestinal barrier integrity

Control of systemic disease requires multiple countering mechanisms that act in concert to reduce the damaging effects of pathogens. The respiratory disease complex that was encountered in the Hanor study involved at least 3 known pathogens (virus, bacteria). Multiple pathogen infection tends to be the norm.

Figure 1: Total isoflavone content for experimental diets based solely on the contribution by SBM.

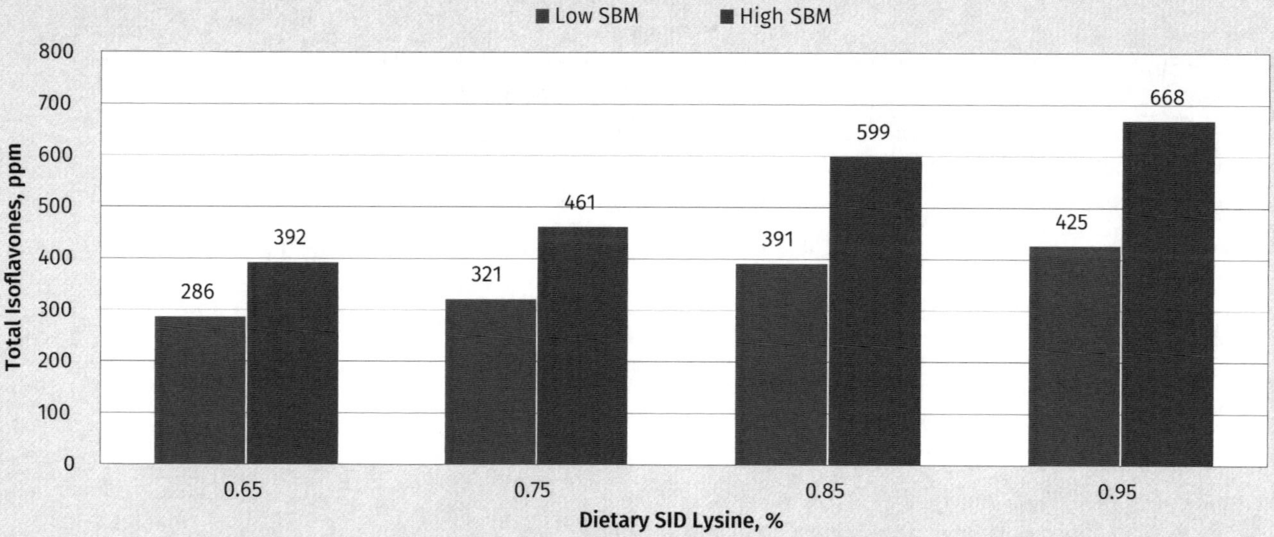

Total isoflavone (ISF) content of diets is based on the contribution by SBM and symbolizes the difference in content of all SBM functional molecules between high and low SBM diets. The value shown is the sum of 3 major ISF, genistein, daidzein and glycitein. Level of each ISF was derived from the review by Smith and Dilger.

Figure 2: Whole-body feed efficiency response for growing pigs receiving Ractopamine (5 ppm) for 21 d: Response to dietary SBM and lysine levels during an active respiratory infection.

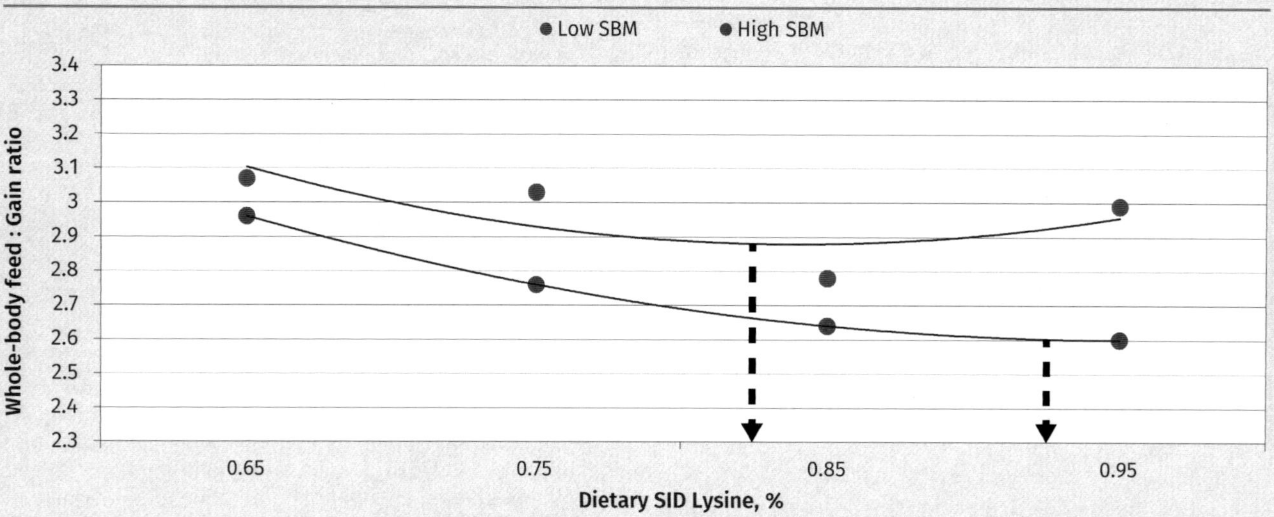

Dietary level of SBM for each lysine level treatment is shown in Table 2. Arrows show the response maximum feed conversion efficiency (FCE) for both SBM levels. The absolute dietary SID lysine requirement to achieve maximum FCE response is 0.84 and 0.94% for low and high SBM curves respectively.

Figure 3: Whole-body daily gain response for growing pigs receiving Ractopamine (5 ppm) for 21 d: Response to dietary SBM and lysine level during an active respiratory disease infection.

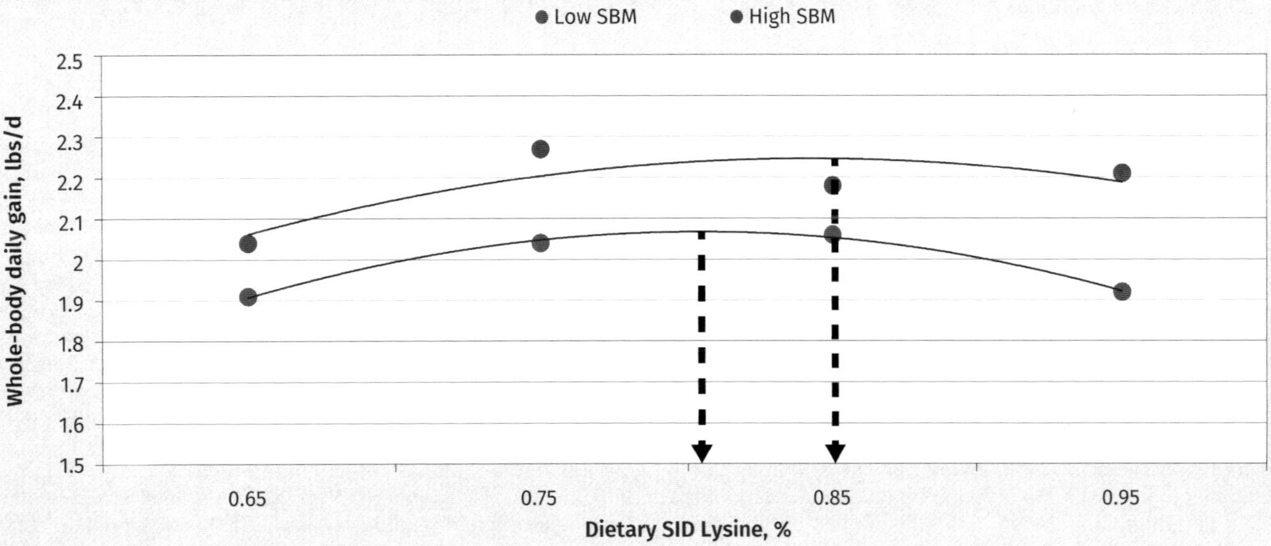

Dietary level of SBM for each lysine level treatment is shown in Table 2. Arrows show the response maximum growth rate (ADG) for both SBM levels. The absolute dietary SID lysine requirement to achieve maximum ADG is 0.80 and 0.85% for low and high SBM curves respectively.

Applying the Hanor Study to other herds

This research provides an answer to the often-asked question, 'can anything be done nutritionally to improve growth when disease is encountered. Diet SBM content, rather than elevated amino acid (or energy) intake, is a means of profoundly mitigating growth suppression by the SRD complex. The financial value of tactical SBM use can be determined by comparing performance for (1) system diets and (2) prescriptive diets (system diets plus 30% more SBM in each diet to finish). Diets need to be equally applied by site and sex, using large numbers and in the time frame specified by the Veterinarian(s). Key measures are tracked to compute the financial outcome (eg, 1000 pig basis), including full-value (FVP) and cull-value pig market revenue, total diet cost and medication cost.

Carcass weight is a priority measure. Although growth rate is reflected in FVP carcass pounds, the diet effect on population carcass weight is important because a decline can begin to emerge even prior to summer heat stress. The motive is to see if the system is experiencing a leading edge to summer carcass weight dip (later article in this series). Carcass weight decline can begin to appear in April when heat is not yet a problem and becomes worse during summer. The reason for a premature decline in carcass weight can be confusing, but it could be due to unchecked SRD impaired growth. Summer carcass weight dip is a very costly problem; one that is preventable provided that pre-summer carcass weight decline is controlled.

Acknowledgments

This paper was adapted from a more comprohensive article published in Feedstuffs in October, 2023. **https://informamarkets.turtl.co/story/feedstuffs-october-2023/page/1**

These materials are (partially) funded by the United Soybean Board.

References

1. Sharma, R., S. Kumar, V. Kumar and A. Thakur. 2019. Comprehensive review on nutraceutical significance of phytochemicals as functional food ingredients for human health management. *J. Pharma. Phytochem.* 8 (5):385-395. **http://doi.org/10.22271/phyto.2019.v8.i5h.9589**

2. Chatterjee, C., S. Gleddie and C-W Xiao. 2018. Soybean bioactive peptides and their functional properties: Review. *Nutrients 10* (9):1211-1227. **http://dx.doi.org/1.3390/nu10091211**

3. Smith, B.N. and R.N. Dilger. 2018. Immunomodulatory potential of dietary soybean-derived isoflavones and saponins in pigs. *J. Anim. Sci.* 96:1288-1304. **http://doi.org/10.1093/jas/skyy036**

4. Johnston, M.E., R.D. Boyd, C. Zier-Rush, and C.E. Fralick. 2010. Soybean meal level modifies the impact of high immune stress on growth and feed efficiency in pigs. *J Anim. Sci.* 88 (E-Suppl. 3):174.

5. Boyd, R.D., M.E. Johnston and C. Zier-Rush. 2010. Soybean meal level modulates the adverse effect of high immune stress on growth and feed efficiency in growing pigs. Proc. 71st Minn. *Nutrition Conf.* pp. 167-174. U. Minn., St. Paul Minnesota. **https://hdl.handle.net/11299/204236x**

6. Cemin, H. S., H. E. Williams, M. D. Tokach, S. S. Dritz, J. C. Woodworth, J. M. Derouchey, K. F. Coble, B. A. Carrender, and M. J. Gerhart. 2020. Estimate of the energy value of soybean meal relative to corn based on growth performance of nursery pigs. *J. Anim. Sci.* Biotech 11:70-78. **https://doi.org/10.1186/s40104-020-00474-x**

7. Boyd, R.D. and A.M. Gaines. 2023. Soybean meal NE value for growing pigs is greater in commercial environments. *Feedstuffs*, August digital edition, page 1. **https://informamarkets.turtl.co/story/feedstuffs-august-2023/page/2/1**

8. Parra, M.D., P. Fuentes, F. Tecles, S. Martinez-Subiela, J.S. Martinez, A. Munoz and J.J. Ceron. 2006. Porcine acute phase protein concentrations in different diseases in field conditions. *J. Vet. Med.* B 53:488-493.

9. Johnson, R.W. 1997. Inhibition of growth by pro-inflammatory cytokines: An integrated view. *J. Anim. Sci.* 75:1244-1255

10. Spurlock, M.E. 1997. Regulation of metabolism and growth during immune challenge: An overview of cytokine function. *J. Anim. Sci.* 75:1773-1783.

https://doi.org/10.54846/am2024/76

Impact of Termin-8® to reduce the risk of transmission of PEDV, PRRSV, and SVA through contaminated feed

S. Dee[1]; J. P. Cano[1]; F. Domingues[2]; O. Lopez[2]
[1] Pipestone Research, Minnesota; [2]Anitox Corporation, Lawrenceville, Georgia

Introduction

In 2014, feed was first identified as a risk factor for porcine epidemic diarrhea viral infection of pigs.[1] This observation was expanded across multiple viruses at both the domestic and transboundary levels.[2] Of special interest are PRRSV (porcine reproductive and respiratory syndrome virus), the most economically significant viral disease of pigs around the world, PEDV (porcine epidemic diarrhea virus), a virus capable of causing 100% mortality in suckling piglets of < 7 days of age, and SVA (Seneca virus A), a picornavirus capable of causing vesicular lesions similar foot and mouth disease virus. To reduce this risk, feed additives were tested to determine their ability to negatively affect survival and quantity of these three viruses.[3] These testing platforms, while informative, have been primarily bench-top level and bioassay-based. Now the need for further evaluation of candidate mitigants under more rigorous, field-based models is needed to better assess their potential efficacy. The goal of the experiment was to determine if 2 kg/ton of Termin-8® (33% formaldehyde from Anitox, Lawrenceville, GA) will protect pigs against infection following consumption of feed contaminated with SVA, PEDV and PRRSV.

Methods

The project involved 3 rooms, each containing 30 pigs. Pigs in rooms 1 and 2 will were fed complete feed treated with Termin8® and challenged with PEDV, PRRSV and SVA via the validated ice block challenge model, which involves a 454g "Ice Block Model", consisting of 100 mL SVA (5 logs TCID50/mL, Ct = 20.72), 100 mL PRRSV 174 (5 logs TCID50/mL, Ct = 21.38), 100 mL PEDV (5 logs TCID50/mL, Ct = 24.25) and balanced with 154 mL MEM.[4] Blocks were frozen at -80C and dropped into each feed bin on days 0 and 13 of the study. The blocks were allowed to melt, with liquid permeating the feed, which were then augured into room for pigs to consume via natural feeding behavior. Room 3 contained 30 pigs that are fed non-mitigated feed that is challenged in a similar manner. Following the challenge, pigs in each room were tested for a period of 30 days and monitored for viral infection via weekly oral fluid testing.

Results and discussion

The results showed both treatment rooms (Termin-8 @ 2 kg/ton) remained free of PRRSV, PEDV, and SVA infection. All oral fluid samples were PCR negative for all 3 viruses and no clinical signs of PRRS, PED and SVA were observed in any animals consuming treated feed. Environmental samples collected from feeders in both rooms were PCR-positive for all 3 viruses indicating successful challenge of the feed through the ice block protocol. In contrast, pigs in the positive control room became infected with all 3 viruses, as documented via PCR-positive oral fluid samples and clinical signs of PRRS (lethargy, dyspnea), PEDV (mild diarrhea) and SVA (vesicular lesions, lameness) were observed consistently throughout the control population. Data is summarized in Table 1. Due to sample size limitations, no statistical analyses were conducted. In closing, the addition of Termin-8 to feed effectively neutralized all three viruses and prevented infection.

References

1. Dee S, Clement T, Schelkopf A, Nerem J, Knudsen D, Hennings J and Nelson E. An evaluation of contaminated complete feed as a vehicle for porcine epidemic diarrhea virus infection of naïve pigs following consumption via natural feeding behavior: Proof of concept. *BMC Vet Res* 2014,10:176.

2. Dee SA, Neill C, Singrey A, Clement T, Cochrane R, Jones C, Patterson G, Spronk G, Christopher-Hennings, J and Nelson E. Modeling the transboundary risk of feed ingredients contaminated with porcine epidemic diarrhea virus. *BMC Vet Res* 2016, 12:51.

3. Dee SA, Bauermann FV, Niederwerder MC, Singrey A, Clement T, et al. (2018) Survival of viral pathogens in animal feed ingredients under transboundary shipping models. PLOS ONE 13(3): e0194509. **https://doi.org/10.1371/journal.pone.0194509**

4. Dee S, Niederwerder MC, Edler Roy, Hanson D, Singrey A, Cochrane R, Spronk G, Nelson E. An evaluation of additives for mitigating the risk of virus-contaminated feed using an ice block challenge model. *Transbound Emerg Dis* **doi:10.1111/tbed.13749**

Table 1: Summary of diagnostic data from Termin-8 trial

Treatment 1	Clinical status	Oral fluids	Feed
PEDV	Normal	0/24 positive	1/24 positive Ct = 37.1
PRRSV	Normal	0/24 positive	1/24 positive Ct = 32.5
SVA	Normal	0/24 positive	1/24 positive Ct = 36.0
Treatment 2	**Clinical status**	**Oral fluids**	**Feed**
PEDV	Normal	0/24 positive	1/24 positive Ct = 36.4
PRRSV	Normal	0/24 positive	1/24 positive Ct = 37.3
SVA	Normal	0/24 positive	1/24 positive Ct = 35.4
Pos control	**Clinical status**	**Oral fluids**	**Feed**
PEDV	Mild diarrhea (1 pen)	2/24 positive Ct = 36.3, 36.8	3/24 positive Ct = 33.6, 29.2, 31.5
PRRSV	Dyspnea, lethargy (5 pens)	5/24 positive Ct = 31.8, 29.5, 36.3, 28.1, 30.9	3/24 positive Ct = 32.5, 33.6, 33.5
SVA	Vesicles, lameness (2 pens)	4/24 positive Ct = 33.2, 33.7, 35.2, 36.8	3/24 positive Ct = 35.6, 33.7, 36.7

https://doi.org/10.54846/am2024/77

Evaluation of anticoccidial therapeutic strategies in nursery piglets

M. J.Nisley[1]; D. Nolan[2]; C. Moura[3]; B. Haberl[3]; E. R. Burrough[4]; N. K. Gabler[1]
[1]Department of Animal Science, Iowa State University, Ames, Iowa; [2]Huvepharma Inc., Peachtree City, Georgia; [3]Iowa Select Farms, Iowa Falls, Iowa; [4]Department of Veterinary Diagnostic and Production Animal Medicine, Iowa State University, Ames, Iowa

Introduction

Enteric pathogenesis stands as one of the most influential factors negatively impacting the livability and performance of suckling and nursery pigs. Bacterial, viral, and parasitic agents collectively contribute to the overall disease status, diarrhea prevalence, and risk of dehydration during this crucial period. In recent years, due to anticoccidial therapeutic restriction, aging facilities, labor shortages, and inadequate cleaning practices in farrowing facilities, the prevalence of coccidiosis has been reemerging. The primary parasite propagating this disease in young pigs is *Cystoisospora suis*, leading to decreased pig performance and a higher incidence of diarrhea in affected pigs.[1] Prior work from our group has shown therapeutic intervention utilizing in-feed salinomycin during the post-wean period can reduce coccidia prevalence, alleviate intestinal lesions, and enhance pig livability in populations burdened by endemic coccidiosis.[2] However, salinomycin is not sanctioned for use in US swine. Amprolium and sulfaquinoxaline emerge as plausible alternatives for anticoccidial treatment in US swine. Amprolium, a thiamine analogue, disrupts cell metabolism by impeding thiamine transport in oocysts,[3] while sulfaquinoxaline inhibits protein translation by blocking tetrahydrofolate biosynthesis in the coccidian parasite.[4] Therefore, our objective was to evaluate amprolium and sulfaquinoxaline on the performance, livability, fecal consistency, and coccidia shedding in nursery pigs facing endemic enteric challenges.

Methods

Three trials were conducted between October 2022 and June 2023 to assess the effectiveness of amprolium and sulfaquinoxaline. In the first trial, there were control and amprolium treatments, while the second and third trials included control, amprolium, and sulfaquinoxaline treatments. Pigs in all three trials were sourced from sow farms known to have coccidia presence and were randomly allocated into mixed-sex pens. These pens were allocated to various treatments: the first trial had 15 pens per treatment with 30 pigs per pen, the second had 12 pens per treatment with 5 pigs per pen, and the third had 16 pens per treatment with 10 pigs per pen. The medication regimen included the administration of amprolium (Amprol) at a concentration of 0.024% for 5 days, followed by 0.006% for 9 days through the water. Sulfaquinoxaline (Sul-Q-Nox) was administered at a concentration of 0.04% for 3 days, followed by a 3-day interval, and then an additional 2 days at 0.04% through the water. In trials two and three, pen bodyweights and feed disappearance were collected at the start and end of the trial to calculate average daily gain (ADG), average daily feed intake (ADFI), and feed efficiency (gain to feed ratio, G:F). Fecal consistency was assessed daily, utilizing a 0 - 3 scale, where 0 represented normal feces, and 3 indicated watery diarrhea during the second and third trials. Additionally, pig mortalities or removals and the detection of coccidia oocysts in fecal samples on days 0, 3 - 4, and 7 were assessed through Sheather's solution floats and microscope analysis within each pen across all three trials. Each trail lasted 42-day in duration starting immediately post-weaning. Performance data, mortality/removals, fecal consistency, and fecal coccidia presence (binary) were analyzed for the fixed effect of medication and trial was utilized as a random effect. The experimental unit was the pen, with significance differences reported when $P < .05$ and tendencies when $.05 \le P < .10$.

Results

Fecal coccidia oocyst presence was reduced by 86% in Sul-Q-Nox treated pigs and 131% in Amprol treated pigs when compared to the control ($P = .045$; Figure 1) at day 3 post-weaning. Similarly, on day 7 fecal coccidia oocyst presence was reduced by 556% in Sul-Q-Nox treated pigs and 340% in Amprol treated pigs when compared to the control ($P < .001$; Figure 1). Over the first two weeks, both Sul-Q-Nox and Amprol treated pigs observed lower proportions of diarrhea incidence compared to untreated pigs ($P < 0.001$). Additionally, overall, Sul-Q-Nox treated pigs observed lower proportions of diarrhea compared to Amprol treated and untreated pigs ($P = .028$). Although not statistically significant, untreated pigs had a 13.2% mortality/removal rate, while Sul-Q-Nox and Amprol treated pigs had rates of 10.1% and 8.5%, respectively. At the culmination of the 42-day trials, body weights were ~9% higher in pigs treated with Sul-Q-Nox or Amprol compared to untreated pigs ($P < .001$; Table 1). Overall, ADFI and ADG was ~10% and ~15% higher in Sul-Q-Nox or Amprol treated pigs compared to untreated pigs ($P < .001$ and $P = .002$, respectively). Feed efficiency did not significantly differ between the treatments.

Discussion

Ideally, proper cleansing practices in farrowing facilities would provide adequate preventative measure in the control of coccidia transmission and proliferation in the suckling period.[5] Unfortunately, due to facility design constraints and cleaning program implantation, coccidiosis remains prevalent among suckling and nursery pigs today. Moreover, with restrictions on current prophylactic anticoccidial medication, intervening during the nursery phase could potentially offer another combative measure against this disease. Our previous findings have demonstrated that unregulated in-feed medication with salinomycin (at 60 - 90 ppm) can enhance nursery pig survival rates and reduce the excretion of coccidia oocysts.[2] These data herein show that amprolium and sulfaquinoxaline administered through water to post-weaned pigs is a viable and promising approach to reducing fecal *Cystoisospora suis* oocyst excretion. These latter compounds also lowered the incidence of diarrhea and enhanced pig performance in a population naturally infected with coccidiosis. In conclusion, the implementation of these compounds might represent a low hanging fruit solution during the nursery phase, although further research is required to address the suckling phase.

Figure 1: Effect of medication with Sul-Q-Nox or Amprol or none (control) on oocyst presence.

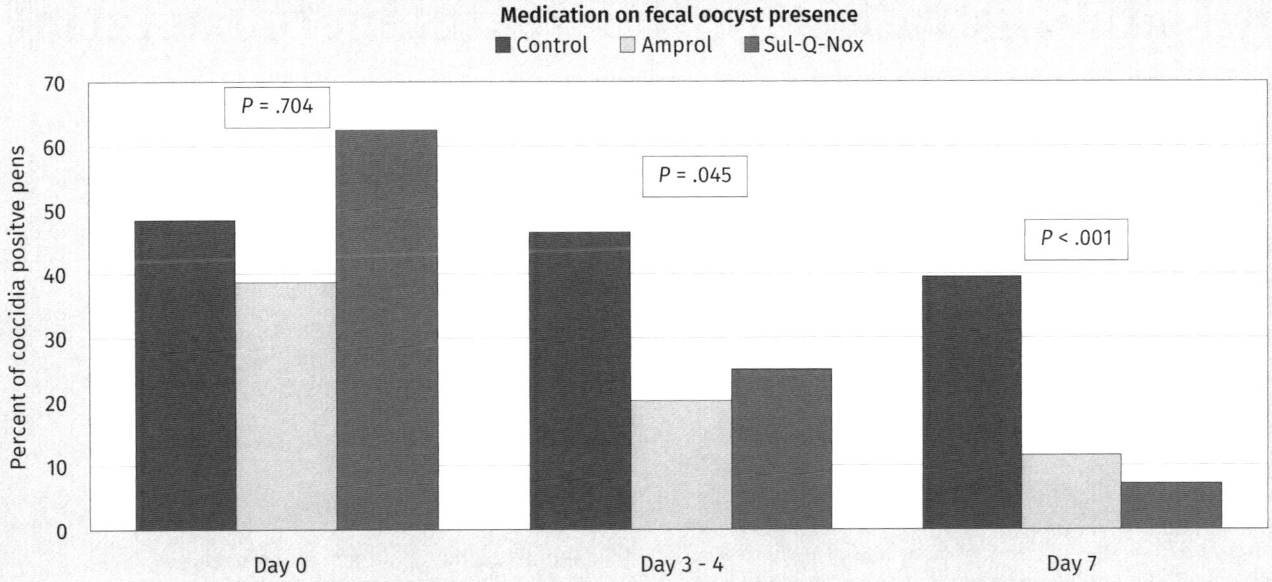

Percent of pens that observed *Cystoisospora suis* oocytes in the pen pooled fecal sample.

Table 1: Growth parameters for pigs treated with Sul-Q-Nox or Amprol or untreated (control)

Parameter	Control	Sul-Q-Nox	Amprol	SEM	P-value
Body weights[1]					
BW d0, kg	4.6[b]	5.0[a]	5.2[a]	0.11	< .001
BW d42, kg	20.4[b]	22.3[a]	22.2[a]	0.37	< .001
Growth performance[1]					
ADG, kg	0.32[b]	0.37[a]	0.36[a]	0.011	.002
ADFI, kg	0.50[b]	0.56[a]	0.55[a]	0.011	< .001
G:F	0.62	0.65	0.65	0.012	.109
Removal/mortality (%)[2]	13.2	10.1	8.5	3.07	.211

[1] Contain observations from trials two and three.

[2] Contain observations from all trials.

[a,b] Means lacking a common superscript are significantly different ($P < .05$).

References

1. Mundt H.-C., A. Cohnen, A. Daugschies, A. Joachim, H. Prosl, R. Schmäschke, and B. Westphal. 2005. Occurrence of *Isospora suis* in Germany, Switzerland and Austria. *Journal of Veterinary Medicine Series B.*

2. Nisley M.J, D. Nolan, C. Sparks, E.R. Burrough, and N.K. Gabler. 2023. Impacts of dietary salinomycin and crude protein on coccidiosis burdened nursery pigs. *Proc American Association of Swine Veterinarians Annual Meeting.*

3. James S. 1980. Thiamine uptake in isolated schizonts of Eimeria tenella and the inhibitory effects of amprolium. *Parasitology.*

4. Looker D.L., J. J. Marr, and R. L. Stotish. 1986. Modes of action of antiprotozoal agents. *In Chemotherapy of parasitic diseases.*

5. Hinney, B., V. Cvjetković, D. Espigares, J. Vanhara, C. Waehner, B. Ruttkowski, R. Selista, D. Sperling, and A. Joachim. 2020. *Cystoisospora suis* control in Europe is not always effective. *Frontiers in Veterinary Science 7.*

https://doi.org/10.54846/am2024/78

PK/PD relationships of Tilmovet® (tilmicosin) oral administration for the treatment of bacterial respiratory diseases in pigs

Ulrich Klein[1]; Wouter Depondt[1]; Thomas Fangman[2]; David Nolan[2]
[1]Huvepharma® NV Belgium [2]Huvepharma® Inc. Peachtree City, Georgia

Introduction

The presence of swine respiratory disease (SRD) is one of the main obstacles for the swine industry. Clinical respiratory disease is often polymicrobial and caused by different respiratory pathogens (*M hyopneumoniae* – Mhyo, *A pleuropneumoniae* – App, *P multocida* – Pm, *G parasuis* – Gp, *B bronchiseptica* – Bb, *S suis* – Ss, PRRSV, SIV).

Tilmovet® (tilmicosin) provides antibacterial and antimycoplasmal activities and is used for pneumonia treatment caused by different respiratory pathogens in pigs.

Tilmovet® (tilmicosin) exhibits broad distribution in respiratory tissues and some unique pharmacokinetic (PK) characteristics. Antimicrobial susceptibility testing (pharmacodynamics PD) is an important component to identify susceptibility patterns of clinical bacteria isolates over time.

The purpose of this paper is to describe the PK properties of Tilmovet® (tilmicosin) based on pharmacokinetic studies, to compare tilmicosin MIC results generated on US and European swine respiratory pathogens and to relate the MICs to the pharmacokinetic`s of tilmicosin. PK/PD relationship data are the basis for the therapeutic use of Tilmovet® (tilmicosin) in the context of prudent use policies.

Materials and methods

Pharmacodynamic (MIC) studies

Antimicrobial susceptibility of respiratory pathogens was done in studies conducted in US and Europe. In the US study the susceptibility of Pm, App, Bb and Ss isolates generated 2019/2020 in the US and Canada to tilmicosin was determined.[15] In the European studies tilmicosin MIC testing of Mhyo, Ap, Pm, Gp, Bb and Ss isolates recovered from SRD cases from eight different European countries was conducted.[3,10] All European strains were generated 2019/2020 with the exception of Mhyo isolates (2015/2016).

In both studies the MICs were determined using broth microdilution method according to CLSI.[1] MICs were assessed following CLSI standards and interpreted using veterinary breakpoints were available.

Clinical veterinary breakpoints are defined for tilmicosin for Pm and App isolates – susceptible MIC ≤ 16 µg/mL, resistant MIC ≥ 32 µg/mL.[1] Veterinary breakpoints for tilmicosin were created based on the outcome clinical treatment studies, pharmacokinetic-pharmacodynamic studies and in vitro susceptibility testing results.[13]

Pharmacokinetic studies

App challenge/PK study: In a study the tilmicosin concentration in the plasma and lung (dosage 16mg tilmicosin/kg bw) were determined after treatment stop following 15 consecutive days of medication (medication start seven days before App serotype 2 challenge). Serum and lung tissue samples were analyzed (HPLC assay) on last day of treatment and 24, 48, 72, 96 hours after the treatment stop. 24 pigs were involved in the study.[5]

PK study: In the study the distribution of tilmicosin was determined (HPLC) in serum, tracheal/bronchial epithelium samples (eight pigs) and in lung and macrophage samples (four pigs) following tilmicosin feed medication of 14 consecutive days (dosage 19.4 mg tilmicosin/kg bw; total number of pigs:10).[14]

Results and discussion

Tilmicosin`s pharmacokinetic behaviour

The pharmacokinetic properties of Tilmovet® (tilmicosin) contribute to its rapid absorption, slow elimination (elimination half-life of 25 hours in pigs), large distribution volume and the broad and consistent tilmicosin concentrations throughout the respiratory tract achieved after oral administration.[12] Tilmicosin has a unique ability to concentrate in, and to be retained within swine phagocytes, over long periods of time. Neutrophils and macrophages migrate preferentially to the sites of infections in the lung.

The uptake of Tilmovet® (tilmicosin) by swine phagocytes facilitates drug delivery to the site of infection and contributes to the clinical efficacy due to release of high concentrations of bioactive tilmicosin at lung infection sites.[11]

Tilmicosin shows substantially higher intracellular accumulation in macrophages and neutrophils in comparison to other macrolides - 2.7 times (tulathromycin), 6 times (tylvalosin), 12.5 times (erythromycin) – and versus other antimicrobial products as shown in Figure 1. The accumulation helps to broadly distribute tilmicosin in the respiratory tract and to provide therapeutic concentrations directly at the infection sites based on prolonged half-lives.

App challenge/PK study

After 15 consecutive days of Tilmovet® (tilmicosin) administration higher tilmicosin lung concentrations versus the plasma concentrations were found (lung:serum ratio 34:1). Lung concentrations increased to peak lung level of 2.712 µg/g 24 hours and maintained on high level (2.063 µg/g) 48 hours after medication cessation. This indicates the intracellular accumulation and prolonged maintenance of higher tilmicosin levels at the site of App infection in the lungs.

PK study

Rapid absorption led to low tilmicosin concentrations in the blood serum. Tilmicosin concentrations in trachea epithelium and lung tissue are in average 20 times higher than the blood concentrations. The accumulation of tilmicosin in alveolar macrophages results in 4.3 times higher concentrations than in lung tissues.

Table 1: Tilmicosin concentrations in plasma and lung after 15 days of consecutive treatment and after medication cessation[5]

	Day 15 of treatment	24 h post treatment	48 h post treatment	72 h post treatment	96 h post treatment
Plasma (µg/mL)	.063[a]	.101[b]	.078[b]	< .050	< .050
Lung (µg/g)	2.183[c]	2.712[c]	2.063[c]	.961[d]	.403[d]

[a,b] $P \leq .01$; [c,d] $P \leq .01$; LOQ < .05 µg/mL

Table 2: Tilmicosin concentrations (Cmax) in different specimen and lung macrophages[14]

	Serum (µg/ml)	Macrophages (bronchial alveolar lavage - µg/ml)	Trachea epithelium (µg/g)	Lung tissue (µg/g)
Mean tilmicosin conc (8 animals in total)	< .1	7.19	2.19	1.69

Figure 1: Intracellular accumulation of tilmicosin in macrophages and neutrophils in comparison to other macrolides and members of other classes of antimicrobials.[2,8,11,15]

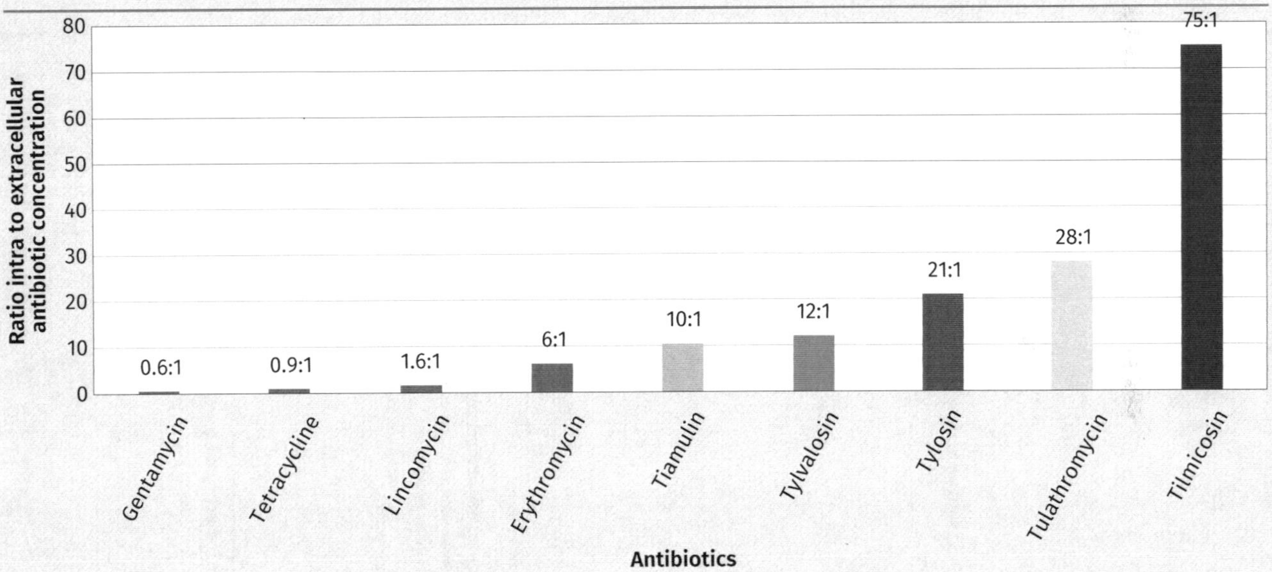

MIC and pharmacokinetic/pharmacodynamic relationship data

MIC`s and MIC frequency distributions as shown in Figures 2 and 3 and their relationship to tilmicosin PK data allow interpretations of PK/PD data and enable efficacy predictions.

The MIC data generated in the US study shows variable susceptibility to tilmicosin (see Figure 2). High levels of susceptibility were determined for Pm and App isolates. The susceptibility patterns for Bb were lower. The majority of Ss isolates show low susceptibility to tilmicosin. For Pm, App and Bb a monomodal MIC distribution was found. In the case of Ss the tilmicosin MICs show a bimodal distribution pattern.

Tilmicosin MIC_{90} values (App 2019 = 16 and 2020 = 8; Pm 2019 = 16 and 2020 = 4) in the US study correspond well with the approved CLSI clinical breakpoints established for Pm and App. The Pm- and App-specific tilmicosin PK/PD relationship data predict a therapeutic success of Tilmovet® (tilmicosin) use in cases of Pm and App infections. In spite of high concentrations in the macrophages the clinical effect in cases of Bb infections will be lower ($MIC_{90} = 32$). Minor clinical effects against Ss are expected in the case of SRD infections caused by this pathogen.

In the EU study very high susceptibility to tilmicosin were determined for Mhyo, Gp and high susceptibility for Pm and App isolates (see Figure 3). The susceptibility patterns for Bb were lower. The majority of the Ss showed low susceptibility to tilmicosin considering a bimodal MIC distribution for this pathogen.

Figure 2: Pharmacokinetic/pharmacodynamic relationships of tilmicosin following feed medication in the case of US/Canada respiratory pathogens.[5,14,16]

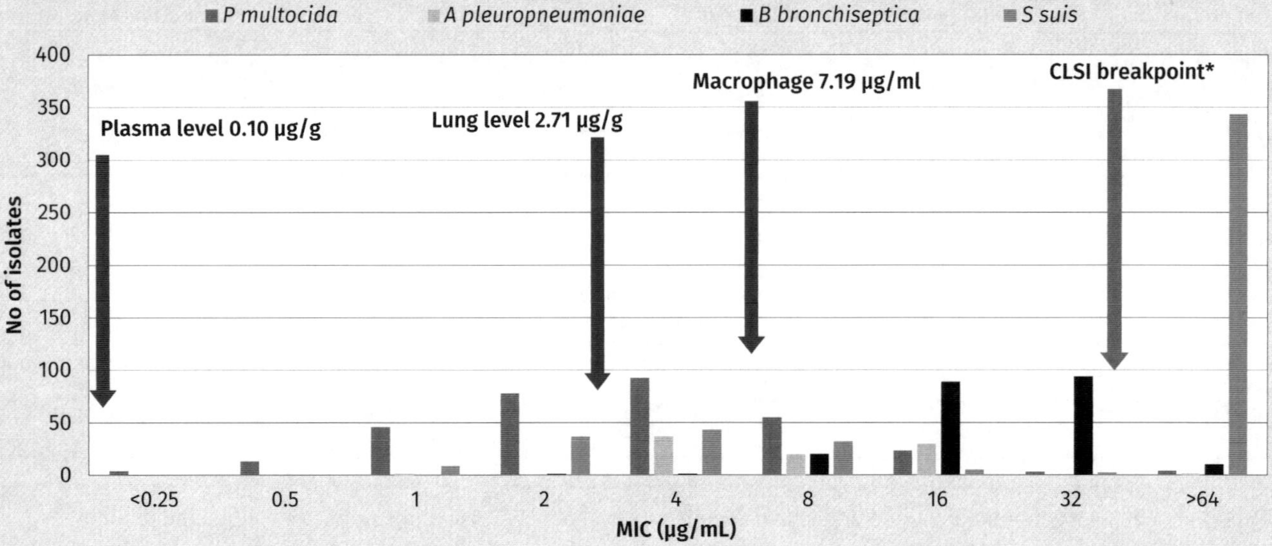

* Tilmicosin CLSI breakpoints: sensitive against Pm, App ≤ 16 µg/mL; resistance ≥ 3

Figure 3: Pharmacokinetic/pharmacodynamic relationships of tilmicosin following feed medication in the case of EU respiratory pathogens.[3,5,10,14]

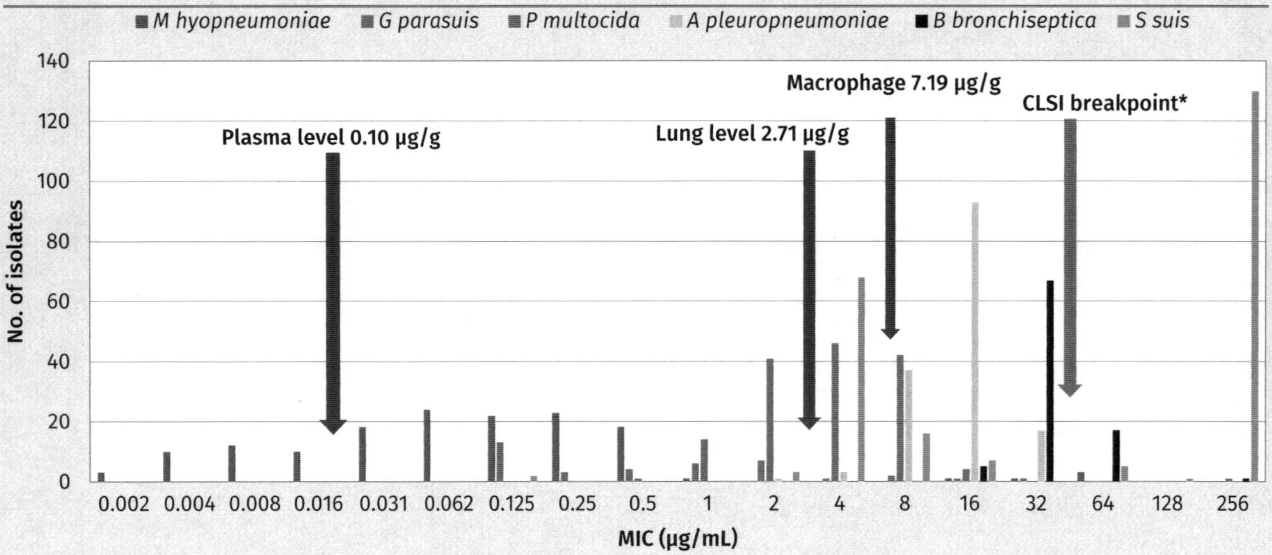

* Tilmicosin CLSI breakpoints: sensitive against Pm, App ≤ 16 µg/mL; resistance ≥ 32

Tilmicosin MIC values for *M hyopneumoniae* isolates (MIC_{90} = 0.5) and for *G parasuis* (MIC_{90} = 4.0) are below tilmicosin concentrations achieved in the lung macrophages. Tilmicosin MICs for Pm ($MIC_{50/90}$ = 4.0/8.0) and App $MIC_{50/90}$ = 16.0/32.0) correspond with the approved CLSI clinical breakpoints. The Mhyo, Gp, App and Pm-specific PK/PD relationship data predict a high therapeutic effect of Tilmovet® (tilmicosin) in pneumonia infections caused by these pathogens.

The pharmacokinetic/pharmacodynamic relationship data presented in this study are in line with the therapeutic outcome of the use of tilmicosin in Mhyo, Gp, App and Pm efficacy studies.[5,6,8] A good therapeutic effect based on the tilmicosin lung concentrations and it`s uptake and accumulation in alveolar macrophages can be expected when treating SRD infections caused by M *hyopneumonie, G parasuis, A pleuropneumonie* and *P multocida.*

Conclusion

Use of pharmacokinetic/pharmacodynamic data including clinical veterinary breakpoints justify the use of Tilmovet® (tilmicosin) for SRD treatment. MIC study data from the US and Europe demonstrate high susceptibility of *M hyopneumoniae, G parasuis, A pleuropneumoniae, P multocida* isolates. Medium susceptibility of *B bronchiseptica* and low susceptibility of *S suis* were determined in both studies.

The pharmacokinetic characteristics contribute to the clinical effectiveness of Tilmovet® (tilmicosin). Effective and high accumulation of tilmicosin in alveolar macrophages and tilmicosin`s presence in lung tissues over long periods of time explain the clinical efficacy of this antimicrobial seen in clinical studies.[4,6,7,9]

The Mhyo-, Gp-, Pm- and App-specific tilmicosin PK/PD relationship data predict a therapeutic success in the case of Tilmovet® (tilmicosin) therapeutic use at treatment dosage in cases of swine respiratory diseases caused by these pathogens.

References

1. Clinical and Laboratory Standard Institute (2018). Performance standards for Antimicrobial Disk and Dilution Susceptibility Tests for Bacteria isolated from Animals. 5th ed. CLSI standard VET01.

2. Blais, J. and S. Chamberland (1994). Intracellular accumulation of tilmicosin in primary swine alveolar macrophages. Proc. 13th IPVS Congress Bangkok, Thailand, p.331.

3. Hawser, S. (2022). European Collection of Veterinary Pathogens (VetPath V) - minimal inhibitory concentration (MIC) testing, IHMA Study Report 3638 September 2022.

4. Ivers, D. et al. (2016). Effectiveness of tilmicosin phospate for controlling respiratory disease associated with M.hyopneumonia in swine in the presence of PRRS. Proc. AASV, 158-160.

5. Karanikolova, M. *et al.* (2012). Maintenance of tilmicosin concentrations post oral administration and clinical effect in pigs experimentally infected with *Actinobacillus pleuropneumoniae*. Proc. 22nd IPVS Congress Jeju, Korea, p.622.

6. MacInnes J.I. *et al.* (2003). Efficacy of prophylactic tilmicosin in the control of experimentally induced *Haemophilus parasuis* infection in pigs. J. Swine Health and Production; 11 (4), 174-180.

7. Moore, G.M. *et al.* (1996) Efficacy dose determination study of tilmicosin phosphate in feed for control of pneumonia caused by APP. Am J Vet Res 57, (2), 220-223.

8. Nielsen, B.H. and Szancer, J. (1998). Uptake and intracellular concentration of tiamulin in human polymorphonuclear leukocytes compared with norfloxacin. Proceedings 15th IPVS Congress, Birmingham, UK, 3, p.241.

9. Paradis, M.A. *et al.* (2004). Efficacy of tilmicosin in the control of experimentally induced *Actinobacillus pleuropneumoniae* infection in swine. Can.J.Vet.Res. 68,7-11.

10. Pridmore, A. (2018). European Collection of *Mycoplasma* Pathogens (MycoPath II): collection, storage, identification and MIC determination for *Mycoplasma* strains isolated from cattle, pigs and poultry. Don Whitley Scientific Study Report 015/16.

11. Scorneaux, B., Shryock, T. (1998). Intracellular accumulation, subcellular distribution and efflux of tilmicosin in swine phagocytes. J. Vet. Pharmacol. Ther.21(4): pp.257-268.

12. Shen, J. *et al.* (2005). Pharmacokinetics of tilmicosin after oral administration in swine. Am J Vet Res, 66; 1071.

13. Shryock, T.R. et al. (2002). MIC breakpoints and disk inhibitory zone interpretation criteria for tilmicosin susceptibility testing against *P.multocida* and *A.pleuropneumoniae* associated with porcine respiratory disease. J.Vet Diagn Invet 14, 389-395.

14. Stoker, J. *et al.* (1996). The concentration of tilmicosin in pig serum and respiratory tissue following administration via feed at level of 400g/tonne. Proc. 14th IPVS Congress, Bologna, p.656.

15. Stuart, A.D. *et al.* (2008). Tylvalosin a macrolide antibiotic inhibits the in-vitro replication of European and American PRRSV. The Pig Journal, 61, 42-48.

16. Sweeney, M.T. *et al.* (2022). Antimicrobial susceptibility of *Actinobacillus pleuropneumoniae, Bordetella bronchiseptica, Pasteurella multocida* and *Streptococcus suis* isolated from diseased pigs in the United States and Canada, 2016-2020. J.Swine Health Prod, 30, 130-144.

https://doi.org/10.54846/am2024/79

Optimizing shipping and receiving of pigs with AI-powered tools for enhanced labor efficiency and animal welfare

J. Labrecque, MSc, P.Eng
Ro-Main, Saint-Lambert-de-Lauzon, Quebec City

Abstract

This comprehensive article delves into the transformative impact of artificial intelligence (AI) on swine industry practices, with a specific focus on the shipping and receiving processes. Examining cutting-edge technologies, exemplified by the smaRt Counting system, the article explores how AI-driven tools are poised to revolutionize the industry, offering solutions to enhance labor efficiency and elevate animal welfare standards. The discussion delves into the precision of AI in pig counting, the application of computer vision to score pig handling, and the advanced features contributing to an industry-wide shift towards a more technology-centric and humane approach.

Introduction

The swine industry, a cornerstone of global agriculture, stands at the cusp of a technological revolution with the integration of AI into its traditional workflows. This article, a detailed exploration of this paradigm shift, focuses on the multifaceted role of AI-powered tools in optimizing the intricate processes involved in the shipping and receiving of pigs. It aims to provide a nuanced understanding of how these technologies, specifically the smaRt Counting system, are reshaping industry practices, fostering heightened efficiency, and advancing animal welfare standards.

AI in precision pig counting

A central challenge in swine logistics is the accurate counting of pigs. The smaRt Counting system, leveraging advanced computer vision and machine learning algorithms, presents a paradigm shift in addressing this challenge. By automating real-time pig counting, this technology eradicates the need for manual enumeration, reducing errors and optimizing time management for workers. The system's ability to adapt to varying environmental conditions and pig sizes underscores its versatility in meeting the dynamic demands of the industry. With a guaranteed minimum accuracy of 99.9%, the smaRt Counting system ensures unparalleled precision, eliminating the need for both manual counting and recounting during shipping and receiving processes.

Computer vision for scoring pig handling

Moving beyond the realm of simple enumeration, the smaRt Counting system incorporates sophisticated features, notably its Usage Compliance module, which scores pig handling through advanced computer vision. This innovative functionality goes beyond mere counting, evaluating worker-pig interactions with precision. By providing real-time insights into stress levels and refining handling techniques, the Usage Compliance feature contributes significantly to superior animal welfare. Additionally, this feature enables the optimization of counting accuracy by promoting handling methods that offer better visibility of the pigs. The system's capacity to analyze and score handling practices allows for immediate adjustments, fostering a continuous improvement cycle in animal welfare standards. This unique aspect sets the smaRt Counting system apart as a comprehensive tool for optimizing both labor efficiency and animal welfare in the swine industry.

Improving labor efficiency

Traditional pig counting methods are characterized by labor-intensive processes prone to errors, contributing to inefficiencies in the overall shipping and receiving workflow. The infusion of AI-driven tools significantly amplifies labor efficiency by automating the counting process. Workers can redirect their focus to other critical aspects of shipping and receiving, assured of the accuracy and speed offered by the AI-powered system. The result is a streamlined workflow that maximizes human resources and minimizes operational bottlenecks.

Stress alleviation for animals and workers

The shipping and receiving processes in the swine industry are inherently stressful for both pigs and workers. AI-powered tools, such as the smaRt Counting system, with their non-intrusive counting and scoring capabilities, serve as a stress-relief solution. The system's automation of tasks that traditionally required manual intervention has a dual benefit – not only does it guarantee precise counting and scoring through advanced computer vision, but it also allows workers to handle pigs more efficiently. This automation fosters a calmer environment for both stakeholders, as workers no longer bear the stress of manually counting pigs while managing other crucial tasks. The positive impact of stress reduction on animal welfare is undeniable, translating into healthier pigs and an overall improvement in the quality of meat produced.

Advanced features of smaRt Counting

While boasting advanced features, the smaRt Counting system is designed with a user-friendly approach, ensuring that even basic users can easily navigate its functionalities. The system's standout features, particularly in user management and permissions, are seamlessly integrated without compromising simplicity. This allows for a tailored approach, ensuring that workers possess permissions aligned with their specific roles and responsibilities. The system's capability to adapt to standard operating procedures of the user enhances its usability and integration into existing workflows. Moreover, the system facilitates robust reporting and analysis, contributing to a continuous enhancement of pig handling practices. The ability to generate detailed reports on pig movements, worker interactions, and system performance provides valuable insights for decision-makers, allowing for data-driven optimizations in logistics and animal management. This adaptability to user-specific procedures ensures a seamless integration of advanced AI tools into the existing operational framework without sacrificing ease of use for basic users.

Adherence to standards and safety

Addressing safety concerns, the smaRt Counting system complies with stringent standards such as EN IEC 61326-1:2021 and FCC 47 CFR Part 15. The device prioritizes safety with explicit warnings and instructions, underscoring the significance of proper handling to prevent risks of electrocution or fire. These safety features, coupled with robust certifications such as IP65, instill confidence in users and stakeholders, establishing the smaRt Counting system as a reliable and secure component of swine industry operations.

Conclusion

The integration of AI-powered tools, exemplified by the smaRt Counting system, marks a pivotal advancement in the swine industry. This comprehensive exploration has underscored the multifaceted role of AI in optimizing the shipping and receiving of pigs, presenting a precise balance between heightened labor efficiency and elevated animal welfare. As the industry embraces these advancements, a future is envisioned where technology not only streamlines processes but also significantly contributes to the well-being of animals and workers alike. The ongoing evolution of AI technologies in the swine industry promises a dynamic landscape, where innovation and sustainability go hand in hand.

https://doi.org/10.54846/am2024/80

Effect of body condition during late gestation on susceptibility to vaginal/uterine prolapse

J. R. Dunkelberger[1,2], PhD; T. Veldhuizen[3]; C. Opschoor[1], MSc; T. Stevens[1], MSc; E. F. Knol[1], PhD
[1]Topigs Norsvin, Den Bosch, The Netherlands; [2]Topigs Norsvin USA, Burnsville, Minnesota; [3]Iowa State University, Ames, Iowa

Introduction

Sow mortality continues to be a major concern for the US swine industry. One of the top reasons for sow mortality is pelvic organ prolapse (POP).[1] Both genetic and non-genetic factors have been associated with susceptibility to POP. Results of genetic analyses show that POP is moderately heritable at 35%.[2] This result suggests that susceptibility to POP can be addressed through the breeding program, but that the majority of phenotypic variation in POP (65%) is due to environmental factors. Therefore, genetic selection alone will not solve POP. Rather, a combination of genetic and non-genetic solutions is needed to address this issue. Several non-genetic factors have already been associated with susceptibility to POP, including: water treatment, bump feeding strategy, perineal score, and body condition during late gestation.[1] The objective of this study was to estimate the effect of one or more of these previously-identified, non-genetic factors on susceptibility to POP in an independent population, with the ultimate goal of identifying non-genetic solutions that can be used to supplement genetic selection for reduced susceptibility to POP.

Materials and methods

Data

Data was collected on crossbred (Landrace x Large White) Topigs Norsvin females from two commercial sow farms (Farm A: n = 1,474; Farm B: n = 2,395) located in the Midwest USA. Data was collected for ten consecutive weeks from May through July 2023. Body condition score (BCS), caliper score, and perineal score were recorded for each female upon entry into the farrowing unit (an average of 4.8 days prior to her due date). Body condition score was assigned using a 5-point scoring system based on evaluation of the pin bones, tail, loin, backbone, and ribs where: "1" = skinny; "2" = thin; "3" = normal; "4" = fat; and "5" = obese. Caliper score was assigned using caliper units on a scale from 5 (skinny) to 24 (obese). Perineal score was assigned based on the 3-point scoring system developed by Kiefer et al[3] where: "1" = little to no risk of prolapse; "2" = moderate risk of prolapse; and "3" = high risk of prolapse. Each instance of POP observed during the study, defined here as a vaginal and/or uterine prolapse, was also recorded.

Statistical analyses

A generalized linear model was used to estimate the effect of BCS and caliper score on perineal score using SAS. Farm (2 levels: A and B), parity class (6 levels; 1, 2, 3, 4, 5, 6+), and sampling month (2 levels; May/June and July) were fitted as fixed effects. May and June were combined into a single category, due to a limited number of sampling days in May. All pairwise interactions were fitted and removed if not significant. Caliper score was analyzed in two different ways: "caliper class" and "caliper range." Levels for "caliper class" were defined using two-unit intervals from the minimum (13) to maximum (22) observed caliper score: 13 to 14, 15 to 16, 17 to 18, 19 to 20, and 21 to 22. Levels for "caliper range" (thin, ideal, fat) were defined based on Topigs Norsvin's recommendations for evaluating body condition using the caliper device: for parity 1 sows: "thin" < 14 caliper units; "ideal" = 14 to 18 caliper units; and "fat" > 18 caliper units. For parity 2+ sows: "thin" < 14 caliper units, "ideal" = 14 to 17 caliper units; and "fat" > 17 caliper units.

Results and discussion

Of the 3,869 females evaluated, 33 demonstrated evidence of POP at some point during the study. Each instance of POP observed occurred while the sow was housed in the farrowing unit within 5 days of her due date. Of the 33 cases, 7 died naturally, 17 were euthanized, and 9 were still alive at the conclusion of the study. The average BCS, caliper score, and perineal score for females that did/did not display POP are listed in Table 1. These results show that sows with POP had a lower average BCS and caliper score upon entry into the farrowing unit than sows that did not display POP. These findings agree with results reported by Ross (2019), where sows with a lower body condition score in late gestation were more likely to display POP. Results from the current study also show that sows with POP had a higher average perineal score than sows that did not display POP (Table 1). Similarly, the higher the perineal score, the higher the number of sows that displayed POP (Figure 1), which supports previous findings that a higher perineal score is associated with increased likelihood of POP.[1]

Due to the limited number of POP observations (n = 33), perineal score was analyzed as the response variable for all analyses presented. The percent of sows per perineal score and farm are listed in Table 2. Perineal score was analyzed as a binary trait, where an ideal perineal score (ie, no risk of developing POP) was coded as "0" (n = 2,554) and a non-ideal perineal score (moderate to high risk of developing POP) was coded as "1" (n = 1,315). Results show that BCS ($P < .0001$), caliper class ($P < .0001$), and caliper range ($P < .0001$) each had a highly significant effect on perineal score. Results for analysis of BCS show that a lower (thinner) body condition was associated with a higher incidence of non-ideal perineal score (Figure 2). These findings are consistent with results for analysis of caliper class, where a lower caliper score was associated with a higher incidence of non-ideal perineal score (Figure 3). Likewise, when caliper score was analyzed using categories (thin, ideal, or fat), thin sows had the highest incidence of non-ideal perineal score (Figure 4). Of the other variables included in the model, "farm" and "parity" were consistently, significantly associated with perineal score. "Farm" had a highly significant effect ($P < .0001$) on BCS, caliper class, and caliper range, where Farm B had a higher incidence of non-ideal score than Farm A. "Parity" also had a highly significant effect ($P < .0001$) on BCS, caliper class, and caliper range, where parity 3 and 4 sows had a higher incidence of non-ideal score than sows of other parities.

Table 1: Average body condition score, caliper units, and perineal score for sows that displayed vaginal and/or uterine prolapse (or not) during the study

Trait	Prolapsed	Did not prolapse
Body condition score[a]	3.34	3.53
Caliper units[b]	16.77	17.27
Perineal score[c]	2.36	1.35

[a] Body condition score: assigned based on visual observation using the following scale: "1" = skinny; "2" = thin; "3" = normal; "4" = fat; or "5" = obese.

[b] Caliper units: range = 5 to 24.

[c] Perineal score: assigned based on visual observation using the following scale: "1" = little to no risk of prolapse; "2" = moderate risk of prolapse; or "3" = high risk of prolapse.

Figure 1: Number of sows that displayed vaginal and/or uterine prolapse by perineal score.*

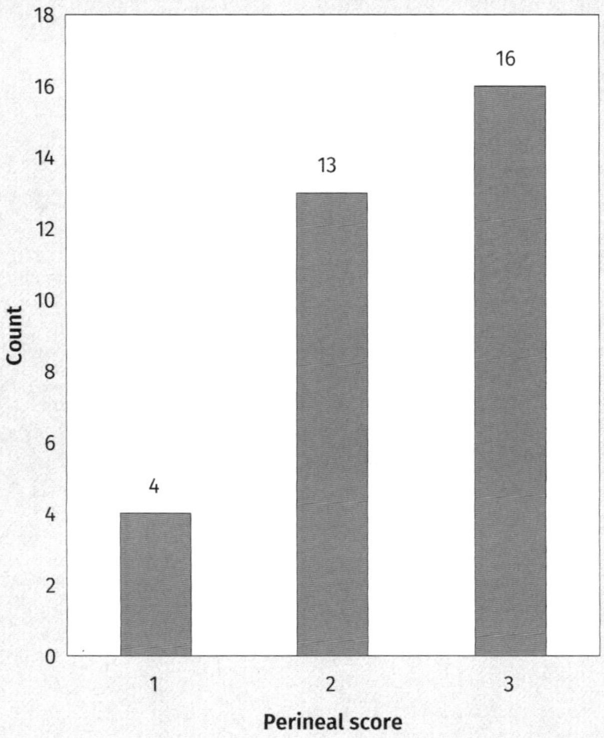

* Perineal score: assigned based on visual observation using the following scale: "1" = little to no risk of prolapse; "2" = moderate risk of prolapse; or "3" = high risk of prolapse.

Table 2: Number of observations per perineal score and farm

Farm		Perineal score[a]		
		1	2	3
A	N =	1,183	281	10
	%	80.26%	19.06%	0.68%
B	N =	1,371	960	64
	%	57.24%	40.08%	2.67%

[a] Perineal score: assigned based on visual observation using the following scale: "1" = little to no risk of prolapse; "2" = moderate risk of prolapse; or "3" = high risk of prolapse.

Figure 2: Effect of body condition score* on perineal score. A higher estimate indicates increased incidence of non-ideal score (moderate to high risk of vaginal and/or uterine prolapse).

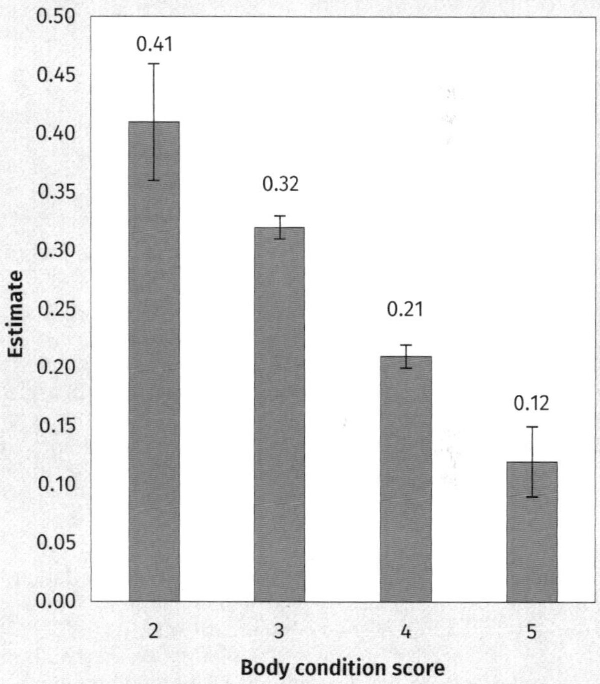

* Body condition score: assigned based on visual observation using the following scale: "1" = skinny; "2" = thin; "3" = normal; "4" = fat; or "5" = obese.

Figure 3: Effect of caliper class (in intervals of two caliper units) on perineal score. A higher estimate indicates increased incidence of non-ideal score (moderate to high risk of vaginal and/or uterine prolapse).

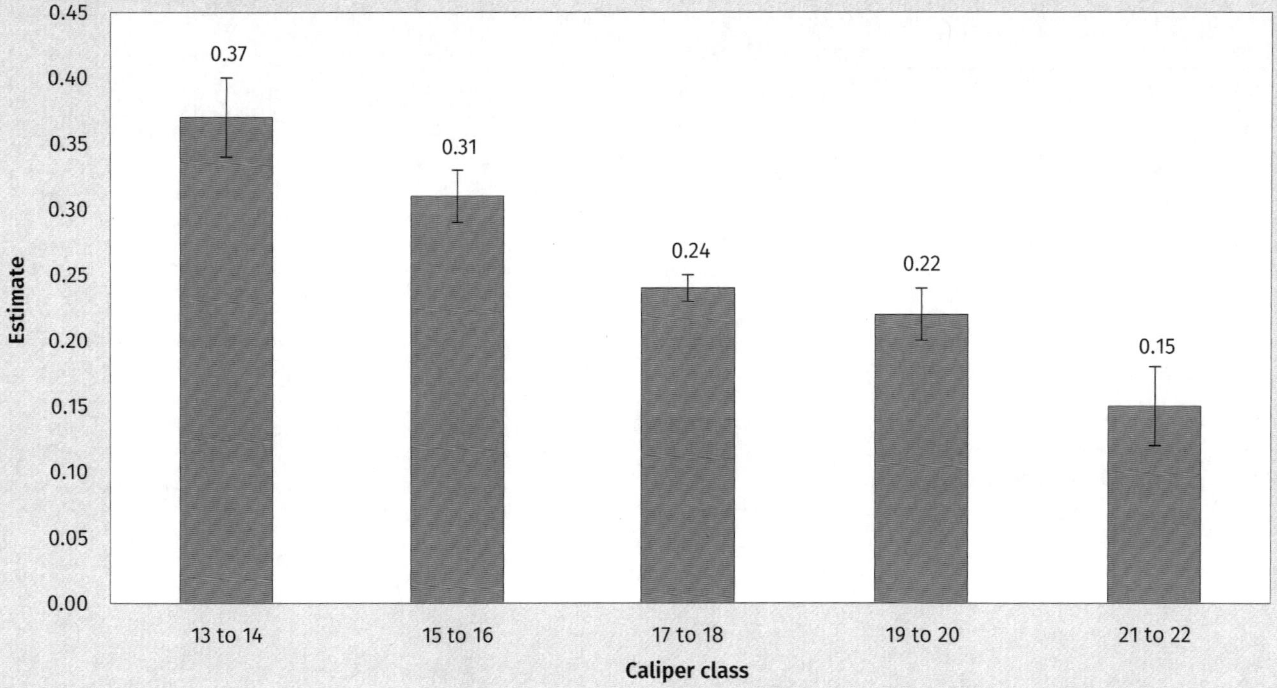

In conclusion, results from this study support previous findings that perineal score is a good indicator of susceptibility to POP, where a higher perineal score was associated with increased risk of developing POP. Results from this study also show that BCS, caliper score (whether analyzed in unit intervals or categories), farm, and parity were highly, significantly associated with perineal score. Thinner sows (whether evaluated based on body condition or caliper measurement) during late gestation had a higher incidence of non-ideal perineal score. This finding supports results from previous work, where thinner sows were more susceptible to POP. Results of previous work conducted by our group show that, although genetics plays a role in susceptibility to POP, POP is mostly influenced by environmental factors. Results from the current study validate the effect of several of these previously identified non-genetic factors. Therefore, genetic selection for reduced incidence of POP, in combination with non-genetic solutions, such as proper management of body condition during late gestation, is needed to address this critical issue in US swine production.

References

1. Ross J. Identification of putative factors contributing to pelvic organ prolapse in sows. *Improving Pig Survivability.* 2019.

2. Bhatia V, Stevens T, Derks MFL, et al. Identification of the genetic basis of sow pelvic organ prolapse. *Front Genet.* 2023; 14.1154713. **https://www.frontiersin.org/articles/10.3389/fgene.2023.1154713/full**

3. Kiefer ZE, Koester LR, Showman L, et al. Vaginal microbiome and serum metabolite differences in late gestation commercial sows at risk for pelvic organ prolapse. *Scientific Reports.* 2021; 11.6189. **https://www.nature.com/articles/s41598-021-85367-3**

Figure 4: Effect of caliper range* on perineal score. A higher estimate indicates increased incidence of non-ideal score (moderate to high risk of vaginal and/or uterine prolapse).

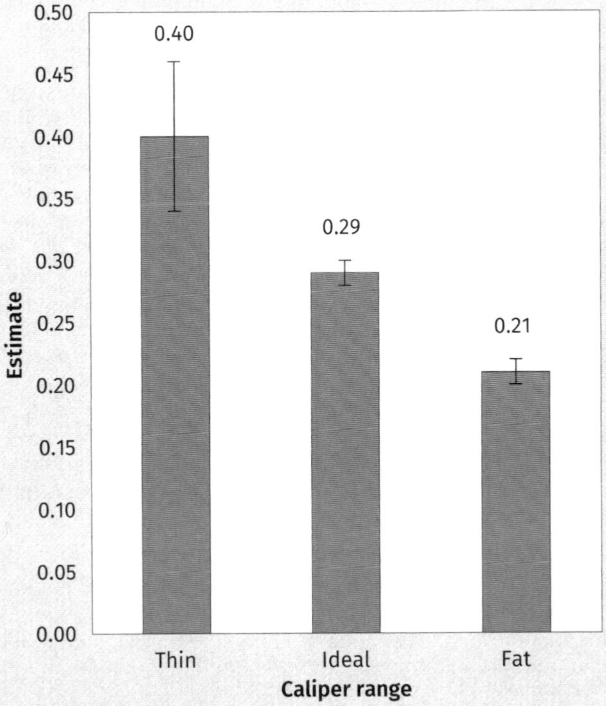

* Caliper range: for parity 1 sows: "thin" < 14 caliper units; "ideal" = 14 to 18 caliper units; and "fat" > 18 caliper units. For parity 2+ sows: "thin" < 14 caliper units, "ideal" = 14 to 17 caliper units; and "fat" > 17 caliper units.

https://doi.org/10.54846/am2024/81

Effects of oral meloxicam in gel and oral drench on pain scores in piglets during processing

C. Bates[1], BS; R. Stika Jensen[2], DVM; C. Haden[2], DVM; R. Edler[2], MS; M. Pairis-Garcia[3], DVM, PhD, DACAW
[1]Iowa State University CVM, Ames, Iowa; [2]Pipestone, Pipestone, Minnesota; [3]North Carolina State University, Raleigh, North Carolina

Introduction

Tail docking and castration, "processing," of young pigs is common in the swine industry to reduce aggression, tail biting and to inhibit boar taint. Currently, the US swine industry has not adopted pain mitigation during processing despite studies demonstrating that pigs present behavioral and physiological behaviors of pain.[1] The main factors influencing this decision is the cost of pain control and additional labor needed to handle pigs a second time. The objective of this study was to evaluate the use of oral meloxicam, an inexpensive NSAID, on pain, average daily gain (ADG) and mortality at processing (5 - 7 days of age) using 2 different administration routes.

Materials and methods

This study was conducted on a 3,000 head sow farm in Iowa. Using parity 2 or 3 sows, 12 litters were selected and randomly assigned to one of the four treatments. Treatment 1 received an oral meloxicam gavage, treatment 2 received no treatment, treatment 3 received an oral gel/meloxicam mixture, and treatment 4 received only oral gel. Since the gel is green, treatments with no gel received green dye in the crate. Litters and video footage were randomly labeled to keep farm staff and scorers blinded. On day 1, 24 hours pre-processing (PP), pigs were weighed and marked with a number (1 - 15). On day 2, treatment 1 was administered (1 mL, 2.4 mg of compounded meloxicam, Veterinary Pharmaceutical Solutions) via oral drench 2 hours prior to processing. Treatment 2 was administered no treatment, but the mat was covered in green dye 2 hours prior to processing. Treatment 3 was administered 30 mL per pig of a compounded meloxicam/gel mixture (Pig-E-Gel, Animal Science Products) 3 hours prior to processing. The mixture contained 4.8 mg of meloxicam/piglet assuming a 50% consumption over 1 hour based on previous observations. Treatment 4 was given 30 mL of gel per pig 3 hours prior to processing. One caretaker performed processing procedures: tail-docking for both sexes and castration of males. Video recording was used to evaluate behaviors. Cameras were placed at the rear of each sow and pigs were observed three times by a blinded observer: 24 hours PP, immediately post-processing (IPP) and three hours post-processing (3PP). At each time point, pigs were scored using UNESP-Botucatu composite pain scale for assessing post-operative pain in piglets.[2] Mortality was tracked and pigs were weighed at weaning. The primary variable was analyzed by a linear mixed model approach for repeated measures where the model considers the fixed effects of treatment, gender, time of observation and the interaction of treatment and time of observation as well as the random effect of pigs.

Results and discussion

There was an increase in pain score from PP to IPP and 3PP across all treatment groups verifying processing is a painful procedure (P-value < .0001). All treatments had similar IPP scores indicating there was no effect on the immediate pain. This may suggest meloxicam administration needs to be earlier than 3 hours PP. However, there was a significant decrease in pain scores 3PP for treatments 1 and 4 respectively, providing evidence that utilizing oral meloxicam drench, an inexpensive NSAID, can decrease pain scores (Figure 1). Treatment 4 was not a significant finding in lowering pain scores due to a low sample number (n = 5) from exclusion of pigs sleeping. Gender of pigs plays a significant role with males having an average score of 1.42 points higher (P-value = .0001) than females (more painful). No notable trends were found between treatments and ADG or mortality, further studies with an ideal sample power to identify the effects is warranted. Use of meloxicam in oral gel, which requires no additional handling of pigs, had no prominent findings. Additional work is needed to increase consumption rate and better understand peak blood concentration. A limitation of this study is that pain scoring is a subjective measure.

References

1. Ison, Sarah H, et al. "A Review of Pain Assessment in Pigs." *Frontiers in Veterinary Science,* U.S. National Library of Medicine, 28 Nov. 2016, **www.ncbi.nlm.nih.gov/pmc/articles/PMC5124671/.**

2. Robles I, Luna SPL, Trindade PHE, Lopez-Soriano M, Merenda VR, Viscardi AV, Tamminga E, Lou ME, Pairis-Garcia MD. Validation of the Unesp-Botucatu pig composite acute pain scale (UPAPS) in piglets undergoing castration. PLoS One. 2023 Apr 13;18(4):e0284218. **doi: 10.1371/journal.pone.0284218.**

Figure 1: A comparison of pain scores by treatment groups at three time points.

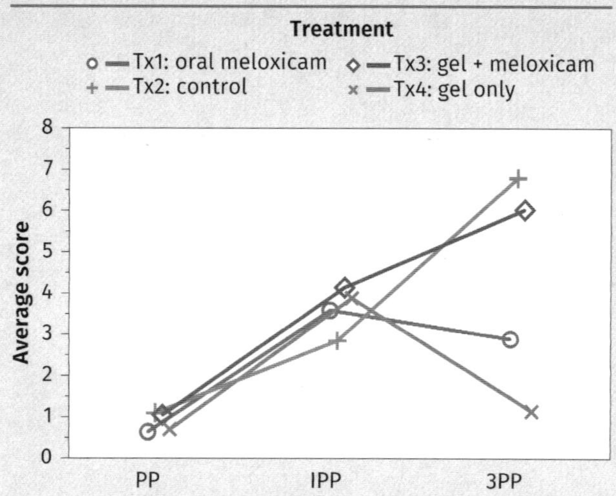

PP = pre-processing; IPP= immediately post-processing; 3PP = 3 hours post-processing.

https://doi.org/10.54846/am2024/82

A comparison of *Mycoplasma hyopneumoniae* infectivity rates of replacement gilts using different fogging techniques

M. Jones[1], BS; B. Whitt[2], DVM; E. Noblett[2], MAS, DVM
[1]College of Veterinary Medicine, North Carolina State University, Raleigh, North Carolina;
[2]Smithfield Hog Production, South Region, Laurinburg, North Carolina

Introduction

Mycoplasma hyopneumoniae (M hyo) is the primary pathogen of enzootic pneumonia in pigs and causes significant economic losses to producers due to reduced feed efficiency and average daily gain.[1] One mitigation strategy commonly used is to acclimate replacement gilts prior to entry to the breeding herd. Intratracheal, intranasal, and aerosol exposure methods have proven successful in a research setting,[2] but aerosol is the most practical for acclimating a large number of animals. Different fogging methods (ie, different equipment, isolates, dosages, exposure intervals, etc) have been tested within the production system used in this study but have previously yielded varied results. The objectives of this project were 1) To compare the replacement gilt infectivity rates when using different inoculum dosages and administration equipment and 2) To determine the earliest time point that infection can be considered successful.

Materials and methods

Entire groups of gilts were exposed via nebulization at 8-9 weeks of age using one stored batch of lung homogenate (average Ct = 20.8/mL). The challenge material used had previously been confirmed infectious in one batch of gilts. Two different dosages (2x and 4x; both at a mixing rate of 40 mLs homogenate/1000 mLs saline) and two different fogging units (Backpack and Hurricane) were evaluated. Gilt groups received one of four treatment groups (T01: 2x dose, administered via Backpack, T02: 4x-dose Backpack, T03: 2x-dose Hurricane, and T04: 4x-dose Hurricane). There were 4 replicates per group. 30 gilts from each group were tagged at treatment administration and deep tracheal sampling was performed at 14 and 28 days post-exposure (DPE). A subset of groups was also tested between 35 - 50 DPE. A catheter was inserted to the bifurcation of the trachea and once removed, roughly 1 inch of the distal end of the catheter was placed in snap top tubes with 2 mLs PBS. All samples were sent to the Iowa State University Veterinary Diagnostic Laboratory and pooled by 5 for M hyo PCR testing. Given a lack of response from aerosol exposure, there was a need to investigate the master seed. Intratracheal inoculations were completed on 10 gilts at 4 sites to test the infectivity of the lung homogenate used. 5 gilts per site received 10 mLs of undiluted homogenate and 5 received 10 mLs diluted homogenate (same mixing rate as the aerosol doses) and were sampled at 14 DPE via tracheal sampling and tested by individual M hyo PCR testing.

Results and discussion

All 14 and 28 DPE swabs (n = 507) from aerosol exposures were negative. There were only 2 groups that tested PCR positive after 45 DPE and while 6 of 12 pools were positive the average Ct was 31.2. These fogging results regardless of dosage and fogger type were unacceptable for the system's needs so further investigation into the master seed was performed. Evaluation of the master seed in gilts that were intratracheally challenged demonstrated a response rate of 46% PCR positive at 14 DPE. Site response rates were quite variable (Site 1 = 90% positive, Site 2 = 0% positive, Site 3 = 55.6% positive, and Site 4 = 40% positive). These poor and highly variable response rates in gilts challenged intratracheally demonstrated the need for a new master seed. Future investigations with a new master seed confirmed infectious are still needed to evaluate fogging techniques which are critical to ensure success of gilt acclimation programs used.

References

1. Figueras Gourgues S, Fano E, Alegre Sabaté A, López Grasa E, Hernández Caravaca I, García Vázquez FA, Rodríguez Vega V, Garcia-Morante B. Assessment of nebulization technology for gilt exposure to *Mycoplasma hyopneumoniae* as an acclimation strategy. *J Swine Health Prod.* 2020;28(6):294-301.

2. Poeta Silva AP, Marostica T, McDaniel A, Arruda B, Alonso C, Derscheid R, Yeske P, Linhares, D, Giménes-Lirola L, Karriker L, Fano E, Zimmerman J, Clavijo M. Comparison of *Mycoplasma hyopneumoniae* response to infection by route of exposure. *Vet Microbiol.* 2021;258:109118.

https://doi.org/10.54846/am2024/83

Minimizing sample cross-contamination for the detection of *Mycoplasma hyopneumoniae*

C. Snow[1]; A. Canturri[1,2], DVM, PhD; M. Pieters[1-3], DVM, PhD
[1]Department of Veterinary Population Medicine, College of Veterinary Medicine, University of Minnesota, St. Paul, Minnesota; [2]Veterinary Diagnostic Laboratory, University of Minnesota, St. Paul, Minnesota; [3]Swine Disease Eradication Center, University of Minnesota, St, Paul, Minnesota

Introduction

Cross-contamination between samples is one of the most common pre-analytical errors and can cause misleading diagnostic results. Recent studies illustrating the issue of sample contamination emphasize DNA's significant resistance to degradation when on environmental surfaces, which is undoubtedly an important problem for PCR testing and posterior result interpretation.[1] Current sampling recommendations include measures that aim at minimizing specimen cross-contamination.[2] For example, materials used to collect tracheal secretion samples for the detection of *Mycoplasma hyopneumoniae,* including mouth speculums and laryngoscopes, should be disinfected between pigs. However, the disinfection method is not standardized, and although most commercially available disinfecting wipes may have bactericidal effects, it is currently unknown if they prevent DNA carryover between samples. Therefore, the objective of this study was to compare various disinfecting compounds on their ability to minimize sample cross-contamination in swine tracheal secretion collection.

Materials and methods

Tracheal secretions were collected with various material sets from nursery pigs that had been experimentally inoculated with *M. hyopneumoniae* (n = 30; strain 232), for natural contamination. In parallel, 1:10 diluted *M hyopneumoniae* strain 232 culture was sprayed onto six separate material sets for artificial contamination. The sets included a mouth speculum, a laryngoscope blade, and a pair of scissors. Swabs were collected from each material. The material sets were thoroughly cleaned with one of the following treatment-based wipes: hydrogen peroxide, alkyl dimethyl benzyl ammonium chloride, bleach, alcohol, or phosphate-buffered saline (PBS). Following the disinfecting wipe, a PBS wipe was used to collect a post-disinfecting sample from each material. The DNA was extracted from the pre-disinfecting swabs, disinfecting wipes, and post-disinfecting samples for both natural and artificial contamination, and a species-specific PCR was performed to detect *M hyopneumoniae*. Ct values were compared with Kruskal-Wallis tests to assess statistical significance.

Results

Prior to disinfection, all materials within each contamination method harbored a similar relative bacterial load of *M hyopneumoniae* ($P = .09$ and $P = .06$ for the natural and artificial contamination, respectively). Thus, data from various materials were grouped as replicates for each treatment group for this study. Starting Ct values were not statistically different between treatment groups for natural ($P = .50$) or artificial ($P = .79$) contamination. Of the naturally contaminated treatment groups, bleach and hydrogen peroxide A were the only disinfectants to eliminate all *M hyopneumoniae* genetic material ($P = .04$). Bleach was the only treatment to completely remove all *M hyopneumoniae* DNA from the artificially contaminated materials ($P = .02$).

Discussion and conclusions

Results of this investigation suggested that bleach was the disinfectant that successfully removed all *M hyopneumoniae* DNA from the sampling materials. Our results seemingly agree with other studies on different pathogens.[3] These results can be further applied for the development of a protocol for minimizing or eliminating sample cross-contamination when collecting swine clinical specimens. It is important to note that current protocols for sample collection could result in cross-contamination from residual DNA between pigs.

References

1. Kampmann, M.-L., Børsting, C., & Morling, N. (2017). Decrease DNA contamination in the laboratories. *Forensic Science International: Genetics Supplement Series,* 6.

2. McDowell, E., Pieters, M., Spronk, T., Nerem, J., Fano, E., Dee, S., & Sponheim, A. (2023). Duration of *Mycoplasma hyopneumoniae* detection in pigs following purposeful aerosol exposure. *Veterinary Microbiology,* 282, 109758.

3. Walia, K., Argüello, H., Lynch, H., Grant, J., Leonard, F. C., Lawlor, P. G., Gardiner, G. E., Duffy, G. (2017). The efficacy of different cleaning and disinfection procedures to reduce salmonella and Enterobacteriaceae in the lairage environment of a pig abattoir. *International Journal of Food Microbiology,* 246, 64–71.

https://doi.org/10.54846/am2024/84

The impact of early life stress on neural plasticity in growing swine

M. Hall; A. Viscardi, PhD
Department of Anatomy and Physiology, College of Veterinary Medicine, Kansas State University, Manhattan, Kansas

Introduction

Within the first three weeks of life, piglets reared in the United States commercial pork industry are exposed to a variety of stressors, including painful processing procedures, social isolation from non-littermates, and a barren environment. Negative early life experiences in this pre-weaning phase, a period of significant brain development in pigs,[1] may be analogous to humans experiencing adverse childhood events (ACEs), which predispose children to developing mental health disorders later in life.[2] The unique parallels between growing pigs and children experiencing ACEs present an opportunity to explore the impact of early life stress on behavior, immune function, and brain development in swine. Reducing early life stress through pre-weaning socialization was hypothesized to result in decreased damaging agonistic behavior, increased immunocompetence, and improved brain development and function later in life.

Methods

Seven sows negative for swine influenza virus (SIV) antibodies—to control for a subsequent immunologic challenge—were randomized into one of two treatment groups: socialized (SOC; n = 4) or unsocialized (CON; n = 3). 10 piglets from each litter were enrolled. Piglets remained confined to gestation pens until day 8 of life, when the barrier between two pens was removed and piglets in the SOC treatment were allowed to intermingle. Pre-weaning behavior was recorded the day prior, the day of, and the day following the socialization treatment using high-definition video cameras mounted on tripods. Videos will be analyzed using scan sampling methodology to evaluate piglet social behavior. Piglets in the CON treatment remained confined to their gestation pen until weaning. At 21 days old, all pigs were weaned and kept within their social treatment groups until a social contest was conducted at 8 weeks of age. Pigs were matched by sex and weight as closely as possible to assess decision-making behaviors when introduced to unfamiliar pigs of the same socialization treatment. Specific behaviors of interest included latency-to-contact, damaging aggression, and a classic head-tilt-retreat from one individual, signaling "loss" of a contest between two pigs. At 11 and 14 weeks, the piglets were challenged with a bivalent SIV vaccination intramuscularly, and antibody responses were assessed periodically through serum hemagglutinin inhibition assays. At 17 weeks, a subset of pigs (n = 24) was humanely euthanized to assess brain structures known in other species to be responsible for decision-making and social behaviors: hippocampus, thalamus, hypothalamus, pre-frontal cortex, and lateral habenula. 22 of the formalin-fixed and sectioned brain specimens were prepared with standard Hematoxylin and Eosin stain and analyzed for their percentage of neuronal cell populations by region of interest.

Results and discussion

Preliminary morphologic data for brain specimens resulted in no gross weight difference between SOC versus CON pigs ($P = .47$). However, pigs deemed "Winners" at the time of the 8-week social contest had significantly heavier brains than "Loser" pigs ($P = .01$). Analyses are ongoing and will be useful to evaluate the downstream effects of early life stress on behavior, immune function, and neurodevelopment in swine. The ultimate goal of this study is to enhance swine welfare in a production setting, by encouraging practices which promote neural plasticity in swine to improve decision-making and social skills.

References

1. Conrad, M.S., R.N. Dilger, and R.W. Johnson. Brain growth of the domestic pig (*Sus scrofa*) from 2 to 24 weeks of age: A longitudinal MRI study. Dev. Neurosci. 34:291-298. 2012.

2. Herzog, J. I., & Schmahl, C. Adverse Childhood Experiences and the Consequences on Neurobiological, Psychosocial, and Somatic Conditions Across the Lifespan. *Frontiers in psychiatry, 9,* 420. 2018.

https://doi.org/10.54846/am2024/85

Evaluation of clinical and immunization sequences of sapovirus and rotavirus in related pig flows using next generation sequencing

Erin Russell[1], MS; Amanda Reever[2], MS, DVM; Brandi Burton[2], DVM
[1]Lincoln Memorial University College of Veterinary Medicine, Harrogate, Tennessee
[2]Suidae Health and Production, Algona, Iowa

Introduction

Sapovirus (SV) and rotavirus (RV) are prevalent causes of diarrhea that impact weight gain in swine. Cross-protection is limited within serogroups and immunization methods are often customized to individual flows. Frequently sequenced proteins are VP7 and VP4 for RV and VP1 for SV. Pigs immunized with homologous VP4s and VP7s have been shown to be protected from shedding and clinical disease in the face of a challenge. Further research is needed on immunization strategies and viral diversity. The objectives of this study were to investigate RV and SV diversity in related flows and to provide an example of next generation sequencing (NGS) analysis to guide immunization strategies, pig flows, sanitation, and more.

Methods

A multiplier flow and two commercial flows (sourced by the multiplier) were enrolled. These farms vaccinate pre-farrow with one type of SV each and utilize a master seed feedback method for RV. Pooled diarrhea samples were collected from each sow farm and one recently weaned nursery in June and July. Feces and feedback were tested by polymerase chain reaction (PCR) and NGS for VP1, VP4, and VP7. Analysis was completed on Disease Bioportal.

Results and discussion

SV VP1 analysis revealed 5 distinct clades (2 or more sequences, 97%+ homology), as shown in the lower box of Figure 1. Two clades only included sequences from one flow, while all flows were represented in the remaining three clades. The majority of sequences were from nursery samples. In this system, pigs are single-sourced but can be placed in any available nursery, meaning that SVs could be spread vertically or through contaminated trailers and nurseries. Since anecdotal evidence suggests that pre-farrow vaccination for SV may benefit post-weaning piglets, immunization against multiple SV serotypes should be considered. Fecal and feedback sequences were included for RV analysis. The upper box of Figure 1 shows VP7 sequences for RVC. The blue clade contains two clinical and one feedback sequence. However, there are three clades (yellow, green, and blue) where sequences found on the sow farm were not present in the feedback. These viruses should be considered for future immunization pre-farrow. Numerous RVAs were found in nursery samples, but there was only a single RVA sequence found on one of the sow farms. The effect of pre-farrow vaccination on post-weaning diarrhea is unknown, therefore future research and better nursery and transport sanitation is likely needed. Limitations of this study include the duration of testing and difficulty interpreting partial sequences. Regardless, this shows that there is both a wide diversity of RVs and SVs, as well as an important overlap within related flows. Using readily available programs and increasingly affordable and efficient sequencing techniques, practitioners can make immediate advances to combat these viruses. Sharing of the additional results of this study, analysis of the number of reads of each sequence, and future sampling of these flows would be beneficial in drawing further conclusions on immunization best practices.

Figure 1: The upper box shows the relationship between VP7 sequences for Rotavirus C, and which farms it was present at. The lower box shows the relationship between Sapovirus sequences, and at which farms they were present at.

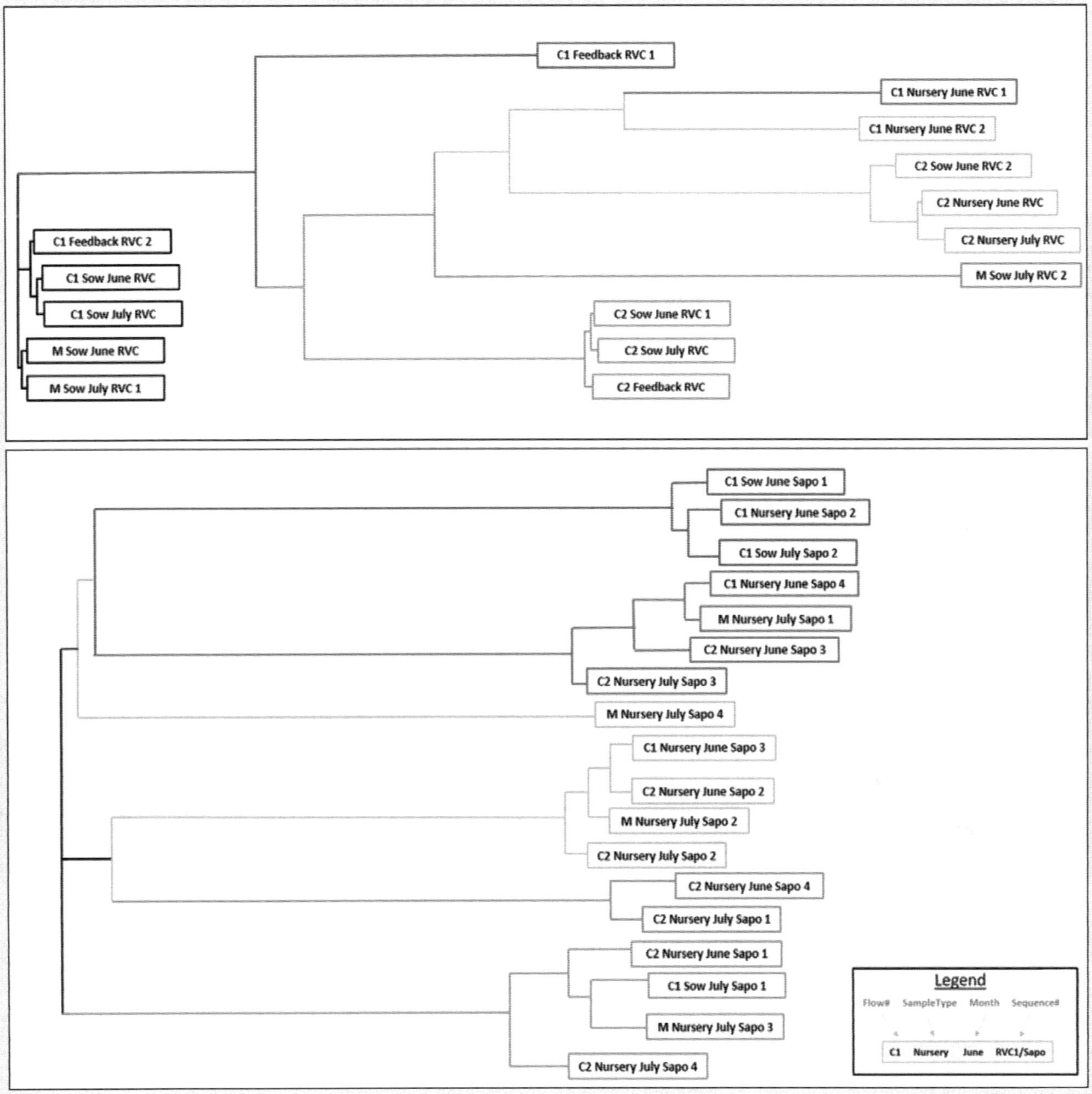

 https://doi.org/10.54846/am2024/86

Exploring the use of ultrasonography at palpable landmarks on sows to measure and track body composition

L. A. Harper[1], BS; S. Thompson[2], BS; M. Benjamin[3], DVM, MS, DACAW
[1]College of Veterinary Medicine, Cornell University, Ithaca, New York; [2]College of Veterinary Medicine, Michigan State University, East Lansing, Michigan; [3]Extension Pork, College of Veterinary Medicine, Michigan State University, East Lansing, Michigan

Introduction

Maintaining sows in healthy body condition throughout their lives can improve reproductive performance, prolong their productive lifetime, maximize farm productivity/profitability, and improve animal welfare. Despite the importance of maintaining sow body condition, measurements such as visual body condition scoring, and sow body calipers (which require consistent and appropriate placement) are the main stays of monitoring sow body condition on farm. Most of the literature on quantitative methods to measure body condition includes ultrasonography at the 10th and last rib landmarks. However, this method is not commonly used in commercial sow systems for body condition monitoring due to cost, specialized training and access to rib locations while sows are standing in stalls.

The problem

Regular monitoring of sow body condition is imperative to minimize sow loss/replacement and maintain sow welfare, yet current monitoring methods are subjective, require specialized training, and are labor intensive.

Our objectives

The goal of this work was to improve ground truth indices on muscle and fat deposition in the sow at palpable landmarks to be used in breeding decisions, further research, and to gain a better understanding of changes in sow body composition. Specifically (1) to identify novel sow ultrasound landmarks and determine intra- and inter-observer repeatability, and (2) use repeatable measurements to observe and understand changes in fat and muscle depth through weekly scanning over a single lactation.

Methods

For objective 1, 18 cull sows, were scanned and measured by two observers daily for 3 consecutive days. Measurements included visual body condition score, height, weight, body caliper measurements at the shoulder, last rib, and hip, and flank-to-flank weight estimates. Ultrasonographic measurements were taken at the last rib, hip, neck, shoulder, tail, flank, and 5 ribs cranial to the last rib (10th rib) (Figure 1). The second objective used repeatable measurements (from objective 1) and the same two observers. Both observers took weekly measurements from pre-farrowing through weaning on a convenience sample of 11 primiparous females and 10 parity 2+ sows. ANOVA and Intraclass Correlation Coefficients (ICC) were used to evaluate intra- and inter- observer repeatability for objective 1. In objective 2, correlation coefficients and correlation test p-values were determined for each pair of measurements and stratified by time and parity.

Results

Several measurements including 10th rib total back fat, last rib middle back fat, and surface to first thoracic vertebrae depth had excellent reliability between and within both observers repeated measurements. For the second objective, hip caliper, 10th rib outer and middle back fat, and last rib total, outer, and middle back fat were significantly associated with days in lactation. Measures using the sow caliper position at the last rib and over the ilium, flank-to-flank, height, weight, and 10th rib L dorsi area were significantly associated with parity groups (P1 vs P2 - 5). Ultrasound measurements of last rib total back fat, an industry standard, and a novel ultrasound location at the hip (skin to the wing of the ilium) were strongly correlated.

Conclusions

Ultrasonography can be used as an alternative to monitoring changes in sow body condition and composition at multiple palpable landmarks including the scapula, neck, and ilium. The ilium location allows for greater access when sows are housed in either gestation or farrowing stalls. This study, using a small sample size, was able to identify palpable landmarks for ultrasound placement and measurement, demonstrate repeatability within and between observers, show correlations in body composition changes to industry standards, and began to illustrate how and from where the sow changes body composition during lactation.

Figure 1: Approximate locations of palpable ultrasound locations used in objective 1.

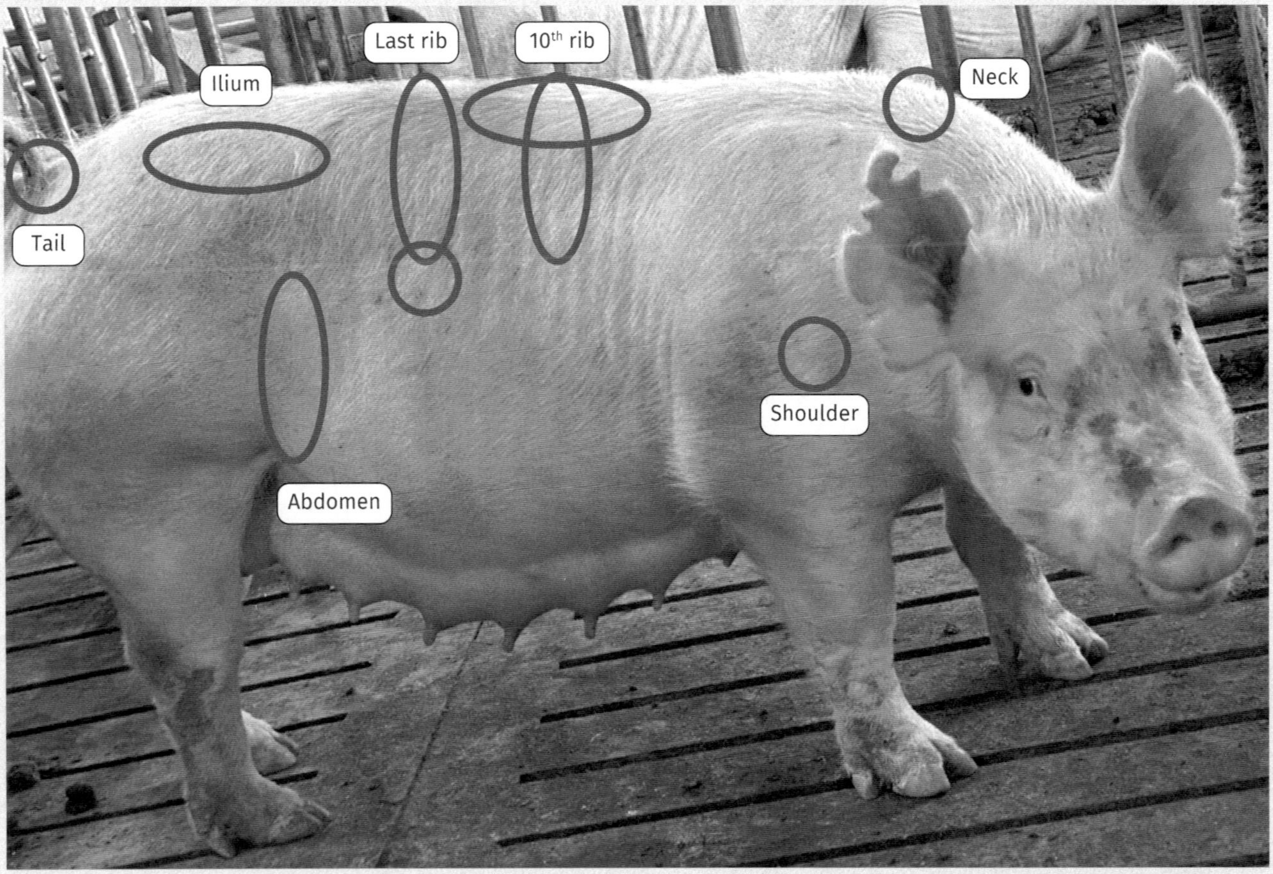

https://doi.org/10.54846/am2024/87

Investigation of oral fluid IgA antibody levels to *Lawsonia intracellularis* and *Salmonella enterica* following vaccination under field conditions

C. Ezell[1], BS; M. Schleper[2], DVM; C. Stewart[2], DVM; A. Sponheim[3], DVM; L. Gimenez-Lirola[4], MS, PhD; R. L. Magtoto[4], PhD; J. C. Mora-Díaz[4], DVM, PhD; D. Zhang[4], BS; F. Leite[3], DVM, MS, PhD
[1]Louisiana State University, Baton Rouge, Louisiana; [2]Christensen Farms, Sleepy Eye, Minnesota; [3]Boehringer Ingelheim Animal Health USA Inc., Duluth, Georgia; [4]Iowa State University, College of Veterinary Medicine, Veterinary Diagnostic Laboratory, Ames, Iowa

Introduction

Lawsonia intracellularis causes porcine proliferative enteropathy, an infectious enteric disease which can lead to decreased weight gain and increased mortality in growing pigs. *Salmonella enterica* is another significant enteric pathogen of swine which also has the potential for causing food-borne illnesses in humans.Attenuated-live vaccinations against these pathogens (Enterisol® Ileitis and Enterisol® Salmonella T/C) are available and can provide immunity to pigs when administered properly. A previous pilot study evaluated the use of Svanova *L intracellularis* competitive ELISA adapted into an indirect format for assessment of anti-*L intracellularis* IgA in oral fluids and serum to measure mucosal immunity following Enterisol® Ileitis vaccination via oral drench. The objective of this project was to investigate if IgA against *L intracellularis* and *Salmonella* could be detected in oral fluids and oral swabs following vaccination with Enterisol® Ileitis and Enterisol® Salmonella T/C through the drinking water as a method to confirm the vaccination status of a herd.

Materials and methods

A 2,500 head wean-to-finish barn with separate water lines in northeast Iowa was chosen. The north barn was randomly selected as treatment Group A, and the south barn as Group B. On Day 0 of the study, three pens were randomly selected in each barn, and ten 45-day-old pigs were randomly identified and tagged within each pen (n = 30 pigs/group). Blood, fecal swabs, and oral swabs were collected from tagged pigs and oral fluids were collected from the six study pens. Following sample collection, Enterisol® Ileitis and Enterisol® Salmonella T/C vaccines were thawed and administered per label instruction via the drinking water to every pen. On Day 21, samples were collected as Day 0, and the oral vaccines were administered through the drinking water exclusively to Group A to elicit a recall immune response. Blood, fecal swabs, and oral swabs were collected from tagged pigs a third time on Day 42 of the study. Oral fluid samples were collected on a weekly basis (Days 0, 7, 14, 21, 28, 35, 42) throughout the study. Oral fluid samples and oral swabs were tested using commercial *L intracellularis* (Svanova) and *Salmonella* ELISAs (IDEXX) both of which were adapted for the assessment of anti-*L intracellularis* and *Salmonella* IgA levels, respectively. Serum samples were tested for *Salmonella* and *L intracellularis* levels by ELISA and fecal swabs were tested by *L intracellularis* PCR to measure field exposure to either pathogen.

Results and discussion

On average, Group A *L intracellularis* IgA S/P value increased in oral fluid (0.73 to 1.19) and decreased in oral swabs (0.18 to 0.01), while *Salmonella* IgA S/P value increased in oral fluid (0.01 to 0.17) and oral swabs (0.04 to 0.14), from the first to the last collection timepoints, respectively. The increase in the average *L intracellularis* IgA S/P value in oral fluid over time was lower than observed values in the previous pilot study. With further validation, *L intracellularis* and *Salmonella* IgA levels in oral fluid samples could allow swine veterinarians to determine immune status of a herd, including the evaluation of vaccination status.

https://doi.org/10.54846/am2024/88

Assessing the impact of a negative air ionization system on particulate matter and gaseous pollutants in swine farm environment

Y.-F. Shen, BS, BA; M. R. Campler, MSc, PhD; B. A. Fonseca, DVM, MSc, PhD; A. G. Arruda, DVM, MSc, PhD
Department of Veterinary Preventative Medicine, The Ohio State University, Columbus, Ohio

Introduction

Poor indoor air quality, such as high levels of airborne particles and residual gases such as ammonia (NH_3) and hydrogen sulfide (H2S), could be a source of respiratory challenges in swine housed in commercial farms.[1,2] Negative air ionization systems (NAIS) have been shown to be effective in reducing airborne particles in human indoor environments,[3] but research is limited for swine-specific farm conditions. The objective of this study is to evaluate NAIS as a technology to reduce airborne particulate matter and gaseous pollutants at two different height levels (2 ft and 5 ft: pig and human respiration levels, respectively) under field conditions.

Methods

Negative air ionization systems (Freshlight LLC) were installed in six farrowing rooms in a 5,000-sow herd. Two treatment groups were assigned: 1) NAIS installed with lamps ON (L-ON), 2) NAIS installed with lamps OFF (L-OFF). Treatment was assigned at the room/batch level for three weeks, following the site's AIAO system; all rooms were cleaned and disinfected between batches. Particulate matter (PM, 2.5 and 10) and gaseous pollutant's (ammonia detector (NH_3), multigas detector (H_2S, CO) measurements were taken in each room at 2 and 5ft heights twice a day (morning and evening), two fixed times per week, using commercially available and scientifically validated monitoring systems. Daily change in particle counts (morning to afternoon) and gas concentrations were analyzed using STATA 17 (College Station, Texas).

Results

During the time of analysis, three replicates for each treatment were completed for a total of 92 time points. At the pig level (2 ft) in L-ON treatment rooms, there was a mean ± SD daily reduction of 126.4 ± 164.1µm for PM 2.5 and 181.2 ± 222.9µm for PM 10, compared to reduction of 61.1 ± 68.7µm for PM 2.5 and 172.1 ± 102.8µm for PM 10 in L-OFF rooms. At the human level (5 ft) there was a reduction of 70.2 ± 156.4µm for PM 2.5 and 149.1 ± 232.1µm for PM10 in the L-ON group, compared to a reduction of 67.1 ± 80.4µm for PM 2.5 and 192.7 ± 130.0µm for PM 10 in the L-OFF group. There was a reduction in H_2S concentrations in the L-ON by 0.3 ± 0.9 ppm (2 ft) and 0.1 ± 0.5 ppm (5 ft). H_2S was not detected in the L-OFF group. CO was not detected throughout the study in any of the rooms. NH_3 concentrations by treatment group are shown in Figure 1.

Discussion

Results showed numerically greater reductions for PM in the L-ON group compared to the L-OFF group except in PM 10 at 5 ft. Greater reductions were observed at the 2 ft level compared to the 5 ft level in the L-ON group. The study found that L-ON successfully kept NH_3 levels below 3 ppm at different heights during the entire study. This is important because recent research indicates that maintaining NH3 levels below 10 ppm can reduce health risks for both pigs and humans. Therefore, the NAIS system shows potential as a practical solution for veterinarians to help producers meet this requirement in real-world field conditions. In conclusion, NAIS appears effective in decreasing airborne particles and residual gases for both swine and production workers. Further analysis will investigate the association between NAIS and swine health (eg, treatment records) and production (e.g., pre-weaning mortality and pigs weaned/sow).

References

1. Ni, Ji-Qin, Albert J. Heber, and Teng-Teeh Lim. "Ammonia and hydrogen sulfide in swine production." *Air quality and livestock farming* 29.47 (2018): 9781315738338-3.

2. Dai, Chunhao, et al. "Concentrations and emissions of particulate matter and ammonia from extensive livestock farm in South China." *Environmental Science and Pollution Research* 26 (2019): 1871-1879.

3. Jiang, Shu-Ye, Ali Ma, and Srinivasan Ramachandran. "Negative air ions and their effects on human health and air quality improvement." *International journal of molecular sciences* 19.10 (2018): 2966.

Figure 1: Ammonia (NH3) concentrations (ppm) by treatment and height measured (2 ft, left graph and 5 ft, right graph). Ammonia measurements taken during morning and afternoon were averaged for each data point (day). Two measurements were taken per week.

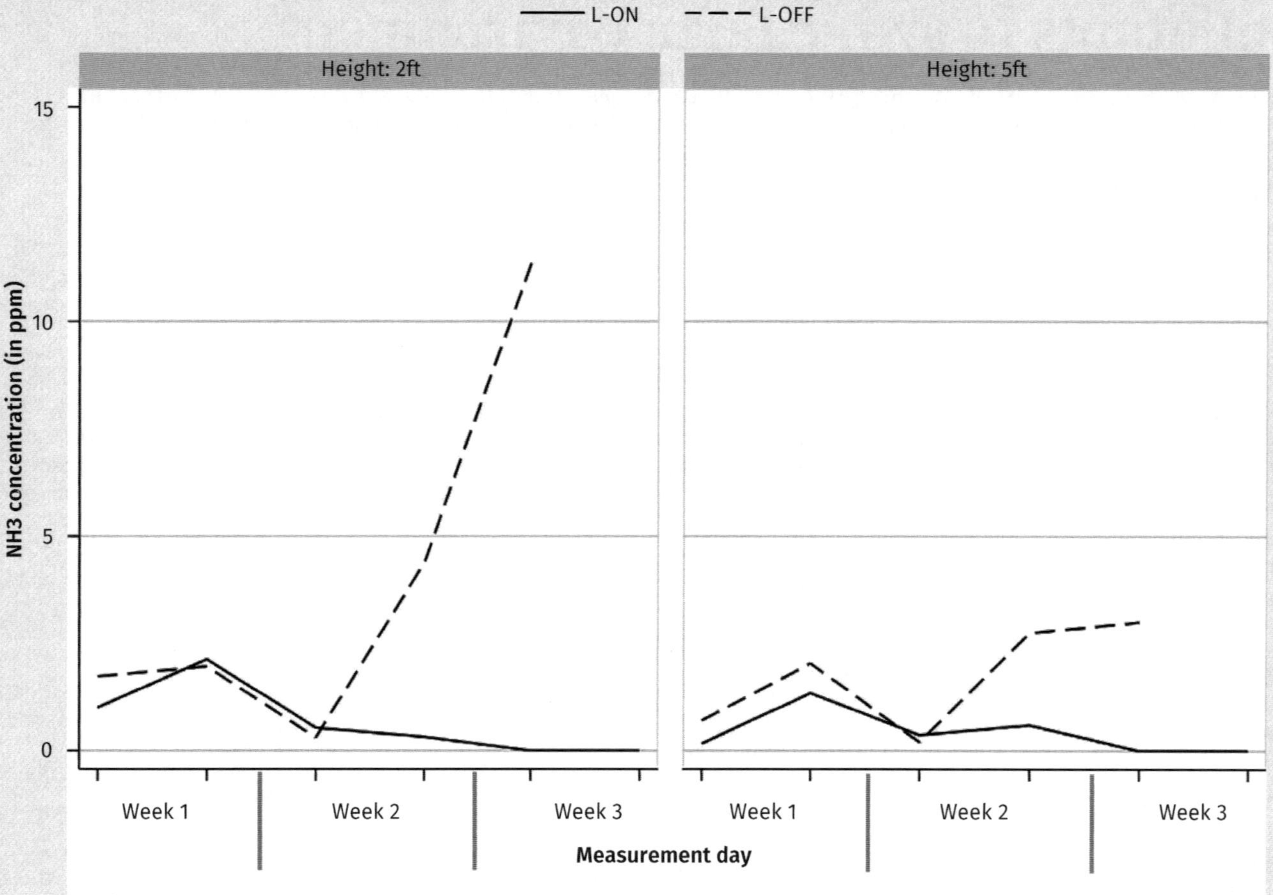

https://doi.org/10.54846/am2024/89

Descriptive study of wild rodent populations as reservoirs for swine dysentery

Kristen Cleaver[1]; Pete Lasley[2], DVM; Robert Evelsizer[3], DVM, Christopher Puls[3], PhD; Eric Burrough[4], DVM, PhD
[1]College of Veterinary Medicine, Iowa State University, Ames, Iowa; [2]Smithfield Hog Production, Princeton, Missouri; [3]Elanco Animal Health, Greenfield, Indiana; [4]Veterinary Diagnostic Laboratory, Iowa State University, Ames, Iowa

Introduction

Swine dysentery was discovered in the 1920s and has reemerged in the US during the last decade. The main pathological agents are strongly beta-hemolytic spirochetes classified as *Brachyspira hyodysenteriae* (B hyo) and *Brachyspira hampsonii* (B hamp). Both bacteria cause mucohemorrhagic diarrhea and malabsorption leading to losses in body condition primarily observed in growing and finishing pigs. *Brachyspira pilosicoli* (B pilo) is a related, weakly beta-hemolytic, zoonotic pathogen causing porcine colonic spirochetosis. Rodents are known reservoirs of infection, and mice have been shown to shed *Brachyspira* sp. in feces for at least 200 days.[1] Previous research in 1982 found *B hyo.* in 4 of 298 wild rodents caught on swine farms.[2] The objective of this descriptive study was to evaluate the presence of rodents carrying pathogenic *Brachyspira* sp at sow farms and finishing locations and to analyze the relationship of new isolates to those previously collected at the locations.

Methods

Dysentery surveillance testing was conducted by fecal swabs at sow farms and finishing locations within the region. Rodent sampling locations were identified from sites that tested positive for at least one agent of swine dysentery. The target was to collect at least 15 rodents from each sample location in 8 weeks. Multi-catch live rodent traps were used. Rodents were euthanized by cervical dislocation, and intestinal tracts were submitted to the Iowa State Veterinary Diagnostic Laboratory for *Brachyspira* sp culture. Environmental samples were collected from recycle water inflow and barn effluent to be sent in for culture. Next-Generation Sequencing (NGS) was performed on new dysentery isolates recovered from mice, and these were compared to previous isolates from pigs in the same locations.

Results

A total of 59 mice were collected from 6 locations. There were 26 mice collected from breeding herds and 33 mice collected from finishing locations. B hyo was recovered from 2 mice. Both were from a breeding herd where active infection was detected in a gilt at the start of sample collection. Both of these mouse isolates showed a resistance to tylosin, but the isolate from the gilt did not. B pilo was isolated from 1 mouse in a different breeding herd and showed a resistance to tiamulin. There were 4 mice and 1 environmental sample of barn effluent that cultured positive for *Brachyspira murdochii,* a related, non-pathogenic commensal. All other mice and environmental samples were negative for agents of swine dysentery. NGS results on the tylosin-resistant B hyo mouse isolates showed that they were different from previous swine isolates in the same location.

Discussion

Resistance to tylosin in agents of swine dysentery is common, and tiamulin resistance is being increasingly reported in the US.[1] Carbadox has been demonstrated to increase lateral gene transfer in *Brachyspira* sp and could serve to increase spread of resistance genes.[3] Endemically infected farms using these antibiotics as swine dysentery control measures could thereby be increasing the frequency of resistance. This study showed that mice can be a reservoir for antibiotic-resistant *Brachyspira* sp that have not been previously detected at a location and that mice can be a source of introduction of new strains. This emphasizes the need for dysentery clean-up programs to include intensive rodent control to help prevent the occurrence and persistence of pathogenic, antibiotic-resistant *Brachyspira* sp.

References

1. Burrough E.R. Swine Dysentery: Etiopathogenesis and Diagnosis of a Reemerging Disease. *Veterinary Pathology,* 2017; 54(1), 22-31.

2. Joens L.A. & Kinyon J.M. Isolation of *Treponema hyodysenteriae* from Wild Rodents. *Journal of Clinical Microbiology,* 1982; 15(6). **https://journals.asm.org/journal/jcm**

3. Stanton T.B., Humphrey S.B., Sharma V.K., & Zuerner R.L. Collateral effects of antibiotics: Carbadox and metronidazole induce VSH-I and facilitate gene transfer among *Brachyspira hyodysenteriae* strains. *Applied and Environmental Microbiology,* 2008; 74(10), 2950-2956. **https://doi.org/10.1128/AEM.00189-08**

https://doi.org/10.54846/am2024/90

A comparative analysis of rectal thermometers and thermal imaging cameras for early detection of disease in pigs

Samantha Nixon[1]; Maite Dias[2] DVM; Matheus O. Costa[2], DVM, PhD, DABVP
[1]University of Saskatchewan, Western College of Veterinary Medicine, SK, Canada; [2]University of Saskatchewan, Western College of Veterinary Medicine, Department of Large Animal Clinical Science, Saskatchewan, Canada

Introduction

Elevated body temperature (fever) is one of the first signs that the body is fighting infection. Identifying the onset of fever is crucial for the detection of septicaemic diseases, early treatment intervention, and for preventing the spread of a pathogen within and between pigs. With timely detection and treatment, the overall efficiency of labor in swine management is improved. Sick animals can be identified and treated more quickly, reducing the workload associated with managing widespread outbreaks and reducing the need for mass use of antibiotics. Avoiding unnecessary and indiscriminate use of antibiotics help mitigate antimicrobial resistance. The use of rectal thermometers is a traditional method for measuring body temperature in pigs. However, it is invasive, labour intensive, and time consuming. Thermal imaging cameras measure the radiated temperature and offer a non-invasive and rapid alternative for monitoring body temperature in swine barns. Cameras can be set up for continuous monitoring, providing real-time data on body temperature variations for individual or groups of animals. This continuous surveillance enhances the chances of early detection of fever or abnormal temperature patterns, allowing for proactive intervention and treatment.

Methods

Five-week-old cross-bred pigs (n = 12) from a high health herd were housed under controlled conditions. All animals were clinically monitored daily over 19 days. An electronic rectal thermometer and a professional handheld thermal imaging camera (SeekThermal, Shotpro, Wisconsin, USA) were used to measure body temperature once daily, in the morning. Rectal thermometer was only used for rectal temperature, whereas radiated temperature was assessed on the right ear, left ear, and whole dorsal body. Data was analyzed using repeated measures ANOVA followed by Tukey's post-hoc.

Results

The average rectal temperature across all animals during the entire study period was 39.25°C (± 0.15°C). The average right ear (38.30°C ± 1.3°C), left ear (37.34°C ± 1.0°C), and whole dorsal body (38.52°C (± 0.5°C) temperatures were lower than the average rectal temperature, and were statistically different among themselves ($P < .001$). Unsurprisingly, rectal temperatures also showed the lowest variation across all the explored methods. A summary of the data is presented in Figure 1.

Discussion

Rectal temperature measurement provides a reliable and consistent method for core temperature assessment. Thermal imaging cameras are becoming increasingly popular as a diagnostic tool for evaluating the health of pigs. They provide the opportunity to reduce stress and curb the spread of disease by functioning as an automated surveillance system.[1] Their quick assessment of skin temperatures allows for the estimation of body temperature without the need for restraint or direct contact. Since fever is one of the first indicators of septicaemic disease, a non-contact method could improve time and labour efficiency in detecting and treating affected animals. However, our data showed that thermal imaging is more susceptible to variation and may artificially provide underestimated data. One of the limitations of thermal imaging is that it measures surface temperature rather than core temperature. While changes in surface temperature can indicate potential health issues, they may not reflect variations in core body temperature. Since skin temperature is constantly influenced by its surroundings, there is a greater potential for variation compared to core temperatures.[2] For instance, variations in barn temperature, drafts, or direct exposure to sunlight may influence surface temperature and impact the readings obtained through thermal imaging cameras. Future trials using animals with fever may shed a light on the usefulness of thermal imaging during disease outbreaks. These trials may help establish correlations between changes in surface temperature captured by thermal imaging and actual increases in core body temperature. Combining thermal imaging with other diagnostic tools, such as rectal thermometry or clinical assessments, may provide a more accurate and nuanced understanding of animal health, in lieu of increased labour.

References

1. Marina Stukelj, Melita Hajdinjak, Igor Pusnik. Stress-free measurement of body temperature of pigs by using thermal imaging – Useful fact or wishful thinking. *Computers and Electronics in Agriculture.* February 2022. Accessed November 13, 2023. **https://doi.org/10.1016/j.compag.2021.106656**

2. Simone M. Schmid, Wolfgang Buscher, Julia Steinhoff-Wagner. Suitability of Different Thermometers for Measuring Body Core and Skin Temperatures in Suckling Piglets. *National Library of Medicine.* April 2021. Accessed November 13, 2023. **https://doi.org/10.3390%2Fani11041004**

Figure 1: Average daily temperature measured using a rectal thermometer (rectal, n = 12/day) or thermal imaging camera aimed at different body sites: whole dorsal body (body, n = 12/day), right ear (R-ear, n = 12/day) and left ear (L-ear, n = 12/day).

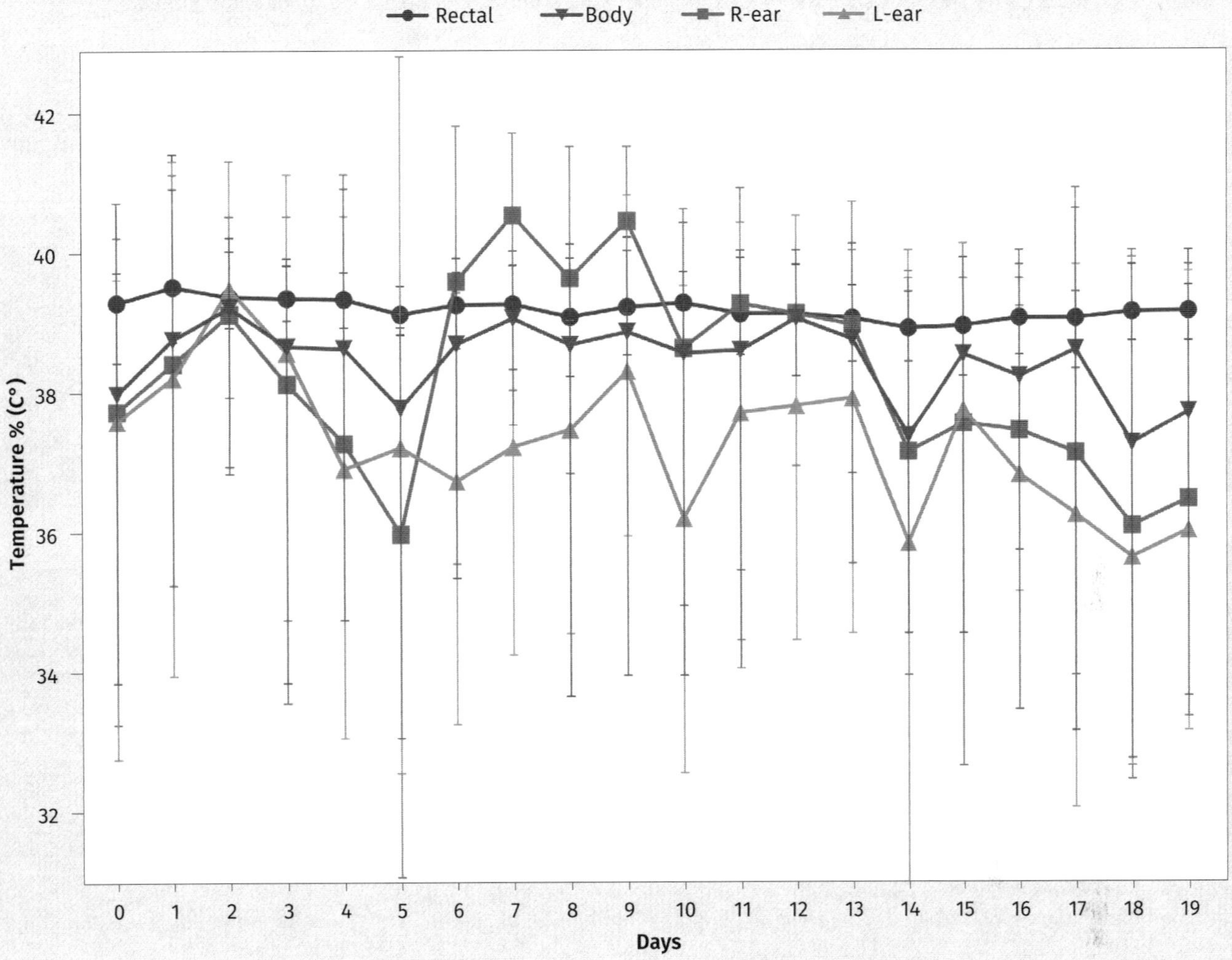

https://doi.org/10.54846/am2024/91

The effect of water enriched with a Nutrient Dense Liquid on weaned pigs to mitigate the effects of post-weaning *E coli*

T. Clark[1]; E. Toohill[2], DVM; D. Shaffer[4]; M. Almeida[3], DVM, PhD; J. E. Guzman[4]; D. McKilligan[4]; K. Coulson[4]; V. Utrera[4], DVM
[1]College of Veterinary Medicine, Iowa State University, Ames, Iowa; [2]Carthage Veterinary Service LTD, Carthage, Illinois; [3]Veterinary Diagnostic and Production Animal Medicine, Iowa State University, Ames, Iowa; [4]TechMix, LLC, Stewart, Minnesota

Introduction

Liquitein® APF is a nutrient dense liquid (NDL) containing energy, electrolytes, and functional proteins. The objective of this study was to determine the effect of Liquitein APF on *E coli* prevalence, fecal scores, average daily gain, and small intestinal morphology of weaned pigs.

Materials and methods

The study was a randomized controlled trial design. Sixty weanage pigs from parity 1 and 2 sows were randomly allotted into two treatment groups: **1)** pigs receiving water enriched with Liquitein APF at 1 oz/gal and **2)** pigs receiving fresh water with no additives. The trial was conducted over seven days. Individual fecal scores were captured daily. Individual pig weights were recorded on days 0, 3, and 7. On days 3 and 7, thirty pigs were weighed and euthanized to collect formalin-fixed samples of proximal jejunum, distal jejunum, and ileum for villous atrophy scoring[1] and submitted to Iowa State University Veterinary Diagnostic Laboratory. Fecal swabs were collected from each pig before euthanasia for Rotavirus PCR and for *E coli* bacterial isolation and genotyping. All statistical analyses were performed using a commercial statistical software package (JMP Pro 17; SAS Institute, Cary, North Carolina). Performance parameters (average daily gain, ADG and body weight, BW), fecal scores, villus height, crypt depth, and villus height:crypt depth ratios were compared using a one-way ANOVA with Tukey-Kramer adjustment for normally distributed data and with nonparametric Steel-Dwass test if normality was not met. Results were considered significant at $P < 0.05$ and P between $> .05$ and ≤ 0.10 was considered a tendency.

Results

Pigs in Liquitein APF group had lower fecal scores on days 3, 4, 5, and 7 after weaning (no significant difference); the lowest fecal score was observed 3 days after weaning ($P = .07$). During the study, the number of watery diarrhea cases detected in pigs in the control group was greater than those observed in the group of pigs receiving Liquitein APF (Table 1). An improvement in ADG was evident in pigs receiving Liquitein APF for 7 days ($P < .05$). These pigs had greater villus height in the distal jejunum at three days ($P < .05$) and at seven days post -weaning, respectively, and greater distal jejunum villous height:crypt ratio ($P < .05$). There was a strong positive correlation ($r = 1.00$) in pigs receiving Liquitein APF between distal jejunum crypt depth and average daily gain ($P < .05$). Additionally, there was a strong negative correlation ($r = 1.00$; $P < .05$) between fecal scores and average daily gain in treated pigs.

Table 1: Number of watery diarrhea (score #3) cases in pigs receiving water enriched with Liquitein APF vs control pigs

Treatment	Day 3	Day 4	Day 5	Day 6	Day 7	Total
Liquitein APF	4	2	3	4	3	16
Control	10	7	6	4	6	33

Discussion

Weaning is usually associated with low feed intake resulting in nutrient disruption. Provision of a nutrient dense liquid at weaning has been shown to improve growth performance and reduce morbidity and mortality in the nursery.[2] The results of this study appear to indicate a positive effect of an Liquitein APF on post-weaning *E coli* incidence, ADG, and gut health. This study only represents 60 pigs and further research would be needed to evaluate a larger sample size.

References

1. Madson et al., 2014. Pathogenesis of porcine epidemic diarrhea virus isolate (US/Iowa/18984/2019) in 3-week-old weaned pigs. *Veterinary Microbiology*, 174(1-2):60-68.

2. Schmitt, R. L., M. Ellis, O. F. Mendoza, C. M. Shull, D. McKilligan, and N. Upah. 2018. The effect of administration of a nutrient dense liquid at weaning on growth performance and morbidity and mortality of pigs during the nursery period under commercial conditions *J. Anim. Sci.* 85(Suppl. 2):115.

https://doi.org/10.54846/am2024/92

Gross lesions in lame versus non-lame finisher pigs

David Buckwalter; Meghann K. Pierdon, BA, VMD; Julie B. Engiles, BA, VMD, DACVP; Nathan Fanzone, VMD
University of Pennsylvania, Kennett Square, Pennsylvania

Statement of the problem

Lameness represents a widespread issue affecting growing pig welfare and has economic implications.[1] Lameness can affect viability and growth, ultimately impeding efficient production and inducing costs to producers.[2] The causes of lameness in growing swine, however, are poorly elucidated and often difficult to diagnose in the field. The objective of this study was to use gross pathological examination to compare lesions in lame and non-lame growing pigs to better understand the etiology of lameness in growing pigs.

Materials and methods

Two production companies enrolled 5 farms each for a total of 10 farms. On each farm 2 pigs were chosen, a single lame pig (L) and a single non-lame control pig (C). Pigs were identified as lame if two observers agreed that the animal was slow to rise, limping, reluctant to walk, or reluctant to place weight on one or more limb. Pigs were euthanized by chemical euthanasia and transported to a diagnostic lab for complete postmortem evaluation. Visceral examination was completed on all pigs, all four legs were excised, and each joint examined grossly. Joints were scored for the presence or absence of synovial hypertrophy, hyperemia, or effusion as well as for lesions consistent with osteochondrosis (OC) and physeal bone lesions. One joint in each L pig that contained synovial lesions was swabbed and a swab was taken from the same location on the C pig from that farm. STATA IC v15.1 was used for statistical analysis. The total count of locations with lesions was calculated for each pig including all assessed locations. Data was checked for normality and a t-test was used to calculate the difference in means for normal data. A Mann Whitney U test was used to calculate the difference in means for non-normal data. Odds ratios were calculated for the odds of OC lesions, visceral lesions, and having multiple types of lesions in the lame versus non-lame pigs using a Fisher's Exact test.

Results

Average farm size was 3,624 pigs and mean age was 14.6 weeks. Eight females and 12 males were selected. Eight sow flows were included with five being comingled. There were 16 times greater odds of having multiple lesion types in the L pigs compared to the C pigs (OR = 16 95% CI 1.3 to 239.5). The odds of having an OC lesion were no different between the L and C pigs (OR = 5.44 95% CI .594 to 55.6). There was no difference in the odds of having a visceral lesion in the L pigs versus the C pigs (OR = 1.71 95% CI .144 to 25.6). All 10 of the L pigs had at least one synovial lesion whereas only 30% of the C pigs had synovial lesions. No odds ratio could be calculated due to the zero value ($P < .01$). None of the C pigs had physeal bone lesions whereas 30% of the L pigs had such lesions so no odds ratio could be calculated ($P > .05$). There was a significant difference in the median number of locations where there were lesions in the L (12.5 ± 7) compared to the C pigs (0 ± 2) ($P < .001$) (Table 1). Lesions were evenly distributed between front and hind limbs with 16% of locations scored in the front limbs having a lesion and 14% of the locations in hind. Only one L pig joint (11 swabs) was found to be positive for *M hyosynoviae* by PCR and none of the C pigs tested positive (9 swabs).

Discussion

L pigs had more lesions, and were more likely to have synovial lesions, than C pigs though only one pig was found positive for *M hyosynoviae* so the cause of the lesions is unclear. Histopathologic investigations are ongoing. In contrast to other studies of pigs this age, the OC lesions found were mild and not more likely to be found in the L pigs, decreasing the likelihood of its involvement in the clinical lameness. Systemic disease was not more prominent in the L pigs indicating that the lameness is likely not secondary to systemic illness with common swine pathogens such as *Streptococcus suis*, especially given that none of the pigs had overt septic lesions (ie, fibrinosuppurative joint or bone lesions). Determining the cause of lameness in these animals remains challenging though bacterial pathogens that cause lesions to the synovium like *M hyosynoviae* may be more likely than other causes based on these findings.[2]

References

1. Zimmerman, J. J. (2019). *Diseases of swine*. Eleventh edition. Hoboken, New JERSEY: John Wiley & Sons, Inc.

2. Zimmerman, J., Karriker, L., Ramirez, A., Schwartz, K., & Stevenson, G. (Eds.). (2012). *Diseases of swine*. John Wiley & Sons, Incorporated.

Table 1: Types of lesions found on gross examination of all joints in the limbs of ten lame and ten non-lame pigs that averaged 14.6 weeks of age and came from 10 commercial finishing farms

	Synovial lesions	OC lesions	Physeal lesions	Multiple lesion types
Non-lame (n = 10)	3 (30%)	3 (30%)	0 (0%)	2 (20%)
Lame (n = 10)	10 (100%)	7 (70%)	3 (30%)	8 (80%)
Total (n = 20)	13 (65%)	10 (50%)	3 (30%)	10 (50%)

https://doi.org/10.54846/am2024/93

In vitro feasibility of bovine and canine blood products as a source of xenotransfusion in swine

V. M. Diaz[1]; DMW; Schaefer[3], DVM, MS, MT; J. S. Smith[2], DVM, MPS, PhD; P. Y. Mulon[2], DVM; L. Giori[3], DVM, PhD; C. E. Hampton[2], DVM, MS
[1]College of Veterinary Medicine, University of Tennessee, Knoxville, Tennessee; [2]Department of Large Animal Clinical Sciences, University of Tennessee, Knoxville, Tennessee; [3]Department of Biomedical and Diagnostic Sciences, University of Tennessee, Knoxville, Tennessee

Introduction

Sourcing porcine blood to perform transfusions in porcine recipients is difficult due to lack of readily available donors, potentially large amounts of blood needed, and anatomical characteristics[1] of swine. Furthermore, standardized pre-transfusion testing is not routinely performed in swine as it is for others.[2] Performing xenotransfusions with bovine whole blood[3] may overcome these difficulties in medium-to-large swine. Also, with companion pigs becoming more commonly treated by small animal practitioners, canine blood products may pose as a potential source. Therefore, both sources of xenotransfusion need investigation. The primary objective of this study was to determine the frequency of incompatible crossmatch (CM), major (CMMa), and minor (CMMi) reactions using bovine whole blood and whole blood from commercial pigs, and between DEA 1.1 negative canine whole blood, DEA 1.1 negative packed red blood cells (pRBCs DEA-), DEA 1.1 positive pRBCs (pRBCs DEA+) and porcine whole blood via standard saline agglutination (SSA). A secondary objective was to determine the agreement between the reference method (SSA) and a quick slide (QS) technique that carries the advantage of being performed quickly with minimal training.

Methods

After determining sample size via power analysis, blood was collected from 12 heifers and 8 commercial-cross pigs. Bovine blood from three subjects was crossmatched and, if compatible, pooled into 8 bags. Whole blood was collected from 3 canine blood donors (DEA 1.1, 5, 7 negative, and DEA 4 positive), and 3 bags of pRBCs DEA- and pRBCs DEA+ were purchased and crossmatched with whole blood from 7 companion pigs. Sixty-four and sixty-three pairs of CM were performed for bovine-to-porcine and canine-to-porcine samples, respectively. Incompatibility was defined as any macroscopic or microscopic agglutination or hemolysis, evaluated by a graded scale and reported in percentage. Mann Whitney U-test was used to elucidate the effect of porcine blood group on incompatibility with bovine blood. Kappa statistics tested the agreement between SSA and QS. Significance was set at $P < .05$.

Results

Agglutination was observed in 9.4% of CMMa and in 100% of CMMi via SSA, and in 9.4% and 98.4% via QS in bovine-to-porcine CMs. All canine-to-porcine CMMa on SSA and QS were incompatible and showed hemolysis severe enough to prevent evaluation of agglutination. CMMa incompatibility of bovine blood was greater with porcine blood type "0" ($P = .0107$) than with type "A" on CMMa, whereas porcine blood group had no effect on CMMi results. Accuracy of QS was limited to 87.5% for detection of agglutination in CMMa and 98.4% in CMMi in bovine-to-porcine samples. Agreement between SSA and QS methods was fair ($\kappa = 0.36$) for bovine-to-porcine CMMa but could not be calculated for CMMi due to lack of compatible matches. For canine-to-porcine samples, the effects of porcine blood group on incompatibility, accuracy of QS, and agreement between SSA and QS could not be calculated for CMMa due to absolute lack of compatible matches. For CMMi, the agreement between tests was poor ($\kappa = 0$).

Discussion

Based on these results, bovine blood holds promising potential as a source of xenotransfusion for pigs. Canine blood is absolutely contraindicated for *in vivo* xenotransfusion to swine based on CMMa incompatibility. The quick slide method cannot be used as a surrogate of SAA in pre-transfusion testing with both blood sources due to increased risk of false compatible results between donor and recipient. For any clinical scenario in which a blood product transfusion is indicated, bovine whole blood appears to be suitable based on *in vitro* pre-transfusion testing as a potential source for xenotransfusion. *In vivo* studies are needed to elucidate the clinical significance of incompatibility of minor crossmatch results in pigs by administering bovine whole blood and bovine pRBCs to swine recipients.

References

1. Clark SG, Coffer N. Normal hematology and hematologic disorders in potbellied pigs. The veterinary clinics of North America Exotic animal practice 2008;11:569-582, vii.

2. Kumar R. Blood transfusion in veterinary medicine. Hematol Transfus Int J. 2017;4(4):116-122.

3. Johnstone JE, MacLaren LA, Doucet J, McAlister VC. In vitro studies regarding the feasibility of bovine erythrocyte xenotransfusion. Xenotransplantation. 2004;11(1):11-17.

https://doi.org/10.54846/am2024/94

Comparison of sample collection techniques to target *Salmonella* on sow farms

E. Gibbs[1], BS; J. Davenport[2], DVM; F. Leite[3], DVM, MS, PhD; G. Cline[3], DVM
[1]University of Missouri College of Veterinary Medicine, Columbia, Missouri; [2]JBS Live Pork, Missouri;
[3]Boehringer Ingelheim Animal Health Inc., Duluth, Georgia

Introduction

Salmonellosis is a disease of pigs caused by a variety of *Salmonella* serotypes. *Salmonella* serovar *Typhimurium* causes enterocolitis with yellow, watery diarrhea and anorexia observed at weaning with high morbidity and moderate mortality.[1] The purpose of this study was to investigate and compare sample collection techniques to identify a *Salmonella* health challenge.

Methods

This study included a 4-week batch breed-to-wean (BTW) farm (S1), a weekly BTW farm (S2), and control (CNT). S1 and S2 herds exhibited clinical *Salmonella* infections in weaned pigs. Feces was collected from 30 P0, 30 P1 - 2, and 30 P3+ sows and pooled into groups of three and a pooled piglet fecal sample was obtained from each crate resulting in 10 total samples per parity for sows and piglets respectively. Piglet feces was collected in tubes filled with buffered peptone water, a medium for pre-enrichment of *Salmonella,* including resuscitation of injured bacteria.[2] One sow from each parity was bled, rectally swabbed and each swab was placed in peptone water. Each piglet in her litter was bled and rectally swabbed. Gestation pens at 30 - 60 days and 90 - 110 days were identified. Each pen was walked, and 10 pooled fecal samples were collected for each time point. At CNT, sows 10 - 14 days post wean were housed in pens so 8 pooled fecal samples were collected. At the other two sites sows were housed in crates post wean and fecal samples were collected by parity. Ten pooled tubes of feces each from three sows across P1, P2 - 3, and P4+ were collected resulting in 30 total samples. All fecal samples were sent to Food Safety Net Services in San Antonio, Texas for *Salmonella* culture with enrichment.[3] Confirmed positive samples were sent to Iowa State University Veterinary Diagnostic Laboratory for serotyping by the Check and Trace method.

Results

There were not significant results in the pooled breeding cultured fecal samples, gestation fecal samples, or rectal swab data. Table 1 shows percent positive results of all sample types evaluated. There was a difference in the piglet pooled feces culture results. Pooled piglet fecal culture samples were highly positive at S1 (57%) and S2 (80%) but were low at CNT (10%).

Discussion

This study investigated which sample types to collect and when for the surveillance of *Salmonella* on sow farms with the intention of controlling infection in weaned pigs. Iowa State VDL concluded that there were 20+ serotypes of *Salmonella* isolated in samples collected. Targeting *Salmonella* infections should begin with pooled piglet fecal samples collected for *Salmonella* culture prior to weaning. Sow fecal samples and serum from both sow and piglets also showed significant results for *Salmonella.* These results may be a better resource only if the pooled piglet fecal culture are investigated and do not produce significant results but the wean pigs are still showing clinical signs in the nurseries.

References

1. Neumann, E. J., Ramirez, A., & Schwartz, K. J. (2020). Salmonellosis. In *Swine disease manual* (5th ed., pp. 46-49). American Association of Swine Veterinarians.

2. Hoorfar J, Baggesen DL. Importance of pre-enrichment media for isolation of *Salmonella* spp. from swine and poultry. FEMS Microbiol Lett. 1998 Dec 1;169(1):125-30. **doi: 10.1111/j.1574-6968.1998.tb13308.x**

3. Love, Brenda C, and Marcos H Rostagno. *Comparison of Five Culture Methods for Salmonella Isolation from Swine ...,* 2008, **journals.sagepub.com/doi: 10.1177/104063870802000514**

Table 1: Positive sample comparison results

	CNT	S1	S2
Processing fluids	50% (2/4)	100% (1)	75% (3/4)
Piglet pooled feces	10% (3/30)	57% (17/30)	80% (24/30)
Farrowing pooled sow feces	17% (5/30)	66% (20/30)	30% (9/30)
Sow serum	66% (2/3)	100% (3/3)	100% (3/3)
Piglet serum	53% (18/34)	68% (25/37)	55% (21/38)
Breeding pooled sow feces	50% (4/8)	0% (0/10)	17% (5/30)
Gestation pooled sow feces	35% (7/20)	35% (7/20)	40% (8/20)
Sow rectal swabs	0% (0/3)	0% (0/3)	0% (0/3)
Piglet rectal swabs	0.05% (2/28)	24% (9/37)	24% (9/38)

CNT = control; SI = 4-week batch breed-to-wean (BTW) farm; S2 = a weekly BTW farm

 https://doi.org/10.54846/am2024/95

Effect of gonadotropin releasing factor analog (Improvest®) timing regimens on immunocastrated barrow behavior in late finishing

S. G. Roach[1], BS; E. A. Jablonski[2], DVM, MS; S. Imwalle, BS; J. Newbold, BS; D. M. Amodie[2], MS
[1]North Carolina State University College of Veterinary Medicine, Raleigh, North Carolina; [2]Zoetis, Parsippany, New Jersey

Introduction

Improvest (IMV), a veterinary prescription product approved in 72 countries, including the US, is labeled for use in boars for temporary immunocastration and in gilts for temporary estrus suppression. Two doses (D1;D2) are administered subcutaneously at least four weeks apart, to boars no younger than nine weeks of age (WOA). A one-hour quality assurance (QA) check is completed two weeks after the second dose to identify immunocastrated barrows exhibiting boar-like behavior (EBB). These animals receive an additional dose of IMV. The primary objective of this study was to compare the current dose timing protocol to three other regimens to potentially reduce EBB and the number of additional IMV injections administered. A secondary objective was to validate that the current one-hour QA time frame was sufficient to identify most EBB, or if an additional hour of observation is warranted.

Materials and methods

A total of 20 barns, with the same barn type, size, and injection crew, were systematically assigned to four treatment groups. Treatment group one (T01; control; n = 2819) was the current IMV dosing regimen with D1 at 11 WOA and D2 at 19 WOA; T02 received D1 at 16 WOA and D2 at 20 WOA (n = 2413), with a shorter interval of four weeks between doses to mimic the recommendation for estrus suppression in gilts;[1] T03 received D1 at 11 WOA and D2 at 18 WOA (n = 3426); a shorter time interval was selected to accommodate the concern of earlier sexual maturity that could possibly occur with the genetic shift that the company is experiencing. T04 is a combination of T02 and T03 and with a later start and a shorter interval between doses; they received D1 at 14 WOA and D2 at 18 WOA (n = 2641). The barn was the experimental unit. Two weeks after D2, QA checks were performed and immunocastrated barrows were observed for a total of two hours with each hour recorded separately.

Results and discussion

Summary statistics were performed for this study. One barn was excluded in the final data set due to an unexpected schedule change. The EBB rate by treatment and by hour are shown in Figure 1. There were no remarkable numerical differences across treatment groups; all were approximately 1%, which aligned with historical rates for the previous 12 months. Overall, the EBB rate for all treatment groups was 1.04%; of these 0.92% were found in the first hour. While the second hour was not zero, immunocastrated barrows that were exhibiting boar-like behavior were not

Figure 1: EBB rate by treatment group and QA hour (n = pigs; percent EBB = number of EBB pigs/total n per barn; experimental unit is barn).

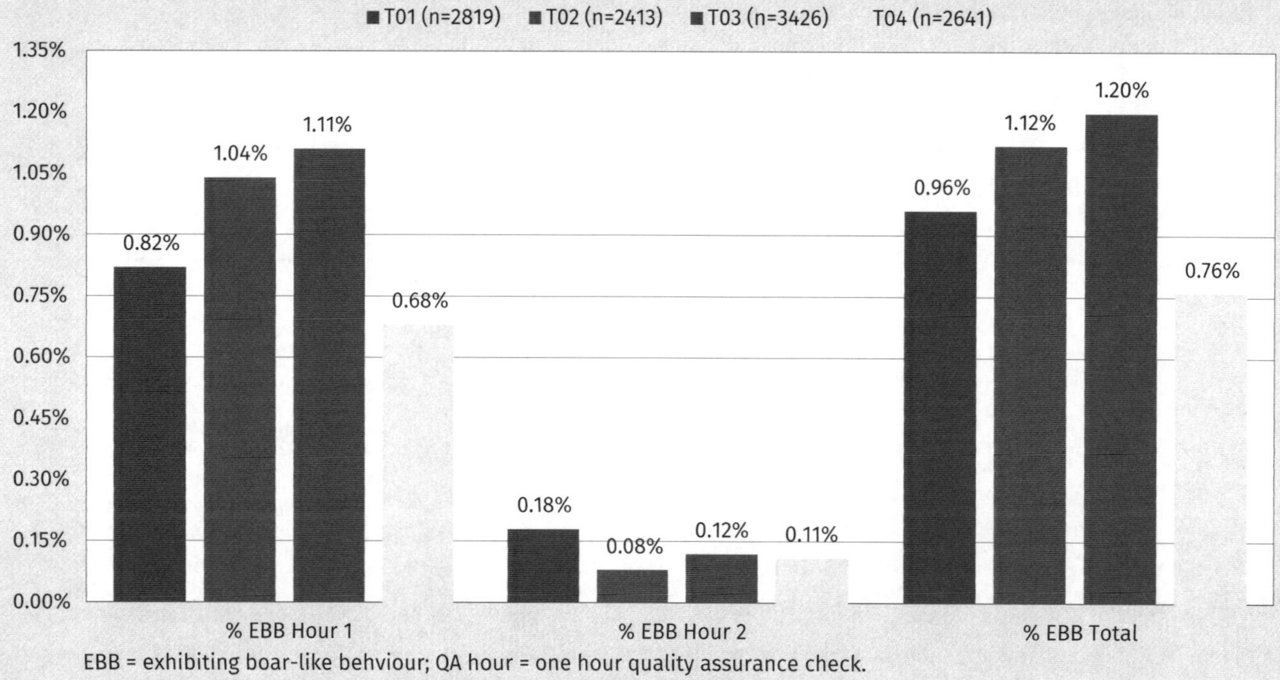

EBB = exhibiting boar-like behviour; QA hour = one hour quality assurance check.

believed to be showing true boar behavior but learned behavior (ie, mounting). Following this study, no process changes were recommended, but further research may be needed with a larger sample size to better evaluate the various dose timing regimens.

Reference

1. Vasquez-Hidalgo, MA, Rueff, L, Pollmann, S, et al. Optimal time interval between Improvest doses in gilts. *Proc Leman Swine Conf.* 2021.

https://doi.org/10.54846/am2024/96

Iron kinetics and tissue storage study

J. Fennell; E. Gruber DVM, PhD; M. Rahe DVM, MPH, PhD; G. Almond DVM, MSc, PhD
College of Veterinary Medicine, North Carolina State University

Introduction

Sow's milk and neonatal piglet iron levels pose a challenge, leading to susceptibility to iron deficiency and subsequent microcytic hypochromic anemia. To counteract this, piglets commonly receive a 200 mg intramuscular iron dextran injection within their first three days of life. While normal pig liver iron content ranges from 100-200 ppm, it's observed that suckling or recently weaned pigs may exhibit higher levels, from 1500 - 2000 ppm. This study aims to establish reference intervals for blood and tissue iron parameters in piglets under 30 days who received injectable iron, addressing the scarcity of data on post-injection liver iron content.

Methods

At Time 0, all pigs received their initial iron injection, with subsequent blood samples taken at Time 0, 7, 14, 21, and 28. Samples were collected only from euthanized pigs, selected randomly. Five pigs euthanized at Time 0 served as "reference standards" for mineral panels. Serum, obtained through centrifugation, was stored for analysis. Serum ferritin and hepcidin were measured via ELISA using porcine-specific reagents. Data were analyzed using one-way ANOVA, and graphs were generated using GraphPad Prism 8.1.1.

Results

A steady weight increase in all groups was observed throughout the study. By Day 28, the 200 and 400 groups exhibited significantly greater body weights than the control group ($P < .05$), indicating no advantage to higher Uniferon® doses (> 200 mg). Basophil concentrations decreased by Day 7, followed by an increase, with all treated pigs surpassing the control group by Day 21. Eosinophil concentrations were erratic, particularly in the 0 and 800 mg groups. Lymphocytes and monocytes increased post-weaning in the control group, with higher concentrations in the 400 and 800 mg groups at Day 28 ($P < .05$). Neutrophil numbers increased post-weaning in all groups. The control group displayed microcytic, hypochromic anemia. Platelet numbers decreased from Day 0 to Day 21 in treated groups, while the control group showed an increase at Day 7 and 14. Serum iron concentrations were higher in the 400 and 800 mg groups at Days 7 and 14, except for the 200 mg group at Day 28. Liver iron concentrations were consistently higher in the 800 mg group. No statistical difference in hepcidin was observed among groups. No significant differences in serum ferritin concentration were noted. Pathology reports indicated no hepatotoxicity at any dose or time point.

Conclusion

This study aimed to determine the amount of iron dextran injections needed by piglets to avoid microcytic, hypochromic anemia. It was found that piglets who did not receive supplemental iron dextran injections showed anemia, with increased hepcidin and reduced growth. Although there were hematological differences among the groups who received 200, 400, and 800 mg of iron dextran, there were no pathological lesions found in the liver of the pigs. The excess hemosiderin and Prussian blue scores simply reflect the iron storage in these pigs (Figure 1). It was expected to see increased serum and liver concentrations of iron following injections of 400 and 800 mg iron dextran, and the pigs thrived and grew without complications. In summary, it is evident that a 200 mg dose of iron dextran was sufficient, and the higher doses of 400 and 800 mg did not cause any pathological issues in the pigs.

Figure 1: Prussian blue scores were used to identify the iron stores through a computer system at Iowa State University Diagnostic Laboratory. The presence of hemosiderin was subjectively scored on a 0 to 3, representing the higher presence of hemosiderin. These values demonstrated the safety of the product.

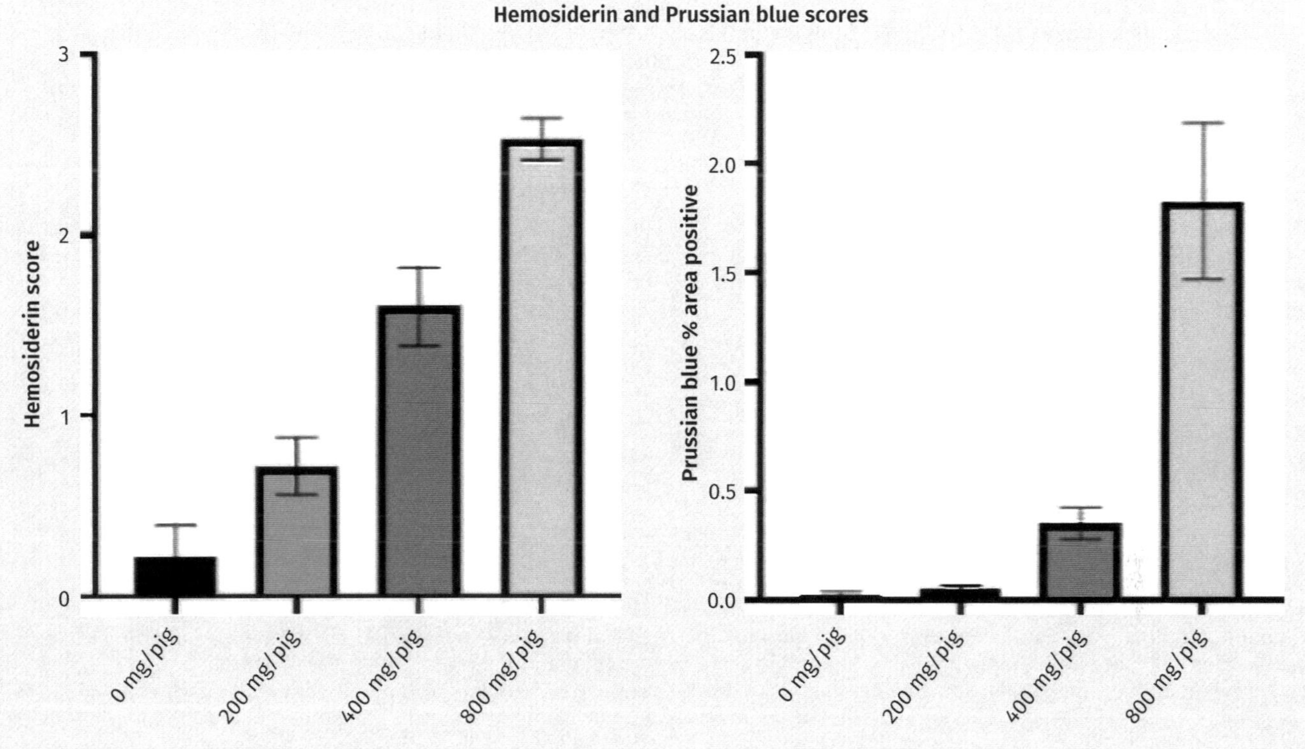

https://doi.org/10.54846/am2024/97

What is “normalization” and why should we normalize PCR results?

B. Armenta-Leyva[1], DVM, MS; B. Munguía-Ramírez[1], DVM, MS; T.-Y. Cheng[2], DVM, PhD; F. Ye[1], BS, PhD; A. Henao-Díaz[3], DVM, MS, PhD; K. Doolittle[4], BS; S. Zimmerman[4], DVM, MS; L. G. Giménez-Lirola[1], BS, MS, PhD; J. J. Zimmerman[1], DVM, PhD
[1]Iowa State University, Iowa; [2]Ohio State University, Ohio; [3]PIC Latin America, Mexico, [4]IDEXX Laboratories, Inc., Maine

Introduction

Benefits of normalized PCRs are major: 1) results within or between labs using the same reference standard are comparable (Cqs are not comparable); 2) all results have an easily interpreted numeric value - the larger the ECq, the more nucleic acid in the sample relative to the reference standard; 3) diagnostic sensitivity and specificity can be derived for normalized PCRs.

Our objective was to explore routine normalization in diagnostic PRRSV RT-qPCR testing.

“Normalization” is re-expressing a test result relative to an agreed-upon reference standard that undergoes the same testing processes as the sample. For example, ELISAs re-express the sample (S) optical density relative to a positive (P) control optical density - which is why ELISA results are reported as “S/Ps”. The purpose of normalization is to account for run-to-run variation, eg, sample handling, pipetting, technicians, etc. Unique to PCR, amplification efficiency (E) is also an important source of variation because E directly impacts the final number of PCR copies. PCR results reported as raw Cqs assume that E is 100%, but in real life, nothing is 100% efficient. Cqs can be easily normalized by converting them to “efficiency standardized Cqs” (ECqs):[1,2]

$$\text{Eq. 1} \quad ECq = E^{-\Delta Cq}$$

where E = amplification efficiency and
ΔCq = (Sample Cq - Reference Standard Cq).

Thus, ECqs represent target concentration relative to the reference standard: The larger the ECq value, the more nucleic acid in the sample relative to the reference standard. Because every Cqs can be converted to a ECq value, there are no “indeterminate” results. This makes it possible to estimate the diagnostic sensitivity and specificity of the test and set cutoffs (this is <u>not</u> possible using raw Cqs).

Materials and methods

Reference standards were created by rehydrating and diluting (1×10^{-4}) a PRRSV MLV (Ingelvac) with serum or oral fluid (OF). That is, a serum reference standard was used when testing serum samples and an OF reference standard when testing OF samples. The test the use of ECqs, serum (n = 132) and individual OF (n = 130) samples were collected from 12 pigs vaccinated with a PRRSV MLV from -7 to 42 DPV. Samples were tested by RT-qPCR and Cqs converted to ECqs. Sample ECqs were calculated using Eq. 1 where the mean E and mean reference standard Cq were calculated from 4 reference standards run on each plate. Testing was done using commercial reagents (IDEXX Laboratories) and the MIC PCR Cycler (Bio Molecular Systems).

Results and discussion

PRRSV RT-qPCR positives were detected on DPV 3 in serum ($\bar{x}$ ECq 0.90) and OF ($\bar{x}$ ECq 1.07). That is, in both serum and oral fluids, the concentration of virus was about the same as in the reference standards. Thereafter, virus concentration peaked on DPV 7 in serum ($\bar{x}$ ECq 2.24) and on DPV 9 in OF ($\bar{x}$ ECq 2.00) (Figure 1). For serum, the diagnostic sensitivity and specificity was estimated as 97.9% and 100% at an ECq cutoff of ≥ 0.20 and, for OF, 82.6% and 100%, respectively, at ECq ≥ 0.45.

Normalization of PCRs has been routine in basic research for 20+ years, but diagnostic laboratories continue to report non-normalized results (raw Cqs). As shown herein, it is not difficult to make the leap from Cqs to ECqs.

References

1. Pfaffl MW. A new mathematical model for relative quantification in real-time RT-PCR. *Nucleic Acids Res* 2001;29:e45.

2. Cheng T-Y, et al. Pseudorabies (Aujeszky’s disease) virus DNA detection in swine nasal swab and oral fluid specimens using a gB-based real-time quantitative PCR. Prev Vet Med 2021;189:105308.

Figure 1: PRRSV RT-qPCR performance by day post-vaccination (n = 12 pigs) for serum and oral fluid. Left y-axis denotes efficiency standardized Cq (ECq) response. Right axis denotes Cq response. Means were calculated as the geometric mean of 12 pigs.

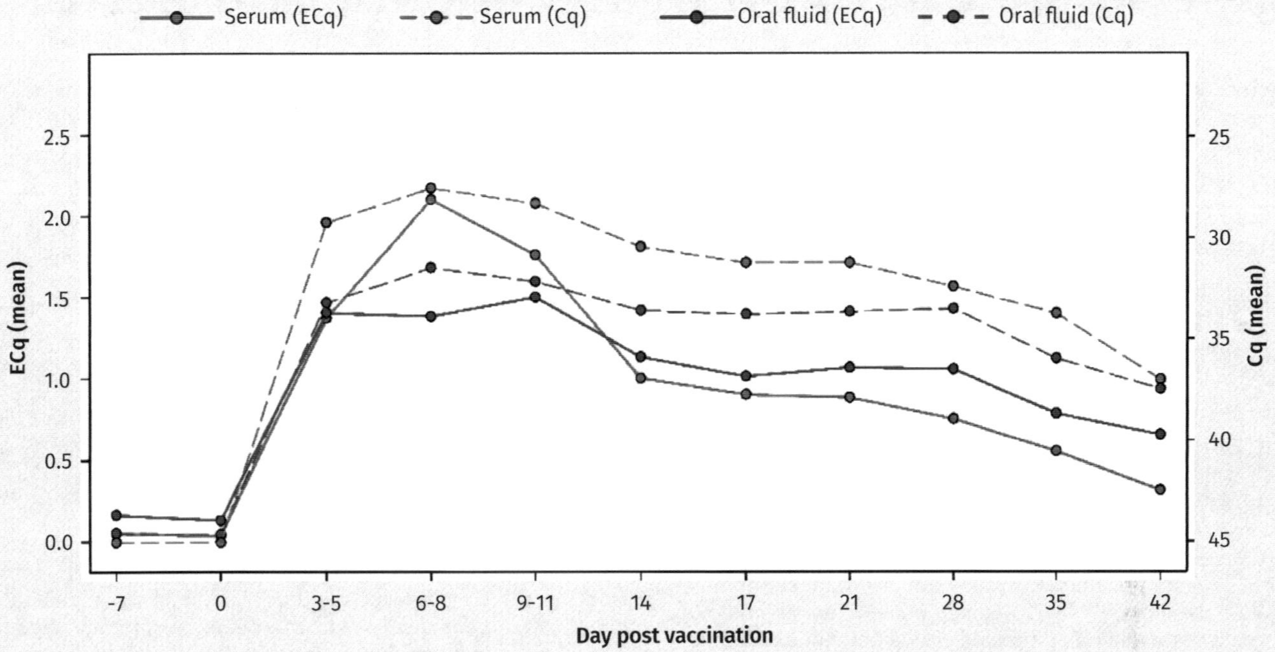

https://doi.org/10.54846/am2024/98

Comparative field evaluation of a new *Cystoisospora suis* PCR in an endemic sow herd

A.M. Betlach[1], DVM, PhD; E. Kettelkamp[1], DVM; S. Rossow[2], DVM, PhD; B. Payne[3], DVM
[1]Swine Vet Center, St. Peter, Minnesota; [2]University of Minnesota, Veterinary Diagnostic Laboratory, St. Paul, Minnesota; [3]Veterinary Pharmaceutical Solutions, St. Peter, Minnesota

Introduction

From a US swine practitioner's perspective, diagnosis of coccidiosis has been limited to clinical observation, fecal flotation, and histopathologic lesions. Based on the limitations of the assays available, a more sensitive method to detect and confirm suspected coccidia in piglet diarrhea cases is desired. The purpose of this paper was to share the work of laboratory validation and initial field evaluation of a specific *Cystoisospora suis* (*C suis*) PCR.

Materials and methods

Validation of PCR

INgene q *C suis* real-time PCR kit (Gold Standard Diagnostics, Madrid, Spain) was evaluated at the UMN-VDL. The test had an analytical sensitivity of 50 DNA copies/mL, a diagnostic sensitivity of 93%, and both analytical and diagnostic specificities of 100%. Sample types validated were feces, fecal swabs, intestines, feedback, and environmental samples. Adequate sample volume for testing was determined to be an important factor for sensitivity.

Field evaluation of *C suis* detection by PCR

A 5,000 commercial sow herd located in Midwestern US with previous and recent diagnosis of coccidiosis was selected. The study was conducted in Spring 2023, in which a 50 - 60% clinical prevalence of coccidiosis in farrowing was observed. Various sample types were collected from 84 farrowing crates over three weekly farrowing groups and submitted to UMN-VDL. Prior to farrowing, environmental Swiffer® samples of crates (ESS, n = 75) and > 50g of feces/sow (SF, n = 76) were collected. The ESS samples were collected by wetting a Swiffer® Dry Sweeping Cloth with 50 mL of PBS and extracting the fluid in a sterile conical tube. On day 17 post-farrow, 2 - 4 randomly tagged pigs in each crate were rectally swabbed, along with the collection of pooled piglet feces from the crates (PF, n = 83 samples). Piglet rectal swabs (PRS) were pooled by litter (n = 84 pooled samples). Fecal flotation exams (n = 159) were conducted on SF and PF for comparative purposes. The validated *C suis* PCR (n = 243) was conducted on ESS, SF, PF and PRS samples. The odds ratio (OR) of obtaining a positive *C suis* result based on sample and diagnostic assay was calculated using a logistic mixed-effects regression model. Data analysis was performed using R statistical software (version 4.2.2). Results were considered statistically significant at $P \leq .05$.

Results

All ESS-PCR, SF-PCR and SF flotation were negative for *C suis*. *Eimeria* species oocysts were found in one PF flotation sample. Twenty-three litters (27.7%) resulted positive for *C suis*, based on parallel testing. The positivity rate was highest in PRS-PCR (22.9%, 19/83) compared to PF-PCR (8.4%, 7/83) and PF-Float (2.4%, 2/83; Table 1). The probability of obtaining a positive *C. suis* result was dependent on sample and diagnostic assay ($P < .001$), with PRS-PCR having 18 times greater odds (OR = 18.25 $P < .001$) and PF-PCR having 4 times greater odds (OR = 4.28, $P = .091$) than PF-Float for *C suis* detection.

Discussion

At the litter level, pooled piglet rectal swabs tested by PCR were 18 times more likely to detect *C suis* than piglet feces via fecal float, with the testing of piglet feces by PCR being intermediate. While the PCR resulted in a high sample positivity rate, false negatives (n = 4) were observed compared to other tests and the utilization of PCR can limit the detection of other pathogens, such as *Eimeria* sp. Additional work should be conducted on the utilization of this *C suis* PCR for herd screening, Ct level interpretation, pooling effect, sample size and age of sample collection. Moreover, an improved understanding of the dynamics of *C suis* transmission between dams and piglets and pathogen shedding is important to help practitioners improve current *C suis* control strategies. To the best of the authors' knowledge, this is the first known study to utilize this *C suis* PCR for coccidiosis detection in the US.

Table 1: *Cystoisospora suis* positivity rate based on sample and diagnostic assay

		PF-PCR[1]		
		Negative	**Positive**	
PF-Float[2]	Negative	89.2%, **(74)**	8.4%, **(7)**	97.6%, (81)
	Positive	2.4%, **(2)**	0%, **(0)**	2.4%, (2)
		91.5%, (76)	8.4%, (7)	100%, (83)
		PRS-PCR[3]		
		Negative	**Positive**	
PF-Float	Negative	74.7%, **(62)**	22.9%, **(19)**	97.6%, (81)
	Positive	2.4%, **(2)**	0%, **(0)**	2.4%, (2)
		77.1%, (64)	22.9%, (19)	100%, (83)
		PRS-PCR[4]		
		Negative	**Positive**	
PF-PCR	Negative	74.7%, **(62)**	16.9%, **(14)**	91.6%, (76)
	Positive	2.4%, **(2)**	6.0%, **(5)**	8.4%, (7)
		77.1%, (64)	22.9%, (19)	100%, (83)

1 Percentage of *C suis* positive samples/total samples collected, (count of positive samples). Samples were collected from 83 litters at approximately 17 days of age.

2 PF-PCR = piglet feces tested by PCR.

3 PF-Float = piglet feces tested by fecal floatation.

4 PRS-PCR = pooled piglet rectal swabs tested by PCR.

 https://doi.org/10.54846/am2024/99

Characterizing the persistence of inactivated *Mycoplasma hyopneumoniae* DNA detection in the respiratory tract of pigs

Albert Canturri[1,2]; Maria Pieters[1-3]
[1]Department of Veterinary Population Medicine, College of Veterinary Medicine, University of Minnesota, St. Paul, Minnesota;
[2]Veterinary Diagnostic Laboratory, College of Veterinary Medicine, University of Minnesota, St. Paul, Minnesota;
[3]Swine Disease Eradication Center, College of Veterinary Medicine, University of Minnesota, St. Paul, Minnesota

Introduction

Molecular diagnostics aiming at determining the presence of *Mycoplasma hyopneumoniae* in clinical specimens are mostly based on the detection of nucleic acids of the pathogen by PCR. However, PCR can detect nucleic acid from non-viable bacteria, potentially misleading diagnostic interpretation. The extensive persistence of DNA *ex-vivo,* in the environment, has been well established in taxonomic and evolutionary biology studies. *In-vivo,* the persistence of nucleic acid detection from non-viable cells has also been reported. For example, microbial cell-free DNA, a highly fragmented nucleic acid released from decomposing cells, can be identified by PCR in various body systems, such as in the respiratory airways of human patients with cystic fibrosis.[1] Similarly, RNA from a bluetongue virus, inactivated vaccine product was detected in the blood of sheep and cattle several days after injection.[2] It is currently unknown if, and for how long, the DNA from non-viable *M hyopneumoniae* can be detectable by PCR *in vivo,* in the respiratory airways of pigs. Therefore, the objective of this study was to assess the PCR detection dynamics of non-viable *M hyopneumoniae* DNA in experimentally inoculated pigs.

Materials and methods

A lung homogenate containing $1x10^{5.5}$CCU/mL of *M hyopneumoniae* strain 232 was combined with Friis medium and administered either in an active or an inactivated form. The active inoculum, containing viable bacteria, was maintained at -80°C and prepared just minutes prior to inoculation. The inactive inoculum, containing non-viable *M hyopneumoniae,* was inactivated via autoclaving at 121°C for 40 minutes. The active and inactive inocula had a Ct value of 28.47 and 29.00, respectively. At time 0 post-inoculation, sixteen, three-week-old pigs were intratracheally inoculated with 10mL of either the active or inactive inoculum (n = 8/group) and transferred to separate rooms. At 6 and 12 hours post-inoculation (hpi), and at 1, 2, 3, 5, 10, and 14 days post-inoculation (dpi), tracheal secretion samples were obtained from each live pig. Additionally, one pig per group was humanely euthanized at each timepoint and post-mortem samples were collected, including tracheal mucosa, broncho-alveolar lavage fluid (BALF), bronchial secretions, thoracic lymph nodes and spleen. Macroscopic and histologic lung lesions were assessed and scored. All samples were tested via a commercial *M hyopneumoniae* species-specific real-time PCR assay.

Results and discussion

Detection of *M hyopneumoniae* DNA by PCR was not obtained at any timepoint post-inoculation in pigs receiving the inactivated inoculum. On the contrary, in pigs that received the active inoculum, there was *M hyopneumoniae* PCR detection starting at 2 dpi in tracheal secretions and BALF, and at 3 dpi in bronchial secretions, tracheal scrapings and thoracic lymph nodes. The results obtained in this study indicated that, in pigs with non-compromised mucociliary apparatuses and immune systems, the DNA of non-viable *M hyopneumoniae* was rapidly sensed and cleared. Thus, a positive PCR result for *M hyopneumoniae* may not be interfered by the residual presence of genetic material from of non-viable cells and suggests that the pig from which the sample was collected was certainly colonized/infected at the time of sampling.

References

1. Barrett SLR, Holmes EA, Long DR, et al. Cell free DNA from respiratory pathogens is detectable in the blood plasma of Cystic Fibrosis patients. *Sci Rep.* 2020;10(1):1-6. **doi:10.1038/s41598-020-63970-0**

2. Steinrigl A, Revilla-Fernández S, Eichinger M, Koefer J, Winter P. Bluetongue virus RNA detection by RT-qPCR in blood samples of sheep vaccinated with a commercially available inactivated BTV-8 vaccine. *Vaccine.* 2010;28(34):5573-5581. **doi: 10.1016/j.vaccine.2010.06.034**

https://doi.org/10.54846/am2024/100

The use of a portable real-time PCR system for point-of-care diagnosis of porcine reproductive and respiratory syndrome virus

T.-Y. Cheng[1], DVM, PhD; S. Ochwo[2], PhD; E. Dalesio[1], BS; R. Schambow[2], DVM, PhD; M. S. Pérez[2], DVM, PhD; A. Rovira[2], DVM, PhD; A. Perez[2], DVM, PhD; A. G. Arruda[1], DVM, PhD
[1]The Ohio State University, Columbus, Ohio; [2]University of Minnesota, St. Paul, Minnesota

Introduction

The development of point-of-care (POC) diagnostic tests has been thriving to fulfill the need for timely test results. POC is a form of diagnostic testing performed where healthcare is provided close to or near the patient. In a swine production setting, POC tests are those tests performed on swine farms or at locations close to farms rather than in dedicated diagnostic laboratories. In this study, a POC real-time PCR system was assessed for the detection of porcine reproductive and respiratory syndrome virus (PRRSV) in swine diagnostic specimens.

Materials and methods

Two trials were conducted to describe the detection of PRRSV RNA in swine serum and oral fluid samples using a POC real-time PCR system (IndiField Portable PCR System, Indical Bioscience GmbH). For Trial 1, the system was evaluated by testing PRRSV presence in 15 pooled serum samples, 9 positives and 6 negatives previously tested by the Veterinary Diagnostic Laboratory of University of Minnesota (UMN-VDL). Nucleic acid extraction was done using a magnetic bead-based kit (IndiMag Pathogen IM48 Cartridge) on an automated, high-throughput system (IndiMag 48s, Indical Bioscience). Diagnostic sensitivity and specificity were calculated to evaluate test performance. For Trial 2, the field application of the POC system was evaluated by performing POC tests on a commercial swine farm using the manual extraction procedure provided by the manufacturer. In April 2023, 7 swine oral fluid (OF) samples were collected from 7 pens within 2 barns in a finisher site that had a confirmed PRRSV outbreak a week before the visit. To further validate result consistency, OF samples were divided into 6 subsets for 6 PCR running conditions: **1)** at the farm right after sample collection; **2)** in the vehicle during commute; **3)** after a 24-hour post-collection incubation at room temperature; **4)** after a 24-hour post-collection incubation on ice; and **5)** after a 48-hour post-collection incubation on ice. The last set was tested by the Animal Disease Diagnostic Laboratory of the Ohio Department of Agriculture (ODA-ADDL).

Results and discussion

The results of Trial 1 showed the PRRSV RNA was detected in 7 of 9 pooled serum samples declared positive by UMN-VDL, whereas all negative samples remained negative by the POC system. The diagnostic sensitivity and specificity were 77.8% and 100%, respectively. For Trial 2, although all 7 samples tested positive by ODA-ADDL, the POC system detected PRRSV RNA in 7 samples after a 24-hour incubation at room temperature and 1 sample after a 48-hour incubation on ice. Overall, the POC system showed promising performance in testing serum samples in a laboratory setting. However, more work toward the validation and implementation of POC OF samples under field conditions is needed.

https://doi.org/10.54846/am2024/101

A mixed method study of beliefs, behaviors and practices of farm biosecurity in the US swine industry

M. C. Chepkwony[1], DVM, MVPH; D. N. Makau[2], DVM, MVEE, PhD; C. Yoder[1,2], DVM, MS; M. Culhane[2], DVM, PhD; M. S. P. Aguirreburualde[1], DVM, PhD; A. M. Perez[1], DVM, PhD; C. A. Corzo[2], DVM, MS, PhD; Michael Mahero[1,3], DVM, MPH, PhD
[1]Center for Animal Health and Food Safety, University of Minnesota; [2]Department of Veterinary Population Medicine, College of Veterinary Medicine, University of Minnesota, Minnesota; [3]Department of Biomedical and Diagnostic Sciences, College of Veterinary Medicine, University of Tennessee Institute of Agriculture, Knoxville, Tennessee

Introduction

Risk perception and behavior can vary in how different production systems handle diseases and implement biosecurity measures on their farms. However, in integrated systems such as the US swine industry a degree of uniformity in biosecurity management may be beneficial in preventing introduction and mitigating disease spread.[1–3] To understand beliefs, behaviors, and practices on farm biosecurity among swine producers in the US swine industry we conducted a mixed methods study.

Methods

We integrated an online survey alongside focus group and key informant interviews (8) among swine producers and practitioners, in the Midwest US region. Development of the survey and qualitative interview guide was informed by the theory of planned behavior and Secure Pork Supply (SPS) framework. Descriptive (quantitative) and qualitative analysis of data from the survey and interviews was done.

Results

Of 54 online questionnaire respondents, 48.1% said they implement some biosecurity measures and 72.2%, felt it was important to always have enhanced biosecurity measures in place. More than half (53.7%) felt that their veterinarians' opinion on biosecurity was most important and 37% were not worried about the global threat of African swine fever. Almost all (90.7%) respondents were sometimes/always confident that in case of an outbreak on their farms, they would be able to contain it. However, based on the recommendations by the SPS biosecurity guidelines, none of the respondents practiced enhanced biosecurity.

Discussion

Focus group and in-depth interviews confirmed the presence of these biosecurity gaps and revealed infrastructural, economic and behavioral drivers of the observed gaps.[2] Also revealed was the differential application of the recommended enhanced biosecurity practices across swine producers depending on size, location and previous production or disease outbreak experience.[4] Regular biosecurity assessments, review of biosecurity practices and tailoring of education packages to address existing cognitive biases would be beneficial to the management and preparedness for both endemic and foreign diseases in the US swine industry.

References

1. Kinsley AC, Perez AM, Craft ME, Vanderwaal KL. Characterization of swine movements in the United States and implications for disease control. Prev Vet Med [Internet]. 2019 Mar;164(April 2018):1–9. **https://doi.org/10.1016/j.prevetmed.2019.01.001**

2. Pudenz CC, Schulz LL, Tonsor GT. Adoption of secure pork supply plan biosecurity by U.S. Swine producers. Front Vet Sci. 2019 May 28;6:146.

3. ISU and UMN. Secure Pork Supply Plan-Biosecurity [Internet]. Secure Pork Supply Plan. 2023 [cited 2023 Jan 30]. Available from: **https://www.securepork.org/pork-producers/biosecurity/**

4. Salvesen HA, Whitelaw CBA. Current and prospective control strategies of influenza A virus in swine. Porc Heal Manag [Internet]. 2021 Dec 28 [cited 2022 Jun 3];7(1):23. **https://porcinehealthmanagement.biomedcentral.com/articles/10.1186/s40813-021-00196-0**

https://doi.org/10.54846/am2024/102

Comparison of PRRSV vaccination strategies on wean-to-finish groups facing a lateral outbreak under field conditions

Swaminathan Jayaraman[1]; Tyler Bauman[2]; Amy Maschhoff[2]; Edison Magalhães[1]; Giovani Trevisan[1]; Daniel C. L. Linhares[1]; Gustavo Silva[1]
[1]Iowa State University, [2]The Maschhoffs Ltd.

Introduction

Porcine reproductive and respiratory syndrome virus (PRRSV) infection in wean-to-finish pigs leads to significant financial losses due to mortality, morbidity, and poor growth. Lateral outbreaks in field settings further exacerbate the problem.[1] Vaccination is a key strategy in managing PRRSV, but a few factors, including dosage and vaccination timing, can impact the effectiveness.[2] The study assessed the effect of different PRRSV MLV vaccination strategies (full vs half dose) using observational data from pigs from stable sow farms that reported a clinical lateral outbreak during the grow-finish phase.

Methods

A retrospective analysis assessed differences in PRRSV vaccine dose (half vs full) in growing pig groups that reported a clinical lateral outbreak. The dataset comprised single flow groups from three different regions in the United States from 2020 to 2023. These groups originated from sow farms that were PRRSV vaccinated. The study groups had a clinical expression of PRRSV post-placement, as defined by the respective vets based on clinical signs and diagnostics. The average age at break was 11 weeks. After cleaning, the data contained 48 nursery groups, of which 11 received half the dose, and 37 received the full dose, becoming 125 finishing groups where 39 received half and 86 received the full dose. A causal diagram was built to understand the relationships between the variables, potential confounders, and the treatment outcome pathway.[3] The nursery and grow-finish datasets were then balanced using the propensity score matching (PSM) technique, using the average treatment effect (ATE) to balance the treatment groups on confounding factors by analyzing standardized mean difference (SMD) and variance.[4] Lastly, linear mixed-effects models were built to assess differences between vaccination strategies (full vs. half-dose), incorporating weights from the PSM and considering factors such as season, year, vaccine type, vaccination timing, farm status, other diseases, and PRRS strain, with a random intercept being the flow. The outcomes were mortality, culling, and grade-A pigs.

Results

There were no significant differences between doses in the nursery phase on the mortality percent, cull rate, and grade A percent. However, in the grow-finish phase, there was a significant difference with $P < .05$ (Table 1) of 3.69% in mortality between the groups receiving the full dose (6.05%) and the groups receiving half dose (9.74%)

Table 1: Ls means and 95% confidence limits from the weighted linear mixed effects model for outcome mortality percentage

Vaccine dose	LS means	95% CL	Group
Full	6.05	4.8 - 7.7	a
Half	9.74	7.1 - 13.3	b

Conclusion

Under the conditions of this production system and balancing for potential confounders, the study revealed that groups that received a full dose of PRRSV MLV and reported a lateral break during the grow-finish phase had lower mortality (3.69 percentage points) compared to groups that received half dose. This finding suggests partial protection in pigs receiving a half-dose and facing a lateral PRRSV outbreak during the grow-finish phase. However, this analysis should be interpreted cautiously when expanded to other regions and production systems.

References

1. Linhares, Daniel Correia Lima. 2013. "Evaluation of Immune Management Strategies to Control and Eliminate Porcine Reproductive and Respiratory Syndrome Virus (PRRSv)." Edited by Montserrat Torremorell. Ann Arbor, United States: University of Minnesota. **https://www.proquest.com/dissertations-theses/evaluation-immune-management-strategies-control/docview/1322974196/se-2**

2. Moura, Cesar A. A., Reid Philips, Gustavo S. Silva, Derald J. Holtkamp, and Daniel C. L. Linhares. 2022. "Comparison of Virus Detection, Productivity, and Economic Performance between Lots of Growing Pigs Vaccinated with Two Doses or One Dose of PRRS MLV Vaccine, under Field Conditions." Preventive Veterinary Medicine 204 (July): 105669.

3. Greenland, S., J. Pearl, and J. M. Robins. 1999. "Causal Diagrams for Epidemiologic Research." Epidemiology 10 (1): 37–48.

4. Austin, Peter C. 2011. "An Introduction to Propensity Score Methods for Reducing the Effects of Confounding in Observational Studies." Multivariate Behavioral Research 46 (3): 399–424.

https://doi.org/10.54846/am2024/103

Integration of PCV2 processing fluids Ct values and serological profiles of sows for assessing grower-finisher performance

Molly Kroeger[1]; Chelsea Stewart[2]; Marjorie Schleper[2]; Amanda Sponheim[3]; Eduardo Fano[3]; Pablo Piñeyro[1]
[1]Department of Veterinary Diagnostic and Production Animal Medicine, Iowa State University, Ames, Iowa; [2]Christensen Farms, Minnesota; [3]Boehringer-Ingelheim Animal Health USA Inc., Georgia

Introduction

The use of processing fluids to detect PRRSV has increased significantly in the past decade, resulting in practical, effective monitoring strategies for PRRSV circulation and shedding in sow herds.[1] However, the applicability and interpretation of this sample matrix regarding PCV2 circulation and herd health status is poorly understood. Therefore, the objective was to assess the correlation between the sow pre-farrow PCV2 immunological status to the nursery viral circulation using PCV2 processing fluid PCR Ct values.

Methods

Sixty sows from two 2000-head commercial sow farms were selected for study. Sows were evenly distributed by parity: P1 (n = 20), P2 - 5 (n = 20), and P > 5 (n = 20). All sows were bled 2 weeks prior to farrowing. No cross fostering occurred between litters. Processing fluids (PF) were collected by litter to assess PCV2 viral levels. Sow antibodies were analyzed by 3 serological methods: total PCV2 IgG by ELISA, specific PCV2a and PCV2d IgG by IFA, and neutralizing antibody (NA) titers against PCV2a and PCV2d by chimeric virus neutralization assay. The correlation between PF Ct values and serological results was assessed by simple linear regression.

Results

For the sow immunological status, the average ELISA S/P ratio was 3.0. The average PCV2a and PCV2d titers were 1:65 and 1:242, respectively. The average PCV1-2a and PCV1-2d VN titers were 1:3.2 and 1:2.3, respectively. Sow immunological status by all methods was not statistically significant between parities (*P* > .05). The average processing fluid Ct was 34.8 and was not statistically significant between parities (*P* > .05). Higher ELISA S/P ratios, IFA titers, and NA titers were not statistically significantly correlated to higher high PF Ct values regardless of sow parity or PCV2 subtype (Figure 1).

Discussion

The PF Ct value was not correlated to sow immunological status by ELISA, IFA, and VN. The high average PF Ct value partially explains the lack of correlation. Identifying a correlation between PF Ct and sow immunological status could result in more accurate timing of sow vaccination to achieve maximal immune status before farrowing. For further interpretation of PF PCR results, future studies should examine the implication of PF Ct on downstream production parameters. Establishing the correlation between PF Ct values and downstream performance could warrant using PF Ct to establish more tailored vaccination programs and as a predictor of grower-finisher performance.

Reference

1. Machado I 2022. Front Vet Sci 9:993442.

Figure 1: Correlation of **A)** ELISA S/P ratio and **B)** PCV1-2a virus neutralization titer to PCV2 PF Ct value.

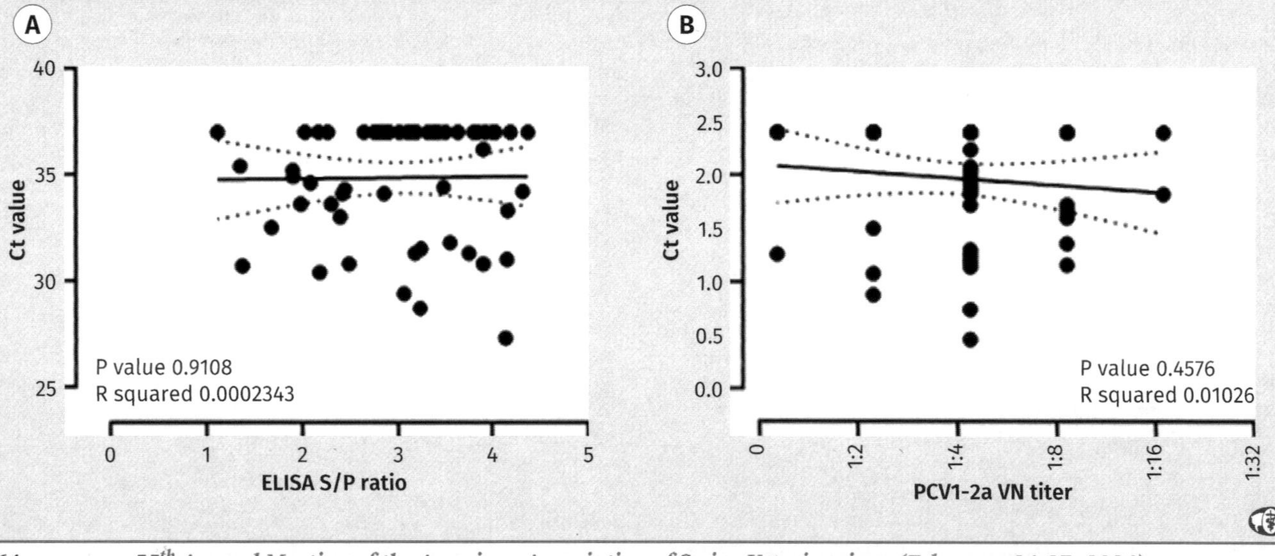

https://doi.org/10.54846/am2024/104

Results of an evaluation of the USDA Swine Hemorrhagic Fevers Integrated Surveillance Plan

C. Lorenz, DVM, MPH, DACVPM; C. Vanicek, DVM, MPH, DACVPM
Center for Epidemiology and Animal Health, Veterinary Services, Strategy and Policy, Animal and Plant Health Inspection Service, United States Department of Agriculture, Fort Collins, Colorado

Introduction

African swine fever (ASF) continues to spread across the globe, most recently to the island of Hispañiola, and classical swine fever (CSF) remains present in the Caribbean, South America, and Asia. The United States Department of Agriculture (USDA) Animal and Plant Health Inspection Service (APHIS) has an integrated surveillance plan targeting these diseases. The plan is designed to accomplish three primary objectives: 1) strengthening detection capabilities by providing timely and consistent surveillance to establish a baseline of disease absence, 2) enhancing outbreak preparedness by testing the capability for high-volume sample collection, laboratory capacity, and data management systems prior to an outbreak, and 3) supporting claims of disease freedom through testing of targeted higher-risk subpopulations. This work summarizes the outcomes of a retrospective evaluation of this plan.

Methods

Veterinary services recently evaluated the functionality, effectiveness, organization, and value of our ASF/CSF surveillance plan in the conterminous United States over the past three years. The evaluation analyzed data from specimens collected from large commercial swine, higher-risk swine, and feral swine between June 1, 2020, to May 31, 2023. The organization, function, and effectiveness of this surveillance plan was evaluated using the RISKSUR framework.[1]

Results

This evaluation revealed many successes of the plan over the past three years, including the addition of sample types to improve flexibility and efficiency of sample collection and the addition of four National Animal Health Laboratory Network laboratories for active surveillance to provide flexible, scalable, and timely diagnostic testing. Consistent sample collection was seen from all surveillance components across time and location with a substantial increase in specimens tested from sick pig submissions at veterinary diagnostic laboratories. Furthermore, improvements were made to data fidelity through automated identification and management of duplicate records, pooled specimens, and other factors impacting data quality.

Several challenges were also identified through this evaluation. These included limitations among data collection systems to consistently capture clinical signs, age/production type, and premises identification number for submitted samples. Accurate and consistent availability of these data would better support surveillance evaluations and other epidemiologic analyses. There were occasional issues with efficiently linking laboratory results with epidemiologic information from specimen submitters. Geographic representativeness could be improved through participation of additional premises and a re-evaluation of state-level sampling targets to best support current needs and animal movement patterns.

Discussion

Overall, the program is meeting its current objectives, strengthening detection capabilities and response preparedness, and supporting disease freedom claims. Continued success of the surveillance system requires ongoing collaboration among federal, state, and industry partners. Focus areas to improve performance include: 1) enhancing risk-based targeting of swine subpopulations through ongoing analyses of risks of disease introduction, 2) regularly re-evaluating state-level sample target numbers, encouraging sample collection from under-represented populations, and identifying resource needs to efficiently execute disease surveillance, and 3) continuing to develop and expand the use of tools to capture and share electronic information though all surveillance components and through all stages of specimen collection, processing, and reporting. The results of this evaluation will inform recommendations for enhancements to the integrated surveillance plan, improving the ability of the USDA to detect, control, and eradicate any ASF introductions and thus mitigate the impact of an ASF introduction to the US.

Reference

1. Peyre M, Hoinville L, Njoroge J, et al. The RISKSUR EVA tool (Survtool): A tool for the integrated evaluation of animal health surveillance systems. *Prev Vet Med.* 2019;173:104777. **doi: 10.1016/j.prevetmed.2019.104777**

https://doi.org/10.54846/am2024/105

Drivers of pre-weaning mortality in a midwestern US producer during 2020 to 2021

K. J. Will[1,2], DVM, MS; E. S. Magalhaes[1], DVM, MS, PhD; C.A. A. Moura[3], DVM, PhD; G. Trevisan[1], DVM, MBA, PhD; G. Silva[1], DVM, MS, PhD; A. P. G. Mellagi[2], DVM, MS, PhD; R. R. Ulguim[2], DVM, MS, PhD; F. Bortolozzo[2], DVM, PhD; D. C. L. Linhares[1], DVM, MBA, PhD

[1]Iowa State University, Ames, Iowa; [2]Setor de Suínos, Universidade Federal do Rio Grande do Sul, Porto Alegre, Brazil; [3]Iowa Select Farms, Iowa Falls, Iowa

Introduction

Piglet pre-weaning mortality (PWM) remains an important issue within the US swine industry, presenting significant economic losses and raising concerns regarding sustainability and animal welfare. To enhance piglet survivability, it is crucial to comprehend the underlying factors and devise effective management strategies. Therefore, this study conducted a comprehensive multivariate analysis to identify the factors associated with PWM in a Midwest US swine production system and aimed to provide valuable insight to implement targeted interventions and improve survival rates.

Methods

Weekly data from 47 sow farms from a US Midwestern production system were captured from January 2020 to December 2021. According to the available literature, 29 variables that could potentially affect PWM were selected for analysis. These variables included farm infrastructure, productivity parameters, health status, and interventions. All numerical variables were categorized into quartiles, and the outcome of this study was the log-transformed weekly PWM. First, a univariate analysis was performed to screen significant risk factors on PWM, retaining those with $P < 0.20$. Thereafter, a multivariate analysis was conducted to further explore the combined effects of 18 variables on PWM. The week served as the unit of analysis and the farm included as a random effect. Variables with $P \leq 0.05$ in the multivariable model were considered statistically significant. Analysis was performed using the GLIMMIX Procedure in SAS 9.4.

Results

The analysis included 4,889 weaning weeks, and the overall average post-weaning mortality (PWM) was 14.19%. There were several risk factors associated with PWM (Table 1).

Discussion and conclusions

This study reports multiple factors affecting PWM, including seasonal variations, specific productivity variables, and farm infrastructure. These findings are strategic in developing targeted interventions to mitigate PWM. For instance, productivity variables allow the identification of high-risk farrowing groups, farm infrastructure characteristics focused on PWM should be evaluated, and strategies applied during remodeling or building new facilities. Additionally, health status results corroborate with the literature regarding the negative effect of PRRSV and PEDV on piglet survival. Therefore, these results provide valuable insights to develop targeted strategies to reduce PWM and enhance the productivity of sow farms.

Table 1: Average weekly pre-weaning mortality (PWM) for each variable in the multivariable model

Variables (PWM)	Category A (%)	Category B (%)	Category C (%)	Category D and E (%)
Year	2020 – 17.36	2021 – 16.36		–
Season	Spring – 16.92^{ab}	Summer – 16.58^{b}	Fall – 16.95^{ab}	Winter – 16.97^{a}
Farm region	Central, Iowa – 17.61^{a}	North, Iowa–15.28^{b}	South, Iowa – 17.40^{ab}	West, Iowa – 15.88^{ab} Other, Iowa – 18.29^{a}
Crate dimensions	< 3.1m^{2} – 17.34^{a}	3.15 to 3.25m^{2} – 15.82^{b}	> 3.9 to 4.9 m^{2} – 17.44^{a}	–
Labor minutes/sow	0.47 min – 16.53^{b}	0.69 min – 16.68^{b}	0.78 min – 16.85^{b}	1.19 min – 17.36^{a}
Farrowing rate	74.0 – 17.88^{a}	82.8 – 16.70^{b}	85.8 – 16.49^{b}	89.3 – 16.38^{b}
Farrowing parity	3.1 – 15.36^{c}	3.9 – 16.81^{b}	4.3 – 16.81^{b}	5.0 – 17.46^{a}
Gestation length	116.2 d – 17.95^{a}	116.7 d – 16.69^{b}	117.0 d – 15.53^{bc}	117.5 d – 15.29^{b}
Stillborn rate	5.9 – 15.59^{c}	7.9 – 15.57^{b}	9.3 – 16.88^{b}	11.5 – 18.53^{a}
Average total born	14.5 – 16.64^{bc}	15.2 – 16.63^{c}	15.5 – 17.02^{ab}	16.1 – 17.12^{a}
Oxytocin use	No – 16.54^{b}	Yes – 17.17^{a}	–	–
PRRSV status	Negative – 14.81^{c}	Endemic – 16.07^{b}	Epidemic – 20.11^{a}	–
PEDV status	Negative – 13.94^{b}	Positive–20.38^{a}	–	–
Feed medication	Before – 17.49^{a}	Feeding – 17.01^{ab}	W1[1] – 16.47^{c}	W2[2] – 16.45^{c} W3[3] – 16.86^{bc}
Rotavirus feedback	No – 17.21^{b}	Yes – 16.50^{a}	–	–

[1] W1 – 1st to 3rd week after feed medication.
[2] W2 – 4th to 6th week after feed medication.
[3] W3 – 7th to 9th week after feed medication.
a,b,c Different superscripts are statistically different ($P \leq 0.05$).

https://doi.org/10.54846/am2024/106

Characterization of farming, management, biosecurity, and marketing practices of alternative pig farms in Minnesota

M. B. Medrano, DVM, MPH; M. R. Culhane, DVM, PhD; C. A. Corzo, DVM, MS, PhD
Department of Veterinary Population Medicine, College of Veterinary Medicine, University of Minnesota, Saint Paul, Minnesota

Objective

The United States (US) swine industry primarily raises hogs indoors., yet there are still farmers who have chosen the alternative to raise their hogs outdoors. These "alternative" pig farms (APFs) include niche, pasture-raised, and USDA organic-certified, among others. As the US swine industry seeks to strengthen its defenses against the introduction and spread of diseases, especially African swine fever, it is therefore important to understand the role APFs play in disease transmission and their connectedness to the broader swine industry. Unfortunately, this information is scarce in the US and to fill this information gap, this research seeks to describe the management, biosecurity, and marketing practices of APFs in Minnesota (MN) and to characterize their interactions with the larger commercial industry.

Methods

An APF database was created and developed to identify the study's target population using processes specific to the state of MN. This target population was emailed an online survey using Qualtrics (Provo, UT) in December 2022 to gather information about their management, biosecurity, and marketing practices. Consenting APFs were visited by the research team to collect blood and/or oral fluid samples from pigs and a herd level prevalence study for porcine reproductive and respiratory syndrome (PRRS) virus was then conducted by testing for the presence of antibodies and virus using ELISA and PCR, respectively.

Results

Over 200 APFs were identified and contacted, of which 54 (27%) responded to the survey and 21/50 (42%) reported having a known, registered Premises ID number. There were 33 breeding herds, of which 32 raised pigs until finishing weight, and the 21 growing pig farms purchased nursery and/or grower pigs. Breeding herds had a median of four sows and ranged between 1 and 60. Twenty-six herds used natural breeding and had a median of two boars but ranged between 1 and 5 while five herds exclusively used artificial insemination. While 67% (35/52) of farms reported having pasture access for their pigs, only 7.7% (4/52) exclusively used hoop barns with deep bedding. Over half (31/54) of the farms have bulk and/or bagged feed delivered by non-farm personnel, compared to those that transport their own feed or produce it on-site.

Of the 67% (36/54) of the farms that used veterinary services, 12 responded they had previous infectious disease diagnoses such as erysipelas, greasy pig, PRRS, influenza, circovirus, bacterial enteritis, and ascariasis. Five of 43 farms reported to have zero biosecurity measures (out of 17 measures in the survey) with the maximum number of measures implemented by a farm being 13. Third-party certification (eg, USDA organic, PQA, and Humane Certified, etc) was reported by 28% (15/53) of the participating farms. Additional survey results can be found in Table 1. The herd PRRS seroprevalence was 33% (ie, 8/24 were PRRS ELISA antibody positive) while 21% (5/24) had at least one PRRS PCR positive result. The PRRS PCR positive farms had a median number of 150 pigs compared with a median number of 24 pigs for PRRS PCR negative farms.

Discussion

The biosecurity practices for APFs vary widely, especially regarding the number of biosecurity? practices implemented by each farm. As with intensively raised pigs in the US commercial swine industry, this variation in biosecurity and other factors may be associated with PRRS prevalence. Additional analysis focusing on risk factors specific to PRRS in APFs is therefore needed. The results of this descriptive study suggest that growing pigs and feed mills may be a potential connection to the commercial swine industry. Additional research is needed to identify the degree to which this occurs, including how geography/location may be a factor in disease transmission.

Table 1: Survey results from select variables

Certificate of Veterinary Inspection (CVI) obtained entering state	10 Yes, 1 No
CVI obtained exiting state	4 Yes, 4 No
Pig breeds	12 Purebred farms, breeds included: Duroc, Red Wattle, Idaho Pasture Pig, Berkshire, Hampshire, Mangalitsa, Chester White
Marketing	27 Direct to consumer
Processing facility type	14 Exempt processing 5 Home slaughter 6 Commercial facility 30 State/USDA facility
Vaccine use	13/23 Sows 1/21 Growers
Mortality disposal method	29 Compost 9 Burial 2 Incinerate/render
Employees	10 Yes, 44 No
Other livestock/poultry on-site	50 Yes, 1 No
Double fencing used	17 Yes, 31 No

https://doi.org/10.54846/am2024/107

Evaluation of the infectivity level of three wild-type PRRSV variants

Claudio Marcello Melini[1]; Mariana Kikuti[1]; Montserrat Torremorell[1]; Kimberly VanderWaal[1]; Stephanie Rossow[2]; Jerry Torrison[2]; Cesar A. Corzo[1]
[1]Department of Veterinary Population Medicine, University of Minnesota, [2]Veterinary Diagnostic Laboratory, University of Minnesota

Introduction

During the fall of 2020 simultaneous PRRSV breaks occurred in both sow and grow-finish sites. Upon investigation highly similar viruses were identified, that appeared to be associated with more severe clinical presentations and a newly emerged variant of PRRSV classified as L1C 1-4-4.[1] The epidemic subsided in early 2021, however, during late spring and summer, another wave of transmission occurred infecting multiple sow and growing pig farms. These cases led the industry to question why this virus was spreading at such a fast pace and breaking through current biosecurity practices. A study was designed to compare this newly emergent variant to be compared to two other variants that caused major epidemics in the past, L1A 1-7-4 and L9 1-4-2, to assess the minimum infectious dose (ID_{50}), shedding pattern, and histopathological lesions in experimentally infected pigs.

Methods

For each variant, thirty-six PRRSV ELISA and RT-PCR negative three-week-old barrows were randomly allocated to five challenge groups and one control group. Pigs were individually housed with no nose-to-nose contact in rooms segregated by treatment groups. All three variants (ie, L1C 1-4-4, L1A 1-7-4, and L9 1-4-2) were grown in MARC-145 cells. Pigs were intranasally challenged (1 mL per nostril) with 10^0 to 10^4 $TCID_{50}$/mL. Blood, nasal and rectal swabs were collected at day post challenge (DPC) 0, 1, 2, 4, 7, 11, 16, 21, 26, and 30 to assess infection and viral shedding through PRRSV RT-PCR. Successful infection was considered if pigs were RT-PCR positive on serum by DPC 4. A probit model was used to obtain the median infectious dose 50 (ID_{50}) (infective dose needed to infect 50% of exposed individuals of each virus, and a logistic regression model was used to obtain the Odds Ratio of the pigs being RT-PCR positive at 4 DPC using the L9 1-4-2 virus as a reference and after adjusting for the infectious dose in $TCID_{50}$/mL. On DPC 11, two pigs from each group were euthanized, and tissues (eg, brain, lung, lymph nodes) were collected for histopathological assessment.

Results

The probit model results for ID_{50} were $10^{1.3}$ (95% CI $10^{0.4}$ – $10^{2.2}$), $10^{2.3}$ (95% CI $10^{1.6}$ – $10^{3.0}$), and $10^{2.6}$ (95% CI $10^{2.0}$ – $10^{3.2}$) $TCID_{50}$/mL for the L1C 1-4-4, L1A 1-7-4, and L9 1-4-2 variants, respectively. When adjusting for the infectious dose in $TCID_{50}$/mL, the pigs challenged with L1C 1-4-4 were 13 (95% CI 3 – 16) times more likely to be infected by DPC 4 compared to pigs challenged with L9 1-4-2. Virus detection in nasal swabs began at 1 DPC and in rectal swabs at 4 DPC with both sample types cycle threshold values decreasing until DPC 7. Intermittent nasal and rectal shedding patterns were similar for all variants with slight differences between inoculum groups during the study period. Lungs had similar microscopic lesions regardless of dose, but pigs infected with the L1C 1-4-4 variant were consistently more severe. Brain lesions were found in at least 1 pig within each concentration group in the L1C 1-4-4 challenged group, whereas this was limited to only 1 pig per variant in either the L1A 1-7-4 or L9 1-4-2 challenged groups. Heart lesions were mostly present in the L1C 1-4-4 variant group.

Discussion

The results of this study indicate that compared to the L9 1-4-2 and L1A 1-7-4 variants, the L1C 1-4-4 virus requires fewer viral particles to successfully infect half of the pigs by DPC 4. In addition, our data showed that the L1C-1-4-4 variant leads to more frequent and more severe tissue lesions in the lung, heart, and brain as compared to the other evaluated PRRSV variants.

Reference

1. Kikuti, M., Paploski, I., Pamornchainavakul, N., Picasso-Risso, C., Schwartz, M., Yeske, P., Leuwerke, B., et al. Emergence of a New Lineage 1C Variant of Porcine Reproductive and Respiratory Syndrome Virus 2 in the United States. *Frontiers in Veterinary Science* 8 (October 18, 2021): 752938. **DOI: 10.3389/fvets.2021.752938**

https://doi.org/10.54846/am2024/108

Genotypic and phenotypic antimicrobial resistance of *Streptococcus suis* strains from North American swine clinical cases

R. Mugabi[1]; X. Hu[1]; N. R. Macedo[1]; O. Sahin[1]; G. Li[1]; P. Harms[2]; B. Martinez-Lopez[3]; M. J. Clavijo[1,2]
[1]Veterinary Diagnostic and Population Animal Medicine, Iowa State University, Ames, Iowa; [2] PIC®, Hendersonville, Tennessee, [3]University of California Davis, Davis, California

Introduction

Streptococcus suis is a Gram-positive coccus that causes various infections in pigs, including meningitis, arthritis, endocarditis, septicemia, and pneumonia, leading to significant economic losses in the swine industry, especially in young pigs. The objective of this study was to the describe the genotypic and phenotypic antimicrobial susceptibility patterns of clinical *S suis* isolates from North America.

Methods

MICs for 19 different antibiotics were determined for 400 whole-genome sequenced isolates using the broth microdilution method.[1] Out of the 19 antibiotics, only 6 have recommended CLSI breakpoints for *S suis* and those were the main focus for this study. Isolates were sequenced in-house using the Illumina MiSeq. The SRT2 program was used to determine the AMR genes, serotypes, and sequence types.

Results

A total of 388/400 (97%) isolates were resistant to tetracycline. Ampicillin, ceftiofur, penicillin, florfenicol, and enrofloxacin resistance rates ranged from 0% to 7.8%. Genotypically, numerous AMR genes were detected in the dataset but mostly in low proportions, with the exception of *tetO* (89%) and the *ermB* (80.1%) genes that were detected in the majority of isolates. Other detected genes included: *tetM* (7%), *tet44* (5%), *tetS* (5%), *ant6-Ia* (5.5%), *lnuC* (4.5%), *linB* (4%), *tet32* (2.5%), and *tetL* (2.5%) were among the most prevalent detected genes. β-lactam resistance genes were not detected. A total of 95.1% of tetracycline resistant isolates were positive for at least one gene conferring resistance to tetracycline (Table 1). This trend was not observed for other drugs evaluated.

Discussion

The high prevalence of *tetO* and *ermB* genes is consistent with previous studies.[1] Phenotypic and genotypic correlation was not evident for β-lactams as there was an absence of known AMR genes in phenotypically resistant isolates. The absence of resistance genes in phenotypically resistant isolates could be attributed to alternative resistance mechanisms or unidentified mechanisms not captured by the currently studied AMR genes. The majority (71%) of the penicillin-resistant isolates are likely opportunistic/commensal based on the absence of virulence-associated markers, not clustering with known pathogenic strains, untypable by serotyping or MLST, and carriage of multiple AMR genes.[2] Only two penicillin-resistant isolates belonged to ST1 or ST28, the most common pathogenic strains in US herds. This suggests that florfenicol, enrofloxacin, and β-lactam drug resistance are likely very low within pathogenic populations. Continued research is crucial for advancing the understanding and effective utilization of whole-genome sequencing data in the context of antimicrobial stewardship programs targeting *S suis*.

References

1. Lunha K, et al., Relationship between Penicillin-Binding Proteins Alterations and β-Lactams Non-Susceptibility of Diseased Pig-Isolated *Streptococcus suis*. *Antibiotics*. 2023 Jan 12;12(1):158.

2. Weinert, L., Chaudhuri, R., Wang, J. et al. Genomic signatures of human and animal disease in the zoonotic pathogen *Streptococcus suis*. Nat Commun 6, 6740 (2015). **https://doi.org/10.1038/ncomms7740**

Table 1: Relationship between phenotypic and genotypic resistance of selected antibiotics

Antibiotic	Phenotype: resistant (R)		Phenotype: susceptible (S)			
	Genotype: R	Genotype: S	Genotype: R	Genotype: S	Percentage[a]	Percentage[b]
Beta-lactam						
Ampicillin	0	0	0	400	0	100
Penicillin	0	31	1	355	0	99.7
Ceftiofur	0	24	0	372	0	100
Quinolone						
Enrofloxacin	0	9		379	0	100
Phenicol						
Florfenicol	0	1	0	390	0	100
Tetracycline	369	19	1	2	95.1	66.7

[a] Percentage of resistant isolates with presence of AMR gene/s; [b] Percentage of isolates susceptible without any detected AMR gene/s for a given drug.

https://doi.org/10.54846/am2024/109

Optimizing PRRSV RNA detection in oral fluids - cheap and easy?

B. Munguía-Ramírez, DVM, MS; B. Armenta-Leyva, DVM, MS; L. Giménez-Lirola, PhD; J. Zimmerman, DVM, PhD
Veterinary Diagnostic and Production Animal Medicine, Iowa State University, Ames, Iowa

Introduction

Oral fluid samples are widely used in swine surveillance for pathogens of economic importance to the swine industry. Surveillance often requires collecting specimens for PCR. Testing results then form the basis of prevention-and-control decisions. Previous data showed a loss in detectable PRRSV RNA in oral fluids as a function of storage temperature × time.[1] Thus, protecting PRRSV RNA from degradation in oral fluids prior to PCR testing is crucial. A variety of commercial products are sold to preserve nucleic acids in diagnostic specimens, eg, RNAlater®, FTA® cards, etc. However, these products are expensive, and there is little peer-reviewed data with sufficient evidence of their efficacy. As an alternative approach, we tested readily available chemical agents in terms of their capacity to stabilize PRRSV RNA in oral fluid samples stored under adverse conditions.

Methods

PRRSV-negative oral fluids collected from the ISU-VDL were pooled and centrifuged (3,300× g for 3 hr) to obtain a 30 ml stock solution. The stock solution was inoculated with 1×10^{-3} of PRRSV MLV (Ingelvac®), aliquoted in triplicates (1 mL), and assigned to one of 7 chemical agent treatments. All samples were stored at 30°C for 24 hours and tested for PRRSV RNA by RT-qPCR (IDEXX Laboratories). For comparison, two untreated controls were included: one held at -80°C and one held at 30°C for 24 hours. Results were expressed in terms of percent (%) recovery relative to the untreated control held at -80°C.

Results

Recovery (%) of PRRSV RNA in oral fluids among treatments were: No stabilizer (68.5%), treatment 1 (83.7%), treatment 2 (31.4%), treatment 3 (48.5%), treatment 4 (92.5%), treatment 5 (91.1%), treatment 6 (91.6%), treatment 7 (90.0%) (Figure 1). This preliminary data strongly suggested that we can stabilize PRRSV RNA in oral fluid samples using readily available chemical agents. Work continues to identify the optimal stabilizer (or combination of stabilizers) and concentration for PRRSV RNA. The best-performing stabilizer will then be evaluated for its capacity to stabilize different swine pathogens of economic importance in oral fluids.

Reference

1. Munguía-Ramírez B, Armenta-Leyva B, Henao-Díaz A, Cheng TY, Zhang J, Rawal G, Ye F, Giménez-Lirola L, Zimmerman J. Effect of extrinsic factors on the detection of PRRSV and a porcine-specific internal sample control in diagnostic specimens tested by RT-rtPCR. *J Vet Diagn Invest.* 2023; 35(4): 375–384. **https://doi.org/10.1177/10406387231174556**

Figure 1: Percent recovery of PRRSV RNA by RT-qPCR in oral fluids stored at 30°C for 24 hours.

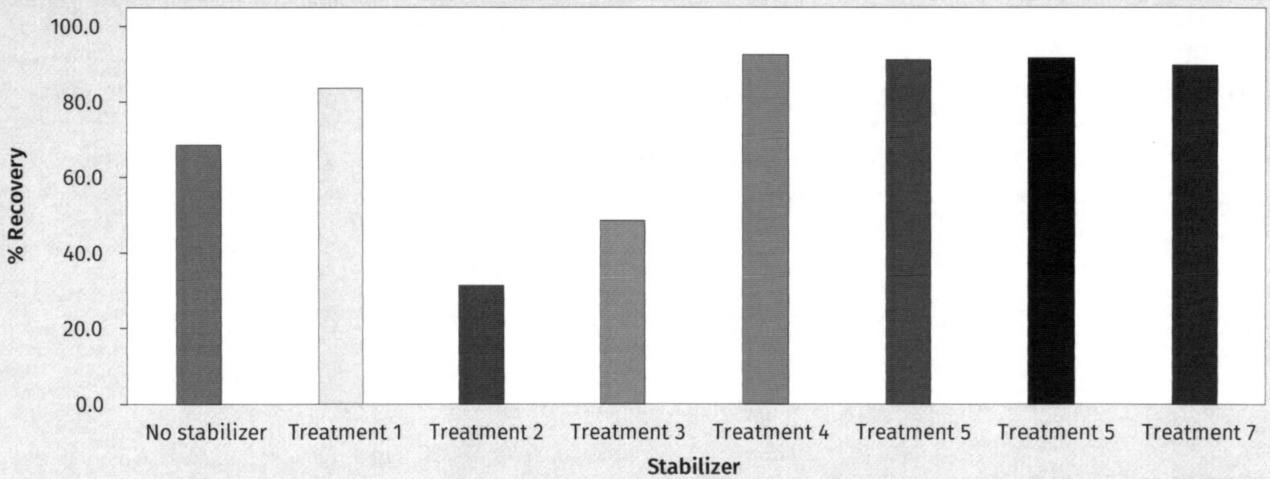

https://doi.org/10.54846/am2024/110

Comparison of the pathology and clinical effects of an F18 enterotoxigenic *Escherichia coli* containing a *tia* adhesin gene against a contemporary F18 *Escherichia coli* strain

Rodrigo C. Paiva[1]; Ethan Aljets[1]; Chen Hsin[1]; Patumporn Jermsutjarit[1]; Parin Watcharavongtip[1]; Diana Bermudez[1]; Hao Tong[1]; Philip Gauger[1]; Deb Murray[2]; Pablo Pineyro[1]; Marcelo Almeida[1]
[1]Department of Veterinary Diagnostic and Production Animal Medicine, Veterinary Diagnostic Laboratory, Iowa State University, Ames, Iowa; [2]New Fashion Pork, Jackson, Minnesota

Introduction

Diarrhea due to enterotoxigenic *Escherichia coli* (ETEC) remains one of the main enteric challenges to the United States (US) swine production.[1,2] The Iowa State University Veterinary Diagnostic Laboratory (ISU-VDL) data shows that ETEC is the primary pathogen involved in post-weaning diarrhea (PWD). Different virulence factors associated with ETEC PWD have been reported. There are two separate chromosomally encoded invasion loci (*tia* and *tib*) described in human-specific classical ETEC strains that have been detected in samples collected for surveillance in market hogs in an abattoir but with unknown clinical relevance in pigs. This study aimed: (a) to assess the clinical impact of ETEC-F18*tia*$^+$; (b) to compare the clinical impact of ETEC-F18*tia*$^+$ with ETEC-F18$^+$/*tia*$^-$; (c) to assess the efficacy of a commercially available F18 *E coli* vaccine as a competitive exclusion product for PWD caused by an ETEC-F18$^+$/*tia*$^+$.

Materials and methods

Seventy-two three-week-old pigs were housed at the Iowa State University (ISU) Livestock Infectious Disease Isolation Facility (LIDIF). Pigs were individually weighed and randomly allocated in five treatment groups, negative control (NC; N = 8), vaccinated, challenged with *E coli* F18:LT:STa:STb:Stx2e (V/C-F18$^+$/*tia*$^-$; N = 16), non-vaccinated, challenged with *E coli* F18:LT:STa:STb:Stx2e (NV/C-F18$^+$/*tia*$^-$; N = 16), vaccinated, challenged with *E. coli* F18:LT:STa:STb:Stx2e:*tia* (V/C-F18$^+$/*tia*$^+$; N = 16), non-vaccinated, challenged with *E coli* F18:LT:STa:STb:Stx2e:*tia* (NV/C-F18$^+$/*tia*$^+$; N = 16). All pigs were confirmed to be susceptible to F18 *E coli* infection by FUT1 gene genotyping by PCR. Pigs were weighed individually on dpi -5, -2, 0, and 7. Rectal swabs were collected on dpi 0, 1, 2, 3, 5, and 7. The hemolytic *E coli* shedding was assessed using a semiquantitative method, and individual colonies' ID was confirmed using Matrix-Assisted Laser Desorption/Ionization time-of-flight mass spectrometry (MALDI-TOF MS) technique. The data were analyzed using a repeated measures model for longitudinal data.

Results

The groups V/C-F18$^+$/*tia*$^+$; NV/C-F18$^+$/*tia*$^+$; V/C-F18$^+$/*tia*$^-$; NV/C-F18$^+$/*tia*$^-$ were significantly different ($P \leq .05$) from the NC. There was a numerical difference in days of diarrhea and weight per pig between NV/C-F18$^+$/*tia*$^+$ group compared with V/C-F18$^+$/*tia*$^-$; NV/C-F18$^+$/*tia*$^-$ and a significant difference compared with NC (Table 1).

The fecal score was significantly different ($P \leq .05$) on treatments V/C-F18$^+$/*tia*$^-$; NV/C-F18$^+$/*tia*$^-$; V/C-F18$^+$/*tia*$^+$; NV/C-F18$^+$/*tia*$^+$) compared to the NC. By 7 dpi, pigs in the NV/C-F18$^+$/*tia*$^+$ group had numerically higher average diarrhea scores (fecal score ≥ 2)

Discussion and conclusions

The ISU VDL cases of *E coli* have been higher than the historical average. In this study, the *E coli* field isolate containing the *tia* gene was capable of causing diarrhea in challenged pigs and should be considered as a potential virulence factor contributing to disease in pigs post-weaning. Further investigation of this gene is needed to better understand its pathogenesis and impact on the swine industry. The results of this study may help to better understand cases of enteric challenges in growing pigs without diagnostic confirmation and consider the novel *E coli tia*$^+$ genotype as a potential differential. Thus, veterinarians should consider the association of this gene with enteric cases of diarrhea and trigger a deep investigation of the field cases.

Table 1: Average days of diarrhea per pig and weight gain (lbs) according to treatment groups

Treatment	Av. days of diarrhea per pig	Weight gain (lbs)
NC	.38 (03/08)	6.1
V/C-F18$^+$/*tia*$^+$	1.56 (25/16)	4.1
NV/C-F18$^+$/*tia*$^+$	1.56 (25/16)	5.8
V/C-F18$^+$/*tia*$^+$	1.69 (27/16)	5.1
NV/C-F18$^+$/*tia*$^+$	2.13 (34/16)	4.5

Acknowledgment

The American Association of Swine Veterinarians Foundation for funding this study.

References

1. Fairbrother, John M., and Éric Nadeau. "Colibacillosis." *Diseases of swine* (2019): 807-834.
2. Gyles, C. L., and J. M. Fairbrother. "Escherichia coli." *Pathogenesis of bacterial infections in animals* 4 (2010): 267-308.

https://doi.org/10.54846/am2024/111

Weathering the storm: Extreme weather events and their association with PED and PRRS occurrence

Igor A. D. Paploski; Gabriela Kiesel; Dennis N. Makau; Nakarin Pamornchainavakul; Julia P. Baker; Mariana Kikuti;Cesar A. Corzo; Kimberly VanderWaal
Department of Veterinary Population Medicine, University of Minnesota, Saint Paul, Minnesota

Introduction

Despite the wealth of knowledge regarding risk factors for porcine epidemic diarrhea (PED) and porcine reproductive and respiratory syndrome (PRRS), weather risk factors historically received little attention to their association with the occurrence of both diseases. Our objective was to investigate the relationship between exposure to extreme weather events and PED or PRRS outbreaks.

Methods

We conducted a case-control study to investigate if the frequency of exposure to extreme weather events was different between cases (farms experiencing new outbreaks of PRRS or PED) and controls (farms not experiencing a new outbreak of PRRS or PED). Outbreaks of both PED and PRRS were captured by the Morrison Swine Health Monitoring Project (MSHMP). County-level extreme weather event data were obtained from the National Oceanic and Atmospheric Administration storm events database for the states in the SE US. We modelled the association between weather events and PED/PRRS occurrence using a 1:4 unmatched logistic regression. For each case (a farm in a given week in which an outbreak occurred), four different controls (farm in weeks in which an outbreak had not occurred) were randomly selected. For each of the analyzed weather event (flood, heavy rain, high wind and tornadoes), we ran ten different models, with each model including a different lag of time between exposure and outbreak occurrence, ranging from lag 1 (if the weather event occurred during the outbreak week) to lag 10 (if the weather event occurred nine weeks before the outbreak). Analyses were run on three periods of time based on trends on PEDV incidence: epidemic (2014 - 2015); endemic (2016 - 2019) and overall (2014 - 2019).

Results

We found that certain extreme weather events are consistently associated with the occurrence of PED. During the endemic period, farms located in counties exposed to floods had three to four times higher odds of having a PED outbreak in the four to nine weeks prior to the flood compared to controls (Figure 1). We also saw increased odds of PED occurrence after heavy rain (endemic period) and high wind (epidemic period). Associations of weather events and PRRS were less clear. We found that PRRS cases only had increased odds of being exposed to high wind in 2016 - 2019. Overall, associations tended to be weak and with wide confidence intervals.

Discussion

The exact mechanism by which each weather event influences the risk of disease occurrence was not possible to investigate, but weather events may potentially influence outbreak occurrence by disrupting aspects of farm management, for example, by increasing the likelihood of biosecurity breaches or changing truck routes and delivery schedules. We suggest that swine-producing companies in the US develop biosecurity protocols to account for and mitigate the impact of extreme weather events. These could potentially include strengthening protocols for visits of maintenance teams assessing physical damages to farms, and more stringent biosecurity practices in the immediate weeks following an extreme weather event.

Figure 1: Odds ratios and 95% confidence intervals for the association of PED outbreaks and floods in the two different analyzed periods (figure columns) for different week-lags after the occurrence of the weather event (rows within graphs). Vertical red lines display the null value for the odds ratio. Confidence intervals with a pipe symbol ("|") on their upper limit display a censored confidence – the true upper limit of the confidence interval extends beyond the x-axis of the graph. An "x" symbol beyond the upper limit of a confidence interval displays associations with a *P*-value < 0.05 (as does a brighter blue color). The association between PED and other extreme weather events or PRRS were less noteworthy and are not shown.

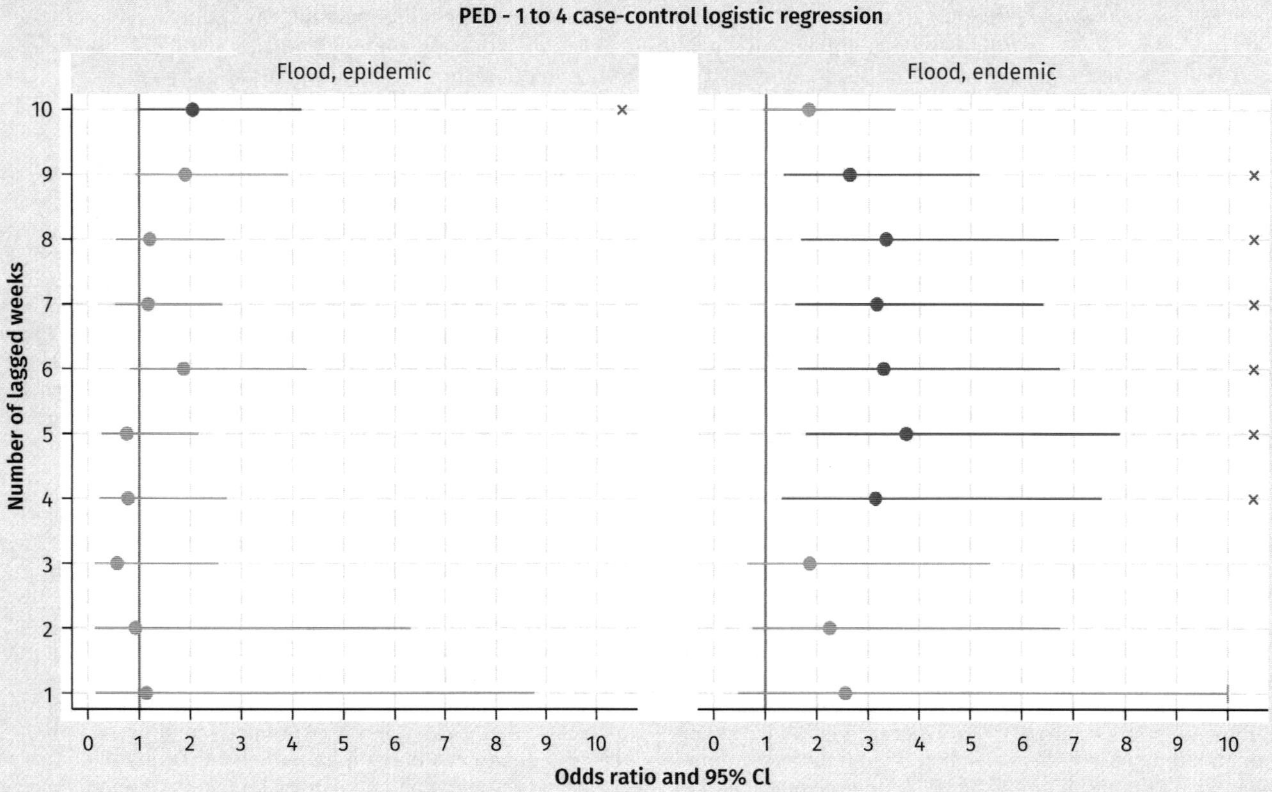

https://doi.org/10.54846/am2024/112

Investigation of *Lawsonia intracellularis* prevalence in suckling piglets

V. Rodriguez-Vega[1,3], DVM; F. L. Leite[2], DVM, PhD; R. García[1], DVM; A. Carvajal[3], DVM, PhD; H. Arguello[3], DVM, PhD
[1]Boehringer Ingelheim España, S.A. Spain [2] Inga Food Extremadura, Almendralejo. Spain[2] Boehringer Ingelheim Animal Health USA Inc. Duluth, Georgia, United States; [3]Grupo de investigación Digespor. Universidad de León

Background and objectives

Porcine proliferative enteropathy (PPE), caused by *Lawsonia intracellularis,* is one of the most prevalent enteric diseases worldwide[1] with a major impact on pig performance. The information about the prevalence of *L intracellularis* in suckling piglets is sparse due to the difficulties of sampling lactating piglets (ie, ileum content). The shedding of *L intracellularis* by peri-parturient sows has been demonstrated and suckling piglets can be potentially exposed to the bacteria,[2] however the infection of suckling piglets has not been assessed yet. The objective of this study was to investigate the prevalence of *L intracellularis* in 3-week-old piglets intended for human consumption (Spanish traditional specialty of oven cooked piglet).

Materials and methods

Fifteen commercial sow herds from the center of Spain were included in this study. Sampling was carried out between March 2022 and July 2022. In these farms piglets never receive any treatment from birth to weaning (ie, no iron injection, no antibiotics...). They are sent to the abattoir for meat at 3 weeks of age and ± 5 kg of live weight.

A total of 375 piglets were sampled, 25 from each farm. The whole intestinal package was collected for this study and submitted to the laboratory for processing. From each intestinal package, the content was collected. These samples were tested individually for *L intracellularis* by quantitative RT PCR.

Results

Lawsonia intracellularis was found in 31 intestinal content samples, 23 considered weak positives, with 10^2 bacteria/gr. Eight samples were strong positives: 5 between 10^3 and 10^4 bacteria/gr and 3 with 10^5 bacteria/gr.

Eight out of 375 ileum samples and 5 out of 15 sampled sow herds were positive (Table 1).

Table 1: Results of *L intracellularis* detection by q PCR in intestinal content

# positive PCR/# samples	8/375
% positive samples	2.1%
# positive PCR/# sow herds	5/15
% positive sow herds	33%

Discussion and conclusions

The results of this study indicate the presence of *L intracellularis* in the intestinal content of piglets. It demonstrates that vertical transmission can occur during the suckling phase from sows to piglets. While the prevalence is low, these piglets are a potential source of infection during the post-weaning period.

The clinical implications of these findings and the preventive measures to prevent the vertical transmission of *L intracellularis* in sow herds must be investigated in further studies.

References

1. Salleras et al. Proc 19th IPVS Congress 2006. Copenhagen P 174.
2. Patton et al 52nd AASV Meeting P74.

https://doi.org/10.54846/am2024/113

Detection of *Mycoplasma hyosynoviae* antibodies in pigs from clinically affected and non-affected herds

Haley K. Schwecke[1]; Emily McDowell[2]; Amanda Sponheim[1,3]; Joel Nerem[2]; Robert Valeris-Chacin[4]; Maria Pieters[1,5,6]
[1] Department of Veterinary Population Medicine, College of Veterinary Medicine, University of Minnesota, St. Paul, Minnesota; [2] Pipestone Veterinary Services, Pipestone, Minnesota; [3] Boehringer Ingelheim Animal Health, Duluth, Georgia; [4]Veterinary Education, Research, and Outreach (VERO), Department of Veterinary Pathobiology, College of Veterinary Medicine and Biomedical Sciences, Texas A&M University, College Station, Texas; [5] Veterinary Diagnostic Laboratory, College of Veterinary Medicine, University of Minnesota, St. Paul, Minnesota; [6] Swine Disease Eradication Center, University of Minnesota, St. Paul, Minnesota

Introduction

Mycoplasma hyosynoviae is a commensal bacterium that can induce lameness in growing pigs.[1] Information on the ecology, epidemiology, pathogenesis and diagnostics of *M hyosynoviae* is limited. However, there is a dire need to generate data for prevention and control of *M hyosynoviae*-associated disease in modern swine production. Therefore, the aim of this study was to determine the natural antibody profile of *M hyosynoviae* in clinically affected and non-affected swine herds in order to optimize the application of preventative measures.

Methods

One clinically affected and one clinically non-affected *M hyosynoviae* sow farm (9,000-head each) and five nursery and/or grow-finish downstream barns, from each sow farm, were enrolled in this study. Blood and colostrum samples were conveniently collected from varying age groups categorized as dams, replacement gilts, and piglets. The following samples were tested via indirect ELISA for *M hyosynoviae* IgG detection"[2] dams at service (n = 132), midgestation (n = 166), farrowing (n = 133), weaning (n = 134), replacement gilts 30 weeks of age (wks; n = 134), 25 wks (n = 133), 20 wks (n=145), 15 wks (n = 137), 10 wks (n = 135), 5 wks (n = 103), newborn piglets (n = 135), weaning piglets (n = 151), colostrum (n = 50), nursery pigs 5 wks (n = 51), 7 wks (n = 110), 8 wks (n = 52), 9 wks (n = 106), 10 wks (n = 110), 12 wks (n = 109), and finisher pigs 13 wks (n = 50), 16 wks (n = 52), 19 wks (n = 103), 20 wks (n = 51). Results (S/P values) were analyzed by herd and age group.

Results

The percent positives pigs for IgG S/P values per age group were: In the clinical herd: dams at service 93.0%, midgestation 97.4%, farrowing 99.0%, colostrum 100.0%, dams at weaning 100.0%, newborn piglets 86.4%, weaning piglets 92.0%; nursery 7 wks 39.1%, 9 wks 28.3%, 10 wks 17.3%, 12 wks 3.7%; finishing 19 wks 9.7%; gilts: 5 wks 66.0%, 10 wks 13.7%, 15 wks 1.0%, 20 wks 11.7%, 25 wks 12.0%, and 30 wks 24.8%. In the non-clinical herd, percent positives pigs were: dams at service 100.0%, midgestation 100.0%, farrowing 100.0%, colostrum 100.0%, dams at weaning 96.9%, newborn piglets 96.9%, weaning piglets 84.3%, nursery: 5 wks 52.9%, 8 wks 33.3%, 13 wks 18.0%, 16 wks 46.2%; gilts: 10 wks 6.1%, 15 wks 9.1%, 20 wks 38.2%, 25 wks 48.5%, and 30 wks 72.7%.

Discussion

The four stages of dams showed the highest percent positives when comparing across the three categories of pigs in both type of herds. The percent positive of IgG antibodies decreased at approximately five weeks of age in the nursery pigs and remained low through the finishing phase in the clinically affected herd. Whereas the percent positive of IgG antibodies in the non-clinical herd increased in growing pigs at approximately 16 weeks of age. Our results appear to support the hypothesis that *M hyosynoviae* antibodies are associated with protection against *M hyosynoviae* clinical signs. Similar serological patterns have been identified in previous investigations.[3,4] However, the commensal nature of the bacterium implies challenges for interpretation of serological profiles. Overall, this information sheds light for the evaluation of application timing for disease prevention and control measures.

References

1. Gomes-Neto JC, Gauger PC, Strait EL, Boyes N, Madson DM, Schwartz KJ. Mycoplasma-associated arthritis: critical points for diagnosis. J Swine Health Prod. 2012; 20: 82-86.

2. Giménez-Lirola LG, Meiroz-De-Souza-Almeida H, Magtoto RL, McDaniel AJ, Merodio MM, Matias Ferreyra FS, Poonsuk K, Gatto IRH, Baum DH, Ross RF, Arruda PHE, Schwartz KJ, Zimmerman JJ, Derscheid RJ, Arruda BL. Early detection and differential serodiagnosis of *Mycoplasma hyorhinis* and *Mycoplasma hyosynoviae* infections under experimental conditions. PLoS One. 2019 Oct 7;14 (10):e0223459.

3. Hagedorn-Olsen NC, Nielsen NC, Friis NF, Nielsen J, Progression of *Mycoplasma hyosynoviae* Infection in Three Pig Herds. Development of Tonsillar Carrier State, Arthritis and Antibodies in Serum and Synovial Fluid in Pigs from Birth to Slaughter. J Vet Med. 1999; 46: 555-564.

4. Nielsen EO, Lauritsen KT, Friis NF, Enøe C, Hagedorn-Olsen T, Jungersen G. Use of a novel serum ELISA method and the tonsil-carrier state for evaluation of *Mycoplasma hyosynoviae* distributions in pig herds with or without clinical arthritis. Veterinary Microbiology. 2005; 11 (1-2): 41-50.

https://doi.org/10.54846/am2024/114

Effects of bioactive feed ingredients on improvement of clinical response to post-weaning F18 enterotoxigenic *Escherichia coli* (ETEC) infection

Corinne Schut[1], MSc; Caroline González-Vega[2], MSc, PhD; Brooke Smith[2], DVM, PhD; Zhenbin Zhang[2], MSc, PhD; Ehsan Khafipour[2], MSc, PhD; Abdolvahab Farzan[1], DVM, MSc, PhD
[1]Department of Population Medicine, University of Guelph, Guelph, Ontario, Canada; [2]Cargill, Minneapolis, Minnesota

Introduction

Enteric diseases caused by enterotoxigenic *Escherichia coli* (ETEC) are one of the most common illnesses in newly weaned pigs and continue to be economically important for the swine industry due to the morbidity and effect on performance. Restrictions on the use of antibiotics and zinc oxide as a prophylaxis for control of enteric pathogens have resulted in development of different bioactive feed additives with varying modes of action.[1] The objective of this study were **1)** to evaluate the efficacy of phytogenics, a postbiotic, and a mix of organic acids with essential oils for inhibition of F18 ETEC, and **2)** to assess the impact of these feed additives on growth performance and fecal shedding of *E coli* in piglets following experimental challenge.

Methods

A total of 144 weaned pigs (Yorkshire/Landrace/Duroc pigs) with initial body weight (BW) of 6.62 ± 1.37 kg were divided in 2 blocks with 12 pens/block and 6 pigs/pen for a total of 24 pigs/treatment adjusted for gender and litter. Pigs were housed in the biosafety level 2 animal isolation facility at the University of Guelph, Canada and randomly assigned to 6 treatments: **1)** a basal diet with no feed additives (negative control, NC), **2)** NC + 0.1% phytogenic feed additive (PFP, Fresta® Protect, Delacon, Austria), **3)** NC + 0.2% postbiotic (PB), **4)** NC + 0.2% new phytogenic prototype (NPP), **5)** NC + 0.2% organic acid and essential oil blend (OA + EO), and **6)** NC + 0.1% phytogenic (Fresta® Protect) + 0.2% postbiotic (PFP + PB). The duration of study was 15 d, with a 7-d pre-challenge and 8 d post-challenge period. On d 8 and 9, after fasting for 5 hours, all pigs were orally inoculated with 5×10^{10} CFU/mL of F18 ETEC (O147:H2, LT:STb:STa:Stx2:F18). Body weight and feed intake were recorded on d 0, 8 and 15. Observed clinical indicators of disease were scored daily pre- and post-challenge. Fecal swabs were collected on d 1, 8, 9, 10, 11, and 15 and cultured on blood agar to quantify *E. coli* shedding. At the end of the trial, all pigs were euthanized, and the consistency of intestinal contents were scored and tissue samples were collected in formalin for histopathological assessment. Data were analyzed using STATA 18 (StataCorp, College Station, Texas, Texas, USA) with a multi-level mixed effects regression method with block and pen as random effects. Effects were considered significant at $P < 0.05$. Trends were discussed at $P < .10$.

Results

Pigs fed PFP, OA + EO, and PFP + PB had reduced ($P < .05$) occurrence of moderate/severe diarrhea (score 2 - 3 out of 3) post-challenge compared to NC pigs (Figure 1A). On d 3 post-challenge, pigs fed NPP, OA + EO, and PFP + PB had reduced (5.44, 5.66, 5.94 log CFU/mL, respectively; $P < .01$) fecal *E coli* shedding compared to NC (7.27 log CFU/mL). On d 7 post-challenge, while no significant difference in fecal *E. coli* shedding was observed among treatments, PFP pigs had numerically the lowest shedding (2.58 log CFU/mL; Figure 1B), reduced number of moderate/severe (score 2 - 3 out of 3) intestinal lesions on the jejunum and ileum, and reduced intestinal content consistency (ie, decreased water content) in the ileum and cecum compared to NC pigs. Pigs fed PFP and PFP + PB had numerically less adherent Bacilli in the jejunum and ileum than NC pigs. Pre-challenge average daily gain (ADG) and BW on d 8 of study were less ($P = .01$) with PFP and PFP + PB, but no differences were observed in growth performance post-challenge or during the total study period.

Discussion

Multiple feed additives evaluated in this study supported the reduction of clinical disease following an acute high dose ETEC challenge; however, the most positive effects were observed with PFP and PFP+PB. Phytogenic feed additives have been reported to control clinical diarrhea associated with ETEC in previous studies, possibly due to their antimicrobial and anti-inflammatory effects reducing the load of ETEC in the gut.[2,3] In the current study, PFP showed reduced rates of diarrhea, ETEC loads on d 3 and 7 compared to NC without negatively impacting overall growth performance making it a promising candidate for control of ETEC infection.

References

1. Castro J, Barros MM, Araújo D, et al. Swine enteric colibacillosis: Current treatment avenues and future directions. *Front Vet Sci.* 2022;9. **https://doi.org/10.3389/fvets.2022.981207**

2. Montoya D, D'Angelo M, Martín-Orúe SM, et al. Effectiveness of two plant-based in-feed additives against an *Escherichia coli* F4 oral challenge in weaned pigs. *Animals.* 2021;11(7). **https://doi.org/10.3390/ani11072024**

3. Sun Y, Kim SW. Intestinal challenge with enterotoxigenic *Escherichia coli* in pigs, and nutritional intervention to prevent postweaning diarrhea. *Anim Nutr.* 2017;3(4):322-330. **https://doi.org/10.1016/j.aninu.2017.10.001**

Figure 1: Effects of various feed additives (phytogenic, postbiotic, and a mix of organic acids with essential oils) on **(A)** diarrhea scores* and **(B)** ETEC fecal shedding in weaned pigs challenged with F18 ETEC.

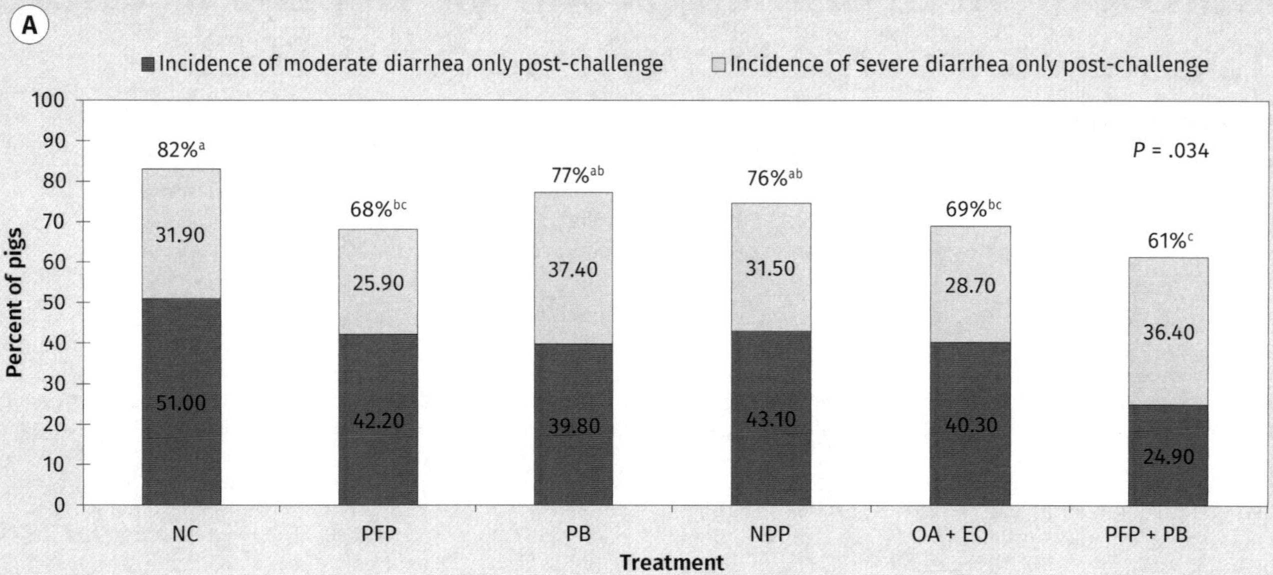

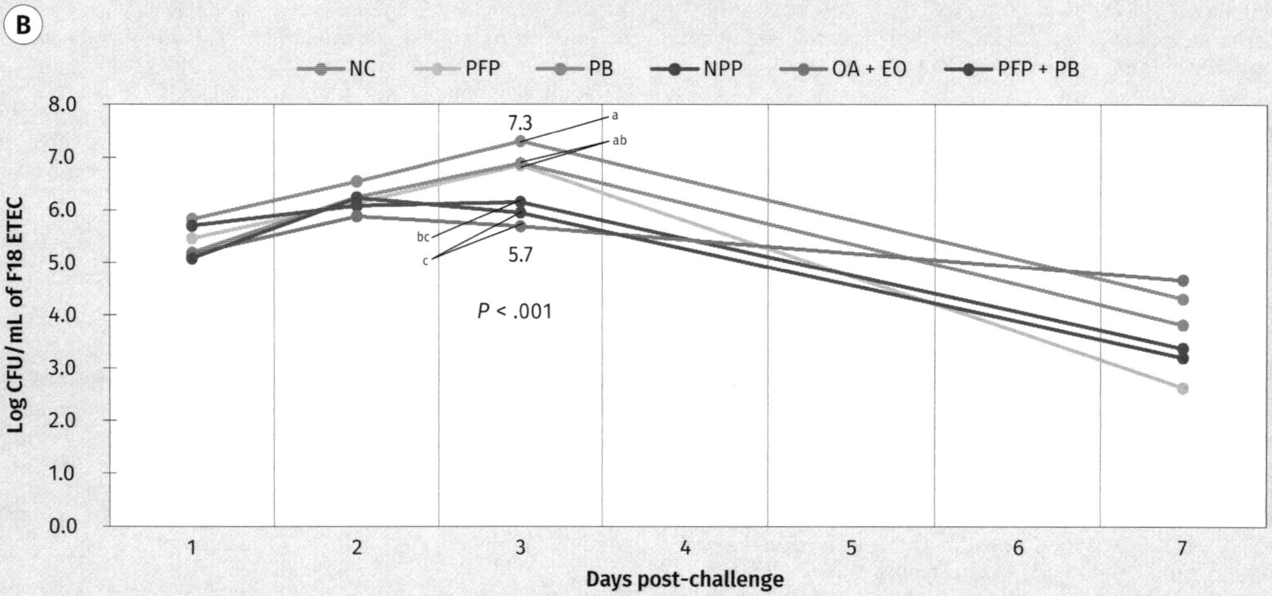

a-c Common superscript indicates no significant difference between treatments.

* Diarrhea was scored as: no diarrhea (0 - 1), moderate diarrhea (score 2), and severe diarrhea (score 3).

NC = negative control; PFP = NC + 0.1% phytogenic feed additive Fresta® Protect, Delacon, Austria); PB = NC + 0.2% postbiotic; NPP = NC + 0.2% new phytogenic prototype OA + EO = NC + 0.2% organic acid and essential oil blend.

https://doi.org/10.54846/am2024/115

A novel method for evaluating environmental and pig-derived targets in pen-based oral fluids

G. Tarasiuk[1]; M. Remmenga[2]; K. C. O'Hara[2]; M. Rotolo[3]; P. Zaabel[4]; L. Gimenez Lirola[1]; J. J. Zimmerman[1]
[1] College of Veterinary Medicine, Iowa State University, Ames, Iowa; [2]USDA: VS: Strategy and Policy, Center for Epidemiology and Animal Health, Fort Collins, Colorado; [3]PIC North America, Hendersonville, Tennessee; [4]National Pork Board, Des Moines, Iowa

Introduction

Oral fluids (OF) are widely used as a surveillance sample for infectious diseases of swine. The detection of specific pathogens is widely researched from the perspective of detection in the lab, but we have an incomplete understanding of the source of the target (pig vs environment) in the oral fluid sample. For example, porcine epidemic diarrhea virus (PEDV) is readily detected in OF, despite the fact that it is not shed orally. This study presents a novel approach, using the detection of fluorescence from food dye, to explore individual and environmental contributions to OF samples.

Experimental design

To avoid the complications of working with an infectious agent in the field, red food dye was used as a surrogate. FD&C Red dye is approved by the FDA for use in drugs, cosmetics, and foods, and has measurable fluorescence that can be quantified using standard laboratory equipment. Pilot studies evaluated: 4 different dye concentrations, treatment volumes, the addition of sugar to entice pigs to interact with the treatment, treatment to sampling intervals, and sample handling requirements, to consistently detect fluorescence in OF samples. The final protocol, used for both pig and environmental "treatments", included a dye solution composed of distilled water (50 mL), white sugar (25 g), and red food dye (50 mL) (McCormick & Co., INC., Hunt Valley, MD), and an extended centrifugation step during processing.

Methods

In brief, the study was conducted in pens holding ~30, ~65, and ~125, 12- to 14-week-old pigs (32 pens per size). Oral fluid samples were collected twice from each pen: one day prior to treatment (negative control) and again immediately after treatment. The negative control sample was used to provide a baseline fluorescence value, ie, to account for naturally occurring fluorescent substances (feces, feedstuffs, etc).

Pens received one of 2 treatments. For pig exposure, a convenience sample of 10% of the pigs in the pen were orally administered 3 mL of the dye solution. For environmental exposure, 20 mL of the dye solution was placed on the floor in the center of the pen. Immediately following treatments, ropes were hung for 30 minutes in pens of ~30 pigs, 45 minutes in the pens of ~65 pigs, and 60 minutes in the pens of 125 pigs. Pre- and post-treatment samples were tested for fluorescence at 530/570 nm (SpectraMax i3x, Molecular Devices LLC, San Jose, California) and the results analyzed for evidence that the dye was transferred from the environment and/or from pigs to the pen-based oral fluid sample.

Results and conclusions

For both treatments (environmental vs pig exposure) and regardless of pen size, specific fluorescence was consistently detected in pen-based oral fluid samples, successfully demonstrating that food dye can be used as a surrogate to evaluate environmental and pig-derived OF contributions.

https://doi.org/10.54846/am2024/116

Medical treatment of pelvic organ prolapse in late-gestation sows

G. Wideman, DVM; G. Greaves, MSc; C. Werth, MSc; K. Vilaca, DVM, MSc; C. Lichty, DVM
South West Ontario Veterinary Services, Stratford, Ontario, Canada

Background

Pelvic organ prolapse (POP) is a common cause of sow removals across the swine industry.[1,2] The welfare impact is severe and surgical treatment options are intensive and of limited value, leading to a high proportion of POP sows being euthanized. Euthanasia of a prolapsed sow leads to the opportunity loss of a cull sow and nurse-off litter, both of which negatively impact management and the cost of production.[3] Studies have sought to identify risk factors for prolapse including nutrition,[2] water treatment,[2] genetics,[4] and perineal score (PS).[5] The perineal scoring system specifically, offers a practical approach for identifying sows at risk for POP; as sows with a PS3 (severe swelling, redness and protrusion) have a higher risk compared to sows with a PS1.[3,5] Theoretically, anesthesia of the pelvis (to reduce the sensation of pain or swelling that may lead to pushing) and use of a diuretic (that reduces swelling via diuresis) may decrease the risk of prolapse in PS3 sows. Furosemide, a loop of Henle diuretic labelled for use in food animals in Canada, is commonly used to relieve tissue swelling in dairy cattle (udder edema) and horses (limb edema); however, little is known about use in swine. The observational study objective was to evaluate a medical POP treatment protocol in herds across Ontario, Canada.

Methods

In June 2023, using a convenience sample of veterinary clinic clients, herds with a POP prevalence of ≥ 2% were identified to implement the following protocol for sows with a PS3: wash tissue with cold water, replace prolapsed tissues (if possible), apply 10 - 20mL of Xylocaine 2% intravaginally once, administer 5mL of Salix® IM (repeat every 12 hrs up to two more times), and administer Dexamethasone-5 [1mL/45kg] IM once. If the treatment was successful (ie, diminished swelling, redness, and protrusion), the sow was culled once the lactation period ended. If the treatment was unsuccessful, the sow was euthanized. Producers were asked to record treatment steps taken and outcome data, per sow.

Results

Herds varied in size (548 - 5370 sows), genetics, and management practices (eg, stall and loose housing). To date, 76 sows were administered the protocol with 67% avoiding prolapse. Some herds were 100% successful at avoiding mortality due to POP in treated sows. Recovered sows lived longer and completed an average 22-day lactation period. Characteristics of sows that recovered versus sows that did not recover will be compared.

Discussion

This nonsurgical treatment for POP in sows with a PS3 may be a feasible option to improve welfare, complete a lactation period, and provide a cull sow opportunity compared to euthanasia; however, the pharmacokinetics of products used are not well understood. Investigation into the biological understanding of POP in sows remains important for informing treatment approaches.

References

1. Supakorn C, Stock JD, Hostetler C, Stalder KJ. Prolapse incidence in swine breeding herds is a cause for concern. *Open Journal of Veterinary Medicine*, 2017;7:85-97.

2. Chipman A, Rademacher CJ, Johnson CD, et al. Pelvic organ prolapse: An industry-wide collaboration to identify putative contributing factors. *ISU James D. McKean Swine Disease Conference, November 1-2*. 2018.

3. Kiefer ZE, Struder J, Chipman A, et al. Circulating biomarkers associated with pelvic organ prolapse risk in late gestation sows. *Journal of Animal Science*, 2021;99(8): 1-8.

4. Bhatia V, Stevens T, Derks MF, et al. Identification of the genetic basis of sow pelvic organ prolapse. *Frontiers in Genetics*, 2023;14:1154713.

5. Kiefer AE, Koester, LR, Showman, L, et al. Vaginal microbiome and serum metabolite differences in late gestation commercial sows at risk for pelvic organ prolapse. *Scientific Reports*, 2021;11: 6189.

https://doi.org/10.54846/am2024/117

Development and implementation of an early warning tool for disease occurrence in the US swine breeding herd population

X. Yue, PhD; M. Kikuti, PhD; C. M. Mellini, DVM; C. A. Corzo, DVM, PhD
Dept. of Veterinary Population Medicine, University of Minnesota, Saint Paul, Minnesota

Introduction

Recognizing the significance of understanding the regional risk of disease introduction for prevention and control in swine farms, tools to understand real-time risk are needed. Overtime, the Morrison Swine Health Monitoring Project (MSHMP) has built the capacity to create such tools through the strong collaborative effort of both practitioners and producers to communicate health events (eg, PRRS breaks) at the national level which ultimately has helped understand risk and translate this information into actionable interventions to reduce risk.[1] Therefore, The Early Regional Occurrence Warning Tool (TEROW) developed in this study aimed to generate an automated near-real-time report to alert participating systems of current regional PRRS occurrences related to their farms.

Methods

MSHMP-participating production systems were invited to participate and enrolled in this study, which currently includes 187 breeding herds representing approximately half a million sows. Upon voluntary weekly reporting, an algorithm was developed to calculate the total number of PRRS-positive sites within the neighborhood of each enrolled farm. The results were then compiled into a report, which is being currently shared weekly with all the participants in the study. Figure 1 shows an anonymous TEROW report with important information obfuscated. To balance the epidemiological relevance and confidentiality, an individual reporting radius for each studied farm was determined in 3 steps. First, the initial reporting radius is set to 25 miles for the farms in pig-dense states (eg, MN, IA) and to 50 miles for farms in other states. Secondly, four variables were generated to count the farms within the initial reporting radius: variable A) Total number of farms that are PRRS positive (AASV PRRS status 1); B) Total number of farms from other production systems that are PRRS positive; C) Total number of farms participating in this study, and D) Total number of MSHMP monitored farms. Subsequently, the reporting radius is automatically increased when there is a confidentiality concern, thus preventing participants from knowing which specific farms from other production systems are PRRS positive. Additionally, changes in the total number of positive farms (variable A) compared to the previous week are indicated by a "trend" variable with an up, down arrow or dash.

Results and conclusion

The report has been shared with participants for 28 weeks. In which participants have learned how to interpret the data but most importantly, there have been participants that are already making adjustments to their biosecurity programs. A clear example is the frequency of supply introduction and transportation route planning. The early PRRS occurrence warning tool developed in this study enhances producers' ability to effectively communicate and provides the opportunity to quickly respond to health threats in the form of infectious diseases to mitigate regional PRRS transmission, ultimately resulting in lower regional disease pressure. Recently, new systems approached us to join this project. More participants, together with the MSHMP project, the developed tool can be particularly useful in the emergence of existing and imminent health threats to the US swine industry allowing stakeholders to timely tailor prevention and control programs to effectively contain these pathogens.

Reference

1. Morrison Swine Health Monitoring Project. Description of the Project. *Morrison Swine Health Monitoring Project*. June, 4, 2023. Accessed November 4, 2023. **https://mshmp.umn.edu/about**

Figure 1: Anonymized report of The Early Regional Occurrence Warning Tool (TEROW).

Swine Health Information Center

October 24, 2023

MSHMP Morrison Swine Health Monitoring Project | University of Minnesota

Dr. Bob Morrison's Swine Health Monitoring Project

The Early Regional Occurrence Warning (TEROW)

A Swine Health Information Center Funded Project

Description of the TEROW report table

11 of the 39 MSHMP participating systems contribute to The Early Regional Occurrence Warning (TEROW). For each site, depending on region and pathogen, a reporting radius that balanced epidemiological relevance and confidentiality was determined. Specifically, the No. positive sites is the total number of PRRS positive sites in the current week (AASV PRRS status 1 – including wild-type introductions into status 1 herds) within the radius; the Trend shows the change in the No. positive sites from previous week, represented by ↑ for an increase, ↓ for a decrease, and — for no change; the No. positive sites from other systems is the number of PRRS positive sites from other TEROW participating systems in the current week within the radius; the No. TEROW sites is the total number of sites participating in TEROW within the radius; and No. MSHMP sites is the total number of sites participating in MSHMP within the radius. Disclaimer: All companies report regularly but at different frequencies. Confidential reminder: this report is generated for System please do not share, forward or distribute to any unauthorized individuals.

System	Farm name	Current PRRS status	Radius miles	No. positive sites	Trend	No. positive sites from other systems	No. TEROW sites	No. MSHMP sites
	[illegible]	4	25	3	—	0	5	7
	[illegible]	4	75	1	↓	0	5	7
	[illegible]	3	25	3	—	0	5	7
	[illegible]	3	25	3	—	0	5	7
	[illegible]	3	30	7	↓	0	8	9
	[illegible]	3	25	6	↓	0	9	9
	[illegible]	3	25	0	—	0	1	1
	[illegible]	3	25	0	—	0	1	1
	[illegible]	1	25	4	—	0	4	7
	[illegible]	1	25	2	—	0	5	10
	[illegible]	1	30	4	—	0	4	7
	[illegible]	1	25	4	—	0	4	7

Page 1 of 2

https://doi.org/10.54846/am2024/118

Factors influencing the PEDV time-to-stability and time-in-positive in US breeding herds: A retrospective analysis based on one decade of monitoring data

X. Yue, PhD; M. Kikuti, PhD; C. M. Melini, DVM; C. A. Corzo, DVM, PhD
Dept. of Veterinary Population Medicine, University of Minnesota, Saint Paul, Minnesota

Introduction

Porcine epidemic diarrhea virus (PEDV) was first identified in the United States (US) in April 2013. The epidemic period of PEDV lasted for approximately 1.5 years, transitioning to endemic levels in 2015. Over the span of 10 years encompassing the epidemic and endemic periods, little published work on how long a sow herd remains PEDV-positive (ie, time-in-positive, TIP) and how long it takes a sow herd to achieve a stable status consistently weaning PEDV-free pigs (ie, time-to-stability, TTS). The objective of this study was to estimate the PEDV TIP and TTS in breeding herds in the US during the epidemic and endemic periods and assess the effects of associated recorded factors.

Materials and methods

PEDV monitoring data originating from 1028 US herds that report PEDV status regularly between May 1, 2013 and June 30, 2023 were obtained from the Morrison Swine Health Monitoring Project, a Swine Health Information Center (SHIC) funded project. After data cleaning, TIP was calculated as the number of weeks that a herd stayed in PED status 1 (ie, positive unstable); TTS as the number of weeks that a herd stayed in status 1 before changing to status 2 (ie, undergoing elimination) or 2fvi (ie, ongoing gilt field virus exposure). Both status 2 and 2fvi require the herd to be absent of clinical signs and no detectable virus for a minimum of 4 weeks. Recorded factors assessed for TIP and TTS included: PEDV epidemic/endemic period, season, PED status before and after the break, whether the site experienced a PED outbreak within the previous 6 months, and the total number of prior breaks. Mixed-effects Cox proportional hazards models were used to assess the association between TIP / TTS and the recorded predictors. Herd ID was involved as the random effect.

Results and conclusion

This study included a total of 466 PEDV breaks in 302 sow herds, in which 236 PEDV breaks from 229 herds correspond to the epidemic period (i.e., from May 1, 2013 to December 31, 2014) and 230 breaks from 158 herds to the endemic period (ie, January 1, 2015 to June 30, 2023). Figure 1 shows the Kaplan-Meier survival curve of TIP (Figure 1A) and TTS (Figure 1B) by the study period in which the PEDV break occurred. Overall, the median TIPs in the PED epidemic and endemic period were 25 weeks (Interquartile range, IQR: 18; 31) and 17 weeks (IQR: 11; 24), respectively. The median TTS of the epidemic period was 24 weeks (IQR: 18, 31), while the median TTS of the endemic period was 14 weeks (IQR: 9; 22). Both TIP and TTS mixed-effects Cox proportional hazards models showed that herds that broke from a PED status 2fvi reached stability or other stable status sooner (hazard ratio 3.4, 95% CI: 2.3; 5.0 in TIP model and 2.5 95% CI: 1.7; 3.8 in TTS model) than

Figure 1: Kaplan-Meier time-in-positive **(A)** and time-to-stability **(B)** survival curve and 95% upper and lower bound confidence intervals (colored area) of 466 PEDV breaks in 302 sow herds in the United States, during the PEDV epidemic (May 1, 2013 to December 31, 2014)) and endemic (January 1, 2015 to June 30, 2023) periods.

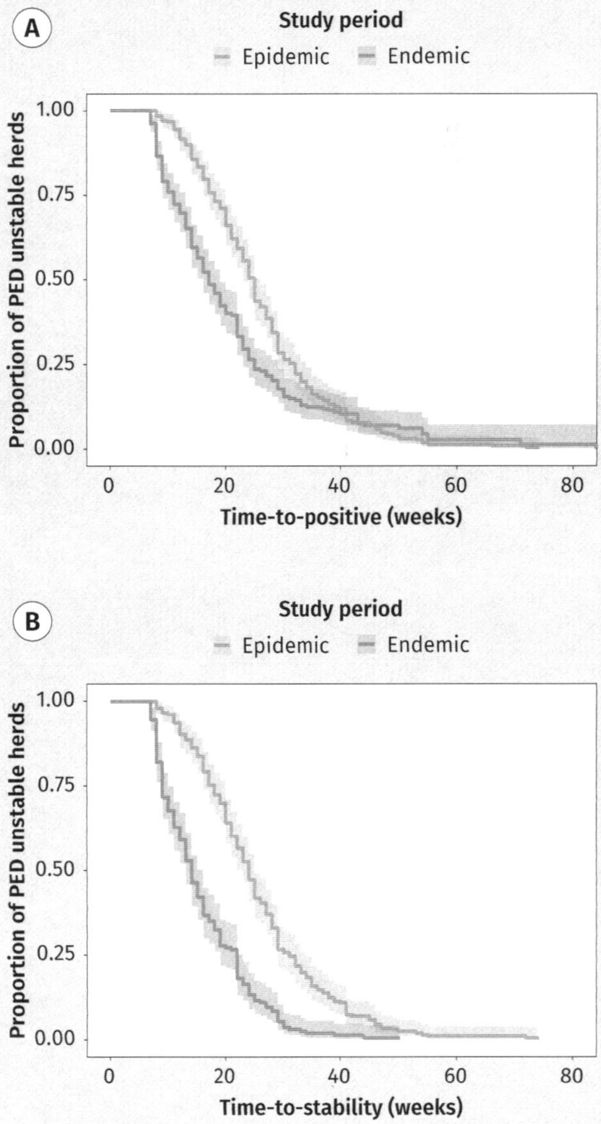

herds breaking from a PED status 4. Whether the site experienced a PED outbreak within the previous 6 months was not associated with TIP and TTS. This study provides valuable and objective insights into TIP and TTS of PEDV breaks in the US breeding herds especially at a time when the industry is considering whether this disease could be eradicated from the US swine population. These support decision-making in PEDV control and elimination strategies (eg, forecasting PEDV elimination timeline, herd closure plan) and farm management such as pig flow management and investment timing.

https://doi.org/10.54846/am2024/119

Risk factors for lameness in finisher pigs

A. Zoltick, DVM; L. Redding, VMD, PhD, DACVPM; M. Pierdon, VMD, DACAW
Clinical Studies New Bolton Center, University of Pennsylvania School of Veterinary Medicine

Introduction

Lameness is a common condition of growing pigs that compromises welfare and production performance due to associated pain, treatment costs, depressed growth rates, culling, and death. Lameness is multifactorial and suspected risk factors may be compounded, complicating diagnosis, treatment, and prevention. Although many studies examine lameness in sows, there is a scarcity of studies that investigate lameness in growing pigs. The three primary objectives of this study are to 1) determine the prevalence of lameness at two different stages of a finisher pig's growing life; 2) investigate both sow and finisher farm-level risk factors for lameness; and 3) quantify antimicrobial use on finisher farms.

Methods

The number of pigs included for lameness evaluation was based on Pork Quality Assurance (PQA) site assessment guidelines. A visual locomotion scoring system was used to evaluate pigs on a 0 - 2 scale ranging from no lameness to severe lameness, with the same observer responsible for each evaluation. Sow farm-level risk factors were collected via questionnaire to obtain information on sow and piglet antibiotic use, vaccination status, disease status, and dam and sire genetics. Finisher farm-level risk factors included housing factors, such as floor quality and cleanliness, stocking density, and facility temperature; nutrition, such as feed type and pan coverage, as well as trace mineral and mycotoxin levels in feed samples; and health, such as disease status with emphasis on oral fluid analysis for *Mycoplasma hyosynoviae* and porcine reproductive and respiratory syndrome virus (PRRSV) via rt-PCR and ELISA. For *M hyosynoviae* testing, two samples were collected: one pooled sample from all pens assessed and one sample from the pen with the highest lameness score. Antimicrobial use was recorded by on-farm treatment records and an antibiotic inventory. Multivariable linear regression analysis was used to examine the association between farm-level factors and lameness prevalence, and correlation analyses were used for the association between pen-level factors and pen-level lameness score.

Results

Preliminary findings were obtained from 6,862 pigs from 4 companies (n = 2,845; n = 1,039; n = 872; n = 2,106) assessed on 26 different farms at 63 - 108 days of age, the first time point of interest. The mean (SD) percentage of lame pigs per farm on the was determined to be 23.8% (10) with a range of 7 - 41% (Figure 1). 26.9% of pooled samples tested positive for *M hyosynoviae* and 42.1% tested positive for PRRSV. 19.2% of samples from pens with the highest combined lameness score tested positive for *M hyosynoviae*. At the farm level, there were no statistically significant associations between lameness prevalence and any farm-level factors. However, statistically significant correlations were observed at the pen level, including a weak correlation between the pen-level lameness score and floor quality (R_S = -0.185, P = .0095) and between the pen-level lameness score and feces content of flooring (R_S = 0.197, P = .006).

Discussion

The results of this study demonstrate a similar lameness prevalence to previous studies for pigs in similar housing conditions (3.9 - 34.3%) though lameness scoring systems varied by study.[1,2]

Figure 1: Proportion of pigs affected by lameness for 6,826 pigs on 26 farms at 63 - 108 days of age where a score of 0 was no lameness, 1 was mild lameness and, 2 was severe lameness. Each bar represents a single farm.

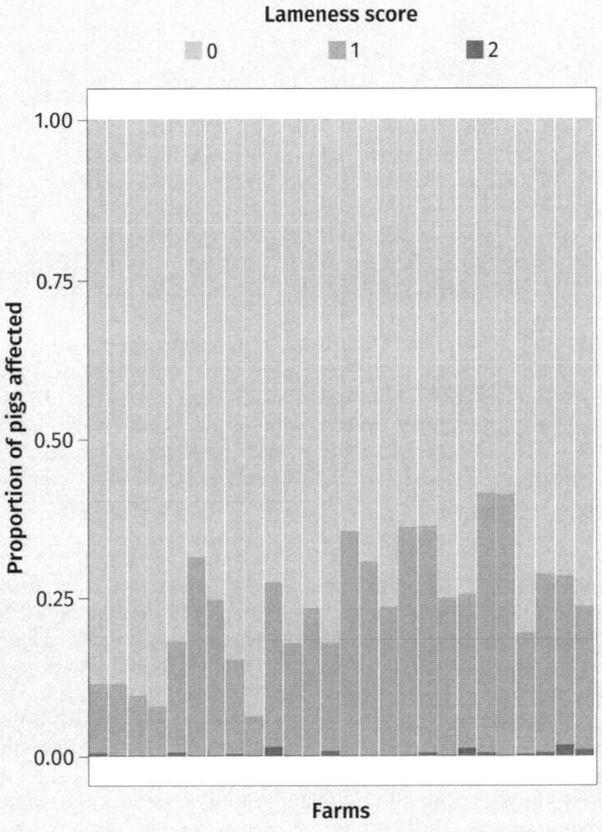

Additional data collection from more farms and from a second time period on each farm is ongoing. Identification of risk factors for lameness may help elucidate disease pathogenesis and direct producers towards beneficial management and husbandry practices. Lameness mitigation may also enhance antibiotic stewardship on swine farms.

References

1. Pillman, D., Nair, M.S., Schwartz, J. and Pieters, M., 2019. Detection of Mycoplasma hyorhinis and Mycoplasma hyosynoviae in oral fluids and correlation with pig lameness scores. *Veterinary microbiology, 239*, p.108448.

2. KilBride, A.L., Gillman, C.E. and Green, L.E., 2009. A cross-sectional study of the prevalence of lameness in finishing pigs, gilts and pregnant sows and associations with limb lesions and floor types on commercial farms in England. *Animal Welfare, 18*(3), pp.215-224.

 https://doi.org/10.54846/am2024/120

A broadly protective vaccine against swine influenza A virus based on the matrix protein 2 (M2)

F. A. Zuckermann[1,3], DVM, PhD; Y. Grionkova[2], PhD; R. J. Husmann[1], DVM, MS; M. Pires-Alvez[1], PhD; S. Storms[1], DVM; S. G. Sligar[2], PhD
Departments of [1]Pathobiology and [2]Biochemistry, University of Illinois at Urbana-Champaign; [3]Vastimmune Biologics, Inc. Champaign, Illinois

Introduction

The syndrome resulting from the infection of swine with influenza A virus (IAV) presents a major economic burden to the pork industry and is one of the top three diseases affecting pigs in all phases of pork production. Effective vaccines are the cornerstone of defense against acute influenza virus infections. Whole inactivated virus (WIV) vaccines are the only type of biologic used in the US to protect pigs against IAV. Experimental and field data indicates a limited protection provided to swine by commercial WIV influenza vaccines against contemporary strains of swine IAV (swIAV). The substantial diversity of the haemagglutinin (HA) among co-circulating IAV viruses in swine herds poses a significant challenge for effective vaccine development. Notably, the matrix protein 2 (M2) of IAV is a highly conserved protein present in the virus envelope. More than 98% of IAV strains circulating in American swine herds share the identical pandemic (pdm) isoform of M2. M2 is a 97 amino acid long transmembrane protein that forms a tetramer in the virus envelope and acts as a viroporin. M2 consists of an intracellular C-terminal domain (positions 47 to 97), a transmembrane domain (positions 24 to 46), and an extracellular N-terminal domain (positions 1 to 23). The ectodomain of M2 (M2e) has been pursued for many years as candidate for a potential universal influenza vaccine for humans, primarily due to the difficulty of assembling the entire M2 protein in its natural transmembrane configuration. For this project, we examined the immunogenicity and protective efficacy of a novel IAV vaccine for swine consisting of recombinant full-length M2 protein displayed in its natural transmembrane configuration in soluble nanoscale membrane assemblies called nanodiscs (NDs).

Methods

To assess the ability of NDs displaying M2 (M2ND) to confer protective immunity against swIAV, groups of pigs (n = 9 each), were immunized at 5 weeks of age via both intranasal and intramuscular routes twice at a 4-week interval with either: M2ND; empty NDs (eND); or empty NDs mixed with M2 (eND + M2). The NDs in all three formulations, prepared as previously described,[1] were adjuvanted by incorporating synthetic oligonucleotides containing the immunostimulatory (ISS) CpG oligodeoxynucleotide (CpG ODNs) ODN19.[2] CpG ODNs act as ISS by triggering cells that express TLR9 to produce interferon alpha. At seventeen days after the booster, the animals in all three groups were challenged intranasally with 10^6 plaque forming units (PFU) of A/swine/MN/A01668936/2016 (H3N2). For five days after virus challenge, pigs were monitored for clinical signs and nasal swabs collected at 0-, 3- and 5-days post challenge (PC). The animals were euthanized at day 5 PC and their lungs scored for the percent of gross lung lesions by a pathologist, who was blinded to the treatment of the animals and bronchoalveolar lavage (BAL) fluids collected. The extent of virus load in the BAL fluids and nasal swabs was determined using a plaque forming unit (pfu) assay in MDCK cells. The presence of the M2- and IAV-specific IFN-g-secreting cells (SC) in PBMC was determined by ELIspot. Virus and M2-specific antibodies were determined by ELISA using as antigen either purified intact IAV virions, recombinant M2 protein produced in *E. coli*, or a synthetic M2e peptide.

Results

Before challenge, none of animals exhibited the presence of virus in nasal swabs samples. The nasal swabs collected at 3- and 5-days PC from animals treated with either eND or eND+M2 exhibited high titers of infectious IAV. In contrast, animals immunized with M2NDs exhibited a >10-fold decrease of infectious IAV titers in nasal swabs collected at 3- and 5-days PC. At the time of euthanasia (5 days PC), animals treated with either eNDs or eND+M2, exhibited extensive pneumonia (Figure 1) and a substantial viral load in BAL samples. In contrast, pigs vaccinated with the M2NDs exhibited the presence of strong protective immunity as determined by the either minimal or complete absence of pneumonic lesions in the lungs in 8 of the 9 pigs in this group (Figure 1), as well as a significantly reduced load of infectious IAV in their BAL samples.

Discussion

This study demonstrates that the immunization of swine with M2 displayed in NDs (M2ND) provides a strong level of protective immunity against swIAV. Since the vaccine formulation consisting of simply mixing M2 with the NDs was unable to elicit protective immunity, the data indicates that for the M2 protein to elicit protective immunity it has to be inserted into the ND assemblies. Future work will aim to ascertain the minimal antigen dose and vaccination schedule necessary to elicit strong protective immunity. This novel vaccine offers great promise as a universal IAV vaccine to solve the challenges presented by the HA antigenically diverse swIAV present in the field.

References

1. Sligar SG, Denisov IG. Nanodiscs: A toolkit for membrane protein science. Protein Sci. 2021 Feb;30(2):297-315.

2. Kayraklioglu N, Horuluoglu B, Klinman DM. CpG Oligonucleotides as Vaccine Adjuvants. Methods Mol Biol. 2021;2197:51-85.

Figure 1: Extent of gross lung pathology 5 days after challenge with H3N2 swIAV. Three groups of pigs (n = 9 each) were immunized at 5 weeks of age via both intranasal and intramuscular routes twice at a 4-week interval with either: Nanodiscs displaying M2 (M2ND); empty NDs (eND); or empty NDs mixed with M2 (eND + M2). At seventeen days after the booster, the animals in all three groups were challenged intranasally with 106 plaque forming units of A/swine/MN/A01668936/2016 (H3N2). The animals were euthanized at day 5 PC and their lungs scored for the percent of gross lung lesions by a pathologist, who was blinded to the treatment of the animals. Each symbol represents the extent of lung consolidation as a % of the total lung surface. Horizontal bars represent the average of the % of lung consolidation for the respective group ± SD. *** *P* = .004; **** *P* < .0001.

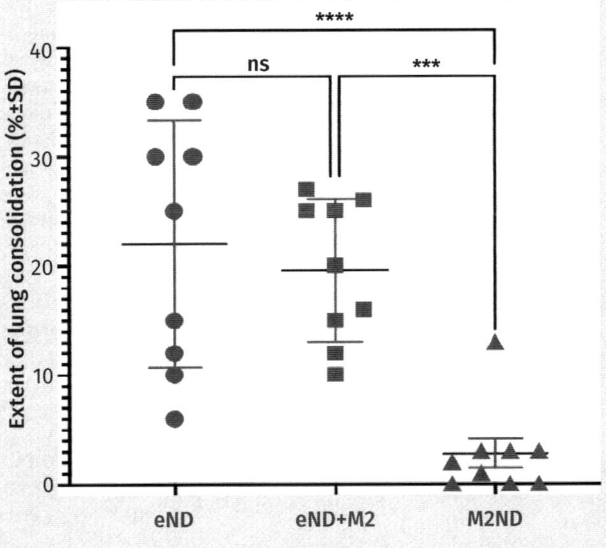

Poster 41 https://doi.org/10.54846/am2024/121

A novel PRRS live virus vaccine with unprecedented safety and efficacy based on the naturally nonpathogenic strain G16X

F. A. Zuckermann[1,2], DVM, PhD; G. Calzada-Nova[2], DVM, PhD; R. J. Husmann[1], DVM, MS
[1]University of Illinois at Urbana-Champaign; [2]Aptimmune Biologics, Inc

Introduction

All commercially available PRRS modified live virus (MLV) vaccines sold in the US are based on virulent viruses whose pathogenicity was attenuated by extensive serial passage on a heterologous cell line. Unfortunately, when administered to a pig, a PRRS MLV will inherently seek to regain its ability to optimally replicate in its natural host cell, ie, a porcine macrophage, by altering its highly mutable RNA genome to regain its virulence.[1] Clearly, the pork industry would benefit from the availability of an effective PRRS live virus vaccine without the predisposition to revert to virulent. Here, we describe the North American PRRSV vaccine strain G16X, which was derived from a naturally non-virulent PRRSV isolated in 1990. Hence, serial passage of the virus in a non-porcine host cell was not necessary to render it nonpathogenic. Because the G16X virus efficiently replicates in porcine alveolar macrophages, vaccine stocks are produced in the porcine macrophage cell line, ZMAC-4,[2] and when administered to pigs, the G16X virus readily replicates. Hence, there is no selective pressure for the virus to mutate to gain replication fitness in the pig.

Methods

Thirty-six PRRSV-free pigs at 6-weeks of age were randomly assigned to a treatment group and housed in a BSL2 containment facility. After 7 days of acclimation, each group of pigs was vaccinated according to their treatment allocation as follows: Group 1 - mock-vaccinated, received an im injection of 2 mL of vaccine diluent (n = 10). Group 2 – vaccinated, received an im injection of 2 mL containing 2×10^4 $TCID_{50}$ of G16X virus grown in ZMAC-4 cells (n = 20). A strict environmental control was not vaccinated nor challenged (n = 6). Twenty-eight days after vaccination, all the animals in groups 1 and 2 were challenged with 8×10^4 $TCID_{50}$ (half the dose administered im and half the dose in) of the Type 2 (NA) Lineage 1 virus strain LTX1, which possess a very high level of virulence, like the current LIC viruses. LTX1 has a < 88% homology in ORF5 amino acid sequence to G16X. The animals were monitored daily for the next 14 days for clinical signs, and body weight gain measured at 0, 7, 11 and 14 days after challenge. Rectal body temperature and SpO2 were measured daily. At 14 days after the challenge, the animals were euthanized, and their lungs scored for gross pathology. The infectious virus ($TCID_{50}$) load in serum samples collected at 0 4, 7, 10 and 14 days after challenge, and BAL fluid determined as described.[3] The GLM Univariate analysis was used to assess differences between groups by using the SAS software.

Results

Pigs immunized with G16X developed a viremia that peaked 6 days post vaccination (PV), began to decline by day 14 PV, and was undetectable in all pigs by day 28 (Figure 1). The injection of G16X did not affect the growth rate of the pigs nor induced any clinical signs. Following challenge with the LTX1 virus, pigs in the mock-vaccinated group exhibited a high level of viremia which lasted > 14 days. In contrast, only 25% (5 of 20) of the G16X-vaccinated pigs had detectable infectious PRRSV by 10 days post challenge and none of the pigs in this group had detectable PRRSV by day 14 (Figure 1). The protective effect of G16X was also evidenced by the ability of the G16X-vaccinated pigs to achieve >75% of the rate of weight gain attained by the unchallenged (strict) control pigs at 14 days after the virulent virus challenge. In contrast, the average weight gain of the mock-vaccinated and challenged pigs reached only 25% of the weight gain attained by the strict control group. Additional parameters indicating strong protective immunity after virulent PRRSV challenge included statistically and clinically significant reduction of the febrile response, a much improved level of SpO2 in G16X-vaccinates (> 93%) at 10 days post challenge, as compared to that exhibited by the mock vaccinates (< 89%) and strict controls (95 -96%), as well as reduced viral load in the lung, and reduced extent and severity of gross and microscopic lung pathology.

Conclusions

Based on all the clinical, virological, production and pathological parameters measured, the G16X virus can confer a strong level of heterologous protective immunity from challenge with a highly virulent Type 2 (North American) PRRSV of Lineage 1. The availability of a effective PRRS vaccine capable of providing heterologous protective immunity, with a high safety profile, ie, very unlikely to become virulent, should prove to be an effective tool to diminish the heavy economic losses due to PRRSV. The G16X vaccine has been approved for use in Mexico (Innovac L-PRRS, Avimex) and in the field has yielded impressive resuts of efficacy and safety.

References

1. Wang J, Zhang M, Cui X, et. al. Attenuated Porcine Reproductive and Respiratory Syndrome Virus Regains Its Fatal Virulence by Serial Passaging in Pigs or Porcine Alveolar Macrophages To Increase Its Adaptation to Target Cells. Microbiol Spectr. 2022 Dec 21;10(6):e0308422.

2. Portugal R, Goatley LC, Husmann R, Zuckermann, FA, Dixon LK. A porcine macrophage cell line that supports high levels of replication of OURT88/3, an attenuated strain of African swine fever virus. Emerg Microbes Infect. 2020 Dec;9(1):1245-1253.

3. Calzada-Nova G, Husmann RJ, Schnitzlein WM, Zuckermann FA. Effect of the host cell line on the vaccine efficacy of an attenuated porcine reproductive and respiratory syndrome virus. Vet Immunol Immunopathol. 2012 Jul 15;148(1-2):116-25.

Figure 1: Extent and frequency of viremia in pigs after vaccination with G16X and after challenge with LTX1. Serum samples were collected from mock-vaccinated (n = 10) and G16X-vaccinated (n = 20) animals immediately prior to and at the indicated days after either vaccination with G16X (day 0) or challenge (day 28) with the wild-type PRRSV LTX1. Samples were also taken from the strict controls (n = 6), which were not vaccinated nor challenged. The virus load in the serum was examined for the presence of infectious virus ($TCID_{50}$) in ZMAC-4 cells. Assay level of detection is ≤ 0.85 log10. Results are presented as the geometric mean for each group of animals. The fractional numbers next to the symbols indicate the number of virus-positive animals over the total number of animals in the group. Asterisks indicate statistically significant differences (P < .01).

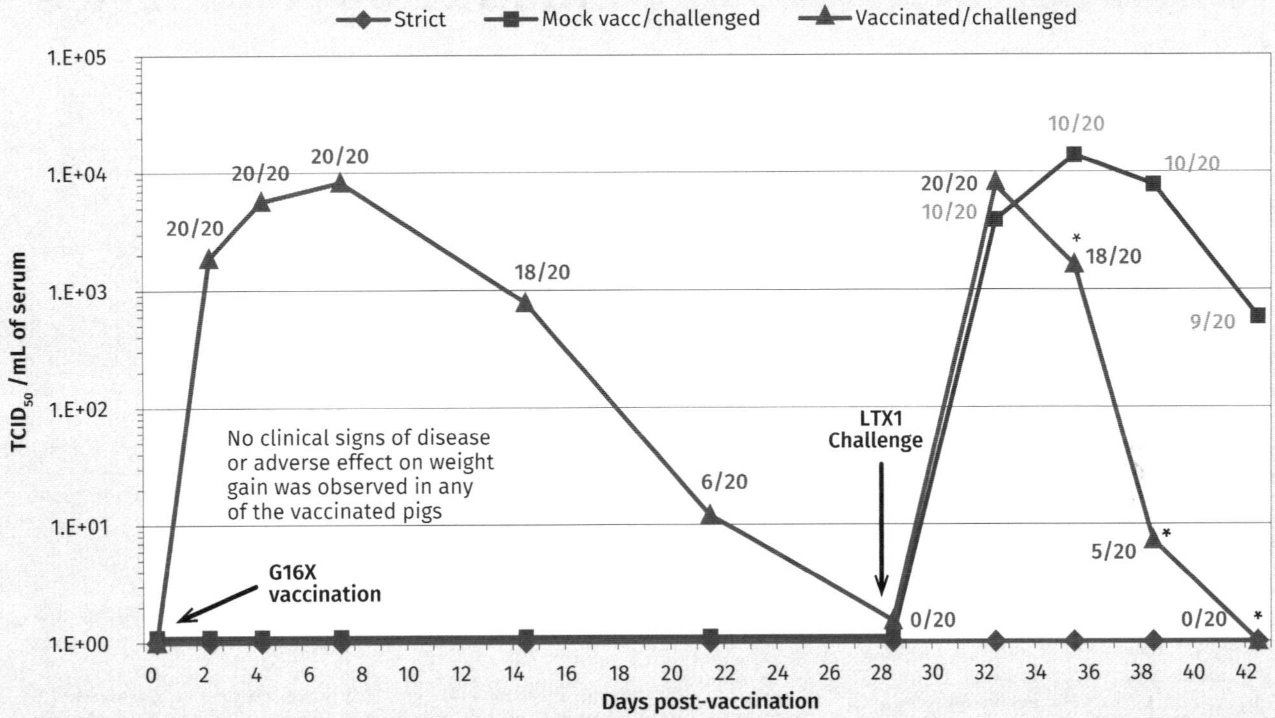

https://doi.org/10.54846/am2024/122

Health benefits to the progeny of sows fed Celmanax™ extend into the nursery phase of production and result in similar overall nursery growth performance as pharmacological levels of zinc oxide

S. Jalukar; E. Davis; J. Christianson; A. Wuethrich; S. Hernandez
Arm and Hammer animal and food production, Princeton, New Jersey

Introduction

Celmanax™ is an enzymatically hydrolyzed yeast product with immunomodulatory[1] and reproductive performance enhancing properties in swine.[2,3] Increases in piglet weaning weights have been observed when Celmanax™ is added to sow lactation diets. The objective of this study was to evaluate if this pre-weaning benefit extends into the nursery phase and compare with pharmacological levels of zinc oxide (ZnO) in nursery pig diets.

Methods

Four treatments were arranged in a 2 x 2 factorial with seven pens/treatment. The two treatment factors included: 1) Celmanax™ (0 or 200 g/MT) in the sow lactation ration; and 2) ZnO (0 or 2500 ppm) in the first two nursery phases. Pigs were weaned at 18 days of age, blocked by initial body weight (BW), and were randomly allocated to treatments using a Randomized Complete Block Design. Pigs were *ad libitum* fed a three-phase mash nursery diet [Ph1, Ph2 and Ph3, 0 - 7, 7 - 21, and 21 - 42, respectively] in a 42-day trial. Individual BW and feed consumption by pen was recorded and used to calculate average daily gain (ADG), average daily feed intake (ADFI) and feed efficiency (FE). Blood samples were obtained on d 42 from two pigs/pen for immune gene expression analysis. The main effects of Celmanax™-sow and nursery ZnO were analyzed using General Linear Model. Least Squares Means were separated using the PDIFF option of SAS (Ver 9.2, Cary NC).

Results

Celmanax™ fed to sows during lactation resulted in heavier pigs entering the nursery ($P < .01$), and these pigs had greater ($P < .01$) ADFI in Ph1, greater ADG in Ph2, and had heavier BW ($P < .01$) at d 42 than pigs from sows fed diets devoid of Celmanax™; Pigs fed ZnO in the nursery also had greater ADFI in Ph1 ($P < .05$), as well as greater ($P < .01$) ADG and FE in Ph2 and heavier ($P < .01$) BW at d 42. ZnO resulted in a lower gene expression of the inflammatory cytokine, TNFα; whereas there was a tendency ($P < .10$) for a higher expression of TNFα in pigs from sows fed Celmanax™. Pigs from Celmanax™-fed sows also had a tendency for greater ($P = .06$) gene expression of the regulatory cytokine, SOC3.

Discussion

Use of pharmacological levels of ZnO in nursery diets to improve nursery pig performance has been banned or being phased out in many countries. Taking proactive measures to improve the pre-weaning health and performance of piglets could potentially reduce the dependency on zinc during the nursery phase. In this study, the heavier weight of pigs from Celmanax™-fed sows at the beginning of the nursery phase and additional ADG and ADFI after weaning resulted in similar overall nursery performance as pharmacological ZnO. Both ZnO and Celmanax™ modulated immune gene expression, however they differed in their immunomodulatory mode of action.

References

1. Hung, I.F. *et. al.* Effect of supplying mannan oligosaccharide (MOS) pig diets on response to an immune challenge (2008) *J. Anim. Sci.* Vol. 86, E-Suppl. 2, Abstract T-168.

2. Hung I. F. and Lindemann M.D. Benefits of mannan oligosaccharide (MOS) for sows and weanling pigs. Mid-west Swine Nutrition Conf., Indianapolis, IN, USA, 2009.

3. Peng Ma, Guozhu C., and Jalukar, S. Evaluation of Celmanax SCP supplementation in sow diets on piglet performance at weaning. *J. Anim. Sci.* 2013; Vol. 91, E-Suppl. 2. Abstract TH 419.

https://doi.org/10.54846/am2024/123

Leading and lagging indicators of growing pig health and performance

D. Polson[1], DVM, MSc, PhD; C. Alonso[2], DVM, MS, PhD
[1]Boehringer Ingelheim Animal Health USA Inc., Duluth Georgia; [2]Boehringer Ingelheim Vetmedica GmbH, Ingelheim, Germany

Introduction

Several different production-related measures are used by pork producers in an effort to detect meaningful deviations from expected performance that would signal the onset of intervals of poor performance attributable to various causal and contributing factors. These combinations of multiple factors may be genetic, nutritional, environmental, management and/or disease related. For at-risk growing pig populations, appropriate interventions must be implemented to mitigate performance problems, and detection of the onset of performance problems is prerequisite to obtaining a diagnosis and optimizing intervention. Producers and veterinarians understand that for any intervention to be optimally effective and cost-efficient, the onset of performance and health problems must be detected and causes with contributing factors diagnosed as soon as possible in at-risk pig populations. Several often-used indicators used to detect the onset and track the course of poor growing pig performance episodes are: Daily mortality, daily treatments, daily water usage and daily feed disappearance. Within the past several years, various precision livestock farming (PLF) technologies have also been utilized for detection of poor health and performance in growing pigs, one example being a sound-based technology that continuously measures respiratory health.[1] Here, we compare the utility of several of these measures.

Materials and methods

For this analysis, we utilized data collected from a growing pig site with three 1,000 head single-room barns. Data represented 24 groups of pigs closed out across a 42-month period from March 2020 to August 2023. The producer utilized a sound-based technology (SoundTalks, SoundTalks NV, Leuven, Belgium) integrated via API with a cloud-based digital platform, dashboard, and phone-based app (Farmera, Boehringer Ingelheim, Ingelheim Germany) for data capture, aggregation, integration, analysis and notification. Each barn was equipped with a cloud-connected water meter and that data was continuously captured and pushed via API to the Farmera platform. Daily water data was standardized to pig inventory and expressed as water (gallons) per pig per day. Mortality was standardized to pig inventory to calculate deads per 1000 pigs at-risk per day. A standardized, indexed respiratory health score (ReHS) was used, as reported by the sound-based technology.

Results and discussion

ReHS index, deads per 1000 per day and water gallons per pig per day results are shown in Figure 1. Mortality is clearly a lagging indicator and as such is generally not suitable to be used for early detection of respiratory health problems. Interestingly, changes in water (gallons) per pig per day also appears to lag the ReSH index by multiple days, which is problematic when attempting to use water data as an indicator of the onset of episodes of respiratory disease.

Conclusions

Based on these data, the sound-based technology appears to be well suited for use as an objective leading indicator and valuable for triggering contingent diagnostic and/or intervention actions. We continue to conduct further research to evaluate the aforementioned as well as other health and performance measures, individually and in combination, using data from many more growing pig sites, to develop measures and methods to more effectively detect and assess pig health, welfare and performance.

Reference

1. Polson D, Playter S, Berckmans D, et al., (2018) Precision livestock farming (PLF) for pig health and production: Sound as a diagnostic sample. *Proceedings of the 2018 American Association of Swine Veterinarians*, San Diego California, USA, 21-24.

Figure 1: ReHS index (Y1 axis), deads per 1000 per day (Y2 axis) and water (gallons) per pig per day (Y2 axis) by consecutive days of classified respiratory episodes. (Note: The "Pre" and "Post" column data represents one single day prior to the beginning of and following the end of each respiratory episode, respectively.)

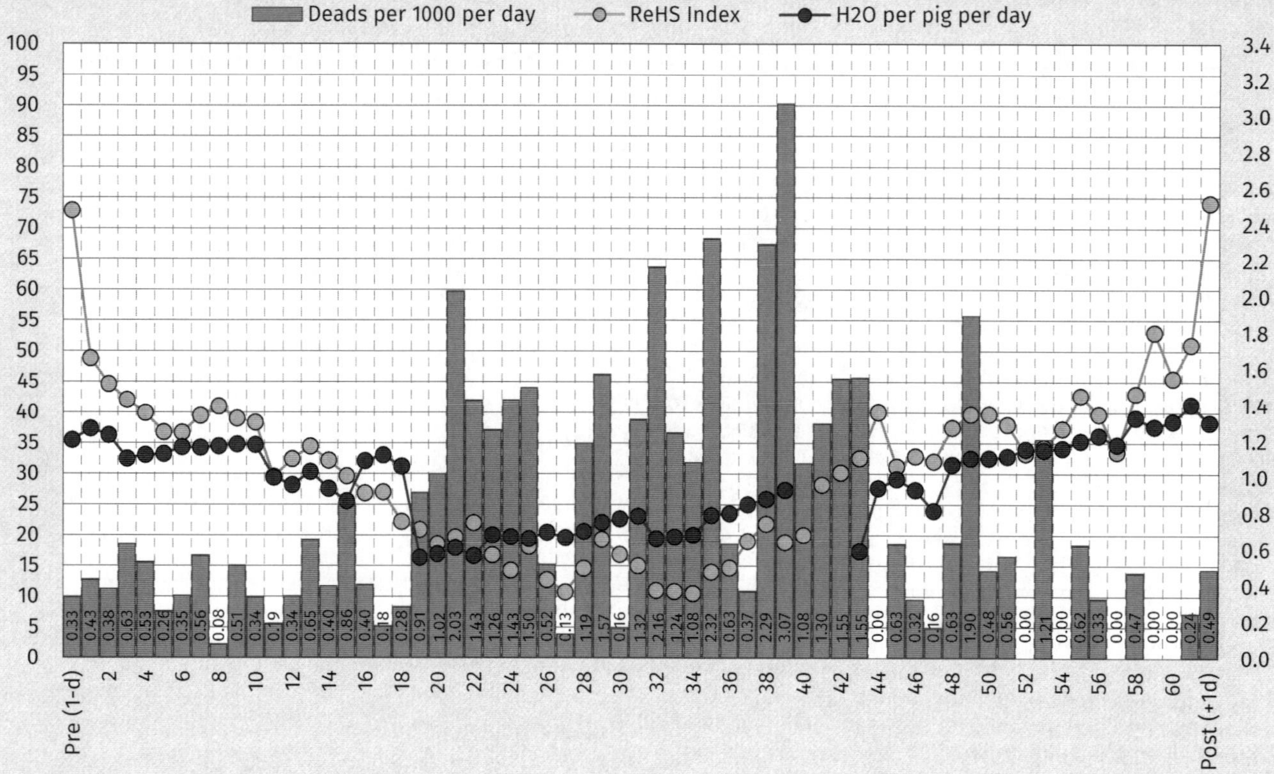

https://doi.org/10.54846/am2024/124

Labor-free fly control ClariFly 267 (diflubenzuron) reduces housefly (*Musca domestica*) populations in pull-plug facilities

Gene Spellman, BS
Central Life Sciences, Schaumburg Illinois

Introduction

Despite extensive investments in biosecurity measures and protocols, farms still have unexplainable cases of viral diseases. Fly control is now recommended as an important plank of swine biosecurity protocols and disease risk reduction.[1-3] Flies can carry and transmit important swine disease viruses such as porcine reproductive and respiratory syndrome virus (PRRSV)[1], porcine epidemic diarrhea virus (PEDV)[2], Senecavirus A (SVA)[3] and now African swine fever virus (ASFV).[4] Flies can carry infectious bacteria,[5] and antibiotic-resistant bacteria.[6] Flies can travel up to four miles.[7] Fly control in swine till now has been labor-intensive, sporadically effective, non-coordinated, using multiple products with harsh chemicals. ClariFly 267 (diflubenzuron) insect growth regulator larvicide was developed and approved for swine as an in-feed fly control product. Earlier research showed ClariFly 267 in-feed larvicide achieved a 99% reduction in housefly populations in swine finishing facilities with deep pit manure handling facilities.[8] Questions remained if ClariFly 267 would perform as well in facilities with pull-plug manure facilities. The purpose of this study is to show similar fly control performance in shallow pit manure facilities with ClariFly 267.

Materials and methods

A 2,500-sow farrow-to-finish operation in the Midwest was selected. ClariFly 267 (diflubenzuron) in-feed larvicide was incorporated within the typical ration at the label level of 0.20 mg ClariFly 267 per kg of body weight per day for farrowing, and 0.10 mg ClariFly 267 per kg body weight in gestation, for the eight-week study. Feed samples were collected at the start and mid-point of the study. The feed samples were assayed for proper levels of ClariFly 267. To count flies and monitor the fly population reduction, Starbar EZ Trap sticky fly traps were hung in each study room. Traps were placed approximately seven feet above the floor. Fans and air vents were avoided for trap placement to avoid air movement interfering with fly activity. Traps were hung for 1 week and replaced with new, unused traps for the next week of the study. Weekly trap collection, fly counting and new trap placement was repeated throughout the study.

Results

After seven weeks on ClariFly 267 in-feed larvicide, the fly populations, as represented by the flies caught and counted on the traps collected weekly during the study, were reduced more than 97% by the end of week seven versus the end of week one, showing that ClariFly 267 was very effective in reducing fly populations in these shallow manure pit facilities. Figure 1.

Discussion

ClariFly 267 in-feed larvicide proved extremely effective in reducing housefly populations in this shallow pit manure study, achieving a 99% reduction in fly counts in each room. ClariFly

Figure 1: 97% reduction in fly counts in week 7.

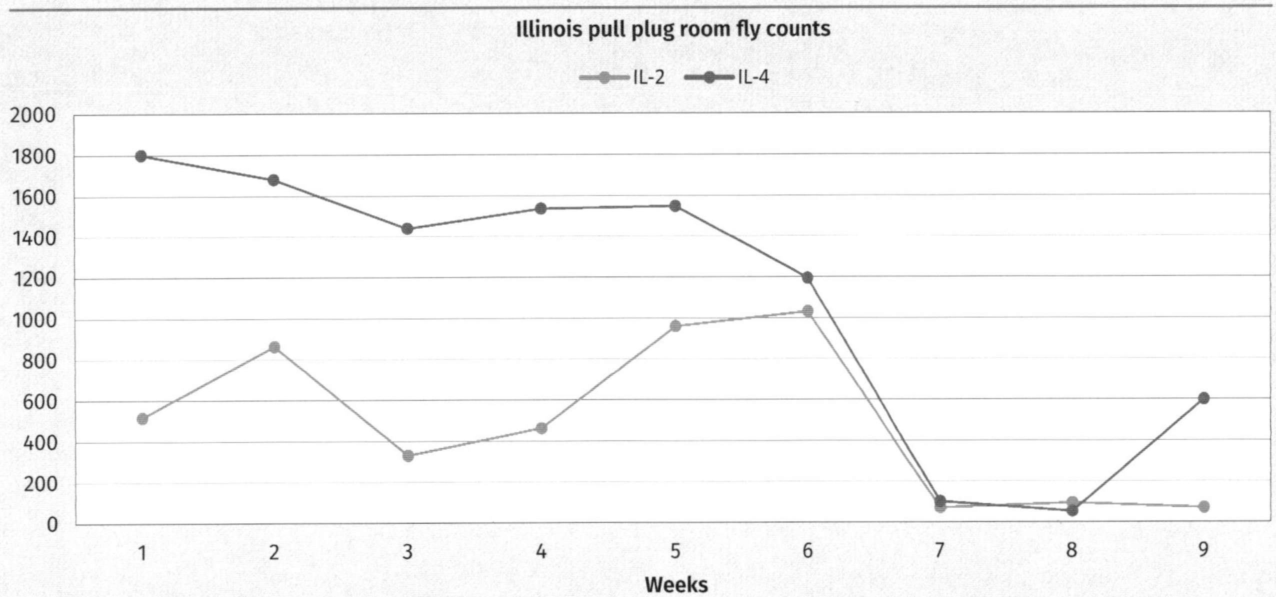

267 in-feed larvicide offers fly control options in a labor-free fashion, since ClariFly 267 is mixed in the swine ration. The pigs do the work, and are effective in delivering ClariFly 267 where it is needed-in the manure. ClariFly 267 is a chitin-synthesis inhibitor. Chitin is a key component of the fly's exoskeleton. Inhibiting chitin synthesis blocks further development of the fly in the larval stage. ClariFly 267 does not kill adult flies. Adult flies are prevented from developing from earlier larval stages. Existing adult flies eventually die off and are not replaced when ClariFly 267 is fed.

It is important to show patience when feeding ClariFly 267, as adult flies are still visible during the initial feeding stages. ClariFly 267 offers a convenient and effective option to achieve fly control in swine biosecurity protocols. Fly control also provided an improved working environment for pigs and employees, as reported by the study cooperator. More research is needed in the area of improved working environment and animal welfare to quantify the benefits resulting from improved fly control.

References

1. Otake S, Dee SA et al. Transmission of porcine reproductive and respiratory syndrome virus by houseflies, (*Musca domestica Linnaeus*). *Vet Rec* 2003. 152: 73-76.

2 Allison G, et al. PEDV positive bioassay reveals houseflies (*Musca domestica*) can transmit infectious PEDV to pigs. *Proceedings 50th AASV 2019.*

3. Allison G, et al. PEDV, PRRSV, and Seneca Valley virus PCR positive flies (*Musca domestica*). What now? *Proceedings 49th AASV 2018.* 394-396.

4. Olesen, A, et al. Infection of pigs with African swine fever virus via ingestion of stable flies (Stomoxys calcitrans) *Transboundary Emerging Diseases 2018,* 65: 1152-1157.

5. Khamesipour F, et al. A systematic review of human pathogens carried by the housefly (*Musca domestica L.*). *BMC Public Health 2018,* 18:1049.

6. Zurek L, et al. Insects Represent a Link between Food Animal Farms and the Urban Environment for Antibiotic Resistance Traits. *Appl Environ Microbiol. 2014 Jun;* 80(12): 3562–3567.

7. Nazni WA, Luke H et al. Determination of the flight range and dispersal of the house fly, *Musca domestica* (L.) using mark release recapture technique. *Trop Biomed 2005 Jun*;22(1):53-61.

8. Spellman, G, et al. New labor-free fly control ClariFly 267 (diflubenzuron) achieves 99% reduction in Housefly *(Musca domestica)* populations in swine facilities. *Proceedings 51st AASV 2020.* 175-176.

https://doi.org/10.54846/am2024/125

An update on the benefits of VVC (a unique combination of benzoic acid and nature-identical flavorings) for feed biosecurity and performance improvement

S. Hough, DVM; J. Bergstrom, PhD
dsm-firmenich, Animal Health and Nutrition, Plainsboro, New Jersey

Introduction

There has been great interest from the veterinary, nutrition and production segments of the swine industry regarding feed biosecurity and mitigation of pathogens. The combination of benzoic acid and nature-identical flavorings from VVC has proven to be beneficial in pig feed by reducing clinical expression of disease when pigs were challenged with PRRSV, SVA and PEDV in Dr. Scott Dee's ice-block challenge model,[1] and by reducing the ASF contamination of feed in an in-vitro study in China.[2] Most products used for feed mitigation are solely used for that purpose and have few, if any, beneficial effects reported on performance in the absence of feed pathogens. Some feed mitigation products have even been reported to reduce feed intake. A recent commercial trial in the Midwest evaluated the use of VVC in diets for nursery pigs and the objective was to determine the effects of VVC on growth performance, morbidity and mortality.

Methods

5,466 pigs were enrolled in the study, using a randomized complete block design. The pigs were allotted to pens, with 72 pigs in each. Thirty-five pens were fed a control diet (C) that was the standard nutrition program without diet acidification. The other 35 pens were fed the control diet plus VVC premix at 7 lb/ton in the first two diet phases and 3.5 lb/ton in the last nursery phase (VVC). Pigs were budgeted to receive 6 pounds of the initial nursery diet, 12 pounds of the second phase, and 46 pounds of phase 3 per pig. Pen weights and feeder measurements were collected at the start, week 2, week 5 and at the end of the study at 8 weeks placed. Body weight, average daily gain, average daily feed intake and gain to feed was recorded by pen. Clinical observations and mortality were captured by pig. All pigs came from PRRSv stable sow farms and received routine vaccinations for porcine circovirus and PRRS at weaning, and ileitis and erysipelas at 5-weeks post weaning.

Results

Results can be found in Table 1. There were numerical improvements in the average daily gain and feed conversion of pigs fed VVC; however, the percentage of total morbidity and mortality was significantly reduced for pigs fed VVC (3.27% vs 4.45%), with pigs observed to have poor body condition being the largest contributing factor for morbidity in the study. The majority of pigs succumbed to morbidity or mortality between weeks 2 and 5.

Table 1: Performance results

	Treatment		
	C	VVC	*P*-value
Number of pens, n	35	35	
Start weight, lb	12.5	12.5	.35
End weight, lb	69.1	69.4	.29
Overall ADG, lb	1.00	1.01	.12
Overall ADFI, lb	1.59	1.60	.62
Overall gain:feed, lb:lb	0.627	0.630	.16
Morbidity %	1.79	1.21	.08
Mortality %	2.66	2.06	.15
Total morbidity and mortality %	4.45[b]	3.27[a]	.03

[a,b] Different superscripts are statistically different ($P < .05$)

Discussion

Other research data from around the globe showcases the same benefits from using VVC in nursery pig diets, with improvements in growth performance resulting from improvements in gut functionality and nutrient digestibility. Improving survivability and robustness in the early nursery is a major focus for the swine industry, so these recent results are promising and showcase benefit beyond pathogen mitigation.

References

1. Dee S, Neiderweider M, Elder R, et al. An evaluation of additives for mitigating the risk of virus-contaminated feed using an ice-block challenge model. *Transbound Emerg Dis*. 2021;68:833-845.

2. Zhai, H, M Walsh, S Hough and J Bergstrom. Evaluation of a combination of benzoic acid and nature identical flavorings to reduce ASF viral contamination of complete feed. Abstract for Poster at AASV Annual Virtual Meeting 2021.

https://doi.org/10.54846/am2024/126

An overview of Microtracers® in Skycis®

Roger A. Arentson, PhD; Michael Shields, PhD; Jenna Seltzer, MS; Gary Fuller; Scott Carr, PhD
Elanco Animal Health, Greenfield, Indiana

Introduction

A recent narasin (Skycis®, Elanco Animal Health, Greenfield, Indiana) meta-analysis done at Kansas State University indicated that pigs fed 15 ppm of narasin for the entire finishing period and achieving normal production parameters have improved ADG, F:G, and carcass yields by 1.4%, 1.4%, and 0.31% units, respectively.[1] These outcomes and lower pig production profitability over that past year have led to increased narasin usage by pig producers in the United States. Because of this increased usage and customer demand for traceability, Elanco has started including Microtracers® in Skycis® 100 premix to improve its safety profile.

Microtracers® are small, iron particles that are coated with food grade dye and added to several common premixes utilized by the feed industry. Microtracers® in diets have no impact on animals, feed, or the environment. In the case of Skycis® 100 premix, these particles are red. Skycis® premix that is currently being manufactured and shipped has "With Microtracer" printed on the label.

Extraction procedure

Microtracers® can be extracted from diets with a simple procedure taking only a few minutes by using a device supplied by Elanco. Depending on the level of precision required, either a qualitative or a semiquantitative method may be used. Qualitative extraction of Microtracers® from feed containing Skycis® is achieved by pouring a feed sample over a piece of filter paper attached to a spinning magnet. The filter paper is then removed, misted with 70% isopropyl alcohol, and warmed slightly. Red dots will appear, and this indicates the presence of Skycis® within the feed (Figure 1). Using another feed sample and a slightly different procedure, semiquantitative analysis of feed containing Skycis® can be achieved. Microtracer analyses do not replace chemical assays for accuracy but do provide a rapid method to access Skycis® in swine diets.

Figure 1: Qualitative analysis of Microtracers® in Skycis®.

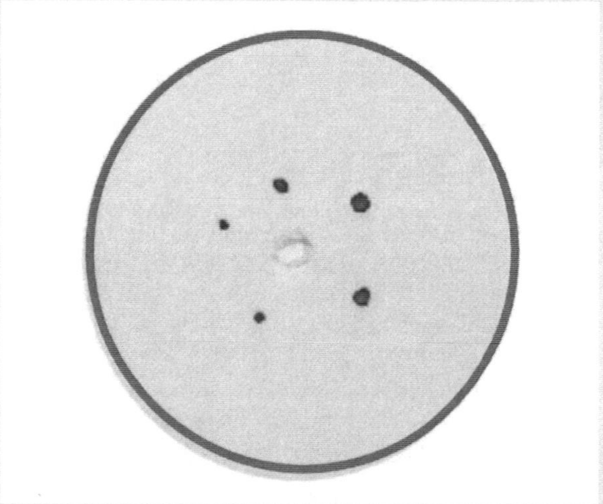

Uses

Microtracer detection may be used in feed mills and on farms. On farm, if a producer observes clinical signs of pigs consistent with ionophore toxicity, the feed from feeders of affected pigs can be assessed for the presence of red Microtracers® and determine if an abnormally high level of Skycis® is present or an "on label" level of Skycis® is present. If the level of Microtracers® indicates the Skycis® inclusion is "on label," producers should investigate if tiamulin is being provided to pigs via the water. By quickly assessing Microtracers®, a producer can quickly verify ionophore toxicity of pigs and then choose the best course of action.

In a feed mill, Microtracer analysis can be incorporated into a feed mill's normal quality control program and standard operating procedures (SOP) for diets containing Skycis® and several other common premixes containing Microtracers®. Firstly, Microtracers® can be used to help confirm that the proper level of Skycis® is added to a diet. Secondly, flush validation of mixers, bins, and trucks can be assessed after diets containing Skycis® have passed through them. Lastly, they can be used to estimate cross-contamination and carry-over of diets.

References

1. Becker LL, Gebhardt JT, Tokach MD, Woodworth JC, Goodband RD, DeRouchey JM, Seltzer JA, Arentson RA, Shields M, and Puls CL. 284 A Meta-Regression Analysis to Evaluate the Effects of Narasin on Grow-Finish Pig Performance, *Journal of Animal Science*, 2023, 101 (2) 176–177, **https://doi.org/10.1093/jas/skad341.195**

2. Elanco PM-US-23-160.

https://doi.org/10.54846/am2024/127

Improved piglet performance after sow vaccination against *C difficile* in Italy

M. Faccenda[1]; S. Rosina[2]; I. Ballarà[3]; O. Boix[3]
[1]Gruppo Chiola, [2]Hipra Italia Srl, [3]Laboratorios HIPRA S.A., Amer, Spain

Introduction

Clostridioides difficile (C difficile) and *Clostridium perfringens* type A (CpA) are significant contributors to the development of diarrhea among piglets aged 1 to 10 days.[1,2] Besides their impact on gastrointestinal health, *C difficile* has been associated with increased pre-weaning mortality rates and subclinical affections such as suboptimal growth patterns and variability in weaning weights.[3] This study aims to assess the health and growth performance of piglets born from sows that received two different vaccination plans under field conditions.

Materials and methods

This study was conducted on a commercial herd comprising of 4,000 sows. The farm exhibited a stable but positive status regarding PRRS, with a recurring issue of neonatal diarrhea during the first two weeks of age in the farrowing unit. The existing vaccination plan included vaccination against *E coli*, CpA, and CpC in combination with individual Florfenicol injections. The experimental design aimed to compare an experimental group (G1) consisting of 25 sows and 326 piglets, vaccinated with a mixture of SUISENG® Coli/C and SUISENG® Diff/A, to a control group (G2) of 23 sows and 321 piglets injected with the previous trivalent vaccine that does not include *C diffcile* in its composition. All vaccination procedures adhered to the manufacturer's instructions. The study period spanned from November 2022 to January 2023, during which the incidence and severity of diarrhea, pre-weaning mortality, and litter weight at weaning were recorded.

Results

The percentage of piglets with clinical scours was reduced in G1 compared to G2, although no significant differences were observed. In terms of average pre-weaning mortality, G1 exhibited an overall mortality rate of 13.6% during the lactation phase, which was significantly lower than the 21.1% overall mortality rate recorded in G2 ($P < .05$). These significant differences persisted when comparing mortality solely within litters affected by neonatal diarrhea (15.3% in G1 vs. 24.5% in G2 [$P < .05$]). Additionally, there were significant improvements in both body weight at weaning (8.3kg ± 1 [G1] vs 8kg ± 1.5 [G2]) and average daily weight gain at weaning (236g ± 28 [G1] vs 228g ± 55 [G2]) (Table 1).

Table 1: Average piglet body weight at weaning and ADG

Group	Piglet weight at weaning	ADG (g/d)
1	8.3kg ± 1[a]	236g ± 28[a]
2	8kg ± 1.5[b]	228g ± 55[b]

[a,b] Show statistically significant differences determined by linear regression model ($P < .05$)

Conclusion

These results demonstrate that the combination of SUISENG® Coli/C and SUISENG® Diff/A significantly reduced the mortality rate and improved piglet performance at weaning. Furthermore, considering that the most significant difference between the experimental treatments was the inclusion of *C difficile* A and B toxins, this study reinforces the importance of immunization against *C difficile* when establishing a vaccination plan in a commercial herd.

References

1. Grzeskowiak LE et al. Developing Gut Microbiota Exerts Colonisation Resistance to Clostridium (syn. Clostridioides) difficile in Piglets. Microorganisms 2019.

2. Chan, G., Farzan, A., Soltes, G. *et al.* The epidemiology of *Clostridium perfringens* type A on Ontario swine farms, with special reference to *cpb2*-positive isolates. *BMC Vet Res* **8**, 156 (2012). **https://doi.org/10.1186/1746-6148-8-156.**

3. Arruda, Paulo H. Elias, "*Clostridium difficile* infection in neonatal piglets: Pathogenesis, risk factors, and prevention" (2014). Graduate Theses and Dissertations. 14096.

 https://doi.org/10.54846/am2024/128

Performance of a newly designed RT-qPCR for multiplex identification of PRRSV-1 and PRRSV-2 RNA: Field study evaluation

M. Angelichio; L. Plourde; L. Gow; H. Webber; A. Hidalgo
IDEXX Laboratories, Inc., Westbrook Maine

Introduction

Porcine reproductive and respiratory syndrome (PRRS) continues to be one of the most economically significant diseases affecting the swine industry worldwide. In the US, the impact of PRRS has being estimated at $4.67 for every pig marketed.[1]

Effective management strategies for controlling PRRS include the use of diagnostic assays for direct identification of PRRSV and testing for evidence of past exposure to the virus, with RT-qPCR and ELISA being the most common techniques, respectively. Nevertheless, PRRSV high mutation rate can lead to changes in key nucleotide positions within a RT-qPCR design region, affecting its detection. As PRRSV continues to evolve in the field, reliable identification of contemporary PRRSV strains depends on RT-qPCR designs adapting to these changes.

This study investigated the performance of a newly designed multiplex RT-qPCR assay for identification and differentiation of PRRSV-1 and PRRSV-2 RNA and compared this to a reference RT-qPCR.

Materials and methods

A total of 370 swine samples, including serum, oral fluids, lung lavage, processing fluids and tongue tip exudates, were obtained from several farms in the US between 2022 and 2023. Samples were extracted using RealPCR DNA/RNA Magnetic Bead Kit (IDEXX Laboratories) and independently tested using two different qPCR assays using the same qPCR instrument. PCR-A refers to the RealPCR PRRSV Type 1 and Type 2 Multiplex RNA Mix, IDEXX Laboratories, while qPCR-B refers to another commercially available qPCR assay for PRRSV-1 and PRRSV-2.

Results and discussion

Identification of PRRSV-1 RNA (Table 1): 101 samples were positive by one or both PCRs and 269 samples were negative by both assays. The percentage of agreement between PCRs was 95.1%. The average Ct value for positive samples was 30.8 for PCR-A and 32.1 for PCR-B. PCR-A identified six more positive samples (confirmed by an alternate method) than PCR-B.

Identification of PRRSV-2 RNA (Table 1): 170 samples were positive by one or both PCRs and 200 samples were negative by both assays. The percentage of agreement between PCR-A and PCR-B was 95.1%. The average Ct value for positive samples was 29.8 for both assays. Discrepant results corresponded to high Ct values (> 36).

Table 1: Comparison of two RT-qPCR assays for identification of PRRSV RNA: **(A)** PRRSV-1 results and **(B)** PRRSV-2 results

A. PRRSV-1 results		**PCR-B (Reference PCR)**		
		Positive	**Negative**	
PCR-A (IDEXX)	Positive	83	12	95
	Negative	6	269	275
		89	281	**370**

B. PRRSV-2 results		**PCR-B (Reference PCR)**		
		Positive	**Negative**	
PCR-A (IDEXX)	Positive	152	9	161
	Negative	9	200	209
		161	209	**370**

Under the conditions of this study, the comparison of two RT-PCR assays for identification of PRRSV-1 and PRRSV-2 RNA of strains circulating in North America showed that both identified the same number of PRRSV-2 RNA positive samples with the same average Ct value. In addition, PCR-A identified more PRRSV-1 RNA positive samples compared to PCR-B and had a lower average Ct value. According to this evaluation, PCR-A showed high performance for RNA identification of PRRSV-1 and PRRSV-2 strains circulating in North America.

Reference

1. Holtkamp D, Kliebenstein J, Neumann E, et al. Assessment of the economic impact of porcine reproductive and respiratory syndrome virus on United States pork producers. *J Swine Health Prod*. 2013;21(2):72–84.

https://doi.org/10.54846/am2024/129

Benefits, differentiation, and results of a chlorine dioxide water sanitation method (PRO-OXINE® AH) in a swine facility[1]

K. Bretey, DVM; V. Iseri, PhD; C. Van Rensberg; E. Rankin
Kemin Industries, Inc., Des Moines, Iowa

Background

Water is an essential nutrient for animals and can have a significantly negative impact on animal health and performance when it is of poor quality. Minerals alone, naturally found in groundwater, can be a concern from a microbiological standpoint. Iron, for instance, can cause proliferation and buildup of *E coli* as well as other pathogens in water lines. In addition, prebiotics, antibiotics, vitamins, and vaccines delivered via water also fuel pathogen buildup resulting in biofilm. The development of biofilm buildup (Figure 1A) in water lines will increase exposure of harmful pathogens to already challenged animals, as well as cause off flavors and reduced water flow. The combination of poor water quality management will impact water intake, feed intake, milk production, and overall health and performance indicators like weight gain, mortality, and the producer's bottom line.

PRO-OXINE® AH

PRO-OXINE® AH is a 5% stabilized chlorine dioxide-based disinfectant by Kemin Industries, a global manufacturer of proprietary chlorine dioxide-based antimicrobial products and application systems for more than 50 years. PRO-OXINE AH® is a high-efficacy, broad-spectrum, stabilized chlorine dioxide antimicrobial that is proven to disinfect drinking water supply of livestock and poultry, by controlling harmful, biofilm-forming pathogens. PRO-OXINE is resistant to neutralization from organic load and is active over a wide pH range (3 - 10). Consequently, superior disinfectant efficacy is observed from the front to the back of the barns, so all animals have access to clean water. PRO-OXINE is continuously applied to water lines to constantly disinfect water supply, resulting in healthier and more efficient animals.

Comprehensive program

Kemin is changing the industry by simplifying water quality management and offering an all-around service program so producers can focus on day-to-day operations. At Kemin, we take ownership of how product is activated and applied to ensure best results. Pro-OXINE is offered with dependable application systems that are designed and manufactured by our Kemin Application Solutions Team, an extensive global network of in-house engineers. The field-proven systems are automated, eliminating the need for manual mixing. Furthermore, our comprehensive program includes support from many other in-house teams with nationwide footprint including 1) water technicians to support proper product application and equipment maintenance; 2) veterinarians and PhDs to support field trials and product assessments for informed decision making; and 3) supports services from our state-of-the-art Customer Laboratory Services team (CLS) to identify product demand, product efficacy and more. To simplify water quality management, turn to PRO-OXINE AH®. Visit Kemin.com/prooxine or contact your Kemin representative for more information on disinfection with PRO-OXINE AH®.

Field study

A pre- and post-PRO-OXINE® AH assessment/evaluation on water quality and wean to finish performance was performed on two 1,250 head rooms of a historically low performing site. PRO-OXINE AH® was applied using the Kemin's EDIE application system (Figure 1B) at 5 ppm level. There was a reduction in total coliform count in the drinking water during the 6-month trial. Performance metrics of 3 previous turns were compared to the PRO-OXINE AH® rooms, which showed an ADG increase of 5.2% (1.73 vs 1.82) and a livability increase of 5.8% (86.1% vs 91.0%), despite a PRRS outbreak and subsequent *E coli* challenge typical for this barn.

Reference

1. Kemin Internal Document: TD-23-8992.

Figure 1: **A)** Biofilm buildup in a water line. **B)** EDIE application system (Kemin Industries).

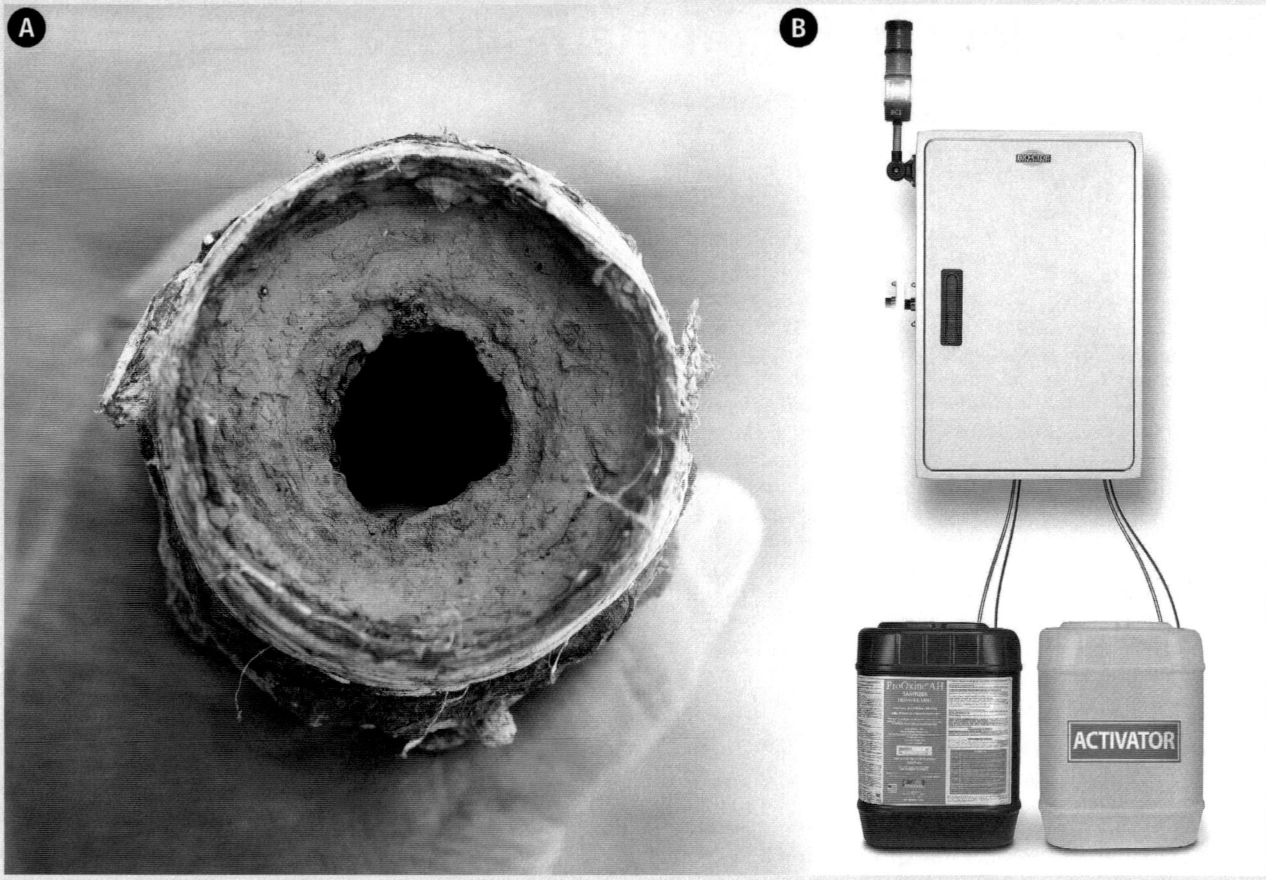

https://doi.org/10.54846/am2024/130

Senecavirus A testing of slaughter swine: A model for foreign animal disease surveillance

R. Anderson[1], DVM, MPVM; J. Torrison[2], DVM, PhD; B. Crossley[3], DVM, MPVM, PhD; S. Aly[4], DVM, MPVM, PhD
[1]California Department of Food and Agriculture, Animal Health Branch, Sacramento, California; [2]Longhorn Vaccines and Diagnostics LLC, Bethesda, Maryland; [3]California Animal Health and Food Safety laboratory, Davis, California; [4]Department of Veterinary Medicine, School of Veterinary Medicine, University of California-Davis, California

Introduction

Senecavirus A (SVA) causes vesicular lesions in swine that are indistinguishable from other agents such as foot-and-mouth virus (FMDV). Finding vesicles on an animal triggers a foreign animal disease (FAD) investigation to rule out a FAD such as FMDV. There was a 14,000% increase in FAD investigations in swine from 2014 to 2019 in the United States,[1] almost exclusively due to the presence (and confirmation) of SVA sampled from lesioned pigs. FAD investigations are costly and may lead to FAD "investigation fatigue" and complacency. For example, there has been a 37% decrease in FAD investigations in swine from 2019 to 2022 in the United States.[1] Surveillance testing for FADs needs to be developed to represent large numbers of animals efficiently that would ideally not involve handling animals individually. Our goal was to evaluate three testing schemes (two involving environmental testing of the livestock trailer, the other testing oral fluids (not specifically addressed here) and two different virus transport media (PrimeStore® MTM and tris-buffered tryptose broth, TBTB) for the detection of SVA in pigs from various sources at a California slaughter facility to determine the optimum testing strategy.

Methods

A total of 17 swine loads were tested, eight "suspect" loads, defined as one in which Food Safety and Inspection Service (FSIS) inspectors identified and held a pig because it that had lesions resembling a FAD, thus triggering a FAD investigation, and nine non-suspect loads. Environmental testing of livestock trailers was accomplished by testing (1) sterile boot swabs (Romer Labs, Inc, Newark, DE) worn while sampling (2) walls and doors of the trailer using sterile gauze sponges. Sterile saline was used to moisten both boot swabs and wall and door samples, the samples squeezed to mix, and the resultant fluid was decanted into conical tubes for testing. Oral fluids (OF) were collected from two areas of the holding pen using ¾" cotton ropes from individual loads of hogs directly after all hogs were penned. The two OF samples were pooled for each pen. Each pooled sample tube was then sampled according to standard sampling guidelines for TBTB and PrimeStore® MTM. All samples were tested using TaqMan-based qRT-PCR for SVA, samples were run for 45 cycles with a positive cut-off value of < 40.00. A Mixed-effects model linear regression was used to determine the best-fit model using Stata15.11 and significance was declared at $P < .05$.

Results

One hundred percent of the samples tested on suspect loads were positive for SVA, regardless of sample or media type. After assessing normality using Q-Q plots, continuous CT results were modeled using general linear mixed models. The best fitting random effect structure was with state of origin specified as a random intercept. Manual variable selection approach identified load status (samples from suspect -13 Ct different from non-suspect, 95% CI: -18.75, -7.24, *P*-value < .001), media type (samples in PrimeStore® MTM -1.8 Ct different from TBTB, 95% CI: -3.50, -0.17, *P*-value .03) and sample type (oral fluids -3.5 Ct different from walls and doors, 95% CI: -5.52, -1.47, *P*-value .001; and floors -2.6 Ct different from walls and doors, 95% CI: -4.75, - 0.53, *P*-value .014) significant variables as assessed using the Bayesian Information Criterion (BIC).

Discussion

There are several viral transport media available for FAD investigations. We evaluated PrimeStore® MTM in this study as an alternative transport media to TBTB which is often used in FAD investigations and found that the qPCR Ct values were significantly lower using PrimeStore® MTM, additionally, PrimeStore® MTM detected more SVA-positive samples. Given that PrimeStore® MTM inactivates virus while maintaining diagnostic efficacy[2] and does not require refrigeration, it seems to be a reasonable choice to be used in a standard FAD surveillance program.

References

1. USDA calendar year 2022 update: FAD investigation report. Accessed October 1, 2023. **https://www.aphis.usda.gov/animal.health/emergency_management/downloads/summary-fad-investigations-cy2022.pdf**

2. Welch JL, Shrestha R, Hutchings, *et al.* Inactivation of Highly Transmissible Livestock and Avian Viruses Including Influenza A and Newcastle Disease Virus for Molecular Diagnostics. bioRxiv. Sept 13, 2023. **doi: https://doi.org/10.1101/2023.09.13.557451**

https://doi.org/10.54846/am2024/131

Uses of an on-farm progesterone strip test to improve gilt management on farms

James Lehman, DVM, MS
Merck Animal Health, Lenexa, Kansas

Introduction

Efficient reproductive performance is an important aspect of successful and profitable swine production. Herds add gilts to replace sows that have poor reproductive performance, die or are culled for other reasons. Most farms expect > 95% of selected gilts to reach puberty and be mated before the gilts are 210 to 240 days of age. When the percent of gilts bred by these targets decreases, an investigation often begins. Two causes are decreased gilt estrus detection and/or decreased rates of puberty. Reduced response rates to hormone intervention (eg, PG600® or MATRIX™) may also occur. PG600 is contraindicated in post-pubertal gilts and conversely, MATRIX is contraindicated in pre-pubertal gilts. In certain situations, diagnostic testing to evaluate progesterone levels in individual animals is indicated to investigate the reason behind reduced gilt breeding targets and to confirm optimal use of hormone treatment plans being used on farm.

Progesterone test description/on-farm applications

Progesterone (P4) is a steroid hormone produced by the corpus luteum (CL), an ovarian structure formed from post-ovulatory follicular tissue. P4 levels are < 5ng/mL in animals that lack functional CL (gilts in the prepubertal, follicular or anestrus phase). Post-pubertal gilts have P4 levels > 10ng/mL during the luteal phase, which in non-pregnant gilts has a duration of approximately 14 days of the 21-day cycle.[1] The Progesterone Strip Test (PST) is an easy-to-use, lateral-flow immunoassay with a detection limit of 10ng/mL. 100µL of serum is required for analysis and the test takes 15 minutes to perform. The accuracy, specificity and sensitivity of the PST assay are 96.8%, 96.7% and 96.8%, respectively when compared to the reference gold standard assay (Chemiluminescence immunoassay, SIEMENS ADVIA Centaur XPT system). Below are a examples of this test's use.

Example 1- Older gilts without a detected estrus are often classified by the herdsman as prepubertal and PG600 is a common treatment. P4 testing (either before or after PG600 treatment) often indicates a subset of gilts have P4 > 10 ng/ml. Gilts with poor response to PG600 and > 10 ng/mL P4 levels 7-20 days following treatment may indicate ovulation occurred following an undetected or silent estrus.[2] Conversely, response to PG600 in animals with confirmed P4 levels < 10 ng/mL at the time of administration has been shown to exceed 92.5% (Table 1).[3]

Example 2- MATRIX is a synthetic progestin that mimics natural P4. Matrix is administered orally to post-pubertal gilts for 14 consecutive days to synchronize their estrus cycle. Prepubertal gilts will have suboptimal response to Matrix treatment. Testing to confirm P4 levels > 10ng/mL ensures gilts are post-pubertal. In a group of randomly cycling animals, approximately 67% should have > 10 ng/mL P4 (~33% in follicular phase, ~66% in luteal phase). It is important to remember that normal, actively cycling animals will have low P4 levels for approximately 7 days out of the 21-day cycle. Response to MATRIX in gilts with P4 levels > 10 ng/mL has been shown to exceed 95% (Table 1).[3]

References

1. Knox R, Vatzias G, Naber C and Zimmerman D. Plasma gonadotropins and ovarian hormones during the estrus cycle in high compared to low ovulation rate gilts. *J. Anim. Sci.* 2003;81:249–260.

2. Estienne and Crawford. *Theriogenology*. 2015; Vol 83, Issue 4, Pages 459-465.

3. Lehman J, Rodibaugh M, Miller D, Thacker B. Utilizing a progesterone ELISA in gilts to augment the use of PG600® and MATRIX™. 2014; Proc 23rd IPVS, p328.

Table 1: Response to hormone treatments in gilts with known progesterone status at time of treatment

Herd A: Progesterone (P4) test results in gilts prior to PG600 use and subsequent estrus response				**Herd B: Progesterone (P4) test results in gilts prior to and after MATRIX use and subsequent estrus response**				
N with P4 < 10 ng/mL on day of PG600 treatment	N treated with PG600	N in heat within 7 days following PG600 treatment	Percent response rate to PG 600	N with P4 > 10 ng/mL on day of first MATRIX dose	N with P4 < 10 ng/mL on day of last MATRIX dose	N mated within 8 days of last MATRIX dose	Avg days from last MATRIX dose to mating	Percent response rate to MATRIX
55	55	51	92.7%	21	21	20	6.35	95.2%
58	58	56	96.6%					

Adapted from Merck Animal Health technical bulletin SW-MAT-015-1; 2014

https://doi.org/10.54846/am2024/133

Machine learning accurately predicts infection in swine farms

A. Halev[1]; B. Martínez-López[2]; M. Clavijo[3]; C. Gonzalez-Crespo[2]; J. Kim[1]; C. Huang[4]; S. Krantz[5]; R. Robbins[6]; X. Liu[4]
[1]Department of Mathematics, University of California, Davis, Davis, California; [2]Department of Medicine and Epidemiology, Center for Animal Disease Modeling and Surveillance (CADMS), School of Veterinary Medicine, University of California, Davis, Davis, California; [3]Department of Veterinary Diagnostic & Production Animal Medicine (VDPAM), Iowa State University, Ames, Iowa; [4]Department of Computer Science, University of California, Davis, Davis, California; [5]Tosh Farms, Henry, Tennessee; [6]Pig Improvement Company, Hendersonville, Tennessee

Harnessing the power of artificial intelligence (AI) and successfully incorporating it into everyday life can be a dilemma. The potential is nothing short of astonishing. Appleby and Basran articulate[1,2] the value of AI and its subfield, machine learning (ML), have to the future of veterinary medicine; boldly stating "veterinarians who use AI will replace veterinarians who don't."

Last year we discussed the need to incorporate web-based AI into our biosecurity tool kit.[3] Specifically, we laid out the potential for ML applied to our common data streams, like diagnostic results and animal movement, to predict infectious disease outbreaks. That information would then inform our biosecurity measures before the event in hopes of preventing, rather than reacting, to an outbreak. The limitations for realizing the potential of ML for swine disease management were data availability and quality. Herein, we summarize a publication, *Infection prediction in swine populations with machine learning,* authored by our multi-disciplinary team.[4]

We consider six ML models with different combinations of the features to determine the model that generalizes and performs best. Specifically, we consider logistic regression, support vector machines, decision trees, gradient boosting, and random forests in both systems; in system B, we also consider a neural network model (auto encoder MLP) to handle distribution shift. There are 5 key contributions:

1. Production data and external factors particularly weather are useful for modeling infection.
   a. In system A, distance of nearby farms, total piglet inventory, the number of pigs weaned per mated female per year (PWMFY), pregnancy rates and average gilt pool inventory (total gilt days per days in period).
   b. In system B, test rates, wind direction and speed, sow gestation feed, incoming swine movements, and biosecurity data are important predictors.
   c. Biosecurity data proves important for PRRS, *Mycoplasma hyopneumoniae* (MHP) and PEDV: the employee access point plan is valuable in PRRS and PEDV prediction, and the carcass disposal plan is valuable in PEDV and MHP prediction.
2. Prediction of positive samples and assigning probability for infection (positive/negative result) at the farm level on a daily basis for PRRSV, PEDV, MHP, IAV-S.
3. Advanced warning (7 and 30 days) for positive sample prediction when using a 60-day historical window.
4. Models used on two independent production systems and are able to generalize to different scenarios of data granularity.
   a. System A with sparse data resulting in low granularity which was used to infer global infection trends. System B with high granularity was used to model detection of PEDV, PRRSV, MHP, and IAV infections.
   b. Movement data aided in generalization determined by the improvement in test scores. There was a limited amount of movement data available for system B thus the disease model performance is likely to improve with addition of movement data. The value of movement data for prediction of PEDV and PRRSV is confirmed by prior investigations.
5. Methods for successfully addressing data availability and distribution shift are described throughout.
   a. Distribution shift can be an unmitigable problem for data gathered in time series.

Model performance was judged by balanced accuracy and is a measure of how well a model correctly classifies cases as positive and negative. The global disease model created with system A data to classify any positive or negative case had 85% balanced accuracy. The disease specific models created with system B data to predict infections of PEDV, PRRSV, MHP and IAV had balanced accuracies ranging from 58 - 74%.

In conclusion, the work considered production systems as a whole, including both sow and nursery/finishing farms in both systems. In addition, analysis and comparison of two distinct farm production systems with differing levels of data availability are presented, and showed that machine learning is able to predict positive samples in both of these contexts.

References

1. Appleby, R. B., & Basran, P. S. (2022). Artificial intelligence in veterinary medicine. *J Am Vet Med Assoc*, 260(8), 819-824. Retrieved Nov 20, 2023, from **https://doi.org/10.2460/javma.22.03.0093**

2. Basran, P. S., & Appleby, R. B. (2022). The unmet potential of artificial intelligence in veterinary medicine. *Am J Vet Res*, 83(5), 385-392. Retrieved Nov 20, 2023, from **https://doi.org/10.2460/ajvr.22.03.0038**

3. Robbins, R., B. Martínez-López, B., Clavijo, M. & Harms, P. (2023). Biosecurity from 30,000 feet: web-based platforms enabling an epidemiological perspective (above the trees to see the forest). *Pre-conference proceedings of the 54th Annual Meeting of AASV.*

4. Halev, A., Martínez-López, B., Clavijo, M. et al. Infection prediction in swine populations with machine learning. *Sci Rep* 13, 17738 (2023). **https://doi.org/10.1038/s41598-023-43472-5**

https://doi.org/10.54846/am2024/134

EnduraPig™ mitigates porcine reproductive and respiratory syndrome and *Streptococcus suis* co-infection in nursery pigs

Huyen Tran[1], PhD; Aileen Joy Mercado[1], PhD; Stacie Crowder[2], PhD; Brenda de Rodas[1], PhD
[1] Purina Animal Nutrition, Gray Summit, Missouri; [2]PMI Additives, Arden Hills, Minnesota

Introduction

Porcine reproductive and respiratory syndrome virus (PRRSV) is one of the most devastating endemic diseases in swine, which is responsible for approximately $664 million annual losses to the US pork production. PRRSV infection induces animal susceptibility to secondary bacterial infections such as *Streptococcus suis,* leading to respiratory disease complex which is costly and difficult to treat. Therefore, the objective of this trial was to evaluate the impacts of EnduraPig™ (PMI Additives), a patent-pending blend of activated medium chain fatty acids (MCFA), phytogenic compounds, a probiotic cocktail, and other compounds on PRRSV viral load, clinical symptoms, lung lesions, and nasal microbiome of nursery pigs experimentally co-challenged with PRRSV and *Streptococcus suis.*

Materials and methods

Twenty-six weaned pigs (average 20 days of age; free of PRRSV, PCV2, SIAV, and *S suis* serotype ½ ST28) were randomly assigned to three treatment groups, including: 1) non-challenged fed control diet (Negative control; NC); 2) PRRSV and *S suis* co-challenged fed control diet (Challenge; Group 2); and 3) PRRSV and *S suis* co-challenged fed EnduraPig™ (Challenge + EnduraPig™; Group 3). Respective diets were fed to the pigs throughout the 29-day trial. After weaning, pigs were acclimated for 8 days (day -8) to adapt to the environment. Subsequently, Group 2 and 3 pigs were intranasally challenged with PRRSV 1-7-4 (2 mL/each nostril;1x10^6 log$_{10}$ TCID50/pig) on day 0 followed by additionally challenged with *S suis* Serotype ½ ST 28 (2 mL/each nostril; 1x10^9 CFU/4 mL) on day 5. Pigs were monitored twice daily for respiratory distress symptoms. On day 10 (Group 1, n = 3 pig; Group 2 and 3, n = 5 pigs each) and 21 (Group 1, n = 3 pig; Group 2 and 3, n = 5 pigs each), pigs were euthanized and necropsied for the evaluation of lung lesions and quantification of viral load in both lung and serum. Microbiome in nasal swabs were characterized by 16S rRNA V3-V4 amplicon sequencing (Illumina MiSeq) on day 21.

Results

Compared to Group 2, Group 3 fed EnduraPig™ had lower PRRS viral copies in both serum (day 21; $P < .05$) and lung (day 10 and 21; $P < .05$), lower % of pigs having pneumonia lesions (day 21; $P < .1$) and % of lung with pneumonia lesions (day 10 and 21). In addition, Linear discriminant analysis Effect Size indicated that PRRSV and *S. suis* co-infection shifted nasal microbiome toward enriching *Mycoplasma*, *Staphylococcus*, and *Pasteurella* (LDA score > 3; $P < .05$); however, feeding EnduraPig™ to the challenged pigs reduced the abundance of these bacteria and *Streptococcus* (LDA score > 3; $P < .05$). In terms of growth performance, EnduraPig™ reduced body weight loss due to the co-infection compared to Group 2.

Discussion

Data from this study indicate that dietary supplementation of EnduraPig™ reduced the severity of PRRSV infection, decreased nasal pathogenic bacterial growth, and decreased the incidence and severity of pneumonia lesions while supporting growth performance of nursery pigs co-infected with PRRSV and *S suis.*

https://doi.org/10.54846/am2024/135

A CSF vaccine based on E2 recombinant glycoprotein and adjuvanted with oil-in-water emulsion induces a full protection in a pig field trial

Marie-Eve Koziol[1]; Sarah Kharief[2]; Juliette Ben Arous[2]; Nicolas Versillé[3]
[1]Seppic - USA - 30 Two Bridges Road, Ste 210 - Fairfield, NJ 07004; [2]Seppic - Paris La Défense - 50 Boulevard National - 92257 La Garenne Colombes, France; [3]Seppic - 127 chemin de la Poudrerie - BP 90228 - 81105 Castres Cedex, France

Introduction

Classical swine fever (CSF) is a highly contagious viral disease which causes important economic losses in the pig industry. Even if many countries have succeeded in eliminating CSF, it remains endemic in South and Central America, Eastern Europe and Asia. Systematic prophylactic vaccination is the most effective strategy to control CSF in the endemic zones. Live attenuated vaccines have been widely used to control the disease but these conventional vaccines fail to identify infected from vaccinated animals (differentiating infected from vaccinated animals). Thus, a new generation of CSF DIVA vaccines are needed. In this study, a recombinant vaccine, based on the highly immunogenic structural glycoprotein E2 of CSF virus, is assessed in a pig trial.

Method

Two groups of 7 pigs are intramuscularly injected in the neck with 2 mL of the E2 recombinant CSF vaccine at day 0 (D0) and D21. The vaccines are either adjuvanted with a carbomer or with an oil-in-water emulsion (Montanide™ ISA 28R VG; 15/85 w/w). A third control group is left unvaccinated. After each vaccination, the body temperature and the local reactions at the injection site are monitored. The body weight gain is also controlled at D21 and D35. Material and Method: Blood samples are taken at different dates in order to assess the E2 specific antibodies by ELISA. At D35, a lethal dose of Shi-Myn strain of the CSF virus is intramuscularly injected into the posterior femoral muscle. After challenge, the vaccine protection is evaluated by calculating the survival rate and by PCR detection of CSF virus in the blood samples and in the spleens and lymph nodes.

Results

In terms of safety, the 2 adjuvanted vaccines were very well tolerated: No abnormality was noticed in the injection site, the body temperatures did not exceed one celsius degree after injections and the body weight gains were normal. Regarding the efficacy, vaccine groups showed a similar antibody profile with a positive threshold reached after the boost at D21. After challenge, only the vaccine adjuvanted with the Montanide™ ISA 28R VG induced a survival rate of 100% while that based on carbomer failed to fully protect the pigs (57%). The monitoring of the viremia also underlined an early and total clearance of the virus with the emulsion-based vaccine compared to the carbomer vaccine.

Discussion

Taken together, these results showed that E2 recombinant CSF vaccine associated with an oily adjuvant as Montanide™ ISA 28R VG is a good vaccine candidate and adapted to protect pigs from the disease while maintaining a good safety profile.

https://doi.org/10.54846/am2024/136

A multi-year quantitative surveillance of nursery pigs reveals increasing carriage of F18+ *Escherichia coli*

K. Pederson[1], BS; E. Galbraith[1], MS; J. Lee[2], MS; S. Son[1], MBA; M. King[1], PhD
[1]Microbial Discovery Group, Oak Creek, Wisconsin; [2]United Animal Health, Sheridan, Indiana

Introduction

The post-weaning period is a critical and stressful phase in the pig's life. A rapid dietary change combined with low feed intake, an immature gastrointestinal microbiota, and a developing immune system give rise to loss of microbial diversity in the gastrointestinal tract, inflammation, and disruption of the intestinal barrier. This dysbiosis allows the proliferation of opportunistic pathogens, often enterotoxigenic *Escherichia coli* (ETEC), resulting in post-weaning diarrhea (PWD) characterized by scouring, reduced growth, or sudden death. PWD is one of the most significant causes of swine health and performance losses globally, with mortality rates reaching 20 to 30% during serious outbreaks.[1,2] Monitoring the incidence and quantities of ETEC and other pathogens through microbial surveillance is one way to reveal underlying pathogen dynamics and allow veterinary health teams to develop proactive approaches for prevention and treatment of enteric disease.

Materials and methods

Rectal swabs were collected from 962 nursery pigs at 26 commercial farms and classified into three groups by time of sampling: 2018 to 2019 (n = 491), 2020 to 2021 (n = 218), and 2022 to 2023 (n = 253). Genomic DNA were extracted from swab samples and *E coli* marker genes were quantified using a panel of quantitative PCR assays for toxin and virulence genes involved in swine enteric disease. *E coli* genes with less than 25 positive results were excluded from this analysis. Gene data were log-transformed and differences in average gene quantity between two-year periods were determined with the Kruskal-Wallis test, followed by pairwise comparisons using Dunn's test with Bonferroni adjustment. Incidence of *E coli* marker genes of interest (number of pigs positive) by two-year period was compared using a Fisher's Exact test with Bonferroni adjustment.

Results and discussion

Surveillance results revealed that a gene marker for the ETEC-associated F18 fimbrial adhesin was detected in increased quantities in rectal swab samples from nursery pigs in 2022 to 2023 compared to the 2018 to 2019 and 2020 to 2021 periods (1.7 vs 1.0 and 0.7 $\log_{10}$ gene copies respectively; $P < 0.01$; Figure 1A). Further, the incidence of *F18* detection was increased in the 2022-2023 period compared to prior years (59% vs 33% and 26% respectively; $P < 0.01$; Figure 1B). Following a similar pattern, an ETEC-associated heat-labile toxin gene (*LT*) was present at highest quantities ($P < 0.01$) and at greatest frequency ($P < 0.01$) in the 2022 to 2023 period. Heat-stable (*STa*) and Shiga toxin (*Stx2*) genes were also found in higher quantities in the 2022 to 2023 period compared to 2018-2019, while intermediate quantities were detected in 2020-2021. In contrast, *K88*, another fimbrial adhesin gene, decreased in quantity from 2020 to 2021 to 2022 to 2023 (0.6 vs 0.3 $\log_{10}$ gene copies, $P < 0.01$). Together these data indicate that disease pressure from F18 + ETEC may be higher in recent years and may require new approaches for prevention and control. Continued surveillance programs for ETEC and other economically important pathogens will provide veterinary health teams with comprehensive tools for development of effective strategies for swine health management during single pathogen insults and co-infection scenarios.

References

1. Castro J, Barros M, Araújo D, et al. Swine enteric colibacillosis: Current treatment avenues and future directions. *Frontiers in Veterinary Science.* 2022; 9:981207.

2. Rhouma M, Fairbrother JM, Beaudry F, et al. Post weaning diarrhea in pigs: Risk factors and non-colistin-based control strategies. *Acta Vet Scand.* 2017; 59(1):1–19.

Figure 1: A) Mean *E coli* virulence gene quantity and **B)** incidence in rectal swabs from nursery pigs over two-year periods.

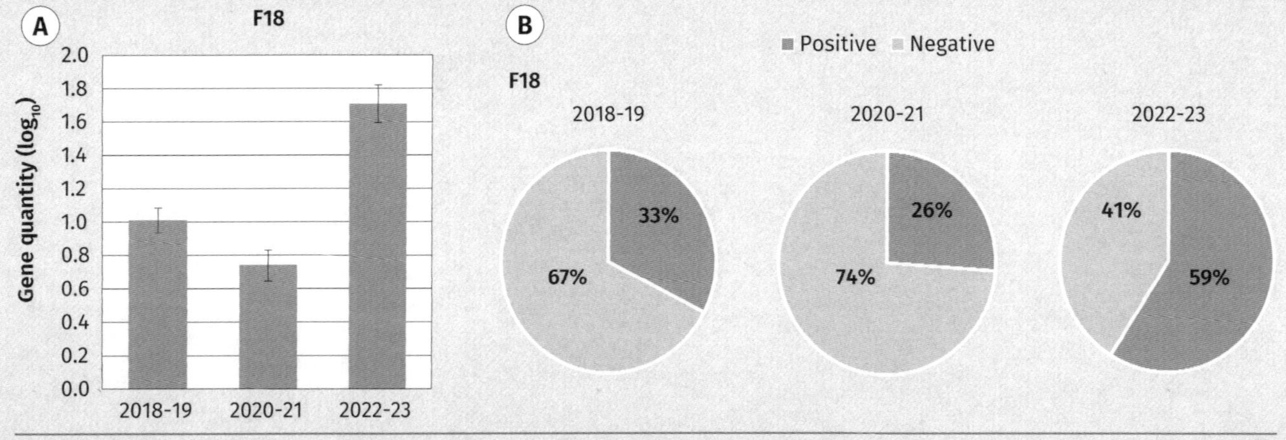

Superscripts with different letters indicate that mean gene quantities differ $P < 0.05$.

https://doi.org/10.54846/am2024/137

Helping people or pigs? Meloxicam vs aspirin impact on antibiotic use and production during weaning, post-vaccination and PRRSV challenge

H. Rotto[1], DVM, MS; D. Madson[2], DVM, PhD, DACVP; G. Silva[2], DVM, MS, PhD;
P. Smith[3], RPh, Pharm D; B. Payne[3], DVM
[1]Innovative Agriculture Solutions, LLC, Ames, Iowa; [2]Iowa State University, Ames, Iowa;
[3]Veterinary Pharmaceutical Solutions (VPS), St. Peter, Minnesota

Introduction

Non-steroidal anti-inflammatories (NSAIDs) delivered through drinking water are used in the swine industry during times of stress, infection, and inflammation. Previous work has demonstrated acetylsalicylic acid (ASA) is unable to achieve therapeutic levels.[1] However, ASA is still prescribed today as a low-cost option for caregivers. A more advanced selective COX-2 inhibitor NSAID, compounded meloxicam (MEL), is available. MEL has been shown to be effective during respiratory challenges.[2-4] This two-phase study adds to the knowledge of NSAIDs' impact when administered to pigs during normal stress of weaning and vaccination and a direct porcine reproductive and respiratory syndrome virus (PRRSV) challenge.

Methods

Three-week-old PRRS-negative pigs (n = 180) were weaned into a two-room facility. Phases 1 and 2 were conducted d0-10 and d21-35, respectively. On d0 and d21, the start of each phase, pigs were allocated into two treatment groups, in thirty-six pens of five pigs each. Blocks were created by weight (mean and variation) in each phase. On d0 all pigs were vaccinated with Circumvent® PCV MH and Ingelvac® PRRS MLV. On d21, all pigs were challenged with 1-8-4 10^4 PRRSV, 2mL, IM. Treatment group 1, ASA, drinking water was equivalent to 4oz ASA/1 gallon of stock solution proportioned at 1:128 rate on d2 and d24. Treatment group 2, meloxicam (MEL) drinking water was equivalent to 2 pints of 6mg/mL compounded (VPS, St. Peter, MN) MEL/5 gallon of stock solution on d2 and d24 proportioned at 1:128 rate. Both treatments rates were cut in half d3-8 and d25-30. Drinking water was made fresh daily and weighed in and out. All feed was weighed. Pigs were weighed on d0, d10, d21 and d35. Individual pig respiratory clinical scoring (0 - 3) was conducted on days 21 - 34 by a blinded observer. Sick pigs in phase 2 were treated with 2mL of enrofloxacin and recorded. Pen was the experimental unit. Individual pig weight differences were calculated. Analysis of Variance (ANOVA) was performed for each outcome to assess statistical significance between treatment groups.

Results and discussion

Production data and treatments are reported in Table 1. Water disappearance (not shown) was lowest on d3 - 4 and d22 - 24 for each phase likely from the impact of the challenges. MEL had the greatest water disappearance during all days. Despite severe respiratory clinical signs, no pigs died. Sum of the clinical scoring was 9% lower with MEL. This study supports MEL's positive impact on clinical picture, production parameters, and reduction of antibiotic treatments during the stress of weaning and vaccination as well as MEL pigs outperforming pigs treated with ASA during a clinical PRRSV challenge. This comparison should be expanded into field production to increase sample size, evaluate further during weaning/vaccination events and PRRSV infections.

References

1. Patterson A, et al. Determination of acetylsalicylic acid and sodium salicylate product stability and achievable serum concentrations under typical nursery conditions. AASV. 2005; 39-42.

2. Georgoulakis I, et al. Meloxicam as adjunctive therapy in treatment and control of porcine respiratory disease complex in growing pigs. JSHAP. 2006; 14(5): 253-7.

3. Lang I, et al. Preliminary findings on the efficacy of meloxicam in pigs with experimental APP infection. Proceedings of 18th IPVS Congress, Germany, 2004; 1: 177.

4. Payne B, et al. Evaluation of NSAIDs on weaning/vaccination stress and a subsequent PRRSV challenge. AASV. 2023; 325-6.

Table 1: Production and treatment parameters during Phase 1 and 2

	Phase 1-Weaning/Vax Stress			Phase 2-Direct PRRSV Challenge		
Parameter	**ASA**	**MEL**	***P*-value**	**ASA**	**MEL**	***P*-value**
Starting mean body weight, per phase, lbs	12.55	12.55	0.911	27.75	27.86	0.297
Ending mean body weight, per phase, lbs	18.10	18.48	0.075	40.27	40.93	0.117
Weight gain per pig, per phase, lbs	5.55	5.93	0.072	12.52	13.08	0.137
Average daily gain (ADG), lbs	0.56	0.59	0.072	0.89	0.93	0.145
Feed conversion ratio, feed/gain, FCR	1.06	0.99	0.417	1.58	1.55	0.358
Number (percent) of pigs treated	-	-	-	19 (21%)	10 (11%)	-
Total antibiotic treatments	-	-	-	24	16	-

https://doi.org/10.54846/am2024/138

Effects of dietary zinc supplementation strategy on immune response, bacterial clearance and disease severity in growing gilts infected with *Mycoplasma hyopneumoniae*

C. B. Carpenter[1,4]; P. E. Urriola[1]; G. C. Shurson[1]; M. Pieters[2]; L. J. Johnston[1,3]; W. P. Schweer[4]; L. A. Rodrigues[4]
[1]University of Minnesota, Department of Animal Science, St Paul, Minnesota; [2]Univeristy of Minnesota, Department of Veterinary Population Medicine, St. Paul, Minnesota; [3]University of Minnesota, West Central Research and Outreach Center, Morris, Minnesota; [4]Zinpro Corporation, Eden Prairie, Minnesota

Introduction

Zinc (Zn) is an essential nutrient[1] that supports respiratory & GI epithelia[2], and cytokine[3] & antibody production.[4] *Mycoplasma hyopneumoniae* infection results in cytokine[5] and IgG antibody production[5], lung lesions[6], and impaired bacterial clearance.[7] Compared with inorganic Zn alone, a blend of inorganic Zn and Zn-amino acid complex reduces the severity of GI[8] and respiratory[9] infections. Few researchers have evaluated the effects of dietary Zn source on disease outcomes during bacterial respiratory infection in pigs. The objective of this study was to assess the impact of dietary Zn supplementation strategy on immune response, bacterial clearance, and disease severity in growing gilts infected with *M hyopneumoniae*.

Methods

Ninety-six, 21-day old gilts were included in a 63-day experiment which consisted of a *M hyopneumoniae* challenge period from d 35-63. On d 0 (-35 days post-inoculation; dpi), gilts were confirmed negative for *M hyopneumoniae* and PRRSV and assigned randomly to 1 of 2 dietary treatments with 8 pens/treatment and 6 gilts/pen. The treatments were: 1) Control; 125 mg/kg Zn from (Zinc Sulfate) and 2) Zn-blend; 125 mg/kg Zn with 50 mg/kg Zn from Zn-amino acid complex (ZnAA) and 75 mg/kg Zn from ZnSO4. At 0 and 1 dpi, all gilts were inoculated intratracheally with 10 mL of a lung homogenate containing 1×10^5 CCU/mL of *M hyopneumoniae* strain 232. Blood and tracheal secretions were collected on -35, 0, 7, 14, 21, 28 dpi. Blood samples and tracheal secretions were collected for *M hyopneumoniae* antibody and bacterial detection, respectively. Number of coughs were recorded daily from 7 to 28 dpi to calculate a coughing index. At 28 dpi, gilts were humanely euthanized, and samples and observations were collected. Gross lung lesions were scored, and lung tissues were fixed for histopathology. Normally distributed data were analyzed in a generalized linear model with the fixed effect of diet and repeated measures for dpi. Non-normal data were analyzed using a Mann Whitney U Test using the NPAR1WAY procedure of SAS with a fixed effect of diet or a Chi-square test.

Results

The mean sample-to-positive ratio and the proportions of gilt seroconversion ($P > .05$) were not different between gilts consuming the control vs. Zn-blend treatments. Gilts consuming the control tended to have less relative bacterial load than those consuming the Zn-blend (21.68 vs 20.90 Ct, $P = .08$). There were no differences in mean macroscopic ($P = .73$) or microscopic ($P = .67$) lung scores, or coughing index score ($P = .25$) over the entire period between gilts consuming the control vs Zn-blend treatment.

Conclusion

Based on these results, growing gilts consuming diets containing different Zn sources exhibited similar signs of infection during the acute phase of a *M hyopneumoniae* infection. Nevertheless, it is important to consider that *M hyopneumoniae* infections are lengthy (months) and the chronic phase of infection was not evaluated in this study.

References

1. National Research Council (NRC). 2012.Nutrient Requirements of Swine, 11th ed.; National Academy Press: Washington, DC, USA.

2. Wessels, I., B. Rolles, & L. Rink. 2020. The potential impact of zinc supplementation on COVID-19 pathogenesis. Front. Immun. 11:1712. **doi: 10.3389/fimmu.2020.01712**

3. Lang, C. J., M. Hansen, E. Roscioli, J. Jones, C. Murgia, M. L. Ackland, P. Zalewski, G. Anderson, and R. Ruffin. 2011. Dietary zinc mediates inflammation and protects against wasting and metabolic derangement caused by sustained cigarette smoke exposure in mice. Bio Met. 24(1). 23–39. **doi:10.1007/s10534-010-9370-9**

4. Prasad, A. S. 2008. Zinc in Human Health: Effect of Zinc on Immune Cells. Mol. Med. 14(5). 353-357. **doi: 10.2119/2008-00033**

5. Summerfield, A. 2020. Immune responses against porcine mycoplasma infections. In: Maes, D., Siblia, M., Pieters, M., Mycoplasmas in Swine. Uitgeverij Acco. 6, 110 – 125.

6. Carr, J., M. Sabila, J. Segales. 2020. *Mycoplasma hyopneumoniae* clinical signs and gross lung lesions, including monitoring. In: Maes, D., Siblia, M., Pieters, M., Mycoplasmas in Swine. Uitgeverij Acco. 5, 97 - 107.

7. Kuhnert, P. & J. Jores. 2020. *Mycoplasma hyopneumoniae* pathogenicity: The known and the unknown. In: Maes, D., Siblia, M., Pieters, M., Mycoplasmas in Swine. Uitgeverij Acco. 3, 73 - 86.

8. Leite, F. L., E. Vasquez, F. A. Vannucci, C. J. Gebhart, A. Rendahl, J. Torrison, A. Mueller, N. L. Winkelman, Z. J. Rambo, and R. E. Isaacson. 2018. The effects of zinc amino acid complex supplementation on the porcine host response to Lawsonia intracellularis infection. Vet. Res. 49:1–9. **doi:10.1186/s13567-018-0581-3**

9. Da Silva, A. P., M. A. Rebollo, & R. A. Gallardo. 2020. Effects of Amino Acid–Bound Zinc and Manganese Feed Additives on MHC Haplotype Chickens Challenged with Infectious Bronchitis Coronavirus. Avian Diseases, 64(4), 451-456. **doi: 10.1637/aviandiseases-D-20-00031**

https://doi.org/10.54846/am2024/139

Metabolomic profile of sows affected by postpartum dysgalactia syndrome

Jamie M. Studer[1], MS; Zoe E. Kiefer[1], MS; Lucas Koester[1], PhD; Erika Johnson[1], MS; Lance H. Baumgard[1], PhD; Stephan Schmitz-Esser[1], PhD; Laura L. Greiner[1], PhD; Aileen F. Keating[1], PhD; Attila Farkas[2], DVM; Lucina Galina Pantoja[3], DVM, PhD; Kimberly A. Vonnahme[3], PhD; Jason W. Ross[1], PhD
[1]Department of Animal Science, Iowa State University, Ames, Iowa;
[2]Carthage Veterinary Services, Carthage, Illinois;[3]Zoetis, Parsippany, New Jersey

Introduction

Genetic selection for prolificacy in modern sows has increased the metabolic burden of lactation due to increased demand for milk production to support larger litters. Additionally, sows undergo substantial immunological and metabolic changes throughout gestation and lactation, with maternal metabolism favoring an anabolic state during early pregnancy and shifting to a catabolic state during late gestation and early lactation.[1] These drastic shifts in metabolism experienced during the transition period may induce pathologies associated with oxidative stress, insulin resistance, and intestinal barrier dysfunction, leading to a state of chronic low-grade inflammation,[2] rendering sows vulnerable to puerperal disorders such as postpartum dysgalactia syndrome (PDS). Described as a metabolic and hormonal disturbance, PDS is a multifactorial disease complex characterized by a pronounced reduction (dysgalactia) in milk secretion. Insufficient milk production limits piglet growth, leading to increased piglet mortality and often removal of affected sows from the herd, ultimately compromising sow longevity and negatively impacting profitability of swine operations. We have previously reported differences in circulating analytes (aspartate aminotransferase, alanine aminotransferase, gamma glutamyl transferase, beta-hydroxybutyrate) and inflammatory markers (lipopolysaccharide binding protein) between healthy sows and sows affected with PDS. However, few differences were observed before clinical signs of PDS were evident. Therefore, the objective of the current study was to utilize metabolomic analysis to identify candidate biomarkers to aid in early detection of PDS.

Methods

Starting within 24 - 36 hours post farrowing, sows (n = 374) were monitored for signs of PDS as determined by elevated rectal temperature (exceeding 103.0 °F), phenotypic evaluation of mammary fill, lying in sternal recumbency to deny piglets access to teats, and piglet wasting. Subsequently, thirty-six sows with clinical symptoms of PDS were identified and matched by parity and day of lactation with thirty-six healthy control (CON) sows that remained unaffected. Blood samples were collected within 24 - 36 hours post farrowing (timepoint 1) and at identification of clinical PDS (timepoint 2). Additionally, piglet mortality and average daily gain (ADG) were measured to validate reduced milk production in sows affected by PDS. Serum samples were subjected to gas chromatography-mass spectrometry (GC-MS) to identify differences in small molecules and alterations to metabolic pathways between healthy sows and sows experiencing PDS.

Results and discussion

As an expected consequence of insufficient milk intake, piglets from PDS sows had lower ADG and higher mortality than piglets from CON sows ($P < .01$). Metabolomic analysis via GC-MS revealed abundance of 16 and 41 molecules differed between CON and PDS sows at timepoint 1 and timepoint 2, respectively ($P < .10$). At timepoint 1, molecules involved in insulin regulation were altered in PDS sows, including increased 2-hydroxybutyric acid (2HB) and propanoic acid ($P = .04$), along with decreased D-pinitol and L-tyrosine ($P \leq .04$) compared to CON. Elevated 2HB is of interest as this metabolite has been identified as a marker for insulin resistance and impaired glucose regulation, and increased levels of 2HB have been reported during periods of deficient energy metabolism, lipid oxidation, and oxidative stress.[3] At timepoint 2, alterations in molecules associated with insulin regulation in PDS sows included decreased L-alanine, glycine, D-glucose, D-pinitol, and L-asparagine, along with increased DL-phenylalanine, myo-inositol, and propanoic acid ($P \leq .05$). Additionally, all three branched chain amino acids (BCAA), valine, leucine, and isoleucine, were increased in PDS sows at timepoint 2 ($P < .01$), and elevated levels of BCAA are highly associated with metabolic diseases including insulin resistance and obesity.[4] Collectively, these data indicate sows affected with PDS experience metabolic dysfunction potentially due to inflammation induced insulin resistance. Further research is needed to determine if metabolites identified at timepoint 1 can be utilized to identify sows at risk for developing PDS, before piglet performance is compromised.

References

1. Collier RJ, McNamara JP, Wallace CR, Dehoff MH. A Review of Endocrine Regulation of Metabolism during Lactation. *J Anim Sci.* 1984;59(2):498-510. **doi:10.2527/jas1984.592498x**

2. Tan C, Zhai Z, Ni X, et al. Metabolomic Profiles Reveal Potential Factors that Correlate with Lactation Performance in Sow Milk. *Sci Rep.* 2018;8(1):10712. **doi:10.1038/s41598-018-28793-0**

3. Sousa AP, Cunha DM, Franco C, et al. Which Role Plays 2-Hydroxybutyric Acid on Insulin Resistance? *Metabolites.* 2021;11(12):835. **doi:10.3390/metabo11120835**

4. White PJ, McGarrah RW, Herman MA, Bain JR, Shah SH, Newgard CB. Insulin action, type 2 diabetes, and branched-chain amino acids: A two-way street. *Mol Metab.* 2021;52:101261. **doi:10.1016/j.molmet.2021.101261**

https://doi.org/10.54846/am2024/140

Eradication of classical swine fever and pseudorabies viruses in the United States ... what worked and why?

Jeff Zimmerman, DVM, PhD
Department of Veterinary Diagnostic and Production Animal Medicine, Iowa State University, Ames, Iowa

Introduction

Classical swine fever virus (CSFV)

Between 1884 and 1913, CSFV killed ~7.5% of the total US swine population annually. In 1903, Bureau of Animal Industry (BAI) scientists Edward A de Schweinitz and Marion Dorset proved the viral etiology of the disease (BAI USDA No. 41). Thereafter, Dorset, WB Niles, and CN McBryde developed a procedure to protect swine against CSFV by simultaneously administering virus and antibody, ie, defibrinated blood from an acutely infected pig and serum from a recovered pig, as previously described by Kolle and Turner (1898) for rinderpest. The simultaneous method remained the standard until the first licensed MLV vaccine appeared on the market in 1951. Even so, a 1961 study estimated annual CSFV loses to US producers at $592.6 million (this value was adjusted to 2023 dollars using an online calculator at **www.bls.gov/data/inflation_calculator.htm**).[19]

Pseudorabies virus (PRV)

In the US, reports suggestive of PRV in cattle date back to 1813.[9] In 1902, Aladàr Aujeszky demonstrated the viral etiology of the disease by producing clinical disease in rabbits injected with tissue homogenate from a cow with clinical signs of PRV.[10] Beginning in the 1960's, clinical PRV in commercial herds was reported with increasing frequency in both North America and Europe. Gustafson and Scherba[8] estimated annual losses to PRV prior to its eradication at $126.0 million (in 2023 dollars).

Leadership, teamwork, and industry support

Animal health organizations and the United States Animal Health Association (USAHA), in particular, have historically played a major role in US animal health initiatives. The first annual meeting of the USAHA (then called the Inter-State Association of Live Stock Boards) was held in in 1897.[2,5] The USAHA first called for the eradication of hog cholera in 1907, just 4 years after the discovery of the virus. This was followed by the organization of a "hog cholera" control committee in 1921 and a renewed call for eradication by the USAHA in 1926. In 1950, the USAHA organized the National Committee on the Eradication of Hog Cholera and, in 1956, the Committee held regional workshops and promoted a ten-point CSFV control plan.[13] Legislation introduced into Congress for a national eradication program was passed and signed by President Kennedy on September 6, 1961 (Public Law 87-209).

In the case of PRV, the USAHA, the Livestock Conservation Institute (now the National Institute for Animal Agriculture), and the National Pork Producers Council led the development and implementation of PRV eradication.[1,5] For example, the 1975 Livestock Conservation Institute's annual meeting featured a symposium on PRV and formed a PRV Committee.[1] Ultimately, a national eradication program was approved by the National Pork Producers Council in 1989 with a start date of January 1.

The tools needed for eradication

Classical swine fever virus (CSFV)

The knowledge base with respect to CSFV at the beginning of the eradication program was generally adequate. That is, a great deal was known regarding the natural history of CSFV including routes of transmission, the presence of persistent carriers, reservoirs, and potential vectors.[15] The one area of noticeable weakness was in diagnostics. Dunne[6] commented that "no completely satisfactory laboratory test is presently available for use in hog cholera diagnosis" and some argued that the eradication program wait until a definitive diagnostic assay was developed.[7] In the program, CSFV-suspect herds were identified on the basis of clinical signs - not routine surveillance. Thereafter, BAI veterinarians evaluated herd history, clinical signs, pathological lesions, and hematology on the farm to arrive at a diagnosis. Hematology was initially an important part of the diagnostic process. That is, leukopenia is one of the cardinal signs of CSFV infection and a total leukocyte count of 9,000 or less in pigs ≥ 8 weeks of age was considered definitive for CSFV infection.[6] The introduction of the tissue direct fluorescent antibody technique in June 1964 (visualization of CSFV antigen in tonsil tissue) was a big step forward in CSFV diagnosis. All states were enrolled in the program by 1966. Vaccination was prohibited in 1969. The last isolation of CSFV was made in 1976 from a herd in New Jersey. On January 31, 1978 (16 years after initiation), the US was declared CSFV-free.

Pseudorabies virus (PRV)

Extensive research on PRV and PRV eradication (pilot projects) conducted in the early 1980s provided a broad knowledge base but, until the mid-1980s, a herd diagnosis of PRV was based on the detection of serum-virus neutralizing antibody. Major improvements in PRV diagnostics and vaccines were achieved in the mid-1980's. In particular, the development of DIVA vaccines based on PRV deletion mutants that did not express specific glycoproteins made it possible to use serum antibody ELISAs to differentiate vaccinated from field virus-infected pigs.[4,11,12,18] The DIVA vaccines were highly effective, ie, protected pigs against clinical disease, reduced the shedding of field virus, and controlled clinical outbreaks, but did not eliminate the PRV infection and spread. The US Federal Government spent $72 million on the project during the first 10 years of the program. Additional costs were incurred by the various state governments and swine producers.[16] Despite some setbacks in the latter 1990s, the National Pseudorabies Control Board announced that all commercial swine herds in the US. were free of pseudorabies virus in 2004.[3]

Herd sampling protocols

In the CSFV program, sampling was done in herds were clinical signs suggestive of CSFV. In suspect herds, all pigs in the herd sampled ("whole herd" or "census" sampling) and tested. Census sampling was feasible in the CSFV program because the average herd held ~50 pigs.

The PRV program differed from the CSFV program by implementing routine testing of herds using statistically-based "representative" sampling.[17] This change was not made easily. At a meeting in Peoria Illinois in 1986, the proponents of census sampling and the advocates of representative sample hotly debated the two approaches, with representative sampling ultimately emerging the winner (thank goodness!). Sampling frequency and sample size were based on herd type and size. For example, commercial herds were sampled annually and seedstock producers sampled quarterly. Sample size varied between 25 to 29 animals, eg, 29 animals were bled herds of ≥ 1000 pigs. The sampling protocol was modified over the course of the eradication program such that each finishing barn was considered a "herd". Thus, a farm with 3 1,100 head finishing barns would bleed 29 pigs per barn for a total of 87 animals.

Reasons for success?

CSFV and PRV eradication were successful because of:

1. Strong, long-term industry support and an effective Federal-State-Industry collaboration.
2. A substantial scientific understanding of the natural histories of CSFV and PRV.
3. The availability of diagnostically sensitive and diagnostically specific tests. In the case of PRV, the crucial role of highly effective DIVA vaccines and antibody ELISAs cannot be exaggerated.
4. Willingness to apply new concepts. Representative sampling was accepted and integrated into the PRV program and the "whole herd" sampling used in the CSFV program abandoned as impractical and unnecessary.

There are are couple of things we should bear in mind. Each of these programs fit the industry of its time because they were designed by people close to the industry. Thus, in 1960, the inception of the CSFV eradication program, there were 640,000 small farms in the US run by family members and mostly raising pigs outdoors.[14] And the program worked for them. By 1994, early in the PRV eradication program, there were 150,000 US hog farms and the industry was in transition to a different production model.[14] That is, the small, outdoor herds of the CSFV eradication era were mostly being replaced with larger, multi-site, specialist (breeder or finisher) operations. These changes in the industry are relevant because the processes that worked on the small-scale farms of the CSFV era would not have worked on the larger farms of the PRV era. And most probably, the PRV eradication design will not work on the 65,000 farms of 2024. Although rarely a winning strategy, "Generals always prepare to fight the last war - especially if they won." We would be wise to keep this caution in mind as we prepare to deal with the issues facing our industry.

References

1. Anderson L, Black N, Hagerty TJ, Kluge JP, Sundberg PL. 2008. Pseudorabies (Aujesxzky's Disease) and its Eradication: A Review of the U.S. Experience. *USDA:APHIS Technical Bulleting* No. 1923.

2. Anon. 1950. History of the United States Livestock Sanitary Association. *Proceedings of the United States Livestock Sanitary Association*, pp. 1-8.

3. Anon. 2005. Succeeding against pseudorabies. *J Am Vet Med Assoc* 226:22.

4. Beran, G.W., 2008. Pseudorabies: A Century of Learning. Trends in Emerging Viral Infections of Swine. *Iowa State Press*, Ames, Iowa, USA, 211-216.

5. Black N. 1998. Animal health: A century of progress. *USAHA Website*. **http://www.usaha. org/history**

6. Dunne HW. 1958. Hog cholera. In: Dunne HW (ed), Diseases of Swine. *The Iowa State College Press*, Ames, Iowa, pp. 111-144.

7. Dunne HW. 1961. The diagnosis of hog cholera. *Proceedings of the United States Livestock Sanitary Association*, pp. 478-484.

8. Gustafson, D.P. and Scherba, G., 1978. National survey of the economic impact of pseudorabies in swine for the period June 1, 1976 - May 31, 1977. *A report to APHIS*, 17 pp.

9. Hanson, RP. 1954. The history of pseudorabies in the United States. *J Am Vet Med Assoc* 124:259-261.

10. Kohler M, Kohler W. 2003. Zentralblatt fur Bakteriologie--100 years ago Aladar Aujeszky detects a 'new' disease--or: It was the cow and not the sow. Int *J Med Microbiol* 292, 423-427.

11. Mettenleiter T. 1994. Pseudorabies (Aujeszky's disease) virus: state of the art. August 1993. *Acta Vet Hung* 42:153-177.

12. Mettenleiter TC. 2000. Aujeszky's disease (pseudorabies) virus: the virus and molecular pathogenesis - state of the art, June 1999. *Vet Res* 31:99-115.

13. Milligan J, Aitkin WA, Campbell CL, et al. 1956. Report of the Committee on Nationwide Eradication of Hog Cholera. *Proceedings of the United States Livestock Sanitary Association*, pp. 270-272.

14. Putman B, Hickman J, Bandekar P, Matlock M, Thoma G. 2018. A Retrospective Assessment of US Pork Productions: 1960 to 2015. *University of Arkansas Resiliency Center*. Retrieved from **https://scholarworks.uark.edu/rescentfs/2**

15. Schwarte LH. 1959. Our present knowledge of reservoirs and vectors of hog cholera virus. *Proceedings of the United States Livestock Sanitary Association*, pp. 317-322.

16. Taft AC. 2000. The eradication of Aujeszky's disease in the United States. *Vet Res* 31:157-158.

17. USDA. 1981. Hog cholera and its eradication. A review of the U.S. experience. *Animal and Plant Health Inspection Service, United States Department of Agriculture, APHIS* 91-55.

18. Van Oirschot JT, Gielkens ALJ, Moormann RJM, Berbs AJM. 1990. Marker vaccines, virus protein-specific antibody assays and the control of Aujeszky's disease. *Vet Microbiol* 23:85-101.

19. Wise GH. 1981. Hog cholera and its Eradication, A Revew of the U.S. Experience. *United States Department of Agriculture (USDA): Animal and Plant Health Inspection Service (APHIS)* 91-55.

https://doi.org/10.54846/am2024/141

Mycoplasma elimination from a regional to national level (why aren't we there yet?)

P. Yeske, DVM, MS
Swine Vet Center, St. Peter Minnesota

Introduction

Mycoplasma hyopneumoniae (*M hyopneumoniae*) is a pathogen that has been in the swine industry for a long time and continues to be a significant problem for the swine industry.[1-3] Disease is primarily observed in the grow-finish phase of production, with effects of decreased average daily gain, poorer feed conversion, increased mortality, lower percent marketed pigs to the primary market, and increased medication cost.[4] Costs attributed to *M hyopneumoniae* disease are significant. When utilizing actual production numbers from 2007-2015 records and conducting an economic model, the cost of *M hyopneumoniae* disease equated to $4.99 per pig for one system.[5] Economic data reported from other production systems have been described to be $2.80 per pig for this disease.[4] Preliminary data suggests an economic impact in the breeding herd, but this has not yet been well documented.[6] It stands to reason that if the sow herd and especially gilts are healthier, then there would be better performance as well.

Health, production, and economic losses are the primary reasons for herd owners and veterinarians to consider *M hyopneumoniae* elimination programs. One of the best ways to guide the decision to eliminate *M hyopneumoniae* is to evaluate performance data from herds that are free from the pathogen or to use numbers from their own experience to demonstrate the value of freedom of disease. Frequently, the decision to eliminate *M hyopneumoniae* is made based on other activities or eradication programs that may take place at the farm at the same time. The diagram in Figure 1 can aid the eradication decision-making process. As depicted in Figure 1, "Stabilization" means that the pathogen level is under control and the herd is still positive but managed to the best of capabilities. Although stabilization in the sow herd may improve herd and grow-finish performance, economic and production losses still occur when compared to *M hyopneumoniae* negative herds. Therefore, the ultimate control method for *M hyopneumoniae* is the elimination of this microorganism from a herd.

Mycoplasma elimination

In the North American industry, *M hyopneumoniae* elimination has been attempted predominately in farrow-to-wean farms. The elimination of *M hyopneumoniae* is also possible from farrow-to-finish farms. However, it is much more difficult and usually

Figure 1: Decision tree for *Mycoplasma hyopneumoniae* disease control options.

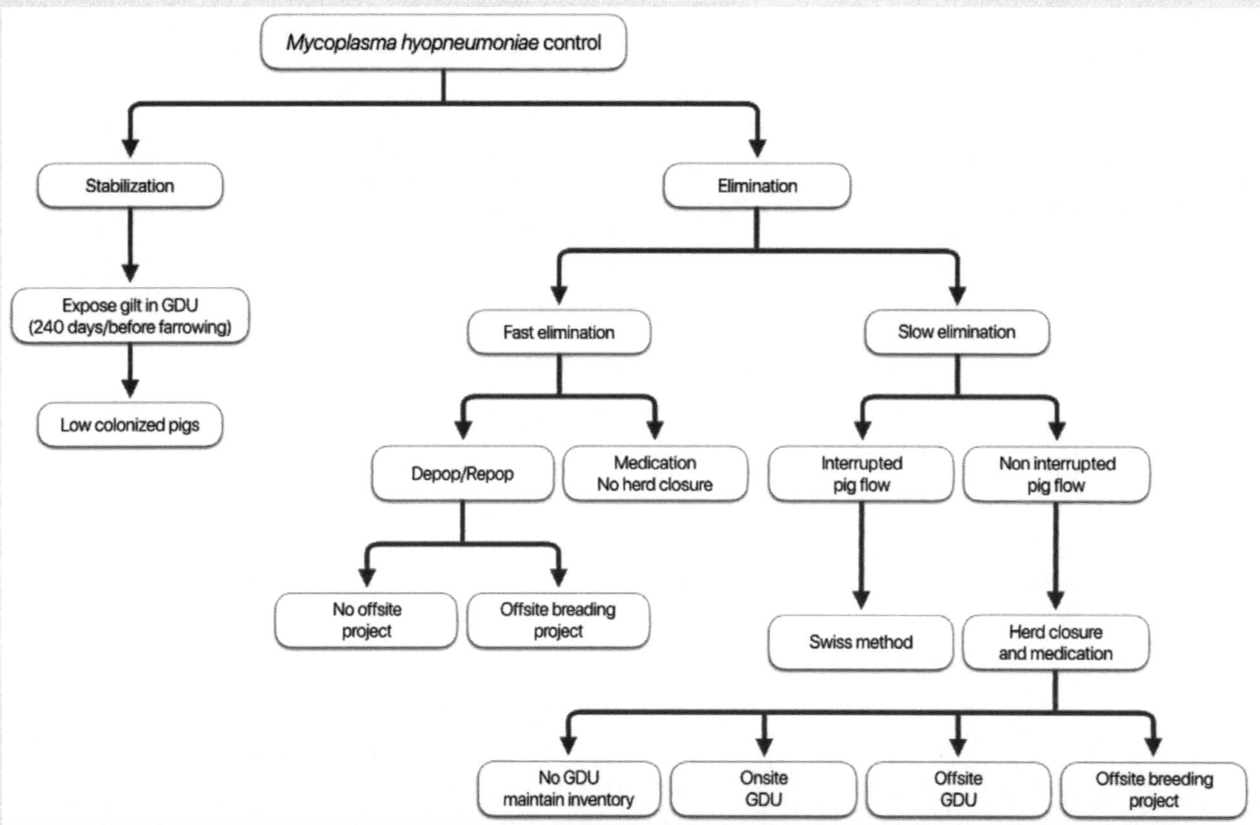

requires a partial depopulation of the nursery and finishing site to be performed successfully in addition to elimination from the sow herd. There is good data supporting that when negative pigs are placed in pig dense areas, a majority (94%) remained negative, allowing for herds to truly capture the advantages of being *M hyopneumoniae* negative.[7]

Several methods have been successfully used for *M hyopneumoniae* elimination.[7,8]

- **Depopulation and repopulation:** This is the time-tested, quickest method to eliminate *M hyopneumoniae,* with the greatest likelihood of success, however, it is generally the most expensive.
- **"Swiss" model:** Has been demonstrated to be very effective but requires a partial depopulation of farrowing, requiring the farrowing house to be completely empty for 1 turn followed by removal of all animals less than 10 months of age.
- **Herd closure and whole herd medication:** One of the most popular methods and can be done with minimal effects on production, depending on the facility and management. This method has been used in combination with other pathogen eliminations, predominantly with PRRSV and with PEDV. Herd closure and whole herd medication method has a moderate cost and has a good success rate. This is the method that I have used for most of my and our clinics clients' eliminations for these reasons. Herd closure and whole herd medication program is detailed in Figure 2 and the long-term success rate of this method in comparison to other methods are shown in Table 1.
- **Whole herd medication without closure:** This method allows for rapid return to negative status but has more cost due to medication programs (there are multiple different ones that have been used) and lower rate of success is the whole herd medication without closure program is detailed in Figure 3 and long-term success rate of this method in comparison to other methods is shown in Table 1.

Every herd will have unique circumstances that will have to be considered. As we move forward, the protocols will continue to evolve, which will be interesting to see if we can reduce the closure time, cost, and impact on the herds' production.

Area control programs

Individual herd elimination plans have been successful for extended periods of time, even in relatively pig dense areas. The addition of filtration to many herds has also reduced the likelihood of area spread to these herds. Herds that have started as negative or have been through depopulation and repopulation have also had good track records of remaining *M hyopneumoniae* negative. All these cases would lead to only seeing additional benefit in regional control with the remarkable success seen at the level of individual herds, extending these efforts to a regional scale could

Figure 2: Herd closure and medication protocol.

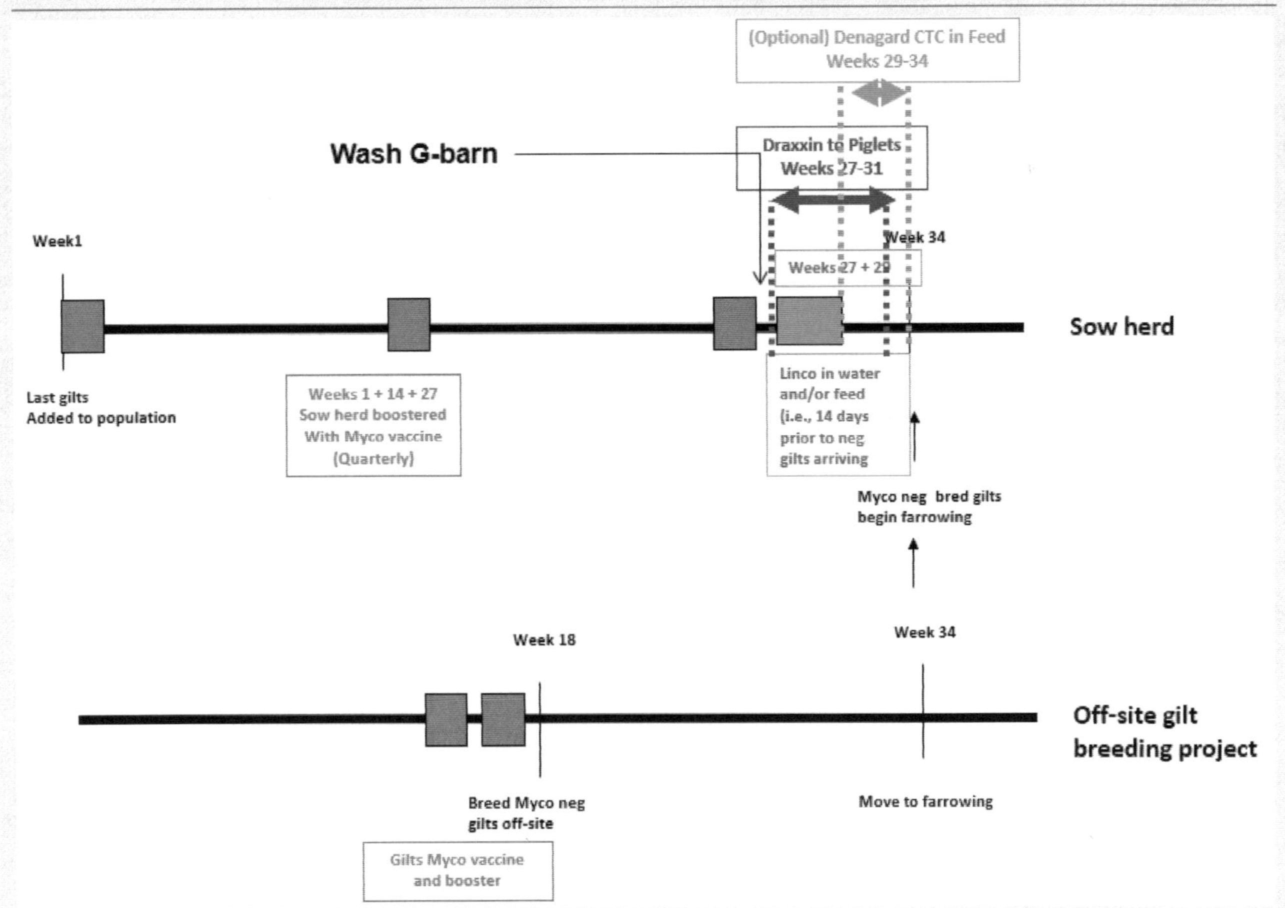

Figure 3: Whole herd medication without closure.

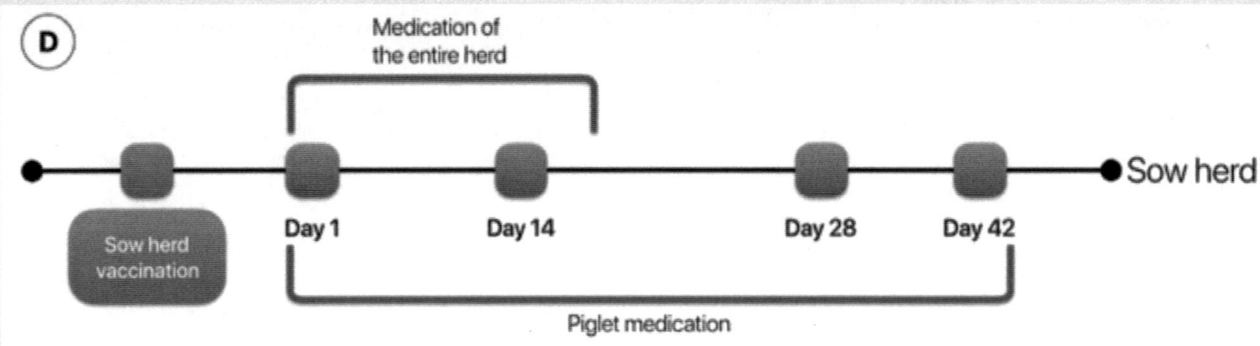

Table 1: Long term success rate with *Mycoplasma hyopneumoniae* elimination

	Herd closure	No herd closure	Total
Number of herds	53	20	73
Number of sows	146388	47450	196388
Percent negative to date	69%	53%	76%
Number of negative herds	45	10	62
Average herd size	2762	2373	2690
Average months negative	143	83	87
Years negative	11.9	6.9	7.2

undoubtedly amplify their effectiveness. Such a strategic approach would not only diminish the incidence of new infections but, considering the identified sources, it also presents a relatively low risk in the present scenario. Regional elimination of *M hyopneumoniae* would reduce the level of new infections without a doubt.

To date, there has been limited activity and focus around *M hyopneumoniae* regional control programs. PRRS area control projects have lost momentum, even though it is still a good idea. The increased number of PRRS regional control failures has made it more difficult to keep these moving forward. I believe this has been a limitation for *M. hyopneumoniae* following the same path, however, I believe that the outcome could be much better based on individual farm experiences. Although it still occurs, area spread of *M hyopneumoniae* is less likely to happen compared to other pathogens, aiding to the success of regional programs and even at a national level.

The timing is right to move forward with a national program as there is momentum in the industry and more systems are moving in that direction already. There just needs to be the drive from producers and veterinarians to move forward on this. The tools are all in place to make this happen and would just need to have a coordinated effort. The vast majority of farms are farrow-to-wean, making *M hyopneumoniae* elimination more practical to accomplish without a major expense. There is also an economic advantage to moving to a national program. Once elimination was achieved routine vaccination that is still done on most commercial farms could be eliminated. Granted vaccination cost is relatively low and included with many combination products there is still a cost and an opportunity to refocus these dollars in other programs.

Summary

Mycoplasma hyopneumoniae is still a concern and costs the swine industry every day. Individual farm data demonstrates the value of elimination and that it can be sustainable. The only thing that is holding us back from moving forward on a regional and even a national program is the political will to make it happen. For any elimination program, there is always risk of failure but at this time, we have good tools for elimination and high success rates have been described. There is a need for easier diagnostic tools, but we have tools that we can use today that work well. With more systems and farms moving in the direction to become *M hyopneumoniae* negative, it is a good time to go to the 'next level' and move forward with elimination on a broader scale.

References

1. Mare, C.J., Switzer, W.P., 1965. New Species: *Mycoplasma hyopneumoniae*; a causative agent of virus pig pneumonia. Vet. Med. *Small Anim. Clin.* 60, 841– 846.

2. Goodwin, R.F.W., Pomeroy, A.P., Whittlestone, P., 1965. Production of enzootic pneumonia in pigs with a mycoplasma. *Vet.* Rec. 77, 1247–1249.

3. Thacker, E.L., Minion, F.C., 2012. Mycoplasmosis. In: Zimmerman, J.J., Karriker, L.A., Ramirez, A., Schwartz, K.J., Stevenson, G.W. (Eds.), *Diseases of Swine*. Iowa State University, 779–797.

4. Schwartz, M., 2015: Cost of M. Hyopneumoniea in growing pigs. Allen D Leman conference.

5. Linhares, D., 2017: A field study on economics of *Mycoplasma hyopneuomiae* elimination. Allen D Leman conference.

6. Britton L., *Personal communication*.

7. Yeske, P., 2016, 2016*: Mycoplasma Hyopneumoniae* Elimination. Proceedings from AASV annual meeting 376-381.

8. Yeske P., Holst S., Betlach A., Pieters M, Chapter 12 Eradication of Mycoplasma hyopneumoniae from Pig Herds. *Mycoplasmas in Swine 2020* pp 229-246.

https://doi.org/10.54846/am2024/142

PRRSV: The NeverEnding Story

A. Maschhoff, DVM, T. Bauman, DVM
The Maschhoffs, LLC, Carlyle Illinois

Introduction

Porcine reproductive and respiratory syndrome virus (PRRSV) continues to be one of the most economically significant pathogens in swine herds across the globe, at $664 million annually in the US, €126 per sow in Dutch herds and ¥822.75 per sow in China.[1-3] Similar to the movies *The NeverEnding Story, Groundhog Day* and the *The Song that Doesn't End*, the economic impact is continuously reported in almost ever paper regarding PRRSV. At the same time, it appears less and less discussion is being had about elimination of this costly disease from our US industry. When looking at the commonly used prevalence reports through the University of Minnesota, the trend for disease prevalence has remained stagnant or slightly increasing regarding vaccinated herds and herds with active wild type virus infection.[4] So, even though its economically important, the industry today appears to be accepting and living with this impactful disease. How does the industry structure, sustainability and mindset about the future play out regarding PRRSV? Is the industry in automatic response mode or will it take on the challenge of eliminating this pathogen from our US herds?

Structural challenges to disease management

The US swine industry has a unique structure when it comes to management of endemic diseases. In the last two decades, the industry has consolidated by more than 70% and specialized operations.[5] In a 2017 report by the USDA 73% of swine farms have an inventory of 5,000 head or more.[6] While it is no surprise this structurally maturity has reduced costs and improved industry efficiency, it does present challenges for disease management. Dense population dynamics within geography and individual unit sizes play a part in the ability to manage PRRSV. With such large herds, sows/gilts can have varying immune statuses at the end of PRRSV closures and create a longer time to stability. Just this past year at the ISU James D. McKean conference, Dr. Jason Kelly presented a project titled "Strategies to shorten time to stability" where the Suidae veterinary team was working on trying to find those last positive sows by completing targeted diagnostic tonsil sampling. Piglets within a large unit or as they get moved to finishing sites also help circulate viruses. When referring to McRebel practices in the farrowing house, one may say "we are trying to manage around the "*Typhoid Mary*" to prevent spread of the wild type virus still active within that herd. But, should the industry have a broader and alterative outlook? Should the US industry start using another human analog reference and create the last exposed and known positive, like Ali Maow Maalin and Rahima Banu (the last two people recorded to have natural exposure to Smallpox)?[7]

The industry structural challenges don't just include site structure, but the relative relationship with the necessary support services to run farms, as well. Interstate movement, for example, is an integral part of the swine industry, however, very segmented. Disease risk is created by connecting sites through this necessary aspect to simply move pigs from site to site. In 2001, APHIS estimated that that 71 percent of pigs in the US enter the growing/finishing phase at a different location from which they were born and that in Iowa alone there are 1.7 million pigs move into the state every month.[8] Several progressive strides to manage transport biosecurity have been implemented like Bio-Dry systems, but it remains a risk factor and important piece of the puzzle when considering industry long term PRRSV management strategies. From transport biosecurity to other technologies in biosecurity like filtration, the industry structure has influenced this RNA virus to evolve around those barriers. There are more and more reports on the lineage L1C variants, regarding viral load and pathogenicity of the strains hitting farms today.[9,10] Though not listed here, feed logistics, supply & distribution of products into farms, workforce dynamics and other supporting function structures have a role to play in PRRSV prevalence within the swine industry. All of which will have to be considered if elimination of the disease will be addressed.

Sustainability

Recently, in an Iowa State seminar, Tom Petznick mentioned a quote alluding to disease being the greatest destroyer of profitability in our industry. While disease may not be the only player, it sure does have an impact on the industry's ability to survive in uncertain financial situations. As the US swine infrastructure continues to mature within the US, the traditional growth by size may not be the only route our industry could take when considering "growth". Growth through mature processes and conversations about sustainability is one area to consider. On the EPA website, they emphasize why sustainability is important[11]:

- Increase operational efficiency by reducing costs and waste
- Respond to or reach new customers and increase competitive advantage
- Protect and strengthen brand and reputation and build public trust
- Build long-term business viability and success
- Respond to regulatory constraints and opportunities

Regarding those principles, should the industry ask itself if it can simply continue to live with PRRSV? For those in attendance at the 2023 US Swine Health Improvement Program (US SHIP) meeting, Dr. Gordon Spronk stood in front of that group and demonstrated the timeline for eradication of Pseudorabies. It was a 30-year effort from 1975 to 2004.[12] Another note he mentioned was the negative impact that disease challenges have on sustainability. A simple graphic of the US swine industry mortality rates from birth to market and breeding female inventory indicates numbers that are less than to be desired.[13] At the same time, can the optimistic person see those graphs and think "This gives the US industry an opportunity to improve".

Mindset of possibility

This isn't the first proceedings paper to mention PRRS elimination and it shouldn't be the last.[14] From PRRS Regional Control and Elimination project papers to the PRRSV heroes in the industry, there are so many leaders that have spent their lives dedicated to better understanding PRRSV.[15,16] In a recent leadership meeting within The Maschhoffs, LLC, Matt Goesling showed an interesting allegory was about telling monkeys over and over not to grab for the banana. For example, even within

The Maschhoffs, after break rates higher than desired the health team has felt obligated to vaccinate most of the breeding herds for PRRSV, so that production and severity of clinical impact can be managed. For good reasons and intentions, I believe this has been the stance at The Maschhoffs and with other producers. The industry recently has managed PRRSV out of frustration, markets and a virus that is rapidly evolving to work around our biosecurity measures. That being said, for the sustainability of the US pork industry, it would be refreshing to hear that a new reinvigorated effort of people and resources has started setting the groundwork for how PRRSV elimination in the US could happen. It would require program standards similar to the pseudorabies eradication playbook.[15] A large group of influential people to would need to lay out that framework, complete the economic models, manage farms during the susceptible phase, establish a diagnostic monitoring plan and as well as maintain the program for producers to be prepared for the threats to come.[16] It may need to include government support and indemnities for herds participating over a certain period of time. In future years, the ability to capitalize on newer research and technologies to help in the process of elimination, as well. Between genetic selection for robustness, gene altering technologies, pen side diagnostics and other biosecurity related solutions, there is potential.

Maybe optimism is too much when it comes to PRRSV and the US industry needs to start with *Mycoplasma hyopneumoniae* or porcine epidemic diarrhea virus. But there are other countries free of this devastating virus and someone or many of us have to start talking about PRRSV elimination more often if we are going to ever step out of "Fantasia."

References

1. Holtkamp DJ, Kliebenstein JB, Neumann EJ, Zimmerman JJ, Rotto HF, Yoder TK, et al.. Assessment of the economic impact of porcine reproductive and respiratory syndrome virus on United States pork producers. J Swine Heal Prod. (2013) 21:72–84. **10.31274/ans_air-180814-28**

2. Nieuwenhuis N, Duinhof TF, Van Nes A. Papers: economic analysis of outbreaks of porcine reproductive and respiratory syndrome virus in nine sow herds. Vet Rec. (2012) 170:225. **10.1136/vr.100101**

3. Zhang Zhendong, Li Zhi, Li Hao, Yang Shuqing, et. al., The economic impact of porcine reproductive and respiratory syndrome outbreak in four Chinese farms: Based on cost and revenue analysis. Frontiers in Veterinary Science (2022) 9, 10.3389/fvets.2022.1024720 **https://www.frontiersin.org/articles/10.3389/fvets.2022.1024720**

4. Morrison Swine Health Monitoring Project. (2023, October 6). PRRS cumulative incidence (last updated April 2023). Morrison Swine Health Monitoring Project. **https://mshmp.umn.edu/reports#Charts**

5. USDA Economic Research Service. **https://www.ers.usda.gov/topics/animal-products/hogs-pork/sector-at-a-glance/** Accessed November 11, 2023.

6. USDA Census of Agriculture. 2017 **https://www.nass.usda.gov/Publications/AgCensus/2017/Full_Report/Volume_1,_Chapter_1_US/usv1.pdf**

7. Centers for Disease Control and Prevention. History of SmallPox. **https://www.cdc.gov/smallpox/history/history.html** Accessed November 11, 2023.

8. USDA Swine Industry Manual. March 2011. **https://www.aphis.usda.gov/animal_health/emergency_management/downloads/documents_manuals/swine_industry_manual.pdf**

9. C.M. Melini, M. Kikuti1, M. Torremorell1, K. VanderWaal1, S. Rossow, J. Torrison, C. A. Corzo. Do all PRRSV variants have the same level of infectivity?

10. Dee, Scott. PRRS 144: A Rapid Response to a National Crisis. **https://www.pipestone.com/prrsv-144-a-rapid-response-to-a-national-crisis/** Accessed October 4, 2023.

11. United State Environmental Protection Agency. Sustainability. **https://www.epa.gov/sustainability/sustainable-manufacturing.** Accessed November 1, 2023.

12. Pseudorabies (Aujeszky's Disease) and Its Eradication: A Review of the U.S. Experience. 2008. Tech. Bul. 1923. APHIS, USDA, VS. 2

13. Derald J. Holtkamp, Dermot J. Hayes, and Lee L. Schulz. Key Performance Indicators in Pork Production: An International Comparison Using 2021 Data

14. Baker, R. B. The industry experience with PRRSv elimination. Thrwarting re-introduction. AASV Annual Conference proceedings. 2006.

15. Wadell, John. PRRS Area Regional Control and Elimination: Why is Nebraska a good place to start? George A Young Swine Conference proceedings. 2010.

16. Dee, Scott. Elimination of PRRSV from North America: Uniting the clans! Allen D. Leman Swine Conference Proceedings. 2010. **https://www.aasv.org/library/swineinfo/item.php?LEMAN/2010/005.pdf**

17. **https://www.aphis.usda.gov/animal_health/animal_dis_spec/swine/downloads/prv_program_standards.pdf**

18. Klepac P, Funk S, Hollingsworth TD, Metcalf CJ, Hampson K. Six challenges in the eradication of infectious diseases. Epidemics. 2015 Mar;10:97-101. **doi: 10.1016/j.epidem.2014.12.001** Epub 2014 Dec 12. PMID: 25843393; PMCID: PMC7612385.

https://doi.org/10.54846/am2024/143

The Zombie Apocalypse approach to biosecurity, biocontainment, and disease control and elimination

Luc Dufresne, DVM

I remember vividly during the early days of PEDV when we were at the height of the first wave of infection in the summer of 2013. We decided to see "World War Z" at the cinema to change our minds. This was an abysmal choice of the movie because all we would do was visit all the similarities between the PEDV epidemic we were seeing and the Zombie apocalypse depicted in the film. After that, I was always drawn to see any of those movies in that genre as a metaphor describing the failure of science and human nature to control and eliminate diseases.

The scenario is usually pretty much the same in most of them.

- Initial contact with zombies is highly traumatic, causing shock, panic, disbelief, and possibly denial, hampering survivors' ability to deal with hostile encounters.
- The response of authorities to the threat is slower than its growth rate, giving the zombie plague time to expand beyond containment.
- This results in the collapse of society. Zombies take complete control while small living groups must fight for their survival.

Sounds eerily familiar (minus the collapse of society, of course)? Even government agencies find similarities between fiction and real life. In 2011, the Centers for Disease Control and Prevention posted a disaster preparation guide called "Preparedness 101: Zombie Apocalypse." The guide. It generated so much traffic on the CDC website that it crashed the server. It also sparked enough confusion that the CDC took it offline; the same year, the US military devised CONPLAN 8888 to train officers in contingency planning. The topic of CONPLAN 8888? The zombie apocalypse.

The CDC and US military weren't planning for an actual zombie apocalypse. However, they found that this hypothetical was a great stand-in for many potential disasters they want people to prepare for. Our readiness for the zombie apocalypse can tell us about our enthusiasm for various catastrophic events. From future pandemics to wildfires to infrastructure collapse, this iswhat disaster preparation experts call "all-hazard preparation."

So, it is fair to ask: Is the swine industry ready for a Zombie apocalypse?

Cliché No. 1: The experiment gone wrong!

The zombie plague is usually an experiment gone wrong or a virus created for nefarious purposes that escapes the laboratory and creates havoc. While we cannot take the blame for the initial apparition of PRRSV, IAV, and ASFV, one can argue that man-made conditions have helped them become more pathogenic and more problematic. The multisite production system evolved into a multistate industry-wide output where we see pigs originating from different regions and with different health statuses moving over significant distances, then comingled or raised in close proximity to each other. As our ability to sequence viruses improves, we realize that recombination is a much more frequent event than previously though for IAV and PRRSV. All you need is two viruses infecting the same cell simultaneously. It is a matter of probability, and this way of raising pigs drastically increases the risk of viable recombination. In China, some new milder ASFV vaccine variants have appeared due to the use of non-approved live vaccines. Those variants make elimination and containment more problematic.

Cliché No. 2: The government's inadequate response

In those movies, we usually see government failure in numerous ways:

- Inadequate crisis management;
- Lack of communication;
- Resource mismanagement;
- Failure to implement and enforce policies;
- Loss of public trust;
- Political infighting; and/or,
- Inability to adapt.

The zombie apocalypse metaphor illustrates the potential consequences of government failure in dealing with real-world crises. This approach can serve as a creative way to discuss the importance of effective governance, crisis preparedness, and the need for strong leadership in times of uncertainty. The same is true of tabletop exercises, which have shown some significant challenges that introducing foreign animal diseases would create. I commend the preparedness effort that state and federal animal health officials have put in place to prepare. However, I still don't think we are ready for the introduction of an FAD in North America or how we will control and eliminate the disease. Most importantly, we must determine how we can compartmentalize production and secure the movement of pork products safely outside the US I firmly believe that the US SHIP program is of the highest importance to be implemented and backed by Government officials to ensure trading partners of the validity of the program.

Cliché No. 3: The large cities

The constant movement of people in and out and high density is a recipe for disaster when zombies strike! Unfortunately, it is the same when a new pathogen comes along in pig-dense areas. The combination of high pig movement, high animal density, low biosecurity in growing pig farms, and high interconnection between farms make disease dissemination extremely difficult to control.

Cliché No. 4: The isolated homesteads

To maintain high health on a farm, swine producers and veterinarians often have a similar mindset to the survivalist and doomsday preppers:

- Geographical isolation
- Minimize contact with the outside world
- Redundancy
- Autonomy

This measure allows the maintenance and protection of the health of the farm. It often does not help to expand the population of healthy animals when they are all sent to be produced in a region where they will most likely get infected with several pathogens. Having improved significantly, the health of the sow herds has not positively impacted overall growing pig performance in the high concentration area of the Midwest. Unfortunately, health is not contagious!

Cliché No. 5: The fortress

This is often the goal in those movies--reach the fortress. A fortified location, a safe haven that allows people inside to live a close to everyday life by keeping the threat outside. There are several advantages of the fortress success:

- Security
- Community defense
- Resources
- Infrastructure
- Location

The fortress, of course, identifies as a region or country that is free of a specific disease. The regional isolation and limited contact with the outside through border control allow the security inside that zone to be very good with minimum effort or cost for the producers inside that zone. What can go wrong? You rely on the group to do the right things and the government to put suitable measures in place to reduce the risk of disease introductions and control it. This can be complicated when a region is not set by a defined border or government structure or when the disease is not considered an FAD per se but is not present and would have disastrous consequences if introduced. Excuse me for introducing yet another metaphor, a mythological one this time, but what could be our Trojan horse that lets the zombies in?

Trojan horse 1: No quarantine for the introduction of new out of region animals

In some areas of North America, there is limited to no growing pig movement inside those regions; the only animals coming in are replacement breeding stock. Although those animals usually come from high-health and biosecure farms, it is not guaranteed that no disease will be introduced with them. I have experienced this three times; isolated but pig-dense regions saw an introduction of a new virus (in that case, new strains of PRRSV) by introducing infected gilts in the area. The gilts were tested before movement started but got infected before the testing regiment could detect it. The fact that no pre-entry quarantines were in place resulted in infection of the receiving farm, and then the virus spread rapidly within the region. Quarantine and proper testing of incoming animals for a virus (or even a specific variant of a virus) that an area is free of and is critical to ensure long-term freedom of that disease.

Trojan horse 2: Disease present in the plant or cull market makes its way to the commercial herd

Slaughter plants and cull markets are at significant risk of creating a reservoir of disease leading to infection of a region free of disease. Some of those sites in Canada are located in areas free of specific pathogens but may receive animal from infected areas. This is considered an acceptable practice since those are considered terminal sites. Several functional risks are present within those operations that everyone agrees are huge risks, but we failed to address them as a North American industry:

- Infected animals can infect other animals or premises in transit.
- Animals enter the packing plant and then are moved to a cull barn because they don't meet plant criteria.
- Animals moving to more than one cull plant before actually being slaughtered.
- Animals spending several days in the cull channel moving to several cull barns in several states.
- Cull barn operating in continuous flow.
- Lack of standardized monitoring at packing plants, cull barns or livestock markets.

There is substantial evidence that the packing plant and cull market creates the risk of dissemination between regions for Seneca Valley virus in the US and PEDV in Canada. Although we are all aware of what that risk is, we, as an industry, fail to find a viable solution, or more likely, we know the answer but do not have the economic or political will to implement it.

Trojan horse 3: Pathogen enters the region on infected products

Historically, a common route of foreign animal disease transmission in free countries is the entry of infected meat products into a country. The likely introduction of PEDV in infected feed ingredients also highlights that disease-causing pathogens can be carried over long distances and time on other vessels than live animals.

Trojan horse 4: Diseases present in wild animals make their way to the commercial herd

The problem of wildlife to serve as a reservoir for the disease is a nightmare for several countries trying to eliminate ASFV. The success of Belgium's eradication of ASFV by systematically eradicating their wild boar population illustrates the importance of controlling wildlife to an eradication process. Unfortunately, the demographic explosion of the feral pig population in the US and Canada makes the control/eradication of specific diseases in this population extremely difficult. Defining commercial production as a compartment with significant containment and testing requirements is probably the most likely way the North American swine industry is considered free of a disease that remains in its wild population.

Trojan horse 5: Movement in and out of the infected zone

Transport of pigs to infected regions represents a risk where a pathogen can contaminate equipment and be brought back to the region. Proper sanitation, downtime, and monitoring measures are critical for all equipment returning from an infected area.

Trojan horse 6: People being people!

It can be ignorance, greed, or sheer stupidity, but people not following the rules can jeopardize the health of an entire region. Failsafe mechanisms should be in place to minimize that risk, but in an area with no legal border, you are at the mercy of people doing the right things.

What can we do to not become another zombie apocalypse cliché?

- Know the threats.
- Know our strengths.
- Know our weaknesses and have the fortitude to address them.
- Don't be complacent.
- Be prepared.

There is nothing earth-shattering about what I describe herein. We know what needs to be done. If we look at controlling disease introduction on a regional basis, we are only as strong as the chain's weakest link. Maintaining a region free of disease can not involve only the sow farm producer that we ask to cover the cost of biosecurity and elimination. We need to improve infrastructure and do a better job:

- This starts with improving biosecurity at the growing pig site and requires coordination and effort to regionalize the flows of pigs better.
- Improve market haul washing and disinfection infrastructure plus continuous monitoring of packing plant and cull market site by developing an inexpensive yet proactive testing protocol.
- When an infection is detected, having the ways and means to clean those premises quickly.
- Think not only of biosecurity but also biocontainment. When disaster strikes, we need to ensure that it does not spread to other farms before the disease is eliminated. Should we move the infected animal, and if yes, what route should be taken? If we depopulate the farm, what is the safest way of doing it? One could argue that in some cases, a slower elimination method (ie, ASFV tooth-extraction approach) is a safer option than a massive depopulation requiring a lot of resources and people.

Prevention demands commitment and some investment from players. We, as veterinarians, are usually not actively involved in the discussion and treat them like political hot potatoes. I think it is time to address those issues constructively with all players involved directly and indirectly. I believe that the US SHIP is a step In the right direction. It shows that all partners and stakeholders want to do what is best for the swine industry. Let us continue to build on this.

https://doi.org/10.54846/am2024/144

Mycoplasma hyopneumoniae vs PRRSV elimination differences

M. C. Rahe
Department of Population Health and Pathobiology, North Carolina State University, Raleigh, North Carolina

Summary

Mycoplasma hyopneumoniae (MHP) and PRRSV are both primary pathogens of the porcine respiratory tract which can be challenging to eliminate from a herd. What are the differences between the biology of these two pathogens, the immune system response to these pathogens, and how can this help in informing elimination efforts?

First, there are the obvious differences between MHP and PRRSV: MHP is a bacterium and PRRSV is a virus. As such, MHP is quite a bit bigger than PRRSV with an approximate diameter of 300 nm compared to PRRSV at 50 to 65 nm and genome size of 892 kb vs PRRSV at 15 kb. It should be noted that MHP has a DNA genome and PRRSV is RNA. Furthermore, MHP's genome has very low GC content which results in complex transcriptional organization and gene expression.[1] PRRSV is regarded as one of the most mutable RNA viruses known to man and has also been shown to recombine with other PRRSV isolates to create even further genome variation. Through horizontal gene transfer, MHP can cause the spread of mobile accessory genes that may play a role specific metabolic pathways or antibiotic resistance.

These two agents also infect and replicate differently. MHP lives on the ciliated surface of the pseudostratified respiratory epithelial cells; whereas, PRRSV infects CD163+ macrophages primarily in the lung but elsewhere as well. It is necessary for PRRSV to replicate within the cytoplasm of host cells, while MHP reproduces through extracellular binary fission. However, it has been noted that Mycoplasmas, including MHP, may undergo an intracellular phase of their life cycle.[2] On that note, MHP infection and pathology appears to be confined to the respiratory tract while PRRSV causes a systemic infection, primarily affecting the lung, and can be shed in feces.

MHP is able to bind to the ciliated surface of epithelial cells via adhesins. This allows for the subversion of the mucociliary apparatus and the colonization of the respiratory epithelium. MHP has over 35 proteins which have been associated with adhesion, including the P97/P102 paralog families. Importantly, MHP has shown the ability to mutate these adhesins as well as through post-translational processing expand protein functions and expose more epitopes. This may decrease immune efficiency. On the immune system side, disruption of the interactions of adhesins with cilia prevents colonization and allows for the eventual clearance of the bacteria. Neutralizing antibodies, both IgG and IgA, likely play an important role here. There has been recent work looking at the T cell response to MHP infection, and it is still unclear if a productive Th1 response is generated and helpful or if a Th2 response predominates.[1]

Neutralizing antibodies against PRRSV are frequently only generated against the strain of PRRSV that the pig was vaccinated/challenged with. Though, rare exceptions do exist.[3] Therefore, mutation of the virus and resulting heterologous challenge render these antibodies useless. Comparatively, T cells to PRRSV have been shown to provide greater latitude of response, as they do not go through the highly specific affinity training that B cells do in the germinal center.[4] However, these cells appear to be helpful for the control and clearance of infection, but not a method for providing protective immunity as neutralizing antibodies would be.

In summary, neutralizing antibodies are generally very effective at protecting a host from infection. However, these can be very challenging to generate against pathogens that are highly mutable or those able to alter their surface proteins which are used for attachment to cells. The T cell response appears very important for heterologous immunity to PRRSV and may play a role in controlling MHP infection as well.

References

1. Li G, Obeng E, Shu J, Chen J, Wu Y, He Y. Genomic Variability and Post-translational Protein Processing Enhance the Immune Evasion of. *Front Immunol.* 2020;11: 510943.

2. Leal Zimmer FMA, Paes JA, Zaha A, Ferreira HB. Pathogenicity & virulence of. *Virulence.* 2020;11: 1600-1622.

3. Kick AR, Amaral AF, Frias-De-Diego A, et al. The Local and Systemic Humoral Immune Response Against Homologous and Heterologous Strains of the Type 2 Porcine Reproductive and Respiratory Syndrome Virus. *Front Immunol.* 2021;12: 637613.

4. Chung CJ, Cha SH, Grimm AL, et al. Recognition of Highly Diverse Type-1 and -2 Porcine Reproductive and Respiratory Syndrome Viruses (PRRSVs) by T-Lymphocytes Induced in Pigs after Experimental Infection with a Type-2 PRRSV Strain. *PLoS One.* 2016;11: e0165450.

https://doi.org/10.54846/am2024/145

Pseudorabies virus elimination versus porcine epidemic diarrhea virus elimination: We did it before, why not do it again?

L. Tokach, DVM, DAVBP; M. Potter, DVM, PhD
Abilene Animal Hospital, Pennsylvania, Abilene, Kansas

Introduction

As a 1990 graduate of the University of Minnesota (Lisa) and a 2007 graduate of Purdue University (Megan), we were not around for the elimination of foot-and-mouth disease in 1929, vesicular exanthema in 1956, or classical swine fever in 1976. Lisa, however, came out as a freshly minted veterinarian in the middle of the elimination of pseudorabies (Aujeszky's disease) from the US swine population which was completed in 2004. It seems that we accomplish industry-level disease elimination of a major swine pathogen on average every 25 years. That means we are coming due.

Since the elimination of pseudorabies virus (PRV), we have made great strides in disease elimination from individual swine herds, but there have been no formal programs for complete elimination of a specific pathogen from the national herd. Perhaps it is time to consider our next step so that we continue to build on the national herd health advancement precedent of those who practiced veterinary medicine before us. Besides the fact that we do not want to be the generation of veterinarians who missed opportunities and failed to accomplish any national disease elimination programs, disease elimination makes sense as it ultimately lowers the cost of production and makes us stronger as net exporters of pork.

The next national disease elimination

When considering what our next move should be, pathogens that have proven potential to be eliminated from individual herds include *Mycoplasma hyopneumoniae*, influenza A, Seneca Valley virus, porcine reproductive and respiratory syndrome virus, porcine epidemic diarrhea virus (PEDV), porcine deltacoronavirus (PDCoV), and transmissible gastroenteritis virus (TGEV). Of those pathogens, porcine reproductive and respiratory syndrome virus elimination has proven to be very difficult on a regional or national level, Seneca Valley virus production loss is not financially significant enough to be a contender, influenza A spreads too easily making prevention of new infections difficult, and *Mycoplasma hyopneumoniae* testing is too inconclusive and elimination plans are very prolonged. The coronavirus group (PEDV, PDCoV, and TGEV), however, seem a possible target for a national elimination program because they are at an all-time low prevalence since their introductions. As Dr. Bill Hollis mentioned in his AASV President's Message, "porcine epidemic diarrhea virus has been driven to the finishers and dark corners of unwashed trucks, buying stations, and slaughter plants."[1]

Let us consider some key points that make successful elimination of the three coronavirus pathogens a likely outcome using the most recent national elimination plan (PRV in 2004) for comparison.

1. **Biosecurity:** Realizing that every biosecurity plan is only as strong as its weakest link, we have a much stronger biosecurity plan than we had in 2004. Many sow farms are now shower-in, shower-out with downtime requirements and entry restrictions for downstream flow personnel. Swine flow tends to be unidirectional. In 1990, when Lisa started veterinary practice, only two of her clients were shower-in, shower-out and a few more required that you wore designated clothes from the farm. Three site production was just getting underway.

2. **Isolation protocols:** Today, we would consider any other facility options before bringing breeding age gilts right into your gestation barn without isolating and testing them, but it was a novel concept back then. When protocols for the PRV elimination program required isolation and testing for PRV, Lisa saw some innovative isolation setups, including four breeding boars that were temporarily housed in the client's garage! Producers are much more aware of and prepared for isolation of replacement breeding stock, boars, feed, and even farm supplies.

3. **Transport:** We have known for a long time that hauling pigs in dirty livestock trailers is a sure way to transmit pathogens, but when Dr. Jim Lowe reported that livestock trailers testing negative prior to delivery to the harvest facility had an increase in testing PEDV positive upon leaving the facility,[2] we had to reconsider our transportation protocols. Not washing transport vehicles in the winter months or not washing between loads when we were dumping a barn suddenly went from cheating the system a little to inviting disease into a barn. Knowing that these coronaviruses are alive and well in finisher systems where they go undetected will be a big factor when considering elimination.

4. **Herd size and numbers:** Since PRV elimination in 2004, herd size has increased dramatically and herd numbers have gone down.[3] Fewer herds likely makes systematic elimination easier because you are not dealing with large numbers of small farms and backyard pigs. However, larger average herd size may make elimination more difficult due to size and complexities of the facilities. The thought of cleaning and disinfecting 1200 farrowing crates vs 120 can be somewhat daunting.

5. **Segregated production:** Keep in mind, a large percentage of the industry was still single-site, continuous flow production during PRV elimination. Producers were moving toward all-in, all-out production by room. We have now moved many flows from all-in, all-out by room, to by barn or by site. This makes elimination plans much easier.

6. **Cost to the industry:** Figuring $1.00 in 1986 has the equivalent value of $2.73 in January 2023,[4] the estimated cost of PRV to the 2023 industry would have been $90.09 to $286.65 per sow.5 Given a $45 per wean pig sale price the estimated cost of a PEDV break for piglet losses (100% for 3 weeks, 50% for 3 weeks) in a 30 pig/sow/year farm would be $116.83 per sow. The losses are still costly even though our industry has moved into an endemic phase with PEDV. The financial incentive is present for elimination.

7. **Diagnostic testing:** The relative availability of rapid PCR testing for coronaviruses makes the ELISA testing used for PRV elimination look like the computers we used to send a man to the moon compared to the smartphones we hold in our pockets today. We have wonderful diagnostic tools that are readily available to us for monitoring purposes.

8. **Vaccine:** PRV elimination was already underway before the widespread availability of a licensed vaccine. However, the biggest differences were the development of the gene marker vaccine that allowed differentiation between field infections and vaccinated animals and that the vaccine was highly effective in the reduction of disease. It did not, however, eliminate the disease, since none of the vaccines prevented latent infection and subsequent reactivation and shedding of virulent field virus later on.[6] Currently we do not have a highly efficacious vaccine in terms of preventing a coronavirus infection.

9. **Survivability and infectivity of the virus:** Pseudorabies (herpesvirus) survivability is variable but often unstable unless in optimal moisture, pH and temperature ranges.[5] The virus does not survive hot or dry conditions well.[7] We believe that PEDV tends to be more environmentally stable which may make elimination more difficult. For instance, it appears that PEDV survives a long time in slurry,[8] longer than we would expect PRV to survive given similar temperature conditions.[5] In addition, PEDV is highly infectious making the environmental survivability of virus more concerning.

10. **Threat to other livestock:** Perhaps a motivator to PRV elimination was the fact that swine were the only natural host species and the virus was deadly to cattle, sheep, cats, dogs, mice, and rats. Plus, there were reported infections in bears, Florida panther, raccoons, coyote, deer, and farm fur animals.[6] This is a big difference from the coronaviruses, however, all significantly cost the industry. Maybe the risk to other livestock was a player in motivating the government's support in PRV elimination, but the swine industry was a large contributor to the overall success of the PRV elimination program. The industry would again need to make disease elimination its priority.

Call to action

Many people have already started a call to action for coronavirus elimination. The launching of the US SHIP (US Swine Health Improvement Plan) is designed to provide some framework for biosecurity, traceability, and disease surveillance. Like PRV elimination, we will likely segregate populations by state, then by regions within states with high swine populations. We have the diagnostic tools, we have the financial incentive to be free of coronaviruses currently affecting our swine population, and we currently have a low prevalence of all three diseases. We eliminated diseases before and we think it is time to do it again.

The biggest challenges will be:

1. Who is going to lead the crusade?
2. How will we control the transport risks of market hogs and culls sows?
3. How will we keep manure handling from breaking herds?
4. Who is going to bear the additional cost of testing to establish boundaries and set goals?

References

1. Hollis W. Disease Elimination. *J Swine Health Prod.* 2023;31(5): 225.

2. Lowe J, Gauger P, Harmon K, et al. Role of Transportation in Spread of Porcine Epidemic Diarrhea Virus Infection, United States. *Emerging Infectious Diseases.* 2014;20(5), 872-874. **https://doi.org/10.3201/eid2005.131628**

3. Davis C, Dimitri C, Nehring R, et al. U.S. Hog Production: Rising Output and Changing Trends in Productivity Growth. ERR-308, U.S. Department of Agriculture, Economic Research Service. 2022; 5-6.

4. U.S. Bureau of Labor Statistics. CPI Inflation Calculator. Accessed November 6, 2023. **https://www.bls.gov/data/inflation_calculator.htm**

5. United States Department of Agriculture (USDA). Pseudorabies (Aujeszky's Disease) and Its Eradication: A Review of the U.S. Experience. 2008. Tech. Bul. 1923. APHIS, USDA, VS. 231 pp.

6. Mettenleiter TC, Ehlers B, Muller T, et al. Herpesviruses. In: Zimmerman JJ, Karriker LA, Ramirez A, et al., eds. *Diseases of Swine.* 11th ed. Wiley Blackwell; 2019: 548-575.

7. Spickler A. Aujeszky's Disease. 2017. Accessed November 6, 2023. **http://www.cfsph.iastate.edu/DiseaseInfo/factsheets.php**

8. Goyal S. Environmental stability of PEDV (porcine epidemic diarrhea virus). NPB Research Update #13-215. 2014.

https://doi.org/10.54846/am2024/146

Current trends in *Mycoplasma hyopneumoniae* eradication

A. Sponheim[1,2], DVM; E. Fano[2,a], DVM, MSc, PhD; M. Pieters[1,3,4], DVM, PhD
[1]Department of Veterinary Population Medicine, College of Veterinary Medicine, University of Minnesota, St. Paul, Minnesota; [2]Boehringer Ingelheim Animal Health Inc., Duluth, Georgia; [3]Veterinary Diagnostic Laboratory, College of Veterinary Medicine, University of Minnesota, St. Paul, Minnesota; [4]Swine Disease Eradication Center, College of Veterinary Medicine, University of Minnesota, St. Paul, Minnesota;
[a]Current affiliation: Pipestone Veterinary Services, Pipestone, Minnesota

Introduction

Respiratory diseases are an important health concern in most pork production areas of the world.[1] *Mycoplasma hyopneumoniae* is regarded as the most notable respiratory bacterial agent affecting pigs in the US swine industry.[2] Eradication of *M hyopneumoniae* from pig flows reduces antibiotic use, improves welfare and production performance of the pig, and provides a competitive advantage to US pork producers, thereby promoting sustainability of the industry. In the following document we will summarize current trends in *M hyopneumoniae* eradication, including diagnostic guidelines to confirm the start and end of a herd closure and medication program, extended detection of *M hyopneumoniae* under herd closure conditions, the importance of sanitary sample collection measures, and the development of a web-based resource for eradication program planning. The following work is the result of collaboration with US field veterinarians to address the most pressing needs for various stages of the *M hyopneumoniae* eradication process.

Background

Herd closure and medication eradication programs are commonly used in the US[3] and are based on two primary components, breeding herd population exposure and the duration of pathogen persistence post-exposure. Exposure of all replacement gilts and breeding females to *M hyopneumoniae* prior to the beginning of the eradication program establishes Day 0.[4] Aerosolization of lung tissue containing *M hyopneumoniae* has been rapidly adopted in the North American swine industry to achieve population exposure,[5-8] although no specific diagnostic protocols were previously available for confirmation of successful exposure to define Day 0. The expected duration of *M hyopneumoniae* persistence, which may last longer than 214 days, defines the end of herd closure.[9] Historically, eradication programs were considered complete based on days post-exposure, rather than by diagnostic confirmation. However, preliminary data following diagnostic schemes for detection of < 5% prevalence[10] showed a small proportion of *M hyopneumoniae* PCR positive pigs beyond 240 days post-exposure in various eradication programs,[11,12] necessitating the extension of herd closures, and thereby straining pig flow and production. Reported success rates from field eradication efforts range from 64 to 83%,[12-14] depending on measurement criteria, indicating the need to further refine disease eradication programs. Here we review recently published diagnostic strategies specific to the two primary components of a herd closure and medication elimination program,[10,15] along with a field case that implemented these strategies to describe the duration of *M hyopneumoniae* detection in sows and replacement gilts following aerosolized lung homogenate exposure, as part of a dual disease elimination program.[16]

Day 0 diagnostic strategies

Sponheim et al[15] addressed Day 0 of herd closure diagnostic strategy by longitudinally sampling gilts from five commercial gilt developer units (GDUs) exposed to *M hyopneumoniae* by natural contact or aerosolization. The study showed detection of *M hyopneumoniae* varied greatly based on sample type with oral fluids having the lowest detection in groups of gilts detected positive by laryngeal and deep tracheal secretions. Furthermore, detection by PCR in deep tracheal secretions was higher than in laryngeal secretions and seroconversion to and PCR detection of *M hyopneumoniae* on oral fluids were delayed compared to PCR detection at the individual level. By two weeks post-exposure, at least 85% of study gilts in three GDUs exposed by aerosolization were PCR positive in deep tracheal secretions. In contrast, natural contact exposure resulted in 100% of study gilts becoming PCR positive by six weeks post-initial exposure on deep tracheal secretions. Deep tracheal secretions required the lowest number of gilts to sample for the earliest detection compared to all other samples evaluated. As observed in one of the GDUs using aerosolization, demonstration of failure to expose gilts to *M hyopneumoniae* allowed for early intervention in the exposure protocol and delay of Day 0, for accurate timing of the eradication protocol. Sampling guidelines proposed in Sponheim et al[15] can be used for verification of *M hyopneumoniae* infection of gilts following exposure to determine Day 0 of herd closure. For example, following aerosolization exposure of 500 gilts, if a low-risk, low-prevalence level of non-infected gilts approach is targeted using mean diagnostic sensitivity for deep tracheal secretions, sample size estimates for a ≤ 5% prevalence level of non-infected gilts, 95% herd sensitivity and specificity would be followed since the target herd sensitivity and specificity cannot be achieved for ≤ 1% design prevalence at the given population size. Three hundred eighteen individuals would be sampled and if < 27 (cut-point) samples were PCR negative, the selected criteria would be achieved. If ≥ 27 samples were PCR negative, the chosen criteria would not be achieved, and contingency plans should be discussed, such as delaying Day 0, resampling, or repeating the exposure method. If a higher prevalence of non-infected gilts is assumed, sample size estimates for ≤ 15% design prevalence of non-infected gilts could be followed. Fifty-five individuals would be sampled and if < 7 (cut-point) samples were PCR negative, the selected criteria would be achieved. If ≥ 7 samples were PCR negative, contingency plans would be followed. Samples could be collected as early as 14 days post-aerosolization exposure.

Day 240 diagnostic strategies

Sponheim et al[10] addressed Day 240 of herd closure diagnostic strategy by first determining individual and pool sensitivities for detection of *M hyopneumoniae* during the chronic phase of infection using deep tracheal secretions, the in vivo sample type with the highest reported diagnostic sensitivity. These sensitivity estimates were applied for detection of *M hyopneumoniae* in a low prevalence scenario, to estimate sample sizes with associated sample collection costs. For instance, following completion of a herd eradication program, if a low-risk approach is targeted, sample size estimates for ≥ 2% prevalence using the lower limit of the diagnostic or pool sensitivity 95% CI may be followed. Additional assumptions included a ≥ 95% population sensitivity, infinite population size, and 100% specificity. If samples were to be tested individually, 167 individuals would be sampled at a cost of $6,012. If pooled by 3, 213 would be sampled (testing cost $3,266), and for pools of 5, 220 individuals would be sampled (testing cost $2,464). These sampling strategies have been applied to eradication programs over multiple sow farms and production systems to determine the success of eradication programs and naïve replacement gilt entry timing. In the event where *M hyopneumoniae* is detected, decision makers are equipped with evidence to extend the length of herd closures, apply additional medications, etc.[11,12,16]

Mycoplasma hyopneumoniae persistence during herd closure

McDowell et al[16] longitudinally sampled a subset of sows and gilts from a commercial sow herd and off-site gilt development unit to collect deep tracheal catheter secretions at various times following confirmation of exposure. Samples were tested for *M hyopneumoniae* using a species-specific real-time PCR. A proportion of 58%, 51%, 52%, 19%, and 2% females were detected positive by PCR at 30, 60, 120, 180 and 240 days post-herd closure (dphc), respectively. In addition, a greater proportion of gilts exposed at the off-site GDU were detected PCR positive for *M hyopneumoniae* at each sampling event (97%, 94%, 90%, 30%, and 3%), compared to sows (19%, 7%, 15%, 9%, and 0%). In McDowell et al,[16] assaying for genetic material in live female pigs showed extended detection of *M hyopneumoniae* until at least 240 dphc. This data suggests persistence of *M hyopneumoniae* longer than previously reported and highlights the importance of performing diagnostic testing to confirm negativity to the bacterium, prior to opening sow herds, especially late in the herd closure timeline.

Diagnostic sample collection considerations

Sanitary sample collection measures should be followed in all of these scenarios, with attention to cleaning of equipment between samples and careful handling of catheters, to avoid false positives due to environmental or cross-contamination.[17,18]

Mycoplasma hyopneumoniae eradication roadmap

In an effort to enhance *Mycoplasma hyopneumoniae* eradication programs, a practical, web-based guide has been developed in collaboration with 29 leaders who have aided the advancement of swine health in North America and beyond and were willing to share their experience. This guide walks step by step through methods for *M hyopneumoniae* eradication in the field, organizing relevant research and resources, that will be enriched and updated as new information is generated.

Discussion

Veterinarian and producer led *M hyopneumoniae* eradication efforts have been increasing in the US over the last 15 years, as a result of the benefits to the pig and the observed program success. Further studies underway are directed at validating the sample collection schemes in the field, adapting a transmission model to estimate time to GDU population exposure based on diagnostic outcome, and developing ongoing surveillance programs for *M hyopneumoniae* negative herds. The combined information summarized here and driven by field partnerships, is our perspective of current trends in *M hyopneumoniae* eradication, which we believe will aid veterinarians and producers to provide certainty and make better-informed decisions when setting a timeline for herd closure, in addition to measuring the success, identifying the need for interventions, and preparing for future eradication programs.[10,15,16]

References

1. Pieters, M. and Maes, D. (2019). Mycoplasmosis in J. J. Zimmermann, L. A. Karriker, A. Ramirez, K. J. Schwartz, G. W. Stevenson, and J. Zhang, Diseases of Swine (11th ed.; pp. 863-883). Hoboken, NJ, USA: John Wiley and Sons, Inc.

2. NAHMS, U.S.D.A. (2012). Swine 2012 Part II: Reference of Swine Health and Management in the United States, 2012.

3. Holst S., Yeske, P., and Pieters, M. (2015). Elimination of *Mycoplasma hyopneumoniae* from breed-to-wean farms: A review of current protocols with emphasis on herd closure and medication. Journal of Swine Health and Production, 23(6), 321-330.

4. Pieters, M. (2018). Achieving day 0 for Mycoplasma hyopneumoniae: What is it? Why should it be done? Retrieved in May 2018, from **www.pig333.com**

5. Toohill, E. (2017). Achieving day zero in large swine operations. Allen D. Leman Swine Conference, St. Paul, Minnesota.

6. Yeske, P. (2017). Intentional *Mycoplasma hyopneumoniae* exposure. James D. McKean Swine Disease Conference.

7. Robbins, R., Betlach, A., Mondragon-Evans, M., and Pieters, M. (2019). Development of a herd-specific lung homogenate for exposure to *Mycoplasma hyopneumoniae* under field conditions. Journal of Swine Health and Production, 27(4), 221-227.

8. Figueras Gourgues, S., Fano, E., Alegre Sabate, A., Lopez Grasa, E., Hernandez Caravaca, I., Garcia Vazquez, F. A., Rodriquez Vega, V., Garcia-Morante, B. (2020). Assessment of nebulization technology for gilt exposure to *Mycoplasma hyopneumoniae* as an acclimation strategy. Journal of Swine Health and Production, 28(6), 294-301.

9. Pieters, M., Pijoan, C., Fano, E., and Dee, S. (2009). An assessment of the duration of *Mycoplasma hyopneumoniae* infection in an experimentally infected population of pigs. Veterinary Microbiology, 134(3-4), 261-266.

10. Sponheim, A., Munoz-Zanzi, C., Fano, E., Polson, D., Pieters, M., 2021. Pooled-sample testing for detection of *Mycoplasma hyopneumoniae* during late experimental infection as a diagnostic tool for a herd eradication program. Prev Vet Med. 189:105313. **https://doi.org/10.1016/j.prevetmed.2021.105313**

11. Chapman, B., Evelsizer, R., Pieters, M. 2018. Putting to test a diagnostic protocol for confirmation of *Mycoplasma hyopneumoniae* elimination in sow farms. Proceedings of the American Association of Swine Veterinarians Meeting, San Diego, CA, pp. 81-82.

12. McDowell et al., data on file.

13. Silva, G., Yeske, P., Morrison, R., Linhares, D., 2019. Benefit-cost analysis to estimate the payback time and the economic value of two *Mycoplasma hyopneumoniae* elimination methods in breeding herds. Prev. Vet. Med. 168, 95-102.

14. Yeske, P., Valeris-Chacin, R., Singer, R.S., Pieters, M., 2020. Survival analysis of two *Mycoplasma hyopneumoniae* eradication methods. Prev Vet Med. 174:104811. **https://dio.org/10.1016/j.prevetmed.2019.104811**

15. Sponheim, A., Alvarez, J., Fano, E., Rovira, A., McDowell, E., Toohill, E., Dalquist, L., Pieters, M. 2023. A diagnostic approach to confirm *Mycoplasma hyopneumoniae* "Day zero" for pathogen eradication. Prev. Vet. Med. 221, 106057. **https://doi.org/10.1016/j.prevetmed.2023.106057**

16. McDowell, E., Pieters, M., Spronk, T., Nerem, J., Fano, E., Dee, S., Sponheim, A., 2023. Duration of *Mycoplasma hyopneumoniae* detection in pigs following purposeful aerosol exposure. Vet. Microbiol. 282, 109758. **https://doi.org/10.1016/j.vetmic.2023.109758**

17. Canturri, A., Pieters, M. 2023. *Mycoplasma hyopneumoniae* diagnostics by PCR: tips from the school of hard knocks. Proceedings of the Allen D. Leman Swine Conference, St. Paul, Minnesota. p.8.

18. Snow, C., Canturri, A., Pieters, M. 2023. Minimizing sample cross-contamination for the detection of *Mycoplasma hyopneumoniae*. Proceedings of the Allen D. Leman Swine Conference, St. Paul, Minnesota. pp. 61-62.

https://doi.org/10.54846/am2024/147

PRRS eradication of pig herds in Hungary 2014 - 2022

I. Szabó[1], DVM, PhD; I. Nemes[1], DVM; L. Bognár[2], DVM, PhD; T. Abonyi[1], DVM, PhD; Zs. Terjék[1], DVM; T. Molnár[1], DVM; Á. Bálint[3] DVM, PhD

[1]National PRRS Eradication Committee, Budapest, Hungary; [2]Ministry of Agriculture, Budapest, Hungary; [3]Department of Virology, National Food Chain Safety Office Veterinary Diagnostic Directorate, Budapest, Hungary

Introduction

Porcine reproductive and respiratory syndrome (PRRS) is a major infectious disease of swine causing one of the highest economic damage in the pork industry worldwide.[1] The average PRRS outbreak could reduce production by about 7.4% compared to disease-free periods, and it resulted in 1.92 fewer piglets per sow.[2] In the US, PRRS results in a loss of $114.71 per sow and $4.67 per slaughter pig.[3] Based on data of 21 infected farms in Germany, Renken et al[4] calculated the losses at €255 per sow (~ $274/sow) per year.

According to these findings, the estimated loss caused by PRRS infection in Hungary - calculating with 4 million slaughtered pigs and 170,000 breeding sows - is close to 5 billion HUF about 14.2 million (USD) per year.[5]

Pig breeding in Hungary has been characterised by the parallel presence of large scale and backyard forms of farming for many decades. Large-scale pig farms produce pigs for the market, while backyard farms keep one or two sows and a limited number of pigs for fattening (1 - 10) intended for on-the-spot processing and consumption. In 1997, the last one's market ratio was 53.7%, in 2004 (the year of Hungary's accession to the EU) it was 41.6%, and in 2019 only 18.4% of pigs were kept in backyard farms.[6]

Hungary has an exceptionally large average herd size (2,933 pigs/herd in 2016) in large-scale pig farms compared to other EU countries, being the second highest after Denmark (3,182 pigs/herd). Large-scale pig farms of Hungary are predominantly (> 85%) farrow-to-finish, leading to the concentration of reproduction, farrowing, pre-fattening, fattening and replacement gilt units in the same farm. There is no uniform housing technology, feeding, disease control, preventive or therapeutic treatment protocols in Hungarian large-scale pig farms.[6]

The majority of slaughtered pigs are produced in farrow-to-finish pig farms. In recent years, large-scale breeding herds have appeared where only weaned piglets or pre-fatteners are raised at the breeding establishments, and either pre-fattening or fattening is performed at other locations which are often run by a different owner (multi-site system). In the case of large-scale fattening farms (with a minimum of 100 pigs/holding), the animals may originate from the same holding, from other herds within Hungary, or from abroad. The large-scale fattening farms operate in two distinct manners: all-in/all-out systems or continuous operation sites.[7]

The majority of PRRS cases were caused by the European type of the virus, but the American type PRRS virus also appeared (mainly of vaccine origin). At the beginning of the program, 4% of the backyard farms, 63.9% of the large-scale fattening farms were PRRS positive, respectively. In 2014, out of 470 large-scale breeding units, 125 were infected. A total of 85% of the large-scale swine farms in Hungary are farrow-to-finish type.[8]

Methods

In Hungary, the national eradication program of PRRS was successfully completed between 2014 and 2022.[5] The national PRRS eradication program in Hungary was based on a territorial principle, meaning that PRRS should be eradicated from the entire pig population of a single administrative unit (district, county and region) within a specified time period. It was obligatory for each swine farm. One of the main objectives of the PRRS eradication program was to increase the competitiveness of the Hungarian pig industry by:

- Eliminating losses due to PRRS.
- Decreasing antibiotic usage level during production.
- Increasing the profitability of producers.

The legal background of the eradication program was set out in Decree 3/2014 (16 January) by the Minister of Agriculture. The relevant committee of the European Union approved the program with state involvement. In order to ensure the uniform implementation of the plan, from the beginning of program implementation the Chief Veterinary Officer of Hungary appointed a National PRRS Eradication Committee to address any professional, epidemiological, administrative or other issues arising during the process. The officials (Ministry of Agriculture, The Ministry, and National Food Chain Safety Office (NÉBIH)) held discussions with the representative bodies of the domestic pig farmers during the eradication planning period. The National PRRS Eradication Committee (NEC) asked all pharmaceutical companies with registered vaccines against PRRS in Hungary to share their own control/eradicating programs, in addition to enlisting their help in the implementation of the program.

The National PRRS Eradication Plan specified the establishment of separate PRRS eradication plans for each large-scale pig holding, including large farrow-to-finish farms. The eradication plan was approved by the competent county veterinary authority on the 31st of May, 2015.

From 2014 onwards, a laboratory serological certification regime was carried out every year as specified by the Decree. The required tests were carried out at expense of the state. Based on the results of the examinations, the herds were officially qualified by the veterinary authority.

If the pig farm was identified as infected with PRRSV, several different steps were taken depending on the type of farm:

a. Backyard farms: The authority ordered the depopulation of the herds with state compensation, but without the need for repopulation.
b. Large-scale fattening only units: At this early stage of eradication, after pigs were transported to slaughter, the premises had to be thoroughly cleaned and disinfected and only PRRSV-free animals were allowed to be used for repopulation.
c. Large-scale breeding farms: Complete freedom to decide

whether eradication of the PRRS virus was to be carried out by complete depopulation-repopulation, or other methods (herd closure, test and removal, etc). The goal was to find an optimal eradication method tailored to every farm according to their technology, production system and management.

The development of the eradication plans by the management of the various swine herds was assisted by the PRRS Expert Veterinary Team delegated by the Chief Veterinary Officer of Hungary and by the technical specialists representing the pharmaceutical companies selling vaccines for PRRS in Hungary. The professional evaluation and approval of every eradication plan was performed by the National PRRS Eradication Committee upon the request of the official approval of the Veterinary Authority.

In 2017 World Organisation for Animal Health (WOAH) declared as non-infection with PRRSV:

1. if vaccine strains of PRRSV isolated,
2. detection of antigen or ribonucleic acid specific to PRRSV, which is a consequence of vaccination,
3. antibodies specific to PRRSV, they are demonstrated to be a consequence of vaccination.

Based on it, the decision of the Chief Veterinary Officer of Hungary made it possible to introduce the concept and category of PRRS 'vaccinated free' (VF) swine herds. This term applies to all three breeding types of large-scale swine farms (farrowing to finish, farrowing to pre-fattening (nursery) and farrowing to weaning). In principle, the breeding stocks can be immunised without any time limit, but the progeny of any age group must be proven to be free from PRRSV by laboratory methods (ELISA and PCR). Vaccination during the growing phase (including lactation, nursery, fattening) is not allowed in these farms, therefore, the tests must not show seropositivity at the end of the fattening period of the progeny, and all animals must be free from PRRSV, including even the vaccine virus.[9]

Results

Backyard farms

In 2012–2013, of the examined backyards, PRRS-seropositive individuals were found at a rate of 4% in both years. Seropositive animals were culled on the basis of an official procedure, so that on 31 December 2015 the PRRS-free status of small-scale of Hungarian swine herds was declared. This free status was confirmed in the 2016, 2017, 2018, 2019, 2020 and 2021 monitoring programmes of PRRS carried out in small-scale swine populations (Table 1).

Blood samples for the PRRS laboratory test from seropositive individuals were positive with two different serological methods (ELISA, IFMA). In the studies, efforts to confirm the presence of the virus by direct virological tests in individuals showing seropositivity were attempted. None of samples originated from breeding sows were PCR positive. The PRRS PCR positivity was observed in small-scale herds when irregular animal transport without veterinary certification was carried out or when infected nursery pigs were placed into backyard farms for fattening purposes. This epizootiology connection was confirmed by the phylogenetic analysis. There was no infection that could have been traced back to using so-called 'hidden boars'. In this period when tonsil samples were collected at the slaughterhouse from earlier seropositive sows, all of them were PRRSV PCR negative.[8]

Table 1: The result of laboratory tests for PRRS carried out in backyard farms

Year	Number of investigated animals	Number of seropositive animals	% of positive/ investigated animals
2012	35253	1650	4.68%
2013	34761	1271	3.66%
2014	42885	1956	4.56%
2015	31744	1082	3.41%
2016	13521	139	1.03%
2017	15582	150	0.96%
2018	13111	147	1.12%
2019	12680	181	1.43%
2020	12264	78	0.64%
2021	9907	15	0.15%
2022	8606	18	0.21%

Large-scale breeding farms

Out of the 470 large-scale breeding farms in Hungary, 345 were classified as PRRSV free (73%) and 125 (27%) as PRRSV infected at the beginning of 2014. The total number of sows of 470 farms was 186 404, of which 68 226 (37%) were kept in infected farms.

The average number of sows in the herds was 396; the average sow number in non-infected and infected farms was 343 and 546, respectively. With the exception of twenty farms, all ones were of the farrow-to-finish type.

During period of eradication process (from 2014 until 2021) 40 large-scale breeding farms, which were free of PRRSV became infected. So altogether 165 large-scale breeding farms had to create its own eradication plan. Finally, 94 out of them had been eradicated by depopulation-repopulation method (with 69,111 sows), six reached Vaccinated Free (VF) status, and others ceased their operations, found to be seronegative during certification process, or turn to producing fatteners.

PRRS eradication on the large-scale swine farms was carried out mainly by the depopulation–repopulation method. It was a major issue of the eradication program from the beginning that it might only be carried out efficiently through depopulation–repopulation of the infected herds, which is a very costly procedure. At the same time, this is the process that allows the utilization of the newest, most productive genetics in the swine sector upon repopulation.

Based on the similarity network analysis we were able to determine the reason of the infection of the above-mentioned 40 large-scale pig farms, infected during 2014 - 2022. Forty percent of them became infected by PRRSV origin of imported nursery pigs into Hungary. Almost 50 % of the cases the mediator of infection was vehicle transport: either to slaughterhouse or dead animals from mortality transport (Table 2).

At 15th of February 2022, we could declare that 99 % of large-scale breeding farms, having 97 % of sows kept on large-scale breeding farms are free of PRRSV. 6 farms, (1 % of total, having 3% of sows kept on large scale breeding farm in Hungary are free of wild virus strain of PRRSV, and the offspring is free of PRRSV.

Table 2: Infected large scale swine farms in Hungary between 2014 and 2022

	All infected large scale swine farms	40
Type of the farm	farrow-to finish	29
	farrow-to-weaning	2
	farrow-to-nursery	4
	farrow-to nursery + separated fattening	3
	farrow-to nursery + a certain fattening capacity	5
Number of sows	below 100	8
	between 101-500	8
	between 501-1000	13
	above 1001	11
Virus origin	nursery pigs imported from abroad, directly or indirectly	16
	virus existed in Hungary before 2014	24
Mode of infection	related to the transport of animals to slaughterhouses	10
	deadstock transport	9
	breeding and fattening farms are too close to each other and imported nursery pigs were infected	7
	irregular personnel traffic on the farm	6
	sperm	2
	even airborne could be	2
	common manure lake for two farms	1
	we could not clarify exactly, but foreign virus	2
	it was only seropositive (PCR negative)	1

Large-scale fattening farms

In 2015, 188 (61.2%) out of 307 large-scale fattening farms registered by the Hungarian veterinary authorities kept PRRS-positive animals. Of the total number of fattening places on these farms, 63.9% was PRRS positive. At national level, there was no significant difference in the number of infected and free stocks per holding. In 2013, 46% of pigs coming to Hungary as fatteners originated from the Netherlands, 39% from Germany, and 12% from Slovakia. The rest arrived from Denmark, Austria, the Czech Republic and Slovenia. In 2016, the Netherlands accounted for 46%, Germany for 33%, Denmark for 10% and Slovakia for 6% of imports of fatteners to Hungary.

In 2017, based on the decision of the National Chief Veterinary Officer:

1. Only pigs coming from PRRS-free farms were allowed to be used for fattening in Hungary;
2. If any diagnostic test after 48 hours of arrival, or after 60 days of quarantine gave even one positive result, a second test [indirect immunofluorescence test (IIF)]; had to be carried out, and if it gave the same result, the stock had to be sold for slaughter within 15 days or placed outside Hungary, so that the infected stock would not compromise the PRRS status of that area.

PRRS eradication, new rules regarding to importation, for large-scale fattening units induced major changes in Hungary for countries exporting pigs to Hungary: Pigs imported from The Netherlands decreased significantly, and at the same time from Denmark, Czech Republic and Slovakia increased exports (Table 3).

Hungarian pig sector

By 2022, the total pig population of the country, including all fatteners, have reached wild type PRRS-free status. The Hungarian Chief Veterinary Officer declared the country free status and reported to the European Union and to the WOAH.[10]

During the national PRRS eradication program in the Hungarian, large-scale, swine herds, the depopulation–repopulation was the most commonly used method to achieve a free status. This approach provided the farms the opportunity to upgrade their biosecurity measures and to get repopulated with animals having the most modern genetics and the highest available animal health status. Therefore, the PRRS eradication program, starting in 2014 and resulting in Hungary declaring freedom from the disease in 2022, has led to a significant improvement in the profitability of the swine sector (Table 4). Based on our calculations, the most decisive production parameter in Hungary was the total number of slaughtered pigs per sow per year being only 17.1 in 2012, but it reached 26.1 by 2022. Based on these results, it can be concluded that the 147,800 sows in Hungary generated almost 131.2 million USD more income in 2022 than in 2014, the starting year of the eradication.

Table 3: Number of prefatteners imported into Hungary between 2013-2022

Country of origin		Number of animals										
		2013	2014	2015	2016	2017	2018	2019	2020	2021	2022	2022/2017
Prefatteners	Austria	5,650	8,832	8,651	10,861	8,053	4,672	2,638	1,180		233	3%
	Czech	3,347	3,878	21,726	34,581	40,663	57,315	94,121	84,646	94,584	87,769	216%
	Denmark	4,591		8,886	74,414	168,370	116,916	143,808	156,401	97,787	29,325	17%
	Germany	236,724	317,390	218,625	249,699	479,782	498,490	440,317	313,210	386,685	383,117	80%
	The Netherlands	279,036	271,224	228,339	352,538	216,316	5,256	1,092		2,050	9,097	4%
	Slovakia	69,551	53,625	24,385	43,288	71,576	42,192	39,877	29,690	98,465	124,928	175%
	Slovenia	3,440	6,250	6,500	0	0	0	0				0%
	Italy	0	0	0	3	0	0	0				0%
	Poland	0	0	0	0	4,800	25,450	21,115	7,000			0%
Total compared to 2013		602,339	661,199	517,112	765,384	989,560	750,291	742,968	592,127	679,571	634,469	64%
		100%	110%	86%	127%	164%	125%	123%	98%	113%	96%	

Table 4: Changes in effectiveness of Hungarian pig industry between 2014-2022

Year	Slaughtered pigs [head]/sow/year	Change (2014 = 100%)	Average live weight of slaughtered pigs [kg]/sow/year	Change (2014 = 100%)	Carcass weight of slaughtered pigs [kg]/sow/year	Change (2014 = 100%)
2014	19.78	100%	2,146.8	100%	1,727.54	100%
2015	19.41	98%	2,088.9	97%	1,730.26	100%
2016	21.71	110%	2,429.3	113%	2,035.31	118%
2017	20.91	106%	2,421.5	113%	2,055.92	119%
2018	21.44	108%	2,424.5	113%	2,036.36	118%
2019	24.91	126%	2,873.1	134%	2,394.08	139%
2020	24.09	122%	2,802.6	131%	2,316.05	134%
2021	25.70	130%	3,247.4	151%	2,687.63	156%
2022	26.12	132%	3,237.7	151%	2,647.23	153%

Discussion

By 2022 Hungary successfully finished its PRRS eradication program of swine sector. During the eradication program the most challenging issue was to enhance the interest of all the players of the swine sector. Implementing of the program was crucial to create a committee of experts who are entitled to take care of all the aspects of the whole program. The main animal health challenge was to improve the biosecurity both at farm level and during live animal transport. Also, a key activity was to disseminate the recent information on the disease and the vaccination possibilities amongst veterinarians. The vaccination is a very good tool to control the disease but in the eradication point of view it could be successful only if very strict farm management and biosecurity measures are in place.

The biggest threat to Hungary's pig population in maintaining the PRRS-free status is the trade of prefatteners. In this connection, due to the strict PRRS regulations, we stimulated trading companies to turn to higher health status, PRRS free prefatteners to transport to and within Hungary. At the same time, we had to increase the level of monitoring of animal transport within the country as well.

References

1. *National Hog farmer.* **https://www.nationalhogfarmer.com/health-diseases/cost-prrs-reaches-millions-0817.** Accessed 4 May 2022.

2. Valdes-Donoso P.; Alvarez J.; Jarvis L.S.; Morrison R.B.; Perez A.M. Production Losses From an Endemic Animal Disease: Porcine Reproductive and Respiratory Syndrome (PRRS) in Selected Midwest US Sow Farms. *Front Vet Sci.* 2018;5:102.

3. Holtkamp D.J.; Polson D.D.; Torremorell M.; Morrison B.; Classen D.M.; Becton L.; Henry S.; Rodibaugh M.T.; Rowland R.R.; Snelson H.; Straw B.; Yeske P.; Zimmerman J. Terminology for classifying swine herds by porcine reproductive and respiratory syndrome virus status. *J Swine Health Prod.* 2011;19:44–56.

4. Renken C.; Nathues C.; Swam H.; Fiebig K.; Weiss C.; Eddicks M.; Ritzmann M.; Nathues H. Application of an economic calculator to determine the cost of porcine reproductive and respiratory syndrome at farm-level in 21 pig herds in Germany. *Porcine Health Management* 2021;7:3

5. Szabó I.; Nemes I.; Búza L.; Polyák F.; Bálint Á.; Fitos G.; Holtkamp D.J.; Ózsvári L. The Impact of PRRS Eradication Program on the Production Parameters of the Hungarian Swine Sector. *Animals* 2023, 13, 1565. **https://doi.org/10.3390/ani13091565**

6. Szabó, I.; Bognár, L.; Molnár, T.; Nemes, I.; Bálint, Á. PRRS eradication from swine farms in five regions of Hungary. *Acta Vet. Hung.* 2020, 68, 257–262.

7 Szabó, I.; Molnár, T.; Nemes, I.; Abonyi, T.; Terjék, Zs.; Bálint, A. PRRSV eradication on large-scale fattening pig farms in Hungary between 2014 and 2019. *Acta Vet. Hung.* 2019, 67, 529–542.

8. Nemes, I.; Molnár, T.; Abonyi, T.; Terjék, Zs.; Bálint, A.; Szabó, I. Eradication of PRRS from backyard swine herds in Hungary between 2012 and 2018. *Acta Vet. Hung.* 2019, 67, 543–552.

9. Fornyos K.; Búza L.; Makkai I.; Polyák F.; Pogácsás I.; Savoia L.; Szegedi L.; Bálint Á.; Jakab S.; Bányai K.; Szabó I. Sampling Strategies in PRRS Elimination in Hungary: An Observational Study Involving Four Farrow-to-Finish Swine Herds. *Vet. Sci.* 2023, 10, 546. **https://doi.org/10.3390/vetsci10090546**

10. Terrestrial Animal Health Code - 26/07/2017 *Chapter 15.3.* Infection with porcine reproductive and respiratory syndrome virus.

https://doi.org/10.54846/am2024/148

Use of a surveillance program and vaccine design to eliminate influenza A virus and improve productivity in an integrated production system

J. Garrido-Mantilla, DVM, PhD; J. Sanhueza, DVM, MS, PhD; J. Hernandez, DVM, MBA; M. Culhane, DVM, PhD

Introduction

Influenza A virus (IAV) is one of the most important pathogens in swine production. It produces important economic losses due to its clinical presentation and predisposition to secondary infections in the host. Vaccination is the most common strategy to control influenza infections in pigs. Influenza-negative systems often experience improvements in production and reduced medication and vaccination costs; however, there is scarce information regarding the returns obtained after implementing an influenza vaccination program to prevent or manage the disease. Consequently, influenza elimination, through the production of negative piglets at weaning, is not a common practice among producers. In addition, there is no solid protocol of influenza elimination as is for other pathogens such as *Mycoplasma hyopneumoniae* or porcine reproductive and respiratory syndrome virus (PRRSV). Our goal is to share the strategy we implemented to eliminate influenza in an integrated swine production system in order to encourage influenza elimination and increase pig health, pork production, and profitability.

Methods

An IAV surveillance program was implemented in a production system with 24 sow farms in Mexico. Most of the flows were PRRS positive stable. Sow farm size ranged between 1800 and 4000 sows. Out of the 24 farms, four were batch farrowing systems (weaning every 3 weeks) and the remaining 20 farms farrowed continuously. Wean to finish population flowed either continuously or all-in all-out according to farm capacity and the use of specific farm management strategies to improve health status. The system had a historically positive status for IAV. However, it did not use any commercial or autogenous IAV vaccine in the two years prior to the implementation of the IAV surveillance program. Every month, 30 oropharyngeal (OP) swabs from 10 litters were randomly collected from piglets around weaning age in all sow farms. Samples were tested by IAV rRT-PCR in pools of three. Positive pools with ct value < 25 were opened and selected individually for IAV isolation in MDCK cells. Positive IAV isolates were sequenced targeting hemagglutinin (HA) and neuraminidase (NA) proteins. The HA protein of the isolated IAV strains were compared to each other and to IAV vaccine strains by pairwise alignment with available reference strains from OctoFLU.[1,2] After alignment, key amino acid comparison at putative antigenic sites were made. A custom-made, trivalent IAV vaccine prototype and was designed to include the IAV strains detected most frequently in the system and that also shared highest HA protein sequence and putative key amino acid identity with the majority of the detected IAV strains. Five sow farms were selected for IAV elimination based in IAV status, production system, alignment with other health strategies (PRRS stabilization), and biosecurity standards. Two mass vaccinations were performed three weeks apart one of each other following the manufacturer's recommendations. At the same time, introduction of gilts was stopped and a protocol to reduce IAV transmission was implemented. Briefly, weaning age was reduced, cross fostering was allowed only the first week of age, use of nurse sows was interrupted in addition of others internal biosecurity practices. OP swab samples were collected before the implementation of the protocol to stablish the IAV prevalence baseline. Six weeks after the elimination program was implemented, 90 OP swabs were randomly collected from piglets around weaning age in 30 litters and tested in pools of three by IAV rRT-PCR for three consecutive weeks. In addition, oral fluids were collected from the GDU. Confirmed negative weaned piglets flowed into its respective site two in an all-in all-out fashion. Production parameters such as weaning age and weight, weight and age at the end of the nursery stage, number of pigs at the beginning of the nursery stage, number of pigs that died at the nursery, and medication costs were collected. The change of daily weight gain, and medication costs in the nursery before and after the elimination program was successfully implemented in the sow farm was assessed using two linear models. The change in mortality was assessed using a Poisson model with the number of pigs death as the outcome and an offset for the total number of pigs in the site.

Results and conclusions

The most frequently detected and isolated IAV strains were H1N1 strains of clade 1A.3.3.2 (pdm). The pdm H1N1 strains were isolated from 24 pig sites within the system and all shared greater than 97% HA protein identity to each other (mean 98.57%, range 97.43 to 98.59%). The pdm H1N1 chosen for the vaccine shared 99.59% HA protein overall similarity to the 23 other pdm H1N1 isolates and had all pdm H1N1 strains had identical amino acid motifs at the Ca2 epitope site of SHAGAKS. Two distinct clades of H3N2 strains were isolated from the system. From three sites, an H3N2 with an H3 of clade 3.1990.4 (Cluster IV) was isolated and all three isolates shared greater than 98.7% HA protein identity to each other with two identical isolates being isolated from two pigs as part of one collection timepoint. The amino acid motifs of the identical H3N2 isolates were NYNNYK and the amino acid motif of the H3N2 that shared 98.7% HA protein overall identity had an KYHNYR amino acid motif. Also isolated from a single site was an H3N2 strain with an H3 of clade HA-human-to-swine-2022. This 2022-human-like H3 was also included in the trivalent custom-made vaccine because spillover from humans to swine of the human seasonal influenza strain is known to occur in swine.[3] In summary, the tri-valent IAV vaccine contain an H1N1 of clade 1A.3.3.2 (pdm) with Ca2 epitope site motif of SHAGAKS, and H3N2 of clade 3.1990.4 with an amino acid motif of NYNNYK, and an H3N2 of clade HA-human-to-swine-2022 (human-like H3 seasonal) with an amino acid motif of STSNNK.

Oropharyngeal swab samples were collected for 16 consecutive months. Throughout the surveillance time, no seasonal pattern was identified. The five farms primary selected for IAV elimination were negative after the implementation of the vaccination and elimination protocol and all of them continued having negative results until the writing of this abstract. Site 2 production parameters were gathered for 21, 20, 16, 8 and 8 weeks in each nursery before the intervention, a period of about 6 weeks during the elimination, and for 20, 17, 13, 6, and 6 weeks after the elimination was completed. After adjusting for the effect of growing-pig site, the average daily gain improved significantly ($P < 0.001$) in site 2 after the elimination of IAV in 36.6 g (95% CI 29.8 g – 43.4g). Mortality was reduced 0.45% (95% CI 0.44% - 0.46%) and treatment costs were reduced USD 0.27 (95% 0.24 – 0.31) per pig raised ($P < 0.001$). Week of the year and farrowing type (continuous v/s batch) did not have a statistically significant impact on the ADG, Mortality or treatment costs. These results from five nursery sites help to quantify the production benefits that could be expected in the nursery after eliminating IAV from the sow farm. The elimination program has been progressively applied to the rest of the system according to the feasibility and system needs.

References

1. octoFLU. **https://github.com/flu-crew**

2. Chang J, Anderson TK, Zeller MA, Gauger PC, Vincent AL. octoFLU: Automated Classification for the Evolutionary Origin of Influenza A Virus Gene Sequences Detected in U.S. Swine. Microbiol Resour Announc. 2019 Aug 8;8(32):e00673-19. doi: 10.1128/MRA.00673-19. PMID: 31395641; PMCID: PMC6687928. **https://doi.org/10.1128/MRA.00673-19**

3. Nelson MI, Vincent AL. Reverse zoonosis of influenza to swine: new perspectives on the human-animal interface. Trends Microbiol. 2015 Mar;23(3):142-53. **doi: 10.1016/j.tim.2014.12.002.** Epub 2015 Jan 4. PMID: 25564096; PMCID: PMC4348213.

https://doi.org/10.54846/am2024/149

Approaches to disease elimination: A cross-species comparison with poultry

Jessica Higgins, DVM
Kalmbach Feeds, Kalmbach Poultry LLC & Kalmbach Swine Management, Upper Sandusky, Ohio

Introduction

The swine and poultry industries share multiple parallels in relation to production strategies, disease challenges, and even consumer pressure to meet specific medication and welfare standards. The last forty to fifty years have allowed for both industries to capitalize on significant technological advancements to efficiently raise more animals within an optimal space; yet, recent market trends have demanded an increase in alternative production strategies focused on perceptions of animal welfare and medication usage. The poultry industry has made significant adjustments to the moving customer demands over the last decade, which has pivoted its approach to disease elimination. In relation to disease elimination, medication has become a less utilized tool by the poultry industry today. Instead, there has been a shift toward certification programs, management adjustments, and vaccination programs to address poultry disease challenges.

Egg production in the United States

As Dr. Craig Rowles explained at the 2023 AASV Annual Meeting, egg production in the United States has followed a similar trend as to what has occurred in the swine industry. From the 1920s up to the 1960s, most of the egg layer industry consisted of small farms with backyard flocks. These flocks produced eggs for that specific farm family and excess was sold locally in small towns and cities. By the 1960s, it was more common to see 10,000-20,000 bird flocks with eggs still being handled manually. Off-site service trucks would then pick up the eggs to be transported to a separate packing and distribution location.[1]

A pivot happened with the poultry industry in the 1970s and 1980s with further advancements in technology which allowed for egg handling to become more automated, and consolidation of the industry became the path moving forward. Large, inline egg grading complexes become the obvious profit choice for the industry, which required larger flock and farm sizes to maximize efficiency and profitability. These sites were often located closer to large population centers to increase the efficiency of transportation and labor needs.[1]

The egg industry experienced another swing within the last 10 to 15 years due to consumer and activist pressure for more cage-free, pasture-raised and free-range, antibiotic-free and organic egg production options from smaller and family-owned farms. This has created an interesting paradigm change for the egg industry with multiple generations of the workforce being accustomed to a more industrialized way of raising poultry in a strictly indoor house setting with caged birds. The recent consumer pressure requesting the perception of a more 'wholesome' and natural product has pushed a significant portion of the industry to move back to smaller flock sizes with outdoor access, including pasture access in many settings, and no use of antibiotics or organic standards for the flocks.[2]

Disease challenge comparison

Foreign animal disease

Foreign animal disease risk is a key concern for both the poultry industry and swine industry. The poultry industry has needed to respond to a US outbreak of highly pathogenic avian influence (HPAI) in 2015 and 2022, and it is a continued concern into the Fall of 2023 at the time of this writing. The 2022 - 2023 outbreak of HPAI has resulted in the depopulation of over 60 million birds[3], which is a depletion of more than a sixth of the entire US poultry inventory on an annual basis.[4,5] The devastating effects of HPAI have prodded the poultry industry to continuously seek improved disease prevention and elimination strategies. Biosecurity and monitoring strategies included in certification programs like the National Poultry Improvement Plan (NPIP) have become cornerstones to disease prevention and elimination for the poultry industry.[6] There is no doubt that the effects of HPAI in the US would have been much worse if the NPIP had not already been put in place. Similarly, the impact of foreign animal disease is not an unforeseen risk to the US swine industry with African swine fever spreading throughout the world in recent years. Foreign animal disease preparedness for the US swine industry has the potential to benefit in multiple ways from the NPIP concepts that have already been implemented nationally for nearly 100 years.[7]

A reappearance of 'old' diseases

The recent shift in the poultry industry toward niche markets focused on antibiotic-free flocks with outdoor access has created a surge in disease challenges over the last decade that the previous generation of workforce had not seen due to differences in rearing approaches. Many of these challenges were known as 'old' diseases for the poultry industry, which were eliminated from flocks as the trend for indoor-only, caged egg production became popular after the 1970s and 1980s. Yet, the move toward customer requests for smaller flocks with access to the outdoors in open group settings created the ideal environment for these 'old' diseases to reoccur within the industry. Diseases like erysipelas, fowl cholera (*Pasteurella multocida*), Egg Drop Syndrome, and infectious coryza reappeared at a significantly increasing rate in recent years, which are diseases previously unseen by an entire generation of the workforce for many companies within the commercial egg industry.[8] Also, customer and activist pressure for increased use of cage-free flocks are expected to continue to increase in the coming years according to the most recent USDA NASS data[9], which means that the trend in disease challenges are also expected to remain a concern (Figure 1).

Figure 1: US cage-free and conventional layers.

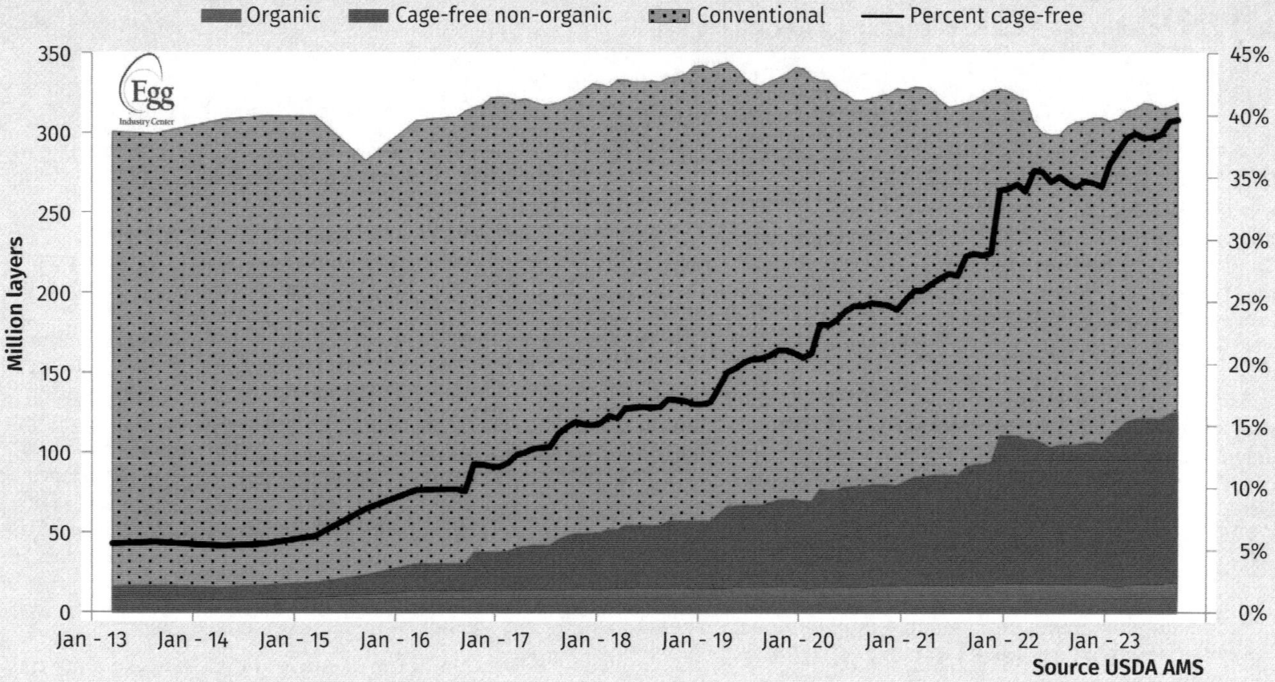

Disease elimination approaches

Certification programs

This trend with diseases in the poultry industry combined with the limited usage of medication in poultry has led to innovative disease management and elimination strategies. As early as the 1930s, the NPIP was as a collaborative effort between industry, state, and federal partners to provide standards for certifying the health status of poultry and egg flocks. It was first initiated to eliminate the spread of Pullorum Disease, caused by *Salmonella pullorum,* that was widespread throughout the poultry industry with devastating mortality greater than 80% in some cases. After its initial success with basically eliminating Pullorum Disease from commercial flocks, the program has expanded to now include monitoring and testing for seven different diseases, including avian influenza.[6] It is also used by greater than 99% of commercial scale poultry and egg operations across the US.[7] The NPIP has also been integral to the poultry industry's development of biosecurity principles that are used as a standard to consistently prevent and reduce the spread of disease between farms.

The US Swine Health Improvement Plan (US SHIP) is an initiative modeled after NPIP. US SHIP is a program sponsored by USDA Veterinary Services with support from state and industry partners. The aim of US SHIP is to safeguard, improve, and represent the health status of swine across participating farm sites, similar to the NPIP platform. The initial objective is to develop and implement an African swine fever (ASF)-classical swine fever (CSF) monitored certification of US pork production operations based on the same principles used for the NPIP H5/H7 avian influenza monitored certification of US commercial poultry operations.[7] This approach would allow for critical continuity of business for swine operations in the face of a foreign animal disease outbreak in the US, which has been modeled in the poultry industry. It also has the potential to provide the format necessary to essentially eliminate a specific disease within the commercial swine industry, as it has been modeled with Pullorum Disease in the US commercial poultry industry.

Disease prevention strategies

The limited utilization of medication in the poultry industry due to regulatory restrictions and customer pressures have naturally led toward a more preventative approach. Vaccination for diseases have become a cornerstone to good layer flock health with the average layer pullet receiving over 15 vaccinations by 16 weeks of age.[10] Also, these layer pullets are hatched with expected maternal antibodies from their breeder flock that will protect them against multiple diseases for the first few weeks of life. In fact, it is becoming a more common practice for chick maternal antibodies to be routinely monitored via serological surveys every few years. The decaying trend in maternal antibody protection is evaluated via serological cross-sectional or time-series surveys for multiple diseases to determine ideal vaccination timing and difference between breeds or varieties. Furthermore, these same serologic approaches are routinely utilized by production companies to understand vaccine administration compliance. Serology is also a commonly utilized tool for the swine industry, but its reliance is not common to the same extent for understanding population immunity over time like it has been within the poultry industry. Advancements in technology have allowed for poultry vaccination strategies to be further fine-tuned with the use of molecular techniques to check vaccination administration compliance via polymerase chain reaction (PCR) testing. Similar approaches have been employed in the swine industry, but there are significant differences between the sample pooling techniques utilized between the two industries. Also, whole genome sequencing techniques have allowed for more extensive isolate comparisons for better vaccine selection, which has similarly become more common within the swine industry. The advancements being utilized within the poultry industry have allowed for more focused activation of the immune system, which results in more rapid disease protection and elimination. These same concepts can be further utilized within the swine industry to aid in disease elimination.

Summary

The US poultry and swine industries have followed similar trajectories over the last 100 years. Arguably, the poultry industry has responded to changing customer and activist pressures more rapidly than the swine industry. The poultry industry's response to change has impacted its approach to disease elimination in both positive and negative ways, especially in recent years. Principles learned by this approach can benefit the swine industry in disease elimination.

References

1. Rowles, C. Lessons learned from the poultry industry for housing and health. In: *Proceedings of the 54th AASV Annual Meeting*. March 6, 2023. American Association of Swine Veterinarians; 2023:335-337.

2. Alonzo A. U.S. organic chicken production grows rapidly. WATTPoultry.com. Published May 5, 2021. Accessed October 31, 2023. **https://www.wattagnet.com/broilers-turkeys/article/15533500/us-organic-chicken-production-grows-rapidly**

3. United States Department of Agriculture Animal and Plant Health Inspection Service. USDA APHIS | 2022/2023 Confirmations of Highly Pathogenic Avian Influenza in Commercial and Backyard Flocks. Published November 3, 2023. Accessed November 4, 2023. **https://www.aphis.usda.gov/aphis/ourfocus/animalhealth/animal-disease-information/avian/avian-influenza/hpai-2022/2022-hpai-commercial-backyard-flocks**

4. Chickens and Eggs Report. National Agricultural Statistics Service, Agricultural Statistics Board, United States Department of Agriculture; 2023. ISSN: 1948-9064. Accessed November 3, 2023. **https://downloads.usda.library.cornell.edu/usda-esmis/files/fb494842n/vq281651h/jq0873504/ckeg1023.pdf**

5. USDA/NASS QuickStats. quickstats.nass.usda.gov. Published 2017. Accessed November 3, 2023. **https://quickstats.nass.usda.gov/**

6. USDA APHIS. USDA APHIS | National Poultry Improvement Plan (NPIP). USDA APHIS. Published June 2, 2020. Accessed October 31, 2023. **https://www.aphis.usda.gov/aphis/ourfocus/animalhealth/nvap/NVAP-Reference-Guide/Poultry/National-Poultry-Improvement-Plan**

7. U.S. Swine Health Improvement Plan Overview. U.S. Swine Health Improvement Plan. Published 2023. Accessed November 3, 2023. **https://usswinehealthimprovementplan.com/about-us/**

8. Gingerich E. 2023 Layer Health Report for October 1, 2022, to October 1, 2023. Paper presented at: 2023 USAHA Annual Meeting, USAHA Committee on Poultry and Other Avian Species (CPAS); October 10, 2023; Online.

9. Ibarburu, M. Flock Trends and Projections for September 2023. Egg Industry Center; 2023:7.

10. Hy-Line International. Technical Update: Vaccination Recommendations. Hy-Line Technical Services; 2019. Accessed November 3, 2023. **https://www.hyline.com/ViewFile?id=7c275f35-64c0-466a-a40f-eb24d1fe0d9f**

https://doi.org/10.54846/am2024/150

Implications of Proposition 12 compliance

Cara Haden, DVM; Lisa Brands
Pipestone Veterinary Services, Pipestone Minnesota

Introduction

In December of 2018 the voters of the state of California passed Proposition 12 (Prop 12), the Farm Animal Confinement Initiative. Prop 12 bans the sale of whole pork meat into the state of California unless pigs are raised according to standards outlined in Prop 12. For pork, these standards include housing sows at a minimum of 24 square feet with exceptions for farrowing and lactation, treatment and for short periods for production purposes.

The mission statement for the California Farm Animal Confinement Initiative is "To serve animal agriculture producers and California consumers by promoting and protecting the welfare and care of animals in agriculture in order for Californians to have access to food that is sourced from humanely and sustainably raised animals".[1] This mission statement makes it clear that improvements in welfare and sustainability are goals of Prop 12.

As more farms become fully compliant and receive official certification, we can begin to understand if the Prop 12 guidelines are meeting the goals put forward by the California Department of Food and Agriculture. In comparison to conventional farms, both stall and pen, has welfare improved? In comparison to conventional farms, both stall and pen, has sustainability improved? The remainder of this paper will be focused on understanding the implications of Prop 12 compliance as it relates to welfare and sustainability metrics.

Methods

The metrics below were gathered as a part of normal production practices. For treatment and lameness metrics, a 10-week span from September to November of 2023 was utilized. Farms were only included in the database if they were PRRSV negative, had no construction projects in the last 6 months, and did not have any other major health outbreaks in the 10-week time period.

Production parameters, including sow death loss, were gathered over a 30-week period from April to November of 2023. The same criteria listed above was used for this data set. The Prop 12 farms had all been Prop 12 compliant for at least 3 months prior April of 2023.

Lameness metrics

Lameness is one of the major health issues seen on farms that contributes to a negative welfare state. Lame sows experience pain, inability to ambulate normally to food and water, and are at risk of being targeted in a pen gestation system by other sows. Because lameness has a large negative impact on welfare, we track numerous lameness-related metrics on our farms.

Table 1 shows a 10-week snapshot of lameness metrics. In comparing our stall, conventional pen, and Prop 12 farms, the percentage of sows treated for a new case of lameness is highest on our Prop 12 farms. Nearly 3 times more animals are treated for a new case of lameness at our Prop 12 compared to our stall barns.

There is a 1.2% cull rate for Prop 12 compared to a 0.45% cull rate in our traditional stall barns; > 2.5 more animals are culled as a result of lameness at our Prop 12 barns. Culling more animals for lameness means the Prop 12 barns have less opportunity to cull for age and production reasons.

The final row in Table 1 compares death loss with lameness as a death loss reason. The majority of these animals are euthanized after failing to respond to treatment after two days. Our stall barns experience a 0.45% rate of lameness related death loss compared to our Prop 12 barns at 1.11%. The lameness related death loss rate is 2.5 times higher at our Prop 12 barns despite treating almost 3 times more animals for lameness.

Treatment rate

Another metric we monitor on farm is an anti-inflammatory treatment rate and antibiotic injection treatment rate. This metric takes the total number of anti-inflammatory or antibiotic injections given on farm in 10 weeks divided by the farm inventory. This metric allows us to compare treatment rates from farm to farm. If we take this rate and standardize for a 5,000-sow farm, we can compare the number of injections given per week in the different housing types using a stall barn as the baseline. Table 2 shows that a 5,000-sow pen barn would give a total of 150 more total injections each week compared to a 5,000-sow stall barn. Eighty of those injections would be anti-inflammatories and 70 antibiotics. A 5,000 sow Prop 12 barn would give a total of 265 injections more per week (approximately 40 per day) than a 5,000-sow stall barn, 105 additional anti-inflammatories and 160 antibiotic injections necessary in Prop 12 compared to the baseline. Despite these increased treatment rates, sow death loss is higher on our Prop 12 farms.

Sow death loss

Sow death loss is an important metric when we are assessing pig welfare on our farms. Table 3 outlines the mortality difference seen across the different housing types. Our healthy stall barns average 10.5% mortality. Our healthy pen barns average 11.3% mortality and our healthy Prop 12 barns average 12.4% mortality. Mortality is consistently approximately 2% higher for Prop 12 when compared to stall barns. In extrapolating to a 5,000-sow barn, this equates to 95 additional dead sows per year.

Production loss

Sustainability can be defined in several ways. A simplified goal for sustainability is to produce more with less. We want to produce more piglets with fewer sows, less land, and fewer medications. We have already discussed the increased medication needs noted on Prop 12 farms, but it is worth noting production metrics as well. Table 4 outlines a few major production parameters we monitor as sustainability measures. Farrow rates are 5.1% lower on our Prop 12 barns. Total born is lower by 0.5 pigs. The pigs per sow per year (PSY) metric shows a decrease in pigs weaned per sow of 1.34 pigs when comparing stall and Prop 12 barns. A reduction of 1.34 pigs at a 5,000-sow farm, results in a 5,000-sow Prop 12 farm would produce 6,700 fewer piglets per year than a stall barn with the same number of sows.

Table 1: Lameness metrics

	Stall	Pen	Prop 12	Change (Stall to Prop 12)
% Treated for lameness	3.50%	7.80%	10.20%	6.70%
% Culled for lameness	0.45%	0.88%	1.20%	0.75%
% Dead for lameness reason	0.45%	0.56%	1.11%	0.66%

Table 2: Injections per week

	Stall	Conventional pen	Prop 12
Anti-inflammatory injections	Baseline	+80 per week	+105 per week
Antibiotic injections	Baseline	+70 per week	+160 per week
Total injections (5,000 sow farm)		+150 injections per week	+265 injections per week

Table 3: Sow death loss

	Stall	Conventional pen	Prop 12	Difference (Stall to Prop 12)
Sow death loss	10.5%	11.3%	12.4%	1.9%
			Total sows (5,000 sow farm)	+ 95 sows

Table 4: Production

	Stall	Conventional pen	Prop 12	Difference (Stall to Prop 12)
Farrow rate	90.6%	89.2%	85.5%	-5.1%
Total born	16.4	16.1	15.9	-0.5 pigs
Live born	14.9	14.7	14.5	-0.4 pigs
PSY	30.35	29.83	29.01	-1.34
			Total fewer piglets (5,000 sow farm)	6,700

PSY = pigs/sow/year

Conclusion

The state of California set a goal to improve pig welfare and pig farm sustainability via their Farm Animal Confinement Initiative. There are positive and negative implications to all initiatives, Proposition 12 included. Unfortunately, in this case, some of the negative implications involve real and significant negative impacts to pig welfare and sustainability. With our current Prop 12 systems, we see a concerning negative impact on many of our welfare related metrics while simultaneously noting negative impacts on sustainability. Our industry has a significant challenge to adapt and improve our Prop 12 systems in order to solve the current welfare and sustainability challenges associated with Prop 12 compliance.

Reference

1. California Department of Food and Agriculture. Animal Care Program. Accessed November 11, 2023. https://www.cdfa.ca.gov/AHFSS/AnimalCare/

https://doi.org/10.54846/am2024/151

Our Proposition 12 journey: Production considerations and lessons we are learning

Carlos Roudergue, DVM, MBA
Production Department, Country View Family Farms

Introduction

In the ever-evolving landscape of food production and agriculture, regulations and policies play a pivotal role in shaping the industry's practices and standards. One such regulation that has generated significant attention and discussion is Proposition 12. Enacted with the aim of improving the welfare of farm animals, Proposition 12 has ushered in a new era for many producers and stakeholders in the agricultural sector. As we embark on our own Proposition 12 journey, we at Country View Family Farms find ourselves at a critical juncture, encountering a multitude of challenges, production considerations, and valuable lessons along the way. This document is intended to offer insights into the multifaceted aspects of production considerations that have arisen as a result of this new legislation. By sharing our experiences and the knowledge gained, our goal is to contribute to a broader understanding of the challenges and opportunities that Proposition 12 presents, lessons that are continuously learned along the way.

Our journey

Following 128 years of company tradition rooted in continuous improvement and innovation, Clemens Food Group (CFG) and its hog production affiliate Country View Family Farms (CVFF) decided to adopt group housing and electronic sow feeding (ESF) systems for gestating sows more than 20 years ago. This strategic decision has been a source of experience and knowledge for the production team, in order to understand the strengths and weaknesses of group housing and the technologies available for it. Additionally, CFG began to solely process hogs from sow farms utilizing group housing at the end of 2022. This includes CVFF and 3rd party suppliers.

Before Proposition 12 legislation started to be in the spotlight, most of the sow inventory in Pennsylvania Operations at CVFF were moved into gestation pens by day 3-5 after 1st service, place where they were already at 22 - 24 sq ft per sow. Because this visionary way to house sows in gestation was already implemented years before, our main challenge in adopting Proposition 12 standards was finding a practical solution to eliminate stalls for wean sows, opportunity sows and gilt pool without compromising performance and welfare.

We started our journey by understanding wean sow behavior and the impact of weaning them into pens by taking one of our sow farms as a trial-and-error unit for our own learning and discovery. At this location we prototyped small pens with different shapes and sizes, evaluating different components including financial, facility maintenance, labor, animal behavior and welfare. Sow behavior and welfare were measured by direct observation through 24/7 video recordings of each pen under our analysis.

After several months of evaluation, it was decided to move forward with a breeding pen design with capacity for 71 sows per pen, with a minimum of 1,704 sq ft. Inside each of these pens there is a set of 10 to 15 individual stalls, which are used only for breeding purposes. Additionally, troughs with ad-libitum feed were located in a well-defined area of the pen, together with water nipples in different areas of the pen at a ratio of 10 sows per nipple.

Taking this design as the new standard, in June 2021 CVFF started the conversion of our first 5,500 sow operation in PA. The implementation of the design, now on a larger scale, provided more lessons learned for the team to consider in the remodeling process of the farms to come. Since June 2021 up today, CVFF has remodeled 8 other sow farms, with an approximate total of 50,000 sows that today are Proposition 12 compliant.

Production considerations & lessons we are learning

All our Proposition 12 sow farms have been audited both internally by the CVFF Animal Care & Compliance Department and externally by independent third-party institutions. Lessons learned from both the operation of these new production standards and the different audits performed over the years, allowed us to outline some production considerations that could be helpful for producers interested on converting facilities to Proposition 12 standards.

1. **Performance:** Overall sow performance and sow inventory could be affected for producers at different magnitudes based their current state of production. In a system where group housing is the standard and sows are moved into pens pre-implantation, the impact may not be as big as in sow farms where sows in gestation are crated from service up to 3 days pre-farrow. In the case of CVFF, performance was measured before and after the remodeling of the first 4 facilities. So far, none of our KPIs has shown significant detrimental changes that we can directly attribute to changes made to our sow housing due to Proposition 12.
2. **Labor:** Breeding into wean pens could be highly demanding on labor. For a 5,500-sow unit, the head count went up by 2 (10%). Based on CVFF experience, producers need to be careful when it comes to weaning schedule and weekend labor availability.
3. **Safety:** Labor safety also becomes relevant, but accidents can still be preventable. It is a matter of having people on staff with good animal husbandry skills and continuously trained, as there is more interaction between animals (sows with sows, sows with boars, boars with boars) and more direct contact between animals and people. So far CVFF has not seen a detectable increase in safety recordable incidents related to new breeding routines.
4. **Compliance is in the details:** After two years of implementation and going through several internal and external audits, two main findings have given us interesting lessons:
   a. Stocking density cannot be calculated assuming barn schematics are accurate. For different reasons, variations from several inches up to a few feet between barn schematics and real measurements by pen have been detected consistently at CVFF operations. The source of these differences usually can be associated but not limited to pen construction, alterations due to the use of the pens over the years, maintenance alteration, etc. This could be a source of confusion for production teams in their efforts to calculate

the right capacity per pen. Based on CVFF experience, there is no way around measuring sq ft of pens one by one using a tape measure or laser measuring tools. Once we measure and record the actual dimensions of each pen, we can calculate the sow capacity per pen. Any decimal obtained from the calculation of the number of sows per pen is rounded down, in order to ensure 24 sq ft per sow in all scenarios. Sow capacity per pen is posted on clearly visible signs in each pen and is monitored on an ongoing basis.

b. A reliable time record-keeping system for sows housed under less than 24 sq ft is needed. The California Proposition 12 law states that sows cannot be under less than 24 sq ft for more than 6 hours in a day, or more than 24 hours in a week. With this in mind, a robust record-keeping process where the number of minutes that sows are in less than 24 sq ft is needed. To address this requirement, CVFF records daily times of sows under less than 24 sq ft every day by using a dedicated form by breeding pen.

Final remarks

As necessity is the mother of invention, Proposition 12 has been creating discussion across the industry about new housing standards. These necessary discussions will create solutions to different challenges as they emerge. As an industry, we are still in a discovery phase where no solutions have emerged as the sole valid solution yet.

Solutions require creativity, but their execution requires discipline from farm level up to upper management. This is considering that along with Proposition 12 are coming new and more sophisticated housing standards with new record needs. At the farm level for example, discipline is required from all Team Members to fill pens to capacity and not place a single additional sow. It is even more challenging when dealing with seasonal infertility or inconsistency in meeting breeding targets. Today we cannot increase the number of sows to be bred as easily as before. Another example of challenges for the farm operation is the combination of sow mass vaccinations, their withdrawal and sow culling schedules. In a Proposition 12 compliant system there is a lot less room for mistakes.

On the other hand, Proposition 12 has given us some back-to-basic lessons. Mixing and truly full-feeding sows at weaning, together with a more intensive boar exposure inside the pens is proving to be highly effective when it comes to reducing wean-to-service interval. The boar exposure process in open pens also gives Team Members the opportunity to make better breeding decisions. They are able to more effectively identify lameness and other quality selection issues in wean sows as those sows stick out in a group of sows walking into the breeding pens, compared to sows standing in breeding stalls next to each other. Another advantage of mixing sows at weaning that we have been able to observe is the reduction of intensity and frequency of fighting when gestation groups are created in pens, which is especially critical in farms where sows are moved pre-implantation.

Finally, continuous, and effective training of Team Members in the new processes is critical for success. But together with training, it is also important to clearly define how to manage the change of processes and routines, since they are significantly different under Proposition 12 standards. Keeping the team involved in the discussions about problems and challenges we are facing, brainstorming ideas and learning from mistakes together are ways to keep the team's morale up, overall engagement elevated and rise the deep desire to be part of something new. Today our Team Members' feedback is clear: If for any reason Proposition 12 is not required at their facility anymore, they would like to keep heat checking and breeding sows as they are doing it today. Farm managers and barn technicians are fully engaged in getting things done right under California's Proposition 12 standards.

https://doi.org/10.54846/am2024/152

Proposition 12: Common findings and suggestions for becoming compliant with the law

Elein Hernandez[1,2]; Alina Porta[2]; Jason McAlister[2]
[1]Department of Clinical Studies and Surgery, Facultad de Estudios Superiores Cuautitlán, Universidad National Autónoma de Mexico, Mexico; [2]CloverLeaf Animal Welfare Systems, St. Joseph, Missouri

Introduction

As the first accredited certifying agent by the California Department of Food and Agriculture (CDFA), CloverLeaf Animal Welfare Systems has been working with producers and has conducted thousands of third-party audits throughout the Americas and Europe. Our findings include many misunderstandings of the law and its exemptions. Most producers have made solid efforts to comply with the law. Still, we find that very few are sufficiently prepared for certification during the audit due to a lack of understanding, misinterpretation of the law, or being told that their current animal welfare certification is sufficient. This abstract describes some common findings in the swine production units in the Americas and Europe, including common enclosure errors, misunderstanding of the law, and suggestions for becoming compliant with current California Proposition 12.[1,2]

National and foreign impact of Proposition 12

There is no exact number of farms or pork producers compliant with California's standards in the USA. Nonetheless, several national and international pork producers have taken an interest in becoming compliant by investing in altering housing structure and production practices. The implications of these changes involve economic investment, training, implementation of new documentation, and an unmeasurable effort to satisfy the client's requests and sows' conditions.

The degree of alterations needed to comply with Proposition 12 directly depends on the individual production conditions before being interested in becoming certified. Pork producers with previous experience in pen gestation, artificial insemination in pen, and provision of limited timing in sow stalls have an advantage, suggesting that infrastructure and experience in this system are needed to succeed as a sustainable farm. In the USA, pork producers with ongoing production systems that did not confine the animals were initially interested in becoming certified with Proposition 12. Still, there has also been an interest from other producers with an intensive production profile.

Similarly, there are no official statements about how many foreign producers seek to be Proposition 12 certified. Nonetheless, some of these foreign producers are in North America (Canada and Mexico), South America (Brazil), and Europe (Belgium, Denmark, Ireland, Netherlands, Poland, Spain, and the UK). Likewise, their changes depend on previous conditions and regulations.

Common findings in production units

Proposition 12 includes animal-based, resource-based, and record-based animal welfare indicators for breeding pigs in gestation areas. Some common audit non-conformities in pork productions seeking Proposition 12 certification include one or more indicators, such as incomplete documentation, non-compliant housing, non-compliant age-related housing conditions, medical exceptions, and insemination in the gilt development unit (GDU). These findings are summarized in Table 1 and described in detail below.

Animal-based and resource-based indicators

Usable floor space non-conformities

According to CDFA, the minimum usable floor space requirement is 24 ft^2 per sow and must allow the sow to turn around freely. However, it has been a common non-conformity finding to see facilities that reflect the 24 ft^2 in blueprints or facilities maps and do not necessarily reflect that during the audit. This is likely because measurements are taken at human eye level, often ignoring feeders and through designs that affect the "pig usable floorspace". It is also common to find enclosures that comply with the 24 ft^2 but do not allow the sow to turn around freely thus rendering the enclosure noncompliant.

The corrections needed regarding housing depend on a case-by-case scenario. The European Union's (EU) minimum floor space allowance is 2.25 m^2 (24.2 ft^2).[3] Therefore, the transition into California's requirement could be easier than in countries with more limited floor allowance legislation. Nonetheless, other non-EU countries have also taken necessary actions to become compliant.

Age-related housing conditions

CDFA describes a breeding pig as a female pig six months of age or pregnant,[4] while a gilt is a female pig that has not yet produced a litter of piglets. There has been confusion over the definitions and when they apply to the Proposition 12 criterion. Therefore, another common non-conformity is to find producers with different housing considerations between sows and gilts. Currently, standard swine industry practices and some international legislation provide different space allowances based on age or body size score. The EU recommendation for floor space allowance is 1.64 m^2 per gilt and 2.25 m^2 per sow; in Canada, it is 1.7 m^2 per gilt and 1.8 to 2.2 m^2 per sow. This has led to confusion while trying to adopt the CDFA criterion, and the corrections needed must be based on California's requirement of six months of age as a six-month-old pig.

Insemination in GDU (gilt development unit) and moving bred sows

An interesting finding during the audits was the certification requirement of the GDU. This unexpected finding required further clarification from the authorities (personal communications). The GDU is where they prepare gilts to enter the breeding herd at a younger age, and it is not commonly discussed compared to other swine production areas. Interestingly, this area has become quite controversial for Proposition 12 compliance. Our findings confirm that many operations practice keeping sows in GDU and then moving those sows to compliant housing after insemination

Table 1: Animal welfare indicators and common non-conformities under Prop 12

Animal welfare indicator	Description	Common non-conformities
Animal-based indicator	"cannot prevent a sow from being able to lie down, stand up, fully extend their limbs (extending all limbs without touching the side of an enclosure, or another animal), and turn around freely (turning in a complete circle without any impediment, including a tether, and without touching the sides or another animal) within their enclosure"[1]	• Some enclosures provide minimum usable floor space, but it does not allow the sow to turn around freely. • Gilts and sows are housed at different stocking densities
	Legislation covers all breeding animals. Therefore, GDU requires certification.	• GDU is not being considered for certification by some producers. This was an unexpected finding.
Resource-based indicator	"One sow is required to have at least 24 square feet of usable floorspace" [1]	• Pig feeders, troughs, and fences often are within the pig's usable space, hence reducing the minimum requirement. • Over stocking pens
Record-based indicator	Documentation that confirms animal confinement standards and exceptions.[2]	• Missing information about pig's expected date of giving birth and temporary confinement in non-compliant enclosures. • Misunderstanding of medical enclosures

GDU = gilt development unit

procedures. This could be considered compliant but would involve the GDU being certified, and the main farm being certified, too. This is based on the definition of a breeding animal as described above, suggesting that when the pig is moved into an enclosure to breed, it is now considered part of the breeding herd and must either be held in compliant housing or under one of the qualified exemptions (see Documentation section).

Record-based indicators

Documentation non-conformities

Due to the recent implementation of this legislation, the lack of documentation is the most common finding across production systems and nations in pork production systems seeking to be Proposition 12 compliant. Producers are expected to eventually improve the lack of documentation as more producers become certified. It is important to mention that although the space requirement is a significant requirement that has caused controversial discussions, documentation is also an auditable criterion commonly neglected by producers. This is likely secondary to the misunderstanding of the requirements or overconfidence based on owning other animal welfare certifications. According to CFDA, recordkeeping should reflect confinement is compliant with Proposition 12 standards exceptions, which include[2]:

1. During medical research;
2. During examination, testing, individual treatment, or operation for veterinary purposes;
3. During transportation;
4. During rodeo exhibitions, state or county fair exhibitions, 4-H programs, and similar exhibitions;
5. During slaughter when performed in accordance with the provisions of Chapter 6 (commencing with Section 19501) of Part 3 of Division 9 of the Food and Agricultural Code, relating to humane methods of slaughter, and other applicable laws and regulations;
6. To a breeding pig during the five-day period prior to the breeding pig's expected date of giving birth, and any day that the breeding pig is nursing piglets; or
7. During temporary periods for animal husbandry purposes, for no more than six hours in any 24-hour period, and no more than 24 hours total in any 30-day period.

Some common non-conformities regarding documentation are the lack of detailed husbandry practices (exceptions 6 and 7) and non-compliant treatment records (exception 2). The corrections needed depend on a case-by-case scenario. It is a common husbandry practice to move sows to the farrowing facility at 110 days after breeding, but if there is no record-keeping evidence of this practice, then it is a non-conformity. Other countries' regulations do not necessarily apply to the present Proposition 12 criterion. Current EU legislation allows sows to be kept in gestation stalls for up to 28 days post-service.[5] However, California's requirements demand "no more than six hours in any 24-hour period". So even if documentation regarding time in gestation stalls is provided, it must be compliant with CFDA's definitions.

In addition, producers have a wide range of understanding of medical treatment exemptions (exemption 2). To comply with this rule, the producer must have the following documentation:

a. A valid veterinarian-client-patient relationship (VCPR)
b. Date of placement into the non-complaint enclosure
c. Treatment plan consistent with veterinary recommendations
d. Have a recorded expected return to compliant enclosure date
e. The space the animal would normally occupy is available for her return to the herd. There must be a space for her in compliant housing.

The common misunderstanding regarding this exemption is in point D. A commonly accepted husbandry practice is to have a sow under clinical observation for up to 28 days in a stall. Placing animals into non-compliant enclosures for medical treatment with extended times, such as "return after farrow date," has been denied as a complaint by officials at the CDFA. Similarly, auditors have been directed by CDFA to ask where the sow would be placed after treatment and confirm there is space available for her return.

Finally, there has been confusion surrounding the certification of other schemes as sufficient to cover regulatory compliance with Proposition 12. Officials at the CDFA have been very clear on this matter that no other scheme certification qualifies a farm or animals produced on that farm as Proposition 12 compliant. Many schemes offer some percentage of compliance, but none are 100% in line with all requirements of California Proposition 12. Regardless of the scheme criteria, all audit tools and certifications must be accredited by the CDFA as compliant and performed by an accredited third-party auditing firm.

Conclusion

There are many resources available to producers for understanding the law. The suggestion is that producers fully understand the law and everything that is required or not required under this law. We have found that producers have made very costly changes to the farms in order to become compliant with the law, regardless of their production size, only to find they misunderstood the law and the costly changes they have made are non-compliant. We encourage producers to review all the publicly available resources and contact CDFA for confirmation before making changes or seeking a certification audit. Several accredited third-party auditing bodies, such as CloverLeaf Animal Welfare Systems, and local pork producer organizations, offer self-assessment guides.

Education, understanding, and compliance with the spirit of the law is the goal for all.

References

1. California Department of Food and Agriculture, Prop 12 – Sow housing Guide, what are the housing requirements for a sow covered under Prop 12? 2023. Accessed December 1, 2023. **https://www.cdfa.ca.gov/AHFSS/AnimalCare/docs/sow_housing_guide.pdf**

2. California Department of Food and Agriculture, Chapter 10 of Division 2 of Title 3 of the California Code of Regulations, 2023. Accessed December 1, 2023. **https://regulations.justia.com/states/california/title-3/division-2/chapter-10/ https://www.cdfa.ca.gov/AHFSS/pdfs/animal_confinement_adopt_text.pdf**

3. Li YZ, Cui SQ, Yang XJ, Johnston LJ, Baidoo SK, Minimal floor space allowance for gestating sows kept in pens with electronic sow feeders on fully slatted floors, J. Anim. Sci., 96:10, 4195–4208, 2018, **doi: 10.1093/jas/sky282**

4. California Department of Food and Agriculture, Animal care program guidance: key terms. 2023. Accessed Dec. 1, 2023. **https://www.cdfa.ca.gov/AHFSS/AnimalCare/docs/Prop12_guidance_Keyterms.pdf**

5. Galli MC, et al. Can we further reduce the time pregnant sows spend in gestation stalls?. Livestock Sci., 264, August, 105049, 2022, **doi: 10.1016/j.livsci.2022.105049.**

Additional resources

Cloverleaf animal welfare systems resource page: **https://www.californiaprop12certified.com/self**

CDFA animal care website: **https://www.cdfa.ca.gov/AHFSS/AnimalCare/**

CDFA resource page: **https://www.cdfa.ca.gov/AHFSS/Animalcare/StakeholderResources/**

California legal code 13.8. Farm Animal Cruelty: **https://leginfo.legislature.ca.gov/faces/codes_displayText.xhtml?lawCode=HSC&division=20.&title=&part=&chapter=13.8.&article=**

CDFA video on prop 12 compliance for pork producers: **https://www.youtube.com/watch?v=PPzlsFMWoDo&t=24s**

https://doi.org/10.54846/am2024/153

The future of farrowing

T. D. Parsons, VMD, PhD, DACAW
University of Pennsylvania School of Veterinary Medicine, Kennett Square, Pennsylvania

Introduction

The United Nation has established seventeen sustainable development goals with an eye towards establishing an inclusive, sustainable and resilient future for people and planet.[1] Agriculture runs though many of the seventeen goals and livestock production often is implicated as an opportunity. While the term sustainability means many things to many people, one facet relevant to animal agriculture is social acceptability,[2] and thus when considering how farrowing systems will look in the future, this is an important point. The other critical aspect of sustainability is economic viability and cannot be neglected.

There have been several recent and extensive reviews in the scientific literature related to questions about farrowing practices[3-5] and are a great resource for a deep dive into this topic. Here, my goal is to summarize recent work related to alternative farrowing systems that will be most relevant to the North American swine industry.

Short- and long-term challenges

One of the challenges in understanding the future of farrowing is the plethora of different farrowing system designs in use and moreover the ballooning terminology to describe these different systems. Goumon et al[5] have recently purposed a standardization of farrowing system terminology which is limited to free farrowing (FF), temporary crating (TC), and permanent crating (PC). Free farrowing consists of zero confinement and describes a system or practice where there is no possibility to confine the sow in a crate during farrowing and lactation. Sows are able to freely turn around at all times. Pens may have feeding stalls or areas to separate the sow from the piglets temporarily for management purposes, but the fixtures would not permit crating for farrowing. Temporary crating allows for temporary confinement and is any system or practice where the sow is confined in a crate (without the possibility to turn around) for a certain period, but not the whole of lactation. Permanent crating describes the farrowing system where the sow is permanently confined in a farrowing crate from entry until weaning.

Conventional housing of farrowing and lactating sows utilizes farrowing crates and remains a public concern given her total confinement during this stage of production. Legislation already restricts permanent crating in Sweden, Switzerland, and Norway, while New Zealand, Austria, and Germany have announced phasing out of farrowing crates by 2025, 2033, and 2036, respectively. Pressure to abolish systems involving close confinement is growing in the rest of Europe. The "End the Cage Age" campaign garnered over 1.4 million signatories from more than 18 member states which was sufficient to trigger the European Commission to draft legislation that would prohibit the use of cage systems on farms for inclusion in the EU Treaties. Outside Europe, consumer expectations are less clear, but surveys in Brazil suggest that citizens have a negative attitude toward farrowing crates and consider that reducing piglet mortality does not fully justify their continued use.[6] Gestation stalls remain the focus of concerns about confinement in the North American swine industry. However, history suggests that welfare concerns in the United States lag behind Europe by about a decade and thus more public scrutiny of farrowing practices can be expected.

Determining how best to maximize farrowing house welfare and productivity is complicated given the unique challenges of the assessment within farrowing systems. In all other phases of swine production, assessments involve pigs at a single stage of their productive life. Within the farrowing room, the sow and her piglets are at two very different stages of their lives and have differing requirements in regard to their thermal, social, and physical environments.[3] Thus, optimizing the sustainability of farrowing systems for the future is complicated by the competing needs that arise from housing sows and piglets together.

The choices

Farrowing crates were introduced in the 1960s and developed to facilitate the ability of a sow to raise large litters. The crate allowed for reduced piglet losses and improved human interventions by limiting the sow's movements during parturition and lactation.[7,8] Decreased piglet preweaning mortality was among the factors driving their adoption in the 1960's and 1970's.[7,9] Today, farrowing crates remain the most common form of sow and litter housing in the United States, Europe, and the United Kingdom.[8,9] Additional economic advantages of the farrowing crate include minimal space requirements for housing the sow and litter (1.23 m^2 crate within 3.6 - 4.6 m^2 pen),[10] and minimal labor requirements for maintaining hygiene given the fully or partially slatted floors above slurry pits. The crate also affords the producer more control around the farrowing period, including a reduction in piglets crushed by the sow.

The sow welfare concerns regarding use of traditional crates and the desire to retain a confinement option to promote piglet survival, have led to the evolution of a hybrid farrowing system. It is designed to accommodate a crate being utilized temporarily to confine the sow within a pen but allows for the sow to be loose for the majority of the time during lactation as determined by the farmer. The systems for temporarily crating a sow often have similar footprint of space utilization in the barn as conventional farrowing systems. This is in contrast to a zero confinement or free farrowing systems that can range in size from 2.66 m^2 to 9.20 m^2 and often utilize more solid flooring and the inclusion of a bedding substrate.[4] Piglet mortality is expected to be higher as well. A recent meta-analysis observed a 14% increase in the relative risk of piglet mortality compared to conventional crates. These factors taken together have been important deterrents to the adoption of zero confinement farrowing systems.[5] Looking forward to future farrowing systems through the lens of the North American swine industry, the temporary confinement systems will likely be the first and most common to be adopted and thus will largely be the focus of this paper.

Piglet mortality

Piglets are extremely vulnerable as neonates as most deaths occur when the piglets are trapped under the sow.[11-13] Trapping can result in painful, often fatal, injuries or suffocation. Suffocation may result in slow death; most piglets trapped less than one minute survive crushing, whereas most of the piglets trapped longer than four minutes die.[14] The majority of piglets die during the first few days post-parturition, particularly within the first 24 hours.[9,13,14] Thus preweaning mortality is both an important production and welfare challenge.

Genetic selection also creates challenges in reducing preweaning mortality, even in crated systems, due to increased litter size as more piglets die in larger litters than small litters.[8] When more piglets are born than the number of functional teats, significant interventions by staff are required to promote survival. Free farrowing systems (zero confinement) impedes the practice of these interventions. Producers see value in using a crate to control a sow's movement to reduce crushing, allow for localized heating during birth, and facilitate interventions by staff to promote piglet survival such as assisted suckling, split suckling, and cross-fostering. Thus in parts of Europe where high rates of preweaning mortality are unacceptable, temporary crating of sows is gaining momentum. Austria and Germany are permitting "crating during the critical period for piglet survival" within their legislation (4[10] and 5[15] days maximum, respectively) and Denmark will also allow the temporary use of crates as part of its pledge to implement alternative farrowing systems.[16]

Sow welfare

Historically, producers adopted a goal of limiting of the sow's movement to increase the piglets' welfare and survivability. However, this restriction prevents the sow from performing species-specific behaviors at a crucial time and thus has the potential to compromise the sow's welfare. Under natural conditions, there is a complex series of behaviors performed in which sows and piglets experience various degrees of isolation followed by community reintegration and living. These behaviors can be divided into six distinct phases: **(i)** isolation and nest site seeking, **(ii)** nest building, **(iii)** farrowing, **(iv)** nest occupation, **(v)** social integration, and **(vi)** weaning.[17] Farrowing crates can preclude access to substrate for nest-building, foraging, and comfort and space needed for the sow to turn-around, form a dedicated dunging area, and the ability to leave the nest.[18] The expression of nesting behavior may be one of the most important behavioral indicator of sow welfare as nesting is a highly motivated behavior,[19] yet these behaviors are largely thwarted by a farrowing crate. The onset and performance of nest building is stimulated primarily by internal hormones as well as external feedback stimuli from the environment. The distinctive motor patterns of nest building behavior in sows have been described in detail for wild sows[20] and feral sows[21] as well as for domestic sows in semi-natural enclosures.[22,23] Compared to a conventional farrowing system where the sows are permanently crated, the temporary confinement opens the door for sows pursue a more natural fulfillment of their different behaviors listed above.

Management of temporary crated systems

Two important decisions need to be made when working with a farrowing system that allows for the temporary confinement of a sow – when to initiate the confinement and when to release the sow from confinement. Both of these decisions can have important implications for the sow, the piglet as well as the staff.

Many producers will want to confine the sow upon entry to the farrowing room; however, this can frustrate the sow's behavioral drive for nest building, especially in the absence of any substrate. Thwarting the performance of nest building activity can result in higher levels of plasma cortisol,[23-25] increased farrowing duration,[26,27] and increased stillbirths[30-32] as well as poor maternal behavior post-farrowing.[28] Thus, some farms will not start the temporary confinement until after the sow is finished farrowing. However, potential benefits of not closing the crate until the conclusion of farrowing may be countered by an increased risk of piglet crushing resulting from the unrestricted sow movement during the immediate neonatal period[29,30] as well as challenges to peri-parturient staff interventions such as managing dystocia. The decision as to when to initiate confinement will likely be determined by farm specific factors including the design of the farrowing system and the skill level of the workers in farrowing.

The decision as when to release the sow from confinement involves weighing the benefits of increased physical and behavioral freedom for the sow and associated benefits to the piglets versus the risks of increased mortality from crushing related to housing the piglets with an unconfined sow. There are also other practical management considerations, which include the ease of maintaining hygienic conditions within pens of limited size or the safety and efficiency of staff when carrying out routine piglet tasks. Upon opening of the crate, sows exhibit increased exploratory behavior,[31-33] however, it is an open question as to whether sows maintain these motivated behaviors for the duration of lactation as most studies have focused only the sow behavior immediate post-crate opening. Sows also exhibited more interactions with their piglets post-opening as well.[31-35] Reports on the impact of crate opening on piglet activity, nursing and suckling behavior, and growth are variable.[5] However, the impact on mortality is clearer. Meta-analysis of several studies where temporary confinement was released 2 to 14 days post-farrowing revealed no advantage (further decrease in mortality) after 4 to 7 days confinement compared to conventional crates (permanent crating).[5]

Other considerations

The number of scientific publications on alternative farrowing systems has expanded greatly in the last 5 or so years. However, there are still several knowledge gaps that exist, often related to the difficulty in studying these factors. The role of sow temperament, and especially expression of maternal behaviors, is poorly understood. The industry has spent the last 40 years selecting for sows to perform in confined systems and it may be that these sows are deficient in the requisite behavioral repertoire needed to be successful in open farrowing systems (FF or TC). Furthermore, the impact of people in these systems is almost completely unstudied. Understanding who are the best people to work in these alternative farrowing systems, what are the optimal standard operating procedures, and exactly what training they need to ensure that these procedures are executed as well as possible will require additional investigation. As an industry, we have over a half-century long collective experience with conventional farrowing systems. What are the opportunities in these alternative systems as we start to gain such experience? Also is there room for improvement in the design of these alternative farrowing systems? Would sows benefit more from the release from confinement if they were then exposed to biologically relevant enrichment and space? Several unanswered questions remain related to the implementation of alternative farrowing systems.

Conclusions

Temporary crated farrowing systems transiently confine sows early in lactation when piglets are most vulnerable. When considering alternatives to conventional farrowing systems, the pendulum in Europe appears to be swinging back from free farrowing (zero confinement) to these temporary crated systems.

Pre-weaning mortality is significantly reduced in the TC systems compared to FF and even with a duration of the confinement period of only 4 to 7 days post-farrowing mortality approaches that of the conventional systems (PC). Motivated behaviors by sows increase prior to and following release from confinement. Taken together TC highlights an improved balance between the competing needs of piglets and sows during farrowing and lactation stages of pig production. Public concern in North America about swine welfare has yet to focus on farrowing systems, but based on experience with gestation stall, one should expect increased scrutiny to arise during the next decade.

References

1. Resolution # 70 of the United Nations General Assembly. 2015. Transforming our world: the 2030 Agenda for Sustainable Development.

2. Broom DM. A history of animal welfare science. *Acta Biotheorectica*. 2011. 59:121-137.

3. Glencorse D, Plush K, Hazel S, D'Souza D, Hebart M. Impact of non-confinement accommodation on farrowing performance: a systematic review and meta-analysis of farrowing crates versus pens. *Animals*. 2019. 9(11):957. **https://doi.org/10.3390/ani9110957**

4. Baxter EM, Moustsen VA, Goumon S, Illmann G, Edwards SA. Transitioning from crates to free farrowing: a roadmap to navigate key decisions. *Front. Vet. Sci.* 2022. 9:998192. **doi: 10.3389/fvets.2022.998192**

5. Goumon S, Illmann G, Moustsen VA, Baxter EM, Edwards SA. Review of temporary crating of farrowing and lactating sows. *Front. Vet. Sci.* 2022. 9:811810. **doi: 10.3389/fvets.2022.811810**

6. Vandresen B, Stadnick ECP, Hötzel MJ. Freedom to move versus piglet crushing: citizens' attitudes towards farrowing housing systems. Proceedings of the Global Virtual Meeting of the International Society for Applied Ethology. 2022.

7. Robertson JB, Laired R, Hall JKS, Forsyth RJ, Thomson JM, Walker-Love J. A comparison of two indoor farrowing systems of sows. *Anim. Prod.* 1966. 8, 171-177.

8. Edwards SA, Fraser D.Housing systems for farrowing and lactation. *Pig J.* 1997. 39, 77–89.

9. Marchant JN, Broom DM, Corning S. The influence of sow behaviour on piglet mortality due to crushing in an open farrowing system. *Anim. Sci.* 2001. 72, 19-28.

10. Baxter EM, Andersen IL, Edwards SA. Sow welfare in the farrowing crate and alternatives. Špinka Advances in Pig Welfare. Cambridge: Elsevier Ltd., Woodhead Publishing. 2018. 27–72. **doi: 10.1016/B978-0-08-101012-9.00002-2**

11. Edwards S. Perinatal mortality in the pig: environmental or physiological solutions? *Livest. Prod. Sci.* 2002. 78, 3-12.

12. KilBride AL, Mendl M, Statham P, Held S, Harris M, Cooper S, Green LE. A cohort study of preweaning piglet mortality and farrowing accommodation on 112 commercial pig farms in England. *Prev. Vet. Med.* 2012. 104:281-291.

13. Marchant JN, Rudd AR, Mendl MT, Broom DM, Meredith MJ, Corning S, Simmins PH. Timing and causes of piglet mortality in alternative and conventional farrowing systems. *Vet. Rec.* 2000. 147:209-214.

14. Cronin GM, Simpson GJ, Hemsworth PH. The effects of the gestation and farrowing environments on sow and piglet behaviour and piglet survival and growth in early lactation. *Appl. Anim. Behav. Sci.* 1996. 46, 175–192.

15. Oliviero C, Heinonen M, Valros A, Hlli O, Peltoniemi OAT. Effect of the environment on the physiology of the sow during late pregnancy, farrowing and early lactation. *Anim Reprod Sci.* 2008. 105:365–77. **doi: 10.1016/j.anireprosci.2007.03.015**

16. Pig Research Centre. Action Plan Better Animal Welfare for Pigs. 2011. Accessed November 15, 2023. **https://en.fvm.dk/fileadmin/user_upload/FVM.dk/Dokumenter/Landbrug/Indsatser/Dyrevelfaerd_og_-transport/Svinehandlingsplan_engelsk_final.docx**

17. Jensen P. Maternal behaviour of free-ranging domestic pigs. I. Results of a three-year study. Swedish University of Agricultural Sciences. 1988.

18. Baxter EM, Lawrence AB, Edwards SA. Alternative farrowing systems: design criteria for farrowing systems based on the biological needs of sows and piglets. *Animal* 2011. 5:580-600.

19. Arey DS. Behavioural observations of peri-parturient sows and the development of alternative farrowing accommodation: a review. *Anim.Welf.* 1997. 6, 217–229.

20. Graves HB. Behavior and ecology of wild and feral swine (Sus scrofa). *J. Anim. Sci.* 1984.58:482-492.

21. Kurz JC, Marchinton RL. Radiotelemetry studies of feral hogs in South Carolina. *J. Wild. Manag.* 1972.1240-1248.

22. Stolba A, Wood-Gush DGM. The identification of behavioural key features and their incorporation into a housing design for pigs. *Ann. Rech. Vet.* 1984. 15, 287-298.

23. Jensen P. Observations on the maternal behaviour of free-ranging domestic pigs. *Appl. Anim. Behav. Sci.* 1986.16, 131-142.

24. Lawrence AB, Petherick JC, McLean KA, Deans LA, Chirnside J, Vaughan A, et al. The effect of environment on behaviour, plasma cortisol and prolactin in parturient sows. *Appl Anim Behav Sci.* 1995. 39:313–30. **doi: 10.1016/0168-1591(94)90165-1**

25. Jarvis S, Lawrence AB, McLean KA, Deans L, Chirnside J, Calvert SK. The effect of environment on behavioural activity, ACTH. b-endorphin and cortisol in pre-farrowing gilts. *Anim Sci.* 1997. 65:465–72. **doi: 10.1017/S1357729800008663**

36. Choi Y, Min Y, Kim Y, Jeong Y, Kim D, Kim J, et al. Effects of loose farrowing facilities on reproductive performance in primiparous sows. *J Anim Sci Technol.* 2020. 62:218–26. **doi: 10.5187/jast.2020.62.2.218**

37. Condous PC, Plush KJ, Tilbrook AJ, van Wettere WHEJ. Reducing sow confinement during farrowing and in early lactation increases piglet mortality. *J Anim Sci.* 2016. 94:3022–9. **doi: 10.2527/jas.2015-0145**

28. Nowland TL, Ernest WH, van Wettere J, Plush KJ. Allowing sows to farrow unconfined has positive implications for sow and piglet welfare. *Appl Anim Behav Sci.* 2019. 221:104872. **doi:10.1016/j.applanim.2019.104872**

29. Hales J, Moustsen VA, Devreese AM, Nielsen MBF, Hansen CF. Comparable farrowing progress in confined and loose housed hyper-prolific sows. *Livest Sci.* 2015. 171:64–72. **doi: 10.1016/j.livsci.2014.11.009**

30. Hales J, Moustsen VA, Nielsen MBF, Hansen CF. The effect of temporary confinement of hyperprolific sows in sow welfare and piglet protection pens on sow behaviour and salivary cortisol concentrations. *Appl Anim Behav Sci.* 2016. 183:19–27. **doi: 10.1016/j.applanim.2016.07.008**

31. Loftus L, Bell G, Padmore E, Atkinson S, Henworth A, Hoyle M. The effect of two different farrowing systems on sow behaviour, piglet behaviour mortality and growth. *Appl Anim Behav Sci.* 2020. 232:105102. **doi: 10.1016/j.applanim.2020.105102**

32. Chidgey KL, Morel PC, Stafford KJ, Barugh IW. Observations of sows and piglets housed in farrowing pens with temporary crating or farrowing crates on a commercial farm. *Appl Anim Behav Sci.* 2016. 176:12–8. **doi: 10.1016/j.applanim.2016.01.004**

33. Ceballos MC, Góis KCR, Parsons TD. The opening of a hinged farrowing crate improves lactating sows' welfare. *Appl Anim Behav Sci.* 2020. 230:105068. **doi: 10.1016/j.applanim.2020.105068**

34. King RL, Baxter EM, Matheson SM, Edwards SA. Temporary crate opening procedure affects immediate post-opening piglet mortality and sow behaviour. *Animal.* 2019. 13:189–97. **doi: 10.1017/S1751731118000915**

35. Baxter EM, Lawrence AB, Edwards SA. Alternative farrowing systems: design criteria for farrowing systems based on the biological needs of sow and piglets. *Animal.* 2011. 5:580–600. **doi: 10.1017/S1751731110002272**

https://doi.org/10.54846/am2024/154

Sustainability in animal agriculture

Dan Thomson, PhD, DVM

Introduction: How did we get here

Sustainability is staying in business. Plain and simple. While many may think that sustainability is about individual topics such as greenhouse gases or carbon or water or labor, the multiple key performance indicators of sustainability are about predicting who will be in business in the future. I have been fortunate to be involved with agriculture and veterinary medicine my entire life. I lived through the 80s when we lost so many farms. We witnessed the consolidation of the swine industry from small herds to large integrators in SW Iowa. While I was an undergraduate, we focused on genetic improvement and as a graduate student we had the advent of growth promoters in the beef industry.

In the 1990s, food safety, *E coli,* Jack in the Box and HACCP in packing plants was a focus but until the big Conagra recall in 2001 or 2002 we did not see pre-harvest food safety action take place in the feedlots. Also, during the early 2000s, the animal welfare era, factory farms, and animal rights groups kicked up. Animal health and disease outbreak with BSE, FMD, avian influenza, porcine epidemic diarrhea and more. Over a decade or so, the natural/organic labeling and specialized grocers increased bringing debate on technology in agriculture. GMO feed and antibiotic usage have been front and center as of late. Human resource issues and keeping rural America's mainstreets and schools open have been topics over the last 40 years. And now, carbon footprint, traceability and plant-based proteins are challenging our industries.

All of these events had an impact on sustainability of livestock systems. They made us balance the issue at hand with all others in concert to figure out how to feed the world. Constant monitoring of all variables is imperative and we cannot a let single agenda, mission or issue to define sustainability.

What is sustainability?

The dictionary says sustainability is the ability to be sustained, supported, upheld or continued. Recently, in conversations sustainability is mostly tied to ecology, the planet's health and environmental indicators. However, sustainability means different things to different people depending on where you sit. Is it the sustainability of humankind? Is it the sustainability of the people in your country? Your individual commodity industry? Is it your corporation's sustainability? Sustainability of livestock production has individual, operation, local, state, national and international definitions.

Sustainability is a complex topic. I have heard it said that for every complex problem there is a simple answer, and it is wrong. Sustainability is more than ecosystem health. The US Roundtable for Sustainable Beef includes land resources, water resources, employee safety and well-being, animal health and well-being, efficiency and yield along with air and greenhouse gas emissions.[1] Being sustainable is about identifying the key indicators of staying in business, and profitability is key.

We must eat. Agriculture is necessary. Livestock systems are necessary. So, ag sustainability is important/necessary for humankind sustainability. Sustainability is measured by outcomes such as profit, performance, mortality, green houses gasses, food security etc, but, agriculture sustainability could also be a key performance indicator for human health.

Balance and monitoring: Nothing lasts forever

"Life is like riding a bicycle. To keep your balance you must keep moving." Albert Einstein

Sustainability and industry evolution are balances. The issues never sit still and they are complex. Antibiotic usage, food safety, food security, environmental stewardship, animal health, human health and so much more must be measured constantly and kept in balance. Sustainability should not be audited as pass/fail but rather constantly tracked for continuous improvement of sustainability key performance indicators which are moving targets.

A quote from a paper from the National Academies of Science written by Mario Herrero and Phillip Thornton[2] says, "Recent global assessments have considered particular elements of livestock and livestock systems, but none addresses such systems and their considerable variations in a comprehensive, integrated way. This has led to inaccurate simplifications of the messages surrounding how to manage the livestock sector's growth in the future. The lack of a systems perspective has also curtailed explorations of more sustainable options for the sector's development. This needs to be rectified. Global change will have highly differentiated impacts on food, livelihoods, and ecosystem goods and services from livestock systems around the world."

The livestock industry must practice brutal honestly. In Wheeler's book,[3] *Understanding variation: The Key to Managing Chaos,* to change an outcome, you must change the process, distort the process or distort the data. What are the real time signals we can use to monitor our industry sustainability beyond supply and demand that help us understand where to improve to remain in business? In other words, which operations are utilizing the correct management practices with the right genetics to remain in business in the future. We can't quit learning. We can't quit improving.

The more I read about sustainability, globalization and climate change, the more I am certain that the environmental changes will have more impact on the production of livestock than livestock production will have on environmental change.

Sustainability could be a holistic view of production systems and technology adaptation. There are many examples of balancing sustainability indicators directly and indirectly related to livestock production. In turn, these indicators can be used by retail to market more beef to our consumers. Sustainability is a balance between playing defense and offense. Our industry has been too defensive minded for too long.

- Animal health and food safety
- Animal growth efficiency and animal welfare – animal housing, factory farming perception, slow growth
- Intensive agriculture and animal health – Bovine respiratory disease, liver abscess and bloat; water
- Extensive agriculture and animal health – Avian influenza, PED, prey
- Extensive agriculture and reliance on weather – drought, blizzard, etc
- AB usage and human health/animal health – antibiotic usage, antibiotic resistance
- Food safety and security – safe, wholesome, nutritious, affordable, available

Marketing of sustainability indicators

Anthony Robins as self-help guru of the 1980s said, "The two things that drive people are fear and pleasure." Today, we see human pharmaceutical companies market disease to get people to use their products ... just ask your doctor for a free trial. Some ads sell drugs from the fear of dying or being in pain and others sell pleasure of better complexion or less pain.

Activists masquerading as consumers are convincing restaurants to market their activist "fear" agendas to sell our beef, poultry, dairy and pork products. David AbiDaoud wrote on a blog[4], "Fear is an interesting emotion, which affects the thought process and reaction of individuals. Therefore, fear can be used as a unique marketing tool to make consumers loyal ... It may not be the safest tactic, but if used correctly, it can create huge impact." Restaurants have fought over a captive 4% of personal income of people in the United States. Restaurants do not feed the poor. They feed those that can afford you to plan the meal, cook the meal and do the dishes. Grocery stores feed the poor. They take food stamps or SNAP coupons. In the end, sustainability has many definitions that can fit many different marketing platforms.

The rich can afford to error on the side of safety and feel pleasure from buying organic food in saving the planet. The poor just need to eat. Retailers must have patience and use sustainability modeling prudently. Getting this right is so important. Most people literally can't afford for us to get it wrong.

Food costs play a major factor in the determination of poverty in this country. Removal or discontinued use of technology or AB or management or housing without evidence-based or outcome-based decision will have lasting effects on society beyond next quarter or next year's sales report. If food prices go up with no changes in incomes, poverty increases. Reliance on food stamps increases. The value of our tax dollar decreases...this is not just an agriculture sustainability issue, it is a societal sustainability issue. This is a human health issue.

We have forgotten how little money most people make and the decisions are being made by people that can afford almost any change. Based on data from USDA Economic Research Service[5], food insecurity in the US occurs at the same rate in urban, suburban and rural areas around 15%. The rate of food insecurity is twice as high Latinos and African Americans than in Caucasian families in the US. We must feed our people. The highest rate of food insecurity occurs in homes where a single mom is raising children (36%). Don't show up on a Saturday to box a meal, or give a meal for so many dollars spent in your store and then remove practices or technology that increases the price of food without proper due diligence.

Bill Gates was quoted to say, "If we can spend the early decades of the 21st century finding approaches that meet the needs of the poor in ways that generate profits and recognition for business, we will have found a sustainable way to reduce poverty in the world."

Conclusion

Farmers and producers should embrace sustainability as the study of how to stay in business. This is not a new concept. We have balanced all of these obstacles and opportunities since the beginning of our industries. Universities need to help our producers sustain their animal agriculture businesses.

Appropriate, honest sustainability studies are necessary for agriculture and livestock producers. Maybe there are changes a person can implement to improve their sustainability or maybe they need to change the crops they raise? Global sustainability, humankind sustainability, national sustainability, local sustainability. We have to feed people. We have to keep food affordable. Sustainability always is tied to economics. Wealth = Food; Poverty = Starvation. It is hard for starving people in poverty to worry about 100 years from now when they are worried about eating tomorrow. Likewise, people who have abundance want to make sure they keep it.

Our globe's climate is going to change. Water availability is going to change. In the future, we will be able to grow crops in different areas of the world, maybe in another world. New disease outbreaks will occur. Population centers of people are going to change. People's tastes are going to change. Therefore, where and how food is produced is going to change over time and with that so will sustainability of agricultural products. In developing countries, sustainability studies are used to determine how, what, when and where to best raise livestock to feed their people. In our developed countries, our sustainability efforts are entwined with so many political agendas because we have an overabundance of food, unbelievably sustainable food production and we can afford it. Local, national and global distribution of food is our downfall but we are getting better. Globalization is here to stay. "Our" developing countries need the livestock systems and the developed countries need to focus more on the mission making sure everyone gets fed. The proper use of sustainability modelling will tell us how to feed the world.

We must have humane leadership that makes sure we do not undo all the good that has been done for so many but look to the future for feeding the planet.

References

1. USRSB High -Priority Indicator Goals and Sector Targets. Accessed at: **https://www.usrsb.org/Media/USRSB/Docs/usrsbgoalsandtargets_supportingdocument_2022.pdf** on October 23, 2023.

2. Herrero, M. and P.K. Thornton. 2013. Livestock and global change: Emerging issues for sustainable food systems. Proc Nat Acad Sci. 110 (52) p. 20878-20881.

3. Wheeler, D. 2000. Understanding variation 2nd Edition: The key to managing chaos. By SPC Press, Inc. First published in 1993.

4. AbiDaoud, D. Why Prank – Marketing. **https://www.linkedin.com/pulse/20140622071033-5339597-why-prank-marketing/**. Accessed April 27, 2021.

5. Prevelance of food insecurity. 2014. USDA Economic Research Service **https://www.ers.usda.gov/webdocs/publications/45425/53740_err194.pdf.** Accessed April 27, 2021.

https://doi.org/10.54846/am2024/155

The veterinarian's role in drug regulation: The views of the National Pork Board

H. Fowler, VMD, MPH, PhD, DACVPM, PMP
National Pork Board, Clive, Iowa

Supporting pork producers and their ability to operate through a holistic, One Health approach

Consumers today are heavily invested in the ethical, nutritional, and sustainable aspects of the food they consume. As result, there has been greater focus put on food safety, animal welfare, and environmental sustainability activities in production agriculture. Therefore, it's important for us as members of the industry to work together to share our sustainability story with them in a way that meets their needs.

Pig farmers across the US have demonstrated a historical and ongoing commitment to the six We Care® Ethical Principles, which represent the foundation for ethical and responsible pork production in the United States. Combined these six principles help drive producers/members of the industry to work in a holistic, One Health way to provide a safe, sustainable product to their consumers (see Figure 1). Furthermore, the Pork Quality Assurance® Plus (PQA Plus) program ensures the implementation of these principles by translating them into good production practices which are readily incorporated into training materials. To date, more than 71,000 members of the industry including farmers, farm employees, veterinarians, etc, have received and/or maintained a certification in this program, firmly grounding them in both the One Health approach and responsible and ethical pork production.

In addition to training and certification programs, the National Pork Board and the industry we represent demonstrates its continued dedication to the One Health approach via the implementation of the We Care® Ethical Principles through our Sustainability efforts. Deliverables from these efforts are shared at numerous venues with key partners and stakeholders throughout the industry. Though as a checkoff organization The National Pork Board is prohibited from influencing government policy; we remain engaged with partners in both the public and private sectors to ensure we are taking an inclusive approach to our work. Funding research, hosting farm tours, and participating in national conferences such as the annual AASV convention allows us to share our story to diverse audiences. During our presentation we will provide an overview of these activities as well as outline how we apply the One Health approach via the We Care® Ethical Principles to everything we do in the swine industry.

Figure 1: We Care® Ethical Principles and the basic One Health framework.

References

1. We Care Ethical Principles. **https://www.porkcares.org/about-we-care/**

2. One Health Definition. OHHLEP. **https://www.who.int/news/item/01-12-2021-tripartite-and-unep-support-ohhlep-s-definition-of-one-health**

3. Pork Quality Assurance® Plus. Pork Checkoff. 2018 ©National Pork Board, Des Moines, USA. **https://www.pork.org/certifications/pork-quality-assurance-plus/**

https://doi.org/10.54846/am2024/155.5

The veterinarian's role in drug regulation: The views of the National Pork Producers Council

A. Johnson, DVM
National Pork Producers Council, Des Moines, Iowa

Timelines and future trends

Public policy decisions constantly affect how pigs can be raised and the abilities of veterinarians to practice. In the past few years, new guidance documents and regulations have changed the way swine veterinarians practice and serve our pork producers. A recent example was the establishment of the Veterinarian Feed Directive (VFD), a regulation that requires the supervision of a licensed veterinarian to issue all animal drugs intended for use in or on animal feed within the context of a veterinarian-client-patient-relationship (VCPR). This final rule changed how medication in swine feed was allowed to be used and required extensive veterinary oversight. This is an example of how regulation can dramatically affect both swine production and veterinarians' roles.

Now, the questions become: What is the regulatory process, and how can veterinarians get involved in these rulemaking processes?

It is important to understand the difference between guidance documents and regulation. Agencies typically create policy with guidance documents, which are non-binding, essentially voluntary, and reflect the agency's policy. On the other hand, agencies can issue regulations to implement laws for state and local governments, industry, other organizations, and the public. Though not laws, regulations have the weight of law. Agencies are given their authority to issue regulations from laws enacted by Congress. When agencies issue regulations, they must follow an open public process.

The open public process starts when agencies publish an annual Regulatory Plan in the fall and an Agenda of Regulatory and Deregulatory Actions in the spring and fall. Combined, this Unified Agenda is used to publicly announce pending and completed regulatory actions. Unified agendas and proposed rules for comment are published on **Regulations.gov** and **RegInfo.gov**.

Before proposing a rule, an agency may publish an Advance Notice of Proposed Rulemaking to invite participation from the public to help the agency shape a proposed rule. The agency may also invite stakeholders to discussion meetings to reach a consensus on a proposed rule, which then becomes an official document that announces and explains the agency's plan to address a problem or accomplish a goal. The proposed rule must be published in the Federal Register to notify the public and give time to comment. Public comment periods can vary but generally last at least 60 days. There can also be submissions to request an extension on comment periods. The agency will then review comments and may revise the document, then issue a final rule.

The current landscape of regulations and guidance documents centers around the principles of One Health. Antimicrobial use and resistance, food safety, and nutrition have been major focus areas, and we expect to see continued emphasis. There is a global push to reduce antibiotic use and resistance. By doing so, the US government wants to understand how antibiotics are used in animal agriculture. Programs that track antibiotic use are continuing to be explored. Policy on antimicrobials in animal agriculture is imminent and could once again affect veterinarians and their customers. Veterinarians are encouraged to understand the guidance documents and regulations and to submit comments. Working within AASV committees on these issues can help drive discussion on these topics to submit comments that will have a meaningful impact on proposed rules and guidance documents.

A call to action

Today's consumers are not only concerned about the health benefits of the food they consume but also the social, environmental, and sustainable impacts of their food choices. As such, it is important for us as veterinarians to not only remain engaged and responsive to consumer needs, but also work to continuously improve in the space of ethical and responsible pork production. As members of the swine industry, veterinarians play an integral role in shaping the future of the industry and ensuring its success. By remaining engaged in national and global conversations, taking an active role in association activities and where feasible submitting comments during an open comment period we can shape the future of this great industry.

References and resources

1. A Guide to the Rulemaking Process. **The Rulemaking Process (federalregister.gov)**

2. Guidance & Regulations: Laws, Regulations, and Guidances; FDA's Center for Veterinary Medicine.

https://doi.org/10.54846/am2024/156

Vaccine platforms: What's currently available?

A. Gill, DVM, PhD
Center for Veterinary Biologics, Ames, Iowa

Industry needs

The Center for Veterinary Biologics (CVB) develops policy to meet challenges of the future. There is urgent need to bring quality, customized products to market quickly to tackle highly mutable agents, emerging diseases, alternatives to antibiotics, and novel technologies.

Vaccine options

The United States Department of Agriculture, Animal and Plant Health Inspection Service, CVB program implements the provisions of the Virus-Serum-Toxin Act (VSTA) to ensure that veterinary biologics are available for the diagnosis, prevention, and treatment of animal diseases and are not worthless, dangerous, contaminated, or harmful. The VSTA mandates that the CVB develop rules and regulations concerning the production and licensing of veterinary biologics, and issue licenses to ensure the veterinary biologics are pure, safe, potent, and effective. The production and sale of veterinary biologic products in the United States, District of Columbia, and United States Territories is regulated by rules codified in the US Code of Federal Regulations, Title 9, Chapter I, Subchapter E, Parts 101 to 124. Guidance documents clarifying the regulations are published in the form of Veterinary Services Memoranda and Center for Veterinary Biologics Public Notices.

CVB grants different types of licenses for veterinary biologics: Fully licensed, conditionally licensed, and autogenous vaccines. In 2018, the CVB published Veterinary Services Memorandum (VSM) 800.213 (initially published in 2015) providing revised guidelines on the production of platform-based, non-replicating, non-viable products. Once licensed, such products offer the flexibility of allowing the gene(s) of interest (GOI) to be swapped out of the platform in an expedited process which may allow for a more rapid response to a (re-)emerging disease situation or other animal health need using an updated vaccine. Upon licensure of the initial product platform, additional flexibility is allowed for GOI swaps with a valid veterinary-client-patient relationship per VSM 800.214 (published in 2018).

Challenges

With advances in technology that allow genetic manipulation of an organism's genome, gene deleted vaccines and vaccines in which a gene(s) is/are inserted into a vector or manipulated into an alternative delivery vehicle have been developed. Even with expedited studies and licensure, it can take several years to bring a product to market. Emerging or re-emerging diseases are a challenge for the producer and veterinarian, especially those that may rapidly change in response to selection pressures such as influenza virus in swine.

Summary

Producers and veterinarians should understand the advantages and limitations of the types of vaccines in order to make informed decisions about vaccination programs.

References

1. Information regarding veterinary biologics may be found on the APHIS website at this link: **https://www.aphis.usda.gov/aphis/ourfocus/animalhealth/veterinary-biologics**

2. Product summaries may be found on the APHIS website at this link: **productdata.aphis.usda.gov**

3. USDA APHIS NVAP Training Module 32: *Regulating Vaccines and Related Biologics* may be found on the APHIS website at this link: **NVAP Training Module 32**

https://doi.org/10.54846/am2024/157

Commercial vaccine development and considerations of vaccine use under field conditions

M. Roof, PhD
Iowa State University, Ames, Iowa

Introduction

Commercial vaccines manufactured and sold within the USA are regulated by APHIS and the Center for Veterinary Biologics (CVB). The focus of the CVB regulatory process is on purity, potency, safety, and efficacy. During vaccine research and development, it is common practice to conduct in vivo testing of vaccines in clinically health and naive pigs. All products licensed must have a defined minimum immunizing dose (MID) or an efficacious dose which is produced according to a CVB approved outline of production. This MID then becomes the baseline that all future serials must meet or exceed in terms of relative antigen levels in the product defined by CVB as potency or relative potency (RP). This also impacts formulation for product release to ensure potency at the end of the products labeled shelf life and dosages used during product safety testing (maximum inclusion in a product). It is important to note that all studies done must use the same age, dose and route of vaccine administration to support the claims on a CVB approved label.

Practical considerations

Testing and development of vaccines in naive and healthy animals is valid from a scientific view but does not always reflect the situation in the field where veterinarians faced multi-factorial challenges. Some important considerations for field use include:

- Vaccine timing – it is important to know when herds become exposed to the pathogen of interest and then place vaccine prior to this exposure. For single dose vaccine general recommendations are to vaccinate 3 - 4 weeks in advance and for a 2-dose vaccine start immunization 6 - 8 weeks in advance of pathogen exposure.
- Vaccination of positive animals – under field conditions it is often difficult if not impossible to vaccinate a naive or negative pig (examples PRRS or Salmonella) which may impact vaccine efficacy or perceived clinical benefit.
- Vaccination in the face of maternal antibody – as noted above vaccine needs to be applied prior to infection and in some cases, this requires vaccination in the presence of maternal antibody which can reduce or slow the vaccine induced immune response.
- Vaccination in the face of co-infections – under field conditions it is common to have pigs exposed to a variety of pathogens concurrently (PRDC) in a disease complex and so vaccination for some but not all relevant pathogens may not fully protect or resolve clinical issues.
- Off-label use may be done but has potential negative implications relative to vaccine safety and efficacy
    - Partial dosing – if practiced may lead to immunization of animals with an antigen level or potency below the MID or efficacious dose. This may lead to no efficacy and perceived vaccine failure.
    - Mixing of non-licensed vaccine fractions – mixing of products not specifically approved by CVB as a combination has risk of host safety (ranging from injection sites to death), adjuvant incompatibility, viricidal activity (live viral vaccines such as PRRS), and antigen interference.
    - Application outside of the label defined age, dose, and route – there are a variety of potential consequences that range from safety (especially in young animals) and lack of efficacy.

Discussion

Vaccines manufactured and sold under CVB requirements provide a solid basis for the industry to have safe and efficacious vaccines using valid scientific methods and study design. Despite this, veterinarians must make conscious decisions how to best apply vaccines under difficult environmental and complex disease situations. It is a best practice to use vaccines as recommended on the CVB approved label for best chances of the desired clinical impact. If or when a vaccine is considered for application off-label it would be wise to review the single tier label data[1] required for all licensed products so that the data from the product can be reviewed prior to use. The CVB provides an electronic web-based location to review all product summaries which show the specific study design and data from critical efficacy studies.[2] Finally, dose-cutting, use by a non-approved route, and incompatible product mixing carry both safety risk and lack of efficacy that may not justify the risk-reward of the effort.

References

1. https://www.federalregister.gov/documents/2014/04/21/2014-08995/viruses-serums-toxins-and-analogous-products-single-label-claim-for-veterinary-biological-products

2. https://www.aphis.usda.gov/aphis/ourfocus/animalhealth/veterinary-biologics/product-summaries

https://doi.org/10.54846/am2024/158

Vaccines and maternal antibody in swine: A brief overview

P. Piñeyro
Department of Veterinary Diagnostic and Production Animal Medicine, Iowa State University, Ames, Iowa

Introduction

Vaccination plays a crucial role in swine health management, aiming to prevent and control diseases that can significantly impact herd productivity. Maternal antibody transfer in swine is a critical aspect of early-life immunity, ensuring piglets receive a degree of protection against pathogens in their environment during the initial stages of life. This process is predominantly facilitated through the transfer of maternal antibodies from sows to piglets via colostrum, the nutrient-rich first milk produced by the sow shortly before and after farrowing.

Colostrum composition

Colostrum is rich in immunoglobulins, particularly IgG, IgA, and IgM, which serve as the primary carriers of maternal antibodies. The concentration of maternal antibodies in colostrum is highest during the first few hours after farrowing, emphasizing the importance of ensuring piglets consume colostrum promptly. The nutritional richness of colostrum extends beyond immunoglobulins. It contains elevated levels of essential nutrients, including proteins, fats, and carbohydrates, surpassing those found in regular milk. This nutrient-dense composition is crucial for supporting the rapid growth and development of neonatal piglets. Colostrum also serves as a reservoir of growth factors and bioactive compounds, contributing to the overall vitality of piglets. The dynamic synergy of these elements underscores the status of colostrum as a nutritional source, laying the foundation for a strong resilient immune system and optimal growth in swine.

Maternal antibodies transfer and duration

Maternal antibody transfer in swine is a pivotal aspect of early immune development by sow-derived antibodies transmitted to piglets via colostrum. The primary immunoglobulins involved in this transfer includes IgG, IgA, and IgM, collectively forming a protective layer of defense against a spectrum of pathogens. The timing of colostrum intake is crucial, as the concentration of these antibodies is highest in the initial hours after farrowing, in addition the absorption of antibodies primarily occurs during the first 24 to 48 hours of life, emphasizing the importance of prompt colostrum consumption by piglets for optimal passive immunity. The mechanism of antibody absorption involves the specialized absorptive cells lining the gut, particularly the small intestine "gut-associated lymphoid tissue" (GALT), where the absorption of antibodies takes place. Immunoglobulins, interact with specific receptors on the surface of these absorptive cells. These receptors facilitate the transfer of antibodies from the colostrum into the bloodstream, providing the piglets with passive immunity against a range of pathogens. This maternal antibody transfer provides piglets with a vital early defense mechanism, offering protection during the critical period before they can mount their own active immune responses through vaccination or exposure to pathogens in the environment. The duration of maternal antibody protection in swine varies based on factors such as antibody type and individual piglet characteristics. While piglets benefit from passive immunity for several weeks to a couple of months, the eventual decline necessitates a timely transition to active immunity through vaccination. Strategic management practices, including careful consideration of vaccination timing, are necessary to optimize the balance between the waning of maternal antibody protection and the onset of piglet-generated immune responses.

Challenges in maternal antibody protection

Despite the essential role that maternal antibodies play in conferring early immunity, there are inherent challenges associated with their interaction with vaccines. Maternal antibodies, particularly immunoglobulins such as IgG, can neutralize vaccine antigens, posing a potential impediment to the piglet's ability to mount a robust immune response. This interference requires a balance in vaccination timing, aiming to administer vaccines when maternal antibody levels have sufficiently waned, yet before piglets become susceptible to infections. The dynamic nature of maternal antibody decay adds complexity to this challenge. The duration of maternal antibody protection varies among piglets within a litter, influenced by factors such as colostrum intake, birth order, and overall piglet vitality. Additionally, the specific pathogens targeted by maternal antibodies can influence their neutralization potential. The interference caused by maternally transferred immunity (MDI) against any vaccine can be characterized as the obstruction of the antigen due to the presence of circulating antibodies or other components, possibly cell-mediated, acquired through colostrum. This phenomenon is well-documented in the context of various swine infectious diseases, including Aujeszky disease virus (ADV), classical swine fever virus, influenza A virus (IAV), and swine parvovirus. For instance, pigs with notable IAV-specific maternally derived antibody (MDA) titers may experience a suppression of adaptive antibody responses during homologous infection or vaccination. Both IAV and ADV show evidence suggesting that the cellular immune response is relatively less affected by MDA. Successful navigation of these challenges requires a good understanding of the swine herd's health status, coupled with good vaccination planning. This intricate approach is vital for optimizing the effectiveness of vaccines and ensuring a seamless transition from maternal antibody protection to independently acquired immunity in swine.

Vaccination strategies

Developing effective vaccination strategies in swine involves integration of timing and vaccine selection to maximize the benefits of maternal immunity. Timely vaccination is crucial to optimize the balance between the waning of maternal antibodies and the piglet's ability to mount its own active immune response. Careful consideration should be given to the specific pathogens prevalent in the swine environment and the optimal window for vaccination to ensure that piglets receive timely protection. Strategic planning, aligned with the sow's reproductive cycle, enables the delivery of vaccines at key stages, such as pre-farrowing and pre-weaning, to boost maternal antibody transfer and subsequent piglet immunity. The selection of appropriate vaccines is equally vital in enhancing maternal immunity. Tailoring vaccines to the specific health challenges faced by the swine herd is essential. For example, vaccines targeting prevalent viruses such as porcine circovirus type 2 (PCV2) or influenza A virus (IAV) can significantly contribute to maternal antibody protection and overall herd health. A comprehensive understanding of the swine herd's health status, together with regular monitoring, allows for the adjustment of vaccination strategies to address emerging threats and improve the efficacy of maternal immunity transfer to piglets.

Examples of vaccines used in swine

Porcine circovirus type 2 (PCV2) vaccines

PCV2 vaccines are essential for preventing diseases associated with this virus, such as porcine circovirus-associated disease (PCVAD). Maternal vaccination against PCV2 helps transfer protective antibodies to piglets through colostrum, providing a foundation for their immune defense. Research in this area has shown that PCV2 vaccine efficacy is influenced more by vaccination age than MDA levels. The research suggested that PCV2 vaccination in the presence of high MDA levels is effective in 3-week-old pigs but not in 1-week-old pigs, indicating that vaccine efficacy is independent of MDA levels. However, other studies suggested that while interference at very high MDA titers cannot ruled out, its impact under field conditions is likely negligible. Thus, an optimal PCV2 vaccination strategy that balances passive immunity levels, management practices, and the timing of infection is imperative.

Porcine reproductive and respiratory syndrome virus (PRRSV) vaccines

PRRSV maternal immunity can be accomplished by natural exposure or a wide vaccination program by MLV on the sow herd. Typically, maternal antibody protection lasts for several weeks to a few months. However, the waning of maternal antibodies emphasizes the need for timely vaccination to ensure a smooth transition to actively acquired immunity in piglets. Thus, MDAs, specifically maternally derived neutralizing antibodies (MDNAs), play a crucial role in protecting suckling piglets from PRRSV infection during their initial weeks and preventing viremia in weaned piglets. Traditionally has been reported that high MDA levels has not impact on vaccine efficacy. However, has been demonstrated that MDNAs have a detrimental effect on PRRSV vaccination in piglets using MLV vaccine at three weeks of age. PRRSV antibodies and IFNγ-secreting cells can be suppressed for four weeks post-vaccination in piglets exhibiting high MDNA levels, reducing protection against PRRSV infection in piglets vaccinated in the presence of elevated MDNA levels. Currently, farms are working toward achieving sow stable herd status characterized by the absence of antibodies and viremia in sows. This strategy can facilitate MDA interference when pigs are vaccinated post-weaning.

Swine influenza A virus (swIAV) vaccines

Sow vaccination represents the primary strategy for managing influenza in breed and grower-finisher pigs. This strategy helps protect sows and gilts from severe clinical disease and enhances the transfer of MDA through colostrum. The role of MDA against IAV is controversial due to viral and host factors that can affect the efficacy of vaccines in the grower-finisher phase. MDAs can play a crucial role in protecting piglets from clinical disease after weaning, mostly in a strain-specific manner, in addition to reducing the transmission of swIAV in weaned pigs. However, elevated MDA levels have been linked with interference with early piglet vaccination, leading to insufficient vaccine-induced humoral response, resulting in more severe clinical signs. Several studies reported the duration of MDA is approximately ten weeks of age, with an average decline lasting 7 to 8.5 weeks. However, experimental and field studies suggested that MDA in piglets from vaccinated sows can last up to 2 - 4 months. Despite the widespread use of IAV vaccines, swIAV continues circulating in pigs, possibly due to challenges in generating adequate immunity against the various co-circulating strains. Furthermore, the complex dynamics of IAV infection in pigs, involving recurrent infections, multiple waves of infection, co-circulation of genetically distinct viruses, and frequent reassortment events observed in growing-finishing pigs, make protection through MDA a complex task. The effect of MDA on weaned pigs and its effect on vaccine efficacy is commonly evaluated from the immunological standpoint based on the presence of protective antibodies on pigs and the effect of viral shedding and risk of infection. Notably, there is a lack of information on the potential impact of MDA interference on vaccine efficacy regarding production parameters, average daily gain (ADG), feed conversion, or coughing incidence.

Mycoplasma hyopneumoniae (Mhp) vaccines

The prevalence and severity of pneumonic lesions associated with Mhp infection have been associated with infection rate at weaning. Therefore, one of the strategies to reduce the effects of Mhp infection is to reduce the vertical transmission and induce strong colostrum-derived humoral and cellular immunity in the piglets by using sow vaccination. Although Mhp is endemic and sow can be re-exposed to the pathogen, levels of MDA may vary based on the management strategies, infectious dose, and pathogenicity of strains circulating in the farm. Although sow vaccination is not commonly included in their vaccination program, it has been reported that at 3 - 4 weeks of age, a considerably higher proportion of piglets from vaccinated sows have higher detectable antibody levels than those from non-vaccinated sows. Amongst gilt acclimatization programs, vaccines are still an important approach to reducing vertical transmission from gilts to their offspring. Thus, vaccination schedules of multiple doses during gilt acclimatization have probed to reduce shedding, and piglets have higher humoral immunity at weaning.

Enterotoxigenic *Escherichia coli* (ETEC) vaccines

Lactogenic immunity acquired through sow vaccination serves as a preventive measure against neonatal enterotoxigenic *Escherichia coli* (ETEC) infections. Various maternal vaccines, employing different approaches such as bacterins, subunit antigens (fimbriae, toxoids), and in some instances, multivalent vaccines targeting common diarrhea-causing pathogens, are available on the market. The shared objective of these vaccines is to stimulate the production of specific antibodies against key adhesion factors and toxins of ETEC strains in the colostrum and milk of sows, ultimately preventing piglet mortality. Given that ETEC infections manifest as non-invasive gastrointestinal issues, the emphasis lies on mucosal, ie, lactogenic immunity, rather than colostral, ie, systemic immunity, to combat the disease. Consequently, the presence of elevated levels of IgA in the milk of vaccinated or environmentally-exposed sows can impede the colonization of the small intestine by ETEC, averting infection. Notably, the success of maternal vaccines, particularly in gilts, centers significantly on these animals' prior mucosal exposure to ETEC as the vaccines are administered parenterally.

Conclusions

In conclusion, a well-managed vaccination program that considers the dynamics of maternal antibodies is vital for maintaining swine health and optimizing productivity. It requires a thoughtful approach to timing and the selection of appropriate vaccines tailored to the specific challenges faced by the swine herd. The vaccine examples presented here showcased the diverse arsenal of vaccines employed in swine production, highlighting the importance of maternal vaccination to confer early immunity against a range of diseases.

References

1. Bandrick, M., Theis, K. & Molitor, T. W. Maternal immunity enhances Mycoplasma hyopneumoniae vaccination induced cell-mediated immune responses in piglets. BMC Veterinary Research 10, 124 (2014). **https://doi.org/10.1186/1746-6148-10-124**

2. Cador, C. et al. Maternally-derived antibodies do not prevent transmission of swine influenza A virus between pigs. Veterinary Research 47, 86 (2016). **https://doi.org/10.1186/s13567-016-0365-6**

3. Chamba Pardo, F. O. et al. Effect of strain-specific maternally-derived antibodies on influenza A virus infection dynamics in nursery pigs. PLOS ONE 14, e0210700 (2019). **https://doi.org/10.1371/journal.pone.0210700**

4. Figueras-Gourgues, S. et al. Effect of Porcine circovirus 2 (PCV-2) maternally derived antibodies on performance and PCV-2 viremia in vaccinated piglets under field conditions. Porcine Health Manag 5, 21 (2019). **https://doi.org/10.1186/s40813-019-0128-7**

5. Keay, S. et al. Does Vaccine-Induced Maternally-Derived Immunity Protect Swine Offspring against Influenza a Viruses? A Systematic Review and Meta-Analysis of Challenge Trials from 1990 to May 2021. Animals 13, 3085 (2023).

6. Martin-Valls, G. E. et al. Levels of neutralizing antibodies against resident farm strain or vaccine strain are not indicators of protection against PRRSV-1 vertical transmission under farm conditions. BMC Veterinary Research 19, 217 (2023). **https://doi.org/10.1186/s12917-023-03785-z**

7. Martínez-Boixaderas, N., Garza-Moreno, L., Sibila, M. & Segalés, J. Impact of maternally derived immunity on immune responses elicited by piglet early vaccination against the most common pathogens involved in porcine respiratory disease complex. Porcine Health Management 8, 11 (2022). **https://doi.org/10.1186/s40813-022-00252-3**

8. Matías, J., Berzosa, M., Pastor, Y., Irache, J. M. & Gamazo, C. Maternal Vaccination. Immunization of Sows during Pregnancy against ETEC Infections. Vaccines 5, 48 (2017).

9. Quesnel, H., Farmer, C. & Devillers, N. Colostrum intake: Influence on piglet performance and factors of variation. Livestock Science 146, 105-114 (2012). **https://doi.org/10.1016/j.livsci.2012.03.010**

10. Renson, P. et al. Maternally-derived neutralizing antibodies reduce vaccine efficacy against porcine reproductive and respiratory syndrome virus infection. Vaccine 37, 4318-4324 (2019). **https://doi.org/10.1016/j.vaccine.2019.06.045**

11. Ryt-Hansen, P. et al. Longitudinal field studies reveal early infection and persistence of influenza A virus in piglets despite the presence of maternally derived antibodies. Veterinary Research 50, 36 (2019). **https://doi.org/10.1186/s13567-019-0655-x**

12. Sibila, M. et al. Effect of sow vaccination against Mycoplasma hyopneumoniae on sow and piglet colonization and seroconversion, and pig lung lesions at slaughter. Veterinary Microbiology 127, 165-170 (2008). **https://doi.org/10.1016/j.vetmic.2007.07.027**

https://doi.org/10.54846/am2024/159

Vaccine immunology and expectations

M. C. Rahe
Department of Population Health and Pathobiology, North Carolina State University, Raleigh, North Carolina

Summary

The immune system is composed of an incredibly complex web of cells and signals that collectively has one common goal – maintain homeostasis largely by combating infectious diseases. To make sense of this complexity, we typically break up the immune system into innate and adaptive immunity. The innate immune response is the first arm of the immune response to react to infection. This non-specific response does its best to reduce infection and clear microbes from the body to prevent tissue damage and resulting disease. The adaptive immune response takes longer to respond to infection than the innate response and is largely composed of lymphocytes and the effector proteins that they produce. The lymphocytes within the adaptive immunity response can be broken down to cell-mediated immunity driven by T cells and humoral immunity driven by B cells and effected through the antibodies that they produce.

T cells are further divided into helper T cells and cytotoxic T cells. Helper T cells are classically thought of as cytokine producing cells which are important for the activation of other white blood cells largely through the cytokines that they secrete. Interferon gamma is the most recognizable cytokine secreted by these cells and has been shown to be important in limiting PRRSV infection.[1] Cytotoxic lymphocytes are significant due to their ability to kill infected cells. Through this, they limit the spread of infection and do this in such a way that does not induce further infection or inflammation. While we tend to compartmentalize T helpers and cytotoxic lymphocytes into their specific functions, it is increasingly recognized that there is functional overlap between these two cell types. For example, some T helper cells are also able to kill cells infected by viruses and cytotoxic lymphocytes may also secrete cytokines.

The B cell response to infection is driven by T helper cells and culminates in the production of antibodies. Although, the antibody isotype that is generated largely depends on where infection or inoculation occurred. Generally, IgG is important for systemic immunity and is generated through subcutaneous (SQ) or intramuscular (IM) vaccination. Whereas IgA protects mucosal surfaces, such as the gastrointestinal tract or upper respiratory tract, and is generated when infection occurs at these sites. Understanding this difference is incredibly important when considering: Does the route of inoculation produce the right type of antibody for the where this pathogen infects? For example, the generation of immunity to PEDV via IM injections may generate a great IgG neutralizing titer against the spike protein, but this will not protect the GI. Immunity against this GI specific pathogen must come from inoculation/infection of the mucosa and development of a neutralizing IgA titer within the gut. Additionally, both cell-mediated and humoral immunity produce a memory response which allows the immune response to rapidly ramp up immunity to infection if the same pathogen comes back.

The adaptive immune response can provide robust protection against infection, and it is the goal of vaccination to induce a strong adaptive immune response against a particular pathogen. Ideally this would result in sterilizing immunity which prevents infection by the pathogen of interest. However, sterilizing immunity is likely impossible to generate/prove and instead the focus of vaccination should be on protective immunity which doesn't prevent infection but will clear the infection before clinical disease occurs. This is achievable for vaccines against many pathogens.

Vaccines against rapidly mutating RNA viruses, such as PRRSV, may only provide partially protective immunity. This means that a previously exposed/vaccinated animal gets infected and displays clinical disease, but it is not as bad as if they hadn't been previously vaccinated or infected. Where this partial protection comes from likely depends on the pathogen. Generally, a strong neutralizing antibody response against any virus is going to result in protective immunity, as long as it is in the right anatomic location, see the PEDV example above. The germinal center affinity maturation that B cells go through tends to make them very specific for certain antigens with little latitude to respond to variation in that antigen. For PRRSV, it has been noted that homologous neutralizing antibodies can be protective but aren't typically associated with heterologous immunity. T cells do not go through germinal center affinity maturation that B cells do, and because of this, they don't have the exquisite specificity of B cells. Furthermore, T cells can be functional against viral non-structural proteins which are much less likely to mutate than surface glycoproteins on variant viruses. The superior cross reactivity of T cells has been shown in PRRSV as well as with coronaviruses such as COVID.[2,3]

In conclusion, with vaccination it is important to have a basic understanding of the immune response as well as the mutability and diversity of the pathogen that you are vaccinating against. This will help inform the type of immunity that you are trying to generate. Where that immunity should be generated (route of inoculation) is indicated by the typical route of infection of the pathogen. There are other variables which must be considered when considering the expectation for success for your vaccination program. While this is not an exhaustive list, pathogen pressure, co-morbidities, vaccine compliance, environment (ex. too cold), and stress may all play a role in the effectiveness of a vaccine. As Terrell Owens once said, "If you align expectations with reality, you will never be disappointed."

References

1. Rowland RR, Robinson B, Stefanick J, et al. Inhibition of porcine reproductive and respiratory syndrome virus by interferon-gamma and recovery of virus replication with 2-aminopurine. *Arch Virol.* 2001;146: 539-555.

2. Chung CJ, Cha SH, Grimm AL, et al. Recognition of Highly Diverse Type-1 and -2 Porcine Reproductive and Respiratory Syndrome Viruses (PRRSVs) by T-Lymphocytes Induced in Pigs after Experimental Infection with a Type-2 PRRSV Strain. *PLoS One.* 2016;11: e0165450.

3. Sette A, Sidney J, Crotty S. T Cell Responses to SARS-CoV-2. *Annu Rev Immunol.* 2023;41: 343-373.

https://doi.org/10.54846/am2024/160

Development of mRNA vaccines for production animals

Davis Verhoeven; David Verhoeven, PhD
Department of Veterinary Medicine and Preventive Medicine, Iowa State University, Ames, Iowa

Introduction

The history of mRNA vaccines dates back to the early 1990s, but it was the global COVID-19 crisis that catapulted these vaccines into the spotlight.[1] However, the use of these vaccines was only possible after the advent and use of pseudouridine[2] made it possible to get protein translation with minimal intrinsic antiviral responses within the transfected cell.

Although not yet licensed, there are several potential advantages to using mRNA vaccines in production animals in the future. First, the potential to rapidly develop these vaccines for new pathogens that arise or cross transboundaries is significant.[3] As evidenced by the COVID-19 vaccines, the development of mRNA vaccines for emerging viruses can be done in a few days rather than the months to years typical of traditional vaccines.[4] Licensure for an mRNA platform rather than individual mRNA vaccines could allow for quick use on farms in response to new outbreaks. Once an immunogen, say a viral receptor on a new virus, is identified and sequenced, then quick DNA synthesis allows for the manufacture of a template to generate mRNA. This approach also allows some biotechs like Genvax and Merck to go to individual farms, sequence all the pathogens on the farm, and develop RNA vaccines targeting each virus found. Thus, rather than using an off-the-shelf flu vaccine that may not fully match the strain circulating on an individual farm, an exact base pair for base pair match can be made and deployed. Although traditionally targeting cancer and viruses, new studies have also shown efficacy against bacterial pathogens.[5] Thus, mRNA against bacteria likely begin to be added to clinical studies against the myriad of bacteria that infect production animals.

Although human COVID mRNA vaccines initially showed significantly higher efficacy over other vaccine approaches,[6] there is no direct evidence that mRNA vaccines in production animals outperform traditional killed, attenuated, or subunit veterinary vaccines. However, there is some evidence that immune responses to mRNA vaccines are slightly more broad than other vaccine formats. For example, mRNA flu vaccines may elicit more antibodies that cross-react with different strains of flu than killed vaccines. In contrast, killed vaccines generally induce very strain-dependent neutralizing antibodies rather than heterosubtypic immunity.[7] The biggest factors that have limited mRNA vaccines are their cost of manufacture and significant technological hurdles in the manufacturing process. The encapsulation efficiency of mRNA into lipid nanoparticles and the transfection efficiency of those particles can be low and thus require higher amounts of RNA. Unfortunately, the raw components and kits needed to make the RNA are expensive, and the entry into RNA vaccine manufacturing is difficult because optimized and consistent protocols, such as in vitro transcription (IVT), require significant time and monetary investment to develop.

Presently, there are two types of RNA-based vaccines that have made the most progress in either animal trials or advancing to licensure. The first is the traditional mRNA format, which is similar to what Moderna and Pfizer developed.[8,9] These have not yet made it to licensure in production animals. The second is a self-replicating RNA construct[9] that, although considered an mRNA vaccine in the human vaccine space, is licensed as an RNA particle for use primarily in swine.[10,11] Thus, these are not licensed as mRNA vaccines. These vaccines, branded Sequivity and made by Merck, have traditionally been used as prescription vaccines, although Merck has recently licensed a broader reactive influenza version for swine.

In this review, we will cover how mRNA vaccines are manufactured, how mRNA carriers are used, and explore some of the preclinical mRNA and licensed RNA particle efficacies that have been published thus far. Given the strong interest in mRNA vaccine approaches, we will likely see increasingly more of these types of vaccines headed to clinical trials and licensure and, eventually, the marketplace.

RNA development process

Traditional mRNA vaccines are synthesized off DNA plasmids in IVT reactions employing standard ribonucleic triphosphates (rNTPs), modified or alternatives to rNTPS (often pseudouracil, N1-methylpseudouridine, or rarely, methylated cytosine), a 5' cap analog (added during IVT or by capping enzymes after IVT), and a poly(A) tail added after the reaction.[12,13] Generally, the IVT reaction consumes the T7 polymerase that catalyzes the RNA synthesis of a linearized plasmid. The T7 polymerase is often a significant problem because: **(a)** the half-life of the enzyme is only about 15 minutes at 37 degrees; **(b)** it is prone to falling off the DNA plasmid early, leading to short spurious RNA products; and **(c)** it may turn around at the end of the plasmid, making double-stranded RNA (dsRNA). Advances in manufacturing new T7 polymerases would increase yields and potentially reduce costs. The synthesized RNA also often contains 5' and 3' untranslated regions (UTR) flanking the coding region. These help to correctly position the ribosomal machinery on the transcript after the ribosome binds to the 5' methyl cap or cap analog.[14] The type of UTR can have profound effects on translation efficiency.[15] Finally, the poly(A) tail directly influences RNA half-life and translation efficiency, with longer tails of 100-150bp being ideal.[16] Polyadenylation occurs after RNA synthesis using yet another enzyme that adds additional costs.

The design of the open reading frame that codes for the desired protein can also be critical.[17] Removal of immunostimulatory sequences (tracts of G/C nucleotides) and codon optimization are essential during plasmid design and construction. Moreover, codon usage should mirror the most common motifs for species for which the vaccine is designed to enhance the level of translation. However, caution must be taken as abundant codon usage, replacing codons naturally encoded by a pathogen, can affect protein folding and change the type of antibodies the immune system makes. Therefore, there can be a trade-off between translational efficiency and the optimization of the types of antibodies induced.

After the IVT process, but before the addition of the poly(A) tail or methyl cap (barring cotranscriptional capping), the RNA needs to be treated with DNase to cleave the plasmid template.

Additionally, alkaline phosphate is often used to remove tri-phosphate bridges on mRNAs that fails to incorporate a 5' cap during the IVT process, as these would set off antiviral sensors such as RIG-I.[18]

The IVT process often leads to many short RNA products and dsRNA species. These must be removed from the mRNA as they are very immunogenic, with many components of the antiviral system in cells designed to find and suppress dsRNAs. To suppress dsRNAs, ribosomal machinery is sequestered away from all cytosolic transcripts, Thus, dsRNA and single-stranded mRNA are not translated if the cell detects dsRNA contamination in the mRNA vaccine. Clean-up of dsRNAs is generally achieved by cellulose binding for small amounts of RNA or chromatography for higher amounts. The DNase step of production can be skipped if chromatography is used, as DNA and dsRNA both eluted from the chromatography column well before single-stranded mRNAs.

The next generation of mRNA vaccines being investigated are circular RNAs.[19] These closed circular RNAs have several advantages over linear mRNAs. For linear mRNA, the exposed ends are often attacked by the cell's exonucleases and endonucleases. These attack the poly(A) tail, then de-cap the RNA, then cleave the body of the RNA. Because circular RNAs lack these exposed ends, the inclusion of 5' methyl caps or poly(A)s is not needed, and nucleases have a difficult time cleaving the RNA. This leads to cost savings and stability within cells. However, strong internal ribosome entry sites (IRESes) and methods to circularize linear RNA are required. These approaches represent the next wave of mRNA vaccines likely to be investigated in humans and animals.

For the generation of self-replicating RNAs, IVT reactions are employed to make an alphavirus transcript that has its capsid and cell receptor genes deleted and replaced with a CDS for a gene of interest. These are transfected into cells with separate plasmids that code for the capsid and cell receptor, generating a virus-like particle (VLP) that can be harvested from the supernatant of transfected cells. VLPs cannot replicate like infectious viruses but still express and replicate the CDS that was included downstream of the coded alphavirus polymerase. These are often easier to manufacture than traditional mRNA constructs.

RNA carrier development

Naked mRNA by itself (very large) or RNA in traditional cationic lipids has not obtained high transfection efficiency outside of cell culture. Instead, lipid nanoparticles have been developed to allow for transfection in the body after injection. The lipid nanoparticles usually contain an ionizable lipid, cholesterol, a phospholipid, and a PEGylated lipid, which are mixed in a microfluidics shaker until the RNA is encapsulated. Several ionizable lipids and phospholipids have been developed and patented, each adding costs to mRNA vaccines. Research into new lipid nanoparticles that can transfect specific tissues or the inclusion of targeting antibodies in the outer lipid shell is ongoing. Research into other carriers, such as polymers or nano/nonstructured lipid carriers, is also ongoing.[20,21]

Mechanism of action

Understanding the mechanism of action is crucial for appreciating the efficacy of mRNA vaccines. Here, we delve into the intricate process of how mRNA vaccines instruct cells to produce viral proteins, triggering an immune response that confers protection against specific pathogens.

After injection, the lipid nanoparticles surrounding the mRNA facilitates its endocytosis into cells. The mRNA is thought to escape from the endosome through pores specialized pores (RNA is often used to communicate between cells), although the lipid nanoparticle may also assist.[22] The host cell then recognizes the mRNA by its poly(A) tail and 5' methyl cap, and its ribosomal machinery is recruited to the mRNA's 5' UTR. Then, ribosomes begin translation of the mRNA into proteins that will be recognized by the immune system. Although mRNA only has a half-life of 6 to 24 hours in mammalian cells, mRNA vaccines, because of their modified bases, can last far longer. Moreover, not all lipid nanoparticles transfect cells immediately. There is some evidence that a subset of vaccinated people still have lipid nanoparticles encasing mRNA payloads for days after COVID-19 vaccination.[23] It is thought that the increased stability of mRNA leads to a depot effect that might contribute to some of the robust immune responses associated with these vaccines. Compared to DNA vaccines, mRNA vaccines elicit stronger cellular and humoral immune responses.

Self-replicating RNA particles simulate a live virus vaccine and elicit robust cellular and humoral immunity. Once the VLP binds with its alpha virus receptor to a permissive cell, the viral genome is endocytosed. Because alphaviruses are positive sense RNA viruses, the leader RNA sequence that codes for the alphavirus polymerase is immediately translated. The polymerase then binds downstream and begins to synthesize new copies of the gene of interest mRNA. Because of its self-amplifying nature, it has been assumed that lower doses of RNA viral particles are needed to elicit immune responses compared to traditional mRNA vaccines. While this approach was shelved as a general vaccine platform in humans since CureVac failed to achieve high efficacy for COVID-19, RNA particles for production animals remain a robust seller as prescription vaccines for swine.

Efficacy and effectiveness

Several pre-licensed vaccines have been developed and are progressing toward efficacy trials or eventual licensure. For example, a foot and mouth vaccine protected mice after a challenge. However, this protection was not due to neutralizing antibodies and the IRES used in the vaccine elicited protection by itself, suggesting the vaccine elicited an innate immune response rather than the favored adaptive.[24] A rabies mRNA vaccine was also developed using the virus' G protein. This vaccine produced neutralizing antibodies comparable to killed rabies vaccines and was thermostable at high temperatures. However, high mRNA levels were needed for protection, making it less cost-effective than killed vaccines. Testing of this vaccine in swine led to protection from challenge.[25] In humans, the number of mRNAs being investigated, with many in mid to late-stage clinical trials, is significant. We will likely see more mRNA vaccines designed and tested for production animals, when cost and other technical hurdles are overcome.

The first RNA particle approved was for PEDV.[26] Recently, Merck developed an RNA particle vaccine for influenza targeting the neuraminidase protein rather than the traditional hemagglutinin.[27] Swine appeared to have significantly fewer lesions than controls and moderately reduced the virus burdens. This vaccine moves the self-replicating RNA particle vaccines out of the prescription model into an off-the-shelf vaccine against many flu strains.

One potential drawback of self-replicating RNA particles that traditional mRNA vaccines in lipid nanoparticles do not face, is that prior exposure to the alphavirus polymerase used in the vaccine can diminish the response to whatever vaccine antigen was included. Furthermore, immunity to the receptors on the VLP that carries the self-replicating RNA can inhibit vaccine entry into cells if booster vaccines are needed.

Most RNA vaccines still suffer from poor thermostability, including the RNA particle ones. Research into stabilizers, UTRs, or RNA carriers to overcome this issue is ongoing. Significant progress is occurring from Moderna's and Pfizer's research into stability, and lessons they learned can be imported to mRNA research for production animals.

Safety profile

Although some safety concerns have been raised with the human COVID-19 vaccines, most of these concerns center around irritation or redness at injection sites. Although we know that COVID-19 vaccines often move around the body and do not just remain at the injection site or go straight to the draining lymph nodes, there is no evidence that these vaccines cause determinantal damage to the organs of human vaccinees. In clinical trials for Sequivity or the newest vaccines from Genvax, clinical trials and ongoing vaccinations did not find any harmful effects on production animals that were injected (Joel Harris, Genvax, personal communication). Most mRNA vaccines are transient, and adaptive immune responses to the immunogen they code for should be enough to elicit cytotoxic T cells to lyse transfected cells, similar to traditional live attenuated virus vaccines.

Although there have been recent studies demonstrating mRNA in breast milk in human vaccinees, those RNAs presented for less than 45 hours post-vaccination. Furthermore, the RNAs were not in their original lipid nanoparticles, but in natural extracellular vesicles that cannot transfect and express proteins in cells after harvest.[28,29]

Indeed, the safety profile of mRNA vaccines can be increased by ensuring that the most minimal effective dose is used, carrier lipid nanoparticles are not designed to last too long in the body, and by determining the withdrawal period for each vaccine. The instability of RNA itself, the prolonged storage of animal products on store shelves, and the hostile gastrointestinal tract environment of people who consume these products should ensure the safety of these vaccines for use in animals or animal products that humans consume.

Future directions

As mRNA vaccine technology continues to evolve, exploring its potential against the myriad of diseases that affect production animals is necessary. This could be especially true for vaccines that have given low to moderate efficacy using transitional vaccines. Long-term safety studies will likely be needed to assuage the public that these vaccines will be safe in food animals. As mRNA vaccines become more common place for humans, public comfort with this vaccine approach should increase and safety fears likely diminish.

References

1. Dolgin E. The tangled history of mRNA vaccines. *Nature*. 2021;597(7876):318-24. **doi: 10.1038/d41586-021-02483-w**

2. Morais P, Adachi H, Yu Y-T. The Critical Contribution of Pseudouridine to mRNA COVID-19 Vaccines. *Frontiers in Cell and Developmental Biology*. 2021;9. **doi: 10.3389/fcell.2021.789427**

3. Chaudhary N, Weissman D, Whitehead KA. mRNA vaccines for infectious diseases: principles, delivery and clinical translation. Nat Rev Drug *Discov*. 2021;20(11):817-38. Epub 20210825. **doi: 10.1038/s41573-021-00283-5**

4. Accelerating vaccine trials. *Bull World Health Organ*. 2021;99(7):482-3. **doi: 10.2471/BLT.21.020721**

5. Kon E, Levy Y, Elia U, Cohen H, Hazan-Halevy I, Aftalion M, Ezra A, Bar-Haim E, Naidu GS, Diesendruck Y, Rotem S, Ad-El N, Goldsmith M, Mamroud E, Peer D, Cohen O. A single-dose F1-based mRNA-LNP vaccine provides protection against the lethal plague bacterium. *Sci Adv*. 2023;9(10):eadg1036. Epub 20230308. **doi: 10.1126/sciadv.adg1036**

6. Zheng C, Shao W, Chen X, Zhang B, Wang G, Zhang W. Real-world effectiveness of COVID-19 vaccines: a literature review and meta-analysis. *Int J Infect Dis*. 2022;114:252-60. Epub 20211117. **doi: 10.1016/j.ijid.2021.11.009**

7. McMahon M, O'Dell G, Tan J, Sarkozy A, Vadovics M, Carreno JM, Puente-Massaguer E, Muramatsu H, Bajusz C, Rijnink W, Beattie M, Tam YK, Kirkpatrick Roubidoux E, Francisco I, Strohmeier S, Kanekiyo M, Graham BS, Krammer F, Pardi N. Assessment of a quadrivalent nucleoside-modified mRNA vaccine that protects against group 2 influenza viruses. *Proc Natl Acad Sci U S A*. 2022;119(45):e2206333119. Epub 20221102. **doi: 10.1073/pnas.2206333119**

8. Gote V, Bolla PK, Kommineni N, Butreddy A, Nukala PK, Palakurthi SS, Khan W. A Comprehensive Review of mRNA Vaccines. *Int J Mol Sci*. 2023;24(3). Epub 20230131. **doi: 10.3390/ijms24032700**

9. Aliahmad P, Miyake-Stoner SJ, Geall AJ, Wang NS. Next generation self-replicating RNA vectors for vaccines and immunotherapies. *Cancer Gene Ther*. 2023;30(6):785-93. Epub 20220222. **doi: 10.1038/s41417-022-00435-8**

10. Opriessnig T, Gerber PF, Xiao CT, Mogler M, Halbur PG. A commercial vaccine based on PCV2a and an experimental vaccine based on a variant mPCV2b are both effective in protecting pigs against challenge with a 2013 U.S. variant mPCV2b strain. *Vaccine*. 2014;32(2):230-7. Epub 20131119. **doi: 10.1016/j.vaccine.2013.11.010**

11. Sandbulte MR, Spickler AR, Zaabel PK, Roth JA. Optimal Use of Vaccines for Control of Influenza A Virus in Swine. *Vaccines* (Basel). 2015;3(1):22-73. Epub 20150130. **doi: 10.3390/vaccines3010022.**

12. Schoenmaker L, Witzigmann D, Kulkarni JA, Verbeke R, Kersten G, Jiskoot W, Crommelin DJA. mRNA-lipid nanoparticle COVID-19 vaccines: Structure and stability. *Int J Pharm*. 2021;601:120586. Epub 20210409. **doi: 10.1016/j.ijpharm.2021.120586**

13. Kim SC, Sekhon SS, Shin WR, Ahn G, Cho BK, Ahn JY, Kim YH. Modifications of mRNA vaccine structural elements for improving mRNA stability and translation efficiency. *Mol Cell Toxicol*. 2022;18(1):1-8. Epub 20210920. **doi: 10.1007/s13273-021-00171-4**

14. Liu A, Wang X. The Pivotal Role of Chemical Modifications in mRNA Therapeutics. *Front Cell Dev Biol*. 2022;10:901510. Epub 20220713. **doi: 10.3389/fcell.2022.901510**

15. Fang E, Liu X, Li M, Zhang Z, Song L, Zhu B, Wu X, Liu J, Zhao D, Li Y. Advances in COVID-19 mRNA vaccine development. *Signal Transduct Target Ther*. 2022;7(1):94. Epub 20220323. **doi: 10.1038/s41392-022-00950-y**

16. Jeeva S, Kim KH, Shin CH, Wang BZ, Kang SM. An Update on mRNA-Based Viral Vaccines. *Vaccines* (Basel). 2021;9(9). Epub 20210829. **doi: 10.3390/vaccines9090965**

17. Zhang H, Zhang L, Lin A, Xu C, Li Z, Liu K, Liu B, Ma X, Zhao F, Jiang H, Chen C, Shen H, Li H, Mathews DH, Zhang Y, Huang L. Algorithm for optimized mRNA design improves stability and immunogenicity. *Nature*. 2023;621(7978):396-403. Epub 20230502. **doi: 10.1038/s41586-023-06127-z**

18. Liu T, Liang Y, Huang L. Development and Delivery Systems of mRNA Vaccines. *Front Bioeng Biotechnol*. 2021;9:718753. Epub 20210727. **doi: 10.3389/fbioe.2021.718753**

19. Niu D, Wu Y, Lian J. Circular RNA vaccine in disease prevention and treatment. *Signal Transduct Target Ther*. 2023;8(1):341. Epub 20230911. **doi: 10.1038/s41392-023-01561-x**

20. Suberi A, Grun MK, Mao T, Israelow B, Reschke M, Grundler J, Akhtar L, Lee T, Shin K, Piotrowski-Daspit AS, Homer RJ, Iwasaki A, Suh HW, Saltzman WM. Polymer nanoparticles deliver mRNA to the lung for mucosal vaccination. *Sci Transl Med*. 2023;15(709):eabq0603. Epub 20230816. **doi:10.1126/scitranslmed.abq0603**

21. Fang CL, Al-Suwayeh SA, Fang JY. Nanostructured lipid carriers (NLCs) for drug delivery and targeting. *Recent Pat Nanotechnol*. 2013;7(1):41-55. PubMed PMID: 22946628.

22. Zheng L, Bandara SR, Tan Z, Leal C. Lipid nanoparticle topology regulates endosomal escape and delivery of RNA to the cytoplasm. *Proc Natl Acad Sci U S A*. 2023;120(27):e2301067120. Epub 20230626. **doi: 10.1073/pnas.2301067120**

23. Krauson AJ, Casimero FVC, Siddiquee Z, Stone JR. Duration of SARS-CoV-2 mRNA vaccine persistence and factors associated with cardiac involvement in recently vaccinated patients. *NPJ Vaccines*. 2023;8(1):141. Epub 20230927. **doi: 10.1038/s41541-023-00742-7**

24. Le T, Sun C, Chang J, Zhang G, Yin X. mRNA Vaccine Development for Emerging Animal and Zoonotic Diseases. *Viruses*. 2022;14(2). Epub 20220215. **doi: 10.3390/v14020401**

25. Schnee M, Vogel AB, Voss D, Petsch B, Baumhof P, Kramps T, Stitz L. An mRNA Vaccine Encoding Rabies Virus Glycoprotein Induces Protection against Lethal Infection in Mice and Correlates of Protection in Adult and Newborn Pigs. *PLoS Negl Trop Dis*. 2016;10(6):e0004746. Epub 20160623. **doi: 10.1371/journal.pntd.0004746**

26. Aida V, Pliasas VC, Neasham PJ, North JF, McWhorter KL, Glover SR, Kyriakis CS. Novel Vaccine Technologies in Veterinary Medicine: A Herald to Human Medicine Vaccines. *Front Vet Sci*. 2021;8:654289. Epub 20210415. **doi: 10.3389/fvets.2021.654289**

27. Kitikoon P, Knetter SM, Mogler MA, Morgan CL, Hoehn A, Puttamreddy S, Strait EL, Segers R. Quadrivalent neuraminidase RNA particle vaccine protects pigs against homologous and heterologous strains of swine influenza virus infection. *Vaccine*. 2023;41(47):6941-51. Epub 20231024. **doi: 10.1016/j.vaccine.2023.10.005**

28. Hanna N, De Mejia CM, Heffes-Doon A, Lin X, Botros B, Gurzenda E, Clauss-Pascarelli C, Nayak A. Biodistribution of mRNA COVID-19 vaccines in human breast milk. *EBioMedicine*. 2023;96:104800. Epub 20230919. **doi: 10.1016/j.ebiom.2023.104800**

29. Hanna N, Heffes-Doon A, Lin X, Manzano De Mejia C, Botros B, Gurzenda E, Nayak A. Detection of Messenger RNA COVID-19 Vaccines in Human Breast Milk. *JAMA Pediatr*. 2022;176(12):1268-70. **doi: 10.1001/jamapediatrics.2022.3581**

https://doi.org/10.54846/am2024/161

Lipid nanoparticle-based DNA vaccines against influenza A virus of swine

T. N. Nguyen[1,2], BS; S. Kumari[1,2], MS; S. Sillman[1,3], DVM, PhD; J. Chaudhari[1,4], BSV, MS, PhD; D. C. Lai[1], BSV; H. L. X. Vu[1,4], BSV, MS, PhD

[1]Nebraska Center for Virology, University of Nebraska-Lincoln, Lincoln Nebraska; [2]School of Veterinary Medicine and Biomedical Sciences and Nebraska Center for Virology, University of Nebraska-Lincoln, Lincoln Nebraska; [3]Nebraska Veterinary Diagnostic Center, University of Nebraska-Lincoln, Lincoln Nebraska ; [4]Department of Animal Science, University of Nebraska-Lincoln, Lincoln Nebraska

Introduction

The rapid evolution nature of viruses, particularly RNA viruses, poses a formidable challenge in developing vaccines with broad protective efficacy. Due to the significant costs and time required for vaccine licensing, veterinary vaccines are not promptly updated to keep up with the continuous evolution of viruses. To address this issue, the Center for Veterinary Biologics (CVB) has issued Memorandum 800.213 dated 03/12/2018, outlining a regulatory framework to expedite the updating process of veterinary vaccines. According to these guidelines, when a veterinary vaccine is developed using a well-established non-replicating, nonviable production platform and the initial vaccine product is fully licensed, the licensure process for new vaccines containing sequence variants of the same vaccine antigens can be accelerated.

DNA plasmids, characterized by their noninfectious and nonreplicating nature, present a safe and cost-effective platform for animal vaccine development. However, the efficacy of DNA vaccines is often limited due to inefficient cellular uptake of the naked DNA plasmids, resulting in suboptimal immune responses. To overcome this challenge, several approaches, such as gene guns, *in vivo* electroporation, and nanoparticle encapsulation, have been developed to enhance their delivery efficacy.

Lipid nanoparticles (LNPs) have recently emerged as a promising nanocarrier for nucleic acid-based vaccines, particularly after the success of the COVID-19 vaccine.[1,2] We have developed different LNP formulations to deliver DNA plasmids in animals. In the present study, we report the immunogenicity and protective efficacy of two LNP-DNA vaccines based on the hemagglutinin (HA) gene of influenza A virus of swine (IAV-S) subtypes H3N2. We demonstrated that pigs receiving a single dose of these LNP-DNA vaccines mounted robust hemagglutination inhibition (HI) antibody titers within 7-14 days post-vaccination and were fully protected against challenge infections with the homologous IAV-S strain. These findings provide compelling evidence that LNP-DNA vaccines can be an attractive platform for veterinary vaccine development.

Materials and methods

Viruses

The IAV-S A/swine/Texas/4199-2/1998 (H3N2-TX98) was obtained from the National Veterinary Services Laboratories (NVSL, Ames, IA) and were propagated and titrated in MDCK cells.

Lipid nanoparticle-based DNA vaccines

The H3N2-TX98 HA gene was codon-optimized and cloned into the pCI plasmid (Promega, Madison, WI). Large-scale DNA plasmid was amplified in *Escherichia coli* DH5a and purified using a plasmid giga prep kit (Zymo Research, Costa Mesa, California). DNA sequencing was performed to confirm the plasmid sequence's authenticity.

Lipid-nanoparticle DNA vaccines were prepared as described previously.[3] Briefly, the lipids (organic phase) and DNA solution (aqueous phase) were mixed by using a mixer-4 chip and the NanoGenerator™ Flex-M Nanoparticle Synthesis System (Precigenome, San Jose, CA). The resulting LNP-DNA products were dialyzed against 100 mM Tris-Cl buffer at pH 7.4 and passed through 0.45 mm filters for sterilization.

Animal experiments

Vaccination and challenge experiments were conducted using 4-week-old IAV-S seronegative pigs. Principal groups were vaccinated intramuscularly with the LNP-DNA vaccines while a control group remained non-vaccinated.

Blood samples were collected weekly after vaccination, and serum was isolated to measure immune responses. At 35 days post-vaccination (dpv), the pigs were subjected to a challenge infection via intratracheal and intranasal inoculation with 2×10^5 TCID50 of the H3N2 TX98 virus. Following the challenge, nasal swabs were taken daily to measure viral shedding. At five days post-challenge (dpc), the pigs were humanely euthanized with an overdose of sodium pentobarbital. The lungs were removed, and bronchioalveolar lavage fluid (BALF) samples were collected using 50 mL of cold PBS to measure viral load in the lungs. Gross lung lesion was visually estimated. The trachea (approximately 1 inch above the carina), left apical, middle, and caudal lung lobes were collected and fixed in 10% buffered formalin for histopathologic examination as previously described.[4]

Results

LNP-DNA vaccines based on the H3N2 elicited systemic and mucosal immune responses

We first developed LNP formulations to encapsulate the HA gene of an H3N2 strain. Using the transfection efficiency *in vitro* as a readout, we screened nearly 100 LNP formulations and found two that showed high transfection efficiency. These formulations are referred to as LNP1-H3 and LNP2-H3.

A study that included three groups of six pigs was conducted to assess the immunogenicity and protective efficacy of LNP1-H3 and LNP2-H3 vaccines. Group 1 was injected intramuscularly (im) with PBS as a control, while Group 2 and 3 were injected im with 500 μg of LNP1-H3 and LNP2-H3, respectively.

Hemagglutinin inhibition (HI) antibodies were detected at 7 dpv and gradually increased. By 35 dpv, the mean titer exceeded 1:640 (Figure 1A). No significant difference in HI titers was observed between the LNP1-H3 and LNP2-H3 groups. At 35 dpv, the number of IFN-g spots in peripheral blood mononucleated cells (PBMCs) collected from LNP1-H3 or LNP2-H3 groups ranged between 50 and 350 spots per 10^6 cells, with no significant difference observed between these two groups (Figure 1B).

To assess the mucosal antibody response, HA-specific IgA and IgG antibodies in BALF samples collected at necropsy were measured using indirect ELISA assays. HA-specific IgG and IgA antibodies were detected in BALF samples collected from both groups of pigs vaccinated with the LNP-DNA vaccines but not from the PBS group (Figure 1C).

The results demonstrate that pigs vaccinated with the LNP-DNA vaccines mounded robust systemic and mucosal immune responses.

LNP-DNA vaccines based on the H3N2 strain protected pigs against challenge infection

On day 35 pv, all pigs were challenged with the homologous H3N2 strain. Nasal swabs were collected daily to measure viral shedding. Viral RNA was not detected in samples collected before the challenge infection (Figure 2A). In the PBS group, viral RNA was detected from all samples starting from 1 dpc, reached a maximal titer of 10^8 copies/100 µL nasal swab at 3 dpc, and declined to 10^6 copies/100 µL at 5 dpc. In the LNP1-H3 group, only one pig had a low copy number of viral RNA in samples collected at 3 and 4 dpc. No pigs in the LNP2-H3 group had detectable levels of viral RNA at any sampling dates (Figure 2A).

At 5 dpc, the pigs were euthanized and necropsied. BALF was collected to evaluate viral loads within the lungs. High viral RNA copies, ranging from 10^6 to 10^8 copies/100 µL, were detected in all pigs from the PBS group (Figure. 2B). In contrast, viral RNA was not detected in pigs from the LNP1-H3 or LNP2-H3 groups.

Lungs from the PBS group presented with purple-red consolidation typical of IAV-S infection, with the total lung surface consolidation ranging from 0.4 to 4.6%. In contrast, only one pig in the LNP2-H3 group exhibited lung consolidation, while no pigs in the LNP1-H3 group displayed visible lung consolidation (Figure 2C).

Lung sections were stained with H&E to evaluate microscopic changes. Pigs in the PBS group exhibited peribronchiolar lymphocytic cuffing with interstitial pneumonia, necrosis, and attenuation of epithelial cells in bronchioles and areas of suppurative bronchiolitis. Conversely, pigs from the LNP1-H3 and LNP2-H3 groups showed only occasional mild interstitial pneumonia and peribronchial lymphocytic infiltration (Figure 2D).

Together, the results demonstrate that pigs vaccinated with either the LNP1-H3 or LNP2-H3 were fully protected against challenge infection with the homologous H3N2 virus.

Discussion

DNA vaccines' immunogenicity and protective capacity have been assessed in various animal models, including the pig model. In general, DNA vaccines often exhibit limited efficacy due to the inefficient cellular uptake of naked DNA plasmids, resulting in subpar immunogenicity. Our current study demonstrates a substantial improvement in the immunogenicity and protective potential of the DNA vaccine through the encapsulation of DNA plasmid molecules within lipid nanoparticles. As a result, our findings highlight the promising potential of the LNP-DNA vaccine platform.

The DNA vaccines used in this study contain only the viral HA gene. Thus, pigs vaccinated with this vaccine will test negative by the commercially available diagnostic ELISA based on the viral nucleoprotein. Therefore, our LNP-DNA vaccines fulfill the requirement of a DIVA vaccine (DIVA: Differentiating Infected from Vaccinated Animals). As a result, swine producers can combine our vaccine with other management methods (such as test-and-removal) to improve the efficacy of controlling IAV-S.

In this study, we demonstrate the efficacy of the LNP-DNA vaccine against IAV-S. However, this can be readily adapted for the rapid development of vaccines against other endemic and emerging swine pathogens, including porcine circovirus type 3 (PCV3) or atypical porcine pestivirus (APPV).

Figure 1: Systemic and mucosal antibody responses following vaccination. **(A)** Reciprocal HI antibody titers against the H3N2 virus. **(B)** Virus-specific IFN-γ-secreting cell responses following vaccination. **(C)** HA-specific IgA and IgG in BALF samples as measured by an indirect ELISA assay. * $P < .05$; *** $P < .001$; **** $P < .0001$.

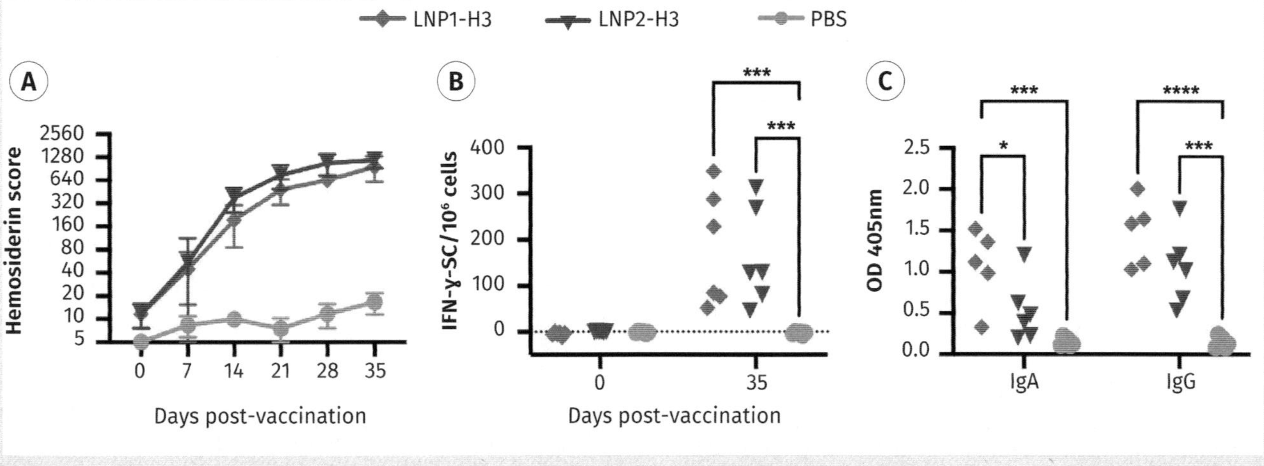

Figure 2: Viral shedding and lung pathology after challenge infection with the H3N2 strain. **(A)** Viral RNA copies in nasal swabs. **(B)** Viral RNA copies in BALF collected on day 5 pc. **(C)** Lung photos taken during necropsy. Black arrows indicate areas with typical consolidation caused by IAV-S. **(D)** Images of lung sections stained with H&E. **** $P < .0001$. Scale bar = 100 μm.

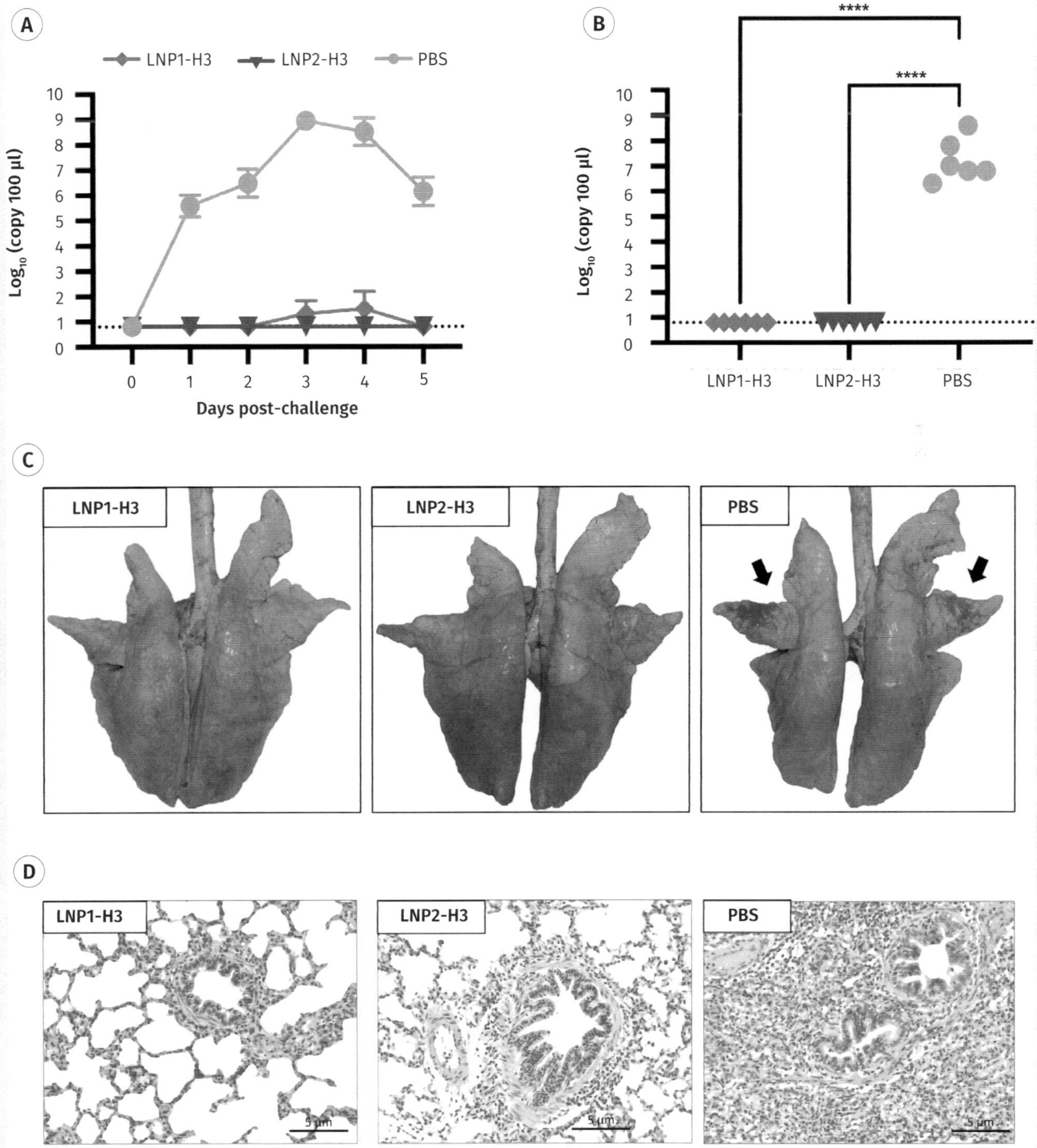

References

1. Mashima, R. & Takada, S. Lipid Nanoparticles: A Novel Gene Delivery Technique for Clinical Application. Curr Issues Mol Biol 44, 5013-5027, **doi:10.3390/cimb44100341** (2022).

2. Corbett, K. S. et al. SARS-CoV-2 mRNA vaccine design enabled by prototype pathogen preparedness. Nature 586, 567-571, **doi:10.1038/s41586-020-2622-0** (2020).

3. Roces, C. B. et al. Manufacturing Considerations for the Development of Lipid Nanoparticles Using Microfluidics. Pharmaceutics 12, **doi:10.3390/pharmaceutics12111095** (2020).

4. Gauger PC, Loving CL, Khurana S, Lorusso A, Perez DR, Kehrli ME, Jr., Roth JA, Golding H, Vincent AL. 2014. Live attenuated influenza A virus vaccine protects against A(H1N1) pdm09 heterologous challenge without vaccine-associated enhanced respiratory disease. Virology 471-473:93-104.

https://doi.org/10.54846/am2024/162

Next-generation nanovaccine platforms for swine health

Balaji Narasimhan, PhD
Department of Chemical and Biological Engineering and Nanovaccine Institute, Iowa State University, Ames, Iowa

Introduction

Respiratory viruses result in devastating economic loss in the swine industry and can be zoonotically transmitted to humans. We have developed a cross-disciplinary approach at the intersection of polymer chemistry, nanotechnology, and immunology for the molecular design of a safe, needle-free, and efficacious nanoparticle-based vaccine (ie, nanovaccine) platform technology. We used this platform technology to design and test multiple formulations of swine influenza nanovaccines.[1,2]

Methods

Inactivated swine influenza virus (H1N2) payloads with or without adjuvants (ie, CpG) were encapsulated within polyanhydride nanoparticles using flash nanoprecipitation and administered intranasally to groups of pigs using a prime-boost regimen. Controls included an inactivated virus vaccine and unvaccinated animals. Both humoral and cell-mediated immune responses induced by the vaccines were characterized. Animals were challenged with a heterologous H1N1 virus and clinical parameters were monitored.

Results

The nanovaccines induced cross-reactive IgA antibody in the nasal cavity, lymphoproliferative responses in peripheral blood mononuclear cells, and IFN-γ secretion upon recall. The nanovaccines also protected animals by reducing fever, nasal virus shedding, and lung virus titers.

Discussion

These studies provide important data on the ability of polyanhydride-based nanovaccines to provide protective immunity against respiratory infections in swine. The robust and long-lived immunity induced by these nanovaccines, when combined with their room temperature shelf life stability and their ability to be administered in a needle free and safe method, provides promise for this platform technology to be widely used to prevent respiratory infections in pigs.

References

1. S Dhakal, JT Goodman, K Bondra, YS Lakshmanappa, J Hiremath, D-L Shyu, K Ouyang, K-I Kang, B Binjawadagi, S Krakowka, MJ Wannemuehler, B Narasimhan, C-W Lee, and GJ Renukaradhya, "Polyanhydride nanovaccine against swine influenza virus in pigs," *Vaccine* **35**, 1124-31 (2017). **doi: 10.1016/j.vaccine.2017.01.019**

2. S Dhakal, S Ghimire, S Renu, KA Ross, YS Lakshmanappa, BT Hogshead, P Bernardo, CW Lee, MJ Wannemuehler, B Narasimhan, and GJ Renukaradhya, "Evaluation of CpG-adjuvanted polyanhydride-based intranasal influenza nanovaccine in pigs," *Vet Microbiol* **237**, 108401 (2019). **doi: 10.1016/j.vetmic.2019.108401**

https://doi.org/10.54846/am2024/163

The international consumer is key to the future of the US pork industry

Erin Borror
US Meat Export Federation

US veterinarians are essential to protecting the industry from foreign animal disease, preparing for the worst-case scenario, and working every day to enhance productivity and sustainability. Pork and edible variety meat exports in 2023 have averaged a record $63 for every hog harvested (Figure 1). This equates to 33% of the hog value, using the pork cutout as the basis. Exports account for close to 30% of US pork and variety meat production, but on a cut-specific basis, exports are even more important. Mexico takes close to half of US hams and if the Mexican market went away, we would only need half of the US swine inventory. In addition to hams, picnics are also heavily exported, and butts are a key item for the top Asian markets as the US exports more end meats and a smaller share of the middle cuts that are popular in the domestic market (Figure 2). Variety meat exports are on a record pace and have averaged nearly $11 per head in 2023, with China and Mexico as the top customers (Figure 3). Top variety meat export items include feet and stomachs.

Exports add value to every animal we produce, by enabling the sale to the highest bidder, worldwide. Worldwide really means in the markets where we have access. Countries like Brazil, South Africa and Australia restrict US pork due to PRRS. Germany, once the largest pork producer and exporter in Europe, has seen its production and exports plummet due in part to African swine fever (ASF) and subsequent market access restrictions. Brazil's pork exports are limited due to their foot-and-mouth disease (FMD) status. China, the world's largest producer, consumer and importer of pork, has had a massive impact on global markets due to its ASF-driven pork shortage followed by its rapid herd rebuild (Figure 4). Animal health is foundational to market access and ASF and FMD must be kept out of the US. Veterinarians are also critical for herd health surveillance and for regionalization and emergency planning.

Market access can also be impacted by regulations on vet drugs, including due to the increased focus on antimicrobial resistance. The first evidence of this is in the European Union's Article 118 and related regulations, requiring countries to prove non-use of the EU-determined list of antibiotics deemed important for human health. The regulation also prevents all antimicrobial use for weight gain and feed efficiency. Veterinarians are needed to help producers manage antibiotic use and to meet export verification requirements.

There has also been increased attention to sustainability and related efficiency. Herd health and productivity are not only needed for industry profitability but also to further improve the industry's benchmarking for sustainability related goals and claims. Sustainability messaging in the international markets is still in its early stages but USMEF has been carrying the industry's story to the markets for several years, including through National Pork Board content customized by USMEF's local staff. Answering consumer questions before they have them helps us to stay ahead of potential negative media.

Overall, the US pork industry has great potential and veterinarians are key to the industry's success. The EU pork industry is undergoing structural change and production has fallen by about 2.8 million metric tons (mt) (or 12%) from its 2021 peak, reaching an estimated 20.8 million mt in 2023. The European Commission expects a continued decline in EU production over the 10-year outlook. This contrasts with the US pork industry, where production growth resumed in 2023, reaching an estimated 12.36 million mt, and modest growth is expected in the coming years. The combined EU-27 has been the largest pork exporter in the world, followed by the US, Canada and Brazil. The decline in EU output and increase in prices has helped the US move toward the top exporter position (Figure 5). But this position and further growth can only happen with herd health and productivity as our foundation.

The EU decline is attributed to a host of factors including: ASF; regulatory costs including regulation at the EU level and by individual member state for animal welfare and environmental factors; regulatory uncertainty; input costs, reliance on imported feed and the war in Ukraine; reliance on China and its massive herd rebuild after adjusting to ASF; and decreasing EU demand.

To avoid the US industry following the path of the EU, we must proactively enhance animal welfare and productivity including through herd health, avoid further state-driven mandates like Proposition 12, maintain consumer trust, diversify our markets including through exports, maintain our animal health status by preventing foreign animal disease, and continue to promote the human health benefits of consuming pork. Veterinarians are at the core of the US pork industry and key to the industry's continued success, including in the international markets which offer added value and growth opportunities (Figure 6).

Figure 1: US pork exports accounting for a record share of production and per-head value.

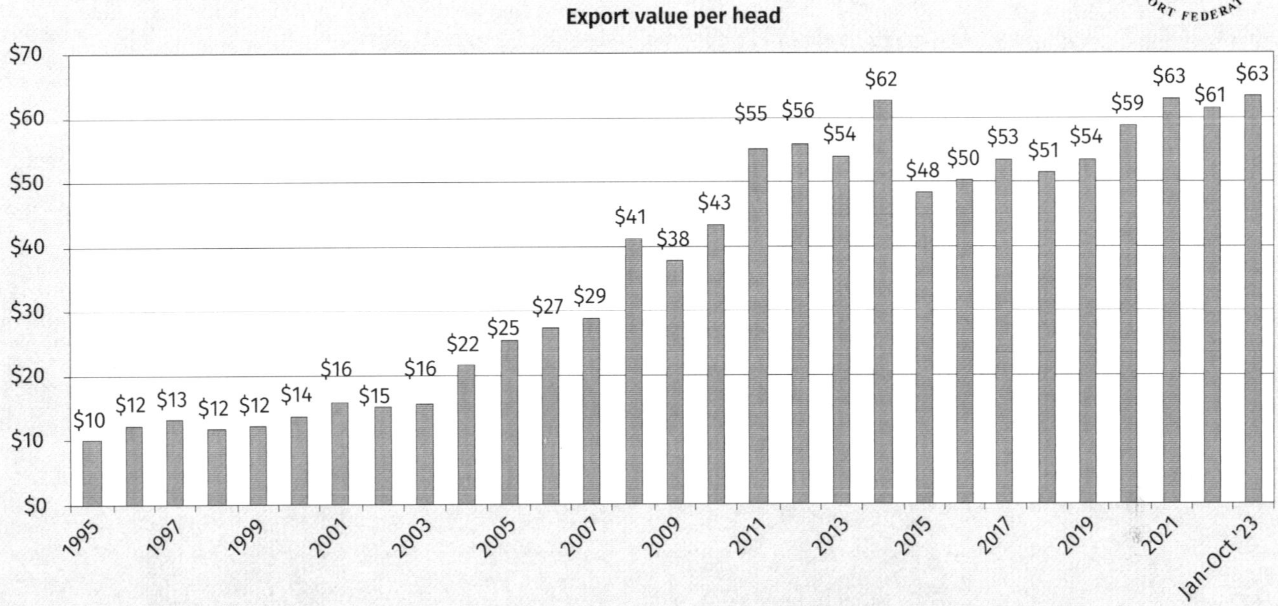

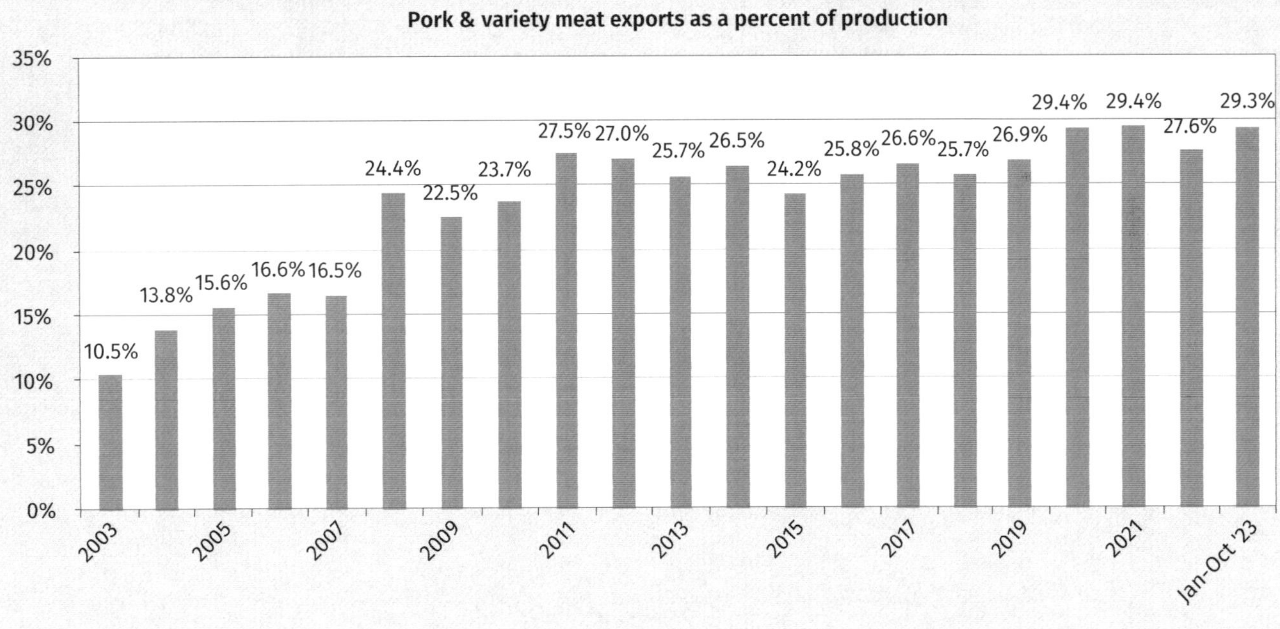

Source: USDA/FAS, USDA/NASS Commercial hog slaughter, USMEF

Figure 2: Pork is complex, exports of end meats allow Americans affordable access to favorite middle cuts and exports optimize overall value of each animal.

2023 Jan-Oct:

- 39.36 lbs. per head exported, 25% of production
- At average of $52 per head
- Pork/pvm export averaged $62.79/head, which was 36% of the hog value of $172.58 and 33% of the per head cutout value of $191.54

Source: USMEF calculations using NASS Commercial hog slaughter, AMS hog and pork cutout values, FAS export data

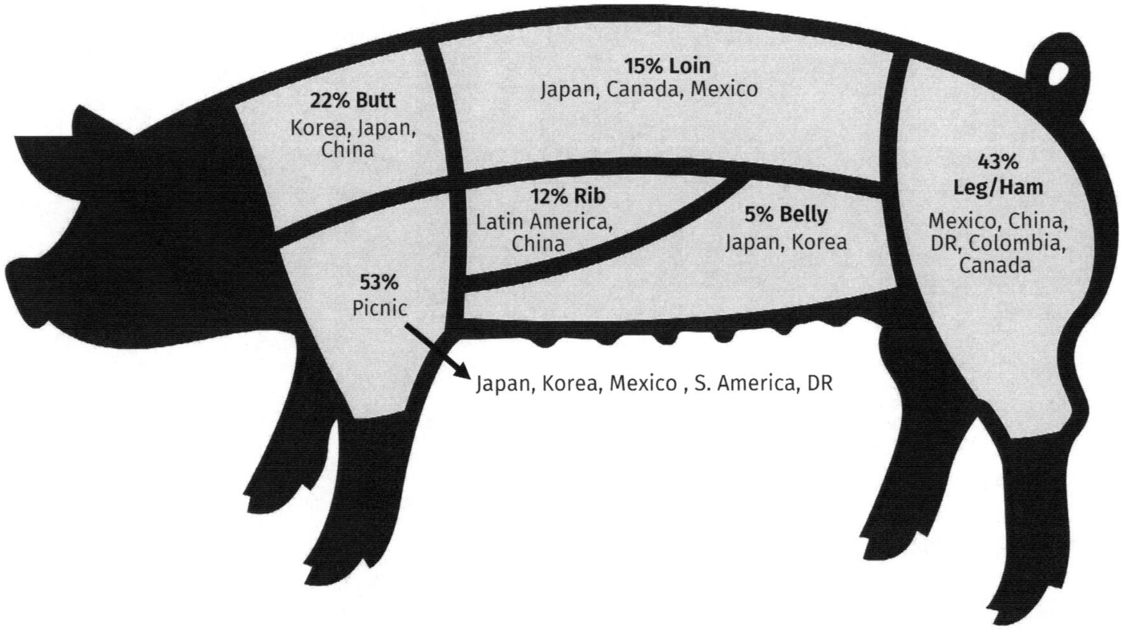

Source: USMEF 2022 estimates for export volume as a share of total U.S. pork primal production

Figure 3: Pork variety meat exports are on a record pace.

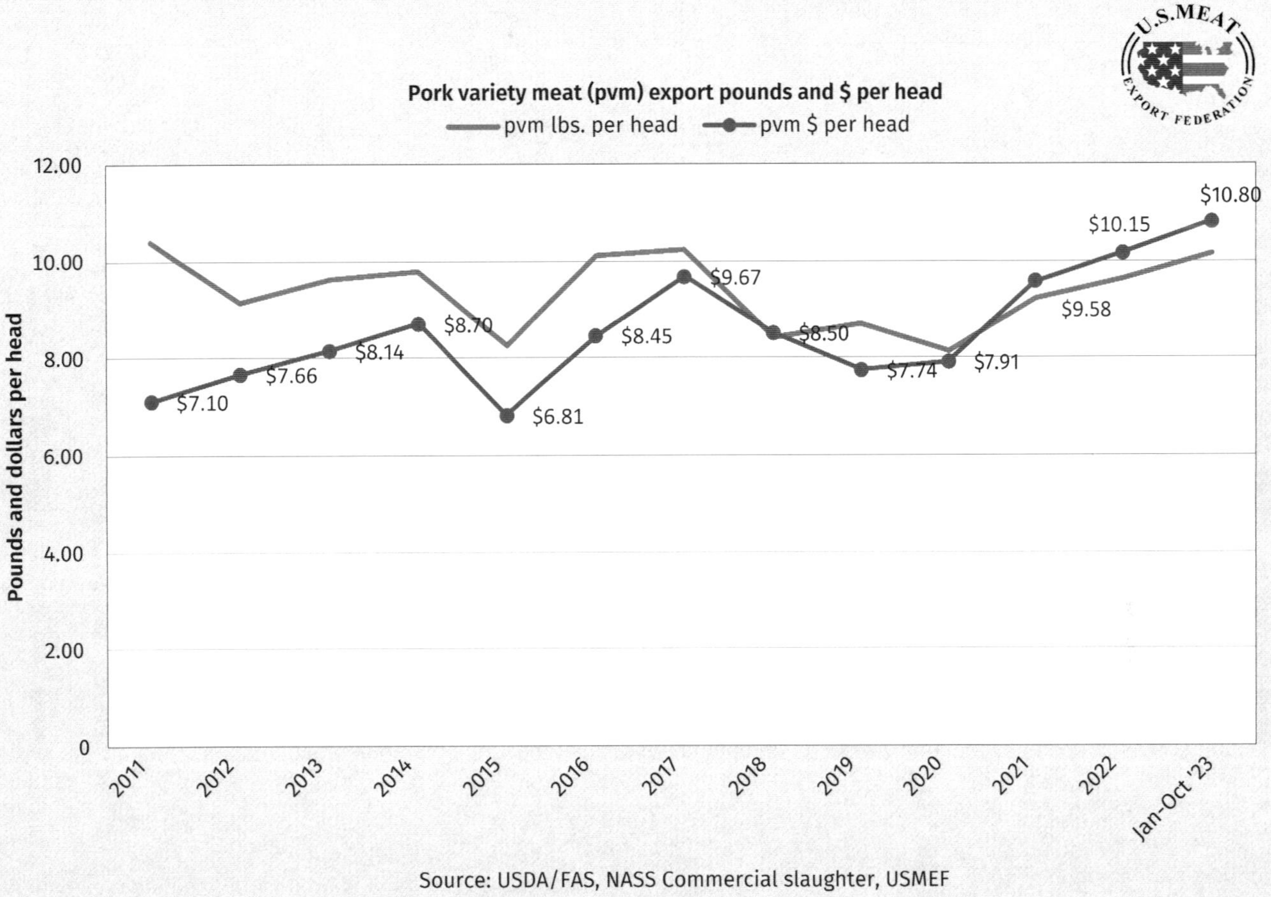

Figure 4: The tremendous impact of ASF; top exporter production to be the smallest since 2017 and down at least 2% year/year.

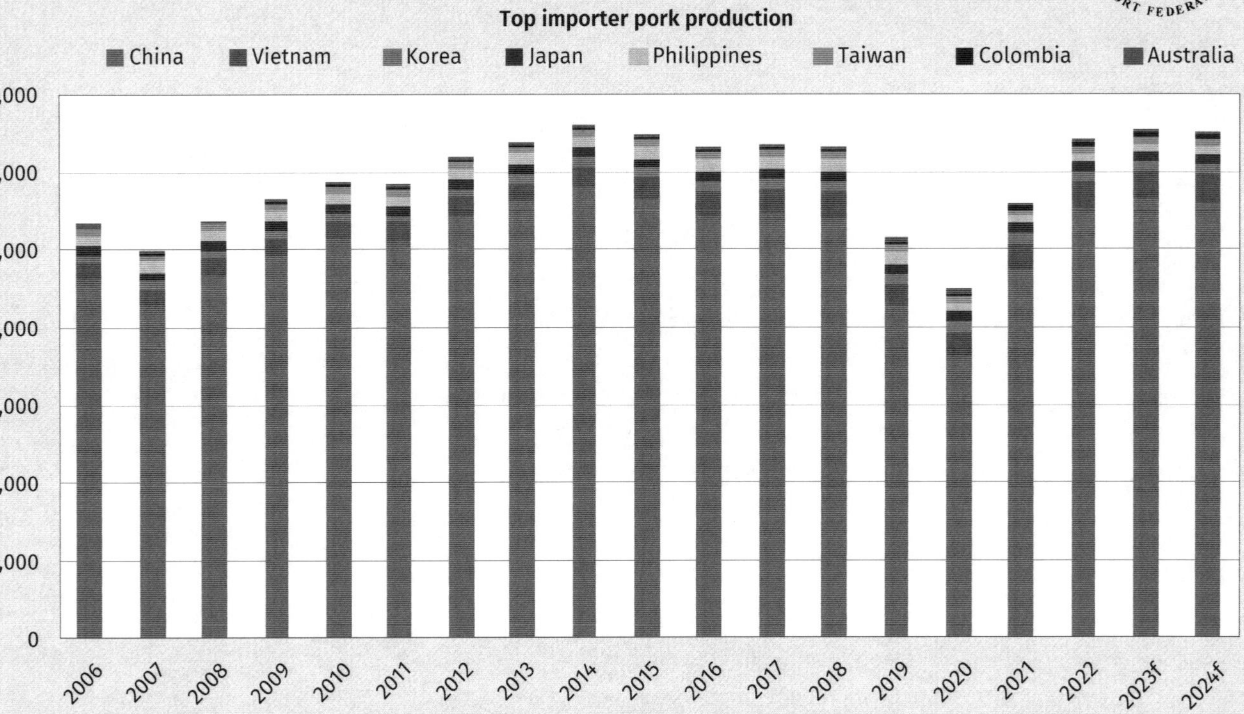

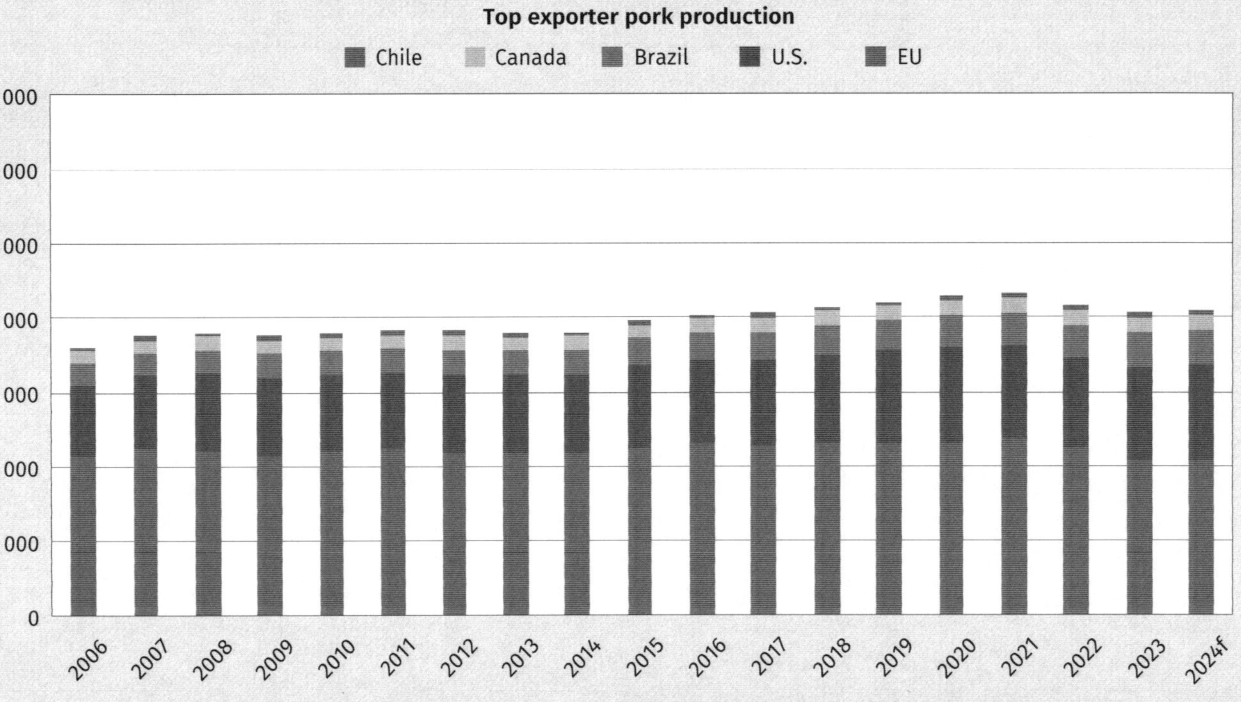

Source: USDA/FAS June PS&D, thousand mt carcass wt.; European Commission and USMEF estimates

Figure 5: Global pork exports down 7% in Jan-Oct, totaling 8.075 million metric tons, but strong 9% rebound in US exports.

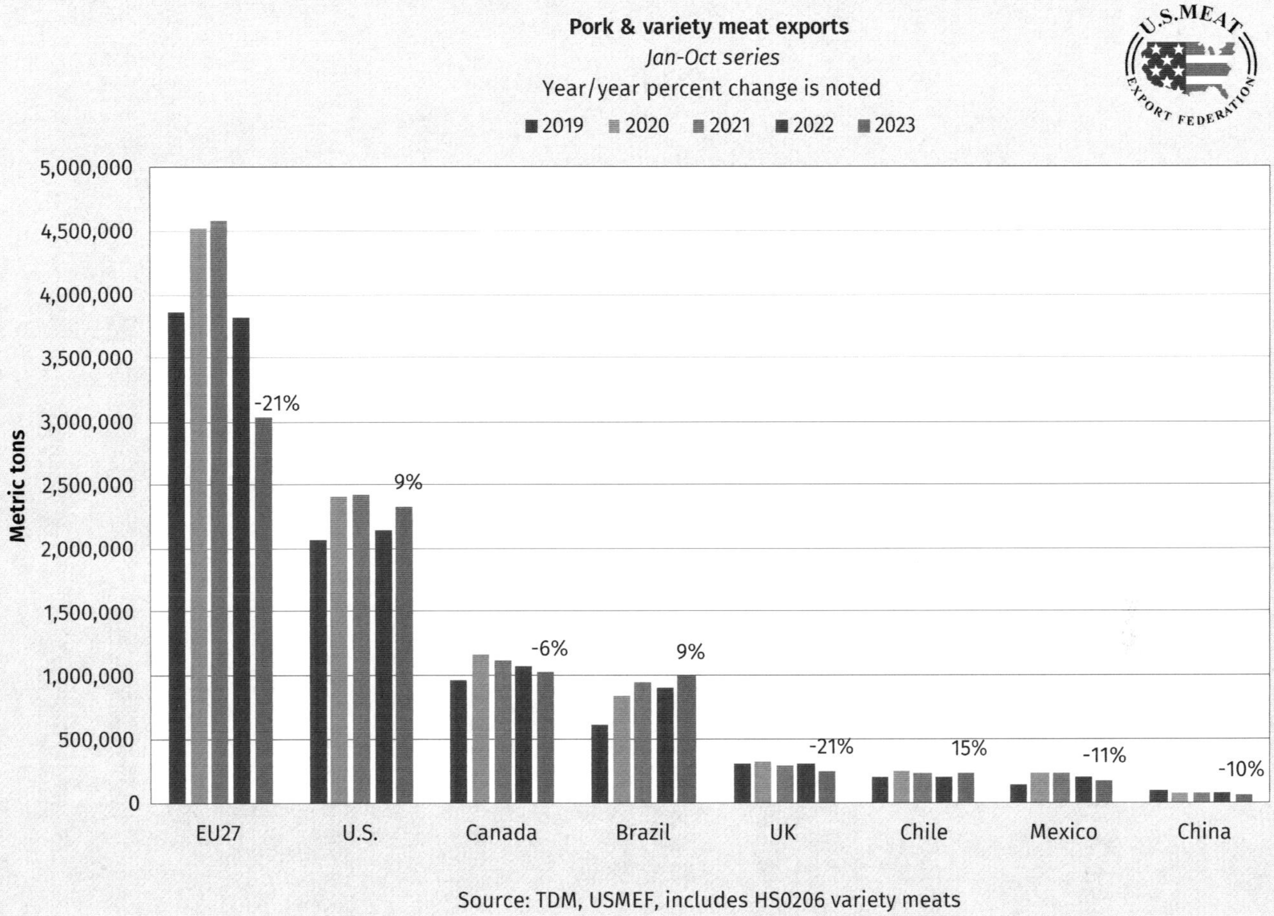

Figure 6: Record US pork export value in 2023, further growth forecast.

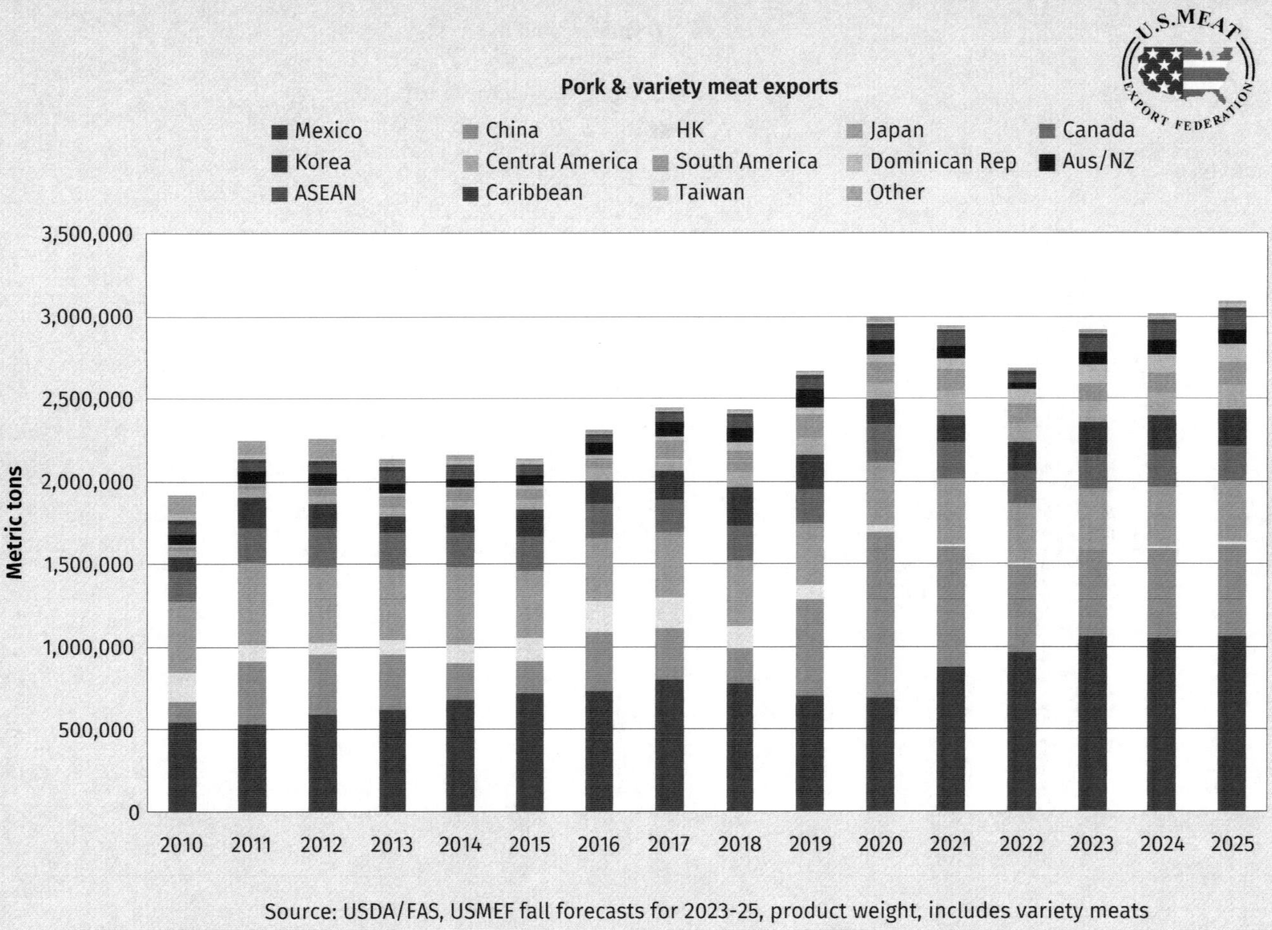

https://doi.org/10.54846/am2024/164

Driving demand: What the Pork Checkoff is doing with your money. What you need to do.

Bill Even
CEO, National Pork Board

Increasing volume and value of US pork domestically

Approximately 70% of total US pork stays in the domestic market. The National Pork Board (NPB) domestic marketing efforts are actively driving consumers to major retailers through a variety of strategic efforts including in-store promotion, e-commerce and digital shopper marketing programs. Overall, the effort is about increasing volume and value of our product.

For pork to remain relevant and ensure future demand, NPB has done significant research to understand the consumer and meet them where they are. By highlighting taste and flavor as the main benefits of pork and implementing strategies like positioning products in the foodservice channel, pork can become relevant to a different generation of consumers. This strategy influences domestic consumer demand for pork, which is why NPB continues building strategic partnerships in the retail and foodservice space, working hard to set pork up for success. NPB has also developed strategic platforms based on consumer research to target multicultural consumers as the US population becomes more diverse.[1] NPB's domestic market demand strategy relies on a quality product, which is influenced by herd health and veterinary relationships with producers.

Expanding international market development with the US meat export federation (USMEF)

The Pork Checkoff has an international market development strategy to build confidence in US pork in a diverse global protein market. In 2022, the value of US pork exports surpassed $7.7 billion. In total, more than 2.5 million metric tons of pork was exported to other markets.[2]

The Checkoff partners with USMEF to maximize pork marketing efforts abroad through cross-industry collaboration. USMEF executes marketing programs on behalf of the Pork Checkoff and positions US pork as a wholesome, high-quality and consistent product. They can further leverage Checkoff dollars with contributions from the USDA Foreign Agriculture Service and other funding opportunities including state associations and other allied industries.

Protecting demand for US pork by preventing animal diseases

Foreign and domestic animal diseases can pose a serious threat to US pork production and marketability, both in the country and around the globe. To protect their herd, pig farmers work with veterinarians to keep their animals healthy and build trust for their product among consumers and exporters. Today's global movements of people and pigs pose greater risks to herd health than ever before.[3] This makes implementing on-farm disease prevention and mitigation strategies critical to maintaining overall herd health and preventing the introduction or spread of disease. The Pork Checkoff leads collaboration with multiple government and industry partners to protect the US from diseases like African swine fever (ASF).

Veterinarians serve an important role in preventing and preparing for foreign animal diseases, as well as protecting herd health on the farm. NPB promotes the use of industry tools and resources like Secure Pork Supply, AgView, Certified Swine Sample Collection, US SHIP and others to help manage the impact of supply chain disruptions. These tools can be used by veterinarians to assist pork producers in keeping their herd healthy.

Building trust with consumers adds value to US pork

We Care® is an overarching philosophy and commitment grounded in the six Ethical Principles that guide America's pig farmers in all that they do. US pork producers recognize their obligation to build and maintain the trust of customers and the public in our products and our practices. The six We Care® Ethical Principles include food safety, animal wellbeing, environment, public health, our people and our communities. Using these principles, American pig farmers can add value to their product by building trust among consumers and ensuring their freedom to operate.[4]

Veterinarians are vital to protecting the health of our animals, and in turn, the public's health. They help guard against disease among pigs and between pigs and people, ensuring only healthy pigs enter the food chain. Veterinarians have helped the pork industry deliver on its three-point antibiotic stewardship pledge to promote research, pig farmer education, and consumer and influencer outreach. Pork Quality Assurance® Plus (PQA® Plus) training and assessment addresses safe and appropriate antibiotic use. Through closer guidance by veterinarians and with the vast majority of pig farmers now certified through PQA® Plus, the occurrence of antibiotic residue exceeding regulatory maximums in pork is extremely rare today.[5] Pig farmers are dedicated to identifying new ways to decrease the overall need for antibiotics. Veterinary relationships with America's pig farmers are an incredibly important piece of US pork production.

References

1. Rabe, M. (2023, October 24). Exploring the racial and ethnic diversity of various age groups. Census.gov. **https://www.census.gov/newsroom/blogs/random-samplings/2023/09/exploring-diversity.html#:~:text=In%202020%2C%20the%20diversity%20index,increased%20between%202010%20and%202020**

2. USDA Foreign Agricultural Service. (2022). *Global Agricultural Trade System (GATS)*. Global Agricultural Trade System. **https://apps.fas.usda.gov/gats/default.aspx?publish=1**

3. Oedekoven, D. (2023, July 17). Foreign animal disease prevention. Pork Checkoff. **https://www.porkcheckoff.org/pork-production-management/animal-disease-prevention/**

4. National Pork Board of Directors

5. Antibiotics. Pork Checkoff. (2021, March 9). **https://porkcheckoff.org/pork-production-management/swine-health/antibiotics/**

https://doi.org/10.54846/am2024/165

Protecting the product: Are your clients participating in price protection?

D. Baker

Introduction

Producing a high-quality, affordable, and nutritious food product is the primary focus of all swine market participants. Swine veterinarians, nutritionists, input suppliers, packers, and processors rely on the economic viability of pork producers to generate economic growth and promote rural sustainability. As such, it is important for all swine market stakeholders to be aware of all the available tools that can be used to offset risk.

Swine producers face volatility in the costs of inputs they purchase as well as the value of the animals they produce. Historically, farmers have been focused on managing the risk of profitability through a price-centric approach. In this way, margin deterioration through eroding sales values on hog prices or input price increases from rising feed costs were treated independent from one another and managed separately. While there is not necessarily anything wrong with this approach and the price movements for both hogs and feed prices like corn and soybean meal are generally separate, there is a different way of thinking about managing profitability risk.

Hog margin management is the process of analyzing financial risk from the perspective of an overall profit margin approach, considering costs and revenues together as a single unit of risk. A margin management process involves modeling forward projected costs and revenues using the futures market as a price discovery mechanism. The futures market is an unbiased estimate of forward value based on a dynamic, daily auction of buyers and sellers discovering hog price and feed price on an organized, regulated exchange of futures contracts. Because there is a strong correlation between futures prices and the cash prices swine producers pay and receive for their feed and hogs, the futures market can effectively be used to discover forward margins through futures contracts.

Building a model

Building an accurate representation of a swine producer's operation by accounting for all costs and revenues is the first step in being able to estimate future profitability using swine margin management. Corn and soybean meal would be the main input costs of producing hogs, along with other feed and non-feed expenses. While there is a strong correlation between cash corn and soybean meal prices with corresponding futures values, the other feed and non-feed expenses do not correlate with a futures price, and thus certain fixed estimated values for these would need to be assumed in the model. On the revenue side, the hog futures market can be used to estimate forward hog sale prices for the swine operation's hog margin management.

Holding certain fixed cost assumptions static and using futures market prices to represent both input costs and revenue values together, forward profit margin opportunities can be identified in deferred time periods to indicate how profitability is projected to change from one period to the next. This is beneficial by itself in providing greater visibility to a swine producer on how the market projects forward profitability. In addition to this though, because the futures market also provides a risk transfer function, a swine producer can contract both input costs and revenue values simultaneously using exchange-traded derivatives to lock-in or protect a profit margin level in a forward time period.

This can significantly add value to an operation by improving planning and budgeting, assisting with lender discussions, and smoothing out volatility in returns over time. Moreover, it may sometimes be the case that a projected forward profit margin opportunity identified in advance may turn into an actual loss in the spot marketing period as market conditions change over time. In other words, should input costs increase from what is projected today and/or revenue values decline it may be beneficial to capture or protect the opportunity in advance.

Objectively assessing opportunities

Once an accurate model of the swine operation is built, the next step would be to evaluate the forward profit margins being represented to determine whether the operation is projecting a profit or a loss in any given period. Assuming the model is projecting a profitable margin opportunity, the producer would want to know how good of an opportunity is being presented. One way to objectively determine the opportunity is to rank the projected margin over a historical period, comparing that profit margin to previous profit margin opportunities in that same period of the calendar year. For example, if one was evaluating a forward profit margin opportunity in upcoming Q4, they could compare that projected margin to the previous ten Q4's to rank that opportunity within the context of the past decade.

Where this objectivity becomes powerful is in helping to guide contracting decisions to protect margin opportunities. As an example, if the current projection for forward margin ranks in the 90th percentile of historical profitability, that means that only 10% of the time have historical margins for that period been better than what is being projected today. It also suggests that there is a much higher probability of the margin deteriorating over time moving into that marketing period than for a further improvement. If by contrast the margin projection ranks at the 50th percentile of historical profitability, that means it's about a coin flip as to whether the margin will be stronger or weaker once we get into that marketing period.

Tools available

There are a variety of different marketing tools at a producer's disposal to manage profit margin opportunities. The cash market is one outlet where the swine operation can contract with local counterparties to secure feed input costs, and/or hog sales prices in advance of a given marketing period. In addition to protecting price risk, these alternatives also allow for physical supply to be secured which can be an important consideration in certain situations. They may also address local basis risk if a delivered price can be established in advance.

Another alternative to protect feed input costs and hog sales revenue around forward margin opportunities is to use exchange-traded derivative contracts. Futures and options allow a swine operation to lock in or protect price levels in a forward period, ahead of realizing actual cash revenues and expenses in that period. Futures are often referred to as "fixed price" tools because they convey both the right and the obligation to purchase or sell a standardized contract at a particular price. Options, on the other hand, allow for flexibility by conveying the right, but not the obligation, to a particular price. The flexibility is oftentimes desirable if the producer perceives there to be opportunity for the prices to move in hir or her favor, but that flexibility comes with a cost. While these contracting alternatives do not address physical delivery or local basis risk, they do offer added flexibility in being able to trade in and out of different structures over time.

Government-sponsored insurance, including Livestock Risk Protection Insurance (LRP) and Livestock Gross Margin Insurance (LGM), is another alternative available to swine producers to protect forward profit margins. These tools offer their own benefits in being subsidized by the government and are cash-flow friendly, as well as offering potential basis applications that may not be as effectively addressed using exchange-traded derivatives or cash contracts. Swaps are also another financial risk management alternative that can be an effective addition to the swine producer's risk management toolbox under certain scenarios.

Discussion

All pork sector participants have a vested interest in the long-term viability and profitability of hog production. Swine margin management may offer significant benefits to a swine operation in helping to create an objective way to evaluate risk and navigate the various contracting tools available in the marketplace today. Learning more about this approach with a commodity price risk consulting service could be well worth the time invested to manage an operation's profitability and help secure it for future generations.

https://doi.org/10.54846/am2024/166

Protecting the product: How I work with my packer

D. Murray, DVM
New Fashion Pork, Jackson, Minnesota

Background

Historically swine veterinarians have been responsible for disease prevention, diagnostics, and treatment. Their job ended when the pigs were loaded on a truck and the animals were hauled to the packer. After harvest, the producer/client received payment for delivered animals. The packer dictated the purchase of the product and product quality, such as percent lean and weight payment matrix. As systems have further integrated, the practicing veterinarian's role not only extends to the packer but also now can follow product attributes that reach the customer. This is further complicated by increased international sales. Many countries do not permit drugs like ractopamine in meat. These countries also have reduced maximum residue levels for tetracycline compared to US. This results in significantly increased withdrawal time for tetracycline from period approved for US markets. In addition, there can also be specific meat quality standards as well as feedstuff limitations and/or requirements.

Relationship building

Relationships between the packer and producer are extremely important to both produce the highest quality product for the end consumer as well as business continuation in the face of foreign animal diseases (FADs). The threat of ASF and other foreign animal diseases is always present. In order for producers to continue to operate through investigations and non-negative events, should they ever arise, communication between the plant and the producer as well as the state and federal animal health officials will be crucial. Existing lines of communication with the plant and USDA officials will alleviate major hurdles when stress is high. Today, for groups of pigs that have had recent FMD negative (Seneca positive diagnostics) through a FAD investigation, communication of the diagnostic findings will help state, APHIS and FSIS veterinarians and plant staff plan and manage harvest schedules.

Observations

Many times, pathology seen at the plant can be a surprise as some groups have had little to no health challenges in the barns but have a high prevalence of significant pneumonia and pleurisy lesions on postmortem examination. Other groups have been challenged with PRRS+/mycoplasma+/-influenza A as well as other secondary pathogens in barns and yet have very few issues on the rail. While we have not developed any new technology for lesion prediction, there have been high rates of condemnation that are cause for concern. In some situations, we see increased preharvest sort offs and loss of full value pigs marketed. This can be an underestimated cost of disease and would be a good area for future study and assessment. Is this due to: Timing of disease? Different disease combinations? Antibiotic use or lack thereof? Most importantly, how could this be prevented for future groups?

Research

The opportunity for conducting research with the plant is extremely valuable. Producers are not paid for end nursery weight or lung lesions (which can be an end point for some research) but rather for the full value product marketed with specific quality characteristics as a product produced in an integrated system as sold by the plant. This relationship and research is a win-win, dual benefit to the producer as well as the plant. If large-scale health concerns arise, product will be lost and line speeds may slow. With help from the plant, we have been able to diagnose the cause of condemnations, conducted research on interventions by testing potential solutions and follow groups through the plant to measure outcomes. Plant staff veterinarians can help collect tissues from condemned carcasses or in our case, lymph nodes. We were then able to culture these lymph nodes and identify a causative pathogen which could be contributing to the lymphadenopathy (other than the normal PRRSV lymphadenopathy that many carcasses would typically have). The cause was *Mycoplasma hyorhinis*. We tested different vaccinations on both healthy and affected groups to compare incidence and corresponding lesions at slaughter. We were able to measure condemnation rate by reason by group and saw a marked decrease in condemns as well as an increase in carcass weights once an effective vaccine strategy was found and implemented.

Future goals

There may be an opportunity for AASV members to hold staff positions in slaughter plants. Ideally, each plant would employ a veterinarian to facilitate health communication between the plant and the field. This individual can both initiate discussion on plant level concerns and actively participate in creating solutions. This would also be an invaluable position should a FAD ever arise in the US.

Recommendations for success

Visit the plant both when there may be problems and when things are running smoothly. This is a great opportunity to familiarize yourself with staff as well as compare any pathology of your pigs to others at the plant. Slaughter checks were another way veterinarians would have interacted with the plant and helped to understand disease prevalence in a herd or flow. They do not seem as popular as they used to be. There are many reasons but if you have never done a slaughter check I'd advise you to try it and see what you can learn. USDA officials see more pathology than most swine vets will ever see and can offer valuable insight with hogs on the rail. Quite often, these concerns can be brought back to improve status at the farm. Utilize veterinarians employed by the plant to help make the farms you work with better and improve the product at the plant for both the producer as well as the plant itself.

https://doi.org/10.54846/am2024/167

Protecting the product: A swine vet's job isn't done when the pig is loaded on the truck

Grace E. Houston, DVM, PhD
Triumph Foods LLC, Saint Joseph, Missouri

Introduction

The goal for any pork processing plant is to take live hogs and turn them into food. With this statement comes many objectives such as animal welfare, food safety, and meat quality. While the producer has spent 6 months raising market hogs, packing plants have approximately 24 hours to turn market hogs into a safe, wholesome meat product. In the process of making this happen, large packing plants work with United States Department of Agriculture Food Safety Inspection Services (USDA FSIS), to make the safest food in the world. As part of this collaboration, these packing plants have responsibilities to ensure safe wholesome pork while USDA FSIS has oversight responsibilities as well. How any packing plant ensures a food safe product is also dependent on the pigs they receive. My goal is to get you thinking about market hogs and safe, wholesome food production after the trailer has left the barn.

Pathway of a pig

When thinking about a packing plant, I consider the receiving dock for loads of pigs as the starting line and shipping docks for packaged product as the finish line. For every load we receive, not every pig will make it to the finish line. So how do these pigs fall out from their cohorts on their way to the finish line?

Transportation and lairage

When receiving a load of pigs, we always consider animal welfare first. Therefore, if pigs are not ambulating off the trailer or have wounds due to exposure to harsh climates during transport, we humanely euthanize these fallout pigs. If we fail to maintain animal welfare, we risk our reputation, customer trust, and the trust of USDA FSIS. Animal welfare infractions are the only reason for immediate plant shut down so we consider anything related to animal welfare a serious matter.

Now that most of the pigs are unloaded, what do they look like upon arrival? We sort for various defects while the pigs walk to the lairage pen, which include flaws like abscesses, hernias, or certain clinical signs. Our USDA FSIS supervisory public health veterinarians and regulations determine the defects we sort for upon arrival.[1] USDA FSIS focuses on these defects because of the food safety risk due to clinical disease contaminating the entire pig or the potential to contaminate the entire slaughter floor. We either resell or euthanize these fallout pigs on a case-by-case basis.

Slaughter floor

As the pigs progress through the various steps on the slaughter floor, how big are these pigs? Large frame pigs can cause backups or downtime on the kill floor. What do you think these pigs look like on the inside? Do they have widespread internal pathology throughout the carcass? USDA FSIS have their own reasons and regulations for what constitutes a carcass condemnation but overall, the theme is to ensure safe food entering the supply chain.[2]

Once carcasses have passed USDA FSIS inspection, carcasses go through some form of chilling and are stored in refrigerated bays. After a certain amount of time, we take carcasses out of the refrigerated bays and the process of breaking them down into primals and sub-primals begins. Again, what do these pigs look like on the inside? Even if these carcasses passed kill floor USDA FSIS inspection, we still can find internal abscesses, needles, and other problems associated with the product because it is not food safe. The carcasses that require excess trimming increase the risk of introducing unsafe food into the food supply and increase the amount of labor needed to make it food safe. Remember: the choices you made on the slats weeks or months ago has an impact on the quality of product that you send to a packing plant.

Behind the scenes work

While the pigs go through the slaughter process, there is a team doing a lot of heavy lifting in the background: The laboratory technicians. Laboratory technicians are responsible for collecting regulatory microbiology and chemistry data, conducting quality testing on byproducts, pathogen testing and shelf life analysis. Generally, our laboratory runs for 20 hours a day. This team provides a lot of the time, skill, and effort that goes into ensuring safe and quality pork products.

Do these pigs have proper drug withdrawal times? Our laboratory technicians also conduct residue testing. Some export markets have lower tolerances for residues so domestic market withdrawal times may not be sufficient. We choose to run drug residue testing as a protectionary measure so we know the quality of product our export customers are receiving.

Products out the door

We use every bit of the pigs you send us but the quality of pigs sent will drive their final destination. Carcasses with minimal pathology and zero foreign material go out the door as primal and subprimal cuts, shelf ready to cook products as well as offal. We can also render fallout hogs into things like choice white grease, meat and bone meal, blood meal, concentrated blood plasma, and heparin.

Conclusion

I hope that this clarifies all of the different products and destinations for a load of pigs. The ability to make these products is largely dependent on the starting material and what is sent to the plant. Remember, the choices you made on the slats weeks or months ago has a huge impact on the quality of product that you send to a packing plant.

References

1. United States Department of Agriculture Food Safety and Inspection Service. FSIS Directive 6100.1 Rev. 3. 2020. **https://www.fsis.usda.gov/sites/default/files/media_file/2020-07/6100.1.pdf.** Accessed 3 November 2023.

2. United State Department of Agriculture Food Safety and Inspection Service. FSIS Directive 6100.2 Rev. 1. 2016. **https://www.fsis.usda.gov/sites/default/files/media_file/2020-07/6100.2.pdf.** Accessed 3 November 2023.

American Association of Swine Veterinarians
830 26th Street
Perry, IA 50220-2328
USA

ISBN 978-1-7138-9240-3

CURRAN ASSOCIATES INC.
proceedings
.com